Springer Series in Translational Stroke Research

Series Editor
John Zhang
Loma Linda, USA

More information about this series at http://www.springer.com/series/10064

Paul A. Lapchak • John H. Zhang
Editors

Cellular and Molecular Approaches to Regeneration and Repair

Springer

Editors
Paul A. Lapchak
Department of Neurology & Neurosurgery
Cedars-Sinai Medical Center, AHSP
Los Angeles, CA, USA

John H. Zhang
Department of Physiology
Loma Linda University
Loma Linda, CA, USA

ISSN 2363-958X ISSN 2363-9598 (electronic)
Springer Series in Translational Stroke Research
ISBN 978-3-319-66678-5 ISBN 978-3-319-66679-2 (eBook)
https://doi.org/10.1007/978-3-319-66679-2

Library of Congress Control Number: 2017957214

Printed on acid-free paper

This Springer imprint is published by Springer Nature
The registered company is Springer International Publishing AG
The registered company address is: Gewerbestrasse 11, 6330 Cham, Switzerland

Contents

Part I
Stem Cells

Chapter 1
Stroke: Cytoprotection, Repair and Regeneration—The Continuum of Patient Care

Paul A. Lapchak

Abstract The ischemic penumbra is now defined as tissue at risk of becoming fully involved in the evolving neurodegenerative process following an embolic stroke. After an ischemic core is quickly developed following vascular occlusion, there is slow spreading of the ischemic injury from the core to areas immediately surrounding the core until full recruitment is achieved. Forty years ago, seminal electrophysiological studies forming the basis of the penumbral hypothesis were conducted in a large animal model, baboons, an animal that is now rarely used in translational stroke research because it can no longer be justified! Thereafter, the rabbit embolic stroke model led the way for approval of Alteplase® (tissue plasminogen activator, tPA, rt-PA) to treat acute ischemic stroke.

Stroke research continues to evolve with the use of rodents primarily mice, rats, Oryctolagus cuniculus (rabbits), and occasionally non-human primates, but recent scientific expert statements have now suggested that non-human primates are not essential for stroke therapy development. One commonality amongst all species used historically is the documented presence of a core and penumbra following vascular occlusion, whether it be an artificial suture or clip occlusion or a blood clot.

This article reviews the historical basis for a few select mechanisms that are currently being targeted for cytoprotection, the rationale for target engagement to arrest penumbral growth and reduce clinical deficits, and it also sets a basis for the future of regeneration strategies to treat stroke patients.

Keywords Translational • Neuroprotection • Neuroprotective • Cytoprotection • Brain • Stroke • Embolic • Hemorrhage • Clinical trial • NIHSS • Stem cell • regeneration

---

P.A. Lapchak, Ph.D., FAHA (✉)
Department of Neurology & Neurosurgery, Cedars-Sinai Medical Center,
Advanced Health Sciences Pavilion, Suite 8318, 127 S. San Vicente Blvd,
Los Angeles, CA 90048, USA
e-mail: Paul.Lapchak@cshs.org

P.A. Lapchak, J.H. Zhang (eds.), *Cellular and Molecular Approaches to Regeneration and Repair*, Springer Series in Translational Stroke Research,
https://doi.org/10.1007/978-3-319-66679-2_1

Abbreviations

ACTION	Effect of Natalizumab on Infarct Volume in Acute Ischemic Stroke
ALS	Amyotrophic lateral sclerosis
ATP	Adenosine triphosphate
AU	Arbitrary units
ESCAPE	Endovascular Treatment for Small Core and Proximal Occlusion Ischemic Stroke
EXTEND-IA	Trial and Extending the Time for Thrombolysis in Emergency Neurological Deficits-Intra-Arterial
FDA	Food and Drug Administration
FRONTIER	Field Randomization of NA-1 Therapy in Early Responders
H_2O_2	Hydrogen peroxide
HO•	Hydroxyl radical
ICB	Intracerebral
IV	Intravenous
M1 or M2	Macrophage
MASTERS	Multipotent adult progenitor cells in acute ischemic stroke
MR CLEAN	Multicenter Randomized Clinical Trial of Endovascular Treatment for Acute Ischemic Stroke in the Netherlands
mRS	Modified Rankin scale
Nd:YaG	Neodymium-doped yttrium aluminum garnet
NADPH	Nicotinamide adenine dinucleotide phosphate
NADH	Nicotinamide adenosine dinucleotide
NEST	NeuroThera® Efficacy and Safety Trial
NIH	National Institutes of Health
NIHSS	National Institutes of Health Stroke Scale
NINDS	National Institute of Neurological Disorders and Stroke
NMDA	N-methyl-D-aspartate
nNOS	Neuronal Nitric oxide synthase
OPB	2-Oxo-3-(phenylhydrazone)-butanoic acid
O_2•−	Superoxide anion radical
PISCES	Human neural stem cells in patients with chronic ischaemic stroke
PSD-95	PSD-95 (postsynaptic density protein 95
PDZ1-2	Postsynaptic density-95, discs large 1, zonula occludens-1
REVASCAT	Endovascular Revascularization With Solitaire Device Versus Best Medical Therapy in Anterior Circulation Stroke Within 8 Hours
SWIFT PRIME	Solitaire With the Intention For Thrombectomy as PRIMary Endovascular Treatment
TCA cycle	Tricarboxylic acid cycle
TIGAR	TP53-inducible regulator of glycolysis and apoptosis

TLT	Transcranial laser therapy
tPA	Tissue plasminogen activator
THRACE	THRombectomie des Arteres CErebrales
TTC	Triphenyl tetrazolium chloride
UV	Ultraviolet
VLA-4	Very late antigen-4

1 Introduction

The existence of an ischemic stroke "penumbra" was first hypothesized by Astrup et al. in 1977 by demonstrating that there was a continuum of "threshold of ischemia" measured by electrical failure (potassium gradient) and reduced cerebral blood flow in the baboon cortical grey matter [1]. The gradient indicated an ischemic core, which is now known to be a mass of dead, unrecoverable tissue at the center of the ischemic infarct. The core is surrounded by **oligemic tissue**, defined as "Oligemia" tissue with reduced blood flow, but function is unaltered, and **ischemic tissue** with reduced blood flow. The **penumbral tissue**, is "at risk" of death tissue with altered potassium release, altered electrical failure and dysfunctional. In 1983, Olsen and colleagues demonstrated that an ischemic penumbra also existed in stroke patients; there was differential distribution of blood flow in non-ischemic, ischemic and hyperemic tissues [2]. The Ischemic Penumbra and Time is Penumbra have been reviewed in some detail by Heiss [3–14] and Donnan [15–26], and two main publications in association with receiving the Johann Jacob Wepfer Award [4, 16].

On this occasion, the 40th anniversary of describing the penumbra, we will take a brief look back at the origin of the stroke penumbra, and forward to review current clinically relevant targets that may arrest penumbral recruitment, and the consequences of such detrimental recruitment. We will also briefly discuss the potential need for multiple forms of therapeutic interventions to maximally promote both short and long term recovery in patients.

2 Cellular Measures of Metabolism and Penumbra

First, we will describe a novel technique to study the penumbral threshold in the rabbit embolic stroke model before addressing drug targets to attempt to arrest penumbra. In Fig. 1.1 we provide a topographical map of the rabbit cerebral cortex following an embolic stroke, comparing unstained fresh brain tissue (Fig. 1.1a) to a Nicotinamide Adenosine dinucleotide (NADH) map (Fig. 1.1c) of the brain followed by standard Triphenyl tetrazolium chloride (TTC) staining (Fig. 1.1b) to demarcate core (umbra) vs. viable tissue [27, 28].

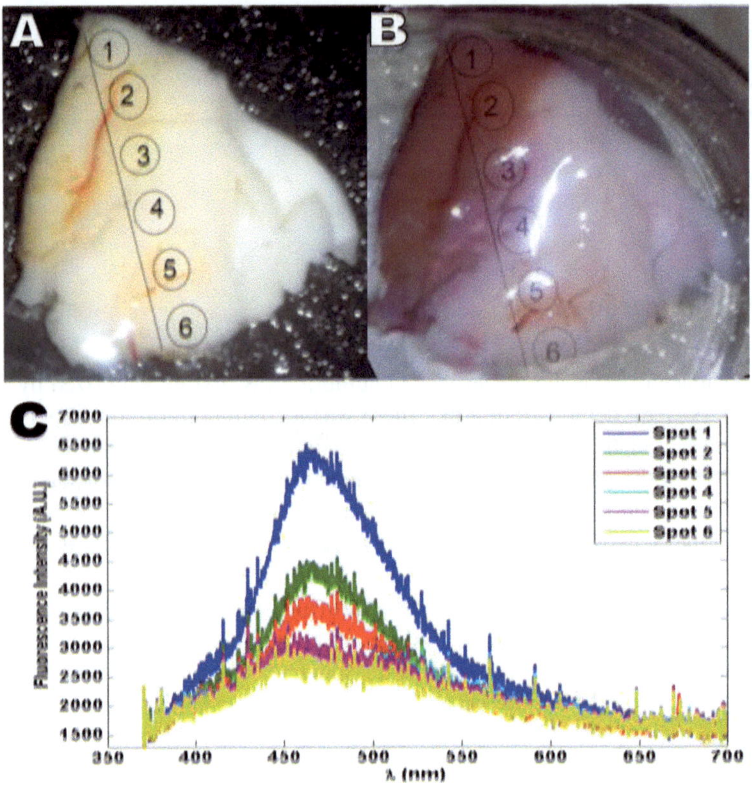

Fig. 1.1 (Credit: USPTO Patent 9404870 B2) Time-resolved laser-induced fluorescence spectroscopy systems and uses thereof [113]. Rabbits were prepared and embolized as described previously [60, 82, 114–118]. Fresh brain tissue was removed 3 h post-embolization for core (umbral) and penumbral NADH measurement. (**a**) Fresh rabbit cortex in oxygenated Krebs medium; (**b**) TTC-stained rabbit cortex; (**c**) Fluorescence measurement of NADH following an embolic stroke

2.1 Nicotinamide Adenosine Dinucleotide (NADH) Penumbra Mapping: A New Technique

NADH is produced in mitochondria from glucose during glycolysis [29]; it is one of the main coenzymes involved in the redox reaction for adenosine triphosphate (ATP) production in aerobic respiration. NADH is oxidized to NAD+ at the mitochondrial membrane by combining the hydrogen to oxygen thus forming H_2O, and in the process, produces ATP. In hypoxia, NADH accumulates inside the cell until there is shut down of tricarboxylic acid cycle (TCA). Alterations in the proton gradient result in further ATP depletion by the catabolic effects of ATP synthase [30]. If the oxygen depletion persists for a long duration, there may be cell death. These variations in NADH level provide us a brief window into the viability and vulnerability of cells under ischemic condition at a fixed time point. One way to evaluate these fluctuations in NADH level is to measure the fluorescence emission from NADH. NAD+ and NADH both have a strong absorption in ultraviolet (UV) spectrum [31–36], but they differ in their fluorescence characteristics NADH [31–36]. Importantly, NADH demonstrates a strong fluorescence in the violet/blue band around 440/460 nm, and this is dependent upon its bound versus free-state. By measuring NADH fluorescence in real time it is possible to monitor the changes in the NADH level and assess the metabolic status of NADH [31–36], as well as the status of the cell.

We have used a Q-switched neodymium-doped yttrium aluminum garnet (Nd:YaG) laser emitting at a wavelength of 350 nm, running at 1 kHz with a pulse width (Full-Width Half-Maximum, FWHM) of 400 ps (Teem Photonics PNVM02510). The total energy per pulse did not exceed 5 μJ to prevent photobleaching of NADH. The excitation light is delivered to the tissue using a custom made trifurcated optical probe, which consists of a central 600μ fiber for delivering the excitation light surrounded by 12 x 200μ fibers to collect the fluorescence. Every other fiber from the twelve collection fibers are bundled together forming two channels. One collection channel/bundle goes to a spectrometer (Ocean Optics, Maya), which measure the fluorescence spectrum every 100 ms.

As shown in Fig. 1.1c, there was a continuum of Fluorescence Intensity in arbitrary units (AU) as the probe was moved from position 1–6 along the surface of the cortex (see Fig. 1.1a, b), and NADH levels were recorded across all wavelengths between 350 and 700 nm. Notably, the peak is at approximately 465 nm, and Position 1, which had the highest levels of NADH was “normal” tissue compared to position 6, which had the lowest NADH levels “infarct core”. Throughout probe placement positions 2–5, various states of the penumbra, there was a gradual decrease in NADH levels measured. This map is reminiscent of the original Astrup et al. [37] electrical failure map with one difference: the current NADH measurement correlates with the ability of cells to synthesize ATP, whereas the Astrup and colleagues measured tissue potassium (K^+) levels. Nevertheless, separated in time by 40 years, we still have a great interest in the stroke penumbra, and both how to measure it and save it!

3 Cytoprotection: What a Time for Intervention?

We have recently reviewed cell death in brain following an ischemic event such as an embolic stroke or ischemic insult [38] (see also [39, 40]) leading us to hypothesize that there are multiple opportunities to provide pharmacological interventions to reduce the primary and secondary consequences of stroke; targets to “attack” in order to reduce widespread cellular death. But when there is widespread cellular death due to inadequate recovery following thrombolysis, embolectomy or new interventions, the stroke patient will move from the neurologist or interventionalist to the transplant neurosurgeon, if intracranial injections of stem cells prove to be superior to other methods of administration.

To recapitulate on recent and current knowledge, basic research demonstrates that neurons are first and foremost affected by ischemia, and die rapidly followed ischemia. Thereafter, brain endothelial cells (i.e.: forming the basis of the blood brain barrier), pericytes (i.e.: contractile cells that wrap around the endothelial cells of capillaries and venules), microglia (i.e.: brain resident macrophage; type M1 or M2), and then astrocytes (microglia) [41–43]. However, all cell types are intricately co-mingled and can directly or indirectly influence each other. This is an important observation and it should form the basis for meaningful and clinically relevant cyto-protection research.

4 The Stroke Cascade and Current Intervention Attempts Toward Efficacy

Recently, stroke interventionalists have advanced endovascular procedures, embolectomy with stentrievers or thromboaspiration catheters to a technical level where statistically significant efficacy was demonstrated in a sextet of positive clinical trials enrolling large vessel-occluded patients with substantial penumbral substrate (See Table 1.1): (1) Multicenter Randomized Clinical Trial of Endovascular Treatment for Acute Ischemic Stroke in the Netherlands (**MR. CLEAN**) [44]; (2) Endovascular Treatment for Small Core and Proximal Occlusion Ischemic Stroke (**ESCAPE**) [45]; (3) Endovascular Revascularization With Solitaire Device Versus Best Medical Therapy in Anterior Circulation Stroke Within 8 Hours (**REVASCAT**) [46]; (4) Solitaire With the Intention For Thrombectomy as PRIMary Endovascular Treatment (**SWIFT PRIME**) [47]; (5) Trial and Extending the Time for Thrombolysis in Emergency Neurological Deficits-Intra-Arterial (**EXTEND-IA**)

Table 1.1 Population efficacy of reperfusion therapies compared to NINDS rt-PA

		No Symptoms----►Death						
Study	**Treatment**	**0**	**1**	**2**	**3**	**4**	**5**	**6**
MR. CLEAN	**Control (267)**	**0**	**6**	**13**	16	30	12	22
	Intervention (233)	**3**	**9**	**21**	18	22	6	21
ESCAPE	**Control (150)**	**7**	**10**	**12**	15	24	12	19
	Intervention (165)	**15**	**21**	**18**	16	13	7	10
REVASCAT	**Control (103)**	**5.8**	**6.8**	**15.5**	19.4	16.5	20.4	15.5
	Intervention (103)	**6.8**	**17.5**	**19.4**	18.4	7.8	11.7	18.4
SWIFT PRIME	**Control (98)**	**9**	**11**	**16**	17	22	26	
	Intervention (98)	**17**	**26**	**17**	12	15	12	
EXTEND-IA	**Control (35)**	**17**	**11**	**11**	11	17	11	20
	Intervention (35)	**26**	**26**	**20**	17	3	0	9
THRACE	**Control (202)**	**11.9**	**16.3**	**13.9**	12.4	27.7	4.5	13.4
	Intervention (200)	**15.5**	**19.5**	**18.0**	12.5	17	5.5	12
NINDS rt-PA	**Control (312)**	**26**		**25**		27		21
	Intervention (312)	**39**		**21**		23		17

Efficacy Analysis: mRS outcome (90 Day shown as (%) per tier; Highlighted Boxes indicat mRS 0-2 functional independence.

[48]; (6) THRombectomie des Arteres CErebrales (**THRACE**) [49], which were completed and published almost 22 years after the "positive" National Institute of Neurological Disorders and Stroke (NINDS) tissue plasminogen activator (tPA) Stroke Study Group trial and publication [50], and subsequent approval of tPA by the Food and Drug Administration (FDA).

Table 1.1 presents the 90-day outcome data for modified Rankin scale (mRS) for 6 endovascular trials compared to the original rt-PA NINDS trial. Every trial indicates that "intervention", thrombolysis plus an endovascular procedure is better than the endovascular procedure alone, when the procedures are done within approximately 8 h of the stroke. Now both reperfusion procedures used independently or in combination are commonplace in US medical centers and hospitals, but recent treatment estimates suggest that only 5–10% of stroke patients are eligible for the procedures, and less than half have full recovery [38, 46, 47, 51–53] (see Fig. 1.2 for study outcome results).

While rational neuroprotection and cytoprotection trials are still evolving after many failed attempts, no specific strategy has realized efficacy in any form of ischemic stroke [54, 55]. Thus, unfortunately, there remains a need for efficacious cytoprotection and regeneration/repair procedures.

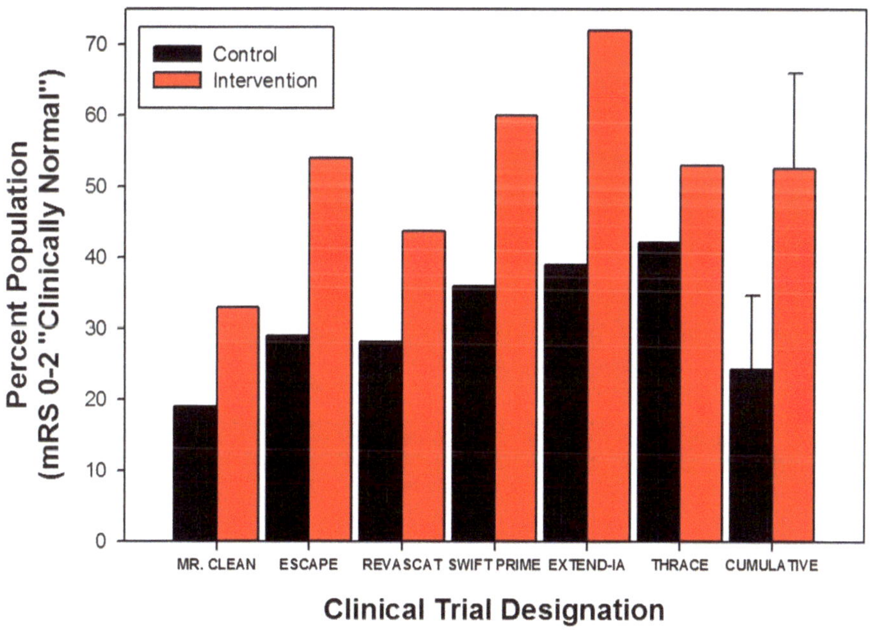

Fig. 1.2 Direct comparison of the patient population that is mRS 0-2 or normal, with functional independence in six recent embolectomy trials [(**MR. CLEAN**) [44]; (**ESCAPE**) [45]; (**REVASCAT**) [46]; (**SWIFT PRIME**) [47]; (**EXTEND-IA**) [48]; and (**THRACE**) [49]. In the trials, there was an a significant improvement in combined thrombolytic/endovascular procedure therapy (intervention) compared to control (endovascular procedure alone). All bars are Mean values, and the CUMULATIVE bars are the Mean ± SD for all patients mRS 0-2 showing that intervention more than doubles the population mRS 0-2. Note: Early enrollment is required for tPA and embolectomy within 8 h

5 Critical Targets for Rapid Intervention

The stroke ischemic cascade has many point of intervention, but there are few targets as important as cellular metabolism and excitotoxicity. As shown in Fig. 1.3, these mechanisms set the stage for cell death because they are the first in the sequence of events.

5.1 Energy Depletion and Cell Death

Early in the 1970s, Siesjo et al. recognized many important features of ischemia [56–59] related to rapid energy depletion by cells and metabolic changes that are either irreversible or reversible. In fact, in Fig. 1.1, we have taken advantage of some of the metabolic changes so that we can map the ubmra and penumbra. Two interesting and potentially important therapeutic approaches at this level have been documented over the last 10–15 years. First, transcranial laser therapy (TLT) was

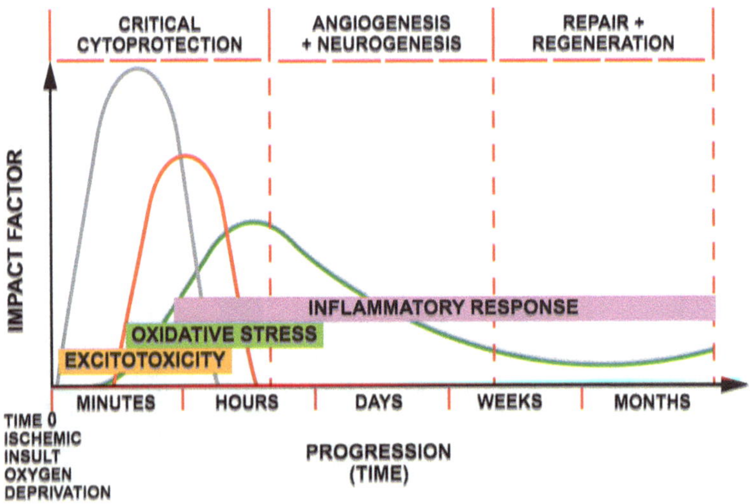

Fig. 1.3 The stroke cascade. Following an ischemic stroke, reduced blood blow and depleted ATP levels cause rapid death of neurons followed by initiation of a sequence of event in cells directly surrounding the core. There are numeours potential intervention points for therapy in the cascade, including reperfusion therapy conducted early after the ischemic event. The temporal profile for therapeutic intervention in stroke patients. The *graph* indicates therapeutic windows for mechanism-based acute pharmacological therapies, potentiating neurogenesis and angiogenesis, and then providing cell restoration or replenishment therapies. None of the interventions are mutually exclusive and optimal cytoprotection may require the use of combined therapies

attempted as an approach to enhance metabolism and increase cerebral blood flow; this worked well in multiple species [60–68], and this was linked to enhanced mitochondrial function and ATP levels [67, 68]. However, in stroke patients, reproducible clinical efficacy could not be achieved [69–71]. Because it has become clear that TLT was not sufficiently optimized before proceeding to the (NeuroThera® Efficacy and Safety Trial) NEST trials, TLT should be revived as a potentially useful approach [65]. New concepts are under development to deliver photons locally to promote cytoprotection.

More recently, Li et al. [72] used nicotinamide adenine dinucleotide phosphate (NADPH) as a therapy based upon initial observations that the TP53-inducible regulator of glycolysis and apoptosis (TIGAR) or fructose-2,6-bisphosphatase TIGAR plays a role in cell survival by increasing the flux of pentose phosphate pathway. The strategy of administering NADPH was successful in animals, where ATP levels were increased, stroke volume decreased and behavior improved statistically.

5.2 *Excitotoxicity*

Excitatory amino acid release (EAA), primarily glutamate is considered the first acute insult as reported by Jorgensen and Diemer in 1982 [73], and reviewed by Schwarz and Meldrum [74]. While many classic attempts at antagonism at EAA receptors with high affinity ligands have failed [54, 55], Tymianski and Cook [75, 76] have made a significant advance in the field by custom designing a therapeutic molecule targeting PSD-95 (postsynaptic density protein 95); the PSD-95 inhibitor Tat-NR2B9c which has shown potential using a non-human primate ischemia model and has advanced to the Field Randomization of NA-1 Therapy in Early Responders (FRONTIER) clinical trial [77]. The article by Bratane et al. referred to this form of neuroprotection as "freezing ischemic penumbra evolution" [78], which if this can be achieved, should significantly attenuate some of the deleterious steps within the temporally-dependent ischemic cascade. Clearly, the problem with the approach is that intervention is required as soon as possible after a stroke to attenuate cascade activation.

Moreover, Avilex Inc. has continued with this line of research and have developed a series of small molecules and peptidomimetics targeting the PSD-95 complex. AVLX-144 is the companies lead compound for the treatment of stroke. PSD-95 inhibitors targeting the postsynaptic density-95, discs large 1, zonula occludens-1 (PDZ1-2) domains of PSD-95 effectively block the formation of the ternary nNOS/PSD-95/NMDA receptor complex and uncouple the NMDA receptor pathway [79–81]. This is a fertile and competitive area of research that may result in identification of a therapeutic to use as a first line of defense in patients, and in patients undergoing thrombectomy.

5.3 *Oxidative Stress*

Free radicals are a valid target for therapeutic intervention for the treatment of AIS [82–85], because oxidative stress is a major early component of the ischemic stroke cascade [84, 86, 87]. Thus, free radical scavengers have been repeatedly tested in stroke and ischemia models, and have been reviewed by leading stroke experts as part of the ischemic cascade [38, 54, 83, 88, 89].

Oxidative stress following an ischemic stroke results in the production of various reactive oxygen species (ROS): hydrogen peroxide (H_2O_2), hydroxyl radical (HO•) and superoxide anion radical (O_2•−) that can induce membrane lipid peroxidation and vascular endothelial cell injury [90–94].

Recently the FDA formally approved edaravone (3-methyl-1-phenyl-2-pyrazolin-5-one; Radicava™) to treat patients with amyotrophic lateral sclerosis (ALS), Lou Gehrig's disease [95]. Edaravone (Radicut™) is a free radical scavenger marketed in Japan by Mitsubishi Tanabe Pharma Corporation to treat acute ischemic stroke patients presenting within 24 h of the attack. For stroke, there are two forms of edaravone currently available. Injectable edaravone ampoules (30 mg b.i.d, i.v. for 14 days) were first approved on May 23, 2001, and then on January 19th 2010, the Radicut BAG was approved by the Japanese Ministry of Health and Welfare and is used as an i.v. infusion. The demonstrated efficacy of Radicut has been limited to the Asian population (China, India and Japan) and there has still not been a randomized clinical trial conducted in North America. The clinical efficacy of Radicut in stroke patients was reviewed in detail by Lapchak [96].

The primary focus of edaravone research has revolved around its potential to scavenge hyrdoxyl, peroxyl and superoxide radicals that mediate neuronal and vascular damage [97–100]. At physiological pH, edaravone is present in an anionic form, and due to electron transfer, electrons released from the edaravone anion can effectively bind radical species containing a free electron. For example, the lipid peroxyl radical (OOL•) that can be formed after free radical extraction of a proton from an unsaturated fatty acid. The transfer of electrons from edaravone to a reactive oxygen species produces an edaravone radical, which then forms a peroxyl radical of edaravone. The result is the formation of a, a 4,5-dione (i.e. 3-methyl-1-phenyl-2-pyrazolin-4,5-dione) and eventually 2-oxo-3-(phenylhydrazone)-butanoic acid (OPB).

With Radicut now given orphan drug status in the United States for ALS, perhaps the near future will see the use of the antioxidant in stroke victims in North America.

5.4 *Immunotherapy*

The immune response following a stroke may present the longest therapeutic window for intervention, but may also be the most problematic pharmacological target because of the diversity of mediators and mechanisms, and the dual role of many mediators. Moreover, the possibility exists that both central and peripheral immune systems can be mediated to modulate brain activity following a stroke [101, 102]. The immune or inflammatory response is normally characterized by the production of a vast array of inflammatory mediators including lymphokines, cytokines, and necrosis factors (i.e.: tumor necrosis factor). There is also recruitment and accumulation of "inflammatory cells", leukocytes, activation of microglia, and distribution of macrophage to the lesion area [103–105]. Natalizumab, a humanized monoclonal antibody against alpha$_4$-integrin (CD49d), specifically targeting α4 integrin within the adhesion molecule very late antigen-4 (VLA-4), is being tested in a clinical trial based upon limited efficacy in animal models [106]. A review of the preclinical literature shows that there is not unanimous neuroprotection or improvement in documented preclinical studies [106–108]. However, since the antibody attenuates the transmigration of leukocytes across the endothelium, it is hypothesized to block lymphocyte and adhesion and subsequent transmigration into the CNS, this attenuating the immune reaction. Preliminary results from the ACTION trial [106] indicate that "Natalizumab administered up to 9 h after stroke onset did not reduce infarct growth. Treatment-associated benefits on functional outcomes might warrant further investigation".

6 Regeneration, Repair and Repopulation

Novel methods of efficacious cytoprotection in stroke patients continues to be a challenge to both the translational and clinical stroke researcher. Based upon the cumulative data from the endovascular trials that were discussed earlier in this chapter, we can estimate that 5–10% of stroke patients will benefit from reperfusion procedures [109], and that only a small fraction of all stroke patients will be eligible for the procedures. Notably, patients with high National Institutes of Health Stroke Scale (NIHSS) scores have reduced probability of returning to normal (mRS 0-2) at 90 days [109], or thereafter. It is further estimated that 16–19% of patients will be dependent for the remainder of their lives [110]. With these assumptions, even if a cytoprotective agent was developed and FDA-approved, the overall shift in the "normal" patient population would be unacceptably small, but it is critical to attempt to intervene at all stages following a stroke.

6.1 What Happens to the Stroke Victim When Penumbra No Longer Exists as a Target for Intervention?

There would also be a need for strategies to regenerate and repair neuronal pathways in some stroke patients, especially those with long-term disability. This section will focus on three first-in-man stem cell clinical trials (Table 1.2).

Table 1.2 Stem-cell based therapy: results of three trials

Study	Cell type	Route	Trial design and result
MASTERS [109]	Multipotent adult progenitor cells	IV	**Design**: Randomized, double-blind, placebo-controlled **Result**: Safe. No difference in global stroke recovery scores at 90 days
PISCES [110]	Immortalized human neural stem-cell line CTX0E03	Intracerebral Stereotaxic	**Design**: Open-label, single center, dose-escalation. **Result**: Safe, no adverse events. ➢ Highly variable changes in clinical function measured over 24 months
SB623 [111]	Bone-marrow-derived mesenchymal cells - SB623 cells transfected with human Notch-1 intracellular domain	Intracerebral Stereotaxic	**Design**: Unblinded, non-randomized **Result**: Safe, but patients did experience treatment emergent adverse events. ➢ Improvement on European Stroke Scale; NIHSS; Fugl-Meter total score; Fugl-Meyer motor function score, but no changes in mRS measured over 12 months

The results of preliminary stem-cell based therapy trials suggest that certain cells can be administered safely, and that a potentially useful intravenous (IV) or intracerebral (ICB) route may be used in patients. Clearly, all published trials were preliminary safety trials, the focus was to be "first", and they were not designed to adequately or comprehensively determine efficacy on standardized stroke scales. More importantly, they were not rigorous [112]. Nevertheless, the three first-in-man clinical trials for stroke will eventually lead to meaningful rigorous blinded, controlled, randomized patient population investigation in this area.

7 Conclusion

In conclusion, translational stroke research continues to evolve from neuroprotection to cytoprotection, and now delves into regeneration and repair. Thrombolysis and embolectomy are now proven efficacious therapeutic interventions, but the procedures can be improved with adjunct therapy. The future of translational stroke research will continue to progress, and patients will be offered multiple forms of therapy depending on the clinical response to initial therapeutic intervention, their initial and follow-up NIHSS and mRS scores, and overall level of functional disability.

Acknowledgments and Funding This article was written without direct financial support from government sources (PAL). PAL is supported by pharmaceutical industry grants and philanthropic donations. P. Butte is thanked for assistance with NADH measurements.

Conflict of Interest Statements: PAL is Editor-in-Chief, Journal of Neurology & Neurophysiology and Associate Editor, Translational Stroke Research.

References

1. Astrup J, Symon L, Branston NM, Lassen NA. Cortical evoked potential and extracellular K+ and H+ at critical levels of brain ischemia. Stroke. 1977;8(1):51–7.
2. Olsen TS, Larsen B, Herning M, Skriver EB, Lassen NA. Blood flow and vascular reactivity in collaterally perfused brain tissue. Evidence of an ischemic penumbra in patients with acute stroke. Stroke. 1983;14(3):332–41.
3. Heiss WD. The ischemic penumbra: how does tissue injury evolve? Ann N Y Acad Sci. 2012;1268:26–34.
4. Heiss WD. The ischemic penumbra: correlates in imaging and implications for treatment of ischemic stroke. The Johann Jacob Wepfer award 2011. Cerebrovasc Dis. 2011;32(4):307–20.
5. del Zoppo GJ, Sharp FR, Heiss WD, Albers GW. Heterogeneity in the penumbra. J Cereb Blood Flow Metab. 2011;31(9):1836–51.
6. Heiss WD. The concept of the penumbra: can it be translated to stroke management? Int J Stroke. 2010;5(4):290–5.
7. Sobesky J, Zaro Weber O, Lehnhardt FG, Hesselmann V, Neveling M, Jacobs A, et al. Does the mismatch match the penumbra? Magnetic resonance imaging and positron emission tomography in early ischemic stroke. Stroke. 2005;36(5):980–5.
8. Umegaki M, Sanada Y, Waerzeggers Y, Rosner G, Yoshimine T, Heiss WD, et al. Peri-infarct depolarizations reveal penumbra-like conditions in striatum. J Neurosci. 2005;25(6):1387–94.
9. Heiss WD, Sobesky J, Hesselmann V. Identifying thresholds for penumbra and irreversible tissue damage. Stroke. 2004;35(11 Suppl 1):2671–4.
10. Heiss WD. Best measure of ischemic penumbra: positron emission tomography. Stroke. 2003;34(10):2534–5.
11. Heiss WD. Imaging the ischemic penumbra and treatment effects by PET. Keio J Med. 2001;50(4):249–56.
12. Heiss WD. Ischemic penumbra: evidence from functional imaging in man. J Cereb Blood Flow Metab. 2000;20(9):1276–93.
13. Heiss WD, Graf R, Wienhard K, Lottgen J, Saito R, Fujita T, et al. Dynamic penumbra demonstrated by sequential multitracer PET after middle cerebral artery occlusion in cats. J Cereb Blood Flow Metab. 1994;14(6):892–902.
14. Heiss WD, Graf R. The ischemic penumbra. Curr Opin Neurol. 1994;7(1):11–9.
15. Campbell BC, Donnan GA, Davis SM. Vessel occlusion, penumbra, and reperfusion - translating theory to practice. Front Neurol. 2014;5:194.
16. Davis S, Donnan GA. Time is Penumbra: imaging, selection and outcome. The Johann jacob wepfer award 2014. Cerebrovasc Dis. 2014;38(1):59–72.
17. Nagakane Y, Christensen S, Ogata T, Churilov L, Ma H, Parsons MW, et al. Moving beyond a single perfusion threshold to define penumbra: a novel probabilistic mismatch definition. Stroke. 2012;43(6):1548–55.
18. Spratt NJ, Donnan GA, McLeod DD, Howells DW. 'Salvaged' stroke ischaemic penumbra shows significant injury: studies with the hypoxia tracer FMISO. J Cereb Blood Flow Metab. 2011;31(3):934–43.
19. Ebinger M, De Silva DA, Christensen S, Parsons MW, Markus R, Donnan GA, et al. Imaging the penumbra - strategies to detect tissue at risk after ischemic stroke. J Clin Neurosci. 2009;16(2):178–87.
20. Guadagno JV, Donnan GA, Markus R, Gillard JH, Baron JC. Imaging the ischaemic penumbra. Curr Opin Neurol. 2004;17(1):61–7.

21. Saita K, Chen M, Spratt NJ, Porritt MJ, Liberatore GT, Read SJ, et al. Imaging the ischemic penumbra with 18F-fluoromisonidazole in a rat model of ischemic stroke. Stroke. 2004;35(4):975–80.
22. Davis SM, Donnan GA. Advances in penumbra imaging with MR. Cerebrovasc Dis. 2004;17(Suppl 3):23–7.
23. Davis SM, Donnan GA. Ischemic penumbra: MRI or PET. Stroke. 2003;34(10):2536.
24. Donnan GA, Davis SM. Neuroimaging, the ischaemic penumbra, and selection of patients for acute stroke therapy. Lancet Neurol. 2002;1(7):417–25.
25. Phan TG, Wright PM, Markus R, Howells DW, Davis SM, Donnan GA. Salvaging the ischaemic penumbra: more than just reperfusion? Clin Exp Pharmacol Physiol. 2002;29(1-2):1–10.
26. Barber PA, Davis SM, Darby DG, Desmond PM, Gerraty RP, Yang Q, et al. Absent middle cerebral artery flow predicts the presence and evolution of the ischemic penumbra. Neurology. 1999;52(6):1125–32.
27. Benedek A, Moricz K, Juranyi Z, Gigler G, Levay G, Harsing LG Jr, et al. Use of TTC staining for the evaluation of tissue injury in the early phases of reperfusion after focal cerebral ischemia in rats. Brain Res. 2006;1116(1):159–65.
28. Bederson JB, Pitts LH, Germano SM, Nishimura MC, Davis RL, Bartkowski HM. Evaluation of 2,3,5-triphenyltetrazolium chloride as a stain for detection and quantification of experimental cerebral infarction in rats. Stroke. 1986;17(6):1304–8.
29. Krebs HA, Johnson WA. Metabolism of ketonic acids in animal tissues. Biochem J. 1937;31(4):645–60.
30. Penna C, Pagliaro P, Rastaldo R, Di Pancrazio F, Lippe G, Gattullo D, et al. F0F1 ATP synthase activity is differently modulated by coronary reactive hyperemia before and after ischemic preconditioning in the goat. Am J Physiol Heart Circ Physiol. 2004;287(5):H2192–200.
31. Marcu L, Jo JA, Butte PV, Yong WH, Pikul BK, Black KL, et al. Fluorescence lifetime spectroscopy of glioblastoma multiforme. Photochem Photobiol. 2004;80:98–103.
32. Butte PV, Pikul BK, Hever A, Yong WH, Black KL, Marcu L. Diagnosis of meningioma by time-resolved fluorescence spectroscopy. J Biomed Opt. 2005;10(6):064026.
33. Yong WH, Butte PV, Pikul BK, Jo JA, Fang Q, Papaioannou T, et al. Distinction of brain tissue, low grade and high grade glioma with time-resolved fluorescence spectroscopy. Front Biosci. 2006;11:1255–63.
34. Butte PV, Fang Q, Jo JA, Yong WH, Pikul BK, Black KL, et al. Intraoperative delineation of primary brain tumors using time-resolved fluorescence spectroscopy. J Biomed Opt. 2010;15(2):027008.
35. Butte PV, Mamelak AN, Nuno M, Bannykh SI, Black KL, Marcu L. Fluorescence lifetime spectroscopy for guided therapy of brain tumors. Neuroimage. 2011;54(Suppl 1):S125–35.
36. Kittle DS, Vasefi F, Patil CG, Mamelak A, Black KL, Butte PV. Real time optical biopsy: time-resolved fluorescence spectroscopy instrumentation and validation. Sci Rep. 2016;6:38190.
37. Astrup J, Siesjo BK, Symon L. Thresholds in cerebral ischemia - the ischemic penumbra. Stroke. 1981;12(6):723–5.
38. Lapchak PA, Zhang JH. The high cost of stroke and stroke cytoprotection research. Transl Stroke Res. 2017;8:307. https://doi.org/10.1007/s12975-016-0518-y.
39. Hall R, Murdoch J. Brain protection: physiological and pharmacological considerations. Part II: The pharmacology of brain protection. Can J Anaesth. 1990;37(7):762–77.
40. Murdoch J, Hall R. Brain protection: physiological and pharmacological considerations. Part I: The physiology of brain injury. Can J Anaesth. 1990;37(6):663–71.
41. Redzic ZB, Rabie T, Sutherland BA, Buchan AM. Differential effects of paracrine factors on the survival of cells of the neurovascular unit during oxygen glucose deprivation. Int J Stroke. 2015;10(3):407–14.
42. Barakat R, Redzic Z. Differential cytokine expression by brain microglia/macrophages in primary culture after oxygen glucose deprivation and their protective effects on astrocytes during anoxia. Fluids Barriers CNS. 2015;12:6.
43. Carmichael ST. The 3 Rs of stroke biology: radial, relayed, and regenerative. Neurotherapeutics. 2016;13(2):348–59.

44. Broderick JP, Berkhemer OA, Palesch YY, Dippel DW, Foster LD, Roos YB, et al. Endovascular therapy is effective and safe for patients with severe ischemic stroke: pooled analysis of interventional management of stroke III and multicenter randomized clinical trial of endovascular therapy for acute ischemic stroke in the Netherlands data. Stroke. 2015;46(12):3416–22.
45. Goyal M, Demchuk AM, Menon BK, Eesa M, Rempel JL, Thornton J, et al. Randomized assessment of rapid endovascular treatment of ischemic stroke. N Engl J Med. 2015;372(11):1019–30.
46. Jovin TG, Chamorro A, Cobo E, de Miquel MA, Molina CA, Rovira A, et al. Thrombectomy within 8 hours after symptom onset in ischemic stroke. N Engl J Med. 2015;372(24):2296–306.
47. Saver JL, Goyal M, Bonafe A, Diener HC, Levy EI, Pereira VM, et al. Stent-retriever thrombectomy after intravenous t-PA vs. t-PA alone in stroke. N Engl J Med. 2015;372(24):2285–95.
48. Campbell BC, Mitchell PJ, Investigators E-I. Endovascular therapy for ischemic stroke. N Engl J Med. 2015;372(24):2365–6.
49. Bracard S, Ducrocq X, Mas JL, Soudant M, Oppenheim C, Moulin T, et al. Mechanical thrombectomy after intravenous alteplase versus alteplase alone after stroke (THRACE): a randomised controlled trial. Lancet Neurol. 2016;15(11):1138–47.
50. NINDS. Tissue plasminogen activator for acute ischemic stroke. The National Institute of Neurological Disorders and Stroke rt-PA Stroke Study Group. N Engl J Med. 1995;333(24):1581–7.
51. Goyal M, Hill MD, Saver JL, Fisher M. Challenges and opportunities of endovascular stroke therapy. Ann Neurol. 2016;79:11.
52. Lapchak PA. Critical early thrombolytic & endovascular reperfusion therapy for acute ischemic stroke victims: a call for adjunct neuroprotection. Transl Stroke Res. 2015;6:345. https://doi.org/10.1007/s12975-015-0419-5(6):345-54.
53. Henninger N, Fisher M. Extending the time window for endovascular and pharmacological reperfusion. Transl Stroke Res. 2016;7(4):284–93.
54. O'Collins VE, Macleod MR, Donnan GA, Horky LL, van der Worp BH, Howells DW. 1,026 experimental treatments in acute stroke. Ann Neurol. 2006;59(3):467–77.
55. Lapchak PA, Boitano PD. Reflections on neuroprotection research and the path toward clinical success. In: Lapchak PA, Zhang JH, editors. Neuroprotective therapy for stroke and ischemic disease, Springer series in transaltional stroke research. Cham: Springer; 2017. p. 1–72.
56. Nordstrom CH, Rehncrona S, Siesjo BK. Effects of phenobarbital in cerebral ischemia. Part II: restitution of cerebral energy state, as well as of glycolytic metabolites, citric acid cycle intermediates and associated amino acids after pronounced incomplete ischemia. Stroke. 1978;9(4):335–43.
57. Nordstrom CH, Siesjo BK. Effects of phenobarbital in cerebral ischemia. Part I: cerebral energy metabolism during pronounced incomplete ischemia. Stroke. 1978;9(4):327–35.
58. Eklof B, MacMillan V, Siesjo BK. The effect of ischemia upon the energy state of the brain. Eur Neurol. 1971;6(1):60–5.
59. Salford LG, Brierley JB, Plum F, Siesjo BK. Energy metabolism and histology in the brain during combined hypoxemia and ischemia. Eur Neurol. 1971;6(1):329–34.
60. Lapchak PA, Wei J, Zivin JA. Transcranial infrared laser therapy improves clinical rating scores after embolic strokes in rabbits. Stroke. 2004;35(8):1985–8.
61. Lapchak PA. Taking a light approach to treating acute ischemic stroke patients: transcranial near-infrared laser therapy translational science. Ann Med. 2010;42(8):576–86.
62. Lapchak PA. Transcranial near-infrared laser therapy applied to promote clinical recovery in acute and chronic neurodegenerative diseases. Expert Rev Med Devices. 2012;9(1):71–83.
63. Lapchak PA, Butte P, Rajput PS. The difficult path to treating acute ischemic stroke patients with transcranial near-infrared laser therapy. In: Hamblin MR, Agrawal T, de Sousa M, editors. Handbook of Low-Level Laser Therapy. Singapore: Pan Stanford Publishing Pte; 2017
64. Lapchak PA, Boitano PD, Butte PV, Fisher DJ, Holscher T, Ley EJ, et al. Transcranial near-infrared laser transmission (NILT) profiles (800 nm): systematic comparison in four common research species. PLoS One. 2015;10(6):e0127580.

65. Lapchak PA, Boitano PD. Transcranial near-infrared laser therapy for stroke: how to recover from futility in the NEST-3 clinical trial. Acta Neurochir Suppl. 2016;121:7–12.
66. Naeser MA, Hamblin MR. Potential for transcranial laser or LED therapy to treat stroke, traumatic brain injury, and neurodegenerative disease. Photomed Laser Surg. 2011;29(7):443–6.
67. Lapchak PA, Boitano PD. A novel method to promote behavioral improvement and enhance mitochondrial function following an embolic stroke. Brain Res. 2016;1646:125–31.
68. Lapchak PA, De Taboada L. Transcranial near infrared laser treatment (NILT) increases cortical adenosine-5′-triphosphate (ATP) content following embolic strokes in rabbits. Brain Res. 2010;1306:100–5.
69. Lampl Y, Zivin JA, Fisher M, Lew R, Welin L, Dahlof B, et al. Infrared laser therapy for ischemic stroke: a new treatment strategy: results of the NeuroThera Effectiveness and Safety Trial-1 (NEST-1). Stroke. 2007;38(6):1843–9.
70. Hacke W, Schellinger PD, Albers GW, Bornstein NM, Dahlof BL, Fulton R, et al. Transcranial laser therapy in acute stroke treatment: results of neurothera effectiveness and safety trial 3, a Phase III clinical end point device trial. Stroke. 2014;45(11):3187–93.
71. Kasner SE, Rose DZ, Skokan A, Walker MG, Shi J, Streeter J, et al. Transcranial laser therapy and infarct volume. Stroke. 2013;44(7):2025–7.
72. Li M, Zhou ZP, Sun M, Cao L, Chen J, Qin YY, et al. Reduced nicotinamide adenine dinucleotide phosphate, a pentose phosphate pathway product, might be a novel drug candidate for ischemic stroke. Stroke. 2016;47(1):187–95.
73. Jorgensen MB, Diemer NH. Selective neuron loss after cerebral ischemia in the rat: possible role of transmitter glutamate. Acta Neurol Scand. 1982;66(5):536–46.
74. Schwarcz R, Meldrum B. Excitatory aminoacid antagonists provide a therapeutic approach to neurological disorders. Lancet. 1985;2(8447):140–3.
75. Cook DJ, Teves L, Tymianski M. Treatment of stroke with a PSD-95 inhibitor in the gyrencephalic primate brain. Nature. 2012;483(7388):213–7.
76. Cook DJ, Teves L, Tymianski M. A translational paradigm for the preclinical evaluation of the stroke neuroprotectant Tat-NR2B9c in gyrencephalic nonhuman primates. Sci Transl Med. 2012;4(154):154ra33.
77. FRONTIER. Field randomization of NA-1 therapy in early responders (FRONTIER). 2015. https://clinicaltrials.gov/ct2/show/study/NCT02315443.
78. Bratane BT, Cui H, Cook DJ, Bouley J, Tymianski M, Fisher M. Neuroprotection by freezing ischemic penumbra evolution without cerebral blood flow augmentation with a postsynaptic density-95 protein inhibitor. Stroke. 2011;42(11):3265–70.
79. Bach A, Clausen BH, Moller M, Vestergaard B, Chi CN, Round A, et al. A high-affinity, dimeric inhibitor of PSD-95 bivalently interacts with PDZ1-2 and protects against ischemic brain damage. Proc Natl Acad Sci U S A. 2012;109(9):3317–22.
80. Bach A, Chi CN, Olsen TB, Pedersen SW, Roder MU, Pang GF, et al. Modified peptides as potent inhibitors of the postsynaptic density-95/N-methyl-D-aspartate receptor interaction. J Med Chem. 2008;51(20):6450–9.
81. Haugaard-Kedström LM, Fernandes EFA, Strømgaard K. Targeting PSD-95 as a novel approach in the treatment of stroke. In: Lapchak PA, Zhang JH, editors. Neuroprotective therapy for stroke and ischemic disease. Cham: Springer; 2017. p. 157–84.
82. Lapchak PA. Translational stroke research using a rabbit embolic stroke model: a correlative analysis hypothesis for novel therapy development. Transl Stroke Res. 2010;1(2):96–107.
83. Dirnagl U, Iadecola C, Moskowitz MA. Pathobiology of ischaemic stroke: an integrated view. Trends Neurosci. 1999;22(9):391–7.
84. Lapchak PA, Araujo DM. Advances in ischemic stroke treatment: neuroprotective and combination therapies. Expert Opin Emerg Drugs. 2007;12(1):97–112.
85. Butterfield JD Jr, McGraw CP. Free radical pathology. Stroke. 1978;9(5):443–5.
86. Facchinetti F, Dawson VL, Dawson TM. Free radicals as mediators of neuronal injury. Cell Mol Neurobiol. 1998;18(6):667–82.
87. Love S. Oxidative stress in brain ischemia. Brain Pathol. 1999;9(1):119–31.

88. Lapchak PA, Boitano PD. Reflections on neuroprotection research and the path toward clinical success. In: Lapchak PA, Zhang JH, editors. Neuroprotective therapy for stroke and ischemic disease. Cham: Springer International Publishing; 2017. p. 3–71.
89. Dirnagl U. Pathobiology of injury after stroke: the neurovascular unit and beyond. Ann N Y Acad Sci. 2012;1268:21–5.
90. Watanabe T, Tahara M, Todo S. The novel antioxidant edaravone: from bench to bedside. Cardiovasc Ther. 2008;26(2):101–14.
91. Yoshida H, Yanai H, Namiki Y, Fukatsu-Sasaki K, Furutani N, Tada N. Neuroprotective effects of edaravone: a novel free radical scavenger in cerebrovascular injury. CNS Drug Rev. 2006;12(1):9–20.
92. Lee BJ, Egi Y, van Leyen K, Lo EH, Arai K. Edaravone, a free radical scavenger, protects components of the neurovascular unit against oxidative stress in vitro. Brain Res. 2010;1307:22–7.
93. Siesjo BK, Katsura K, Zhao Q, Folbergrova J, Pahlmark K, Siesjo P, et al. Mechanisms of secondary brain damage in global and focal ischemia: a speculative synthesis. J Neurotrauma. 1995;12(5):943–56.
94. Siesjo BK, Siesjo P. Mechanisms of secondary brain injury. Eur J Anaesthesiol. 1996;13(3):247–68.
95. FDA. FDA approves drug to treat ALS. 2017. https://www.fda.gov/NewsEvents/Newsroom/PressAnnouncements/ucm557102.htm?source=govdelivery.
96. Lapchak PA. A critical assessment of edaravone acute ischemic stroke efficacy trials: is edaravone an effective neuroprotective therapy? Expert Opin Pharmacother. 2010;11(10):1753–63.
97. Higashi Y. Edaravone for the treatment of acute cerebral infarction: role of endothelium-derived nitric oxide and oxidative stress. Expert Opin Pharmacother. 2009;10(2):323–31.
98. Kono H, Woods CG, Maki A, Connor HD, Mason RP, Rusyn I, et al. Electron spin resonance and spin trapping technique provide direct evidence that edaravone prevents acute ischemia-reperfusion injury of the liver by limiting free radical-mediated tissue damage. Free Radic Res. 2006;40(6):579–88.
99. Banno M, Mizuno T, Kato H, Zhang G, Kawanokuchi J, Wang J, et al. The radical scavenger edaravone prevents oxidative neurotoxicity induced by peroxynitrite and activated microglia. Neuropharmacology. 2005;48(2):283–90.
100. Shichinohe H, Kuroda S, Yasuda H, Ishikawa T, Iwai M, Horiuchi M, et al. Neuroprotective effects of the free radical scavenger Edaravone (MCI-186) in mice permanent focal brain ischemia. Brain Res. 2004;1029(2):200–6.
101. Skaper SD, Pollock M, Facci L. Mast cells differentially express and release active high molecular weight neurotrophins. Brain Res Mol Brain Res. 2001;97(2):177–85.
102. Skaper SD, Giusti P, Facci L. Microglia and mast cells: two tracks on the road to neuroinflammation. FASEB J. 2012;26(8):3103–17.
103. Seifert HA, Pennypacker KR. Molecular and cellular immune responses to ischemic brain injury. Transl Stroke Res. 2014;5(5):543–53.
104. Becker K. Autoimmune responses to brain following stroke. Transl Stroke Res. 2012;3(3):310–7.
105. Czlonkowska A, Korlak J. The immune response during aging. J Gerontol. 1979;34(1):9–14.
106. Elkins J, Veltkamp R, Montaner J, Johnston SC, Singhal AB, Becker K, et al. Safety and efficacy of natalizumab in patients with acute ischaemic stroke (ACTION): a randomised, placebo-controlled, double-blind phase 2 trial. Lancet Neurol. 2017;16(3):217–26.
107. Tatlisumak T. Can natalizumab be beneficial in acute ischaemic stroke? Lancet Neurol. 2017;16(3):176–7.
108. Simats A, Garcia-Berrocoso T, Montaner J. Natalizumab: a new therapy for acute ischemic stroke? Expert Rev Neurother. 2016;16(9):1013–21.
109. Hess DC, Wechsler LR, Clark WM, Savitz SI, Ford GA, Chiu D, et al. Safety and efficacy of multipotent adult progenitor cells in acute ischaemic stroke (MASTERS): a randomised, double-blind, placebo-controlled, phase 2 trial. Lancet Neurol. 2017;16(5):360–8.

110. Kalladka D, Sinden J, Pollock K, Haig C, McLean J, Smith W, et al. Human neural stem cells in patients with chronic ischaemic stroke (PISCES): a phase 1, first-in-man study. Lancet. 2016;388(10046):787–96.
111. Steinberg GK, Kondziolka D, Wechsler LR, Lunsford LD, Coburn ML, Billigen JB, et al. Clinical outcomes of transplanted modified bone marrow-derived mesenchymal stem cells in stroke: a phase 1/2a study. Stroke. 2016;47(7):1817–24.
112. Lapchak PA, Zhang JH, Noble-Haeusslein LJ. RIGOR guidelines: escalating STAIR and STEPS for effective translational research. Transl Stroke Res. 2013;4(3):279–85.
113. Butte P, Lapchak PA, Kittle DS. Time-resolved laser-induced fluorescence spectroscopy systems and uses thereof. 2016. USPTO Patent 9404870 B2.
114. Lapchak PA, Daley JT, Boitano PD. A blinded, randomized study of L-arginine in small clot embolized rabbits. Exp Neurol. 2015;266:143–6.
115. Lapchak PA. A cost-effective rabbit embolic stroke bioassay: insight into the development of acute ischemic stroke therapy. Transl Stroke Res. 2015;6(2):99–103.
116. Lapchak PA, Schubert DR, Maher PA. Delayed treatment with a novel neurotrophic compound reduces behavioral deficits in rabbit ischemic stroke. J Neurochem. 2011;116(1):122–31.
117. Lapchak PA. A clinically relevant rabbit embolic stroke model for acute ischemic stroke therapy development: mechanisms & targets. In: Lapchak PA, Zhang JH, editors. Translational stroke research: from target selection to clinical trials. New York, NY: Springer; 2011. p. 541–84.
118. Lapchak PA, Kirkeby A, Zivin JA, Sager TN. Therapeutic window for nonerythropoietic carbamylated-erythropoietin to improve motor function following multiple infarct ischemic strokes in New Zealand white rabbits. Brain Res. 2008;1238:208–14.

Chapter 2
Interdisciplinary Advances Towards Understanding and Enhancing the Therapeutic Potential of Stem Cell-Based Therapies for Ischaemic Stroke

Pascal Gervois, Yörg Dillen, Tim Vangansewinkel, Petra Hilkens, Ronald B. Driesen, Greet Merckx, Melissa Lo Monaco, Jessica Ratajczak, Annelies Bronckaers, Ivo Lambrichts, and Esther Wolfs

Abstract Worldwide, stroke is the second single most common cause of death and is a major cause of permanent disability. Moreover, the highest incidence of these pathologies is observed in the elderly, increasing the socioeconomic burden in an aging population. Current available therapies lead to insufficient functional improvement or are not applicable to all patients. This stresses the urgent need for alternative strategies in treating stroke patients, for example cell-based therapies. These cells showed great preclinical potential although the underlying therapeutic mechanisms, preferential route of administration and most suitable stem cell-subtype are unknown. Mechanisms of action include neuroprotection, cell replacement, neurogenesis, immunomodulation and the promotion of both neuroplasticity and angiogenesis in damaged central nervous system regions. Moreover, stem cells have been genetically engineered to enhance their beneficial effects after transplantation. Additionally, noninvasive imaging can be used to provide detailed spatial and functional information on the donor cell fate and the response of the host microenvironment. This chapter provides an overview of recent advances in (bio-)medical research using or manipulating stem cell-based therapies for ischaemic stroke with a focus on their neuroprotective, neuroregenerative and immunomodulatory properties. Additionally, the use of noninvasive imaging to allow temporospatial evaluation of stem cell fate following transplantation in animal stroke models will be discussed.

P. Gervois (✉) • Y. Dillen • T. Vangansewinkel • P. Hilkens • R.B. Driesen • G. Merckx
M.L. Monaco • J. Ratajczak • A. Bronckaers • I. Lambrichts • E. Wolfs
Morphology Research Group, Biomedical Research Institute, Hasselt University,
Agoralaan Building C, Diepenbeek 3590, Belgium
e-mail: pascal.gervois@uhasselt.be; yorg.dillen@uhasselt.be; tim.vangansewinkel@uhasselt.be; petra.hilkens@uhasselt.be; ronald.driesen@uhasselt.be; greet.merckx@uhasselt.be; melissa.lomonaco@uhasselt.be; jessica.ratajczak@uhasselt.be; annelies.bronckaers@uhasselt.be; ivo.lambrichts@uhasselt.be; esther.wolfs@uhasselt.be

P.A. Lapchak, J.H. Zhang (eds.), *Cellular and Molecular Approaches to Regeneration and Repair*, Springer Series in Translational Stroke Research,
https://doi.org/10.1007/978-3-319-66679-2_2

Keywords Stem cells • Ischaemic stroke • Noninvasive imaging • Gene therapy • Regenerative medicine • Immunomodulation

Abbreviation

AMPA	α-amino-3-hydroxy-5-methyl-4-isoxazole propionic acid
Ang-1	Angiopoietin-1
ASC	Adipose-derived stem cell
ATP	Adenosine triphosphate
BBB	Blood-brain barrier
BDNF	Brain-derived neurotrophic factor
bFGF	Basic fibroblast growth factor
BLI	Bioluminescence imaging
BM-MNC	Bone marrow-derived mononuclear cells
BMMSC	Bone marrow-derived MSC
CCR2	C-C chemokine receptor type 2
CT	Computed tomography
CXCR4	C-X-C chemokine receptor type 4
DAMPs	Danger associated molecular pattern molecules
DPSC	Dental pulp stem cell
EC	Endothelial cells
ECM	Extracellular matrix
EGF	Epidermal growth factor
ESC	Embryonic stem cell
EVs	Extracellular vesicles
FGF	Fibroblast growth factor
FLI	Fluorescence imaging
GDNF	Glial-derived neurotrophic factor
GFAP	Glial fibrillary acid protein
hESC	Human embryonic stem cell
HGF	Hepatocyte growth factor
ICAM-1	Intercellular Adhesion Molecule 1
IDO	Indoleamine 2,3-dioxygenase
IFN-γ	Interferon-gamma
IGF-1	Insulin-like growth factor 1
IL	Interleukin
iPSC	Induced pluripotent stem cell
MCAO	Middle cerebral artery occlusion
MCP-1	Monocyte chemotactic protein 1
MHC	Major histocompatibility complex
MLR	Mixed lymphocyte reaction
MMP	Matrix metalloproteinase

MRI	Magnetic resonance imaging
MSC	Mesenchymal stem cell
NF-κB	Nuclear factor kappa B
NGF	Nerve growth factor
NK cells	Natural killer cells
NMDA	N-methyl-D-aspartic acid
NO	Nitric Oxide
NSC	Neural stem cell
OGD	Oxygen-glucose deprivation
PDGF-BB	Platelet-derived growth factor BB
PET	Positron emission tomography
PGE2	Prostaglandin E2
ROS	Reactive oxygen species
SDF-1α	stromal cell-derived factor 1 α
SGZ	Subgranular zone
SPECT	Single-photon emission computed tomography
SPIO	Superparamagnetic iron oxide
STAT3	Signal transducer and activator of transcription 3
SVZ	Subventricular zone
TGF-β	Transforming growth factor beta
TIMP	Tissue inhibitor of metalloproteinase
TNF-α	Tumour necrosis factor alfa
Treg	Regulatory T cell
VEGF	Vascular endothelial growth factor

1 Introduction

The pathophysiology of stroke is defined as a neurologic dysfunction of vascular origin with the rapid occurrence of symptoms and signs corresponding to the involvement of focal areas in the brain [1]. Two different types of stroke can occur: ischaemic stroke (80–85%) and haemorrhagic stroke (15–20%). Ischaemic stroke is most frequently caused by thromboembolisms while haemorrhagic stroke most often results from vessel wall pathology associated with hypertension and microaneurysms [2]. This chapter will only focus on ischaemic stroke as the main pathology.

Worldwide, stroke is the second most common cause of death and is a major cause of permanent disability [3, 4]. Moreover, the highest incidence of these pathologies is observed in the elderly, increasing the socioeconomic burden in an ageing population [4]. In ischaemic stroke, the blood supply to certain areas of the brain is compromised which triggers a cascade of deleterious events ultimately leading to neuronal cell death [5]. This in turn triggers the acute immune response which can have a persistent and detrimental effect on stroke outcome [6]. The resulting severe neurological dysfunction is clinically translated into symptoms such as paralysis, sensory disturbances, aphasia, urinary incontinence and cognitive impairment. Limited stroke-induced endogenous neurogenesis can be observed in patients but adequate functional recovery is not achieved [7]. Recombinant tissue plasminogen activator is the only FDA-approved pharmacological treatment for stroke but comes with many restrictions. Administration should be started within a time window of 4.5 h post-ischaemia, limiting its use to merely 2–4% of the patients and leading to an insufficient functional improvement [8]. These indications highlight the urgent need for alternative strategies in treating stroke patients.

Stem cell therapy is a promising approach to minimize neurological damage and enhance functional recovery after stroke. Preclinical studies in animal stroke models using for instance neural stem cells (NSC) [9], mesenchymal stem cells (MSC), induced pluripotent stem cell (iPSC)-derived cells delivered encouraging results (See Table I and Table II in [10]). However, the optimal stem cell-source, mechanisms of action, cell fate and optimal treatment protocol remain to be elucidated. Mechanisms of action include neuroprotection, cell replacement, immunomodulation and the promotion of both neuroplasticity and angiogenesis in damaged central nervous system regions [10]. Moreover, stem cells that were genetically engineered to overproduce growth factors such as brain-derived neurotrophic factor (BDNF), nerve growth factor (NGF), Noggin and angiopoietin-1 (Ang-1) have been previously shown to enhance post-stroke recovery [11–13]. Additionally, noninvasive imaging can be used to provide detailed spatial and functional information on the donor cell fate and the response of the host microenvironment following cell-transplantation into animal stroke models [10].

This chapter aims to provide an up to date overview of current interdisciplinary advances in preclinical stroke research, focussing on neuroregeneration, neuroprotection and immunomodulation supported by noninvasive imaging opportunities.

2 Therapeutic Approaches and Evaluation of the Post-stroke Microenvironment

The multiple mechanisms that have been proposed for stem cell-mediated therapies include brain protection, cell replacement, immunomodulation and promoting both brain plasticity and angiogenesis in damaged brain regions (Fig. 2.1) [10]. Interestingly, these mechanisms are mainly thought to be mediated by the paracrine effect of the transplanted cells on endogenous stem cells and on the host microenvironment instead of directly replacing the lost cells, although encouraging results have also been achieved with cell-replacement studies. Therefore, the engrafted cells can be seen as a vehicle for persistent growth factor delivery at the stroke lesion which can also respond dynamically to changes in the local microenvironment. In addition, the transplanted cells directly or indirectly influence extracellular matrix (ECM) remodelling and glial scar formation.

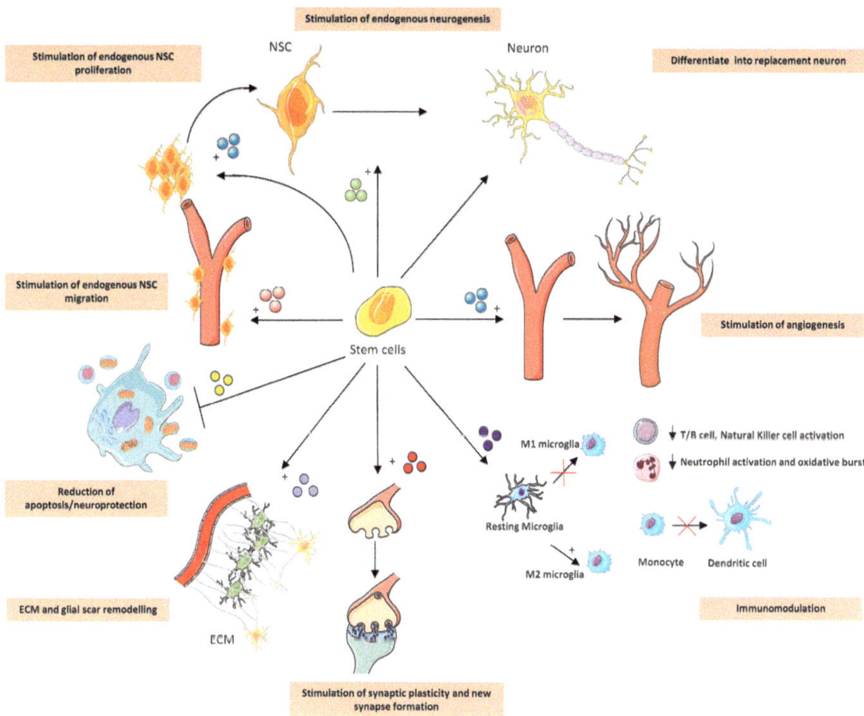

Fig. 2.1 Mechanisms of action of cell-based therapies in ischaemic stroke. The poststroke microenvironment can be influenced by exogenously delivered stem cells by multiple mechanisms to trigger tissue repair. Stem cells contribute to poststroke recovery by stimulating endogenous NSC migration toward the stroke lesion, where proliferation and differentiation toward replacement neurons can be triggered. Additionally, transplanted stem cells are thought to be able to replace the lost neurons themselves. Moreover, the formation and attraction of novel blood vessels toward the ischaemic lesion and the stimulation of synaptogenesis and synaptoplasticity contribute to brain repair. In addition to directly stimulating the formation of new brain tissue, the degradation of resident cells is inhibited by neuroprotective mechanisms and transplanted cells can influence the extent of glial scar formation. Immunomodulatory effects are also observed and include the inhibition of neutrophil activation and migration, effector T-cell and B-cell inhibition, reducing the activation and attraction of peripheral dendritic cells, and stimulating the M2 microglial phenotype. These effects are predominantly caused by the soluble factors released by the stem cells, but also cell–cell interactions appear to play a role. Image was adapted from [10] with permission

2.1 *The Neuroprotective Effect of Stem Cell-Based Therapies*

The ideal therapeutic approach for stroke would be to prevent neuronal cell death induced by the ischaemic insult, thereby minimizing neurological damage and stroke severity. Any strategy that aims to inhibit or antagonize the pathophysiological cascade of biochemical events resulting in irreversible cell damage and neuronal cell death is considered a neuroprotective approach [14].

2.1.1 The Complexity of the Ischaemic Cascade

Neuronal cells located in the ischaemic core die within minutes after stroke onset, whereas peripheral cells residing in the penumbra provided with limited collateral blood flow become dysfunctional but do not undergo acute cell death. However, if left untreated, the neuronal cells in the penumbra are likely to progress into delayed neuronal cell death hours to days after the ischaemic insult [15]. Therefore, a restricted time window exists wherein reversibly damaged neuronal cells can be salvaged from cell death in order to limit infarct size and improve functional outcome after stroke.

In the ischaemic core and penumbra, a series of neurochemical events occur, described as the ischaemic cascade [16]. The brain depends on oxygen and glucose to assure normal neuronal cell function and maintain ionic homeostasis. Impairment of cerebral blood flow causes disturbances in these vital energetic processes [5]. Ischaemia leads to dysfunction of ATP-dependent ion pumps, including the Na^+/K^+ pump, which results in alterations in the membrane potential and depolarisation of neurons and glial cells [17]. Subsequently, voltage-dependent Ca^{2+} channels are activated and excitatory neurotransmitters, including glutamate, are released into the extracellular space. This accumulation of glutamate in the extracellular space leads to the stimulation of α-amino-3-hydroxy-5-methyl-4-isoxazole propionic acid (AMPA) and N-methyl-D-aspartic acid (NMDA)-type glutamate receptors on adjacent neurons. Consequently, these neurons become depolarized, which results in additional Ca^{2+} influx and glutamate release, causing an exacerbation of the initial ischaemic insult. Due to the ionic imbalance, intracellular water accumulation occurs, which is responsible for the formation of cytotoxic oedema [5, 16]. Furthermore, the increase in intracellular Ca^{2+} leads to the activation of Ca^{2+}-dependent enzymes, including proteases, endonucleases, phospholipases and cyclooxygenases. These enzymes cause extensive cell damage and are partially responsible for the generation of reactive oxygen species (ROS). ROS are important mediators of cellular damage by inducing DNA damage, lipid peroxidation and protein denaturation, which ultimately results in mitochondrial failure and membrane disruption [5, 16, 17]. The outcome of these detrimental molecular events is cell death via necrosis or apoptosis, depending on the degree of ischaemic injury [17, 18].

2.1.2 Stem Cell-Mediated Neuroprotection

Numerous neuroprotective agents which target different components of the pathophysiological cascade have been investigated in the context of stroke (reviewed by Moretti et al. [19]). These molecular targets comprise various mechanisms of calcium influx, glutamate excitotoxicity, ROS scavenging, NO metabolism and apoptosis. Despite promising preclinical results, no neuroprotective agent has passed clinical trials in stroke patients. Alternatively, approaches like therapeutic hypothermia and decompressive craniectomy have been explored to minimize neurological damage. Both these approaches seem to result in mild improvements in neurological scores in stroke patients. However, these patients still have a poor functional outcome after 6 months [20, 21].

Stem cell-based therapy is considered a promising treatment strategy to minimize neurological damage and enhance functional recovery following stroke. By secreting neurotrophic and anti-apoptotic factors, stem cells can provide support to reversibly damaged neurons present in the penumbra. In this way, the stem cells can exert neuroprotective effects and rescue neurons which otherwise would progress into delayed cell death, thereby preventing additional neuronal damage [22, 23]. It has been demonstrated that MSC, including bone-marrow-derived MSC (BMMSC) and dental pulp stem cells (DPSC), secrete a plethora of paracrine factors comprising BDNF, NGF, neurotrophin-3 (NT-3) and glial cell-derived neurotrophic factor (GDNF), which are considered hallmark neurotrophic and neuroregulatory factors [24–27].

Neuroprotective effects mediated by stem cells have been observed in numerous *in vitro* studies performed by independent research groups. It has been shown that MSC protect neuroblastoma cells against hypoxia and glutamate excitotoxicity [28–30]. Furthermore, MSC are able to rescue primary cortical neurons which are exposed to oxygen-glucose deprivation and trophic factor withdrawal [25, 31]. Additionally, MSC can also prevent cell death in ischaemic human astrocytes [32]. Extensive evidence of neuroprotective effects mediated by MSC has also been provided in *in vivo* stroke models. Injection of the conditioned medium of stem cells as well as the stem cells as such have proven to exert neuroprotective effects in experimental stroke models [28, 33–37]. The paracrine factors secreted by the stem cells are believed to be responsible for this neuroprotective effect, since no or very limited integration of the stem cells in the lesion site is observed. These studies comprised the use of MSC derived from different sources, including adipose-derived stem cells, umbilical tissue-derived stem cells, BMMSC and DPSC. Furthermore, different routes of administration, cell dosages and time points of transplantation have been used in these studies.

2.2 *Regeneration of Endogenous Tissue by Stem Cells After Ischaemic Stroke*

In the 1990–2000s, MSC therapy gained a lot of attention to be used as therapy after stroke because of their assumed ability to transdifferentiate into neuronal cells, endothelial cells and glial cells. Numerous *in vitro* experiments demonstrating this so-called transdifferentiation potential sparked hope to use these stem cells for neuronal tissue replacement. However, these properties are merely induced under certain artificial cell culture conditions, not representing their endogenous properties. Another disadvantage is that this processes turned out to be time-consuming and very inefficient. Despite of this, transplantation of undifferentiated MSC has been explored with the rationale that they would locally replace the neural tissue. Intracranial transplantation of MSC showed that the cells migrated towards the infarct region, survived in the host brain and stimulated functional recovery [38]. However as already mentioned above, numerous studies indicated that only a small percentage of the transplanted MSC survived and locally differentiated preferentially towards astroglial cells instead of the desired neurons [10, 34, 39]. It is generally accepted that functional replacement of the lost neurons is not the main mode of action of MSC.

Embryonic stem cells (ESC) and iPSC have the tremendous potential to differentiate in all possible cells from the nervous system including glial cells and neurons. However, achieving mature neurons from these pluripotent stem cells is even *in vitro* very time-consuming, costly and inefficient. Also their possible tumorigenicity when used in undifferentiated state is a major disadvantage [40, 41]. Therefore, iPSC are irreversibly pre-differentiated *in vitro* in order to minimize tumour formation and improve the functional outcome as only the undifferentiated iPSC form teratomas. The true potential of iPSC and ESC in stroke research is to apply them *in vitro* as a patient-in-a-dish model, trying to understand disease mechanisms and for drug discovery [42]. Especially the novel method culturing cerebral organoids resembling the 3D brain structure is a big step forward closing the gap between 2D cell cultures and animal models [43, 44].

When it comes to the use of stem cells for neuronal tissue regeneration as well as for neuroprotection, the current focus lays on the angiogenic and neurotrophic factors, cytokines and chemokines that are able to enhance the endogenous repair mechanisms after cerebral ischaemia, i.e. vascular remodelling, activation of endogenous neuroregeneration and remodelling of the extracellular matrix.

2.2.1 Vascular Remodelling After Stroke

Cerebral ischaemia leads to increased vascular remodelling in both the acute and chronic phase [45]. During acute blood flow obstruction, arteriogenesis also referred to as collateralisation can occur. This development of a functional blood flow from pre-existing arterial anastomoses is induced by mechanical forces such as shear stress and is independent of a hypoxia state [46]. By contrast, angiogenesis, the development of new capillaries sprouting from existing small blood vessels, is a key endogenous process induced by chronic hypoxia. The angiogenic process is a complex cascade of events, involving breakdown of the extracellular matrix, activation of endothelial cells (EC), followed by the proliferation, migration of EC. In a final step, pericytes are recruited towards the formed tubular network of EC resulting in mature blood vessels. A multitude of angiogenic factors and signalling molecules such as vascular endothelial cell growth factor (VEGF), fibroblast growth factor (FGF), angiopoietin-1 (Ang-1), Platelet-derived growth factor BB (PDGF-BB), nitric oxide (NO) have to co-operate in concert with spatiotemporal precision [47, 48]. Although distinct triggering mechanisms induce either collateralisation or angiogenesis, similar growth factors, cytokines and signalling molecules are shared by both modes of vascular remodelling [45]. A lot of angiogenic therapies applying recombinant proteins gained disappointing results in clinical trials [49, 50]. This can be partly explained by the fact that the most studied angiogenic factor VEGF, is linked to the generation of immature and unstable vessels leading to oedema and vessel regression over time, aggravating stroke progression. In addition, the majority of angiogenic therapy regimens only involved administration of a single angiogenic protein. In that respect, MSC which have been showed to secrete numerous cytokines and angiogenic factors and as a surplus can act as a pericyte, can create the right angiogenic microenvironment [51]. Furthermore, numerous animal stroke studies support the pro-angiogenic properties of MSC obtained from different tissues and that the reported improved recovery is associated with increased blood vessel density. For example, increased VEGF and VEGFR2 expression was observed after intravenous injection of human BMMSC in a rat model of ischaemic stroke [52]. Another research group reported elevation of Ang-1 and Ang-2 mRNA levels in BMMSC treated rats [53]. Wakabayashi et al. showed that intravenous injection human MSC in a rat middle cerebral artery occlusion (MCAO) model induced functional improvement and reduced infarct volume by producing angiogenic factors. Moreover, MSC locally secreted IGF-1 in the ischaemic core and interestingly, this IGF-1 production was only detected *in vivo*, suggesting its specific induction by the ischaemic environment. In addition, the transplanted MSC affected the host cells, as endogenous VEGF, EGF, and bFGF levels were significantly elevated in stem cell-treated rats 7 days after injury [54]. Not only the stem cells themselves have a significant therapeutic potential in ischaemic stroke but also their extracts were shown to have a therapeutic effect. Intraperitoneal injection of a cell-free extract derived from MSC was shown to dramatically decrease the ischaemic volume and improve motor function after stroke [55].

2.2.2 Endogenous Regeneration After Stroke

For a long time, cells of the adult central nervous system were considered to be incapable of regeneration. However, it was demonstrated that human adult NSC reside in the dentate gyrus of the hippocampus and in the subventricular zone (SVZ) [56, 57]. Under normal physiologic conditions, adult NSC predominantly produce neurons, interneurons of the olfactory bulb for SVZ-derived cells, and dentate granule cell neurons for SGZ-derived cells. Ischaemic stroke enhances proliferation of the SVZ cell population and these cells migrate towards the lesion and differentiate into mature striatal neurons and replace damaged neurons. SDF-1α/CXCR4 and MCP-1/CCR2 receptor signalling has been shown to regulate the directed migration of these NSC to the injured area [58, 59]. NSC themselves have been shown to produce matrix metalloproteinase (MMP)-3 and MMP-9 in response to these extrinsic signals. Blocking the expression of MMP-3 or MMP-9 in NSC interferes with both their differentiation and migration [60], suggesting a prominent role of these MMPs in the endogenous NSC response. The leading fraction of the migrating NSC is closely associated with blood vessels, suggesting that this interaction provides a scaffold to direct the NSC towards the damaged brain region [61]. However, the amount of endogenous cells generated is considered to be too low to have a significant impact on functional recovery after stroke. Nevertheless, various preclinical trials have been performed to enhance neurogenesis after stroke using EGF, VEGF, erythropoietin and statins [62, 63]. These investigations have also formed the basis to investigate the ability of cell transplantation to activate NSC and to mediate their differentiation towards neurons.

Neurotrophins/growth factors found in the MSC secretome include GDNF, NT-3, NGF and BDNF [10, 24, 30, 64]. To our knowledge, the effect of MSC on NSC migration and/or differentiation has not yet been tested *in vitro*. CM of DPSC, BMMSC and Wharton Jelly MSC have shown to enhance neuronal maturation of a pre-differentiated neuroblastoma cell line SH-SY5Y cells [65, 66]. In addition, the MSC secretome has been demonstrated to enhance neurite outgrowth in various types of primary neurons including, dopaminergic, primary cortical neurons and neurons derived from dorsal root and the retinal ganglia [10, 24, 30, 64]. In contrast to the overwhelming preclinical evidence on the induction of angiogenesis by MSC transplantation, only few reports on activation of endogenous NSC proliferation, migration and maturation are available. Munoz et al. showed that hMSC injected stereotactically induced migration of BrdUrd-labeled endogenous cells throughout the dorsal hippocampus, which were doublecortin-positive, and expressed markers for astrocytes as well as for neural or oligodendrocyte progenitors 7 days after treatment. In addition, 30 days after implantation, the newly generated NSC expressed markers for more mature neurons and astrocytes [67]. Another study in a rat stroke model, demonstrated that human MSC transplanted intracranially induce proliferation of endogenous NSC and subsequent migration as shown by double staining of BrdU and doublecortin at 1 and 2 weeks after MCAO induction [63]. Recent work in the setting of traumatic brain injury (TBI) showed that transplanted exogenous MSC are able to guide the migration of endogenous cells from the neurogenic site to the area of injury in the cortex via the formation of a 'biobridge' between the neurogenic and ischaemic site. This biobridge, visualized immunohistochemically and laser captured, corresponded to an area between the neurogenic SVZ and the injured cortex and consists of an altered endogenous expression of MMPs and ECM [68]. Despite the fact that MSC transplantations have been shown to induce both proliferation and differentiation of SVZ-derived NSC, neuronal differentiation rates were very low. As a consequence, there is a controversy on the fact whether or not MSC-induced enhancement of endogenous neurogenesis significantly contributes to an enhanced post stroke recovery [69, 70].

2.2.3 ECM and Scar Tissue Remodelling

After ischaemia, gliosis also referred to as scar formation is strongly induced at the infarct boundary. Damaged neurons initially interact with the adjacent astrocytes, which become activated and show increased expression GFAP, musashi-1 and secrete inflammatory cytokines [71, 72]. These triggered astrocytes in turn rapidly surround the infarct with fibrils [73]. The possible role of this demarcation consisting of activated astrocytes and ECM, but also microglia and oligodendrocytes is to separate the necrotic tissue from viable brain and avoid further spreading of damage. Furthermore, this seal has also shown to play a role in maintenance of ion and fluid balance, preventing further inflammation, free radical scavenging and increasing tropic and metabolic support of the nerve tissue and for blood vessel ingrowth. On the other hand, it has a devastating effect on functional recovery as it impedes axonal regeneration [71, 72]. The ECM compound represents a physical barrier for new regenerating axons to cross. In addition, the reactive astrocytes secrete growth-inhibitory molecules such as Nogo [74]. According to this rationale, therapies that are able to reduce gliosis would thus be beneficial and enhance stroke recovery [72]. Several studies indicate that MSC secrete MMPs that cleave the ECM and would play a role in scar tissue destruction. BMMSC were able to produce active MMP-2, MMP-3 and also membrane-bound MT1-MMP [75, 76]. MSC are also able to activate exogenous proMMP-2 and proMMP-13. Interestingly, a recent study showed that the majority of the MMP activity is associated with the MSC cell surface while they secrete high levels of TIMPs, which strongly inhibits soluble MMPs. Since they bind and activate MMPs at their surfaces, the net result is a very controlled pericellular localization of MMP activities by MSC [77]. However, in the context of stroke, the contribution of this MMP production and the beneficial effects of MSC treatment remains to be elucidated. Only a few reports are available that studied the effect of stem cells on scar tissue formation. MSC treatment reduced the thickness of the scar wall and reduced the number of microglia/macrophages within the scar wall 4 months after surgery in a rat MCAO stroke model [71]. The same research group reported that long-term follow up (more than 1 year) of BMMSC injected in the carotid artery 1 day after MCAO significantly reduced thickness of the lesion scar wall and the number of Nogo-positive cells [78]. The exact molecular mechanisms behind this reduction of the scar wall thickness remain to be elucidated.

Although several studies emphasize that reactive astrocytes after CNS injury induce glial scar formation, which inhibits axon regeneration and impedes functional recovery, others indicated a neuroprotective role of astrocytes in CNS injury [72]. In that respect, it is worth to study the impact of stem cells on astrocyte survival. Indeed, MSCs suppress astrocyte apoptosis induced by OGD *in vitro*, an effect that has been attributed through the MSC-induced activation of IL-6 signalling in injured astrocytes [32, 79]. In addition, Song et al. showed that both BMMSCs as wells as DPSCs attenuated OGD-induced GFAP, nestin, and musashi-1 expression and inhibited OGD-induced ROS and interleukin-1β production in activated astrocytes *in vitro* [32].

2.3 Immunomodulatory Properties of Candidate Stem Cell-Based Therapies for Ischaemic Stroke

Whereas the neuroregenerative and neuroprotective effect of transplanted stem cells on the stroke-affected microenvironment has been studied thoroughly as described in the previous sections, the effect of the transplanted cells on the immune system and the infiltrating immune cells remains to be fully characterized.

2.3.1 Introduction to Stroke Immunology

The immune system and inflammation play a key role in the pathophysiology of stroke and can greatly influence stroke outcome [80]. Moreover, as a response to the ischaemic insult, the brain exerts a suppressive effect on the systemic immune system which leads to systemic lymphocytopenia [81]. This makes patients more susceptible to infections and is a major cause of stroke-associated morbidity and mortality [82, 83].

The various elements of the immune system are involved in all stages of ischaemia-induced brain loss. Early vascular events after arterial occlusion initiate inflammation where hypoxia, the production of ROS and changes in blood flow trigger the coagulation cascade, blood platelets and complement [84–86]. These events are followed by the upregulation of adhesion molecules on the platelet—and EC surface such as P-selectin, E-selectin and intercellular adhesion molecule 1 (ICAM-1) [87]. Moreover, the production of pro-inflammatory signals/cytokines is increased as well as the production of the vasodilator nitric oxide (NO) [86, 87]. Ultimately, EC junctions are weakened which allows protein and cellular extravasation into the perivascular space where mast cells and macrophages are activated and secrete proteases and pro-inflammatory mediators leading to blood-brain barrier (BBB) damage and leukocyte infiltration [86, 88].

In the subsequent phase of ischaemic cell death, the dying neuronal cells send out danger signals that activate the immune system [89]. These so-called danger associated molecular pattern molecules (DAMPs) include extracellular ATP or other nucleotides [90], heat-shock proteins, ECM breakdown proteins [91] and the high mobility group box 1 protein (HMGB1) [92] which are released from dying brain tissue following stroke [89, 93]. These DAMPs activate ionotropic purine receptors and scavenger—or pattern recognition receptors on inflammatory cells, leading to the production of pro-inflammatory mediators by resident brain cells and infiltrating leukocytes (for in-depth DAMP signalling, see review Gelderblom et al. [93]).

Stroke-induced inflammation eventually diminishes and triggers several pathways needed for the repair and reorganisation of the injured brain. This switch from a tissue-damaging pro-inflammatory stroke microenvironment to an anti-inflammatory, repair-stimulating environment remains poorly understood, but is coordinated by an intertwined cascade of inter- and intracellular signalling [86]. In this transition, macrophages and microglia switch from a pro-inflammatory M1 phenotype to an M2 phenotype that stimulates repair processes and attenuates the inflammatory response [94, 95]. Dead cells and debris attract and activate infiltrating macrophages and microglia which subsequently phagocytize the lost tissue. Phagocytosis induces the production of cytokines such as transforming growth factor beta (TGF-β), IGF-1 and IL-10 which were shown to have a neuroprotective and/or an anti-inflammatory effect [96–98].

In addition to the innate immune system, the adaptive immune system was also shown to contribute to inflammation-induced neuronal damage. DAMPs from damaged cells can also function as antigens that are presented to cells of the adaptive immune system, leading to immunity against these antigens [99]. Although the damage to the post-ischaemic brain does not appear to be caused by an autoimmune response, the observed injury also does not fit the profile of classical adaptive immunity due to the temporal profile of the cellular infiltrate [86]. Blocking postischaemic trafficking of T cells 24–48 h after ischaemia provides a neuroprotective effect [100], whereas the classical adaptive immune response takes up to 1 week to develop and damage the ischaemic tissue. Interestingly, B cells do not significantly contribute to brain injury [101] and T-cell mediated damage is associated with γδ T cells which release the pro-inflammatory cytokine interleukin-17 (IL-17) [102, 103] and blocking the IL-17 signalling axis was shown to decrease neutrophil infiltration and ameliorate stroke outcome [103]. The contribution of natural killer (NK) cells and NK T cells to stroke injury remains to be elucidated [100].

This brief introduction in post-ischaemic inflammation provides several targets for stem cell-directed therapies for immunomodulation. Starting at the early onset, for example ROS scavenging can reduce the initial ischaemia-reperfusion injury, whereas stem cell-mediated therapies can also influence stroke outcome by modulating other aspects of the inflammatory cascade, as will be discussed next.

2.3.2 Mechanisms of Stem Cell-Mediated Immunomodulation

Although stem cell survival can be influenced by the host immune system, the transplanted cells themselves are believed to possess immunomodulatory properties [104, 105]. When considering immunomodulation as a stem cell-based therapy for ischaemic stroke, the post-stroke systemic immunosuppression needs to be taken into account. Additional systemic immunosuppression by cell-based therapies could worsen stroke outcome. Fortunately, no adverse effects on systemic cytokine levels were observed following syngeneic BMMSC transplantation in a mouse model of ischaemic stroke [105]. Moreover, the majority of completed clinical pilot studies with autologous MSC showed no adverse effects and improved clinical outcome, although post-stroke immunosuppression was not investigated [106, 107].

When considering mechanisms of stem cell-mediated immunomodulation, several *in vitro* reports are available. BMMSC and their extracellular vesicles (EVs) were shown to suppress T-cell proliferation [108, 109], BMMSC and adipose-derived stem cells (ASC) suppressed lymphocyte proliferation and the mixed lymphocyte reaction (MLR) [110]. Similarly, DPSC possess immunomodulatory properties [111]. Demircan et al. showed that the suppressive actions on T cells were mediated via paracrine effects by means of a transwell and MLR assay [112]. Increased levels of hepatocyte growth factor (HGF), TGF-β, ICAM-1, IL-6, IL-10, VEGF and human leukocyte antigen-G were found in DPSC/T cell co-cultures [112], the latter factor was additionally shown to suppress T cell and NK cell function and induce regulatory T cell function when secreted by BMMSC [113]. Moreover, the expression of pro-inflammatory cytokines by T-cells such as interferon-gamma (IFN-γ), IL-2, IL-12, IL-17A and tumour necrosis factor alpha (TNF-α) were decreased in the transwell system whereas the expression of the anti-inflammatory cytokine inducible protein-10 was upregulated [112]. Interestingly, the expression of the regulatory T cell (Treg) markers CD4, CD25 and Foxp3 was increased. T cell apoptosis was increased after 24 h incubation with DPSC [112]. A similar paracrine mediated immunosuppression was exerted by the secretome of porcine and human DPSC [114–116]. In addition to T cells, BMMSC were shown to inhibit NK cell activation by producing prostaglandin E2 (PGE2) and indoleamine 2,3-dioxygenase (IDO) [117]. The proliferation, activation, maturation and antigen presentation of dendritic cells was also inhibited by MSC subtypes [118–122] and macrophage/microglia polarization was shifted towards an M2 phenotype after exposure to MSC, their secretome or EVs [120–125]. This effect was presumably mediated by PGE2 [124] and by inhibiting the nuclear factor kappa B (NF-κB) pathway and stimulating the signal transducer and activator of transcription 3 (STAT3) pathway [126]. These reports include both paracrine-mediated immunomodulation [124–126] as direct co-cultures [120, 121, 123, 124, 126]. M2 polarized macrophages increased the production of the anti-inflammatory cytokines IL-10 and IL-6 and of arginase-1 while production of TNF-α was decreased [120, 123, 126]. Interestingly, pro-inflammatory cytokines were shown to stimulate MSC to secrete PGE2 and IDO and upregulate the expression of cell adhesion molecules, thereby stimulating the immunomodulatory capacity of MSC [127–129].

In addition to MSC, iPSC were also shown to modulate the immune system. Naïve major histocompatibility complex (MHC)-matched and/or -mismatched iPSC were reported to have superior immunomodulatory properties compared to MHC-matched and/or -mismatched MSC [130]. iPSC- or ESC-derived MSC acquired similar immunomodulatory properties as adult MSC and were able to inhibit lymphocyte proliferation and function [131–133] and NK cell function [132]. Soluble factors secreted by bone marrow-derived mononuclear cells (BM-MNC) were able to prevent macrophage-microglia induced neuronal cell death and ROS induced neurotoxicity [134]. Conversely, NSC were shown to upregulate arginase-1 production when exposed to inflammatory cytokines [135]. Whereas most *in vitro* proof for the immunomodulatory properties of stem cell-based therapies comes from MSC-centred research, proof for NSC-mediated immunomodulation mostly comes from *in vivo* data [136] and is mainly thought to be an effect of the transplanted cells on peripheral immunosuppression after intravenous administration [137, 138].

Studies that specifically focused on stem cell-based immunomodulation in stroke after transplantation *in vivo* are scarce. Similar to NSC, systemically injected BMMSC home towards the spleen, where TNF-α production is diminished. Moreover, the percentage of MHC-II-activated immune cells in the brain is reduced [139]. Intracranial ASC administration decreased the number of Iba-1$^+$ cells in the brain [140] and IL-10 production was increased after stem cell transplantation in a monkey stroke model [141]. Interestingly, intravenous delivered BMMSC-derived EVs had a similar functional outcome than BMMSC transplantation and attenuated postischaemic immunosuppression in the peripheral blood without altering the number of brain-infiltrating immune cells [142].

Although several encouraging results have been achieved and underlying mechanisms of stem cell-mediated immunomodulation were elucidated *in vitro*, the systemic and local effect of stem cells or stem cell-derived therapies such as EVs on immunomodulation *in vivo* remains elusive. In accordance with previous sections, noninvasive imaging modalities can be applied to acquire insight in underlying immune mechanisms in ischaemic stroke.

3 Noninvasive Monitoring of Stem Cells in the Stroke Microenvironment

For many years, medical imaging focused on the anatomical changes taking place following ischaemic stroke. During the last decade, however, there has been a major shift in medical imaging towards the molecular processes underlying these anatomical changes. A large number of noninvasive imaging methods have now been developed to study molecular processes such as metabolic changes, gene expression and cell migration. Visualizing these processes can be of great benefit in the diagnosis and treatment follow-up in ischaemic stroke.

These noninvasive imaging methods comprise magnetic resonance imaging (MRI), computed tomography (CT), positron emission tomography (PET), single-photon emission computed tomography (SPECT) and optical methods such as bioluminescence imaging (BLI) and fluorescence imaging (FLI).

In order to noninvasively track stem cell therapy *in vivo*, cells need to be labelled with a specific imaging probe. This can be done by means of incubation of the cells *in vitro* with contrast agents or radioactive tracer molecules prior to injection, which is referred to as direct cell labelling. Contrast dilution and leakage of these agents from the cells hampers long-term imaging, which has led to the development of indirect cell labelling methods. Hence, so-called "imaging reporter genes" are introduced into the cells and their expression enables the accumulation of imaging probes on a cellular level. This enables repeated stem cell visualization *in vivo* over time within the same subject [143].

Imaging stem cells in the field of ischaemic stroke research is mainly focused on determining the optimal injection route, cell dosing, engraftment, survival and effect on the lesion volume. ESC-derived NSC were imaged with BLI and MRI after labelling with superparamagnetic iron oxide (SPIO) particles, and migration as well as differentiation of the NSC towards neural lineages was confirmed [144]. MRI has also been used in a clinical setting of stem cell transplantation. For example, Zhu et al. have tracked autologous SPIO-labelled NSC transplanted in a patient with brain trauma. Cells migrated towards the lesion site but the signal disappeared 7 weeks after the transplantation [145].

ESC-derived NSC were genetically engineered to express the herpes simplex virus type 1 thymidine kinase (HSV1-tk) reporter gene for PET and labelled with SPIO for MRI. 3 months after stroke, PET and MRI showed a decrease infarct size and functional engraftment of the transplanted cells [146].

Intra-arterial injection of stem cells for the treatment of ischaemic stroke seems to be a favourable injection route. MRI tracking of ASC in a rat MCAO model has shown that the neuroprotective effect might be due to the secretion of trophic factors. Intra-arterially transplanted cells actively migrated towards the lesion sites, but only a low number of cells survived 8 weeks post transplantation [147]. Human umbilical cord blood-derived stem cells were labelled for MRI and injected intra-arterially 60 min after stroke. Researchers found an improved cerebrovascular function, a reduced infarct size and improvement in behavioural deficits [148]. Furthermore, Grudzenski et al. have shown using MRI that not only the injection route, but also the number of cells injected is important in the treatment of ischaemic stroke [149].

As all imaging modalities have their specific strengths and weaknesses, modern molecular imaging often combines several modalities. Multimodal imaging of MSC transplanted in rats with ischaemic stroke has combined MRI together with SPECT and FLI, using one single tri-modal probe (^{125}I-fSiO4@SPIOs). MSC transplanted intracerebrally or intravenously both improved neurobehavioral outcomes of these stroke animals [150].

4 Conclusion

Multiple advances have been made to understand stem cell-mediated mechanisms of brain regeneration following ischaemic stroke. Stem cell-based therapies applied via various administration routes have shown great promise in *in vitro* and *in vivo* models of ischaemic stroke focussing on a plethora of regenerative mechanisms including neuroregeneration, ECM and vascular remodelling and/or angiogenesis, stimulating endogenous repair and immunomodulation. However, these studies were not able to pierce the veil and pinpoint precise mechanisms of action of the transplanted cells. Nonetheless, harmonized stroke- and stem cell research will continue to contribute to the discovery of new targets and modulable pathways potential therapeutic approaches could be directed at. Moreover, noninvasive imaging methods allow changes in host microenvironment caused by the transplanted cells or cell-derived therapies to be connected with functional improvement.

Acknowledgements Pascal Gervois, Esther Wolfs, Jessica Ratajczak, Tim Vangansewinkel, Petra Hilkens, Yörg Dillen and Annelies Bronckaers are funded by Fonds Wetenschappelijk Onderzoek Vlaanderen by grants 12U7718N, G0A7514N, G089213N, G029112N, 12D8516N, 1134717N and 1508015N respectively. Greet Merckx and Melissa Lo Monaco are funded by Bijzonder Onderzoeksfonds grants BOF16DOC06 and BOF16DOCNA02 respectively.

Conflict of Interest: The authors declare that they have no conflict of interest.

References

1. Goldstein M, Barnett HJM, Orgogozo JM, Sartorius N, Symon L, Vereshchagin NV. Stroke—1989: Recommendations on stroke prevention, diagnosis, and therapy. Report of the WHO Task Force on Stroke and other Cerebrovascular Disorders. Stroke. 1989;20(10):1407–31.

2. Donnan GA, Fisher M, Macleod M, Davis SM. Stroke. Lancet. 2008;371(9624):1612–23. PubMed PMID: 18468545.
3. Mozaffarian D, Benjamin EJ, Go AS, Arnett DK, Blaha MJ, Cushman M, et al. Heart disease and stroke statistics-2016 update: a report from the American Heart Association. Circulation. 2016;133(4):e38–360. PubMed PMID: 26673558.
4. Feigin VL, Forouzanfar MH, Krishnamurthi R, Mensah GA, Connor M, Bennett DA, et al. Global and regional burden of stroke during 1990-2010: findings from the Global Burden of Disease Study 2010. Lancet. 2014;383(9913):245–54. PubMed PMID: 24449944. Pubmed Central PMCID: 4181600.
5. Dirnagl U, Iadecola C, Moskowitz MA. Pathobiology of ischaemic stroke: an integrated view. Trends Neurosci. 1999;22(9):391–7. PubMed PMID: 10441299.
6. Chamorro A, Meisel A, Planas AM, Urra X, van de Beek D, Veltkamp R. The immunology of acute stroke. Nat Rev Neurol. 2012;8(7):401–10. PubMed PMID: 22664787.
7. Jin K, Wang X, Xie L, Mao XO, Zhu W, Wang Y, et al. Evidence for stroke-induced neurogenesis in the human brain. Proc Natl Acad Sci U S A. 2006;103(35):13198–202. PubMed PMID: 16924107. Pubmed Central PMCID: 1559776.
8. Del Zoppo GJ, Saver JL, Jauch EC, Adams HP Jr. American Heart Association Stroke C. Expansion of the time window for treatment of acute ischemic stroke with intravenous tissue plasminogen activator: a science advisory from the American Heart Association/American Stroke Association. Stroke. 2009;40(8):2945–8. PubMed PMID: 19478221. Pubmed Central PMCID: 2782817.
9. Barkho BZ, Zhao X. Adult neural stem cells: response to stroke injury and potential for therapeutic applications. Curr Stem Cell Res Ther. 2011;6(4):327–38. PubMed PMID: 21466483. Pubmed Central PMCID: 3199296.
10. Gervois P, Wolfs E, Ratajczak J, Dillen Y, Vangansewinkel T, Hilkens P, et al. Stem cell-based therapies for ischemic stroke: preclinical results and the potential of imaging-assisted evaluation of donor cell fate and mechanisms of brain regeneration. Med Res Rev. 2016;36(6):1080–126. PubMed PMID: 27439773.
11. Nomura T, Honmou O, Harada K, Houkin K, Hamada H, Kocsis JD. I.V. infusion of brain-derived neurotrophic factor gene-modified human mesenchymal stem cells protects against injury in a cerebral ischemia model in adult rat. Neuroscience. 2005;136(1):161–9. PubMed PMID: 16229956. Pubmed Central PMCID: 2605391.
12. Onda T, Honmou O, Harada K, Houkin K, Hamada H, Kocsis JD. Therapeutic benefits by human mesenchymal stem cells (hMSCs) and Ang-1 gene-modified hMSCs after cerebral ischemia. J Cereb Blood Flow Metab. 2008;28(2):329–40. PubMed PMID: 17637706. Pubmed Central PMCID: 2605394.
13. Ding J, Cheng Y, Gao S, Chen J. Effects of nerve growth factor and Noggin-modified bone marrow stromal cells on stroke in rats. J Neurosci Res. 2011;89(2):222–30. PubMed PMID: 21162129.
14. Ginsberg MD. Neuroprotection for ischemic stroke: past, present and future. Neuropharmacology. 2008;55(3):363–89. PubMed PMID: 18308347. Pubmed Central PMCID: 2631228.
15. Mergenthaler P, Dirnagl U, Meisel A. Pathophysiology of stroke: lessons from animal models. Metab Brain Dis. 2004;19(3-4):151–67. PubMed PMID: 15554412.
16. Brouns R, De Deyn PP. The complexity of neurobiological processes in acute ischemic stroke. Clin Neurol Neurosurg. 2009;111(6):483–95. PubMed PMID: 19446389.
17. Durukan A, Tatlisumak T. Acute ischemic stroke: overview of major experimental rodent models, pathophysiology, and therapy of focal cerebral ischemia. Pharmacol Biochem Behav. 2007;87(1):179–97. PubMed PMID: 17521716.
18. Doyle KP, Simon RP, Stenzel-Poore MP. Mechanisms of ischemic brain damage. Neuropharmacology. 2008;55(3):310–8. PubMed PMID: 18308346. Pubmed Central PMCID: PMC2603601.

19. Moretti A, Ferrari F, Villa RF. Neuroprotection for ischaemic stroke: current status and challenges. Pharmacol Ther. 2015;146:23–34. PubMed PMID: 25196155.
20. Kamal Alam B, Bukhari AS, Assad S, Muhammad Siddique P, Ghazanfar H, Niaz MJ, et al. Functional outcome after decompressive craniectomy in patients with dominant or non-dominant malignant middle cerebral infarcts. Cureus. 2017;9(1):e997. PubMed PMID: 28286721. Pubmed Central PMCID: PMC5338989.
21. van der Worp HB, Macleod MR, Bath PM, Demotes J, Durand-Zaleski I, Gebhardt B, et al. EuroHYP-1: European multicenter, randomized, phase III clinical trial of therapeutic hypothermia plus best medical treatment vs. best medical treatment alone for acute ischemic stroke. Int J Stroke. 2014;9(5):642–5. PubMed PMID: 24828363.
22. Gutierrez-Fernandez M, Fuentes B, Rodriguez-Frutos B, Ramos-Cejudo J, Vallejo-Cremades MT, Diez-Tejedor E. Trophic factors and cell therapy to stimulate brain repair after ischaemic stroke. J Cell Mol Med. 2012;16(10):2280–90. PubMed PMID: 22452968. Pubmed Central PMCID: PMC3823421.
23. Li G, Yu F, Lei T, Gao H, Li P, Sun Y, et al. Bone marrow mesenchymal stem cell therapy in ischemic stroke: mechanisms of action and treatment optimization strategies. Neural Regen Res. 2016;11(6):1015–24. PubMed PMID: 27482235. Pubmed Central PMCID: PMC4962565.
24. Mead B, Logan A, Berry M, Leadbeater W, Scheven BA. Paracrine-mediated neuroprotection and neuritogenesis of axotomised retinal ganglion cells by human dental pulp stem cells: comparison with human bone marrow and adipose-derived mesenchymal stem cells. PLoS One. 2014;9(10):e109305. PubMed PMID: 25290916. Pubmed Central PMCID: 4188599.
25. Wilkins A, Kemp K, Ginty M, Hares K, Mallam E, Scolding N. Human bone marrow-derived mesenchymal stem cells secrete brain-derived neurotrophic factor which promotes neuronal survival in vitro. Stem Cell Res. 2009;3(1):63–70. PubMed PMID: 19411199.
26. Nosrat IV, Smith CA, Mullally P, Olson L, Nosrat CA. Dental pulp cells provide neurotrophic support for dopaminergic neurons and differentiate into neurons in vitro; implications for tissue engineering and repair in the nervous system. Eur J Neurosci. 2004;19(9):2388–98. PubMed PMID: 15128393.
27. Gervois P, Struys T, Hilkens P, Bronckaers A, Ratajczak J, Politis C, et al. Neurogenic maturation of human dental pulp stem cells following neurosphere generation induces morphological and electrophysiological characteristics of functional neurons. Stem Cells Dev. 2015;24(3):296–311. PubMed PMID: 25203005. Pubmed Central PMCID: 4303022.
28. Egashira Y, Sugitani S, Suzuki Y, Mishiro K, Tsuruma K, Shimazawa M, et al. The conditioned medium of murine and human adipose-derived stem cells exerts neuroprotective effects against experimental stroke model. Brain Res. 2012;1461:87–95. PubMed PMID: 22608076.
29. Hau S, Reich DM, Scholz M, Naumann W, Emmrich F, Kamprad M, et al. Evidence for neuroprotective properties of human umbilical cord blood cells after neuronal hypoxia in vitro. BMC Neurosci. 2008;9:30. PubMed PMID: 18312640. Pubmed Central PMCID: 2294131.
30. Crigler L, Robey RC, Asawachaicharn A, Gaupp D, Phinney DG. Human mesenchymal stem cell subpopulations express a variety of neuro-regulatory molecules and promote neuronal cell survival and neuritogenesis. Exp Neurol. 2006;198(1):54–64. PubMed PMID: 16336965.
31. Scheibe F, Klein O, Klose J, Priller J. Mesenchymal stromal cells rescue cortical neurons from apoptotic cell death in an in vitro model of cerebral ischemia. Cell Mol Neurobiol. 2012;32(4):567–76. PubMed PMID: 22290155.
32. Song M, Jue SS, Cho YA, Kim EC. Comparison of the effects of human dental pulp stem cells and human bone marrow-derived mesenchymal stem cells on ischemic human astrocytes in vitro. J Neurosci Res. 2015;93(6):973–83. PubMed PMID: 25663284.
33. Ikegame Y, Yamashita K, Hayashi S, Mizuno H, Tawada M, You F, et al. Comparison of mesenchymal stem cells from adipose tissue and bone marrow for ischemic stroke therapy. Cytotherapy. 2011;13(6):675–85. PubMed PMID: 21231804.

34. Leong WK, Henshall TL, Arthur A, Kremer KL, Lewis MD, Helps SC, et al. Human adult dental pulp stem cells enhance poststroke functional recovery through non-neural replacement mechanisms. Stem Cells Transl Med. 2012;1(3):177–87. PubMed PMID: 23197777. Pubmed Central PMCID: 3659845.
35. Toyoshima A, Yasuhara T, Kameda M, Morimoto J, Takeuchi H, Wang F, et al. Intra-arterial transplantation of allogeneic mesenchymal stem cells mounts neuroprotective effects in a transient ischemic stroke model in rats: analyses of therapeutic time window and its mechanisms. PLoS One. 2015;10(6):e0127302. PubMed PMID: 26075717. Pubmed Central PMCID: 4468176.
36. Zhang L, Yi L, Chopp M, Kramer BC, Romanko M, Gosiewska A, et al. Intravenous administration of human umbilical tissue-derived cells improves neurological function in aged rats after embolic stroke. Cell Transplant. 2013;22(9):1569–76. PubMed PMID: 23127976.
37. Gutierrez-Fernandez M, Rodriguez-Frutos B, Ramos-Cejudo J, Otero-Ortega L, Fuentes B, Vallejo-Cremades MT, et al. Comparison between xenogeneic and allogeneic adipose mesenchymal stem cells in the treatment of acute cerebral infarct: proof of concept in rats. J Transl Med. 2015;13:46. PubMed PMID: 25637958. Pubmed Central PMCID: 4322805.
38. Zhao LR, Duan WM, Reyes M, Keene CD, Verfaillie CM, Low WC. Human bone marrow stem cells exhibit neural phenotypes and ameliorate neurological deficits after grafting into the ischemic brain of rats. Exp Neurol. 2002;174(1):11–20. PubMed PMID: 11869029.
39. Yasuhara T, Matsukawa N, Hara K, Maki M, Ali MM, Yu SJ, et al. Notch-induced rat and human bone marrow stromal cell grafts reduce ischemic cell loss and ameliorate behavioral deficits in chronic stroke animals. Stem Cells Dev. 2009;18(10):1501–14. PubMed PMID: 19301956.
40. Liu Z, Tang Y, Lu S, Zhou J, Du Z, Duan C, et al. The tumourigenicity of iPS cells and their differentiated derivates. J Cell Mol Med. 2013;17(6):782–91. PubMed PMID: 23711115. Pubmed Central PMCID: 3823182.
41. Seminatore C, Polentes J, Ellman D, Kozubenko N, Itier V, Tine S, et al. The postischemic environment differentially impacts teratoma or tumor formation after transplantation of human embryonic stem cell-derived neural progenitors. Stroke. 2010;41(1):153–9. PubMed PMID: 19940279.
42. Brennand KJ, Marchetto MC, Benvenisty N, Brustle O, Ebert A, Izpisua Belmonte JC, et al. Creating patient-specific neural cells for the in vitro study of brain disorders. Stem Cell Rep. 2015;5(6):933–45. PubMed PMID: 26610635. Pubmed Central PMCID: PMC4881284.
43. Lancaster MA, Knoblich JA. Generation of cerebral organoids from human pluripotent stem cells. Nat Protoc. 2014;9(10):2329–40. PubMed PMID: 25188634. Pubmed Central PMCID: PMC4160653.
44. Lancaster MA, Renner M, Martin CA, Wenzel D, Bicknell LS, Hurles ME, et al. Cerebral organoids model human brain development and microcephaly. Nature. 2013;501(7467):373–9. PubMed PMID: 23995685. Pubmed Central PMCID: PMC3817409.
45. Liu J, Wang Y, Akamatsu Y, Lee CC, Stetler RA, Lawton MT, et al. Vascular remodeling after ischemic stroke: mechanisms and therapeutic potentials. Prog Neurobiol. 2014;115:138–56. PubMed PMID: 24291532. Pubmed Central PMCID: PMC4295834.
46. Troidl K, Schaper W. Arteriogenesis versus angiogenesis in peripheral artery disease. Diabetes Metab Res Rev. 2012;28(Suppl 1):27–9. PubMed PMID: 22271719.
47. Potente M, Gerhardt H, Carmeliet P. Basic and therapeutic aspects of angiogenesis. Cell. 2011;146(6):873–87. PubMed PMID: 21925313.
48. Stapor P, Wang X, Goveia J, Moens S, Carmeliet P. Angiogenesis revisited - role and therapeutic potential of targeting endothelial metabolism. J Cell Sci. 2014;127(Pt 20):4331–41. PubMed PMID: 25179598.
49. Liu J. Poststroke angiogenesis: blood, bloom, or brood? Stroke. 2015;46(5):e105–6. PubMed PMID: 25813191. Pubmed Central PMCID: PMC4414877.
50. Said SS, Pickering JG, Mequanint K. Advances in growth factor delivery for therapeutic angiogenesis. J Vasc Res. 2013;50(1):35–51. PubMed PMID: 23154615.

51. Caplan AI, Correa D. The MSC: an injury drugstore. Cell Stem Cell. 2011;9(1):11–5. PubMed PMID: 21726829. Pubmed Central PMCID: PMC3144500.
52. Chen J, Zhang ZG, Li Y, Wang L, Xu YX, Gautam SC, et al. Intravenous administration of human bone marrow stromal cells induces angiogenesis in the ischemic boundary zone after stroke in rats. Circ Res. 2003;92(6):692–9. PubMed PMID: 12609969.
53. Ma XL, Liu KD, Li FC, Jiang XM, Jiang L, Li HL. Human mesenchymal stem cells increases expression of alpha-tubulin and angiopoietin 1 and 2 in focal cerebral ischemia and reperfusion. Curr Neurovasc Res. 2013;10(2):103–11. PubMed PMID: 23469950.
54. Wakabayashi K, Nagai A, Sheikh AM, Shiota Y, Narantuya D, Watanabe T, et al. Transplantation of human mesenchymal stem cells promotes functional improvement and increased expression of neurotrophic factors in a rat focal cerebral ischemia model. J Neurosci Res. 2010;88(5):1017–25. PubMed PMID: 19885863.
55. Jeon D, Chu K, Lee ST, Jung KH, Ban JJ, Park DK, et al. Neuroprotective effect of a cell-free extract derived from human adipose stem cells in experimental stroke models. Neurobiol Dis. 2013;54:414–20. PubMed PMID: 23376682.
56. Curtis MA, Kam M, Nannmark U, Anderson MF, Axell MZ, Wikkelso C, et al. Human neuroblasts migrate to the olfactory bulb via a lateral ventricular extension. Science. 2007;315(5816):1243–9. PubMed PMID: 17303719.
57. Eriksson PS, Perfilieva E, Bjork-Eriksson T, Alborn AM, Nordborg C, Peterson DA, et al. Neurogenesis in the adult human hippocampus. Nat Med. 1998;4(11):1313–7. PubMed PMID: 9809557.
58. Thored P, Arvidsson A, Cacci E, Ahlenius H, Kallur T, Darsalia V, et al. Persistent production of neurons from adult brain stem cells during recovery after stroke. Stem Cells. 2006;24(3):739–47. PubMed PMID: 16210404.
59. Yan YP, Sailor KA, Lang BT, Park SW, Vemuganti R, Dempsey RJ. Monocyte chemoattractant protein-1 plays a critical role in neuroblast migration after focal cerebral ischemia. J Cereb Blood Flow Metab. 2007;27(6):1213–24. PubMed PMID: 17191078.
60. Barkho BZ, Munoz AE, Li X, Li L, Cunningham LA, Zhao X. Endogenous matrix metalloproteinase (MMP)-3 and MMP-9 promote the differentiation and migration of adult neural progenitor cells in response to chemokines. Stem Cells. 2008;26(12):3139–49. PubMed PMID: 18818437. Pubmed Central PMCID: PMC2758553.
61. Kojima T, Hirota Y, Ema M, Takahashi S, Miyoshi I, Okano H, et al. Subventricular zone-derived neural progenitor cells migrate along a blood vessel scaffold toward the post-stroke striatum. Stem Cells. 2010;28(3):545–54. PubMed PMID: 20073084.
62. Wang L, Zhang Z, Wang Y, Zhang R, Chopp M. Treatment of stroke with erythropoietin enhances neurogenesis and angiogenesis and improves neurological function in rats. Stroke. 2004;35(7):1732–7. PubMed PMID: 15178821.
63. Yoo SW, Kim SS, Lee SY, Lee HS, Kim HS, Lee YD, et al. Mesenchymal stem cells promote proliferation of endogenous neural stem cells and survival of newborn cells in a rat stroke model. Exp Mol Med. 2008;40(4):387–97. PubMed PMID: 18779651. Pubmed Central PMCID: PMC2679267.
64. Kingham PJ, Kolar MK, Novikova LN, Novikov LN, Wiberg M. Stimulating the neurotrophic and angiogenic properties of human adipose-derived stem cells enhances nerve repair. Stem Cells Dev. 2014;23(7):741–54. PubMed PMID: 24124760.
65. Gervois P, Wolfs E, Dillen Y, Hilkens P, Ratajczak J, Driesen RB, et al. Paracrine Maturation and Migration of SH-SY5Y Cells by Dental Pulp Stem Cells. J Dent Res. 2017;96(6):654–62. PubMed PMID: 28141971.
66. Pires AO, Neves-Carvalho A, Sousa N, Salgado AJ. The Secretome of Bone Marrow and Wharton Jelly Derived Mesenchymal Stem Cells Induces Differentiation and Neurite Outgrowth in SH-SY5Y Cells. Stem Cells Int. 2014;2014:438352. PubMed PMID: 25132857. Pubmed Central PMCID: PMC4124228.
67. Munoz JR, Stoutenger BR, Robinson AP, Spees JL, Prockop DJ. Human stem/progenitor cells from bone marrow promote neurogenesis of endogenous neural stem cells in the

hippocampus of mice. Proc Natl Acad Sci U S A. 2005;102(50):18171–6. PubMed PMID: 16330757. Pubmed Central PMCID: PMC1312406.
68. Tajiri N, Kaneko Y, Shinozuka K, Ishikawa H, Yankee E, McGrogan M, et al. Stem cell recruitment of newly formed host cells via a successful seduction? Filling the gap between neurogenic niche and injured brain site. PLoS One. 2013;8(9):e74857. PubMed PMID: 24023965. Pubmed Central PMCID: PMC3762783.
69. Doeppner TR, Hermann DM. Mesenchymal stem cells in the treatment of ischemic stroke: progress and possibilities. Stem Cells Clon. 2010;3:157–63. PubMed PMID: 24198521. Pubmed Central PMCID: PMC3781740.
70. Sullivan R, Duncan K, Dailey T, Kaneko Y, Tajiri N, Borlongan CV. A possible new focus for stroke treatment - migrating stem cells. Expert Opin Biol Ther. 2015;15(7):949–58. PubMed PMID: 25943632. Pubmed Central PMCID: PMC4465850.
71. Li Y, Chen J, Zhang CL, Wang L, Lu D, Katakowski M, et al. Gliosis and brain remodeling after treatment of stroke in rats with marrow stromal cells. Glia. 2005;49(3):407–17. PubMed PMID: 15540231.
72. Pekny M, Wilhelmsson U, Pekna M. The dual role of astrocyte activation and reactive gliosis. Neurosci Lett. 2014;565:30–8. PubMed PMID: 24406153.
73. Aldskogius H, Kozlova EN. Central neuron-glial and glial-glial interactions following axon injury. Prog Neurobiol. 1998;55(1):1–26. PubMed PMID: 9602498.
74. Chen MS, Huber AB, van der Haar ME, Frank M, Schnell L, Spillmann AA, et al. Nogo-A is a myelin-associated neurite outgrowth inhibitor and an antigen for monoclonal antibody IN-1. Nature. 2000;403(6768):434–9. PubMed PMID: 10667796.
75. Annabi B, Lee YT, Turcotte S, Naud E, Desrosiers RR, Champagne M, et al. Hypoxia promotes murine bone-marrow-derived stromal cell migration and tube formation. Stem Cells. 2003;21(3):337–47. PubMed PMID: 12743328.
76. Iohara K, Zheng L, Wake H, Ito M, Nabekura J, Wakita H, et al. A novel stem cell source for vasculogenesis in ischemia: subfraction of side population cells from dental pulp. Stem Cells. 2008;26(9):2408–18. PubMed PMID: 18583536.
77. Lozito TP, Jackson WM, Nesti LJ, Tuan RS. Human mesenchymal stem cells generate a distinct pericellular zone of MMP activities via binding of MMPs and secretion of high levels of TIMPs. Matrix Biol. 2014;34:132–43. PubMed PMID: 24140982.
78. Shen LH, Li Y, Chen J, Cui Y, Zhang C, Kapke A, et al. One-year follow-up after bone marrow stromal cell treatment in middle-aged female rats with stroke. Stroke. 2007;38(7):2150–6. PubMed PMID: 17525391.
79. Gu Y, He M, Zhou X, Liu J, Hou N, Bin T, et al. Endogenous IL-6 of mesenchymal stem cell improves behavioral outcome of hypoxic-ischemic brain damage neonatal rats by supressing apoptosis in astrocyte. Sci Rep. 2016;6:18587. PubMed PMID: 26766745. Pubmed Central PMCID: PMC4725911.
80. Moskowitz MA, Lo EH, Iadecola C. The science of stroke: mechanisms in search of treatments. Neuron. 2010;67(2):181–98. PubMed PMID: 20670828. Pubmed Central PMCID: 2957363.
81. Liesz A, Hagmann S, Zschoche C, Adamek J, Zhou W, Sun L, et al. The spectrum of systemic immune alterations after murine focal ischemia: immunodepression versus immunomodulation. Stroke. 2009;40(8):2849–58. PubMed PMID: 19443795.
82. Dirnagl U, Klehmet J, Braun JS, Harms H, Meisel C, Ziemssen T, et al. Stroke-induced immunodepression: experimental evidence and clinical relevance. Stroke. 2007;38(2 Suppl):770–3. PubMed PMID: 17261736.
83. Urra X, Cervera A, Villamor N, Planas AM, Chamorro A. Harms and benefits of lymphocyte subpopulations in patients with acute stroke. Neuroscience. 2009;158(3):1174–83. PubMed PMID: 18619524.
84. Carden DL, Granger DN. Pathophysiology of ischaemia-reperfusion injury. J Pathol. 2000;190(3):255–66. PubMed PMID: 10685060.

85. Eltzschig HK, Carmeliet P. Hypoxia and inflammation. N Engl J Med. 2011;364(7):656–65. PubMed PMID: 21323543. Pubmed Central PMCID: 3930928.
86. Iadecola C, Anrather J. The immunology of stroke: from mechanisms to translation. Nat Med. 2011;17(7):796–808. PubMed PMID: 21738161. Pubmed Central PMCID: 3137275.
87. Yilmaz G, Granger DN. Leukocyte recruitment and ischemic brain injury. Neuromolecular Med. 2010;12(2):193–204. PubMed PMID: 19579016. Pubmed Central PMCID: 2878882.
88. Engelhardt B, Sorokin L. The blood-brain and the blood-cerebrospinal fluid barriers: function and dysfunction. Semin Immunopathol. 2009;31(4):497–511. PubMed PMID: 19779720.
89. Shichita T, Ito M, Yoshimura A. Post-ischemic inflammation regulates neural damage and protection. Front Cell Neurosci. 2014;8:319. PubMed PMID: 25352781. Pubmed Central PMCID: 4196547.
90. Eltzschig HK, Sitkovsky MV, Robson SC. Purinergic signaling during inflammation. N Engl J Med. 2013;368(13):1260. PubMed PMID: 23534573.
91. Al'Qteishat A, Gaffney J, Krupinski J, Rubio F, West D, Kumar S, et al. Changes in hyaluronan production and metabolism following ischaemic stroke in man. Brain. 2006;129(Pt 8):2158–76. PubMed PMID: 16731541.
92. Qiu J, Nishimura M, Wang Y, Sims JR, Qiu S, Savitz SI, et al. Early release of HMGB-1 from neurons after the onset of brain ischemia. J Cereb Blood Flow Metab. 2008;28(5):927–38. PubMed PMID: 18000511.
93. Gelderblom M, Sobey CG, Kleinschnitz C, Magnus T, et al. Ageing Res Rev. 2015;24(Pt A):77–82. PubMed PMID: 26210897.
94. Hu X, Li P, Guo Y, Wang H, Leak RK, Chen S, et al. Microglia/macrophage polarization dynamics reveal novel mechanism of injury expansion after focal cerebral ischemia. Stroke. 2012;43(11):3063–70. PubMed PMID: 22933588.
95. Kacimi R, Giffard RG, Yenari MA. Endotoxin-activated microglia injure brain derived endothelial cells via NF-kappaB, JAK-STAT and JNK stress kinase pathways. J Inflamm. 2011;8:7. PubMed PMID: 21385378. Pubmed Central PMCID: 3061894.
96. Spera PA, Ellison JA, Feuerstein GZ, Barone FC. IL-10 reduces rat brain injury following focal stroke. Neurosci Lett. 1998;251(3):189–92. PubMed PMID: 9726375.
97. Dobolyi A, Vincze C, Pal G, Lovas G. The neuroprotective functions of transforming growth factor beta proteins. Int J Mol Sci. 2012;13(7):8219–58. PubMed PMID: 22942700. Pubmed Central PMCID: 3430231.
98. Kooijman R, Sarre S, Michotte Y, De Keyser J. Insulin-like growth factor I: a potential neuroprotective compound for the treatment of acute ischemic stroke? Stroke. 2009;40(4):e83–8. PubMed PMID: 19197073.
99. Kono H, Rock KL. How dying cells alert the immune system to danger. Nat Rev Immunol. 2008;8(4):279–89. PubMed PMID: 18340345. Pubmed Central PMCID: 2763408.
100. Kleinschnitz C, Schwab N, Kraft P, Hagedorn I, Dreykluft A, Schwarz T, et al. Early detrimental T-cell effects in experimental cerebral ischemia are neither related to adaptive immunity nor thrombus formation. Blood. 2010;115(18):3835–42. PubMed PMID: 20215643.
101. Hurn PD, Subramanian S, Parker SM, Afentoulis ME, Kaler LJ, Vandenbark AA, et al. T- and B-cell-deficient mice with experimental stroke have reduced lesion size and inflammation. J Cereb Blood Flow Metab. 2007;27(11):1798–805. PubMed PMID: 17392692. Pubmed Central PMCID: 2592689.
102. Shichita T, Sugiyama Y, Ooboshi H, Sugimori H, Nakagawa R, Takada I, et al. Pivotal role of cerebral interleukin-17-producing gammadeltaT cells in the delayed phase of ischemic brain injury. Nat Med. 2009;15(8):946–50. PubMed PMID: 19648929.
103. Gelderblom M, Weymar A, Bernreuther C, Velden J, Arunachalam P, Steinbach K, et al. Neutralization of the IL-17 axis diminishes neutrophil invasion and protects from ischemic stroke. Blood. 2012;120(18):3793–802. PubMed PMID: 22976954.
104. Uccelli A, Moretta L, Pistoia V. Mesenchymal stem cells in health and disease. Nat Rev Immunol. 2008;8(9):726–36. PubMed PMID: 19172693.

105. Scheibe F, Ladhoff J, Huck J, Grohmann M, Blazej K, Oersal A, et al. Immune effects of mesenchymal stromal cells in experimental stroke. J Cereb Blood Flow Metab. 2012;32(8):1578–88. PubMed PMID: 22549620. Pubmed Central PMCID: 3421097.
106. Lee JS, Hong JM, Moon GJ, Lee PH, Ahn YH, Bang OY, et al. A long-term follow-up study of intravenous autologous mesenchymal stem cell transplantation in patients with ischemic stroke. Stem Cells. 2010;28(6):1099–106. PubMed PMID: 20506226.
107. Honmou O, Houkin K, Matsunaga T, Niitsu Y, Ishiai S, Onodera R, et al. Intravenous administration of auto serum-expanded autologous mesenchymal stem cells in stroke. Brain. 2011;134(Pt 6):1790–807. PubMed PMID: 21493695. Pubmed Central PMCID: 3102237.
108. Glennie S, Soeiro I, Dyson PJ, Lam EW, Dazzi F. Bone marrow mesenchymal stem cells induce division arrest anergy of activated T cells. Blood. 2005;105(7):2821–7. PubMed PMID: 15591115.
109. Del Fattore A, Luciano R, Pascucci L, Goffredo BM, Giorda E, Scapaticci M, et al. Immunoregulatory effects of mesenchymal stem cell-derived extracellular vesicles on T lymphocytes. Cell Transplant. 2015;24(12):2615–27. PubMed PMID: 25695896.
110. Puissant B, Barreau C, Bourin P, Clavel C, Corre J, Bousquet C, et al. Immunomodulatory effect of human adipose tissue-derived adult stem cells: comparison with bone marrow mesenchymal stem cells. Br J Haematol. 2005;129(1):118–29. PubMed PMID: 15801964.
111. Pierdomenico L, Bonsi L, Calvitti M, Rondelli D, Arpinati M, Chirumbolo G, et al. Multipotent mesenchymal stem cells with immunosuppressive activity can be easily isolated from dental pulp. Transplantation. 2005;80(6):836–42. PubMed PMID: 16210973.
112. Demircan PC, Sariboyaci AE, Unal ZS, Gacar G, Subasi C, Karaoz E. Immunoregulatory effects of human dental pulp-derived stem cells on T cells: comparison of transwell co-culture and mixed lymphocyte reaction systems. Cytotherapy. 2011;13(10):1205–20. PubMed PMID: 21905956.
113. Selmani Z, Naji A, Zidi I, Favier B, Gaiffe E, Obert L, et al. Human leukocyte antigen-G5 secretion by human mesenchymal stem cells is required to suppress T lymphocyte and natural killer function and to induce CD4+CD25highFOXP3+ regulatory T cells. Stem Cells. 2008;26(1):212–22. PubMed PMID: 17932417.
114. Ishizaka R, Hayashi Y, Iohara K, Sugiyama M, Murakami M, Yamamoto T, et al. Stimulation of angiogenesis, neurogenesis and regeneration by side population cells from dental pulp. Biomaterials. 2013;34(8):1888–97. PubMed PMID: 23245334.
115. Murakami M, Horibe H, Iohara K, Hayashi Y, Osako Y, Takei Y, et al. The use of granulocyte-colony stimulating factor induced mobilization for isolation of dental pulp stem cells with high regenerative potential. Biomaterials. 2013;34(36):9036–47. PubMed PMID: 23988014.
116. Tang R, Ding G. Swine dental pulp stem cells inhibit T-cell prolifcration. Transplant Proc. 2011;43(10):3955–9. PubMed PMID: 22172878.
117. Spaggiari GM, Capobianco A, Abdelrazik H, Becchetti F, Mingari MC, Moretta L. Mesenchymal stem cells inhibit natural killer-cell proliferation, cytotoxicity, and cytokine production: role of indoleamine 2,3-dioxygenase and prostaglandin E2. Blood. 2008;111(3):1327–33. PubMed PMID: 17951526.
118. Ramasamy R, Fazekasova H, Lam EW, Soeiro I, Lombardi G, Dazzi F. Mesenchymal stem cells inhibit dendritic cell differentiation and function by preventing entry into the cell cycle. Transplantation. 2007;83(1):71–6. PubMed PMID: 17220794.
119. Nauta AJ, Kruisselbrink AB, Lurvink E, Willemze R, Fibbe WE. Mesenchymal stem cells inhibit generation and function of both CD34+-derived and monocyte-derived dendritic cells. J Immunol. 2006;177(4):2080–7. PubMed PMID: 16887966.
120. Jose S, Tan SW, Ooi YY, Ramasamy R, Vidyadaran S. Mesenchymal stem cells exert anti-proliferative effect on lipopolysaccharide-stimulated BV2 microglia by reducing tumour necrosis factor-alpha levels. J Neuroinflammation. 2014;11:149. PubMed PMID: 25182840. Pubmed Central PMCID: 4156657.
121. Rahmat Z, Jose S, Ramasamy R, Vidyadaran S. Reciprocal interactions of mouse bone marrow-derived mesenchymal stem cells and BV2 microglia after lipopolysaccharide stim-

ulation. Stem Cell Res Ther. 2013;4(1):12. PubMed PMID: 23356521. Pubmed Central PMCID: 3706938.
122. Zhang B, Liu R, Shi D, Liu XX, Chen Y, Dou XW, et al. Mesenchymal stem cells induce mature dendritic cells into a novel Jagged-2-dependent regulatory dendritic cell population. Blood. 2009;113(1):46–57. PubMed PMID: WOS:000262162800011. English.
123. Zhang QZ, Su WR, Shi SH, Wilder-Smith P, Xiang AP, Wong A, et al. Human gingiva-derived mesenchymal stem cells elicit polarization of m2 macrophages and enhance cutaneous wound healing. Stem Cells. 2010;28(10):1856–68. PubMed PMID: 20734355. Pubmed Central PMCID: 3114043.
124. Vasandan AB, Jahnavi S, Shashank C, Prasad P, Kumar A, Prasanna SJ. Human Mesenchymal stem cells program macrophage plasticity by altering their metabolic status via a PGE2-dependent mechanism. Sci Rep. 2016;6:38308. PubMed PMID: 27910911. Pubmed Central PMCID: 5133610.
125. Lo Sicco C, Reverberi D, Balbi C, Ulivi V, Principi E, Pascucci L, et al. Mesenchymal stem cell-derived extracellular vesicles as mediators of anti-inflammatory effects: endorsement of macrophage polarization. Stem Cells Transl Med. 2017;6:1018. PubMed PMID: 28186708.
126. Gao S, Mao F, Zhang B, Zhang L, Zhang X, Wang M, et al. Mouse bone marrow-derived mesenchymal stem cells induce macrophage M2 polarization through the nuclear factor-kappaB and signal transducer and activator of transcription 3 pathways. Exp Biol Med (Maywood). 2014;239(3):366–75. PubMed PMID: 24500984.
127. Ren G, Zhao X, Zhang L, Zhang J, L'Huillier A, Ling W, et al. Inflammatory cytokine-induced intercellular adhesion molecule-1 and vascular cell adhesion molecule-1 in mesenchymal stem cells are critical for immunosuppression. J Immunol. 2010;184(5):2321–8. PubMed PMID: 20130212. Pubmed Central PMCID: 2881946.
128. Raicevic G, Najar M, Najimi M, El Taghdouini A, van Grunsven LA, Sokal E, et al. Influence of inflammation on the immunological profile of adult-derived human liver mesenchymal stromal cells and stellate cells. Cytotherapy. 2015;17(2):174–85. PubMed PMID: 25455740.
129. Crop MJ, Baan CC, Korevaar SS, Ijzermans JN, Pescatori M, Stubbs AP, et al. Inflammatory conditions affect gene expression and function of human adipose tissue-derived mesenchymal stem cells. Clin Exp Immunol. 2010;162(3):474–86. PubMed PMID: 20846162. Pubmed Central PMCID: 3026550.
130. Schnabel LV, Abratte CM, Schimenti JC, Felippe MJ, Cassano JM, Southard TL, et al. Induced pluripotent stem cells have similar immunogenic and more potent immunomodulatory properties compared with bone marrow-derived stromal cells in vitro. Regen Med. 2014;9(5):621–35. PubMed PMID: 24773530. Pubmed Central PMCID: 4352342.
131. Yen BL, Chang CJ, Liu KJ, Chen YC, Hu HI, Bai CH, et al. Brief report--human embryonic stem cell-derived mesenchymal progenitors possess strong immunosuppressive effects toward natural killer cells as well as T lymphocytes. Stem Cells. 2009;27(2):451–6. PubMed PMID: 18988708.
132. Tan Z, Su ZY, Wu RR, Gu B, Liu YK, Zhao XL, et al. Immunomodulative effects of mesenchymal stem cells derived from human embryonic stem cells in vivo and in vitro. J Zhejiang Univ Sci B. 2011;12(1):18–27. PubMed PMID: 21194182. Pubmed Central PMCID: 3017412.
133. Fu QL, Chow YY, Sun SJ, Zeng QX, Li HB, Shi JB, et al. Mesenchymal stem cells derived from human induced pluripotent stem cells modulate T-cell phenotypes in allergic rhinitis. Allergy. 2012;67(10):1215–22. PubMed PMID: 22882409. Pubmed Central PMCID: 3555482.
134. Sharma S, Yang B, Strong R, Xi X, Brenneman M, Grotta JC, et al. Bone marrow mononuclear cells protect neurons and modulate microglia in cell culture models of ischemic stroke. J Neurosci Res. 2010;88(13):2869–76. PubMed PMID: 20629187. Pubmed Central PMCID: 3401573.

135. Drago D, Basso V, Gaude E, Volpe G, Peruzzotti-Jametti L, Bachi A, et al. Metabolic determinants of the immune modulatory function of neural stem cells. J Neuroinflammation. 2016;13(1):232. PubMed PMID: 27590826. Pubmed Central PMCID: 5009670.
136. Kokaia Z, Martino G, Schwartz M, Lindvall O. Cross-talk between neural stem cells and immune cells: the key to better brain repair? Nat Neurosci. 2012;15(8):1078–87. PubMed PMID: 22837038.
137. Einstein O, Fainstein N, Vaknin I, Mizrachi-Kol R, Reihartz E, Grigoriadis N, et al. Neural precursors attenuate autoimmune encephalomyelitis by peripheral immunosuppression. Ann Neurol. 2007;61(3):209–18. PubMed PMID: 17187374.
138. Pluchino S, Zanotti L, Rossi B, Brambilla E, Ottoboni L, Salani G, et al. Neurosphere-derived multipotent precursors promote neuroprotection by an immunomodulatory mechanism. Nature. 2005;436(7048):266–71. PubMed PMID: 16015332.
139. Acosta SA, Tajiri N, Hoover J, Kaneko Y, Borlongan CV. Intravenous bone marrow stem cell grafts preferentially migrate to spleen and abrogate chronic inflammation in stroke. Stroke. 2015;46(9):2616–27. PubMed PMID: 26219646. Pubmed Central PMCID: 4542567.
140. Zhou F, Gao S, Wang L, Sun C, Chen L, Yuan P, et al. Human adipose-derived stem cells partially rescue the stroke syndromes by promoting spatial learning and memory in mouse middle cerebral artery occlusion model. Stem Cell Res Ther. 2015;6:92. PubMed PMID: 25956259. Pubmed Central PMCID: 4453264.
141. Li J, Zhu H, Liu Y, Li Q, Lu S, Feng M, et al. Human mesenchymal stem cell transplantation protects against cerebral ischemic injury and upregulates interleukin-10 expression in Macacafascicularis. Brain Res. 2010;1334:65–72. PubMed PMID: 20353760.
142. Doeppner TR, Herz J, Gorgens A, Schlechter J, Ludwig AK, Radtke S, et al. Extracellular vesicles improve post-stroke neuroregeneration and prevent postischemic immunosuppression. Stem Cells Transl Med. 2015;4(10):1131–43. PubMed PMID: 26339036. Pubmed Central PMCID: 4572905.
143. Wolfs E, Verfaillie CM, Van Laere K, Deroose CM. Radiolabeling strategies for radionuclide imaging of stem cells. Stem Cell Rev. 2015;11(2):254–74. PubMed PMID: 25534590.
144. Daadi MM, Li Z, Arac A, Grueter BA, Sofilos M, Malenka RC, et al. Molecular and magnetic resonance imaging of human embryonic stem cell-derived neural stem cell grafts in ischemic rat brain. Mol Ther. 2009;17(7):1282–91. PubMed PMID: 19436269. Pubmed Central PMCID: 2835224.
145. Zhu J, Zhou L, XingWu F. Tracking neural stem cells in patients with brain trauma. N Engl J Med. 2006;355(22):2376–8. PubMed PMID: 17135597.
146. Daadi MM, Hu S, Klausner J, Li Z, Sofilos M, Sun G, et al. Imaging neural stem cell graft-induced structural repair in stroke. Cell Transplant. 2013;22(5):881–92. PubMed PMID: 23044338.
147. Oh SH, Choi C, Chang DJ, Shin DA, Lee N, Jeon I, et al. Early neuroprotective effect with lack of long-term cell replacement effect on experimental stroke after intra-arterial transplantation of adipose-derived mesenchymal stromal cells. Cytotherapy. 2015;17(8):1090–103. PubMed PMID: 26031742.
148. Huang L, Liu Y, Lu J, Cerqueira B, Misra V, Duong TQ. Intraarterial transplantation of human umbilical cord blood mononuclear cells in hyperacute stroke improves vascular function. Stem Cell Res Ther. 2017;8(1):74. PubMed PMID: 28330501. Pubmed Central PMCID: PMC5361847.
149. Grudzenski S, Baier S, Ebert A, Pullens P, Lemke A, Bieback K, et al. The effect of adipose tissue-derived stem cells in a middle cerebral artery occlusion stroke model depends on their engraftment rate. Stem Cell Res Ther. 2017;8(1):96. PubMed PMID: 28446216. Pubmed Central PMCID: PMC5407025.
150. Tang Y, Zhang C, Wang J, Lin X, Zhang L, Yang Y, et al. MRI/SPECT/fluorescent tri-modal probe for evaluating the homing and therapeutic efficacy of transplanted mesenchymal stem cells in a rat ischemic stroke model. Adv Funct Mater. 2015;25(7):1024–34. PubMed PMID: 26290659. Pubmed Central PMCID: PMC4539160.

Chapter 3
Stem Cell Transplants in the Aged Stroke Brain: Microenvironment Factors

Aurel Popa-Wagner and Mario Di Napoli

Abstract In aged humans, stroke is a major cause of disability for which no neuroprotective measures are available. The incidence of stroke increases significantly with age both in men and women with incidence rates accelerating above 70 years. Since stroke afflicts mostly the elderly comorbid patients it is highly desirable to test the efficacy of cell therapies in an appropriate animal stroke model. It has been noted that the potential for neurogenesis is also preserved in aged, stroke-injured brains and the environment of the aged brain is not hostile to cell therapies. However, there remain significant developmental and translational issues that remain to be resolved in future studies such as (1) Understanding the differentiation into specific phenotypes. Upon transplantation, the differentiated cells often de-differentiate; (2) Tumorigenesis remains a significant concern; (3) Anti-neuroinflammatory therapies is a potential target to promote regeneration and repair after brain injury and neurodegenerative conditions by stem cell therapy; (4) Efficacy of cell therapy can be enhanced by physical rehabilitation; (5) One potential weakness of the preclinical dataset is, however, the lack of proof in aged subjects. It is in fact a general drawback of preclinical evaluations of candidate stroke drugs that due to cost effectiveness and practicability most studies were done in young animals. A lack of data from aged subjects in preclinical studies may at least in part explain the failure of candidate neuroprotective drugs in clinical trials. The aged brain has compared to the young brain, an enhanced susceptibility to stroke and displays a limited recovery from an ischemic injury. Finally, a better understanding of potential risks of stem

A. Popa-Wagner (✉)
Department of Neurology, University of Medicine Essen, Essen, Germany

Department of Functional Sciences, University of Medicine and Pharmacy of Craiova, Craiova, Romania

Griffith University School of Medicine,
Gold Coast Campus, Gold Coast, QLD 4222, Australia
e-mail: aurel.popa-wagner@geriatrics-healthyageing.com

M. Di Napoli
Neurological Service, San Camillo de' Lellis General Hospital, Rieti; and the Neurological Section, Neuro-epidemiology Unit, SMDN, Centre for Cardiovascular Medicine and Cerebrovascular Disease Prevention, Sulmona, L'Aquila, Italy

P.A. Lapchak, J.H. Zhang (eds.), *Cellular and Molecular Approaches to Regeneration and Repair*, Springer Series in Translational Stroke Research,
https://doi.org/10.1007/978-3-319-66679-2_3

cell therapies in stroke shall make the translation of cell therapies safer. Likewise, awareness of may help improve their efficacy to achieve therapeutic success.

Keywords Aging • Stroke • Therapies • Stem cells • G-CSF • BM-MSC • BM-MNC

Abbreviations

BBB	Blood–brain barrier
BMECs	Brain microvascular endothelial cells
BM-MNC	Bone marrow-derived mononuclear cells
BM-MSC	Bone marrow mesenchymal cells
ECA	External carotid artery
EPC	Endothelial progenitor cells
ESCs	Embryonic stem cells
G-CSF	(Granulocyte-Colony Stimulating Factor) Hematopoietic factor
hBMMSCs	Mesenchymal cells of human origin
HSPC	Hematopoietic stem/progenitor cells
iPSC	Human-derived inducible pluripotent cells
MCA	Middle cerebral artery
MCAO	Middle cerebral artery occlusion
MRI	Magnetic Resonance Imaging
MSCs	Mesenchymal stem cells
NSCs	Neural stem cells
SVZ	Subventricular zone
UCB	Umbilical-cord blood
VEGF	Vascular endothelial growth factor
SHR	Spontaneously hypertensive rat model

1 Introduction

Stroke is globally the second cause of death by 2015, just 1.5% less prevalent than ischemic heart disease (13.2%). From 1990 to 2010, the age-standardised incidence of stroke significantly decreased by 12% in high-income countries, while in low- and middle-income countries, it increased, although non-significantly, by about 12%, while mortality decreased significantly in both (mean 37% vs. 20%, respectively), of which 31% (with about 80% of it in low- and middle-income countries) were in children and young adults (below 65 years) [1].

Stroke patients are at the highest risk of death in the first week after the event, and between 20% and 50% die within the first month depending on type, severity, age, co-morbidity and effectiveness of treatment of complications. Stroke burden as measured by the disability-adjusted life-years has risen from the fifth place in 1990 to the third in [2]. Considerable spontaneous recovery occurs up to 6 months. However, patients who survive may be left with mild, moderate or severe disability [1]. Moreover, patients with a history of stroke are at risk of a subsequent event of around 10% in the first year and 5% per year thereafter [2].

The negative consequences of stroke extend well beyond the victims themselves, ultimately including families, caregivers, social networks and employers. The consequences of stroke are often devastating and a significant proportion of survivors may suffer from disabilities requiring a temporary or lifelong assistance. The proportion of patients achieving independence in self-care by 1 year after a stroke ranges from around 60% to 83%. This wide variation relates to whether the studies are community based or hospital based, which activities are considered in estimating independence, and the methods used to rate ability. In established marked economies, depending on the organization of hospital services, between 10% and 15% of survivors are resident in an institution at 1 year [3].

Primary prevention does not diminish the risk of cerebrovascular events, but only postpones the onset of the first stroke in later ages [4]. In demographically developed countries, the average age at which stroke occurs is around 73 years reflecting the older age structure of these countries. This could explain why after a first decrease of stroke incidence in developed countries, there is now a stabilization and a new increase due to ageing of the population (for an increase of at risk people). The increase in life expectancy after stroke mainly due to stroke medicine to reduce fatal outcomes and its progression through increasingly disabled states will produce an expansion of stroke morbidity worldwide. Stroke survivors will survive for longer, and therefore, the period of time that they spend in a state of chronic ill-health and disability at the end of life will increase. In less developed regions, the lower average age of stroke onset will increase further this period of ill-health and disability in the population if age-specific prevalence rates remain constant as the population ages.

2 Age Is the Principal Risk Factor for Stroke

The incidence of stroke increases significantly with age both in men and women with incidence rates accelerating exponentially above 70 years [5]. However, there are gender differences in the incidence by age subgroups. Men aged up to 75 years old are more likely to be hit by stroke than women. The risk to have a stroke then becomes higher in women than men aged 85 years or older [5]. This may be attributed to sex-related differences in life expectancy of women and the development of age-related atherosclerosis. It should be noted that the age-associated decline in functional reserve is most pronounced after the age of 85, and implies an impaired response to stressors and illnesses. Importantly, age-associated changes show great variability among individuals, which may be modulated by genetic and long-term lifestyle factors [5–7].

3 Stroke Commorbidities

Since stroke afflicts mostly the elderly comorbid patients, it is highly desirable to test the efficacy of stroke therapies in an appropriate animal stroke model. Animal models of stroke often ignore age and comorbidities frequently associated with senescence, and this could be one of the explanations for unsuccessful bench-to-bedside translation of neuroprotective strategies.

Worldwide stroke is increasing in parallel with modernization, changes in life-style, and the growing elderly population. In particular, rates in Eastern Europe have been increasing, such that currently the highest rates are found in countries such as Bulgaria, Romania, and Hungary. Women and men individuals with a low-risk life-style (smoking, exercising daily, consuming a prudent diet including moderate alcohol) and having a healthy weight during mid-life had a significantly lower risk of stroke than individuals without a low-risk lifestyle. Therefore, the relatively high incidence of stroke may be due, in part, to the impact of numerous known risk factors: arterial hypertension, diabetes, high cholesterol, smoking, alcoholism, obesity, stress, and a sedentary lifestyle [8].

Comorbidities such as diabetes, arterial hypertension or hypercholesterolemia, are common in elderly persons and are associated with a higher risk of stroke, increased mortality and disability [9]. Moreover, simultaneous presence of vascular diabetic complications-associated comorbidities like hypertension and chronic diabetes, significantly increase the level of ischemic damage in humans and animal models [10, 11].

High blood pressure is a major risk factor for stroke. Large clinical trials have shown that ACE inhibitors reduce the incidence of stroke by up to 43% [12]. However, because normotensive patients also benefit from ACE inhibition it has been suggested that these effects may also be independent of the blood pressure-lowering effects of ACE inhibition [13]. Indeed, neither short (7 days) nor long-term (42 days) administration of ACE inhibitors to SHR prior to stroke, reduced the infarct size despite lowering the blood pressure while WKY normotensive rats showed, paradoxically, marked reductions in infarct volume [14].

By Magnetic Resonance Imaging MRI, hyperglycemia was also shown to accelerate infarct progression in cortical areas [15]. However, the mechanisms of hyperglycemia-associated infarct progression remain unclear. It could be that hyperglycemia aggravates brain infarction by hemorrhagic transformation that leads to blood–brain barrier (BBB) disruption and neuronal cell death [16, 17].

Our knowledge about the molecular and cellular mechanisms underlying accelerated infarct progression in subjects with metabolic syndrome is still poor. Some studies report a strong connection between nutrition and body weight, on one hand, and increased oxidative stress or pro-inflammatory changes in the brain, which promote neural imbalance and glucose level elevation, on the other hand [18]. Zhang and colleagues suggested that metabolic inflammatory changes in the brain are linked to the inflammatory IKK/NF-kB signaling pathway [19]. Moreover, the patient's weight may show a complex relation with stroke outcomes, leading to more deaths among individuals with dramatically lower BMIs (obesity paradox) [22]. However, this may reflect undernourishment by consciously avoiding essential nutrients in daily meals [23].

Observational studies have shown a strong correlation between increased blood lipid levels and stroke [20]. In animal models, it could be shown that vascular endothelial growth factor (VEGF)-induced angiogenesis is compromised by hyperlipidemia and provided an explanation of the poor efficacy of pro-angiogenic therapies is animal models of hyperlipidemia [21]. Over-nutrition and hypercholesterolemia may not only be responsible for metabolic inflammation of the brain but can also induce mitochondrial dysfunction and increased oxidative stress [18, 22]. In this light, a better understanding of molecular factors and signaling pathways underlying the metabolic syndrome as well as the contribution of comorbidities to stroke-induced sequelae, may be translated into more successful treatments or prevention therapies against age-associated diseases which, in turn, would extend lifespan and improve lifespan quality.

The effect of age and gender on stroke incidence, functional recovery and mortality has not only been shown in humans but also in animal models [23, 24]. Indeed, the age-dependent increase in the infarct volume strongly suggests that age accounts for the variability in tissue outcome in acute human stroke [25].

4 Stroke Models Using Aged Animals Are Clinically More Relevant

Over the past 10 years, a variety of models of middle cerebral artery occlusion (MCAO) have been established in rodents [26]. MCAO in aged rodents has been produced with permanent or transient occlusion for 30–120 min using (1) MCA ligation after cranioectomy [27]; intraluminal thread occlusion [28]; using a hook attached to a micromanipulator [29]; cauterization [30, 31]; photothrombosis [32]; endothelin injection [33, 34], injection of a thrombus via external carotid artery (ECA) [35], or intraluminal thrombus formation by thrombin injection using occlusion of distal branches of the middle cerebral artery (MCA).

Since focal cerebral ischemia is technically difficult to perform in very old rats and since based on epidemiological studies human stroke occurs more often in late middle aged (60–70 years old) subjects [36], it is advisable to use middle aged instead of very old animals for stroke research [37].

4.1 Spontaneous Stroke Recovery in Aged Patients and Animals

Stroke patients regain some of their lost neurological functions during the first weeks or months after the stroke. In clinical practice recovery is thought to occur via recruitment of neighbouring neuronal circuitries and physical therapy is widely

Table 3.1 Recent therapies for acute ischemic stroke in humans

Therapy	Effects on recovery	Effect on survival	Reference
Endovascular thrombectomy	1. Lower degrees of disability at 3 months	1. Increases survival rate	[44]
Recombinant tissue plasminogen activator	1. Improves learning and memory 2. Promotes neurite outgrowth, synaptic plasticity	1. Increases survival rate	[45]
CTX0E03 (human neural stem-cells)	1. Improved neurological function	1. Improves quality of life 2. Prevents reoccurrence of stroke	[46]
Cortexin	1. Normalization of the quality of life in the early rehabilitation petriod		[47]

used to exploit this phenomenon for stimulating post stroke brain recovery [38–41].

In animal models of stroke, complete spontaneous recovery may occur in young rats, depending on the size and location of the ischemic lesion. Under normal conditions, young rats begin to show improvements of neurological function starting by day 2 post stroke, whereas in aged rats, neurological recovery is hardly detectable before days 4–5, with about 75% of the functional improvement observed in young rats. However, stroke recovery is delayed and often incomplete in aged rats [42]. Therefore, unlike stroke therapies in humans (Table 3.1) most of recent therapies involving young animals are reportedly successful (Table 3.2).

Housing experimental animals in an enriched environment enhances the recovery from brain damage both in young and aged animals [42]. When aged rats were allowed to recover in an enriched environment, the delay period was shortened and behavioural performance was significantly improved. The improvement in task performance positively correlated with slower infarct development, fewer proliferating astrocytes and smaller size of the glial scar [42]. It has been hypothesized that older brains may be more vulnerable to stroke because of decreased rates of compensatory oligodendrogenesis due to an age-related decline in cyclic AMP response element-binding protein (CREB)-mediated oligodendrogenesis after brain injury [43]. Even more effective rehabilitation of the contralateral forelimb could be achieved by Corbett and colleagues by combining enriched environment with training [44].

Spontaneous recovery is common if the infarct is located in striatum, a subcortical structure that exhibits activity-dependent plasticity and is important for controlling movement and motor learning. The enhanced recovery was associated with structural and synaptic plasticity in the contralesional striatum [45]. This may explain why patients with subcortical lacunar stroke are more likely to have early functional recovery after stroke [46, 47]. Other studies suggest that the beneficial

Table 3.2 Recent therapies for ischemic stroke in animal models

Therapy	Molecular targets	Effects on recovery	Effect on stroke area	Reference
Thiamet G	1. Inhibits β-N-acetylglucosaminidase (OGA) 2. Modulates the expression of pro-inflammatory and anti-inflammatory cytokines 3. Decreases the expression of iNOS and COX2 mainly by suppressing NF-κB p65 signaling	Improved outcome in neurobehavioral tests	Reduces the infarct volume	[48]
Ruscogenin	1. Down-regulation of intercellular adhesion molecule-1 (ICAM-1) 2. Involved in nuclear factor-κB (NF-κB) activation in anti-inflammatory pathways 3. Suppresses inducible nitric oxide synthase (iNOS), cyclooxygenase (COX-2), tumor necrosis factor-α (TNF-α) and interleukin-1β (IL-1β)	Improved neurological deficits	Decreased the infarct volume Reduced brainwater content	[49]
Curcumin	1. Inhibits the activation of TLR2/4-NF-κB signaling pathway 2. Attenuates the release of TNF-α and IL-1β in blood	Reduced neurological deficit scores	Decreased the infarct volume Reduced brainwater content	[50]
Xanthotoxol	1. Inhibits the neutrophil infiltration, 2. Decreases the expression of ICAM-1 and E-selectin 3. Attenuates Brain Blood Barrier disruption 4. Reduces the IL-1β, TNF-α, IL-8 and NO level, and attenuates the iNOS activity	Reduced neurological deficit scores	Reduced brain edema	[51]
Leonurine	1. Attenuates mitochondrial membrane swelling, 2. Restores the mitochondrial membrane potential and content of cytochrome c (Cyt-C) in mitochondria from ischemic cortex 3. Decreases the expression of Bax and increases the expression of Bcl-2 4. Decreases reactive oxygen species (ROS) level	Improved neurological outcome	Decreased the infarct volume	[52]

(continued)

Table 3.2 (continued)

Therapy	Molecular targets	Effects on recovery	Effect on stroke area	Reference
Liraglutide	1. Prevents apoptosis by increasing the anti-apoptotic protein Bcl-2 expression and decreases pro-apoptotic protein Bax expression 2. Decreases oxidative stress.	Attenuated neurological deficit	Decreased the infarct volume	[53]
Scutellarin	1. Has an inhibitory effect on the ACE/Ang II/AT1 axis 2. Proinflammation inhibition	Attenuated neurological deficit	Decreased the infarct volume	[54]
Endothelial cells-derived microvesicles	1. Improves local cerebral blood flow 2. Modulates astrocyte functions, Brain Blood Barier integrity	Attenuated neurological deficit	Reduced infarct volume	[55]

effect could be due to in situ secretion of neuroprotective factors by the transplanted cells. For example, human-derived inducible pluripotent cells (iPSC) implanted into striatum of young animals at 1 week after MCAO protected substantia nigra from atrophy, probably through a trophic effect [48].

5 Stroke Therapy in Aged Subjects Using G-CSF

The hematopoietic factor G-CSF (Granulocyte-Colony Stimulating Factor) effectively reduces infarct size and improves functional outcome after various types of experimental stroke [49–52]. G-CSF exerts a wide range of potential effects and can reduce the number of fatal hemorrhages after experimental thrombolysis in young animals models of stroke [55]. Under ischemic conditions, G-CSF inhibits programmed neuronal cell death [53] and stimulates neural progenitor cell differentiation. These mechanisms and others, including immunomodulation, anti-apoptotic properties, blood vessel plasticity, and by reducing excitotoxicity-driven penumbral apoptosis are currently thought to be responsible for infarct size reduction and improved functional outcome in young-adult rodent stroke models treated with G-CSF [54, 56–58]. Indeed, G-CSF was once viewed as one of the best preclinically studied candidate stroke drugs of the recent years that was translated into clinical development [59].

One potential weakness of the preclinical dataset is, however, the lack of proof in aged subjects. A lack of data from aged subjects in preclinical studies may at least in part explain the failure of candidate neuroprotective drugs in clinical trials. The aged brain has compared to the young brain an enhanced susceptibility to stroke and displays a limited recovery from an ischemic injury [29, 31, 42, 60].

G-CSF treatment was recently shown to increase substantially the number of neural progenitor cells and immature neurons in subcortical regions adjacent to the infarcted area. G-CSF also increased neurogenesis in the dentate gyrus of the hippocampus. This cell-regenerative effect in young adult animals could be reconfirmed to some extent in the aged rats that have been treated with G-CSF for 14 days after stroke [61]. Although G-CSF treatment in aged rats increased the number of proliferating cells in the dentate gyrus and in the subventricular zone (SVZ), in aged rats there were more newborn neurons only in the SVZ of the damaged hemisphere. Likewise, G-CSF treatment in aged rats after stroke enhances survival, functional neurological recovery [61, 72].

6 Cell Therapy of Stroke Using a Combination Therapy

Cellular therapy can enhance the endogenous restorative mechanisms of the injured brain by supporting processes of neovascularization, neurogenesis, neural reorganization and functional recovery (Fig. 3.1).

Despite some positive impact of G-CSF on the post-stroke aged brain, a monotherapy may have, nevertheless, limited effects on tissue and functional recovery after ischemic stroke. Therefore, treating post-stroke aged rats with a combination of bone marrow-derived mononuclear cells (BM-MNC) and G-CSF might improve the long term (56 days) functional outcome by compensating the delay before G-CSF comes to full effect. To this end, 1×10^6 syngeneic BM-MNC per kg bodyweight in combination with G-CSF (50 μg/kg, intraperitoneal application, continued for 28 days) were administered via the jugular vein to aged Sprague-Dawley rats at 6 h post-stroke. Infarct volume was measured by magnetic resonance imaging at 3 and 48 days post-stroke and additionally by immunohistochemistry at day 56 (Fig. 3.2). Functional recovery was tested during the entire post-stroke survival period. Daily G-CSF treatment led to robust and consistent improvement of neurological function, but did not alter final infarct volumes. This result was unexpected since benefits of G-CSF and BM-MNC treatment paradigms in stroke, independently

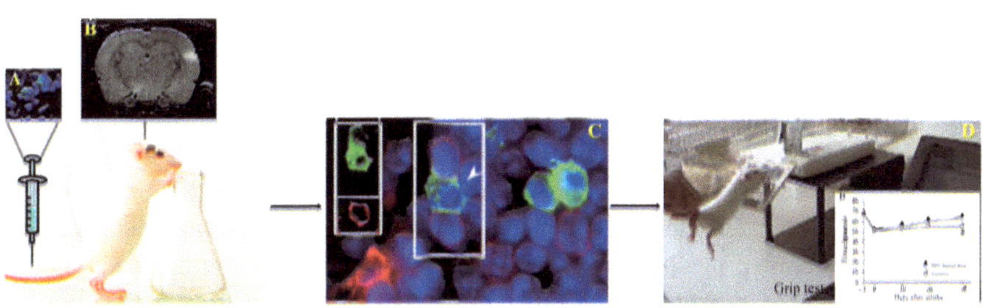

Fig. 3.1 Cellular therapy can enhance the endogenous restorative mechanisms of the injured brain by supporting processes of neovascularization, neurogenesis, neural reorganization and functional recovery

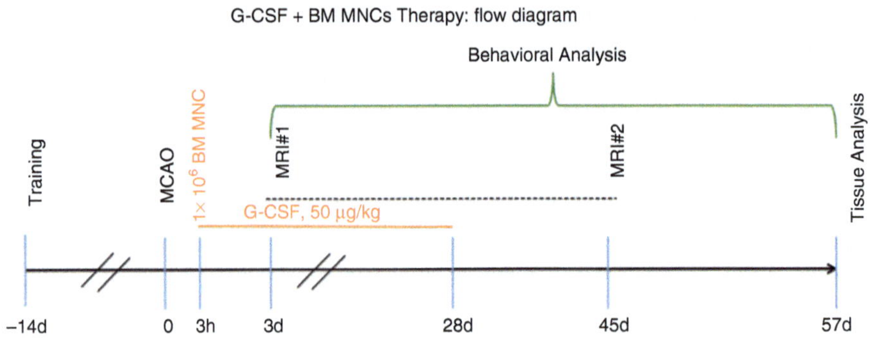

Fig. 3.2 Bone marrow derived mononuclear cells (BM-MNC) and G-CSF might improve the long term (56 days) functional outcome by compensating the delay before G-CSF comes to full effect. To this end, 1×10^6 syngeneic BM-MNC per kg bodyweight in combination with G-CSF (50 μg/kg, intraperitoneal application, continued for 28 days) were administered via the jugular vein to aged Sprague-Dawley rats at 6 h post-stroke. Infarct volume was measured by magnetic resonance imaging at 3 and 48 days post-stroke and additionally by immunohistochemistry at day 56

from each other, have been repeatedly reported by independent experiments and groups were hypothesized to work synergistically especially in the aged, stroke-lesioned brain. The lack of an additional benefit may be due to an hitherto not well investigated interaction between both approaches and, to a minor extent, to the insensitivity of the aged brains to regenerative mechanisms. Also considering recent findings on other tandem approaches involving G-CSF in animal models featuring relevant co-morbidities, we conclude that such combination therapies are not the optimal approach to treat the acutely injured aged brain.

Current knowledge suggests that administered BM-MNC provide indirect neuroprotection leading to infarct size reduction after ischemic damage in a time window of up to 1 month [62]. G-CSF in turn, induces BM MNC mobilization while the SDF-1/CXCR4 mechanisms, causing BM-MNC to invade the ischemic brain [51, 63], where they are believed to exert therapeutic effects. However, the initiation of this potentially beneficial action may take simply too much time: although a granulocyte boost is seen after about 48 h, peaking G-CSF-based mobilization can take up to 9 days [64], which is beyond the therapeutic time window for BM-MNC. Since endogenous G-CSF is not available in sufficient concentrations directly at the lesion site after the ischemic event [57], a combination therapy providing (1) G-CSF in sufficient amounts to act neuroprotectively and (2) exogenous BM-MNC early enough to bridge the time gap until G-CSF-based endogenous BM-MNC mobilization comes to full effect, seemed promising—but failed to fulfill the expectations. One may assume that either the lesioned and aged rat brain environment was insensitive to regenerative mechanisms by BM-MNC or cell treatment has been mainly ineffective. Indeed, the aggravated impact of ischemic damage on the aged brain is well known while potential detrimental effects of ageing on BM-MNC have been anticipated [37, 65, 66]. Moreover, technical complications may come into play as well: a limited influence of long-term cryopreservation on the therapeutic efficacy of umbilical cord blood MNC, a population being very similar to BM-MNC, has been discussed recently [67]. However, deriving syngeneic cells from young animals and limiting cryopreservation to no more than 4 weeks might have limited effects on the donor side due ageing and cryopreservation process itself. An alternative explanation for the reduced efficacy of the combination treatment could be interference between both treatment regimes.

A recent study in hypertensive animals demonstrated that intravenously administered BM-MNC challenges the splenic granulocyte clearance capacities for apoptotic cells [81]. This clearance system usually removes apoptotic granulocytes from the circulation, which represents an important anti-inflammatory mechanism [82]. Being already compromised by externally administered BM-MNC, the newly generated granulocyte boost from the BM by G-CSF may have completely exhausted the clearance system after treatment. This detrimental interaction may have caused a sustained systemic and central pro-inflammatory bias, leading to subtle additional damage, not enhancing, but partly reducing the neuroprotective G-CSF effect.

7 Co-administration of G-CSF and BM-MSC in the Microenvironment of the Post-stroke Aged Rats

Due to the ethical concerns and limited availability of using pluripotent embryonic stem cells (ESCs) and iPS in the clinic, the emphasis was placed on mesenchymal stem cells (MSCs), which are free of both ethical concerns and teratoma formation.

Recent studies suggest that modified (MSCs) are able to form a "biobridge" between neurogenic subventricular zone (SVZ) and the ischemic cortex area (penumbra). Using this road, endogenous stem cells can migrate from the neurogenic area to the site of lesion and may ameliorate outcome in experimental models of cerebral ischemia [39, 68–70]. Several studies showed that grafting bone marrow derived stem cells in the peripheral circulation improved functional neurological outcome and reduced the infarct volume. Most of these studies used bone marrow mesenchymal cells (BM-MSC) [38, 71].

Mesenchymal stem cells and hematopoietic stem/progenitor cells (HSPC) that are most frequently used in preclinical and clinical neurorestorative studies after stroke, augments this endogenous response. MSC can also be obtained from adipose tissue [72]. HSPC can be isolated from bone marrow or from umbilical-cord blood (UCB), or can be mobilized into the blood by the administration of G-CSF.

Amniotic fluid has been investigated as a new cell source for mesenchymal stem cells in the development of cell-based transplantation. Earlier studies have demonstrated the ability of amniotic fluid-derived stem cells to differentiate along a neurogenic pathway [73].

The combination of mesenchymal stem cells and neural stem cells (NSCs) could improve also functional recovery after stroke if given prior stroke. To this end, a mix of MSCs isolated from the femurs and tibias of rats and NSCs isolated from rat embryo ganglion eminence were labeled with PKH26-GL and administered 1 day before stroke into the lateral ventricle and neurological recovery evaluated for 28 days after stroke. The results suggest that this combination cell therapy is more efficient in promoting brain recovery after stroke than each stem cell alone [74].

More recently, triple cells co-transplantation with a mixture of rat NSCs, astrocytes and brain microvascular endothelial cells (BMECs) have been attempted. After grafting these cells into the ischemic brain it was found that the learning and memory ability of these rat improved to some extend. Moreover, rats with triple cells transplantation did perform better than those who grafted with two cells only. Rats grafted only one cell showed least improvement. From this experiment, it was concluded that co-transplantation of NSCs with astrocyte and BMECs can improve learning and memory in the water maze test, probably due to the microenvironment improvement by the transplanted astrocytes and BMECs [75].

In another approach, the combination therapy was given in sequence. First, mesenchymal stem cells were transplanted during the acute phase after stroke (1 day) in an attempt to diminish the inflammation and provide an appropriate microenvironment for regeneration after ischemia. Then, the neural stem cells were transplanted at 7 days after stroke to help regeneration by differentiation into neurons, oligodendrocytes or astrocytes [74].

Finally, experiments aimed at improving long term functional outcome in aged rodents by grafting pre-differentiated BM-MSC in G-CSF-treated animals, have been performed. To this end, rat BM-MSC isolated from young Sprague-Dawley rats were administered a single dose of BM-MSCs (10^6/kg) given in combination with G-CSF (50 μg/kg) via the jugular vein or intrathecally at 6 h post-stroke. The phenotypes of BMSCs used in this study were positive for CD105, CD166, CD29, and CD44 [92]. Cells tested negative for CD14, CD34 and CD45. Prior to transplantation the cells were tested for purity by flow cytometry and for their ability to differentiate into osteogenic, chondrogenic and adipogenic lineages. The control groups received daily injections of either G-CSF 50 μg/kg or vehicle (5% glucose) for 28 days. To investigate the localization of injected cells, a separate group of aged animals were injected with mesenchymal cells of human origin (hBMMSCs). Although hBMMSCs are poorly immunogenic [76] the animals were given cyclosporine A (s.c., Sandimmun, Novartis, 10 mg/kg) diluted in Chremophor EL, Sigma, to prevent graft rejection.

Infarct volume was measured by MRI at 3- and 48 days post-stroke and additionally by immunhistochemistry at day 56. Functional recovery was tested during the entire post-stroke survival period of 56 days. Daily treatment of post-stroke aged rats with G-CSF led to a robust and consistent improvement of neurological function. The combination therapy, GG-CSF + BM MSC in aged rats showed, surprinsingly, no additional improvement in recuperation of the sensory function (adhesive tape), although recuperation of more complex motor (rotating pole) and spatial reference-memory tasks was improved both by G-CSF and the combination. Paradoxically, MCAO rats swam slightly faster than unoperated animals, probably due to a post-stroke excitatory state, an observation confirmed in previous studies [77].

Of the treated groups, the best recovery rate was seen for the G-CSF group which showed significant improvement in the watermaze spatial reference-memory task between days 21 and 42, suggesting that the beneficial effect of the G-CSF treatment is restricted to the G-CSF treatment period [61]. Probably, the improved functional recuperation of the G-CSF group may have been helped by the stimulation of endogenous neurogenesis by G-CSF as previously reported [61]. In the combined treatment study it was found an increased cellularity in the formerly infarct core of the G-CSF + BM MSC group at day 56 post-stroke and intact neurogenesis in the lateral ventricle region. However, there was a clear regional separation of the DCX^+ cells which emanated from the ventricular wall, and BrdU-labeled nuclei which were localized mainly in the vascular network of the lateral ventricle. Since BrdU was administered for the first 14 days after stroke, it seems likely that at 2 months post-stroke, DCX^+ cells with BrdU nuclei did not survive. Instead, $BrdU^+$ nuclei survived most likely in endothelial cells of the vascular wall [78]. Further, the combination therapy significantly improved recuperation and microvessel density in the formerly infarct core and beyond. Finally, it was found that the aged brain environment is permissive for the migration of human BMMSCs toward the lesion site. Finally, it was suggested that, in a real clinical situation involving older post-stroke patients, successful regenerative therapies may have to be delivered throughout a prolonged period, perhaps for 6–12 months.

8 Functional Neurological Recovery and Tissue Repair After Neural Tissue Transplantation

Although rehabilitation is important for improving functional recovery in the early stages after stroke, it does not provide a replacement of lost tissue. Moreover, if and how the aged brain responds to grafted neural tissue is largely unknown. For example mouse fetal hippocampal NSCs implanted into the injured hippocampus of 24-months-old rats, exhibited limited neuronal plasticity, robust astrocytic differentiation, and impaired migration [96].

Most clinical studies conducted so far used neural cells derived from human fetal donors. The techniques to achieve effective survival and growth of neuronal tissues transplanted into the CNS are meanwhile will established [79]. Even though effective, neural grafting has, however, not become a standard treatment for several reasons including the limited supply of fetal tissue of human origin and the beneficial effects have also been controversial [80]. Of the various options, stem cell therapy presents us with a viable alternative [81]. In order to enable the replacement of lost tissue, cell replacement strategies were used in human stroke patients. However, these early clinical studies lacked appropriate control groups [82, 83].

9 Stem Cell Therapy in Subcortical Stroke. Role of Endogenous Neurogenesis and Aging

The ultimate goal of stroke treatment is restoration of neurological function. Stroke is a heavily undertreated disease demanding a vigorous search for new therapies. Despite of improving knowledge about stroke pathology, therapeutical benefits for stroke patients are limited. Distinguished by a necrotic core surrounded by the ischemic area (penumbra), stroke is still the largest cause of disability in stroke survivors. Crucial for recovery phase are the first days and weeks after stroke. Studies before showed that even years after stroke is possible "remodeling" of the brain by neuroplasticity. However, despite the recent progression in stroke research, the major problems to be solved for stroke survivors remains the restorative process.

Spontaneous recovery is common, whenever the infarct is located in the striatum, a subcortical structure that exhibits activity-dependent plasticity and is important for controlling movement and motor learning. Neurological recovery is associated with structural dendritic and synaptic plasticity in the contralesional striatum [45] and axonal plasticity in contralesional motor cortex [84], which may explain why patients with subcortical stroke are likely to exhibit functional neurological recovery [46, 47].

Cell-based therapy augments this endogenous response. Thus, human iPSCs implanted into striatum of young-adult animals at 1 week after MCAO protected substantia nigra from atrophy, probably through a trophic effect via release of survival-promoting growth factors [48]. However, how cells are transplanted and where they are placed after stroke are important issues in graft survival and efficacy in promoting behavioural recovery. Data from many groups have shown that stroke increases proliferation of neuronal progenitors in the ipsilateral subventricular region of young-adult rodents with a maximum at 1–2 weeks, and the newly generated neuroblasts migrate to the damaged area in the peri-infarcted striatum over a period of several months. Eventually the neuroblasts differentiate into medium size spiny neurons and may become part of the neuronal network [85–89]. It seems that the injected cells itself can also stimulate neurogenesis in the SVZ [32, 88].

It has been shown that intracerebral transplantation of NSI-566RSC, a spinal cord-derived NSC line, at two sites within the striatum reduced behavioral deficits associated with ischemic stroke. Significant improvements in both motor and neurological tests were detected in the NSI-566RSC-treated stroke animals. In addition, the results revealed significant dose-dependent differences in the behavioral improvement across treatment groups at post-transplantation periods with the highest NSI-566RSC dose showing the most significant improvement in both motor and neurological tests. These results have demonstrated the safety and efficacy of NSI-566RSC in a subacute model of ischemic stroke in rats [70].

However, the proportion of surviving neurons is discouragingly low [85, 90, 91]. In animal models, the number of new striatal neurons in aged rodents after stroke was similar to that in young animals [92, 93] despite 50% decline in neurogenesis in the subventricular zone of elderly rodents compared to young adult animals [94, 95]. Similar findings have been reported in humans [87, 96–98]. Earlier studies on postmortem human brains provided evidence that there might be SVZ cell proliferation and

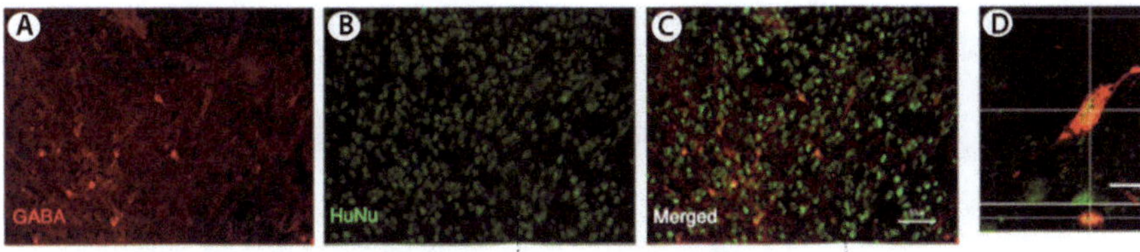

Fig. 3.3 Human iPSC survived and differentiated into neurons after intracortical transplantation in aged rats with cortical stroke and also improved functional recovery in cylinder test at 4 and 7 weeks

neuroblast formation after stroke even in aged patients [87, 97, 99]. The finding that new neurons are continuously added in the adult human striatum along with the presence of an increased number of putative neuroblasts in the human striatum after stroke lends support to this hypothesis [97]. However, whether endogenous neurogenesis contributes to spontaneous recovery after stroke has not yet been established. In addition, age, co-morbidities, physical condition of the patient and severity of disease could substantially influence these steps and, therefore, the outcome of the healing process.

The establishment of iPSCs offers new prospects for stroke treatment. iPSCs can be generated from a patient, avoiding both ethical problems and immune rejection and a limited differentiation potential of adult stem cells [100]. However, if and how the aged brain responds to grafted cells is largely unknown. Experiments done so far yielded conflicting results. For example, mouse fetal hippocampal NSCs implanted into the injured hippocampus of 24-months-old rats, exhibited limited neuronal plasticity, robust astrocytic differentiation, and impaired migration. Still another study using NSCs transplanted into young-adult (3-month-old) and aged (24-month-old) rat brains at 1 day after stroke reportedly reduced ischemic brain injury in aged rats [101]. In stroke models, hiPSC-lt-NES cells derived from a young adult male have the potential to survive, differentiate into immature and mature neurons, and migrate to the peri-infarct area of aged rats. The treated aged rats showed improved behavioral recovery after implantation into the stroke-injured striatum and cortex of adult rats [102, 103]. In a recent study, it could be shown that human iPSC survived and differentiated into neurons after intracortical transplantation in aged rats with cortical stroke and also improved functional recovery in cylinder test at 4 and 7 weeks (Fig. 3.3) [104].

Recent studies indicate that inducible pluripotent cells (iPSCs) can also be generated from aged humans and differentiate into specific cell types [104–106]. Moreover it seems that the re-differentiation efficiency of human fibroblasts via iPSCs into functional motor neurons is the same as in 29–82 year old individuals [107].

10 Mesenchymal Cells Can Be Used as Drug Carriers for Stroke Therapy

Cyclin-dependent kinase-5 (Cdk5) is over-expressed in both neurons and microvessels in hypoxic regions of stroke tissue and has a significant pathological role following hyper-phosphorylation leading to calpain-induced cell death. Recently, the

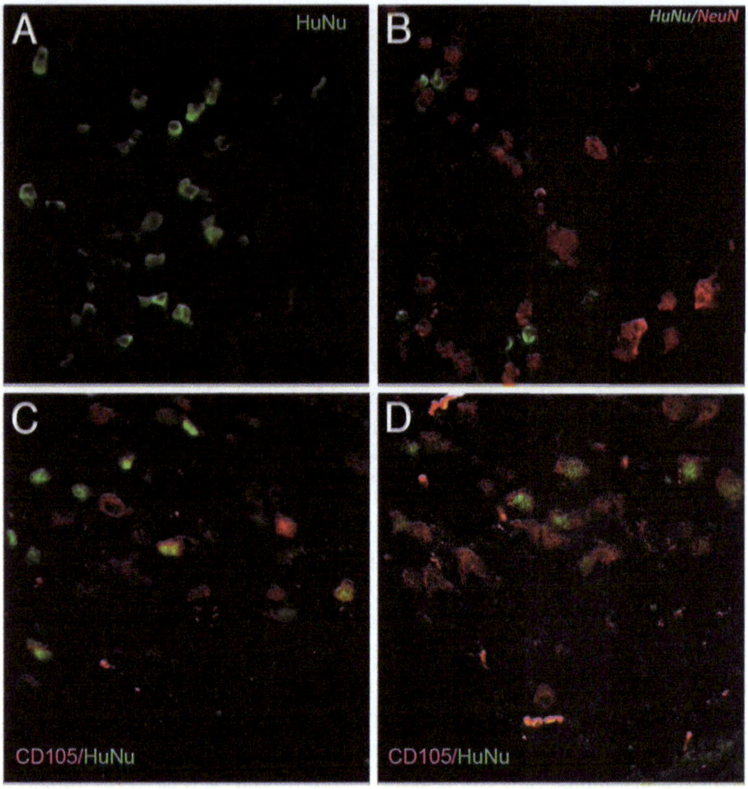

Fig. 3.4 Intracortical administration of mesenchymal harboring the CIP-peptide to middle aged rats with stroke increased survival of transplanted cells in the perinfarct area as shown by double immunofluorescence

neuronal cyto-protective potential of a natural small peptide (CIP-peptide) was demonstrated after neurotoxic stress. CIP is a derived-p35 cleavage peptide, which selectively targets Cdk5/p25 activity without affecting Cdk5/p35 signalling. In hypoxia, insertion of Cdk5/p25-inhibitory peptide (CIP) vector preserved and enhanced in vitro angiogenesis [108]. Indeed, intracortical administration of mesenchymal harboring the CIP-peptide to middle aged rats with stroke increased survival of transplanted cells in the perinfarct area as shown by double immunofluorescence (Fig. 3.4).

11 Angiogenesis After Stroke in the Aged Brains

Recuperative therapeutic strategies for stroke are focused on revascularization, neuroprotection and neuroregeneration, but most of the strategies that have been clinically tested failed to show benefit in humans. Post-stroke vascular remodeling is an essential event with crucial importance for neuroregeneration, but unfortunately this process is still incompletely understood and therefore not exploited for therapeutic purposes [109–112].

Impaired neovascularization was described in elderly, but the effect of aging on angiogenesis and vascular remodeling after stroke has not been studied in detail. Previous studies from our group showed that, following insult to the brain, old rats are still capable of upregulating genes that are active during development, but the response is often blunted and temporarily uncoordinated [113].

Understanding mechanisms underlying angiogenesis and vascular remodeling after stroke in the elderly is crucial for developing new treatment strategies to improve the functional outcome after stroke in aged patients. Unfortunately, the molecular mechanisms regulating angiogenesis and vascular remodeling in aging brains are still poorly understood. Recently the regenerative potential of endothelial progenitor cells (EPC) has been under intense investigation. Endothelial progenitor cells are likely to promote vasculogenesis after cerebral ischemia. Therefore the regenerative potential of endothelial progenitor cells (EPCs) has been under intense investigation [114]. Many angiogenic factors, such as VEGF, IGFs or FGFs, are involved in the mobilization of EPCs and the differentiation and increased levels of EPCs were correlated with increased plasma VEGF levels in stroke patients [108]. However, currently there is no effective and safe stem-cell based therapy for stroke [115].

Other studies have established that bone marrow-derived (EPCs) are present in the systemic circulation, and they are able to differentiate into mature endothelial cells in the ischemic area, but the number of these cells is reduced by aging [116, 117]. The cytokine-induced generation of bone marrow-derived EPCs can be enhanced by the administration of G-CSF, and leads to improved functional outcome after stroke in aged rats [61].

Few studies have investigated human post-stroke angiogenesis at the molecular level. Thus, Krupinski and colleagues [118] noted active angiogenesis in the penumbral areas of patients who survived from several days to weeks after cerebral stroke, as well as a positive correlation between microvessel density and patient survival. In subsequent studies, the authors demonstrated an increased synthesis of angiogenic growth factors such as FGF-2, PDGF, VEGF and their receptors within hours of stroke that correlated with blood vessel growth in the penumbra [119, 120].

The literature on gene expression profiles after stroke in humans also is limited. In this regard, Vikman and Edvinsson [121] have shown similarities in gene expression profiles between human strokes and those in animal models, and reported new genes that support the dynamic changes that occur in the middle cerebral artery branches supplying the ischemic region. Also, promising results of blood genomic profiling in human stroke have been obtained in pilot studies [122–124].

These results argue for the utility of proangiogenic therapies in stroke, given the potential effects consisting of increasing blood flow, decreasing infarct size and supporting the restoration and recovery of neurovascular networks after ischemia [61].

Despite the obvious clinical significance of post-stroke angiogenesis in aged subjects, a detailed transcriptomic analysis of post-stroke angiogenesis has not yet been undertaken in an aged experimental model. By combining stroke transcriptomics with immunohistochemistry in aged rats and post-stroke patients, an attempt was made to identify an age-specific gene expression pattern that may characterize the angiogenetic process after stroke. It was found that both young and old infarcted rats initiated vigorous angiogenesis. However, the young rats had a higher vascular density by day 14 post-stroke. "New-for-stroke" genes that were linked to the increased vasculature density in young animals included *Angpt2, Angptl2, Angptl4, Cib1, Ccr2, Col4a2, Cxcl1, Lef1, Hhex, Lamc1, Nid2, Pcam1, Plod2, Runx3, Scpep1, S100a4, Tgfbi* and *Wnt4*, which are required for sprouting angiogenesis, reconstruction of the basal lamina and the resolution phase. The vast majority of genes involved in sprouting angiogenesis (*Angpt2, Angptl4, Cib1, Col8a1, Nrp1, Pcam1, Pttg1ip, Rac2, Runx1, Tnp4, Wnt4*); reconstruction of a new basal lamina (*Col4a2, Lamc1, Plod2*) *or* tube formation and maturation (*Angpt1, Gpc3, Igfbp7, SPARC, Tie2, Tnfsf10*), had however, a delayed upregulation in the aged rats. The angiogenetic response in aged rats was further diminished by the persistent upregulation of "inflammatory" genes (*Cxcl12, Mmp8, Mmp12, Mmp14, Mpeg1, Tnfrsf1a, Tnfrsf1b*) and vigorous expression of genes required for the build-up of the fibrotic scar (*Cthrc1, Il6ra, Il13ar1, Il18, Mmp2, Rassf4, Tgfb1, Tgfbr2, Timp1*). Beyond this barrier angiogenesis in the aged brains was similar to that in young brains. It was also reported that the aged human brain is capable of mounting a vigorous angiogenic response after stroke, which most likely reflects the remaining brain plasticity of the aged brain [125].

12 Conclusions

Recent results using a variety of drug, cell therapy or combination thereof suggest that, (1) G-CSF in aged rats has primarily a beneficial effect on functional outcome most likely via supportive cellular processes such as neurogenesis; (2) the combination therapy, G-CSF with mesenchymal cells (G-CSF + BM-MSC or G-CSF + BM-MNC) did not further improve behavioral indices, neurogenesis or infarct volume as compared to G-CSF alone in aged animals; (3) better results with regard to integration of transplanted cells in the aged rat environment have been obtained using iPS of human origin; (4) mesenchymal cells may be used as drug carriers for the aged post-stroke brains. Finally, while the middle aged brain does not seem to impair drug and cell therapies, in a real clinical practice involving older post-stroke patients, successful regenerative therapies would have to be carried out for a much longer time.

Acknowledgments This study has been supported by a UEFISCDI grant PN-II-RU-TE-2014-4-0705, "The impact of microglia phagocytosis of live neurons on the efficacy of stem cell therapy of stroke "No 165/2015.

Conflict of Interest: The author confirms that he has no conflict of interest to declare for this publication.

References

1. Bonita R, Beaglehole R. Recovery of motor function after stroke. Stroke. 1988;19(12):1497–500.
2. Burn J, Dennis M, Bamford J, Sandercock P, Wade D, Warlow C. Long-term risk of recurrent stroke after a first-ever stroke. The Oxfordshire Community Stroke Project. Stroke. 1994;25(2):333–7.
3. Appelros P, Nydevik I, Viitanen M. Poor outcome after first-ever stroke: predictors for death, dependency, and recurrent stroke within the first year. Stroke. 2003;34(1):122–6.
4. Tacutu R, Budovsky A, Fraifeld VE. The NetAge database: a compendium of networks for longevity, age-related diseases and associated processes. Biogerontology. 2010;11(4):513–22.
5. Roger VL, Go AS, Lloyd-Jones DM, Benjamin EJ, Berry JD, Borden WB, et al. Heart disease and stroke statistics--2012 update: a report from the American Heart Association. Circulation. 2012;125(1):e2–e220.
6. Tacutu R, Budovsky A, Yanai H, Fraifeld VE. Molecular links between cellular senescence, longevity and age-related diseases - a systems biology perspective. Aging (Albany NY). 2011;3(12):1178–91.
7. Wolfson M, Budovsky A, Tacutu R, Fraifeld V. The signaling hubs at the crossroad of longevity and age-related disease networks. Int J Biochem Cell Biol. 2009;41(3):516–20.
8. Donnan GA, Davis SM. Breaking the 3 h barrier for treatment of acute ischaemic stroke. Lancet Neurol. 2008;7(11):981–2.
9. Goldstein LB, Bushnell CD, Adams RJ, Appel LJ, Braun LT, Chaturvedi S, et al. Guidelines for the primary prevention of stroke: a guideline for healthcare professionals from the American Heart Association/American Stroke Association. Stroke. 2011;42(2):517–84.
10. Rewell SS, Fernandez JA, Cox SF, Spratt NJ, Hogan L, Aleksoska E, et al. Inducing stroke in aged, hypertensive, diabetic rats. J Cereb Blood Flow Metab. 2010;30(4):729–33.
11. McCabe C, Gallagher L, Gsell W, Graham D, Dominiczak AF, Macrae IM. Differences in the evolution of the ischemic penumbra in stroke-prone spontaneously hypertensive and Wistar-Kyoto rats. Stroke. 2009;40(12):3864–8.
12. Yusuf S. (After the HOPE Study. ACE inhibitor now for every diabetic patient?. Interview by Dr. Dirk Einecke). MMW Fortschr Med. 2000;142(44):10.
13. Sleight P. The role of angiotensin-converting enzyme inhibitors in the treatment of hypertension. Curr Cardiol Rep. 2001;3(6):511–8.
14. Porritt MJ, Chen M, Rewell SS, Dean RG, Burrell LM, Howells DW. ACE inhibition reduces infarction in normotensive but not hypertensive rats: correlation with cortical ACE activity. J Cereb Blood Flow Metab. 2010;30(8):1520–6.
15. Martin A, Rojas S, Chamorro A, Falcon C, Bargallo N, Planas AM. Why does acute hyperglycemia worsen the outcome of transient focal cerebral ischemia? Role of corticosteroids, inflammation, and protein O-glycosylation. Stroke. 2006;37(5):1288–95.
16. Kumari S, Anderson L, Farmer S, Mehta SL, Li PA. Hyperglycemia alters mitochondrial fission and fusion proteins in mice subjected to cerebral ischemia and reperfusion. Transl Stroke Res. 2012;3(2):296–304.
17. Soejima H, Ogawa H, Morimoto T, Nakayama M, Okada S, Sakuma M, et al. Aspirin possibly reduces cerebrovascular events in type 2 diabetic patients with higher C-reactive protein level: subanalysis from the JPAD trial. J Cardiol. 2013;62(3):165–70.
18. Cai D, Liu T. Inflammatory cause of metabolic syndrome via brain stress and NF-kappaB. Aging (Albany NY). 2012;4(2):98–115.
19. Zhang X, Zhang G, Zhang H, Karin M, Bai H, Cai D. Hypothalamic IKKbeta/NF-kappaB and ER stress link overnutrition to energy imbalance and obesity. Cell. 2008;135(1):61–73.
20. Iso H, Jacobs DR Jr, Wentworth D, Neaton JD, Cohen JD. Serum cholesterol levels and six-year mortality from stroke in 350,977 men screened for the multiple risk factor intervention trial. N Engl J Med. 1989;320(14):904–10.

21. Zechariah A, ElAli A, Doeppner TR, Jin F, Hasan MR, Helfrich I, et al. Vascular endothelial growth factor promotes pericyte coverage of brain capillaries, improves cerebral blood flow during subsequent focal cerebral ischemia, and preserves the metabolic penumbra. Stroke. 2013;44(6):1690–7.
22. Herz J, Hagen SI, Bergmuller E, Sabellek P, Gothert JR, Buer J, et al. Exacerbation of ischemic brain injury in hypercholesterolemic mice is associated with pronounced changes in peripheral and cerebral immune responses. Neurobiol Dis. 2014;62:456–68.
23. Bergerat A, Decano J, Wu CJ, Choi H, Nesvizhskii AI, Moran AM, et al. Prestroke proteomic changes in cerebral microvessels in stroke-prone, transgenic(hCETP)-Hyperlipidemic, Dahl salt-sensitive hypertensive rats. Mol Med. 2011;17(7-8):588–98.
24. Gokcay F, Arsava EM, Baykaner T, Vangel M, Garg P, Wu O, et al. Age-dependent susceptibility to infarct growth in women. Stroke. 2011;42(4):947–51.
25. Ay H, Koroshetz WJ, Vangel M, Benner T, Melinosky C, Zhu M, et al. Conversion of ischemic brain tissue into infarction increases with age. Stroke. 2005;36(12):2632–6.
26. Bacigaluppi M, Pluchino S, Peruzzotti-Jametti L, Kilic E, Kilic U, Salani G, et al. Delayed post-ischaemic neuroprotection following systemic neural stem cell transplantation involves multiple mechanisms. Brain. 2009;132(Pt 8):2239–51.
27. Wang LC, Futrell N, Wang DZ, Chen FJ, Zhai QH, Schultz LR. A reproducible model of middle cerebral infarcts, compatible with long-term survival, in aged rats. Stroke. 1995;26(11):2087–90.
28. Sutherland GR, Dix GA, Auer RN. Effect of age in rodent models of focal and forebrain ischemia. Stroke. 1996;27(9):1663–7. discussion 8.
29. Popa-Wagner A, Schroder E, Walker LC, Kessler C. beta-Amyloid precursor protein and ss-amyloid peptide immunoreactivity in the rat brain after middle cerebral artery occlusion: effect of age. Stroke. 1998;29(10):2196–202.
30. Katsman D, Zheng J, Spinelli K, Carmichael ST. Tissue microenvironments within functional cortical subdivisions adjacent to focal stroke. J Cereb Blood Flow Metab. 2003;23(9):997–1009.
31. Rosen CL, Dinapoli VA, Nagamine T, Crocco T. Influence of age on stroke outcome following transient focal ischemia. J Neurosurg. 2005;103(4):687–94.
32. Zhang S, Boyd J, Delaney K, Murphy TH. Rapid reversible changes in dendritic spine structure in vivo gated by the degree of ischemia. J Neurosci. 2005;25:5333–8.
33. Soleman S, Yip P, Leasure JL, Moon L. Sustained sensorimotor impairments after endothelin-1 induced focal cerebral ischemia (stroke) in aged rats. Exp Neurol. 2010;222(1):13–24.
34. Trueman RC, Harrison DJ, Dwyer DM, Dunnett SB, Hoehn M, Farr TD. A critical re-examination of the intraluminal filament MCAO model: impact of external carotid artery transection. Transl Stroke Res. 2011;2(4):651–61.
35. DiNapoli VA, Huber JD, Houser K, Li X, Rosen CL. Early disruptions of the blood-brain barrier may contribute to exacerbated neuronal damage and prolonged functional recovery following stroke in aged rats. Neurobiol Aging. 2008;29(5):753–64.
36. Krishnamurthi RV, Moran AE, Feigin VL, Barker-Collo S, Norrving B, Mensah GA, et al. Stroke prevalence, mortality and disability-adjusted life years in adults aged 20-64 years in 1990-2013: data from the global burden of disease 2013 study. Neuroepidemiology. 2015;45(3):190–202. https://doi.org/10.1159/000441098.
37. Popa-Wagner A, Carmichael ST, Kokaia Z, Kessler C, Walker LC. The response of the aged brain to stroke: too much, too soon? Curr Neurovasc Res. 2007;4(3):216–27.
38. Hermann DM, Chopp M. Promoting brain remodelling and plasticity for stroke recovery: therapeutic promise and potential pitfalls of clinical translation. Lancet Neurol. 2012;11(4):369–80.
39. Honmou O, Onodera R, Sasaki M, Waxman SG, Kocsis JD. Mesenchymal stem cells: therapeutic outlook for stroke. Trends Mol Med. 2012;18(5):292–7.
40. Liepert J, Hamzei F, Weiller C. Lesion-induced and training-induced brain reorganization. Restor Neurol Neurosci. 2004;22(3-5):269–77.

41. Hallett M. Plasticity of the human motor cortex and recovery from stroke. Brain Res Brain Res Rev. 2001;36(2-3):169–74.
42. Buchhold B, Mogoanta L, Suofu Y, Hamm A, Walker L, Kessler C, et al. Environmental enrichment improves functional and neuropathological indices following stroke in young and aged rats. Restor Neurol Neurosci. 2007;25(5-6):467–84.
43. Miyamoto N, Pham LD, Hayakawa K, Matsuzaki T, Seo JH, Magnain C, et al. Age-related decline in oligodendrogenesis retards white matter repair in mice. Stroke. 2013;44(9):2573–8.
44. Hicks AU, Hewlett K, Windle V, Chernenko G, Ploughman M, Jolkkonen J, et al. Enriched environment enhances transplanted subventricular zone stem cell migration and functional recovery after stroke. Neuroscience. 2007;146(1):31–40.
45. Qin L, Jing D, Parauda S, Carmel J, Ratan RR, Lee FS, et al. An adaptive role for BDNF Val66Met polymorphism in motor recovery in chronic stroke. J Neurosci. 2014;34(7):2493–502.
46. Bejot Y, Catteau A, Caillier M, Rouaud O, Durier J, Marie C, et al. Trends in incidence, risk factors, and survival in symptomatic lacunar stroke in Dijon, France, from 1989 to 2006: a population-based study. Stroke. 2008;39(7):1945–51.
47. Rothrock JF, Clark WM, Lyden PD. Spontaneous early improvement following ischemic stroke. Stroke. 1995;26(8):1358–60.
48. Polentes J, Jendelova P, Cailleret M, Braun H, Romanyuk N, Tropel P, et al. Human induced pluripotent stem cells improve stroke outcome and reduce secondary degeneration in the recipient brain. Cell Transplant. 2012;21(12):2587–602.
49. Han JL, Blank T, Schwab S, Kollmar R. Inhibited glutamate release by granulocyte-colony stimulating factor after experimental stroke. Neurosci Lett. 2008;432(3):167–9.
50. Lee ST, Chu K, Jung KH, Ko SY, Kim EH, Sinn DI, et al. Granulocyte colony-stimulating factor enhances angiogenesis after focal cerebral ischemia. Brain Res. 2005;1058(1-2):120–8.
51. Shyu WC, Lin SZ, Yang HI, Tzeng YS, Pang CY, Yen PS, et al. Functional recovery of stroke rats induced by granulocyte colony-stimulating factor-stimulated stem cells. Circulation. 2004;110(13):1847–54.
52. Xiao BG, Lu CZ, Link H. Cell biology and clinical promise of G-CSF: immunomodulation and neuroprotection. J Cell Mol Med. 2007;11(6):1272–90.
53. Komine-Kobayashi M, Zhang N, Liu M, Tanaka R, Hara H, Osaka A, et al. Neuroprotective effect of recombinant human granulocyte colony-stimulating factor in transient focal ischemia of mice. J Cereb Blood Flow Metab. 2006;26(3):402–13.
54. Minnerup J, Heidrich J, Wellmann J, Rogalewski A, Schneider A, Schabitz WR. Meta-analysis of the efficacy of granulocyte-colony stimulating factor in animal models of focal cerebral ischemia. Stroke. 2008;39(6):1855–61.
55. de la Pena IC, Yoo A, Tajiri N, Acosta SA, Ji X, Kaneko Y, et al. Granulocyte colony-stimulating factor attenuates delayed tPA-induced hemorrhagic transformation in ischemic stroke rats by enhancing angiogenesis and vasculogenesis. J Cereb Blood Flow Metab. 2015;35(2):338–46.
56. Schneider A, Kruger C, Steigleder T, Weber D, Pitzer C, Laage R, et al. The hematopoietic factor G-CSF is a neuronal ligand that counteracts programmed cell death and drives neurogenesis. J Clin Invest. 2005;115(8):2083–98.
57. Bratane BT, Bouley J, Schneider A, Bastan B, Henninger N, Fisher M. Granulocyte-colony stimulating factor delays PWI/DWI mismatch evolution and reduces final infarct volume in permanent-suture and embolic focal cerebral ischemia models in the rat. Stroke. 2009;40(9):3102–6.
58. Baltan S. Excitotoxicity and mitochondrial dysfunction underlie age-dependent ischemic white matter injury. Adv Neurobiol. 2014;11:151–70.
59. Philip M, Benatar M, Fisher M, Savitz SI. Methodological quality of animal studies of neuroprotective agents currently in phase II/III acute ischemic stroke trials. Stroke. 2009;40(2):577–81.

60. Badan I, Buchhold B, Hamm A, Gratz M, Walker LC, Platt D, et al. Accelerated glial reactivity to stroke in aged rats correlates with reduced functional recovery. J Cereb Blood Flow Metab. 2003;23(7):845–54.
61. Popa-Wagner A, Stocker K, Balseanu AT, Rogalewski A, Diederich K, Minnerup J, et al. Effects of granulocyte-colony stimulating factor after stroke in aged rats. Stroke. 2010;41(5):1027–31.
62. Komatsu K, Honmou O, Suzuki J, Houkin K, Hamada H, Kocsis JD. Therapeutic time window of mesenchymal stem cells derived from bone marrow after cerebral ischemia. Brain Res. 2010;1334:84–92.
63. Sharma S, Yang B, Strong R, Xi X, Brenneman M, Grotta JC, et al. Bone marrow mononuclear cells protect neurons and modulate microglia in cell culture models of ischemic stroke. J Neurosci Res. 2010;88(13):2869–76.
64. Hill QA, Buxton D, Pearce R, Gesinde MO, Smith GM, Cook G. An analysis of the optimal timing of peripheral blood stem cell harvesting following priming with cyclophosphamide and G-CSF. Bone Marrow Transplant. 2007;40(10):925–30.
65. Popa-Wagner A, Buga AM, Kokaia Z. Perturbed cellular response to brain injury during aging. Ageing Res Rev. 2011;10(1):71–9.
66. Wagner DC, Bojko M, Peters M, Lorenz M, Voigt C, Kaminski A, et al. Impact of age on the efficacy of bone marrow mononuclear cell transplantation in experimental stroke. Exp Transl Stroke Med. 2012;4(1):17.
67. Weise G, Lorenz M, Posel C, Maria Riegelsberger U, Storbeck V, Kamprad M, et al. Transplantation of cryopreserved human umbilical cord blood mononuclear cells does not induce sustained recovery after experimental stroke in spontaneously hypertensive rats. J Cereb Blood Flow Metab. 2014;34(1):e1–9.
68. Duncan K, Gonzales-Portillo GS, Acosta SA, Kaneko Y, Borlongan CV, Tajiri N. Stem cell-paved biobridges facilitate stem transplant and host brain cell interactions for stroke therapy. Brain Res. 2015;1623:160–5.
69. Kocsis JD, Honmou O. Bone marrow stem cells in experimental stroke. Prog Brain Res. 2012;201:79–98.
70. Tajiri N, Quach DM, Kaneko Y, Wu S, Lee D, Lam T, et al. Behavioral and histopathological assessment of adult ischemic rat brains after intracerebral transplantation of NSI-566RSC cell lines. PLoS One. 2014;9(3):e91408.
71. Moniche F, Gonzalez A, Gonzalez-Marcos JR, Carmona M, Pinero P, Espigado I, et al. Intra-arterial bone marrow mononuclear cells in ischemic stroke: a pilot clinical trial. Stroke. 2012;43(8):2242–4.
72. Otero-Ortega L, Gutierrez-Fernandez M, Ramos-Cejudo J, Rodriguez-Frutos B, Fuentes B, Sobrino T, et al. White matter injury restoration after stem cell administration in subcortical ischemic stroke. Stem Cell Res Ther. 2015;6:121.
73. De Coppi P, Bartsch G Jr, Siddiqui MM, Xu T, Santos CC, Perin L, et al. Isolation of amniotic stem cell lines with potential for therapy. Nat Biotechnol. 2007;25(1):100–6.
74. Hosseini SM, Farahmandnia M, Kazemi S, Shakibajahromi B, Sarvestani FS, Khodabande Z. A novel cell therapy method for recovering after brain stroke in rats. Int J Stem Cells. 2015;8(2):191–9.
75. Cai Q, Chen Z, Song P, Wu L, Wang L, Deng G, et al. Co-transplantation of hippocampal neural stem cells and astrocytes and microvascular endothelial cells improve the memory in ischemic stroke rat. Int J Clin Exp Med. 2015;8(8):13109–17.
76. Chamberlain G, Fox J, Ashton B, Middleton J. Concise review: mesenchymal stem cells: their phenotype, differentiation capacity, immunological features, and potential for homing. Stem Cells. 2007;25(11):2739–49.
77. Bingham D, Martin SJ, Macrae IM, Carswell HV. Watermaze performance after middle cerebral artery occlusion in the rat: the role of sensorimotor versus memory impairments. J Cereb Blood Flow Metab. 2012;32(6):989–99.

78. Popa-Wagner A, Dinca I, Yalikun S, Walker L, Kroemer H, Kessler C. Accelerated delimitation of the infarct zone by capillary-derived nestin-positive cells in aged rats. Curr Neurovasc Res. 2006;3(1):3–13.
79. Dunnett SB. Neural tissue transplantation, repair, and rehabilitation. Handb Clin Neurol. 2013;110:43–59.
80. Morizane A, Li JY, Brundin P. From bench to bed: the potential of stem cells for the treatment of Parkinson's disease. Cell Tissue Res. 2008;331(1):323–36.
81. Posel C, Scheibe J, Kranz A, Bothe V, Quente E, Frohlich W, et al. Bone marrow cell transplantation time-dependently abolishes efficacy of granulocyte colony-stimulating factor after stroke in hypertensive rats. Stroke. 2014;45(8):2431–7.
82. Bang OY, Lee JS, Lee PH, Lee G. Autologous mesenchymal stem cell transplantation in stroke patients. Ann Neurol. 2005;57(6):874–82.
83. Kondziolka D, Wechsler L, Goldstein S, Meltzer C, Thulborn KR, Gebel J, et al. Transplantation of cultured human neuronal cells for patients with stroke. Neurology. 2000;55(4):565–9.
84. Reitmeir R, Kilic E, Reinboth BS, Guo Z, ElAli A, Zechariah A, et al. Vascular endothelial growth factor induces contralesional corticobulbar plasticity and functional neurological recovery in the ischemic brain. Acta Neuropathol. 2012;123(2):273–84.
85. Arvidsson A, Collin T, Kirik D, Kokaia Z, Lindvall O. Neuronal replacement from endogenous precursors in the adult brain after stroke. Nat Med. 2002;8(9):963–70.
86. Hou SW, Wang YQ, Xu M, Shen DH, Wang JJ, Huang F, et al. Functional integration of newly generated neurons into striatum after cerebral ischemia in the adult rat brain. Stroke. 2008;39(10):2837–44.
87. Jin K, Wang X, Xie L, Mao XO, Zhu W, Wang Y, et al. Evidence for stroke-induced neurogenesis in the human brain. Proc Natl Acad Sci U S A. 2006;103(35):13198–202.
88. Mine Y, Tatarishvili J, Oki K, Monni E, Kokaia Z, Lindvall O. Grafted human neural stem cells enhance several steps of endogenous neurogenesis and improve behavioral recovery after middle cerebral artery occlusion in rats. Neurobiol Dis. 2013;52:191–203.
89. Thored P, Arvidsson A, Cacci E, Ahlenius H, Kallur T, Darsalia V, et al. Persistent production of neurons from adult brain stem cells during recovery after stroke. Stem Cells. 2006;24(3):739–47.
90. Lindvall O, Kokaia Z. Neurogenesis following stroke affecting the adult brain. Cold Spring Harb Perspect Biol. 2015;7(11):a019034.
91. Parent JM, Vexler ZS, Gong C, Derugin N, Ferriero DM. Rat forebrain neurogenesis and striatal neuron replacement after focal stroke. Ann Neurol. 2002;52(6):802–13.
92. Ahlenius H, Visan V, Kokaia M, Lindvall O, Kokaia Z. Neural stem and progenitor cells retain their potential for proliferation and differentiation into functional neurons despite lower number in aged brain. J Neurosci. 2009;29(14):4408–19.
93. Darsalia V, Heldmann U, Lindvall O, Kokaia Z. Stroke-induced neurogenesis in aged brain. Stroke. 2005;36(8):1790–5.
94. Enwere E, Shingo T, Gregg C, Fujikawa H, Ohta S, Weiss S. Aging results in reduced epidermal growth factor receptor signaling, diminished olfactory neurogenesis, and deficits in fine olfactory discrimination. J Neurosci. 2004;24(38):8354–65.
95. Tropepe V, Craig CG, Morshead CM, van der Kooy D. Transforming growth factor-alpha null and senescent mice show decreased neural progenitor cell proliferation in the forebrain subependyma. J Neurosci. 1997;17(20):7850–9.
96. Shetty AK, Hattiangady B, Rao MS. Vulnerability of hippocampal GABA-ergic interneurons to kainate-induced excitotoxic injury during old age. J Cell Mol Med. 2009;13(8B):2408–23.
97. Macas J, Nern C, Plate KH, Momma S. Increased generation of neuronal progenitors after ischemic injury in the aged adult human forebrain. J Neurosci. 2006;26(50):13114–9.

98. Marti-Fabregas J, Romaguera-Ros M, Gomez-Pinedo U, Martinez-Ramirez S, Jimenez-Xarrie E, Marin R, et al. Proliferation in the human ipsilateral subventricular zone after ischemic stroke. Neurology. 2010;74(5):357–65.
99. Minger SL, Ekonomou A, Carta EM, Chinoy A, Perry RH, Ballard CG. Endogenous neurogenesis in the human brain following cerebral infarction. Regen Med. 2007;2(1):69–74.
100. Yuan T, Liao W, Feng NH, Lou YL, Niu X, Zhang AJ, et al. Human induced pluripotent stem cell-derived neural stem cells survive, migrate, differentiate, and improve neurologic function in a rat model of middle cerebral artery occlusion. Stem Cell Res Ther. 2013;4(3):73.
101. Liu X, Ye R, Yan T, Yu SP, Wei L, Xu G, et al. Cell based therapies for ischemic stroke: from basic science to bedside. Prog Neurobiol. 2014;115:92–115.
102. Oki K, Tatarishvili J, Wood J, Koch P, Wattananit S, Mine Y, et al. Human-induced pluripotent stem cells form functional neurons and improve recovery after grafting in stroke-damaged brain. Stem Cells. 2012;30(6):1120–33.
103. Tornero D, Wattananit S, Gronning Madsen M, Koch P, Wood J, Tatarishvili J, et al. Human induced pluripotent stem cell-derived cortical neurons integrate in stroke-injured cortex and improve functional recovery. Brain. 2013;136(Pt 12):3561–77.
104. Tatarishvili J, Oki K, Monni E, Koch P, Memanishvili T, Buga AM, et al. Human induced pluripotent stem cells improve recovery in stroke-injured aged rats. Restor Neurol Neurosci. 2014;32(4):547–58.
105. Mohamad O, Drury-Stewart D, Song M, Faulkner B, Chen D, Yu SP, et al. Vector-free and transgene-free human iPS cells differentiate into functional neurons and enhance functional recovery after ischemic stroke in mice. PLoS One. 2013;8(5):e64160.
106. Phanthong P, Raveh-Amit H, Li T, Kitiyanant Y, Dinnyes A. Is aging a barrier to reprogramming? Lessons from induced pluripotent stem cells. Biogerontology. 2013;14(6):591–602.
107. Boulting GL, Kiskinis E, Croft GF, Amoroso MW, Oakley DH, Wainger BJ, et al. A functionally characterized test set of human induced pluripotent stem cells. Nat Biotechnol. 2011;29(3):279–86.
108. Rafii S, Lyden D, Benezra R, Hattori K, Heissig B. Vascular and haematopoietic stem cells: novel targets for anti-angiogenesis therapy? Nat Rev Cancer. 2002;2(11):826–35.
109. Haruchika M, Takayuki A. Post-natal endothelial progenitor cells for neovascularization in tissue regeneration. Cardiovasc Res. 2003;58:390–8.
110. Caiado F, Dias S. Endothelial progenitor cells and integrins: adhesive needs. Fibrogen Tissue Repair. 2012;5:4. https://doi.org/10.1186/1755-1536-5-4.
111. LimanTG EM. New vessels after stroke: postischemic neovascularization and regeneration. Cerebrovasc Dis. 2012;33(5):492–9.
112. Hayashi T, Deguchi K, Nagotani S, Zhang H, Sehara Y, Tsuchiya A, Abe K. Cerebral ischemia and angiogenesis. Curr Neurovasc Res. 2006;3(2):119–29.
113. Buga AM, Sascau M, Pisoschi C, Herndon JG, Kessler C, Popa-Wagner A. The genomic response of the ipsilateral and contralateral cortex to stroke in aged rats. J Cell Mol Med. 2008;12(6B):2731–53.
114. Peichev M, Naiyer AJ, Pereira D, Zhu Z, Lane WJ, Williams M, Oz MC, Hicklin DJ, Witte L, Moore MA, Rafii S. Expression of VEGFR-2 and AC133 by circulating human CD34(+) cells identifies a population of functional endothelial precursors. Blood. 2000;95(3):952–8.
115. Lees JS, Sena ES, Egan KJ, Antonic A, Koblar SA, Howells DW, Macleod MR. Stem cell-based therapy for experimental stroke: a systematic review and meta-analysis. Int J Stroke. 2012;7(7):582–8.
116. Zhang W, Zhang G, Jin H, Hu R. Characteristics of bone marrow-derived endothelial progenitor cells in aged mice. Biochem Biophys Res Commun. 2006;348(3):1018–23.
117. Mikirova NA, Jackson JA, Hunninghake R, Julian K, Kenyon J, Chan KW, Swindlehurst CA, Minev B, Patel AN, Murphy PM, Smith L, Alexandrescu DT, Ichim TE, Riordan NH. Circulating endothelial progenitor cells: a new approach to anti-aging medicine. J Transl Med. 2009;7:106. https://doi.org/10.1186/1479-5876-7-106.

118. Krupinski J, Kaluza J, Kumar P, Kumar S, Wang JM. Role of angiogenesis in patients with cerebral ischemic stroke. Stroke. 1994;25(9):1794–8.
119. Krupinski J, Kumar P, Kumar S, Kaluza J. Increased expression of TGF-beta 1 in brain tissue after ischemic stroke in humans. Stroke. 1996;27(5):852–7.
120. Krupinski J, Issa R, Bujny T, Slevin M, Kumar P, Kumar S, Kaluza J. A putative role for platelet-derived growth factor in angiogenesis and neuroprotection after ischemic stroke in humans. Stroke. 1997;28(3):564–73.
121. Vikman P, Edvinsson L. Gene expression profiling in the human middle cerebral artery after cerebral ischemia. Eur J Neurol. 2006;13(12):1324–32.
122. Moore DF, Li H, Jeffries N, Wright V, Cooper RA Jr, Elkahloun A, Gelderman MP, Zudaire E, Blevins G, Yu H, Goldin E, Baird AE. Using peripheral blood mononuclear cells to determine a gene expression profile of acute ischemic stroke: a pilot investigation. Circulation. 2005;111(2):212–21.
123. Tang Y, Xu H, Du X, Lit L, Walker W, Lu A, Ran R, Gregg JP, Reilly M, Pancioli A, Khoury JC, Sauerbeck LR, Carrozzella JA, Spilker J, Clark J, Wagner KR, Jauch EC, Chang DJ, Verro P, Broderick JP, Sharp FR. Gene expression in blood changes rapidly in neutrophils and monocytes after ischemic stroke in humans: a microarray study. J Cereb Blood Flow Metab. 2006;26(8):1089–10102.
124. Tan KS, Armugam A, Sepramaniam S, Lim KY, Setyowati KD, Wang CW, Jeyaseelan K. Expression profile of MicroRNAs in young stroke patients. PLoS One. 2009;4(11):e7689. https://doi.org/10.1371/journal.pone.0007689.
125. Buga AM, Margaritescu C, Scholz CJ, Radu E, Zelenak C, Popa-Wagner A. Transcriptomics of post-stroke angiogenesis in the aged brain. Front Aging Neurosci. 2014;6:44. https://doi.org/10.3389/fnagi.2014.00044.

Chapter 4
Modulating Endogenous Adult Neural Stem Cells to Improve Regeneration in Stroke Brain

Fucheng Luo and Yu Luo

Abstract Stroke is a major cause of death and disability globally. Experimental and clinical stroke studies have demonstrated that endogenous brain repair processes could be activated in the brain following stroke. However, the spontaneous brain repair process is constrained with limited improvement of neurological outcome. Neurogenesis, oligodendrogenesis, angiogenesis, axonal outgrowth, and synaptogenesis are major brain repair processes during stroke recovery. In adult rodents and human, there are endogenous neural stem cells that generate new neurons, astrocyte, oligodendrocyte, and NG2-glia under physiological or pathological conditions. Much progress has been made in preclinical studies on the roles of endogenous neural stem cells in brain repair processes in response to stroke. In this review, we will summarize recent progress on the cellular and molecular mechanisms underlying how endogenous adult neural stem cells contribute to neurogenesis and oligodendrogenesis, and their modulatory effects on angiogenesis and inflammation, which may play critical roles in brain repair and leads to improvement of neurological function after stroke.

Keywords Stroke • Neural stem cells • Neurogenesis • Oligodendrogenesis • Brain repair

Abbreviations

3V	Third ventricle
ANg	Angiopoietin
BDNF	Brain-derived neurotrophic factor
BMP	Bone morphogenetic protein

F. Luo, Ph.D. • Y. Luo, Ph.D. (✉)
Department of Neurological Surgery, Case Western Reserve University, Cleveland, OH, USA

Department of Neurosciences, Case Western Reserve University, Cleveland, OH, USA
e-mail: yxl710@case.edu

P.A. Lapchak, J.H. Zhang (eds.), *Cellular and Molecular Approaches to Regeneration and Repair*, Springer Series in Translational Stroke Research,
https://doi.org/10.1007/978-3-319-66679-2_4

CB2R	Cannabinoid type-2 receptor
CCR2	C-C chemokine receptor type 2
ChAT	Choline acetyl-transferase
CNS	Central nervous system
CNTF	Ciliary neurotrophic factor
CSPGs	Chondroitin sulfate proteoglycans
CX3CR1	CX3C chemokine receptor 1
CXCL12	C-X-C motif chemokine 12
CXCR4	C-X-C chemokine receptor type 4
DARPP-32	cAMP-regulated neuronal phosphoprotein
DCX	Doublecortin
DG	Dentate gyrus
ECM	Extracellular matrix
EGF	Epidermal growth factor
EGFR	Epidermal growth factor receptor
FGF10	Fibroblast growth factor 10
FGF2	Fibroblast growth factor 2
GABA	Gamma aminobutyric acid
GAD67	Glutamic acid decarboxylase
GAP43	Growth Associated Protein 43
GSK-3β	Glycogen synthase kinase-3β
HDACs	Histone deacetylases
IGF-1	Insulin-like growth factor 1
MCAO	Middle cerebral artery occlusion
MCP-1	Monocyte chemoattractant protein 1
MMPs	Matrix metalloproteases
mTORC1	Mechanistic target of rapamycin complex 1
Nf1	Neurofibromatosis type 1
NPCs	Neural progenitor cells
NSCs	Neural stem cells
OB	Olfactory bulb
P57kip2	Cyclin-dependent kinase inhibitor 1C
PDGF	Platelet-derived growth factor
PDGFR	Platelet-derived growth factor receptor α
Ptc-1	Patched 1
PV	Parvalbumin
RMS	Rostral migratory stream
Robo	Roundabout
ROCK	Rho-associated kinase
SDF-1	Stromal cell-derived factor 1
SGZ	Subgranular zone
Shh	Sonic hedgehog
siRNA	Short interfering ribonucleic acid
Smo	Smoothened

SVZ	Subventricular zone
TGF-α	Transforming growth factor-alpha
TIA	Transient ischemic attack
Tregs	Regulatory T cells
Usp9x	Ubiquitin-specific peptidase 9, X-linked
VEGF	Vascular endothelial growth factor

1 Introduction

Globally, stroke is the second leading cause of death and the third most common cause of disability [1]. There are three types of stroke: ischemic stroke, hemorrhagic stroke, and transient ischemic attack (TIA, also called a "mini-stroke"). Ischemic stroke is caused by obstruction within a blood vessel and accounts for 87% of all stroke cases, while hemorrhagic stroke occurs when blood vessel rupture. TIAs are caused by a transient clot or blockage in the brain. Although TIAs last only a few minutes and causes no permanent damage to the brain, they are indicative of the likelihood of a coming stroke and should be taken seriously. Only a small percentage of stroke patients benefit from thrombolysis and endovascular thrombectomy treatments due to the short window (4.5–6 h) of these treatments. As a result, a large population of stroke patients still suffer from permanent severe neurological deficits in stroke survivors. Thus, there is an urgent need to develop new therapies for stroke to enhance functional recovery.

Studies from experimental stroke and patients with stroke show that some degree of spontaneous neurological recovery occurs after stroke. However, this endogenous brain self-repair is not sufficient to restore neurological function after stroke [2, 3]. Endogenous brain repair involves a set of highly interactive processes during stroke recovery, such as neurogenesis and oligodendrogenesis, which is induced mostly by endogenous neural stem cells (NSCs). Coupling of neurogenesis and angiogenesis has been implicated in some recent stroke studies [2]. In addition, stroke-induced inflammation, which is characterized by the activation of resident microglia and infiltration of monocytes and lymphocytes, is a major causative factor for neurological deficits [4]. Recent studies also suggest that there is cross-talk between neural stem cells and immune cells in response to brain injury [5]. Therefore, a promising field of investigation is to focus on modulating endogenous adult neural stem cells and their interactions with other cellular processes such as angiogenesis and neuroinflammation to improve functional recovery following stroke. Understanding how endogenous stem cells are activated, differentiate, migrate, integrate, and restore neuronal circuitry will help us develop less invasive therapeutic interventions. Elucidation of the interactions of neurogenesis with other cellular processes such as angiogenesis and inflammation after stroke will provide additional information needed to modulate this process to improve brain recovery after stroke. In this review, we will provide an update on the recent findings on the mechanisms underlying endogenous NSC-mediated neurogenesis and oligodendrogenesis and their modulatory effects on angiogenesis and inflammation after stroke.

2 NSCs Responses in Adult Brain Following Stroke

NSCs are multipotent stem cells that can self-renew, divide, and differentiate into new mature neurons, astrocytes, and oligodendrocytes. In the adult brain, there are three main neurogenic niches containing NSCs: the subventricular zone (SVZ) of the lateral ventricle, the subgranular zone (SGZ) in the dentate gyrus of the hippocampus, and the recently discovered hypothalamic stem cell niche [6] (Fig. 4.1). In these regions, there is a basal rate of neurogenesis in normal conditions. In response to stroke, endogenous NSCs are activated and participate in brain repair processes.

2.1 *Radial Glial Cells (Type B Cells) in SVZ*

The NSCs in the SVZ are termed as Type B cells. They divide slowly to generate transit-amplifying type C cells, which proliferate actively and further differentiate into neuroblasts (also named type A cells). Finally, these neuroblasts form chains and migrate via the rostral migratory stream into the olfactory bulb (OB), where they differentiate into granule cells or periglomerular interneurons. Adult NSCs in the SVZ also generate NG2-glia that migrates toward the gray and white matter. Focal cerebral ischemia stimulates SVZ NSC proliferation and neurogenesis in adult rodent, monkeys, and even human brains [7–10]. Augmented neuroblasts could migrate from the SVZ to ischemic sites and differentiate into neurons in

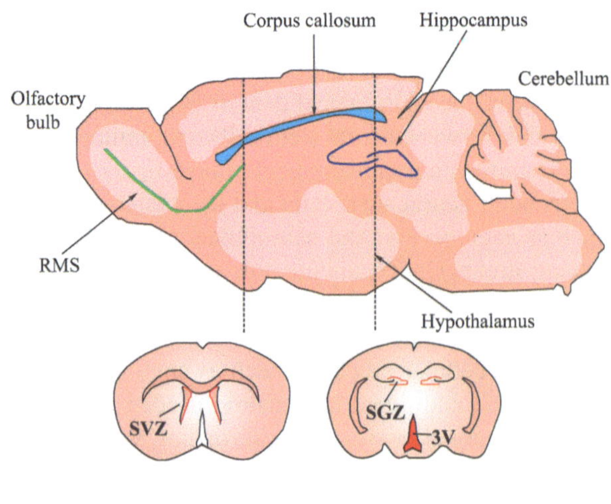

Fig. 4.1 Sagittal and coronal view of the adult rodent brain, illustrating the three niches where adult NSCs reside: the subventricular zone (SVZ), the subgranular zone (SGZ), and the recently described hypothalamic NSCs niche around the third ventricle

rodent middle cerebral artery occlusion (MCAO) models [11, 12]. In addition, stroke also induces oligodendrogenesis in the SVZ and the generated NG2-glia can migrate to the lesion site and differentiate into myelinating oligodendrocytes [13, 14]. Furthermore, activated SVZ NSCs give rise to a subpopulation of reactive astrocytes in the cortex that contribute to astrogliosis and scar formation [15]. Altogether, these data indicate that SVZ NSCs are a major therapeutic target for improving functional recovery after stroke.

2.2 *Radial Glia-Like Cells (Type-1 Cells) in SGZ*

SGZ NSCs are also known as type-1 cells or radial glia-like cells. These cells divide slowly and give rise to type-2 cells or transit amplifying progenitors that could differentiate into neurons and astrocytes [16]. However, it is still a matter of debate whether these cells can spontaneously, that is without any exogenous manipulation, give rise to oligodendroglial cells. Indeed, either ectopical and elevated Ascl1 expression or inactivation of p57kip2, Nf1, Drosha, or Usp9x induce oligodendrogenesis in SGZ NSCs [17–21]. The function of neurogenesis derived from SGZ NSCs is associated with learning, memory, and cognition. Following a stroke, there is significantly enhanced proliferation of NSCs and neurogenesis in the SGZ of many species, such as rats, mice, monkeys, and humans [22]. Generally, the increased proliferation starts bilaterally at 3–4 days post-ischemia, peaks at 7–10 days, and returns to control levels by 3–5 weeks after the ischemia [22]. Recent studies show that hippocampal neurogenesis is responsible for some aspects of recovery following brain ischemia, such as learning and memory [23]. These data suggest that target SGZ NSCs might help to improve functional recovery after stroke.

2.3 *Tanycytes*

It has been recently demonstrated that NSCs also reside in the adult hypothalamus. The NSCs/NPCs of this region are termed as tanycytes, which express classical markers of neural precursor cells and multipotent cell markers, such as nestin, Sox2, UGS148, and FGF10 [6]. These tanycytes belong to ependymal glial cells and surround the lateral walls of the infundibular recess of the third ventricle. In response to peripheral signaling (i.e. CNTF, Leptin and high-fat diet), tanycytes are able to proliferate, migrate, and differentiate into neurons, such as arcuate pro-opiomelanocortin neurons and orexigenic and anorexigenic neurons [6, 24–27]. Importantly, tanycytes exhibited increased proliferation on the infarcted side on day 4 after ischemic stroke injury (MCAO model in rats) [28]. However, the functional role of tanycyte proliferation after stroke is still largely unknown.

3 Promoting Neurogenesis of Endogenous NSCs

Neurogenesis is a multistep process that includes proliferation, fate determination, migration, maturation, and survival of NSCs. Understanding the molecular mechanisms regulating these processes is essential for developing therapies to improve neurological recovery (Fig. 4.2). Many factors are involved in the regulation of adult NSCs, including growth factors, neurotransmitters, and chemokines. We will briefly summarize them in this review.

3.1 *Proliferation*

The initial response of NSCs following stroke is to increase proliferation, a process that is regulated by various intrinsic and extrinsic factors. The mechanism underlying stroke-induced proliferation of NSCs is unclear. Several hypotheses have been suggested as potential mechanisms to regulate proliferation of NSCs. Adult rodent stroke studies have shown that quiescent adult neural stem cells can be activated and recruited to an active pool to increase neurogenesis. As a response to stroke, an increased neurogenesis might result from transiently switching neural progenitors division from asymmetric to symmetric and from a reduction of the length of the cell cycle [29, 30]. Stroke can trigger the early expansion of the progenitor cell pool by shortening the cell-cycle length and retaining daughter cells within the cell cycle at an early stage after stroke. At a later stage, lengthening the cell cycle and the G1 phase leads to the daughter cells exiting the cell cycle and differentiating into neurons [31]. Several important pathways that may regulate the proliferation of NSCs and their early progeny have been identified.

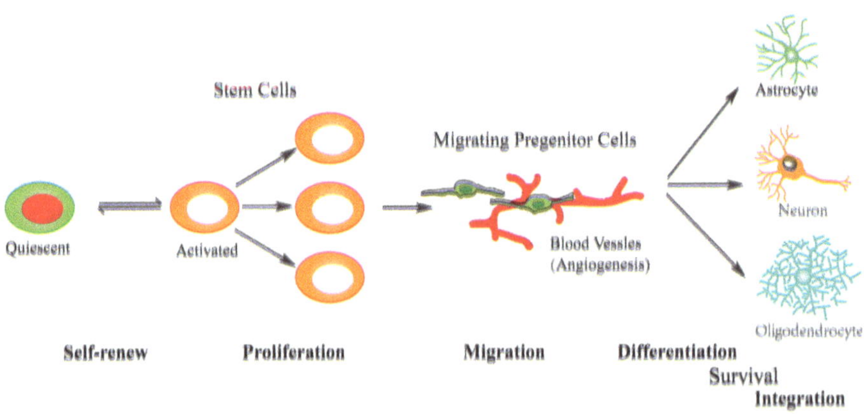

Fig. 4.2 Cellular and molecular processes that are involved in the maintenance of adult NSCs, generation of different lineages of cells and their integration in the brain after stroke

3.1.1 Sonic Hedgehog (Shh)

Shh is a secreted glycoprotein. It binds to its receptor Patched (Ptc-1) to de-repress Smoothened (Smo) and activate transcription factors of the Gli family. Shh signaling is required for SVZ NSC maintenance as conditional deletion of smoothened gene in adult SVZ NSC leads to decreased BrdU-positive cells and DCX+ neuroblasts in the SVZ [32]. Studies have found that stroke upregulates Shh signal in multiple cell types, such as neurons, reactive astrocyte, and SVZ neural progenitor cells [33, 34]. In vivo, blockage of the Shh signaling pathway with cyclopamine, a specific inhibitor of Smo, suppressed ischemia-induced proliferation of subgranular NPCs in the hippocampus [34]. Conditional deletion of shh genes in nestin-expressing cells leads to significantly more severe behavioral deficits in a cortical ischemic model [33]. Administration of cyclopamine also abolished carbamylated erythropoietin-induced neurogenesis [35]. These data suggests that Shh signaling is a key factor for NSC self-renewal or proliferation. Interestingly, at a lower dosage, delayed post-stroke treatment of Shh agonist improves functional recovery by enhancing survival of newly born neurons and angiogenesis [36] but not by increasing BrdU-positive cells at the NSC niche, suggesting that Shh signaling might play multiple roles in ischemia-induced neurogenesis and whether it enhances the proliferation or survival of the newly generated NSC progeny is dose-dependent.

3.1.2 Epidermal Growth Factor (EGF)/Fibroblast Growth Factors 2 (FGF2)

Studies have reported that FGF-2 and EGF expression in the brain increased significantly after ischemic stroke [37, 38]. Importantly, cerebral ischemia resulted in an increase in the number of EGF receptor (EGFR)-positive transit-amplifying cells (type C cells) in the SVZ [39]. Overexpression of FGF-2 significantly increased the proliferation of progenitor cells after ischemic stroke in both FGF-2-deficient mice and wild-type mice [40]. Meanwhile, in vivo infusion of EGF into adult mouse forebrain for 6 consecutive days resulted in a dramatic increase in the proliferation and the total number of subependymal cells and induced their migration away from the lateral ventricle walls into adjacent parenchyma [41]. Furthermore, infusion of EGF together with FGF-2 into the brain of adult rats was found to promote dentate gyrus (DG) and SVZ NPC proliferation after focal ischemic stroke [42, 43].

3.1.3 Insulin-Like Growth Factor 1 (IGF-1)

The progenitors in both the SVZ and DG show IGF-1 receptor expression [44]. In vitro studies demonstrate that IGF-1 stimulated the proliferation of cultured NPCs via activating the PI-3-kinase/Akt signaling pathway [45]. Following ischemic stroke, IGF-1 expression is increased in the activated astrocytes in the ischemic penumbra [44]. Inhibiting IGF-1 activity by intracerebroventricular infusion of

IGF-1 antibody significantly blocked the ischemia-induced neural progenitor proliferation [44]. *Exogenous* IGF-1 injection after ischemic stroke promoted neurogenesis [46]. Meanwhile, post-ischemic IGF-1 gene transfer in the peri-infarct region potently promoted neural and vascular regeneration in the chronic stage of cerebral infarction [47]. These results suggest that IGF-1 formed in the ischemic penumbra might be one of the endogenous diffusible factors that mediate post-ischemic neural progenitor proliferation.

3.1.4 Notch Signaling Pathway

Notch signaling is an evolutionarily conserved pathway that regulates cell-fate determination during development and maintains adult tissue homeostasis. Recent studies have shown that stroke increases the expression of Notch1 and Hes1 in SVZ cells [48]. Transient administration of Notch ligands to the brain of adult rats increases the numbers of newly generated precursor cells and improves motor skills after ischemic injury [49], while the blockage of the Notch pathway by short interfering ribonucleic acid (siRNA) against Notch or a gamma secretase inhibitor significantly blocked ischemia-induced cell proliferation in the SVZ [50]. These data suggest that the Notch signaling pathway mediates adult SVZ neural progenitor cell proliferation after stroke. Interestingly, it has recently been shown that striatal astrocytes can turn on nestin expression and generate neurons in stroke model through downregulation of Notch signaling, suggesting that Notch signaling might also suppress "NSC status" in mature astrocytes [51].

3.1.5 Other Regulators

Finally, other potential mediators of stroke-induced proliferation of NSCs in the neurogenic niches have been described. These include vascular endothelial growth factor (VEGF) [52], glial cell-derived neurotrophic factor (GDNF) [53], brain-derived neurotrophic factor (BDNF) [54], Wnt signaling, retinoic acid [55], bone morphogenetic protein [56], and microRNA [57, 58]. In addition, the communication between NSCs and other cell types also affects NSC proliferation after stroke. It has been reported that M2 phenotype microglia-derived transforming growth factor-alpha (TGF-α) is one of the key factors to enhance proliferation and neural differentiation of NSPCs after ischemic stroke [59]. Activated regulatory T cells (Tregs) enhanced SVZ NSC proliferation in normal and ischemic mice; blockage of IL-10 abolished Tregs-mediated NSC proliferation in vivo and in vitro [60]. Furthermore, astrocytic calcium waves are long-range signals capable of transmitting the occurrence of a brain injury to the SVZ, where they stimulate NSC proliferation and self-renewal and increase the migratory potential of NSPCs. It is shown that the Notch signaling pathway mediates effects of elevated calcium levels on NSPCs [61].

3.2 Migration

After stroke, following NSC proliferation, another critical biological process is the migration of these NSCs from neurogenic niches to the ischemic region. In the normal adult brain, SVZ neuroblasts migrate along the rostral migratory stream to the olfactory bulb. Lateral migration into the striatum and parenchyma is not observed in the rodent brain under normal condition. However, in the ischemic damaged brain, neuroblasts will migrate laterally into the ischemic injury region [11]. Although little is known about the molecular mechanisms underlying stroke-induced redirected migration, several potential mediators have been identified. These include stromal cell-derived factor 1 (SDF-1), monocyte chemoattractant protein 1 (MCP-1/CCL2), matrix metalloproteases (MMPs), cannabinoid type-2 receptor (CB2R), and β1 integrin. Further, the neurovascular niche within the SVZ and SDG is also a key regulator of neuroblast migration following stroke.

3.2.1 Stromal Cell-Derived Factor 1 (SDF-1)

SDF-1, also known as C-X-C motif chemokine 12 (CXCL12), is a chemokine protein that exerts biological functions by binding to its receptors CXCR4 and CXCR7. SDF-1 (CXCL12) is a member of the alpha (CXC) chemokine family which are involved in inflammatory responses [62]. SDF-1 and its receptor CXCR4 have been demonstrated to play an important role in the mobilization and homing of stem cells to bone marrow [63, 64]. Neuroblasts are reported to express CXCR4 [65]. During adult neurogenesis, SDF-1 is secreted by vascular endothelial cells and plays a role in the directional migration of neuroblasts in the CNS [65]. Following stroke, SDF-1 is upregulated by activated astrocyte and endothelial cells, subsequently guiding neuroblast migration toward the injured tissue [66–68]. In contrast, CXCR4 blockade blocks this pathology-directed chain migration [69].

3.2.2 Monocyte Chemoattractant Protein-1 (MCP-1)

MCP-1 is a member of the C-C chemokine family that regulates migration and infiltration of monocytes/macrophages [70]. Following cerebral ischemia, MCP-1 is induced in activated astrocytes and neurons within the injured tissue [71, 72]. The migrating neuroblasts in the ischemic brain express MCP-1 receptor CCR2. Infusion of MCP-1 into the normal striatum induced neuroblast migration to the infusion site [73]. In knockout mice that lacked either MCP-1 or its receptor CCR2, there was a significant decrease in the number of migrating neuroblasts from the SVZ to the ischemic striatum [73].

3.2.3 Matrix Metalloproteinases (MMPs)

Matrix metalloproteinases (MMPs) are members of the metzincin group of proteases that participate in several physiological processes, such as bone remodeling, angiogenesis, immunity, and wound healing [74]. Recent studies suggest that MMPs are involved in guiding neuroblast migration from the neurogenic region to the ischemic boundary [75]. Neuroblasts express MMP-3 and MMP–9. Inhibition of MMPs diminishes neuroblast migration after stroke [76, 77]. Moreover, MMP2 and MMP9 secreted by endothelial cells are also associated with neuroblast migration after stroke [78].

3.2.4 CB2R

The endocannabinoids (eCBs) 2-arachidonoylglycerol and anandamide are lipid signaling messengers involved in the homeostatic control of a large variety of functions of the nervous system through binding cannabinoid type-1 receptor (CB1R) and cannabinoid type-2 receptor (CB2R) [79]. CB2R is expressed in resident microglia, NG2-glia, and NSCs. CB2R is neuroprotective in acute experimental stroke by anti-inflammatory mechanisms [80]. *In vitro* studies show that CB2R promotes NSC proliferation via mTORC1 signaling [81]. Furthermore, in stroke, CB2R is required for neurogenesis by promoting neuroblast migration toward the injured brain tissue [82].

3.2.5 β1 Integrin

β1-class integrins are transmembrane receptors for several extracellular matrix (ECM) proteins such as laminin [83]. Under normal conditions, migrating neuroblasts generated in the adult SVZ express β1 integrin, which is required for their chain formation during RMS migration [84, 85]. Following stroke, laminin-β1 integrin signaling enables neuroblasts to form chains and migrate efficiently along vascular scaffolding in the post-stroke brain [86].

3.2.6 Neurovascular Niche

Stroke-induced directional migration of neuroblasts is closely associated with thin astrocytic processes and blood vessels, suggesting that blood vessels may act as a scaffold for neuroblast migration [87, 88]. Virally labeled SVZ NPCs were observed to migrate along both newly formed and pre-existing blood vessels toward the ischemic injured area. Live imaging showed that migrating SVZ NPCs have their leading process closely associated with blood vessels, suggesting that this interaction provides directional guidance for the NPCs [89]. In addition, vasculature promotes neuroblast migration via secreting various growth and chemotactic factors, including BDNF, MMPs, angiopoietins, and SDF-1 [22].

3.2.7 Other Regulators

Wnt3a, Angiopoitin (ANg)-1 and its receptor Tie 2, and Slit and its receptor (ROBO) also promote post-stroke neuroblast migration and behavioral recovery [66, 90, 91]. It also should be noted that stroke also induces inhibitory molecules to block the migration of neuroblasts. Glycogen synthase kinase-3β (GSK-3β) inhibition promoted proliferation of neural stem cells (NSCs) and migration of nascent doublecortin (DCX+) neuroblasts from the SVZ to the lesioned cortex [92]. Inhibition of Na^{+}-K^{+}-Cl^{-}-co-transporter can increase migration of neuroblasts in the SVZ towards the infarct areas and improve sensorimotor recovery [93]. Stroke also induces chondroitin sulfate proteoglycans (CSPGs), which could block neuroblast migration through Rho-associated kinase (ROCK) activation [94, 95].

3.3 Survival, Differentiation, and Integration of Newborn Neurons

The long-term survival and functional maturation of newborn neurons following stroke are also crucial for neurological recovery. However, there are fewer studies that have examined the survival, differentiation, and integration of newborn neurons in the ischemic brain. The migration of SVZ neuroblasts to the lesion sites may persist for up to 1 year after ischemia [96], thus offering a long-term window for therapeutic manipulations. Ischemia-induced newly generated cells in the damaged areas express medium-size spiny neuronal marker dopamine- and cAMP-regulated neuronal phosphoprotein (DARPP-32) or neurotransmitter synthesizing enzymes such as glutamic acid decarboxylase (GAD67) and choline acetyl-transferase (ChAT) [11, 97, 98]. Moreover, ischemia-induced newly formed striatal GABAergic and cholinergic neurons could exhibit electrophysiological activity and functional synapses [97]. These data indicate that proliferating neuroblasts that migrate into the damaged areas following stroke are able to differentiate into a variety of functional neuronal cells. Compared to our knowledge of factors that promote adult neural precursor cell proliferation or migration, there is comparatively little known about factors that promote newborn neuron survival and integration in stroke. Intraventricular administration of EGF and albumin enhance the differentiation of newly born immature neurons into mature PV-expressing neurons, replacing more than 20% of PV+ interneurons lost after cerebral ischemia [99]. Complement-derived peptide C3a regulates neural progenitor cell migration and differentiation in vitro and C3a receptor signaling stimulates neurogenesis in unchallenged adult mice. Daily intranasal treatment of wild-type mice with C3a beginning 7 days after stroke induction robustly increased synaptic density and expression of Growth Associated Protein 43 (GAP43) in the peri-infarct cortex [100]. Post-stroke p53 inhibitor enhances the survival of NSCs and their progeny by inhibition of apoptosis in these cells through PUMA gene regulation [101]. Similarly, Shh agonist, delivered after stroke at a lower dose that did not affect BrdU-positive cells in SVZ and SGZ improved the long-term survival of YFP-labeled NSCs and their progeny in stroke model [36]. Some knowledge has been gained regarding factors that enhance newborn neuron integration and survival under normal physiological conditions, including the RhoA family of small GTPases, suppressor of cytokine signaling-2, neurotrophins, neurotransmitters (GABA and glutamate), and semaphorins [102]. Logically, we might get some implication from those factors and explore their roles after ischemic injury. However, it should be noted that the stroke-affected CNS environment is quite an inhibitory environment for newborn cell survival. Experimental studies have shown that only a small proportion of cells survive long enough to integrate into the damaged parenchyma after stroke [46, 68, 102, 103]. Thus, neuroprotective or anti-inflammatory strategies might need to be included with therapy to improve newborn neuron maturation, integration, or plasticity in stroke treatment.

4 Endogenous NSCs and Oligodendrogenesis

NG2-glia, also called oligodendrocyte precursor/progenitor cells and polydendrocytes, characterized by expression of chondroitin sulfate proteoglycan NG2 and platelet-derived growth factor receptor α (PDGFR α). It is widely distributed in the brain and can continuously produce differentiating and mature oligodendrocytes in the central neural system throughout the lifespan of the animals. Myelination of axons in the adult brain is critical for saltatory conduction, axonal integrity, neural plasticity, and circuitry function, which are important for functional recovery after stroke [104]. Stroke acutely leads to mature oligodendrocyte damage, resulting in myelin loss, which is associated with a loss of axons. Oligodendrogenesis is induced in the regions surrounding the lateral ventricles and peri-infarct areas during stroke recovery [105–107]. Studies demonstrate that stroke not only activates resident NG2-glia in white and gray matter, but also increases NG2-glia generation in the SVZ and attracts them to the ischemic area [108–111]. The newly generated NG2-glia could differentiate into mature myelinating oligodendrocytes in the peri-infarct areas, which is involved in the brain repair process [106]. Therefore, these studies suggest that NG2-glia generated by adult NSCs contribute to oligodendrogenesis after stroke. The process of SVZ NSC-mediated oligodendrogenesis is regulated by many intrinsic and extrinsic factors, therefore offering many pathways for potential therapeutic interventions to promote functional recovery following stroke.

4.1 *Extrinsic Factors for Oligodendrogenesis*

4.1.1 Shh

In addition to neurogenesis, Shh signaling regulates oligodendrogenesis by inducing transcription factor olig2 expression [112]. In the SVZ, there is a dorsal Shh-dependent domain producing many oligodendroglial lineage cells [113]. The blockage of Shh signal with cyclopamine could abolish cerebrolysin-enhanced oligodendrogenesis in stroke [114]. Bone marrow stromal cell transplantation stimulates oligodendrogenesis by activation of Shh/Gli1 pathway, which mediates subsequent functional recovery after stroke [115]. Thus, these data suggest that Shh signaling in SVZ plays an important role in mediating oligodendrogenesis in the ischemic brain.

4.1.2 Stromal-Derived Factor 1 (SDF-1)

SDF-1, has been shown to be able to promote neurogenesis and angiogenesis, leading to functional recovery in ischemic mice [116, 117]. Through binding with CXCR4 in NG2-glia, SDF-1 activates their proliferation, migration, and differentiation [118–120]. SDF-1 gene therapy at 1 week after ischemia promotes NG2-glia proliferation in the SVZ and further enhances their migration to the ischemic lesion area [121]. These data support that in addition to enhancing neurogenesis, SDF-1 promotes oligodendrogenesis as well after stroke.

4.1.3 Vascular Endothelial Growth Factor (VEGF)

VEGF is a signaling protein that is important for vasculogenesis and angiogenesis. The administration of VEGF improves neurological performance through mediating angiogenesis and survival of newborn neurons in the rat MCAO model [122]. Studies have shown that VEGF-C stimulates NG2-glia proliferation [123] while VEGF-A can induce NG2-glia migration via ROS and FAK-dependent mechanisms, but did not affect their proliferation and differentiation [124]. In the neonatal hypoxia-ischemia rat model, VEGF-A and VEGF-C are induced in the SVZ. Moreover, VEGF-C promotes the proliferation of both early and late oligodendrocyte progenitors through VEGFR-3 receptor [108]. These data suggest that besides promoting angiogenesis and neurogenesis, VEGF signaling is also involved in oligodendrogenesis after stroke.

4.1.4 Brain-Derived Neurotrophic Factor (BDNF)

BDNF is a well-known member of a neurotrophin family that regulates neuronal survival, synaptic plasticity, learning, and memory. Recent studies show that BDNF could promote the proliferation and differentiation of NG2-glia and is required for normal CNS myelination [125–127]. Astrocyte-derived BDNF supports oligodendrogenesis and regeneration after white matter ischemic injury or cuprizone-induced demyelination [128, 129]. BDNF administration improves functional recovery and promoting oligodendrogenesis and remyelination in rats subjected to ischemic stroke [130]. These data suggest that in addition to neuroprotective effects, BDNF plays important roles in white matter protection and remyelination after stroke.

4.1.5 Other Factors

There are many other factors regulating oligodendrogenesis under normal and pathological conditions [131]. For example, neuregulin-1 promotes NG2-glia survival and maintains NG2-glia in an immature state [132]. Platelet-derived growth factor (PDGF) is an important factor for maintaining NG2-glia proliferation and stimulating their differentiation into mature oligodendrocytes [133]. PDGF signaling in the SVZ promotes oligodendrocyte generation [134]. Insulin-like growth factor (IGF)-1 could promote the differentiation of adult NSCs into oligodendrocyte lineage cells through inhibiting BMP signaling [135]. Epidermal growth factor (EGF) induces the progeny of SVZ NSCs to migrate and differentiate into oligodendrocytes [136]. It has been reported that these above growth factors play positive roles in functional recovery after stroke [137]. However, the contribution of ischemia-induced oligodendrogenesis to functional recovery in stroke by these growth factors remains to be established.

4.2 Epigenetic Modulators and Stroke-Induced Oligodendrogenesis

Epigenetics is defined as the heritable changes in gene expression without a change in the DNA sequence [138]. Recent studies have shown that the multiple steps of oligodendrocyte generation (i.e., specific cell fates, proliferation, differentiation, and myelination) can be regulated through epigenetic mechanisms [139–141]. The epigenetic modulators of gene expression include post-translational modulations of nucleosomal histones, histone modification, chromatin remodeling enzymes, DNA methylation, and microRNAs [142]. Among them, we will mainly focus on miRNA and histone deacetylases (HDACs) in this review.

4.2.1 microRNAs

A number of miRNAs have been found to play a critical role in the proliferation or differentiation of OPCs into mature oligodendrocytes as well as myelination [143, 144]. miR-219 and miR-338 could promote NG2-glia differentiation into mature oligodendrocytes through suppressing the expression of PDGFRa, Sox6, Zfp238, FoxJ3, and NeuroD1 [145]. Stroke considerably increased miR-146a density in the corpus callosum and SVZ of the lateral ventricle of the ischemic hemisphere. In vitro, overexpression of miR-146a in neural progenitor cells (NPCs) significantly increased their differentiation into O4+ NG2-glia [146]. During development, miR17-92 cluster can regulate proliferation and survival of NG2-glia in the brain. In stroke, the miR17-92 cluster was significantly up-regulated in SVZ neural progenitor cells [147]. It could mediate the proliferation and survival of SVZ NPCs in the ischemic brain [148]. miR17-92 cluster-enriched exosomes could increase neural plasticity and functional recovery after stroke [149]. In addition, miR-23a, miR-9, and miR-200b are also likely involved in stroke-induced oligodendrogenesis by regulating serum response factor [150, 151]. Collectively, these findings suggest that miRNAs play an important role in stroke-induced oligodendrogenesis.

4.2.2 Histone Deacetylases (HDACs)

The administration of HDACs inhibitor suberoylanilide hydroxamic acid or TSA can confer protection against ischemia-induced brain injury [152, 153]. HDAC1 and HDAC2 are associated with oligodendrocyte differentiation and remyelination during brain development and disease [154–157]. In ischemic brains, there is increased expression of HDAC1 and HDAC2 proteins in NG2-glia [158]. In addition, blockage of HDACs with valproic acid considerably increased OPCs and new oligodendrocytes after stroke [159]. HDACs clearly play important roles in stroke-induced oligodendrogenesis.

5 The Implicating Effects of NSC-Mediated Neurogenesis and Oligodendrogenesis on Angiogenesis and Inflammation

Stroke continuously induces neuroblasts, which migrate to peri-infarct regions for at least 1 year [160]. The ablation of neuroblasts after stroke reduces ischemic brain repair and exacerbates functional recovery [161]. Experimental studies show that only a small fraction of neuroblasts derived from endogenous NSCs in the peri-infarct regions differentiate into mature neurons and survive [162–164]. Meanwhile, there is increased production of NG2-glia and some of them in the peri-infarct regions generate into mature myelinating oligodendrocytes after stroke [165–167].

These data suggest that stroke-induced neurogenesis and oligodendrogenesis might provide additional beneficial effects that are independent of cellular replacement of dead neurons and myelinating oligodendrocyte production to re-wire neuronal circuitry.

5.1 Angiogenesis

Angiogenesis is characterized by the formation of new vessels from existing blood vessels. Coupling and bi-directional regulation of neurogenesis and angiogenesis have been implicated both under normal and pathological conditions [2]. Both SVZ and SGZ niches have unique vasculature characteristics compared to non-neurogenic regions and adult NSCs extend their long processes to directly contact blood vessels, which enables the easy access of NSCs to molecules and factors in the blood [168]. Under the ischemic condition, it has been shown that angiogenic genes are upregulated rapidly after the onset of cerebral ischemia and the increased expression of angiogenic proteins can be sustained in the ischemic area for a prolonged period of time after stroke [169]. Both neurogenesis and angiogenesis have been suggested to contribute to the functional recovery after stroke [170] and the two critical biological processes might have synergistic effects and influence each other. Co-culture of ischemic neural progenitor cells with non-ischemic endothelial cells increases angiogenesis in vitro [171, 172] and co-culture of ischemic endothelial cells with non-ischemic NSCs increases progenitor cell proliferation and neuronal differentiation. On one hand, neuroblasts induced by stroke in the SVZ migrate along cerebral blood vessels to peri-infarct regions where angiogenesis occurs [96]. On the other hand, it is possible that NSC-derived progeny cells (neuroblasts and astrocytes) can regulate angiogenesis and help maintain the function and integrity of the newly generated blood vessels. Importantly, NG2-glia are also in close proximity to astrocyte, pericytes, or endothelial cells [173, 174]. It is an important component of the neurovascular unit in cerebral white matter [174]. NG2-glia and oligodendrocytes can act as a critical source of trophic factors [175, 176]. In addition, NG2-glia can support blood-brain barrier integrity by upregulating tight junction proteins via TGF-β1 signaling [177]. NG2-glia-specific TGF-β1-deficient mice exhibited cerebral hemorrhage and loss of BBB function [177]. It has been shown that signaling from NG2-glia to ECs plays an important role in angiogenesis during development. Wnt7a and Wnt7b secreted by hypoxic NG2-glia could increase the proliferation of endothelial cells and angiogenesis [178]. These data suggest close interaction and potentially synergistic effects of endogenous neurogenesis, oligodendrogenesis, and angiogenesis in stroke recovery.

5.2 *Inflammation Modulation*

Inflammation plays an important role in the pathogenesis of stroke, which contributes to neuronal death and impairs functional recovery. In the ischemic brain, there is activation of microglia, production of pro-inflammatory factors, and immune cell infiltration (i.e. neutrophils, monocyte/macrophages, T cells and B cells). Recent studies have shed new light on the interaction between endogenous NSCs and immune cells, such as microglia, T cells, and natural killer cells [179–182]. Both in vitro and after transplantation in vivo, NSCs can directly change inflammatory responses through releasing immunomodulatory factors [183–185]. However, it is still unknown whether endogenous NSCs in their native location have similar capacities under stroke conditions. Endocannabinoids are reported to play an important role in maintaining immune homeostatic balance within the host [186]. Anandamide, an endogenous cannabinoid, contributes to immune tolerance in the gut by promoting the presence of CX3C chemokine receptor 1 ($CX3CR1^{hi}$) macrophages, which are immunosuppressive [187]. In response to the excitotoxic damage occurring in stroke and epilepsy, SVZ NSCs can release endogenous endocannabinoids to exert a protective role for striatal neurons [188]. In the EAE model of multiple sclerosis, SVZ NSCs produce interleukin-15 and sustain functionally competent natural killer cells [180]. Studies have shown that there is an accumulation of natural killer cells in ischemic brain tissues [189–191]. These data suggest that endogenous NSCs maybe regulate stroke-induced inflammation through releasing immunomodulatory factors. In addition, NG2-glia and oligodendrocytes express a wide range of immunomodulatory molecules [192, 193], suggesting that endogenous NSCs might indirectly affect immune cell function and inflammation through regulating oligodendrogenesis. Further studies are needed to understand whether, when, and how endogenous NSCs can take over and locally manifest an immunomodulatory effect. It will help to develop novel therapies to promote functional recovery in stroke through modifying the immunomodulatory effects of endogenous NSCs.

6 Conclusion and Discussion

Brain repair processes after stroke are regulated by multiple cellular pathways, which include neurogenesis, oligodendrogenesis, angiogenesis, axonal sprouting, and synaptogenesis. The presence of endogenous NSCs in the adult brain and their capacity to generate new neurons, oligodendrocytes, and astrocytes raises hope that new therapeutic strategies can be designed based on appropriate modulation of endogenous NSCs in stroke. Over the past five decades, since its discovery, adult neurogenesis and NSCs have evolved into an established research field that has made substantial and promising progress as regenerative medicine for neurological disease. However, there are still many critical questions that need to be addressed.

The defining characteristics of stem cells are their ability to self-renew and to differentiate into various cell types. We have just started to appreciate the complexity and heterogeneity of adult NSCs. Balance and integration are important themes to consider when trying to modulate this process to improve brain recovery after injury. For example, adult NSCs have quiescent and activated states. Adult NSCs are largely quiescent in vivo, a state that recently has been recognized as not a passive state but rather maintained by active transcriptional regulation [194]. The mechanisms that trigger the activation of NSCs by entering multiple rounds of proliferation followed by potential terminal differentiation after brain injury are still unknown. Since the quiescent state is actively maintained by NSCs and might serve important roles to preserve these cells from metabolic stress and maintain genome integrity over a long lifetime, strategies that only target to enhance the activation and proliferation of NSCs might need to take cautions as these might have the risk of depleting quiescent NSCs over a prolonged period of time. In this regard, it is possible that treatment strategies that target the enhanced survival of NSCs and their progeny might be a better strategy as the majority of the newly born cells derived from NSCs fail to survive at weeks to months after stroke.

Similarly, the precise mechanisms that trigger differentiation of NSCs to different types of cells in vivo after brain injury are largely unknown. When cultured in vitro, adult NSCs are able to self-renew and differentiate into all three neuronal lineages [195]. However, under normal conditions, SVZ and SGZ cells generate different types of neurons and non-neuronal cells, suggesting that the microenvironment of the NSC niche might limit their differentiation potential. Adult NSCs are also capable of responding to a variety of brain injury by altered differentiation phenotypes as well as migration into the injured area instead of their "original path". What are the precise molecules and signals that direct the differentiation of these cells under the pathological condition? Knowledge in these areas will help us modulate the fate of these cells and help guide them to targeted areas to repair the brain. Substantial interests in the field have been focused on the neuronal differentiation of NSCs after injury; however, neuroblasts have been shown to play important roles through non-neuronal replacement mechanisms [2]. SVZ NSCs have also been reported to generate astrocytes that migrate to the injured cortex. Defects in this astrogenic process, which resulted in a shift in SVZ NSCs fate from glial cells to neuroblasts, resulted in abnormal glial scar formation and increased microvascular hemorrhage in stroke animals. In addition, although glial scar formation was previously considered as an inhibitory factor for axonal outgrowth, there is evidence indicating that the glial scar aids axonal outgrowth in spinal cord injury [196]. Therefore, strategies that aim to guide NSC differentiation towards a single cell type (neurons) might not provide desired effects in brain recovery. Considering the heterogeneity of astrocytes and their role in synapse formation and glial scar formation, whether reactive astrocyte derived from NSCs in stroke could affect axonal outgrowth and synaptogenesis also needs to be investigated. In addition, besides the role of producing new neurons and myelinating oligodendrocytes, it remains to be defined whether and how NSCs, NSC-derived neuroblasts, and NG2-glia contribute to angiogenesis and immunomodulation after stroke. Overall, understanding the fundamental mechanisms underlying the endogenous NSC-mediated brain repair process will provide the basis for future endogenous NSC therapy for stroke. By elucidating the relationship and interactions of neurogenesis with other cellular and molecular processes such as angiogenesis, glial scar formation, and inflammation responses, it is possible that more effective therapies could be developed in the future to improve regeneration and functional recovery of the ischemic brain.

References

1. Feigin VL, Norrving B, Mensah GA. Global burden of stroke. Circ Res. 2017;120(3):439–48.
2. Zhang R, Zhang Z, Chopp M. Function of neural stem cells in ischemic brain repair processes. J Cereb Blood Flow Metab. 2016;36(12):2034–43.
3. Koh SH, Park HH. Neurogenesis in stroke recovery. Transl Stroke Res. 2017;8(1):3–13.
4. Ahmad M, Graham SH. Inflammation after stroke: mechanisms and therapeutic approaches. Transl Stroke Res. 2010;1(2):74–84.
5. Tobin MK, et al. Neurogenesis and inflammation after ischemic stroke: what is known and where we go from here. J Cereb Blood Flow Metab. 2014;34(10):1573–84.
6. Lee DA, et al. Tanycytes of the hypothalamic median eminence form a diet-responsive neurogenic niche. Nat Neurosci. 2012;15(5):700–2.
7. Tonchev AB, et al. Enhanced proliferation of progenitor cells in the subventricular zone and limited neuronal production in the striatum and neocortex of adult macaque monkeys after global cerebral ischemia. J Neurosci Res. 2005;81(6):776–88.
8. Jin K, et al. Evidence for stroke-induced neurogenesis in the human brain. Proc Natl Acad Sci U S A. 2006;103(35):13198–202.
9. Macas J, et al. Increased generation of neuronal progenitors after ischemic injury in the aged adult human forebrain. J Neurosci. 2006;26(50):13114–9.
10. Minger SL, et al. Endogenous neurogenesis in the human brain following cerebral infarction. Regen Med. 2007;2(1):69–74.
11. Arvidsson A, et al. Neuronal replacement from endogenous precursors in the adult brain after stroke. Nat Med. 2002;8(9):963–70.
12. Jin K, et al. Directed migration of neuronal precursors into the ischemic cerebral cortex and striatum. Mol Cell Neurosci. 2003;24(1):171–89.
13. Zhang RL, et al. Ascl1 lineage cells contribute to ischemia-induced neurogenesis and oligodendrogenesis. J Cereb Blood Flow Metab. 2011;31(2):614–25.
14. Li L, et al. Focal cerebral ischemia induces a multilineage cytogenic response from adult subventricular zone that is predominantly gliogenic. Glia. 2010;58(13):1610–9.
15. Faiz M, et al. Adult neural stem cells from the subventricular zone give rise to reactive astrocytes in the cortex after stroke. Cell Stem Cell. 2015;17(5):624–34.
16. Suh H, et al. In vivo fate analysis reveals the multipotent and self-renewal capacities of Sox2+ neural stem cells in the adult hippocampus. Cell Stem Cell. 2007;1(5):515–28.
17. Braun SM, et al. Programming hippocampal neural stem/progenitor cells into oligodendrocytes enhances remyelination in the adult brain after injury. Cell Rep. 2015;11(11):1679–85.
18. Jadasz JJ, et al. p57kip2 regulates glial fate decision in adult neural stem cells. Development. 2012;139(18):3306–15.
19. Oishi S, et al. USP9X deletion elevates the density of oligodendrocytes within the postnatal dentate gyrus. Neurogenesis (Austin). 2016;3(1):e1235524.
20. Rolando C, et al. Multipotency of adult hippocampal nscs in vivo is restricted by drosha/NFIB. Cell Stem Cell. 2016;19(5):653–62.
21. Sun GJ, et al. Latent tri-lineage potential of adult hippocampal neural stem cells revealed by Nf1 inactivation. Nat Neurosci. 2015;18(12):1722–4.

22. Marlier Q, et al. Mechanisms and functional significance of stroke-induced neurogenesis. Front Neurosci. 2015;9:458.
23. Ratan RR. Beyond neuroprotection to brain repair: exploring the next frontier in clinical neuroscience to expand the therapeutic window for stroke. Transl Stroke Res. 2010;1(2):71–3.
24. Gouaze A, et al. Cerebral cell renewal in adult mice controls the onset of obesity. PLoS One. 2013;8(8):e72029.
25. Lee DA, Blackshaw S. Functional implications of hypothalamic neurogenesis in the adult mammalian brain. Int J Dev Neurosci. 2012;30(8):615–21.
26. Kokoeva MV, Yin H, Flier JS. Neurogenesis in the hypothalamus of adult mice: potential role in energy balance. Science. 2005;310(5748):679–83.
27. Recabal A, Caprile T, Garcia-Robles MLA. Hypothalamic neurogenesis as an adaptive metabolic mechanism. Front Neurosci. 2017;11:190.
28. Lin R, et al. Neurogenesis is enhanced by stroke in multiple new stem cell niches along the ventricular system at sites of high BBB permeability. Neurobiol Dis. 2015;74:229–39.
29. Zhang R, et al. Stroke transiently increases subventricular zone cell division from asymmetric to symmetric and increases neuronal differentiation in the adult rat. J Neurosci. 2004;24(25):5810–5.
30. Zhang RL, et al. Reduction of the cell cycle length by decreasing G1 phase and cell cycle reentry expand neuronal progenitor cells in the subventricular zone of adult rat after stroke. J Cereb Blood Flow Metab. 2006;26(6):857–63.
31. Zhang RL, et al. Lengthening the G(1) phase of neural progenitor cells is concurrent with an increase of symmetric neuron generating division after stroke. J Cereb Blood Flow Metab. 2008;28(3):602–11.
32. Balordi F, Fishell G. Mosaic removal of hedgehog signaling in the adult SVZ reveals that the residual wild-type stem cells have a limited capacity for self-renewal. J Neurosci. 2007;27(52):14248–59.
33. Jin Y, et al. The shh signaling pathway is upregulated in multiple cell types in cortical ischemia and influences the outcome of stroke in an animal model. PLoS One. 2015;10(4):e0124657.
34. Sims JR, et al. Sonic hedgehog regulates ischemia/hypoxia-induced neural progenitor proliferation. Stroke. 2009;40(11):3618–26.
35. Wang L, et al. The Sonic hedgehog pathway mediates carbamylated erythropoietin-enhanced proliferation and differentiation of adult neural progenitor cells. J Biol Chem. 2007;282(44):32462–70.
36. Jin Y, et al. Poststroke sonic hedgehog agonist treatment improves functional recovery by enhancing neurogenesis and angiogenesis. Stroke. 2017;48(6):1636–45.
37. Tanaka R, et al. Neurogenesis after transient global ischemia in the adult hippocampus visualized by improved retroviral vector. Stroke. 2004;35(6):1454–9.
38. Naylor M, et al. Preconditioning-induced ischemic tolerance stimulates growth factor expression and neurogenesis in adult rat hippocampus. Neurochem Int. 2005;47(8):565–72.
39. Ninomiya M, et al. Enhanced neurogenesis in the ischemic striatum following EGF-induced expansion of transit-amplifying cells in the subventricular zone. Neurosci Lett. 2006;403(1-2):63–7.
40. Yoshimura S, et al. FGF-2 regulation of neurogenesis in adult hippocampus after brain injury. Proc Natl Acad Sci U S A. 2001;98(10):5874–9.
41. Craig CG, et al. In vivo growth factor expansion of endogenous subependymal neural precursor cell populations in the adult mouse brain. J Neurosci. 1996;16(8):2649–58.
42. Tureyen K, et al. EGF and FGF-2 infusion increases post-ischemic neural progenitor cell proliferation in the adult rat brain. Neurosurgery. 2005;57(6):1254–63. discussion 1254-63
43. Nakatomi H, et al. Regeneration of hippocampal pyramidal neurons after ischemic brain injury by recruitment of endogenous neural progenitors. Cell. 2002;110(4):429–41.
44. Yan YP, et al. Insulin-like growth factor-1 is an endogenous mediator of focal ischemia-induced neural progenitor proliferation. Eur J Neurosci. 2006;24(1):45–54.
45. Kalluri HS, Vemuganti R, Dempsey RJ. Mechanism of insulin-like growth factor I-mediated proliferation of adult neural progenitor cells: role of Akt. Eur J Neurosci. 2007;25(4):1041–8.

46. Dempsey RJ, et al. Stroke-induced progenitor cell proliferation in adult spontaneously hypertensive rat brain: effect of exogenous IGF-1 and GDNF. J Neurochem. 2003;87(3):586–97.
47. Zhu W, et al. Postischemic IGF-1 gene transfer promotes neurovascular regeneration after experimental stroke. J Cereb Blood Flow Metab. 2009;29(9):1528–37.
48. Wang X, et al. Involvement of Notch1 signaling in neurogenesis in the subventricular zone of normal and ischemic rat brain in vivo. J Cereb Blood Flow Metab. 2009;29(10):1644–54.
49. Androutsellis-Theotokis A, et al. Notch signalling regulates stem cell numbers in vitro and in vivo. Nature. 2006;442(7104):823–6.
50. Wang L, et al. The Notch pathway mediates expansion of a progenitor pool and neuronal differentiation in adult neural progenitor cells after stroke. Neuroscience. 2009;158(4):1356–63.
51. Magnusson JP, et al. A latent neurogenic program in astrocytes regulated by Notch signaling in the mouse. Science. 2014;346(6206):237–41.
52. Wang Y, et al. VEGF-overexpressing transgenic mice show enhanced post-ischemic neurogenesis and neuromigration. J Neurosci Res. 2007;85(4):740–7.
53. Kobayashi T, et al. Intracerebral infusion of glial cell line-derived neurotrophic factor promotes striatal neurogenesis after stroke in adult rats. Stroke. 2006;37(9):2361–7.
54. Schabitz WR, et al. Intravenous brain-derived neurotrophic factor enhances poststroke sensorimotor recovery and stimulates neurogenesis. Stroke. 2007;38(7):2165–72.
55. Plane JM, et al. Retinoic acid and environmental enrichment alter subventricular zone and striatal neurogenesis after stroke. Exp Neurol. 2008;214(1):125–34.
56. Chou J, et al. Neuroregenerative effects of BMP7 after stroke in rats. J Neurol Sci. 2006;240(1-2):21–9.
57. Liu XS, et al. MicroRNAs in cerebral ischemia-induced neurogenesis. J Neuropathol Exp Neurol. 2013;72(8):718–22.
58. Volvert ML, et al. MicroRNAs tune cerebral cortical neurogenesis. Cell Death Differ. 2012;19(10):1573–81.
59. Choi JY, et al. M2 phenotype microglia-derived cytokine stimulates proliferation and neuronal differentiation of endogenous stem cells in ischemic brain. Exp Neurobiol. 2017;26(1):33–41.
60. Wang J, et al. Activated regulatory T cell regulates neural stem cell proliferation in the subventricular zone of normal and ischemic mouse brain through interleukin 10. Front Cell Neurosci. 2015;9:361.
61. Kraft A, et al. Astrocytic calcium waves signal brain injury to neural stem and progenitor cells. Stem Cell Rep. 2017;8(3):701–14.
62. Matthys P, et al. AMD3100, a potent and specific antagonist of the stromal cell-derived factor-1 chemokine receptor CXCR4, inhibits autoimmune joint inflammation in IFN gamma receptor-deficient mice. J Immunol. 2001;167(8):4686–92.
63. Hattori K, Heissig B, Rafii S. The regulation of hematopoietic stem cell and progenitor mobilization by chemokine SDF-1. Leuk Lymphoma. 2003;44(4):575–82.
64. Petit I, et al. G-CSF induces stem cell mobilization by decreasing bone marrow SDF-1 and up-regulating CXCR4. Nat Immunol. 2002;3(7):687–94.
65. Kokovay E, et al. Adult SVZ lineage cells home to and leave the vascular niche via differential responses to SDF1/CXCR4 signaling. Cell Stem Cell. 2010;7(2):163–73.
66. Ohab JJ, et al. A neurovascular niche for neurogenesis after stroke. J Neurosci. 2006;26(50):13007–16.
67. Robin AM, et al. Stromal cell-derived factor 1alpha mediates neural progenitor cell motility after focal cerebral ischemia. J Cereb Blood Flow Metab. 2006;26(1):125–34.
68. Thored P, et al. Persistent production of neurons from adult brain stem cells during recovery after stroke. Stem Cells. 2006;24(3):739–47.
69. Imitola J, et al. Directed migration of neural stem cells to sites of CNS injury by the stromal cell-derived factor 1alpha/CXC chemokine receptor 4 pathway. Proc Natl Acad Sci U S A. 2004;101(52):18117–22.
70. Deshmane SL, et al. Monocyte chemoattractant protein-1 (MCP-1): an overview. J Interferon Cytokine Res. 2009;29(6):313–26.

71. Deng YY, et al. Monocyte chemoattractant protein-1 (MCP-1) produced via NF-kappaB signaling pathway mediates migration of amoeboid microglia in the periventricular white matter in hypoxic neonatal rats. Glia. 2009;57(6):604–21.
72. Che X, et al. Monocyte chemoattractant protein-1 expressed in neurons and astrocytes during focal ischemia in mice. Brain Res. 2001;902(2):171–7.
73. Yan YP, et al. Monocyte chemoattractant protein-1 plays a critical role in neuroblast migration after focal cerebral ischemia. J Cereb Blood Flow Metab. 2007;27(6):1213–24.
74. Loffek S, Schilling O, Franzke CW. Series "matrix metalloproteinases in lung health and disease": biological role of matrix metalloproteinases: a critical balance. Eur Respir J. 2011;38(1):191–208.
75. Grade S, et al. Brain-derived neurotrophic factor promotes vasculature-associated migration of neuronal precursors toward the ischemic striatum. PLoS One. 2013;8(1):e55039.
76. Barkho BZ, et al. Endogenous matrix metalloproteinase (MMP)-3 and MMP-9 promote the differentiation and migration of adult neural progenitor cells in response to chemokines. Stem Cells. 2008;26(12):3139–49.
77. Lee SR, et al. Involvement of matrix metalloproteinase in neuroblast cell migration from the subventricular zone after stroke. J Neurosci. 2006;26(13):3491–5.
78. Wang L, et al. Matrix metalloproteinase 2 (MMP2) and MMP9 secreted by erythropoietin-activated endothelial cells promote neural progenitor cell migration. J Neurosci. 2006;26(22):5996–6003.
79. Katona I, Freund TF. Endocannabinoid signaling as a synaptic circuit breaker in neurological disease. Nat Med. 2008;14(9):923–30.
80. Hosoya T, et al. In vivo TSPO and cannabinoid receptor type 2 availability early in post-stroke neuroinflammation in rats: a positron emission tomography study. J Neuroinflammation. 2017;14(1):69.
81. Palazuelos J, et al. CB2 cannabinoid receptors promote neural progenitor cell proliferation via mTORC1 signaling. J Biol Chem. 2012;287(2):1198–209.
82. Bravo-Ferrer I, et al. Cannabinoid type-2 receptor drives neurogenesis and improves functional outcome after stroke. Stroke. 2017;48(1):204–12.
83. Hallmann R, et al. Expression and function of laminins in the embryonic and mature vasculature. Physiol Rev. 2005;85(3):979–1000.
84. Belvindrah R, et al. Beta1 integrins control the formation of cell chains in the adult rostral migratory stream. J Neurosci. 2007;27(10):2704–17.
85. Emsley JG, Hagg T. alpha6beta1 integrin directs migration of neuronal precursors in adult mouse forebrain. Exp Neurol. 2003;183(2):273–85.
86. Fujioka T, et al. beta1 integrin signaling promotes neuronal migration along vascular scaffolds in the post-stroke brain. EBioMedicine. 2017;16:195–203.
87. Yamashita T, et al. Subventricular zone-derived neuroblasts migrate and differentiate into mature neurons in the post-stroke adult striatum. J Neurosci. 2006;26(24):6627–36.
88. Le Magueresse C, et al. Subventricular zone-derived neuroblasts use vasculature as a scaffold to migrate radially to the cortex in neonatal mice. Cereb Cortex. 2012;22(10):2285–96.
89. Kojima T, et al. Subventricular zone-derived neural progenitor cells migrate along a blood vessel scaffold toward the post-stroke striatum. Stem Cells. 2010;28(3):545–54.
90. Young CC, et al. Cellular and molecular determinants of stroke-induced changes in subventricular zone cell migration. Antioxid Redox Signal. 2011;14(10):1877–88.
91. Wei ZZ, et al. Neuroprotective and regenerative roles of intranasal Wnt-3a Administration after focal ischemic stroke in mice. J Cereb Blood Flow Metab. 2017: 271678X17702669.
92. Zhao Y, et al. GSK-3beta inhibition induced neuroprotection, regeneration, and functional recovery after intracerebral hemorrhagic stroke. Cell Transplant. 2017;26(3):395–407.
93. Xu W, et al. Chloride co-transporter NKCC1 inhibitor bumetanide enhances neurogenesis and behavioral recovery in rats after experimental stroke. Mol Neurobiol. 2017;54(4):2406–14.
94. Carmichael ST. Rodent models of focal stroke: size, mechanism, and purpose. NeuroRx. 2005;2(3):396–409.

95. Galindo LT, et al. Chondroitin sulfate impairs neural stem cell migration through ROCK activation. Mol Neurobiol. 2017;
96. Thored P, et al. Long-term neuroblast migration along blood vessels in an area with transient angiogenesis and increased vascularization after stroke. Stroke. 2007;38(11):3032–9.
97. Hou SW, et al. Functional integration of newly generated neurons into striatum after cerebral ischemia in the adult rat brain. Stroke. 2008;39(10):2837–44.
98. Parent JM, et al. Rat forebrain neurogenesis and striatal neuron replacement after focal stroke. Ann Neurol. 2002;52(6):802–13.
99. Teramoto T, et al. EGF amplifies the replacement of parvalbumin-expressing striatal interneurons after ischemia. J Clin Invest. 2003;111(8):1125–32.
100. Stokowska A, et al. Complement peptide C3a stimulates neural plasticity after experimental brain ischaemia. Brain. 2017;140(Pt 2):353–69.
101. Luo Y, et al. Delayed treatment with a p53 inhibitor enhances recovery in stroke brain. Ann Neurol. 2009;65(5):520–30.
102. Turnley AM, Basrai HS, Christie KJ. Is integration and survival of newborn neurons the bottleneck for effective neural repair by endogenous neural precursor cells? Front Neurosci. 2014;8:29.
103. Gu W, Brannstrom T, Wester P. Cortical neurogenesis in adult rats after reversible photothrombotic stroke. J Cereb Blood Flow Metab. 2000;20(8):1166–73.
104. Jiang L, et al. Oligogenesis and oligodendrocyte progenitor maturation vary in different brain regions and partially correlate with local angiogenesis after ischemic stroke. Transl Stroke Res. 2011;2(3):366–75.
105. Eda H, et al. Ischemic damage and subsequent proliferation of oligodendrocytes in hippocampal CA1 region following repeated brief cerebral ischemia. Pathobiology. 2009;76(4):204–11.
106. Mandai K, et al. Ischemic damage and subsequent proliferation of oligodendrocytes in focal cerebral ischemia. Neuroscience. 1997;77(3):849–61.
107. Tanaka K, et al. Activation of NG2-positive oligodendrocyte progenitor cells during postischemic reperfusion in the rat brain. Neuroreport. 2001;12(10):2169–74.
108. Bain JM, et al. Vascular endothelial growth factors A and C are induced in the SVZ following neonatal hypoxia-ischemia and exert different effects on neonatal glial progenitors. Transl Stroke Res. 2013;4(2):158–70.
109. Kim HJ, Chuang DM. HDAC inhibitors mitigate ischemia-induced oligodendrocyte damage: potential roles of oligodendrogenesis, VEGF, and anti-inflammation. Am J Transl Res. 2014;6(3):206–23.
110. Zhang L, et al. Erythropoietin amplifies stroke-induced oligodendrogenesis in the rat. PLoS One. 2010;5(6):e11016.
111. Zhang RL, et al. Sildenafil enhances neurogenesis and oligodendrogenesis in ischemic brain of middle-aged mouse. PLoS One. 2012;7(10):e48141.
112. Ferent J, et al. Sonic Hedgehog signaling is a positive oligodendrocyte regulator during demyelination. J Neurosci. 2013;33(5):1759–72.
113. Tong CK, et al. A dorsal SHH-dependent domain in the V-SVZ produces large numbers of oligodendroglial lineage cells in the postnatal brain. Stem Cell Rep. 2015;5(4):461–70.
114. Zhang L, et al. Sonic hedgehog signaling pathway mediates cerebrolysin-improved neurological function after stroke. Stroke. 2013;44(7):1965–72.
115. Ding X, et al. The sonic hedgehog pathway mediates brain plasticity and subsequent functional recovery after bone marrow stromal cell treatment of stroke in mice. J Cereb Blood Flow Metab. 2013;33(7):1015–24.
116. Li M, et al. Chemokine CXCL12 in neurodegenerative diseases: an SOS signal for stem cell-based repair. Trends Neurosci. 2012;35(10):619–28.
117. Li Y, et al. Postacute stromal cell-derived factor-1alpha expression promotes neurovascular recovery in ischemic mice. Stroke. 2014;45(6):1822–9.
118. Dziembowska M, et al. A role for CXCR4 signaling in survival and migration of neural and oligodendrocyte precursors. Glia. 2005;50(3):258–69.

119. Kadi L, et al. Differential effects of chemokines on oligodendrocyte precursor proliferation and myelin formation in vitro. J Neuroimmunol. 2006;174(1-2):133–46.
120. Maysami S, et al. Modulation of rat oligodendrocyte precursor cells by the chemokine CXCL12. Neuroreport. 2006;17(11):1187–90.
121. Li Y, et al. CXCL12 gene therapy ameliorates ischemia-induced white matter injury in mouse brain. Stem Cells Transl Med. 2015;4(10):1122–30.
122. Sun Y, et al. VEGF-induced neuroprotection, neurogenesis, and angiogenesis after focal cerebral ischemia. J Clin Invest. 2003;111(12):1843–51.
123. Le Bras B, et al. VEGF-C is a trophic factor for neural progenitors in the vertebrate embryonic brain. Nat Neurosci. 2006;9(3):340–8.
124. Hayakawa K, et al. Vascular endothelial growth factor regulates the migration of oligodendrocyte precursor cells. J Neurosci. 2011;31(29):10666–70.
125. Cellerino A, et al. Reduced size of retinal ganglion cell axons and hypomyelination in mice lacking brain-derived neurotrophic factor. Mol Cell Neurosci. 1997;9(5-6):397–408.
126. Xiao J, et al. Brain-derived neurotrophic factor promotes central nervous system myelination via a direct effect upon oligodendrocytes. Neurosignals. 2010;18(3):186–202.
127. Tsiperson V, et al. Brain-derived neurotrophic factor deficiency restricts proliferation of oligodendrocyte progenitors following cuprizone-induced demyelination. ASN Neuro. 2015;7(1):1759091414566878.
128. Miyamoto N, et al. Astrocytes promote oligodendrogenesis after white matter damage via brain-derived neurotrophic factor. J Neurosci. 2015;35(41):14002–8.
129. Fulmer CG, et al. Astrocyte-derived BDNF supports myelin protein synthesis after cuprizone-induced demyelination. J Neurosci. 2014;34(24):8186–96.
130. Ramos-Cejudo J, et al. Brain-derived neurotrophic factor administration mediated oligodendrocyte differentiation and myelin formation in subcortical ischemic stroke. Stroke. 2015;46(1):221–8.
131. Itoh K, et al. Mechanisms of cell-cell interaction in oligodendrogenesis and remyelination after stroke. Brain Res. 2015;1623:135–49.
132. Canoll PD, et al. GGF/neuregulin is a neuronal signal that promotes the proliferation and survival and inhibits the differentiation of oligodendrocyte progenitors. Neuron. 1996;17(2):229–43.
133. Deierborg T, et al. Brain injury activates microglia that induce neural stem cell proliferation ex vivo and promote differentiation of neurosphere-derived cells into neurons and oligodendrocytes. Neuroscience. 2010;171(4):1386–96.
134. Woodruff RH, et al. Platelet-derived growth factor regulates oligodendrocyte progenitor numbers in adult CNS and their response following CNS demyelination. Mol Cell Neurosci. 2004;25(2):252–62.
135. Hsieh J, et al. IGF-I instructs multipotent adult neural progenitor cells to become oligodendrocytes. J Cell Biol. 2004;164(1):111–22.
136. Gonzalez-Perez O, et al. Epidermal growth factor induces the progeny of subventricular zone type B cells to migrate and differentiate into oligodendrocytes. Stem Cells. 2009;27(8):2032–43.
137. Chan SJ, et al. Endogenous regeneration: engineering growth factors for stroke. Neurochem Int. 2017;107:57.
138. Chuang JC, Jones PA. Epigenetics and microRNAs. Pediatr Res. 2007;61(5 Pt 2):24R–9R.
139. Liu J, et al. Epigenetic control of oligodendrocyte development: adding new players to old keepers. Curr Opin Neurobiol. 2016;39:133–8.
140. Purger D, Gibson EM, Monje M. Myelin plasticity in the central nervous system. Neuropharmacology. 2016;110(Pt B):563–73.
141. Yu Y, et al. Olig2 targets chromatin remodelers to enhancers to initiate oligodendrocyte differentiation. Cell. 2013;152(1-2):248–61.
142. Liu J, Casaccia P. Epigenetic regulation of oligodendrocyte identity. Trends Neurosci. 2010;33(4):193–201.

143. Barca-Mayo O, Lu QR. Fine-tuning oligodendrocyte development by microRNAs. Front Neurosci. 2012;6:13.
144. Maki T, et al. Mechanisms of oligodendrocyte regeneration from ventricular-subventricular zone-derived progenitor cells in white matter diseases. Front Cell Neurosci. 2013;7:275.
145. Dugas JC, et al. Dicer1 and miR-219 are required for normal oligodendrocyte differentiation and myelination. Neuron. 2010;65(5):597–611.
146. Liu XS, et al. MicroRNA-146a promotes oligodendrogenesis in stroke. Mol Neurobiol. 2017;54(1):227–37.
147. Liu XS, et al. MicroRNAs in cerebral ischemia-induced neurogenesis. J Neuropathol Exp Neurol. 2013;72(8):717–21.
148. Liu XS, et al. MicroRNA-17-92 cluster mediates the proliferation and survival of neural progenitor cells after stroke. J Biol Chem. 2013;288(18):12478–88.
149. Xin H, et al. MicroRNA cluster miR-17-92 cluster in exosomes enhance neuroplasticity and functional recovery after stroke in rats. Stroke. 2017;48(3):747–53.
150. Buller B, et al. Regulation of serum response factor by miRNA-200 and miRNA-9 modulates oligodendrocyte progenitor cell differentiation. Glia. 2012;60(12):1906–14.
151. Siegel C, et al. miR-23a regulation of X-linked inhibitor of apoptosis (XIAP) contributes to sex differences in the response to cerebral ischemia. Proc Natl Acad Sci U S A. 2011;108(28):11662–7.
152. Baltan S, et al. Histone deacetylase inhibitors preserve white matter structure and function during ischemia by conserving ATP and reducing excitotoxicity. J Neurosci. 2011;31(11):3990–9.
153. Faraco G, et al. Pharmacological inhibition of histone deacetylases by suberoylanilide hydroxamic acid specifically alters gene expression and reduces ischemic injury in the mouse brain. Mol Pharmacol. 2006;70(6):1876–84.
154. Shen S, Casaccia-Bonnefil P. Post-translational modifications of nucleosomal histones in oligodendrocyte lineage cells in development and disease. J Mol Neurosci. 2008;35(1):13–22.
155. Shen S, et al. Epigenetic memory loss in aging oligodendrocytes in the corpus callosum. Neurobiol Aging. 2008;29(3):452–63.
156. Shen S, et al. Age-dependent epigenetic control of differentiation inhibitors is critical for remyelination efficiency. Nat Neurosci. 2008;11(9):1024–34.
157. Ye F, et al. HDAC1 and HDAC2 regulate oligodendrocyte differentiation by disrupting the beta-catenin-TCF interaction. Nat Neurosci. 2009;12(7):829–38.
158. Kassis H, et al. Histone deacetylase expression in white matter oligodendrocytes after stroke. Neurochem Int. 2014;77:17–23.
159. Liu XS, et al. Valproic acid increases white matter repair and neurogenesis after stroke. Neuroscience. 2012;220:313–21.
160. Kazanis I, et al. The late response of rat subependymal zone stem and progenitor cells to stroke is restricted to directly affected areas of their niche. Exp Neurol. 2013;248:387–97.
161. Sun F, et al. Ablation of neurogenesis attenuates recovery of motor function after focal cerebral ischemia in middle-aged mice. PLoS One. 2012;7(10):e46326.
162. Bonfanti L. Adult neurogenesis 50 years later: limits and opportunities in mammals. Front Neurosci. 2016;10:44.
163. Liu F, et al. Brain injury does not alter the intrinsic differentiation potential of adult neuroblasts. J Neurosci. 2009;29(16):5075–87.
164. Obernier K, Tong CK, Alvarez-Buylla A. Restricted nature of adult neural stem cells: re-evaluation of their potential for brain repair. Front Neurosci. 2014;8:162.
165. Gregersen R, et al. Focal cerebral ischemia induces increased myelin basic protein and growth-associated protein-43 gene transcription in peri-infarct areas in the rat brain. Exp Brain Res. 2001;138(3):384–92.
166. Ueno Y, et al. Axonal outgrowth and dendritic plasticity in the cortical peri-infarct area after experimental stroke. Stroke. 2012;43(8):2221–8.
167. Zhang R, Chopp M, Zhang ZG. Oligodendrogenesis after cerebral ischemia. Front Cell Neurosci. 2013;7:201.

168. Mirzadeh Z, et al. Neural stem cells confer unique pinwheel architecture to the ventricular surface in neurogenic regions of the adult brain. Cell Stem Cell. 2008;3(3):265–78.
169. Hayashi T, et al. Temporal profile of angiogenesis and expression of related genes in the brain after ischemia. J Cereb Blood Flow Metab. 2003;23(2):166–80.
170. Arenillas JF, et al. The role of angiogenesis in damage and recovery from ischemic stroke. Curr Treat Options Cardiovasc Med. 2007;9(3):205–12.
171. Zhang Z, Chopp M. Neural stem cells and ischemic brain. J Stroke. 2016;18(3):267–72.
172. Teng H, et al. Coupling of angiogenesis and neurogenesis in cultured endothelial cells and neural progenitor cells after stroke. J Cereb Blood Flow Metab. 2008;28(4):764–71.
173. Maki T, et al. Potential interactions between pericytes and oligodendrocyte precursor cells in perivascular regions of cerebral white matter. Neurosci Lett. 2015;597:164–9.
174. Shindo A, et al. Subcortical ischemic vascular disease: roles of oligodendrocyte function in experimental models of subcortical white-matter injury. J Cereb Blood Flow Metab. 2016;36(1):187–98.
175. Sakry D, et al. Oligodendrocyte precursor cells synthesize neuromodulatory factors. PLoS One. 2015;10(5):e0127222.
176. Wilkins A, et al. Oligodendrocytes promote neuronal survival and axonal length by distinct intracellular mechanisms: a novel role for oligodendrocyte-derived glial cell line-derived neurotrophic factor. J Neurosci. 2003;23(12):4967–74.
177. Seo JH, et al. Oligodendrocyte precursor cells support blood-brain barrier integrity via TGF-beta signaling. PLoS One. 2014;9(7):e103174.
178. Yuen TJ, et al. Oligodendrocyte-encoded HIF function couples postnatal myelination and white matter angiogenesis. Cell. 2014;158(2):383–96.
179. Butovsky O, et al. Microglia activated by IL-4 or IFN-gamma differentially induce neurogenesis and oligodendrogenesis from adult stem/progenitor cells. Mol Cell Neurosci. 2006;31(1):149–60.
180. Liu Q, et al. Neural stem cells sustain natural killer cells that dictate recovery from brain inflammation. Nat Neurosci. 2016;19(2):243–52.
181. Ribeiro Xavier AL, et al. A distinct population of microglia supports adult neurogenesis in the subventricular zone. J Neurosci. 2015;35(34):11848–61.
182. Drago D, et al. Metabolic determinants of the immune modulatory function of neural stem cells. J Neuroinflammation. 2016;13(1):232.
183. Ben-Hur T. Immunomodulation by neural stem cells. J Neurol Sci. 2008;265(1-2):102–4.
184. Pluchino S, et al. Neurosphere-derived multipotent precursors promote neuroprotection by an immunomodulatory mechanism. Nature. 2005;436(7048):266–71.
185. Kokaia Z, et al. Cross-talk between neural stem cells and immune cells: the key to better brain repair? Nat Neurosci. 2012;15(8):1078–87.
186. Cabral GA, Ferreira GA, Jamerson MJ. Endocannabinoids and the immune system in health and disease. Handb Exp Pharmacol. 2015;231:185–211.
187. Acharya N, et al. Endocannabinoid system acts as a regulator of immune homeostasis in the gut. Proc Natl Acad Sci U S A. 2017;114(19):5005–10.
188. Butti E, et al. Subventricular zone neural progenitors protect striatal neurons from glutamatergic excitotoxicity. Brain. 2012;135(Pt 11):3320–35.
189. Gan Y, et al. Ischemic neurons recruit natural killer cells that accelerate brain infarction. Proc Natl Acad Sci U S A. 2014;111(7):2704–9.
190. Liu Q, et al. Brain ischemia suppresses immunity in the periphery and brain via different neurogenic innervations. Immunity. 2017;46(3):474–87.
191. Zhang Y, et al. Accumulation of natural killer cells in ischemic brain tissues and the chemotactic effect of IP-10. J Neuroinflammation. 2014;11:79.
192. Zeis T, Enz L, Schaeren-Wiemers N. The immunomodulatory oligodendrocyte. Brain Res. 2016;1641(Pt A):139–48.
193. Fitzner D, et al. Selective transfer of exosomes from oligodendrocytes to microglia by macropinocytosis. J Cell Sci. 2011;124(Pt 3):447–58.

194. Shin J, et al. Single-cell RNA-seq with waterfall reveals molecular cascades underlying adult neurogenesis. Cell Stem Cell. 2015;17(3):360–72.
195. Reynolds BA, Weiss S. Generation of neurons and astrocytes from isolated cells of the adult mammalian central nervous system. Science. 1992;255(5052):1707–10.
196. Anderson MA, et al. Astrocyte scar formation aids central nervous system axon regeneration. Nature. 2016;532(7598):195–200.

Chapter 5
Mobilization of Endogenous Neural Stem Cells to Promote Regeneration After Stroke

Monika Rabenstein and Maria Adele Rueger

Abstract Endogenous neural stem cells (eNSC) in the adult brain mainly reside in two stem cell niches, the subventricular zone (SVZ), and the hippocampal dentate gyrus. Following cerebral insults, they are mobilized from their niches to engage in regeneration and mediate functional recovery. After cerebral ischemia, eNSC generate new neurons in a process called neurogenesis, but also indirectly mediate regeneration via pleiotropic functions including neuroprotection, reduction of neuroinflammation, revascularization, and induction of plasticity. However, the physiological capacity of the brain for self-repair after stroke is insufficient in mammals. Thus, a promising therapeutic approach in stroke constitutes the targeted activation of eNSC by pharmacological substances, e.g. osteopontin or FGL, and by non-pharmacological approaches, such as transcranial direct current stimulation (tDCS). Since treatments based on the transplantation of stem cells harbor several disadvantages including poor long-term cell survival and a lack of integration into the host circuitry, mobilizing the eNSC niche for therapeutic purposes constitutes a most promising approach in stem cell research.

Keywords Osteopontin • FGL • Ar-tumerone • Transcranial direct current stimulation (tDCS) • Recovery • Neurogenesis • Neuroprotection • Plasticity • Functional recovery

Abbreviations

BrDU	Bromodeoxyuridine
CNS	central nervous system
DCX	Doublecortin

M. Rabenstein
Department of Neurology, University Hospital of Cologne,
Kerpener Strasse 62, 50937 Cologne, Germany

M.A. Rueger, M.D. (✉)
Neural Stem Cell Laboratory, Department of Neurology, University Hospital of Cologne,
Kerpener Str. 62, 50924 Cologne, Germany
e-mail: adele.rueger@uk-koeln.de

P.A. Lapchak, J.H. Zhang (eds.), *Cellular and Molecular Approaches to Regeneration and Repair*, Springer Series in Translational Stroke Research,
https://doi.org/10.1007/978-3-319-66679-2_5

EAE	Experimental autoimmune encephalomyelitis
EGF	Epithelial growth factor
eNSC	Endogenous neural stem cells
FGF 2	Fibroblast growth factor 2
FGL	Neural cell adhesion molecule FG Loop
NCAM	Neural cell adhesion molecule
OPN	Osteopontin
PET	Positron-Emission-Tomography
RMS	Rostral migratory stream
SGZ	Subgranular zone of the hippocampus
Shh	Sonic hedgehog
SVZ	Subventricular zone
tDCS	Transcranial direct current stimulation

1 Introduction

For a long time it has been assumed that neurogenesis does not occur in the postnatal brain. This paradigm was opposed by Altman et al. [1], who first described the ability of the postnatal mammalian brain to generate new neurons in the postnatal rat hippocampus, and by Kaplan et al., who essentially discovered neurogenesis in the adult dentate gyrus [2]. They found labeled neurons using radioactive thymidine to label all dividing cells and histological examination of postmortem brains. This proved that new neurons in the adult brain are generated following cell division. The extend of neurogenesis declines in adulthood, but remains in a significant steady-state. Subsequent studies showed that in addition to the hippocampal dentate gyrus, immature precursor cells also persist in the subventricular zone of the lateral ventricles [3, 4] and are capable of differentiating into all three cell fates of the central nervous system (CNS): neurons, astrocytes and oligodendrocytes [5–9].

2 Endogenous Neural Stem Cells Under Physiological Conditions

2.1 Neural Stem Cells in Animals Models

Endogenous neural stem cells (eNSC) in the adult brain mainly reside in two stem cell niches: the subventricular zone (SVZ) adjacent to the lateral ventricles, and the subgranular zone of the hippocampus (SGZ). These stem cell niches are part of two distinct neuronal networks: eNSC from the SVZ can migrate via the rostral migratory stream (RMS) to the olfactory bulb. ENSC from the SGZ give rise to neurons and glia cells in the hippocampal dentate gyrus (reviewed by [10, 11]).

Under physiological circumstances, eNSC are mostly in a quiescent state. Interestingly, driven by Sonic hedgehog (Shh) signalling, they are able to self-renew over the course of a year and generate multiple cell types in vivo such as (inter) neurons, astrocytes or oligodendrocytes [12]. Furthermore, neurogenesis seems not be restricted to the SVZ and SGZ. There have been several reports of adult neurogenesis in the neocortex, the striatum, the amygdala, the hypothalamus, the substantia nigra, in the brain-stem, olfactory tubercle and piriform cortex. However, data about neocortical neurogenesis remains conflicting (reviewed by [13]). Inside the stem cell niches, eNSC reside in close connection to blood vessels. Extending this anatomical connection, endothelial vascular cells stimulate the self-renewal of neural stem cells. Even activated eNSC maintain their connection to blood vessels. This microenvironment is referred to as the neurovascular niche [11, 14].

Hippocampal neurogenesis seems to play an important role in learning and memory. Exercise and an enriched environment increase hippocampal neurogenesis and thus improves learning abilities [15–17]. Importantly, while eNSC numbers and neurogenesis decline during aging [18, 19], the capacity of the remaining eNSC to respond to cerebral insults seems stable over most of the life-span [20].

Moreover, eNSCs secrete trophic factors supporting neuroprotection such as glial-derived neurotrophic factor, vascular endothelial growth factor, or Shh [21, 22]. Additionally, eNSCs promote other regenerative processes including remyelination, angiogenesis, remodeling, and immunomodulation [23, 24].

2.2 *Neural Stem Cells in Humans*

The knowledge of eNSC derived neurogenesis in humans is restricted by the limited amount of detection methods (reviewed by [25]). The first study reporting about neurogenesis in the human hippocampus examined human brain tissue that was obtained postmortem from patients who had been treated during a cancer treatment with the thymidine analog, bromodeoxyuridine (BrdU), that labels DNA during the S phase [26]. Another method to estimate human neurogenesis was established by Jonas Frisén's group by measuring the concentration of nuclear bomb test-derived 14C in genomic DNA [27–29]. Though, limitations of this technique are the demanding infrastructure and the natural decline of 14C levels [28].

In humans, the rostral migratory stream is organized around a tubular extension of the lateral ventricle that reaches the olfactory bulb [30]. However, neurogenesis in the olfactory bulb in humans seems not to reach relevant levels [31]. On the other hand, relevant human adult hippocampal neurogenesis with an estimated number 700 new neurons are added per day was described [27]. Moreover, integration of newborn neurons into the striatum was observed in humans [32]. Notably, in most cases, the intrinsic response of eNSCs is obviously not sufficient to lead to detectable neocortical neurogenesis after stroke [33]. Of note, all of the latter results were obtained by 14C measuring. This method can only detect larger numbers of new cells that were generated at a given time point with a detectable limit at about 1% of the total population of neurons [29]. Therefore, neurogenesis occurring at low levels, and new neurons that are not permanently integrated into the circuitry, may not be tracked using this method. This may explain contradictory data about neocortical neurogenesis occurring after stroke, which was reported to be present in immunocytochemical analyzes of human postmortem brain slices [34]. Additionally, some reports found cortical neurogenesis in small numbers in stroke animal models [35, 36].

3 Neural Stem Cells After Ischemia

Stroke is one of the major causes of adult disability [37]. To date, re-perfusion treatment is only possible in a narrow time window, and there is no neuroprotective or even regenerative treatment for the subacute or chronic phase after stroke yet. Thus, current treatment in this phase is limited to functional treatment such as physiotherapy. From the pathophysiological point of view, after the initial ischemic damage with disruption of the blood flow that leads to necrotic cell death, brain resident immune cells such as microglia and astrocytes are rapidly activated, and blood-borne immune cells (granulocytes, T-cells, monocytes/macrophages) are recruited from the blood stream to the lesion site [38–43]. This process is called neuroinflammation. There are many beneficial effects of neuroinflammation such as containment of necrotic damage, trophic support, support of neurons and mobilization of endogenous stem cells [44]. But on the other hand, persistent neuroinflammation can also cause secondary tissue damage by excessive release of proinflammatory cytokines and reactive oxygen species [45].

The immune cells attract eNSC to the site of the lesion by secretion of various inflammatory cytokines such as stromal cell-derived factor-1, tumor necrosis factor-alpha, and interferon-γ [46–49]. This attraction of eNSC is also referred to as mobilization of eNSCs. In various models of cerebral ischemia in experimental animals, including transient global ischemia, transient focal ischemia, or permanent focal ischemia, a mobilization of eNSC was demonstrated [50–53]. However, this eNSC mobilization is not sufficient to provide functional recovery, because the majority of newly generated neuroblasts in ischemic stroke models die by the time they have reached the peri-infarct area [52]. Moreover, in humans, no relevant neocortical neurogenesis in humans was detected after stroke [33].

4 Mobilizing the Endogenous NSC Niche

Since the endogenous neural stem cell response after stroke is not strong enough for sufficient repair processes, boosting the eNSC response by pharmacological or non-pharmacological methods constitutes a promising therapeutic approach for stroke. In contrast, the transplantation approach of "exogenous" cells is associated with certain disadvantages like poor long-term cell survival, a lack of integration into the host circuitry, immune reactions against the transplants, and limited availability of appropriate cells (reviewed by [54]). Mobilizing the endogenous neural stem cell niche overcomes those difficulties and is additionally less invasive.

4.1 Pharmacological Mobilization of Endogenous Neural Stem Cells

ENSCs can be mobilized for therapeutic purposes by different types of drugs. One group of substances consists of stem cell growth- and regulation factors that target specific intracellular signaling pathways: Early studies reported that intraventricular treatment with fibroblast growth factor 2 (FGF 2) and epithelial growth factor (EGF) stimulates the proliferation of eNSCs in vivo [55, 56]. Intraventricular co-treatment with FGF 2 and EGF increases the number of hippocampal pyramidal neurons after cerebral ischemia by enhancing eNSC proliferation, and their differentiation into neurons [57]. Likewise, augmenting long-term FGF2 expression in rats after stroke increases SVZ and cortical neurogenesis and behavioral outcome [35]. Notch signaling is an important signaling pathway in eNSC and evokes pleiotropic effects in stem cells. Notch receptor activation promotes the survival of neural stem cells. Transient administration of Notch ligands to the brain of adult rats increases the numbers of newly generated precursor cells and improves motor skills after ischemic injury [21]. Moreover, angiopoietins are significant regulators of endothelial and hematopoietic stem cells. Angiopoietin2 rescues injured dopamine neurons with motor behavioral improvement in an experimental model of neurodgeneration [58]. The neural cell adhesion molecule (NCAM) enhances neurite outgrowth, synaptogenesis, and neuronal differentiation. Its mimetic peptide FG Loop (FGL) induces NSC mobilization in vitro and in vivo, and supports oligodendroglial differentiation [59]. After focal cerebral ischemia, FGL mobilizes eNSC from the niches and enhances regeneration by amplifying remyelination and modulating neuroinflammation via affecting microglia [60]. Another important eNSC signaling pathway is initiated by the ligand sonic hedgehog. Jin et al. showed that oral administration of a sonic hedgehog agonist increased functional recovery, neurogenesis and angiogenesis after experimental stroke [61].

In a second pharmacological approach, eNSC mobilization can also be induced by certain nutrition ingredients: Curcumin and ar-turmerone are the major bioactive compounds of the herb Curcuma longa. Ar-turmerone induces NSC proliferation in vitro and promotes neuronal differentiation of eNSC. Concordantly, there was also increased proliferation and mobilization of eNSC in vivo as shown by Positron-Emission-Tomography (PET) [62].

A third group of drugs mobilizing eNSC are endogenous or exogenous factors that are involved in (neuro-)inflammation: Osteopontin (OPN) is an endogenous phosphoglycoprotein with important roles in tissue homeostasis, wound healing, immune regulation, and stress responses. OPN increases survival, proliferation, migration, and neuronal differentiation of eNSC. Increased survival and migration are mediated via the chemokine receptor CXCR4. After cerebral ischemia, OPN increases neurogenesis [63] (Fig. 5.1). Additionally, OPN seems to polarize microglia to a neuroprotective subtype in an inflammation setting [64]. The tetracycline antibiotic minocycline is commonly used to treat bacterial infections. Additionally, it has pleotropic effects on immune processes [65, 66]. In stem cells, minocycline enhances cell survival in vitro, and increases eNSC activity in both the SVZ as well as the hippocampus in animals after experimental stroke [67]. Additionally, minocycline antagonizes the rapid glial differentiation induced by proinflammatory cytokines in vitro, and restores the neurogenic and oligodendrogenic potential [68].

4.2 *Non-pharmacological Mobilization of Endogenous Neural Stem Cells*

Clinical data suggest that transcranial direct current stimulation (tDCS) may facilitate rehabilitation after stroke [69, 70]. However, the neurobiological mechanisms underlying tDCS remain poorly explored. TDCS can be applied with either with an anodal or a cathodal current polarity, and with various current densities. In the healthy rat brain, certain polarities and current densities of tDCS increase neural stem cell migration and activate microglia [71, 72]. Under specific conditions, tDCS accelerates functional recovery in animals after experimental stroke. Moreover, both anodal and cathodal tDCS at different current densities induce neurogenesis (Fig. 5.2). Only cathodal tDCS recruits oligodendrocyte precursors towards the lesion, but also supports a proinflammatory M1-polarization of microglia. In contrast, anodal tDCS leads to downregulation of the constitutive expression of Iba1 by microglia [73, 74]. In conclusion, the different tDCS polarities seem to exert different effects on eNSC as well as on migroglia. TDCS acts through multifaceted mechanisms that far exceed its primary neurophysiological effects, encompassing proliferation and migration of stem cells, their neuronal differentiation, and modulation of microglia responses.

5 Future Perspectives

ENSC can be targeted by pharmacologial or non-pharmacological approaches. Thereby, enhancement of eNSC proliferation, migration and differentiation to neurons and oligodendrocytes is possible in order to promote neuroregeneration and functional recovery. As a second step, a translational approach to establish these therapies in clinical treatment for humans is needed. Such clinical trials could include osteopontin (OPN) that modulates eNSC as well as immune cells, and can be applied via a nose spray [75]. As for non-pharmacological approaches, tDCS is already applied experimentally in the clinical setting [69, 70]. With more knowledge

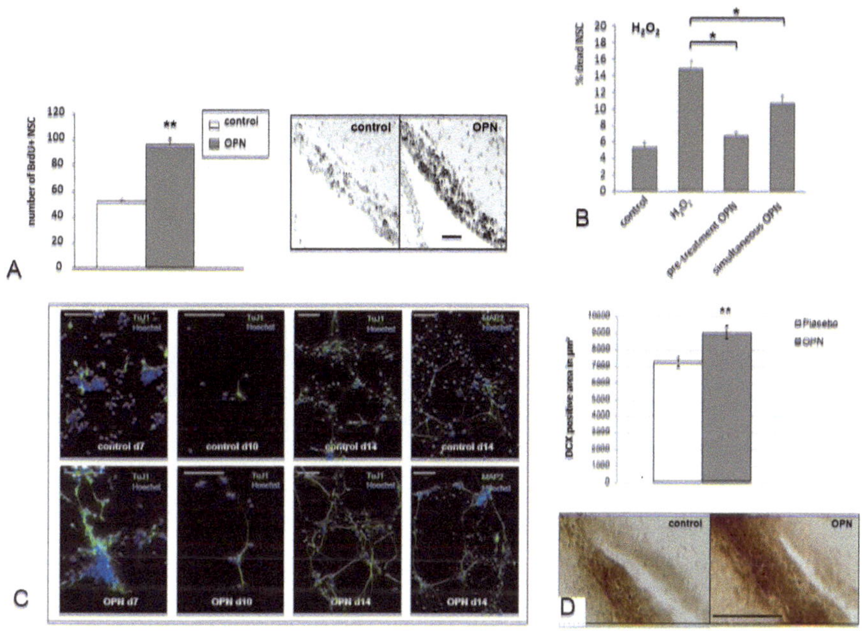

Fig. 5.1 Osteopontin (OPN) increases survival, proliferation and neurogenesis of neural stem cells (NSC). (**a**) Adult male rats injected with a single dose of 500 µg OPN i.c.v. displayed a significantly higher number of proliferating NSC in the SVZ as the major NSC niche, corroborating the effects of OPN on NSC proliferation in vivo (values displayed as means ± SEM; $^{**}p < 0.01$). Representative images from the SVZ of rats treated with either placebo (*left*) or OPN (*right*; scale bar represents 200 µm). (**b**) NSC cultures were exposed to oxidative stress by H_2O_2 (300 nM for 24 h), increasing cell death as assessed by propidium iodide staining. Pre-treatment of NSC with 6.25 µg/ml OPN 24 h prior to oxidative stress completely rescued NSC from this toxicity, while simultaneous addition of OPN and H_2O_2 prevented about half the cells from dying (values displayed as means ± SEM; $^{*}p < 0.05$). (**c**) Generation of TuJ1-positive neurons (*green*) during differentiation was increased by OPN treatment (*lower row*) as compared to control (*upper row*) at days 7, 10, and 14 after mitogen withdrawal. During that period, the axon length grew notably and neurons began to form networks; both observations were more pronounced in OPN-treated cells. By day 14, mature MAP2+-positive neurons had formed (*right column*; scale bars represent 100 µm). (**d**) Osteopontin (OPN) promoted neurogenesis after stroke in vivo. In adults rats that underwent photothrombosis, a single i.c.v. injection of 500 µg OPN significantly increased the area covered by DCX-positive neuroblasts in the SVZ (values are displayed as means ± SEM, $^{**}p < 0.01$). Representative, DCX-stained images from the SVZ of rats subjected to cerebral ischemia, treated with either placebo (*left*) or OPN (*right*). OPN treatment led to an increase of neuroblasts in the SVZ (scale bar represents 100 µm). Adapted from Rabenstein et al. [63] with permission

about the neurobiological and polarity-dependent effects, a better targeted use of tDCS in the clinic could be archived. Currently, in an animal model of experimental autoimmune encephalomyelitis (EAE), inhibition of Gli1—a transcriptional effector of the sonic hedgehood pathway—improves the functional outcome and offers neuroprotection. This inhibition can be achieved by intraventricular application of

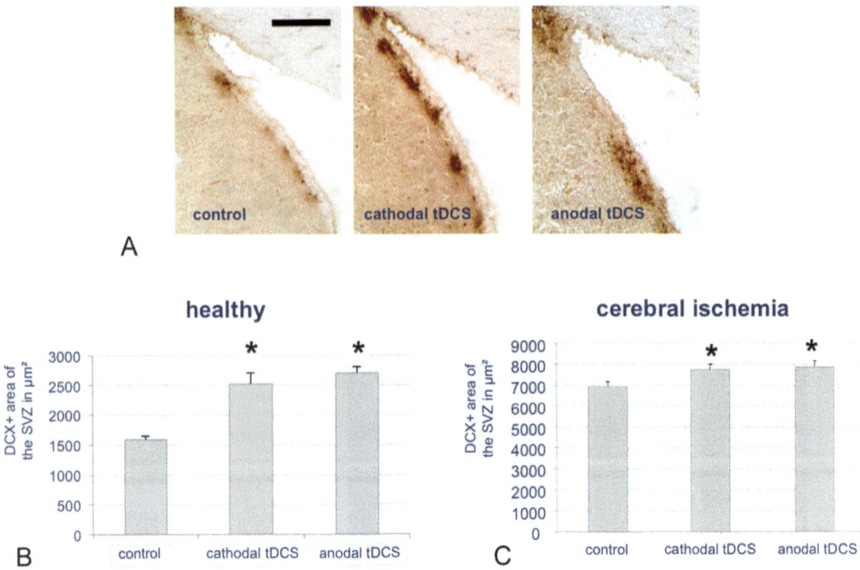

Fig. 5.2 Multisession tDCS induced neurogensis in the subventricular zone (SVZ). (**a**) Representative, DCX-stained images from the SVZ of mice, treated with either sham (*left*), cathodal tDCS (*middle*) or anodal tDCS (*right*). Multisession cathodal or anodal tDCS at 99 kC/m^2 increased DCX immunoreactivity in the SVZ (scale bar represents 100 µm). Multisession cathodal or anodal tDCS increased the number of DCX+ neuroblasts in the SVZ of control animals (**b**) and rats subjected to cerebral ischemia (**c**) (values are displayed as means ± SEM, $*p < 0.05$). Adapted from Pikhovych et al. [74] and Braun et al. [73] with permission

GANT61, a small molecule inhibitor of Gli12 [76]. This pathway might be interesting for targeted eNSC activation after stroke as well. However, for clinical application a less invasive application methods needs to be found.

Most importantly, in order to translate experimental findings into the clinical setting, translational read-outs need to be advanced to non-invasively monitor treatment effects. Moreover, it is crucial to learn more about physiological human neurogenesis, to then evaluate the treatment efficacy of eNSC mobilization in humans. In this, MRI detection methods require invasive labelling to specifically detect eNSC: This can be achieved either by direct intraventricular injection of labels or viral- or antibody-coupled labels, thus all of these methods are not applicable in humans. Another option is PET-imaging with the radiotracer 3′-deoxy-3′-[^{18}F]fluoro-L-thymidine that labels proliferating cells. This approach offers a promising method to noninvasively quantify eNSC in the live brain [77, 78].

Taken together, targeted activation of eNSC by pharmacological substances, e.g. stem cell regulating factors or osteopontin, and by non pharmacological approaches such as transcranial direct current stimulation (tDCS), constitute a promising approach to facilitate regeneration and enhance recovery after stroke.

References

1. Altman J, Das GD. Autoradiographic and histological evidence of postnatal hippocampal neurogenesis in rats. J Comp Neurol. 1965;124(3):319–35.
2. Kaplan MS, Bell DH. Mitotic neuroblasts in the 9-day-old and 11-month-old rodent hippocampus. J Neurosci. 1984;4(6):1429–41.
3. Luskin MB. Restricted proliferation and migration of postnatally generated neurons derived from the forebrain subventricular zone. Neuron. 1993;11(1):173–89.
4. Morshead CM, Reynolds BA, Craig CG, McBurney MW, Staines WA, Morassutti D, et al. Neural stem cells in the adult mammalian forebrain: a relatively quiescent subpopulation of subependymal cells. Neuron. 1994;13(5):1071–82.
5. Doetsch F, Caillé I, Lim DA, García-Verdugo JM, Alvarez-Buylla A. Subventricular zone astrocytes are neural stem cells in the adult mammalian brain. Cell. 1999;97(6):703–16.
6. Reynolds BA, Weiss S. Generation of neurons and astrocytes from isolated cells of the adult mammalian central nervous system. Science. 1992;255(5052):1707–10.
7. Hockfield S, McKay RD. Identification of major cell classes in the developing mammalian nervous system. J Neurosci. 1985;5(12):3310–28.
8. Lois C, Alvarez-Buylla A. Proliferating subventricular zone cells in the adult mammalian forebrain can differentiate into neurons and glia. Proc Natl Acad Sci U S A. 1993;90(5):2074–7.
9. Temple S. Division and differentiation of isolated CNS blast cells in microculture. Nature. 1989;340(6233):471–3.
10. Ming GL, Song H. Adult neurogenesis in the mammalian brain: significant answers and significant questions. Neuron. 2011;70(4):687–702.
11. Silva-Vargas V, Crouch EE, Doetsch F. Adult neural stem cells and their niche: a dynamic duo during homeostasis, regeneration, and aging. Curr Opin Neurobiol. 2013;23(6):935–42.
12. Ahn S, Joyner AL. In vivo analysis of quiescent adult neural stem cells responding to Sonic hedgehog. Nature. 2005;437(7060):894–7.
13. Gould E. How widespread is adult neurogenesis in mammals? Nat Rev Neurosci. 2007;8(6):481–8.
14. Masjkur J, Rueger MA, Bornstein SR, McKay R, Androutsellis-Theotokis A. Neurovascular signals suggest a propagation mechanism for endogenous stem cell activation along blood vessels. CNS Neurol Disord Drug Targets. 2012;11(7):805–17.
15. van Praag H, Kempermann G, Gage FH. Running increases cell proliferation and neurogenesis in the adult mouse dentate gyrus. Nat Neurosci. 1999;2(3):266–70.
16. Garthe A, Roeder I, Kempermann G. Mice in an enriched environment learn more flexibly because of adult hippocampal neurogenesis. Hippocampus. 2016;26(2):261–71.
17. Gonçalves JT, Schafer ST, Gage FH. Adult neurogenesis in the hippocampus: from stem cells to behavior. Cell. 2016;167(4):897–914.
18. Bouab M, Paliouras GN, Aumont A, Forest-Bérard K, Fernandes KJ. Aging of the subventricular zone neural stem cell niche: evidence for quiescence-associated changes between early and mid-adulthood. Neuroscience. 2011;173:135–49.
19. Signer RA, Morrison SJ. Mechanisms that regulate stem cell aging and life span. Cell Stem Cell. 2013;12(2):152–65.
20. Adamczak J, Aswendt M, Kreutzer C, Rotheneichner P, Riou A, Selt M, et al. Neurogenesis upregulation on the healthy hemisphere after stroke enhances compensation for age-dependent decrease of basal neurogenesis. Neurobiol Dis. 2017;99:47–57.
21. Androutsellis-Theotokis A, Leker RR, Soldner F, Hoeppner DJ, Ravin R, Poser SW, et al. Notch signalling regulates stem cell numbers in vitro and in vivo. Nature. 2006;442(7104):823–6.
22. Ourednik J, Ourednik V, Lynch WP, Schachner M, Snyder EY. Neural stem cells display an inherent mechanism for rescuing dysfunctional neurons. Nat Biotechnol. 2002;20(11):1103–10.
23. Chopp M, Li Y, Zhang ZG. Mechanisms underlying improved recovery of neurological function after stroke in the rodent after treatment with neurorestorative cell-based therapies. Stroke. 2009;40(3 Suppl):S143–5.

24. Einstein O, Ben-Hur T. The changing face of neural stem cell therapy in neurologic diseases. Arch Neurol. 2008;65(4):452–6.
25. Jessberger S, Gage FH. Adult neurogenesis: bridging the gap between mice and humans. Trends Cell Biol. 2014;24(10):558–63.
26. Eriksson PS, Perfilieva E, Björk-Eriksson T, Alborn AM, Nordborg C, Peterson DA, et al. Neurogenesis in the adult human hippocampus. Nat Med. 1998;4(11):1313–7.
27. Spalding KL, Bergmann O, Alkass K, Bernard S, Salehpour M, Huttner HB, et al. Dynamics of hippocampal neurogenesis in adult humans. Cell. 2013;153(6):1219–27.
28. Spalding KL, Bhardwaj RD, Buchholz BA, Druid H, Frisén J. Retrospective birth dating of cells in humans. Cell. 2005;122(1):133–43.
29. Bhardwaj RD, Curtis MA, Spalding KL, Buchholz BA, Fink D, Björk-Eriksson T, et al. Neocortical neurogenesis in humans is restricted to development. Proc Natl Acad Sci U S A. 2006;103(33):12564–8.
30. Curtis MA, Kam M, Nannmark U, Anderson MF, Axell MZ, Wikkelso C, et al. Human neuroblasts migrate to the olfactory bulb via a lateral ventricular extension. Science. 2007;315(5816):1243–9.
31. Bergmann O, Liebl J, Bernard S, Alkass K, Yeung MS, Steier P, et al. The age of olfactory bulb neurons in humans. Neuron. 2012;74(4):634–9.
32. Ernst A, Alkass K, Bernard S, Salehpour M, Perl S, Tisdale J, et al. Neurogenesis in the striatum of the adult human brain. Cell. 2014;156(5):1072–83.
33. Huttner HB, Bergmann O, Salehpour M, Rácz A, Tatarishvili J, Lindgren E, et al. The age and genomic integrity of neurons after cortical stroke in humans. Nat Neurosci. 2014;17(6):801–3.
34. Jin K, Wang X, Xie L, Mao XO, Zhu W, Wang Y, et al. Evidence for stroke-induced neurogenesis in the human brain. Proc Natl Acad Sci U S A. 2006;103(35):13198–202.
35. Leker RR, Soldner F, Velasco I, Gavin DK, Androutsellis-Theotokis A, McKay RD. Long-lasting regeneration after ischemia in the cerebral cortex. Stroke. 2007;38(1):153–61.
36. Kreuzberg M, Kanov E, Timofeev O, Schwaninger M, Monyer H, Khodosevich K. Increased subventricular zone-derived cortical neurogenesis after ischemic lesion. Exp Neurol. 2010;226(1):90–9.
37. Mackay J, Mensah G. Atlas of heart disease and stroke. Geneva: World Health Organization; 2004.
38. Schroeter M, Jander S, Witte OW, Stoll G. Local immune responses in the rat cerebral cortex after middle cerebral artery occlusion. J Neuroimmunol. 1994;55(2):195–203.
39. Schroeter M, Franke C, Stoll G, Hoehn M. Dynamic changes of magnetic resonance imaging abnormalities in relation to inflammation and glial responses after photothrombotic cerebral infarction in the rat brain. Acta Neuropathol. 2001;101(2):114–22.
40. Schroeter M, Jander S, Witte OW, Stoll G. Heterogeneity of the microglial response in photochemically induced focal ischemia of the rat cerebral cortex. Neuroscience. 1999;89(4):1367–77.
41. Wang Q, Tang XN, Yenari MA. The inflammatory response in stroke. J Neuroimmunol. 2007;184(1-2):53–68.
42. Mabuchi T, Kitagawa K, Ohtsuki T, Kuwabara K, Yagita Y, Yanagihara T, et al. Contribution of microglia/macrophages to expansion of infarction and response of oligodendrocytes after focal cerebral ischemia in rats. Stroke. 2000;31(7):1735–43.
43. Hallenbeck JM, Dutka AJ, Tanishima T, Kochanek PM, Kumaroo KK, Thompson CB, et al. Polymorphonuclear leukocyte accumulation in brain regions with low blood flow during the early postischemic period. Stroke. 1986;17(2):246–53.
44. Belmadani A, Tran PB, Ren D, Miller RJ. Chemokines regulate the migration of neural progenitors to sites of neuroinflammation. J Neurosci. 2006;26(12):3182–91.
45. Stoll G, Jander S, Schroeter M. Detrimental and beneficial effects of injury-induced inflammation and cytokine expression in the nervous system. Adv Exp Med Biol. 2002;513:87–113.
46. Imitola J, Raddassi K, Park KI, Mueller FJ, Nieto M, Teng YD, et al. Directed migration of neural stem cells to sites of CNS injury by the stromal cell-derived factor 1alpha/CXC chemokine receptor 4 pathway. Proc Natl Acad Sci U S A. 2004;101(52):18117–22.

47. Robin AM, Zhang ZG, Wang L, Zhang RL, Katakowski M, Zhang L, et al. Stromal cell-derived factor 1alpha mediates neural progenitor cell motility after focal cerebral ischemia. J Cereb Blood Flow Metab. 2006;26(1):125–34.
48. Thored P, Arvidsson A, Cacci E, Ahlenius H, Kallur T, Darsalia V, et al. Persistent production of neurons from adult brain stem cells during recovery after stroke. Stem Cells. 2006;24(3):739–47.
49. Widera D, Mikenberg I, Elvers M, Kaltschmidt C, Kaltschmidt B. Tumor necrosis factor alpha triggers proliferation of adult neural stem cells via IKK/NF-kappaB signaling. BMC Neurosci. 2006;7:64.
50. Jin K, Minami M, Lan JQ, Mao XO, Batteur S, Simon RP, et al. Neurogenesis in dentate subgranular zone and rostral subventricular zone after focal cerebral ischemia in the rat. Proc Natl Acad Sci U S A. 2001;98(8):4710–5.
51. Liu J, Solway K, Messing RO, Sharp FR. Increased neurogenesis in the dentate gyrus after transient global ischemia in gerbils. J Neurosci. 1998;18(19):7768–78.
52. Arvidsson A, Collin T, Kirik D, Kokaia Z, Lindvall O. Neuronal replacement from endogenous precursors in the adult brain after stroke. Nat Med. 2002;8(9):963–70.
53. Schroeter M, Dennin MA, Walberer M, Backes H, Neumaier B, Fink GR, et al. Neuroinflammation extends brain tissue at risk to vital peri-infarct tissue: a double tracer [11C] PK11195- and [18F]FDG-PET study. J Cereb Blood Flow Metab. 2009;29(6):1216–25.
54. Lindvall O, Kokaia Z. Stem cells in human neurodegenerative disorders--time for clinical translation? J Clin Invest. 2010;120(1):29–40.
55. Martens DJ, Seaberg RM, van der Kooy D. In vivo infusions of exogenous growth factors into the fourth ventricle of the adult mouse brain increase the proliferation of neural progenitors around the fourth ventricle and the central canal of the spinal cord. Eur J Neurosci. 2002;16(6):1045–57.
56. Kuhn HG, Winkler J, Kempermann G, Thal LJ, Gage FH. Epidermal growth factor and fibroblast growth factor-2 have different effects on neural progenitors in the adult rat brain. J Neurosci. 1997;17(15):5820–9.
57. Nakatomi H, Kuriu T, Okabe S, Yamamoto S, Hatano O, Kawahara N, et al. Regeneration of hippocampal pyramidal neurons after ischemic brain injury by recruitment of endogenous neural progenitors. Cell. 2002;110(4):429–41.
58. Androutsellis-Theotokis A, Rueger MA, Park DM, Mkhikian H, Korb E, Poser SW, et al. Targeting neural precursors in the adult brain rescues injured dopamine neurons. Proc Natl Acad Sci U S A. 2009;106(32):13570–5.
59. Klein R, Blaschke S, Neumaier B, Endepols H, Graf R, Keuters M, et al. The synthetic NCAM mimetic peptide FGL mobilizes neural stem cells in vitro and in vivo. Stem Cell Rev. 2014;10(4):539–47.
60. Klein R, Mahlberg N, Ohren M, Ladwig A, Neumaier B, Graf R, et al. The neural cell adhesion molecule-derived (NCAM)-peptide FG loop (FGL) mobilizes endogenous neural stem cells and promotes endogenous regenerative capacity after stroke. J Neuroimmune Pharmacol. 2016;11(4):708–20.
61. Jin Y, Barnett A, Zhang Y, Yu X, Luo Y. Poststroke sonic hedgehog agonist treatment improves functional recovery by enhancing neurogenesis and angiogenesis. Stroke. 2017;48(6):1636–45.
62. Hucklenbroich J, Klein R, Neumaier B, Graf R, Fink GR, Schroeter M, et al. Aromatic-turmerone induces neural stem cell proliferation in vitro and in vivo. Stem Cell Res Ther. 2014;5(4):100.
63. Rabenstein M, Hucklenbroich J, Willuweit A, Ladwig A, Fink GR, Schroeter M, et al. Osteopontin mediates survival, proliferation and migration of neural stem cells through the chemokine receptor CXCR4. Stem Cell Res Ther. 2015;6:99.
64. Rabenstein M, Vay SU, Flitsch LJ, Fink GR, Schroeter M, Rueger MA. Osteopontin directly modulates cytokine expression of primary microglia and increases their survival. J Neuroimmunol. 2016;299:130–8.
65. Brown A. Osteopontin: a key link between immunity, inflammation and the central nervous system. Transl Neurosci. 2012;3(3):288–93.

66. Denhardt DT, Noda M, O'Regan AW, Pavlin D, Berman JS. Osteopontin as a means to cope with environmental insults: regulation of inflammation, tissue remodeling, and cell survival. J Clin Invest. 2001;107(9):1055–61.
67. Rueger MA, Muesken S, Walberer M, Jantzen SU, Schnakenburg K, Backes H, et al. Effects of minocycline on endogenous neural stem cells after experimental stroke. Neuroscience. 2012;215:174–83.
68. Vay SU, Blaschke S, Klein R, Fink GR, Schroeter M, Rueger MA. Minocycline mitigates the gliogenic effects of proinflammatory cytokines on neural stem cells. J Neurosci Res. 2016;94(2):149–60.
69. Sparing R, Thimm M, Hesse MD, Küst J, Karbe H, Fink GR. Bidirectional alterations of interhemispheric parietal balance by non-invasive cortical stimulation. Brain. 2009;132(Pt 11):3011–20.
70. Hummel F, Celnik P, Giraux P, Floel A, WH W, Gerloff C, et al. Effects of non-invasive cortical stimulation on skilled motor function in chronic stroke. Brain. 2005;128(Pt 3):490–9.
71. Rueger MA, Keuters MH, Walberer M, Braun R, Klein R, Sparing R, et al. Multi-session transcranial direct current stimulation (tDCS) elicits inflammatory and regenerative processes in the rat brain. PLoS One. 2012;7(8):e43776.
72. Keuters MH, Aswendt M, Tennstaedt A, Wiedermann D, Pikhovych A, Rotthues S, et al. Transcranial direct current stimulation promotes the mobility of engrafted NSCs in the rat brain. NMR Biomed. 2015;28(2):231–9.
73. Braun R, Klein R, Walter HL, Ohren M, Freudenmacher L, Getachew K, et al. Transcranial direct current stimulation accelerates recovery of function, induces neurogenesis and recruits oligodendrocyte precursors in a rat model of stroke. Exp Neurol. 2016;279:127–36.
74. Pikhovych A, Stolberg NP, Jessica Flitsch L, Walter HL, Graf R, Fink GR, et al. Transcranial direct current stimulation modulates neurogenesis and microglia activation in the mouse brain. Stem Cells Int. 2016;2016:2715196.
75. Topkoru BC, Altay O, Duris K, Krafft PR, Yan J, Zhang JH. Nasal administration of recombinant osteopontin attenuates early brain injury after subarachnoid hemorrhage. Stroke. 2013;44(11):3189–94.
76. Samanta J, Grund EM, Silva HM, Lafaille JJ, Fishell G, Salzer JL. Inhibition of Gli1 mobilizes endogenous neural stem cells for remyelination. Nature. 2015;526(7573):448–52.
77. Rueger MA, Backes H, Walberer M, Neumaier B, Ullrich R, Simard ML, et al. Noninvasive imaging of endogenous neural stem cell mobilization in vivo using positron emission tomography. J Neurosci. 2010;30(18):6454–60.
78. Rueger MA, Schroeter M. In vivo imaging of endogenous neural stem cells in the adult brain. World J Stem Cells. 2015;7(1):75–83.

Chapter 6
Transcriptional and Genomic Advances on the Pathophysiology of Stem Cell Repairment After Intracerebral Hemorrhage

Sheng Zhang, Yongjie Zhou, and Yujie Chen

Abstract Intracerebral hemorrhage is a life-threatening disease characterized by a sudden rupture of cerebral blood vessels, and it is widely believed that neural cell death occurs after exposure to blood metabolites or subsequently damaged cells. Based on these disappointing results of 1026 neuroprotective agents, researchers turned their interests on neurogenesis, which is traditionally considered as an endogenous neuroprotective mechanism after acute central nervous system injuries. However, because of complexity in stem cell survival, migration, differentiation, and maturation, current strategies have either been proved unsatisfactory or resulted in serious side effects during clinical translation. It is well known that transcriptional and genomic pathways play important roles in ensuring the normal functions of stem cells, including proliferation, migration, differentiation and neural reconnection. And reprogramming technology and other non-invasive electromagnetic stimulation were recently employed and proved effective for the stem cell characteristics. Therefore, in the present chapter, we sought to summarize the advances in the pathophysiology and strategies of stem cell repairment after ICH at the level of transcription and genome, hoping to provide potential sparks for better stem cell repairment for ICH patients.

Keywords Stem cell • Intracerebral hemorrhage • Transcriptome • Genome • Bioinformatics • Reprogramming technology • Neurological recovery

S. Zhang • Y. Chen, M.D., Ph.D. (✉)
Department of Neurosurgery, Southwest Hospital, Third Military Medical University, 29 Gaotanyan Street, Shapingba District, Chongqing 400038, China
e-mail: yujiechen6886@foxmail.com

Y. Zhou
Department of Neurosurgery, The 184th Hospital of People's Liberation Army, Yingtan, City, Jiangxi Province, China

P.A. Lapchak, J.H. Zhang (eds.), *Cellular and Molecular Approaches to Regeneration and Repair*, Springer Series in Translational Stroke Research,
https://doi.org/10.1007/978-3-319-66679-2_6

Abbreviations

ICH	Intracerebral hemorrhage
LncRNA	Long non-coding RNA
miRNA	MicroRNA
NSC	Neural stem cell
SPIO	super-paramagnetic iron oxide
STICH	Surgical Trial in Intracerebral Hemorrhage
SVZ	subependymal ventricular zone
US FDA	United States Food and Drug Administration

1 Introduction

Intracerebral hemorrhage (ICH) often leads to high mortality and morbidity [1, 2]. Despite long-standing and worldwide efforts, no effective neuroprotective strategies are available to improve the neurological outcomes in patients with stroke, including ICH. Although surgical decompression for cerebral hemorrhage benefits the survival of patients, but the long-term outcome is still not improved. A famous clinical trial, Surgical Trial in Intracerebral Hemorrhage (STICH, Phase I and II) has failed to provide enough evidence to support early surgical hematoma removal comparing to conservative therapy [3, 4]. And the recombinant activated factor VII significantly reduces hematoma volume without improving survival or functional outcomes in ICH patients [5]. Therefore, potential therapeutic strategies targeting secondary brain injury are attracting a lot of attentions in translational studies of intracerebral hemorrhage [6, 7].

Secondary brain injury is triggered by blood metabolites which subsequently activate cytotoxic, excitotoxic, reactive oxygen species-related, and inflammatory-mediated pathways, and so on. Nevertheless, according to a systemic review, 1026 neuroprotective agents works for the neurological outcomes in animal stroke model, have failed to exhibit convincing clinical benefits [8, 9]. Based on these disappointing results, researchers turned their interests on neurogenesis, which is traditionally considered as an endogenous neuroprotective mechanism after acute central nervous system injuries. Previous studies indicate neurogenesis occurs after ICH to repair the brain lesions and restore brain connections [10–12]. In addition, researchers have made great efforts to transplant exogenous stem cells to the brain lesions from different sources. However, because of complexity in stem cell survival, migration, differentiation, and maturation, current strategies have either been proved unsatisfactory or resulted in serious side effects during clinical translation [13–15]. Therefore, in the present chapter, we sought to summarize the advances in the pathophysiology and strategies of neurogenesis after ICH at the level of transcription and genome, hoping to provide potential sparks for better stem cell repairment for ICH patients.

2 Current Situation of Stem Cell Repairment After Intracerebral Hemorrhage

2.1 Potential Neuroprotective Effects of Stem Cell Repairment

Back to 1998, Eriksson PS, et al. found the nascent neurons in hippocampus tissue from post mortem cancer patients [16, 17], providing the first evidence for human neurogenesis [16]. More recently, Spalding, et al. retrospectively marked the hippocampal cells in post mortem patients exposed to nuclear testing before death with the ratio of 14^{C} to 12^{C} in DNA, and found that the turnover rate of newborn neurons in the dentate gyrus could be as high as 700 per day [18]. Later, by using two-photon microscopy, Shen J, et al. obtained specimens from patients with hypertensive ICH and found that neural stem cell (NSC)-specific proteins and cell proliferation markers were localized in the perihematomal areas of basal ganglia and the parietal lobe [10]. These data suggest that ICH could induce de novo neurogenesis in the adult human brain, potentially with the ability to repair brain lesions and improve neurological outcome in patients suffered with ICH.

Similar with other acute central nervous system injuries, the preclinical studies of stem cell repairment strategy were gradually carried out to support the beneficial role of stem cell after intracerebral hemorrhage including proliferation, migration, and differentiation. Back in 2004, Tang T, et al. found Nestin-stained or BrdU-labeled cells were mainly located in the basal ganglion and nearby SVZ around hematoma and ependyma after ICH in rats. Additionally, no cells positive for these markers were found in control or sham groups or in non-lesioned parenchyma [19]. Masuda T, et al. injected BrdU for 2 weeks after ICH in rats and found that BrdU-labeled cells significantly increased in both the contralateral and ipsilateral SVZs. Meanwhile, doublecortin-positive, immature, and migratory neurons were also seen in the dorsal striatum and peri-hematoma area 2 weeks post-ICH. In addition, they also noticed clusters of doublecortin-stained cells in the striatum surrounding the hemorrhagic lesion 4 weeks post-ICH. These findings implicate that experimental ICH induces the proliferation and migration of endogenous NSCs to repair the hemorrhagic lesion [12]. In addition to endogenous NSCs, exogenous NSC transplantation also exhibits the potential to attenuate neurological deficits after hemorrhagic stroke. In 2003, Jeong SW, et al. intravenously transplanted human NSCs into experimental ICH rats. Their results indicated that NSCs can cross blood brain barrier and enter the rat brain with ICH. Interestingly, those surviving NSCs in the rat brain helped with the functional recovery [20]. Another investigation transplanted alltrans retinoic acid-induced NSCs into the contralateral ventricle up to 7 days after ICH, and found new neurons and astrocytes surrounding the hematoma lesions of the brain 4 weeks later in all rats receiving the transplantations [21]. Moreover, these results were confirmed by super-paramagnetic iron oxide (SPIO)—labeled human NSCs detected by 3T Magnetic Resonance Imaging, which indicated the presence of prominent NSCs in the periventricular region at 4 and 6 weeks post-transplantation [22]. Most importantly, compared with the control group, the NSC-transplanted rats exhibited excellent functional performance on neurofunctional tests after 2–8 weeks, which indicates that the exogenously supplied NSCs may be used for the functional recovery after hemorrhagic stroke [23].

In addition to NSCs repairment strategies, other types of stem cell were also demonstrated positive in neuroprotective after ICH. In 2009, Liao W, et al. human umbilical cord-derived mesenchymal stem cells were tested for the ICH rat model by intracerebrally transplantation, they found the neurological recovery was enhanced, potentially due to neuroinflammation inhibition and angiogenesis [24]. Similarly, Liu, AM. et al. also demonstrated the umbilical cord-derived mesenchymal stem cells could enhance remyelination and functional recovery in ICH rat model [25]. After them, adipose-derived stem cells [26], bone marrow stromal cells [27], human bone marrow-derived mesenchymal stem cells [28–30], and bone marrow-derived mononuclear cells [31] were proved to be benefit for ICH models one and another. More recently, induced pluripotent stem cells were transplanted into ICH animal models, which could improve the neurological function as expected [32, 33]. According to a meta-analysis, 30 studies using five different type of stem cells showed consistent improvements in neurobehavioral function, but the extensive potential of stem cells repairment for ICH should be further evaluated with more high-quality preclinical studies and clinical trials [34].

2.2 Limitations of Stem Cell Therapy

Despite the potential neuroprotective effects of different stem cells, a lot of factors could influence the efficacy of stem cell repairment for ICH patients, such as intervention timepoint, administration routes, microenvironment of stem cell, the source and status of stem cells, and possible immune responses. Other factors including but not limited to metabolism regulators, epigenetic modifiers, vascular constrictors or dilators, modulators of immune response, activators or inhibitors of signal transduction pathways. Moreover, proliferation, differentiation, maintenance and self-renewal of stem cells in niche are controlled by a network of intrinsic and extrinsic regulators, such as neurotrophins, cyclins and cyclin-dependent kinases, transcription factors. These factors act in concert within their biological network during the establishment and maintenance of neural connections. Epigenetic modulations during hippocampal development can also have impacts on one's learning and memorizing abilities. Genetic polymorphism in genes involving neurogenesis may have essential roles in variations of stem cell differentiation between individuals in adult neural regeneration [35]. Thus, analysis and manipulation of favorable genetic variations for neurogenesis may have the vital practical significance for ICH therapeutic implications [35].

3 Transcriptional and Genomic Advances on the Pathophysiology of Stem Cell Repairment

3.1 Genomic Analysis for the Stimulation of Endogenous NSCs

In mammals, new neurons are constantly generated in the subependymal ventricular zone and subgranular zone of the dentate gyrus throughout developmental stage and adult life. This continuous neurogenesis after birth may be important in processing information, daily learning and memorization etc. During hippocampal neurogenesis, doublecortin-positive immature neurons and neuronal precursor cells mature into neurons. In the immature stage, cells are sensitive and susceptible to extrinsic stimuli. However, knowledge on the dynamics which lead to neuron maturation is limited. Moreover, to date, purification of NSCs in vitro proves to be a challenging task to allow for investigation of their biology and application in clinical medicine.

Due to the development of "omics" technology, emerging evidence has demonstrated that both transcriptional and genomic pathways play important roles in ensuring the normal function of stem cells. At the transcriptional level, sequence-specific transcription factors and coregulators work together to orchestrate the transcriptional landscape of stem cells, which determines the on/off state of target genes, thereby controlling the cell fate of stem cells. At the genomic level, the replication and repair machineries maintain the genomic stability of stem cells. By examining gene expression at single cell level using RNA-seq technology, Gao Y, et al. found that there existed two subgroups among immature neurons with distinct gene expression profiles and different molecular markers. Comparisons of the two subgroups indicated that Notch and Sonic Hedgehog and the Hippo pathways are all important in neuron maturation and NSC activity [36, 37].

Factors which form a regulatory network to support NSC self-renewal has not been fully elucidated up to now. Understanding of the key transcription factors, the promoter region and other non-coding regions that they bind, and the target genes that they regulate, will be essential in unleashing the full potential of these cells for therapeutic use. At the center of this regulatory network are SOX family and FOX family transcription factors, nuclear factor I, basic helix-loop-helix transcription factor family. Coordinated action of these factors to promote proliferation and at the same time prevent untimely differentiation and quiescence is crucial to NSC self-renewal [38]. By analyzing the region specific regulatory networks based on available published databases on subependymal ventricular zone and subgranular zone, Mateo JL, et al. discovered the potential microenvironment associated differences based on membrane and nuclear receptors via HIF-1α, Ar, and NR3C1. They also performed cell fate determinant test for NSCs from subependymal ventricular zone to the interneurons of olfactory bulb and NSC populations from subgranular zone to the granule cells of the granular cell layer. The existence of membrane and nuclear receptors in this region-specific regulatory network shows the importance of niche-derived extracellular molecules and region-specific factors for the neurogenesis in subependymal ventricular zone and subgranular zone [39].

Genomic approaches in modern time have facilitated unprecedented advances in our understanding of the development, function and evolution of central nervous system. By contrast, little is recorded or published about the possible interplay between different genetic factors, epigenetic modulators, non-coding RNAs and environmental factors in causing or modulating neurological disorders in populations from under-developed countries [40]. Both pharmacological intervention and genetic manipulation of epigenetic modulators can trigger profound changes in molecular expression, neuron identity and complex behavioral and cognitive phenotypes. Apparently, epigenetics plays a non-trivial role in the pathogenesis of neurological disorders. Emerging paradigms in possible connections between epigenetics and hemorrhagic stroke include the followings: how gene mutations of epigenetic factors induce hemorrhagic stroke; how is genetic polymorphism of epigenetic factors linked to disease risk of hemorrhagic stroke; how changes in the expression, localization, or function of epigenetic factors affect hemorrhagic stroke; how epigenetic factors modulate disease-linked genomic loci, protein expression and cellular pathways; and how differential epigenetic profiles from patient-derived tissue samples affect disease outcome [41].

3.2 Microenvironment Complexity for the Exogenous Stem Cell

Since the sole neuron protective strategies could not exhibit satisfied outcome in clinical trials, neural vascular coupling that is accepted as Neural Vascular Unit and Vascular Neural Network provides crucial guiding direction for exploring latent mechanisms of other participators near neuron and their associating with numerous central nervous system diseases [42]. A complex network of elements, consisting of macromolecules of the extracellular matrix, glial cells, astrocytes, oligodendrocytes, adhesion molecules for cell-cell and cell-extracellular matrix connections, blood vessels, neurotrophins, and so on. All of them have an impact on tissue homeostasis and maintenance of a homing microenvironment for stem cells in central nervous system. Among these components, extracellular matrix derived from stem cells provides a unique and indispensable microenvironment that helps with stem cell differentiation and neural regeneration. Analysis of protein expression by two-dimensional gel electrophoresis and liquid chromatography-tandem mass spectrometry provided proteomic profiles that corresponded to unique niche properties for each group tested. Proteomic results demonstrated that NSC-derived extracellular matrix can impact the decision-making process of stem cell fate by offering microenvironment for specialized stem cell niches in the process of tissue development and regeneration [43].

Metabolites and nutrients in the bloodstream fulfil the energy demands of adult stem cells. However, more context-specific roles have been shown for several circulating factors. An array of diffusible, non-diffusible and circulating cues influences stem cell quiescence, proliferation, self-renewal and differentiation. A key question is how stem cells integrate all of these signals, particularly since they can have overlapping or antagonistic effects. Some ligands, such as PEDF and BTC, have multiple sources [44–46]—does the source influence the effect on stem cells? Most of neurovascular studies have focused on endothelial cells. The difficulty in identifying pericytes in vivo means that their contributions to NSC behavior are only just emerging [47]. How do pericytes and astrocytes interact with the endothelial cell-NSC crosstalk? Advances in single cell transcriptional profiling, as well as cell type-specific tools for manipulating cells in vivo, will no doubt shed light on these questions.

MicroRNA (miRNA) is a recently discovered group of small, genome-encoded endogenous RNAs that are transcribed but are not translated into proteins [48]. It was reported to play an key regulatory role in many cellular functions, including cell growth, proliferation, differentiation, lineage determination and metabolism [49]. In central nervous system, miRNA was well established to be vital in neurogenesis [50], neural development [51], differentiation [52] and synaptic plasticity [53].

Exosomes are types of nano extracellular vesicles, naturally released from different kinds of living cells, can be taken up by recipient cells [54]. Faure et al. described the release of exosomes by neuron in vitro [55]. And Taylor et al. reported the secretion of exosomes by astrocytes, also in vitro [56]. Furthermore, exosomes can also be found in cerebral spinal fluid in animal models [57, 58], which might be a reminder of its vital role in brain physiologies. Exosome has been recently known as intermediate link in cell-cell communication, with the capability of transferring the proteins, DNAs, microRNAs, non-coding RNAs and lipids with or without direct contact [59]. Xin et al. demonstrated exosomes secreted by mesenchymal stem cells can transfer microRNA-133b into neurons, resulting in the induction of synaptic growth [60]. Morel et al. showed that neuron secretes microRNA-124a via exosomes, which subsequently transport into astrocytes, then indirectly increasing protein expression [61]. Taken together, these finds support the hypothesis that exosome could mediate cell–cell communication in central nervous system. But, emerging evidences suggest exosome based interventions and technologies maybe promising in central nervous system disease diagnostic [62–65]. And their simple structure and ability to cross the blood-brain barrier makes exosome-based cell and gene therapies a bright prospects in the future [66–69].

Long non-coding RNA (LncRNA) is a classic non-coding RNA with transcript frame longer than 200 nucleotide, usually in cytoplasm or nuclear, being well transcripted but lack of protein-coding capability [70]. At the beginning, LncRNA was considered to be the accessory substance of RNA polymerase II transcription, with no meaningful biological function. But, growing evidences suggests LncRNA maybe prevalent in cerebral cortex, hippocampus and olfactory bulb in central nervous system, involve in cell differentiation, cellular transportation, cell-cycle regulation, stem cell reprogramming and other molecular genetical and cellular functions [71], and play an important role in various pathologies including ICH and other central nervous system diseases [72, 73].

4 Assistant Approaches for Stem Cell Repairment of Intracerebral Hemorrhage

4.1 Reprogramming Technology

Cellular reprogramming technology has created new opportunities in understanding human disease, drug discovery, and regenerative medicine. While a combinatorial code was initially found to reprogram somatic cells to pluripotency, a "second generation" of cellular reprogramming involves lineage-restricted transcription factors and microRNAs that directly reprogram one somatic cell to another. This technology was enabled by gene networks active during development, which induce global shifts in the epigenetic landscape driving cell fate decisions. A major utility of direct reprogramming is the potential of harnessing resident support cells within damaged organs to regenerate lost tissue by converting them into the desired cell type in situ.

Reprogramming technology enables the production of neural progenitor cells from somatic cells by direct trans-differentiation. However, little is known on how neural programs in these induced neural stem cells differ from those of alternative stem cell populations in vitro and in vivo. It was reported that transcription factor-mediated reprogramming can efficiently convert differentiated cells into induced pluripotent stem cells, then theoretically possible to obtain a large number of neural stem cells in vivo or in vitro for the transplantation. However, the mechanisms related to the amenability of these cell types to be reprogrammed are still unknown, and tumors are likely to form due to the presence of residual undifferentiated cells following transplantation of the induced cells. Therefore, future clinical application of reprogramming technique for stem cell replacement after ICH are still needed to be improved with transcriptional and genomic analysis of proper artificial manipulations.

Although there were not much practices of transcriptional and genomic analysis for the reprogrammed stem cells for ICH patients. Hallmann AL, et al. performed comparison on murine brain-derived NSCs and pluripotent stem cell-derived neural progenitor cells, which revealed distinct global, neural, metabolic and cell cycle-associated marks in these two populations, with significant implications for the applications of induced NSCs [74]. And the genome-wide gene expression analyses reveal unique cellular characteristics related to the amenability of hematopoietic stem cells into high-quality induced pluripotent stem cells [75]. In addition, chromatin connectivity networks analysis revealed the pivotal genes of reprogramming functions, which linked the chromatin architecture to coordinated gene expression in embryonic stem cells [76]. Furthermore, the network-based approaches could also be used to identify potential targets and drugs for neuroprotection and neurorepair after stroke [77].

4.2 Non-invasive Electromagnetic Field Stimulation

Despite of these novel technologies, traditionally physical methods were employed and demonstrated benefit for the neurological recovery in central nervous diseases. Francis, et al. find that adult mice exposed to electromagnetic field can produces a significant enhancement in the number of new-born neurons in dentate gyrus [78–80]. BrdU and nestin-Corporation method shows us that electromagnetic field also can promote the number of and BrdU and nestin-positive cells within the area between subependymal ventricular zone and lesion 1 week after brain injury, which indicating that electromagnetic field exert a positive effect on proliferation and migration of neural stem cell [81]. Cuccurazzu, et al. also indicated electromagnetic field stimulation could promote adult hippocampal neurogenesis [82]. Meanwhile, Arias-Carrión, et al. demonstrated that transcranial magnetic field stimulation promoted neurogenesis in cells of subependymal ventricular zone for the nigrostriatal lesions [83]. More importantly, the repeated transcranial magnetic stimulation was already approved by US FDA for the treatment of depression. It may conventionally be used for the ICH patients with enough evidences for its neuroprotective effects and underlying mechanisms.

A number of miRNAs are the key factor that involved in determination of the stem cell fate, for instance, NSCs differentiation and proliferation [84–86]. Given to the significant effects of repeated transcranial magnetic stimulation exert on the expression of gene, this technology reminds us it is of great possibility to modulate miRNA in theory. Guo, et al. found that after 10 Hz repeated transcranial magnetic stimulation, cerebral ischemia rat model exhibited a remarkably enhancement of miR-25. Meanwhile, Brett, et al. demonstrated that the miR-106b/25 cluster mainly promoted the adult NSCs proliferation [87, 88]. In addition, Liu, et al. also performed experiments in the focal cerebral ischemia rat model, which demonstrated the miR-106b/25 cluster could increase the NSC proliferation in vitro after high frequency repeated transcranial magnetic stimulation in a dose dependent manner [89].

Meanwhile, more and more proof suggests that epigenetic mechanism especially chromatin modifications may act as a critical role in the modulation of differentiation and proliferation of NSCs [90, 91]. Leone, et al. demonstrated a marked increasing of the pro-proliferative gene Hes-1 and neuronal determination genes NeuroD1 and Neurogenin1 after exposing to electromagnetic fields [92]. And Hes1 is a repressive type of bHLH transcriptional factor that sustain the stemness for NSCs by the means of repressing pro-neural gene expression [93]. In contrast, inactivation of Hes1 means the effect repression of pro-neural genes weakened and correspondingly upregulates the expression of pro-neural genes, including Mash1,

Neurogenin1, and NeuroD1, result in acceleration of neuronal differentiation [94–96].

5 Perspective and Conclusion

In the past 20 years or so, multiple technologies have been developed to utilize the regenerative potential of stem cells and the plasticity of neural cells in central nervous system to repair brain lesions and improve structural and functional recovery after acute central nervous system injuries, including ICH [97]. Based on the pathophysiology of secondary brain injury after ICH, targets regarding prediction, diagnosis, treatment strategies, and neurofunctional recovery need to be further identified and verified in large cohorts of patients, especially those controlling stem cells at both the transcription and genomic levels (Fig. 6.1). Bioinformatics methods is a new field that focuses on the acquisition, storage and analysis of physiological and other data relevant to the bedside care of patients [98]. Advanced statistical and mathematical tools are now being applied to the large volume of clinical and physiological data routinely monitored in neurocritical care with the goal of identifying better markers of secondary brain injury and providing clinicians with an improved ability to target specific parameters in the management of ICH patients [99], which may provide much more information about therapeutic strategies for endogenous neurogenesis and exogenous stem cell transplantation for ICH patients.

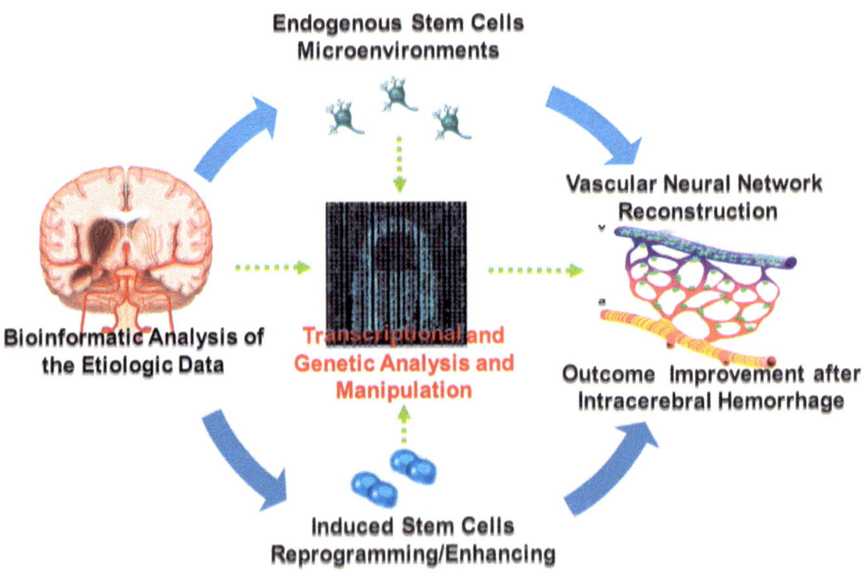

Fig. 6.1 Diagram of transcriptional and genetic analysis for applications of stem cells in intracerebral hemorrhagic

Acknowledgements This work was supported by Incubation Foundation of Interdisciplinary Laboratory of Physics and Biomedicine (Grant No. WSS-2015-08), Basic Science and Advanced Technology Research Project of Chongqing (Grant No. cstc2016jcyjA1730), National Natural Science Foundation of China (Grant No. 81501002, 81220108009) and National Basic Research Program of China (973 Program, Grant No. 2014CB541600).

Conflicts of Interest: The authors declared no potential conflicts of interest.

References

1. Keep RF, Hua Y, Xi G. Intracerebral haemorrhage: mechanisms of injury and therapeutic targets. Lancet Neurol. 2012;11(8):720–31.
2. Macdonald RL, Schweizer TA. Spontaneous subarachnoid haemorrhage. Lancet. 2016;
3. Mendelow AD, Gregson BA, Fernandes HM, Murray GD, Teasdale GM, Hope DT, et al. Early surgery versus initial conservative treatment in patients with spontaneous supratentorial intracerebral haematomas in the International Surgical Trial in Intracerebral Haemorrhage (STICH): a randomised trial. Lancet. 2005;365(9457):387–97.
4. Mendelow AD, Gregson BA, Rowan EN, Murray GD, Gholkar A, Mitchell PM, et al. Early surgery versus initial conservative treatment in patients with spontaneous supratentorial lobar intracerebral haematomas (STICH II): a randomised trial. Lancet. 2013;382(9890):397–408.
5. Mayer SA, Brun NC, Begtrup K, Broderick J, Davis S, Diringer MN, et al. Efficacy and safety of recombinant activated factor VII for acute intracerebral hemorrhage. N Engl J Med. 2008;358(20):2127–37.
6. Adeoye O, Broderick JP. Advances in the management of intracerebral hemorrhage. Nat Rev Neurol. 2010;6(11):593–601.
7. Morgenstern LB, Hemphill JC, 3rd, Anderson C, Becker K, Broderick JP, Connolly ES Jr., et al. Guidelines for the management of spontaneous intracerebral hemorrhage: a guideline for healthcare professionals from the American Heart Association/American Stroke Association. Stroke 2010;41(9):2108-2129.
8. O'Collins VE, Macleod MR, Donnan GA, Horky LL, van der Worp BH, Howells DW. 1,026 experimental treatments in acute stroke. Ann Neurol. 2006;59(3):467–77.
9. Aronowski J, Hall CE. New horizons for primary intracerebral hemorrhage treatment: experience from preclinical studies. Neurol Res. 2005;27(3):268–79.
10. Shen J, Xie L, Mao X, Zhou Y, Zhan R, Greenberg DA, et al. Neurogenesis after primary intracerebral hemorrhage in adult human brain. J Cereb Blood Flow Metab. 2008;28(8):1460–8.
11. Sgubin D, Aztiria E, Perin A, Longatti P, Leanza G. Activation of endogenous neural stem cells in the adult human brain following subarachnoid hemorrhage. J Neurosci Res. 2007;85(8):1647–55.
12. Masuda T, Isobe Y, Aihara N, Furuyama F, Misumi S, Kim TS, et al. Increase in neurogenesis and neuroblast migration after a small intracerebral hemorrhage in rats. Neurosci Lett. 2007;425(2):114–9.
13. Bang OY. Clinical trials of adult stem cell therapy in patients with ischemic stroke. J Clin Neurol. 2016;12(1):14–20.
14. Lin R, Iacovitti L. Classic and novel stem cell niches in brain homeostasis and repair. Brain Res. 2015;1628(Pt B):327–42.
15. Kalladka D, Muir KW. Brain repair: cell therapy in stroke. Stem Cells Clon. 2014;7:31–44.
16. Lazarov O, Hollands C. Hippocampal neurogenesis: learning to remember. Prog Neurobiol. 2016;138-140:1–18.
17. Eriksson PS, Perfilieva E, Bjork-Eriksson T, Alborn AM, Nordborg C, Peterson DA, et al. Neurogenesis in the adult human hippocampus. Nat Med. 1998;4(11):1313–7.

18. Spalding KL, Bergmann O, Alkass K, Bernard S, Salehpour M, Huttner HB, et al. Dynamics of hippocampal neurogenesis in adult humans. Cell. 2013;153(6):1219–27.
19. Tang T, Li XQ, Wu H, Luo JK, Zhang HX, Luo TL. Activation of endogenous neural stem cells in experimental intracerebral hemorrhagic rat brains. Chin Med J (Engl). 2004;117(9):1342–7.
20. Jeong SW, Chu K, Jung KH, Kim SU, Kim M, Roh JK. Human neural stem cell transplantation promotes functional recovery in rats with experimental intracerebral hemorrhage. Stroke. 2003;34(9):2258–63.
21. Nonaka M, Yoshikawa M, Nishimura F, Yokota H, Kimura H, Hirabayashi H, et al. Intraventricular transplantation of embryonic stem cell-derived neural stem cells in intracerebral hemorrhage rats. Neurol Res. 2004;26(3):265–72.
22. Chang NK, Jeong YY, Park JS, Jeong HS, Jang S, Jang MJ, et al. Tracking of neural stem cells in rats with intracerebral hemorrhage by the use of 3T MRI. Korean J Radiol. 2008;9(3):196–204.
23. Lee HJ, Kim KS, Kim EJ, Choi HB, Lee KH, Park IH, et al. Brain transplantation of immortalized human neural stem cells promotes functional recovery in mouse intracerebral hemorrhage stroke model. Stem Cells. 2007;25(5):1204–12.
24. Liao W, Zhong J, Yu J, Xie J, Liu Y, Du L, et al. Therapeutic benefit of human umbilical cord derived mesenchymal stromal cells in intracerebral hemorrhage rat: implications of anti-inflammation and angiogenesis. Cell Physiol Biochem. 2009;24(3-4):307–16.
25. Liu AM, Lu G, Tsang KS, Li G, Wu Y, Huang ZS, et al. Umbilical cord-derived mesenchymal stem cells with forced expression of hepatocyte growth factor enhance remyelination and functional recovery in a rat intracerebral hemorrhage model. Neurosurgery. 2010;67(2):357–65. discussion 65–6.
26. Chen J, Tang YX, Liu YM, Chen J, XQ H, Liu N, et al. Transplantation of adipose-derived stem cells is associated with neural differentiation and functional improvement in a rat model of intracerebral hemorrhage. CNS Neurosci Ther. 2012;18(10):847–54.
27. Liang H, Yin Y, Lin T, Guan D, Ma B, Li C, et al. Transplantation of bone marrow stromal cells enhances nerve regeneration of the corticospinal tract and improves recovery of neurological functions in a collagenase-induced rat model of intracerebral hemorrhage. Mol Cells. 2013;36(1):17–24.
28. Bao XJ, Liu FY, Lu S, Han Q, Feng M, Wei JJ, et al. Transplantation of Flk-1+ human bone marrow-derived mesenchymal stem cells promotes behavioral recovery and anti-inflammatory and angiogenesis effects in an intracerebral hemorrhage rat model. Int J Mol Med. 2013;31(5):1087–96.
29. Ding R, Lin C, Wei S, Zhang N, Tang L, Lin Y, et al. Therapeutic benefits of mesenchymal stromal cells in a rat model of hemoglobin-induced hypertensive intracerebral hemorrhage. Mol Cells. 2017;40(2):133–42.
30. Zhang Q, Shang X, Hao M, Zheng M, Li Y, Liang Z, et al. Effects of human umbilical cord mesenchymal stem cell transplantation combined with minimally invasive hematoma aspiration on intracerebral hemorrhage in rats. Am J Transl Res. 2015;7(11):2176–86.
31. Suda S, Yang B, Schaar K, Xi X, Pido J, Parsha K, et al. Autologous bone marrow mononuclear cells exert broad effects on short- and long-term biological and functional outcomes in rodents with intracerebral hemorrhage. Stem Cells Dev. 2015;24(23):2756–66.
32. Qin J, Ma X, Qi H, Song B, Wang Y, Wen X, et al. Transplantation of induced pluripotent stem cells alleviates cerebral inflammation and neural damage in hemorrhagic stroke. PLoS One. 2015;10(6):e0129881.
33. Qin J, Gong G, Sun S, Qi J, Zhang H, Wang Y, et al. Functional recovery after transplantation of induced pluripotent stem cells in a rat hemorrhagic stroke model. Neurosci Lett. 2013;554:70–5.
34. Ma X, Qin J, Song B, Shi C, Zhang R, Liu X, et al. Stem cell-based therapies for intracerebral hemorrhage in animal model: a meta-analysis. Neurol Sci. 2015;36(8):1311–7.
35. Horgusluoglu E, Nudelman K, Nho K, Saykin AJ. Adult neurogenesis and neurodegenerative diseases: a systems biology perspective. Am J Med Genet B Neuropsychiatr Genet. 2017;174:193.
36. YH Y, Narayanan G, Sankaran S, Ramasamy S, Chan SY, Lin S, et al. Purification, visualization, and molecular signature of neural stem cells. Stem Cells Dev. 2016;25(2):189–201.

37. Gao Y, Wang F, Eisinger BE, Kelnhofer LE, Jobe EM, Zhao X. Integrative single-cell transcriptomics reveals molecular networks defining neuronal maturation during postnatal neurogenesis. Cereb Cortex. 2017;27:2064.
38. Mateo JL, van den Berg DL, Haeussler M, Drechsel D, Gaber ZB, Castro DS, et al. Characterization of the neural stem cell gene regulatory network identifies OLIG2 as a multifunctional regulator of self-renewal. Genome Res. 2015;25(1):41–56.
39. Ertaylan G, Okawa S, Schwamborn JC, Del Sol A. Gene regulatory network analysis reveals differences in site-specific cell fate determination in mammalian brain. Front Cell Neurosci. 2014;8:437.
40. Karikari TK, Aleksic J. Neurogenomics: an opportunity to integrate neuroscience, genomics and bioinformatics research in Africa. Appl Transl Genom. 2015;5:3–10.
41. Qureshi IA, Mehler MF. Understanding neurological disease mechanisms in the era of epigenetics. JAMA Neurol. 2013;70(6):703–10.
42. Zhang JH, Badaut J, Tang J, Obenaus A, Hartman R, Pearce WJ. The vascular neural network--a new paradigm in stroke pathophysiology. Nat Rev Neurol. 2012;8(12):711–6.
43. Yan Y, Martin LM, Bosco DB, Bundy JL, Nowakowski RS, Sang QX, et al. Differential effects of acellular embryonic matrices on pluripotent stem cell expansion and neural differentiation. Biomaterials. 2015;73:231–42.
44. Gomez-Gaviro MV, Scott CE, Sesay AK, Matheu A, Booth S, Galichet C, et al. Betacellulin promotes cell proliferation in the neural stem cell niche and stimulates neurogenesis. Proc Natl Acad Sci U S A. 2012;109(4):1317–22.
45. Ramirez-Castillejo C, Sanchez-Sanchez F, Andreu-Agullo C, Ferron SR, Aroca-Aguilar JD, Sanchez P, et al. Pigment epithelium-derived factor is a niche signal for neural stem cell renewal. Nat Neurosci. 2006;9(3):331–9.
46. Andreu-Agullo C, Morante-Redolat JM, Delgado AC, Farinas I. Vascular niche factor PEDF modulates Notch-dependent stemness in the adult subependymal zone. Nat Neurosci. 2009;12(12):1514–23.
47. Chou CH, Sinden JD, Couraud PO, Modo M. In vitro modeling of the neurovascular environment by coculturing adult human brain endothelial cells with human neural stem cells. PLoS One. 2014;9(9):e106346.
48. Lagos-Quintana M, Rauhut R, Lendeckel W, Tuschl T. Identification of novel genes coding for small expressed RNAs. Science. 2001;294(5543):853–8.
49. Zhang C. Novel functions for small RNA molecules. Curr Opin Mol Ther. 2009;11(6):641–51.
50. Li X, Jin P. Roles of small regulatory RNAs in determining neuronal identity. Nat Rev Neurosci. 2010;11(5):329–38.
51. Cochella L, Hobert O. Diverse functions of microRNAs in nervous system development. Curr Top Dev Biol. 2012;99:115–43.
52. Lau P, Hudson LD. MicroRNAs in neural cell differentiation. Brain Res. 2010;1338:14–9.
53. Goldie BJ, Cairns MJ. Post-transcriptional trafficking and regulation of neuronal gene expression. Mol Neurobiol. 2012;45(1):99–108.
54. Qin J, Functions XQ. application of exosomes. Acta Pol Pharm. 2014;71(4):537–43.
55. Faure J, Lachenal G, Court M, Hirrlinger J, Chatellard-Causse C, Blot B, et al. Exosomes are released by cultured cortical neurones. Mol Cell Neurosci. 2006;31(4):642–8.
56. Taylor DD, Gercel-Taylor C. Exosomes/microvesicles: mediators of cancer-associated immunosuppressive microenvironments. Semin Immunopathol. 2011;33(5):441–54.
57. Vella LJ, Greenwood DL, Cappai R, Scheerlinck JP, Hill AF. Enrichment of prion protein in exosomes derived from ovine cerebral spinal fluid. Vet Immunol Immunopathol. 2008;124(3-4):385–93.
58. Bachy I, Kozyraki R, Wassef M. The particles of the embryonic cerebrospinal fluid: how could they influence brain development? Brain Res Bull. 2008;75(2-4):289–94.
59. Ailawadi S, Wang X, Gu H, Fan GC. Pathologic function and therapeutic potential of exosomes in cardiovascular disease. Biochim Biophys Acta. 2015;1852(1):1–11.
60. Xin H, Li Y, Buller B, Katakowski M, Zhang Y, Wang X, et al. Exosome-mediated transfer of miR-133b from multipotent mesenchymal stromal cells to neural cells contributes to neurite outgrowth. Stem Cells. 2012;30(7):1556–64.

61. Morel L, Regan M, Higashimori H, Ng SK, Esau C, Vidensky S, et al. Neuronal exosomal miRNA-dependent translational regulation of astroglial glutamate transporter GLT1. J Biol Chem. 2013;288(10):7105–16.
62. Kucharzewska P, Christianson HC, Welch JE, Svensson KJ, Fredlund E, Ringner M, et al. Exosomes reflect the hypoxic status of glioma cells and mediate hypoxia-dependent activation of vascular cells during tumor development. Proc Natl Acad Sci U S A. 2013;110(18):7312–7.
63. Colombo E, Borgiani B, Verderio C, Furlan R. Microvesicles: novel biomarkers for neurological disorders. Front Physiol. 2012;3:63.
64. Rao P, Benito E, Fischer A. MicroRNAs as biomarkers for CNS disease. Front Mol Neurosci. 2013;6:39.
65. Felder RA, White MJ, Williams SM, Jose PA. Diagnostic tools for hypertension and salt sensitivity testing. Curr Opin Nephrol Hypertens. 2013;22(1):65–76.
66. Braccioli L, van Velthoven C, Heijnen CJ. Exosomes: a new weapon to treat the central nervous system. Mol Neurobiol. 2014;49(1):113–9.
67. Li Y, Liu Z, Xin H, Chopp M. The role of astrocytes in mediating exogenous cell-based restorative therapy for stroke. Glia. 2014;62(1):1–16.
68. Xin H, Li Y, Chopp M. Exosomes/miRNAs as mediating cell-based therapy of stroke. Front Cell Neurosci. 2014;8:377.
69. O'Loughlin AJ, Woffindale CA, Wood MJ. Exosomes and the emerging field of exosome-based gene therapy. Curr Gene Ther. 2012;12(4):262–74.
70. Ponting CP, Oliver PL, Reik W. Evolution and functions of long noncoding RNAs. Cell. 2009;136(4):629–41.
71. Clark MB, Mattick JS. Long noncoding RNAs in cell biology. Semin Cell Dev Biol. 2011;22(4):366–76.
72. Qureshi IA, Mattick JS, Mehler MF. Long non-coding RNAs in nervous system function and disease. Brain Res. 2010;1338:20–35.
73. Guennewig B, Cooper AA. The central role of noncoding RNA in the brain. Int Rev Neurobiol. 2014;116:153–94.
74. Hallmann AL, Arauzo-Bravo MJ, Zerfass C, Senner V, Ehrlich M, Psathaki OE, et al. Comparative transcriptome analysis in induced neural stem cells reveals defined neural cell identities in vitro and after transplantation into the adult rodent brain. Stem Cell Res. 2016;16(3):776–81.
75. Gao S, Tao L, Hou X, Xu Z, Liu W, Zhao K, et al. Genome-wide gene expression analyses reveal unique cellular characteristics related to the amenability of HPC/HSCs into high-quality induced pluripotent stem cells. Stem Cell Res Ther. 2016;7:40.
76. Zhang Y, Wong CH, Birnbaum RY, Li G, Favaro R, Ngan CY, et al. Chromatin connectivity maps reveal dynamic promoter-enhancer long-range associations. Nature. 2013;504(7479):306–10.
77. Wang Y, Liu H, Lin Y, Liu G, Chu H, Zhao P, et al. Network-Based Approach to Identify Potential Targets and Drugs that Promote Neuroprotection and Neurorepair in Acute Ischemic Stroke. Sci Rep. 2017;7:40137.
78. Francis F, Koulakoff A, Boucher D, Chafey P, Schaar B, Vinet MC, et al. Doublecortin is a developmentally regulated, microtubule-associated protein expressed in migrating and differentiating neurons. Neuron. 1999;23(2):247–56.
79. Gleeson JG, Lin PT, Flanagan LA, Walsh CA. Doublecortin is a microtubule-associated protein and is expressed widely by migrating neurons. Neuron. 1999;23(2):257–71.
80. Nacher J, Crespo C, McEwen BS. Doublecortin expression in the adult rat telencephalon. Eur J Neurosci. 2001;14(4):629–44.
81. Sherafat MA, Heibatollahi M, Mongabadi S, Moradi F, Javan M, Ahmadiani A. Electromagnetic field stimulation potentiates endogenous myelin repair by recruiting subventricular neural stem cells in an experimental model of white matter demyelination. J Mol Neurosci. 2012;48(1):144–53.
82. Cuccurazzu B, Leone L, Podda MV, Piacentini R, Riccardi E, Ripoli C, et al. Exposure to extremely low-frequency (50 Hz) electromagnetic fields enhances adult hippocampal neurogenesis in C57BL/6 mice. Exp Neurol. 2010;226(1):173–82.

83. Arias-Carrion O, Verdugo-Diaz L, Feria-Velasco A, Millan-Aldaco D, Gutierrez AA, Hernandez-Cruz A, et al. Neurogenesis in the subventricular zone following transcranial magnetic field stimulation and nigrostriatal lesions. J Neurosci Res. 2004;78(1):16–28.
84. Zhao C, Sun G, Li S, Shi Y. A feedback regulatory loop involving microRNA-9 and nuclear receptor TLX in neural stem cell fate determination. Nat Struct Mol Biol. 2009;16(4):365–71.
85. Cremisi F. MicroRNAs and cell fate in cortical and retinal development. Front Cell Neurosci. 2013;7:141.
86. Perruisseau-Carrier C, Jurga M, Forraz N, McGuckin CP. miRNAs stem cell reprogramming for neuronal induction and differentiation. Mol Neurobiol. 2011;43(3):215–27.
87. Brett JO, Renault VM, Rafalski VA, Webb AE, Brunet A. The microRNA cluster miR-106b~25 regulates adult neural stem/progenitor cell proliferation and neuronal differentiation. Aging. 2011;3(2):108–24.
88. Peck B, Schulze A. A role for the cancer-associated miR-106b~25 cluster in neuronal stem cells. Aging. 2011;3(4):329–31.
89. Liu H, Han XH, Chen H, Zheng CX, Yang Y, Huang XL. Repetitive magnetic stimulation promotes neural stem cells proliferation by upregulating MiR-106b in vitro. J Huazhong Univ Sci Technolog Med Sci. 2015;35(5):766–72.
90. Luo J, Hu X, Zhang L, Li L, Zheng H, Li M, et al. Physical exercise regulates neural stem cells proliferation and migration via SDF-1alpha/CXCR4 pathway in rats after ischemic stroke. Neurosci Lett. 2014;578:203–8.
91. Morris DC, Chopp M, Zhang L, Lu M, Zhang ZG. Thymosin beta4 improves functional neurological outcome in a rat model of embolic stroke. Neuroscience. 2010;169(2):674–82.
92. Leone L, Fusco S, Mastrodonato A, Piacentini R, Barbati SA, Zaffina S, et al. Epigenetic modulation of adult hippocampal neurogenesis by extremely low-frequency electromagnetic fields. Mol Neurobiol. 2014;49(3):1472–86.
93. Bai G, Sheng N, Xie Z, Bian W, Yokota Y, Benezra R, et al. Id sustains Hes1 expression to inhibit precocious neurogenesis by releasing negative autoregulation of Hes1. Dev Cell. 2007;13(2):283–97.
94. Hatakeyama J, Kageyama R. Retinal cell fate determination and bHLH factors. Semin Cell Dev Biol. 2004;15(1):83–9.
95. Ishibashi M, Ang SL, Shiota K, Nakanishi S, Kageyama R, Guillemot F. Targeted disruption of mammalian hairy and Enhancer of split homolog-1 (HES-1) leads to up-regulation of neural helix-loop-helix factors, premature neurogenesis, and severe neural tube defects. Genes Dev. 1995;9(24):3136–48.
96. Tomita K, Nakanishi S, Guillemot F, Kageyama R. Mash1 promotes neuronal differentiation in the retina. Genes Cells. 1996;1(8):765–74.
97. Jessberger S. Neural repair in the adult brain. F1000Res. 2016;5:F1000 Faculty Rev-169.
98. Hemphill JC, Andrews P, De Georgia M. Multimodal monitoring and neurocritical care bioinformatics. Nat Rev Neurol. 2011;7(8):451–60.
99. White TE, Ford BD. Gene interaction hierarchy analysis can be an effective tool for managing big data related to unilateral traumatic brain injury. In: Kobeissy FH, editor. Brain neurotrauma: molecular, neuropsychological, and rehabilitation aspects. Boca Raton, FL: Frontiers in Neuroengineering; 2015.

Chapter 7
Modulation of Post-Stroke Plasticity and Regeneration by Stem Cell Therapy and Exogenic Factors

Lukas Andereggen, Raluca Reitmeir, Stefano Di Santo, Raphael Guzman, Hans R. Widmer, Serge Marbacher, and Robert H. Andres

Abstract Revascularization therapy in the acute post-stroke phase nowadays is reducing the grade of disability and mortality after cerebral ischemia. Post-acute to chronic therapeutic strategies in the phase of irreversible brain parenchyma damage showed until now controversial results in pre-clinical studies: currently there are no effective treatment strategies apart from neurological rehabilitation aiming at restoration of functional post-ischemic deficits.

Spontaneous functional recovery appears immediately after stroke and was proven to correlate with the endogenous regeneration potential represented by rewiring of

L. Andereggen
Department of Neurosurgery, University of Berne, Inselspital, Berne, Switzerland

Department of Neurosurgery and F.M. Kirby Neurobiology Center, Harvard Medical School, Boston, MA, USA

R. Reitmeir • S. Di Santo • H.R. Widmer
Department of Neurosurgery, University of Berne, Inselspital, Berne, Switzerland

R. Guzman
Department of Neurosurgery, University Hospital Basel, Basel, Switzerland

Department of Neurosurgery and Stanford Stroke Center, Stanford University School of Medicine, Stanford, CA, USA

S. Marbacher
Department of Neurosurgery, University of Berne, Inselspital, Berne, Switzerland

Department of Neurosurgery, Cantonal Hospital of Aarau, Aarau, Switzerland

R.H. Andres, M.D. (✉)
Department of Neurosurgery, University of Berne, Inselspital, Berne, Switzerland

Department of Neurosurgery and Stanford Stroke Center, Stanford University School of Medicine, Stanford, CA, USA

Department of Clinical Research, University of Berne, Berne, Switzerland

Department of Neurosurgery, Research Laboratory, Inselspital,
Pavillon 47, Freiburgstrasse 10, 3010 Berne, Switzerland
e-mail: robert.andres@dkf.unibe.ch

P.A. Lapchak, J.H. Zhang (eds.), *Cellular and Molecular Approaches to Regeneration and Repair*, Springer Series in Translational Stroke Research,
https://doi.org/10.1007/978-3-319-66679-2_7

neuronal circuits through promotion of dendritic and axonal sprouting, improving axonal function, synaptogenesis, neurogenesis, and angiogenesis. These observations have led to numerous preclinical studies investigating a new therapeutic direction after stroke, the neurovascular restoration impacting stroke recovery potential.

This chapter summarizes achievements to date, current challenges and ongoing research in the field of regenerative processes after ischemic stroke, focusing on the formation of functional anatomical pathways responsible for enhanced recovery.

Keywords Stroke • Regeneration • Plasticity • Repair • Stem cells • Neural progenitors • Neurogenesis • Neuroprotection • Endothelial progenitors • Trophic factors

Abbreviations

AMPA	Alpha-amino-3-hydroxy-5-methyl-4-isoxazolepropionic acid
AQP	Aquaporin
ATSC	Adipose-tissue stem cell
BBB	Blood-brain barrier
BDNF	Brain-derived neurotrophic factor
BMSC	Bone-marrow derived stem cell
BrdU	5-Bromo-2′-deoxyuridine
CCL2	C-C chemokine ligand 2
CCR2	C-C chemokine receptor type 2
CM	Conditioned medium
CNS	Central nervous system
DCX	Doublecortin
EC	Endothelial cell
EPC	Endothelial progenitor cell
Epo	Erythropoietin
ERK	Extracellular signal-regulated kinase
ESC	Embryonic stem cell
GABA	Gamma aminobutyric acid
GFAP	Glial fibrillary acidic protein
HIF-1	Hypoxia-inducible factor-1
HuNu	Human nuclear antigen
IL	Interleukin
LTD	Long term depression
LTP	Long term potentiation
LV	Lateral ventricle
MAPK	Mitogen activated protein kinase
MMP	Matrix metalloproteinase
NMDA	N-methyl-D-aspartate
NPC	Neural progenitor cell

NO	Nitric oxide
NSC	Neural stem cell
PI3K	Phosphoinositide 3-kinase
Robo	Roundabout protein
SEM	Standard error of the mean
SGZ	Subgranular zone
STDP	Spike timing depending plasticity
SVZ	Subventricular zone
TPEN	*N,N,N′,N′*-tetrakis-(2-pyridylmethyl)ethylenediamine
TSP	Thrombospondin
TuJ1	Neuron-specific class III beta-tubulin
UCBC	Umbilical cord blood stem cell
VCAM-1	Vascular cell adhesion molecule 1
VEGF	Vascular endothelial growth factor
VLA-4	Very large antigen-4
Zn	Zinc
Zn^{2+}	Ionic zinc

1 Introduction

The controversial principle of *diaschisis* introduced 100 years ago by *Constantin von Monakow* represents the beginning of understanding the general model of plasticity underlying functional recovery after central nervous system (CNS) damage nowadays. *Diaschisis*, meaning 'shocked throughout' in Greek was defined by an 'interruption of function' in an intact brain region which will lead to a 'struggle for the preservation of the disrupted nervous function, and the CNS is always prepared for a struggle' [1, 2].

Ischemic stroke results from a sudden impairment of blood supply in specific parts of the brain, being the leading cause of long-term disability in adults in industrialized countries [3]. Sensorimotor and cognitive impairment after stroke are often severe with little chance of complete rehabilitation, which is associated with high socio-economic costs. Therefore, there is high demand for the development of new, effective treatment strategies to improve the functional outcome after ischemic brain damage.

The scientific efforts from the last decade showed that not only the brain itself has an intrinsic potential for reorganization and repair after stroke [4], but also that these processes of regeneration can be successfully stimulated by means of extrinsic factors. This plasticity potential is represented at the anatomical level by: 1.) recruitment of pathways that sustain the same function as the destroyed ones but have a different anatomical form, 2.) synaptogenesis, 3.) dendritic arborization, 4.) fortification of functionally silent synaptic connections, 5.) long distance fiber sprouting and branching, and 6.) endogenous neurogenesis [5–8]. These events take place in the first days up to weeks after the ischemic lesion, and the susceptibility to external therapeutic influence is negatively correlated with time. However, post-stroke plasticity is challenged by unexpected and sudden onset of ischemic damage [9], and by the high complexity of the damaged neural structures [10]. Novel therapeutic approaches like transplantation of stem and progenitor cells or administration of factors influencing endogenous repair capabilities of the post-ischemic brain have been investigated in the last decade. The purpose of this chapter is to provide an overview of the current research on neuroregenerative strategies after stroke focusing on underlying mechanisms of action, therapeutic window and possible implications for targeted neurorehabilitation.

2 Clinical Aspects of Stroke Treatment Research

The human brain detains its own rescue mechanisms in case of acute ischemic injury such as: 1.) recruitment of existing collateral blood vessels and induction of angiogenesis, preparing them for takeover in case of sudden obstruction [11], 2.) glial scar formation in the close vicinity of the ischemic core with neuroprotective potential, 3.) self-regeneration by reactivation of ontogenetic repair mechanisms [12]. These three pathways observed after stroke were further investigated in preclinical and clinical studies, giving rise to three therapeutic directions: re-establishing cerebral blood flow (revascularization), neuroprotection and neuroregeneration.

2.1 Revascularization

Mechanical thrombectomy with stent retrievers after large artery occlusion has been proven in recently published randomized studies [13] to re-establish blood flow and to reduce the functional disabilities, being nowadays the gold standard of acute stroke therapy (for review see *Balami et al.* [14]). The reperfusion of the ischemic tissue in the therapeutic window is meant to save the penumbra, limiting the ischemic damage, as well as to prevent vasogenic edema. It also sets the basic conditions for regenerative processes in the peri-infarct zone after stroke.

2.2 Neuroprotection

Neuroprotection is a broad term for mechanisms and strategies aiming at preventing neuronal cell death, therefore reducing deleterious effects of ischemic injury. This terminology is being used in preclinical research with regard to treatments that prevent or interrupt the molecular injury cascade in the penumbra and preventing secondary neuronal death [15, 16].

Neuroprotective strategies were developed on all progression pathways of ischemic injury described earlier: molecular injury, brain edema, inflammation, excitotoxicity, apoptosis, and spreading depression [17, 18].

Glutamate antagonists were studied with regard to their inhibitory effect upon peri-infarct depolarization and proved to reduce the size of ischemic lesion [19, 20]. Trying to reverse or stop the cascade of molecular injury after stroke lead to the development of different strategies like: stopping neuronal death by excitotoxicity by glutamate antagonists, using antioxidant substances to stop the formation of reactive species of oxygen or of peroxynitrite, antiapoptotic substances meant to stop the delayed neuronal death [17].

Formation of cytotoxic edema was considered as a target for aquaporin (AQP) channels, which are located in the plasma membrane and facilitate water transport. Inhibition of AQP water conductance was demonstrated to reduce the severity of ischemic brain edema [21]. Later studies proved that an intrinsic mechanism of early induction of AQPs may decrease cytotoxic edema formation after stroke but has no influence upon blood-brain barrier (BBB) disruption and therefore has a limited time effect after stroke [22].

The cellular inflammatory response after ischemia was proven to have both detrimental effects contributing to lesion expansion but also to play an important role in the orchestration of lesion repair, the outcome after stroke being seen as a result of the interaction between the injured brain and the immune system [23].

The most active inflammatory pathway after stroke is lead by cytokines and their answer after stroke. Especially the cytokine interleukin (IL)1-beta was for a long time considered a strong neuroprotective target, since administration of IL1-beta receptor antagonists reduces infarct size [24].

The translation of these therapies failed repeatedly, despite the convincing preclinical and phase IIb available data. The SAINT-II (Stroke Acute Ischemic NXY Treatment) study investigating the antioxidative agent NXY-059 as neuroprotective therapy after stroke in patients had to be stopped due to lack of efficiency in the beginning of the phase III trial [25, 26].

2.3 *Neurovascular Restoration*

Since the main clinical impact of stroke is due to its long time disability effect and because neuroprotective studies did not succeed in the clinical trials, the focus of stroke research changed in the last years on neuroregenerative approaches. The observation of endogenous regeneration potential after stroke, by means of neurogenesis [8, 27, 28] angiogenesis [29], axonal and dendritic sprouting potential [30] and synaptogenesis [30] started a new therapeutic direction after stroke, called neurovascular restoration.

2.3.1 Endogenous Neural Stem Cells as a Possible Pool for Regeneration After Stroke

Formation of neural stem cells (NSCs) starts in the gastrulation phase and continues throughout the embryonic brain by a continuous proliferation of NSCs and subsequent differentiation and migration of neural progenitor cells (NPCs) [31, 32]. After birth there are still neurogenic niches situated in the subventricular zone (SVZ) and the subgranular zone (SGZ) of the dentate gyrus [33]. Accordingly, proliferation of residential NSCs is observed in the adult brain in the SVZ, SGZ and the posterior periventricular area [34–37]. This represents an endogenous pool of NSCs which was proven to be activated by focal ischemia [38] both in the ipsilesional and in the contralesional hemisphere [39, 40], presenting a well determined timing following transient focal ischemia by reaching the peak point 1–2 weeks after stroke and returning to sham levels by 3–4 weeks [41, 42].

The process of neurogenesis includes three major anatomical steps: proliferation, migration and differentiation [43], which have to be followed by functional integration of the newborn neurons, including integration in the extracellular matrix environment and electrophysiological integration in neuronal circuits.

Proliferation of neurons after stroke was intensively studied until now with regards to different growth factors, some of the most promising being erythropoietin (Epo) and vascular endothelial growth factor (VEGF). Ischemia was shown to stimulate the hypoxia-inducible factor-1 (HIF-1) pathway as a main player in the signal cascade after stroke. The smallest reduction in oxygen partial pressure in the brain leads to a strong activation of HIF-1. Both VEGF and Epo are responsible for downstream effects of the transcription factor HIF-1 cascade. Epo knock-out mice have deficiencies in post-ischemic neurogenesis [44, 45] and VEGF was proven to promote neurogenesis both *in vitro* and *in vivo* [46]. Further neuroregeneration-specific aspects of these two growth factors are going to be discussed later in detail.

Migration and differentiation of endogenous NPCs in the normal brain was demonstrated to follow the route of the rostral migratory stream towards the olfactory bulb, whereas in the ischemic preconditioned brain, the NPCs migrate towards the injured areas in the brain [47, 48]. Important mediators in this process of migration and maturation are represented by matrix metalloproteinases (MMPs), especially the MMP9 molecule, which is upregulated in the infarcted cortex at 7–14 days in rats and was shown to colocalize with the NPC marker doublecortin (DCX) and proliferating 5-bromo-2′-deoxyuridine (BrdU) positive cells migrating from the SVZ [49]. *Wang et al.* proved in their studies that conditioned medium from Epo-treated epithelial cell cultures significantly promoted NPC migration, which was blocked by specific MMP inhibitors [50].

Even if there is ample evidence for migration and maturation of NPCs to the ischemic lesion, an aspect that still causes controversies involves the functionality of these neurons, their long-time survival and their integration in the peri-neural and angiogenetic milieu in order to sustain the beneficial recovery after stroke. Among other molecules, VEGF and Epo are thought to be promising candidates to facilitate this functional integration.

2.3.2 Angiogenesis and Neurovascular Remodeling After Stroke

A strong intercellular orchestration is needed to create the permissive conditions for functionally relevant axonal and dendritic sprouting after ischemic injury of the brain. The vascular and the nervous system share multiple similarities in their development, both of them using long-distance projections to reach their targets, being guided by gradients of chemokines and growth factors. Especially in the peripheral nervous tissue, the parallel tracking of blood vessels and nerves is obvious. In the CNS, neurogenesis takes place in the embryological vascular niches where endothelial cells (ECs) proliferate. This is why the two systems have to be taken into consideration as a homeostatic unit, especially in neuropathological conditions such as stroke.

Angiogenesis in the adult brain is the hypoxia-driven sprouting of new capillaries from postcapillary venules [51]. Tissue hypoxia in the adult brain stimulates the activation of HIF-1α expression, which then stimulates the transcription of VEGF, VEGF receptors flt-1 and neuropilin-1, and angiopoietin [52]. Besides the molecular aspect of angiogenesis activation, there are two further systems that are implicated immediately after stroke: loss of vascular integrity and cell matrix degradation [53]. These two processes activate growth factors, their receptors and the guidance molecules which were until then in a dormant phase by being incorporated in the cellular matrix. One of the most important activated growth factors is VEGF, which induces endothelial cell proliferation and their migration [54, 55]. The VEGF family comprises 5 related genes: VEGF-A, -B, -C, -D and PIGF (placenta induced growth factor). VEGF-A is the vascular permeability factor and is known in several isoforms (VEGF-A_{204}, -A_{189}, -A_{165}, -A_{145}, -A_{121}), with different amino acid length and has the capacity of binding to heparin sulfates. VEGF-A_{165} has some degree of heparan sulfate binding which reduces its diffusibility, but at the same time increases its ability to stimulate VEGF receptors [56].

As discussed for neurorestoration, an important aspect of angiogenesis is its functionality translated either by a significant increase in overall blood flow to the tissue that suffered an ischemic damage or by creating the foundation for late restoration processes in the ischemic tissue together with the neural network. The formation of new blood vessels after stroke seems to develop parallel to neurogenesis, being initiated rather late at 48 h after stroke [57, 58]. Because of this delay, there are no reasons to assume that angiogenesis influences brain hemodynamics during an acute ischemic stroke in a relevant way [53]. However, the observed timing hints towards a coupling of neurovascular remodeling after stroke in order to prepare the necessary background for long term restorative processes, e.g. axonal sprouting.

Another important aspect in the process of new blood vessel formation is the VEGF induced disruption of the BBB immediately after stroke, leading to edema formation. Early post-ischemic administration of VEGF in rats increased BBB leakage and infarction volume, whereas its late administration (48 h) enhanced angiogenesis and decreased BBB leakage, resulting in improved recovery volume [59].

2.3.3 Axonal Sprouting and Plasticity

An interesting general observation in the maturation of the corticospinal tract was that the early widespread distribution reaches the specific mature distribution by means of collateral selection. Neuronal cell death is not known to take place in the developing brain, so the hypothesis of postnatal reorganization could be explained just by means of collateral elimination. Using retrograde fluorescent tracer injections into the pyramidal decussation at the spinomedullary junction in adult versus postnatal rats, *Stanfield et al.* could prove that beside the frontal and parietal cortex, the occipital cortex was involved in building the corticospinal tract in postnatal rats [60]. These studies lead to the conclusion that transient pyramidal tract axons are eliminated during development, e.g. being found as projections to the superior colliculus and/or the pons [61, 62]. The understanding of developmental sculpturing of cortical efferent systems is important in further perception of remodeling processes in the adult brain after stroke or other types of injury.

It is now well accepted that the CNS has an intrinsic recovery capacity after stroke, by means of reactivating the ontogenetic machinery stimulating gene expression, protein synthesis, and cellular genesis, reconstructing the needed environment for recovery [63]. Preclinical and clinical studies on unilateral ischemic brain damage demonstrate an increased amount of corticospinal projections and shift of cortical sensorimotor functions to the intact hemisphere. Whether the intact hemisphere increases functionality after contralateral stroke just by means of increasing pyramidal corticospinal projections is not clear. A series of recent studies could identify stimulation of interhemispheric, cortico-reticular or cortico-thalamic pathways [5, 6, 64]. The involvement of the intact pyramidal tract in taking over functions of the damaged contralateral pyramidal system requires large–scale reorganization and a competition between the two cerebral hemispheres for spinal synaptic space. This implies that the degree of abnormality of these corticospinal projections following unilateral lesions might not reflect simply the extent of the initial lesion but also the consecutive competitive disadvantage of the surviving corticospinal projections. This competitive disadvantage would lead to cortico-spinal projections from the intact hemisphere progressively replacing a part of the surviving cortico-spinal projections from the damaged hemisphere and thus to a progressive degeneration.

Post-ischemic endogenous responses of the CNS go in line with an enhanced sensitivity to rehabilitative [65] and plasticity-promoting [66, 67] treatments, opening a time window in which ontogenetic brain repair mechanisms might successfully be reactivated [12, 68]. Stroke recovery is associated with reorganization of neuronal circuits both at the cortical and subcortical level. A series of events set the stage for brain reorganization in the intact hemisphere, such as increased angiogenesis [69] and axonal sprouting [66, 70].

Recruitment of contralesional brain areas correlated with a better recovery from stroke in animal studies [66, 70]. By administering anterograde tract tracers into the contralesional motor cortex, these authors suggested that contralateral projections may be recruited by plasticity-promoting therapies, underlining the relevance of contralesional reorganization for neurological recovery. However, models of permanent focal cerebral ischemia were used in the latter studies, in which motor cortex tissue was irreversibly destroyed. Brain plasticity ipsilateral to the stroke was not systematically assessed in these studies.

The vascular system is strongly linked to the neural system due to common ontogenic developing pathways. When investigating the circulatory system in cases of neurovascular pathologies of the brain, a series of dynamic processes were identified, which modulate development, survival and differentiation of neurons, rewriting the embryologic developmental phase in a restricted time and space manner [71]. This interconnected developmental network also depends on an important common regulator: the VEGF protein family and its receptor system.

2.3.4 Dendritic Elaboration and Dendritic Spine Proliferation

Whereas in the uninjured adult brain dendritic branching and spines are considered to be stable entities [72], important changes in density of dendritic spines were observed in ischemic situations [73].

By means of two-photon imaging techniques, *Brown et al.* demonstrated an increase in dendritic spine formation with a peak around 1–2 weeks and lasting around 6 weeks in the peri-infarct region after cortical ischemic injury [74]. Until now, no direct link between the rate of dendritic spine formation and functional recovery after stroke has been shown. Due to the spatial and temporal coincidence between dendritic branching and spine reorganization and changes in functional representation of the peri-infarct region [75] and functional recovery [76], restorative therapies focused also on stimulation of dendritic branching and spine density in the early phase after stroke.

After an ischemic injury that affects the axons of pyramidal neurons in the cortex without having a direct effect upon dendrites, significant dendritic spine loss was observed, which was proven to be a result of profound deafferentation. With time, this is followed by an increase in dendritic spine densities and neuritic outgrowth. There are two main sprouting directions involved: the horizontal cortico-cortical connections [6, 77] and the vertical connections from the contralesional hemisphere that travel through the corpus callosum [78].

Large-scale dendritic plasticity was proven to depend on the balanced elongation-retraction of pyramidal dendrites in the peri-infarct cortex after a small photo-thrombotic stroke [79]. Layer V of pyramidal neurons in the cortex is considered the main excitatory zone, therefore differences in dendritic branching and connectivity in this region will be expected to have a major impact upon cortical circuits. Still the evidence that this branching takes place is controversial, and some studies done with the MCAO stroke model found no differences in dendritic branching and spine densities between lesioned and control animals [80, 81]. By repetitive imaging *in vivo* over 3 months after ischemia, another study failed to prove evidence of *de novo* branching formation in the surviving L5 pyramidal neurons in the peri-infarct cortex [82].

The failure of spontaneous regeneration to induce functional recovery after stroke was proved not to be a failure of forming new connections [83], but mainly to be dependent on the non-permissive environment. The main players in the inhibition of the sprouting are the growth-inhibitory proteins, parts of the CNS myelin [84].

2.3.5 Synaptic Plasticity

Synaptic plasticity during the developing phase experiences a burst early in the postnatal period especially in the occipital cortex, reaching a density that is approximately twice the density in the adult brain [85]. Waves of synaptic plasticity appear also in parieto-temporal and then frontal regions during development, reaching their peaks around early adolescence [86, 87].

In physiological situations in the brain, synapses are modulated (strengthened or weakened) and shaped by activity-induced mechanisms called also Hebbian plasticity. Whereas regulation of individual synapses by means of long term potentiation (LTP) or long term depression (LTD) is the main mechanism responsible for learning and memory processes. In addition, mechanisms regulating levels of activity are considered important in synaptic plasticity for network function [88]. The Hebbian plasticity describes positive-feedback mechanisms responsible either for strengthening of effective synapses or for weakening of passive synapses. Regulation of neuronal activity during synaptic modulation was theoretically and experimentally devised in three possible underlying processes: synaptic scaling, spike timing depending plasticity (STDP) and synaptic redistribution.

Synaptic scaling refers to the competitive interaction between synapses coupling on the same neuron. The biological substrates that lead to synaptic scaling are dependent of glutamate receptors number. This leads to specialization of neuronal pathways depending on their synaptic stimulus. Whereas synaptic scaling is non-Hebbian plasticity forming active neuronal pathways by means of mainly postsynaptic firing rate, STDP is thought to respect Hebbian plasticity by considering both pre- and postsynaptic activity. Synaptic redistribution was observed in some forms of cortical neurons in which short-term plasticity of synapses induced by LTD can be modified by LTP, which increases the presynaptic elimination of the neurotransmitter. The mechanism of synaptic redistribution is not completely understood. For further information please see the review of *Abbott and Nelson* [88].

In the ischemic cortex, synaptogenesis is progressively increased in the peri-infarct region together with markers of axonal sprouting such as GAP-43, underlining the importance of the penumbral region in synaptic reorganization after stroke [30].

2.3.6 Modulation of the Immune Response and Inflammation

MMPs represent a family of zinc endopeptidases with a major role during development of the CNS and distinct functions in pathological states. Recent data showed specific roles of MMPs in different time periods after stroke. Whereas acute post-ischemic activation of MMPs leads to increased ischemic injury by enhancing the neuroinflammatory response [89], the postacute activation was shown to contribute to neurovascular remodeling and promote stroke recovery [90].

3 Cell Therapy and Brain Plasticity After Stroke

In the last decade, extensive research efforts have been carried out to establish cell-based therapies, e.g. transplantation of NSCs and/or NPCs, bone-marrow derived stem cells (BMSCs), umbilical cord blood stem cells (UCBCs), adipose-tissue derived stem cells (ATSCs), and embryonic stem cells (ESCs) as a possible experimental therapeutic avenue for ischemic stroke and other disorders of the CNS (for review see *Lindvall et al.* [91]). A series of clinical studies has already demonstrated the feasibility and safety of this approach in clinical practice (for review see *Bliss et al.* [92]).

Transplanted NPCs have been shown to survive, migrate and integrate in the post-ischemic host brain, thereby acquiring adequate neuronal and glial phenotypes and display functional electrophysiological integration into neuronal circuitry (for review see *Hermann et al.* [93]).

Investigations on the possible underlying mechanisms of cell transplantation have revealed that processes other than direct replacement of neurons and glial cells by the grafted NSCs or NPCs are involved in promoting neurological recovery after stroke [94]. The grafted cells orchestrate tissue plasticity of the host brain [95, 96]. These effects include secretion of neurotrophic factors, immunomodulation, and angiogenesis [97–102].

We have previously shown that transplantation of human NPCs improved functional outcome after experimental stroke in rats with temporal coincidence of increased dendritic arborization of layer V pyramidal neurons and promotion of cortico-cortical, cortico-striatal and cortico-spinal axonal projections [103].

Extensive migration of grafted cells towards the perilesional area was observed (Fig. 7.1). Only a very small percentage (about 6%) of the grafted NPCs showed differentiation into the neuronal lineage, most of them stayed in an undifferentiated state. At the same time, grafted animals showed robust functional recovery demonstrated with a battery of neurological tests, as compared to sham-operated controls (Fig. 7.1). Impaired axonal transport processes were partially restored in grafted animals, as demonstrated by reduced amyloid precursor protein accumulation. Using *in vitro* assays with indirect co-culture of human NPCs and cortical neurons, we demonstrated that increased dendritic and axonal plasticity depends on molecules secreted by NPCs. In a further step, some of these mediating factors were identified as VEGF, thrombospondins 1 and 2, and slit using immunodepletion assays (Fig. 7.2). Endogenous remapping of the ipsi- and contralesional hemispheres is a well-known phenomenon in recovery after ischemic damage [104–109]. It is reasonable that the above-mentioned soluble factors secreted by the grafted NPCs and their progeny, among many others, influence the host cells during this process by promoting dendritic and axonal regeneration. Other types of stem cells,

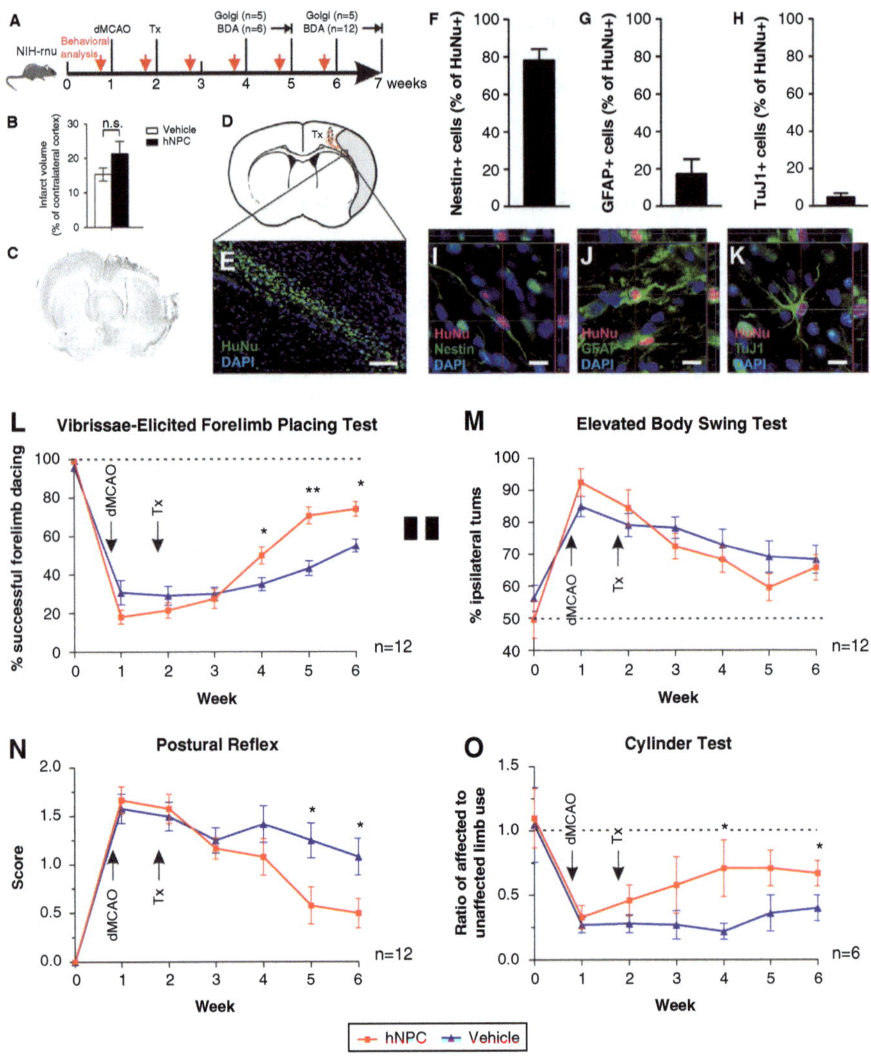

Fig. 7.1 Fate of transplanted human NPCs (hNPCs) and behavioral recovery after experimental stroke in rats. Experimental setup (**a**). Transplanted animals (hNPC) showed a tendency to smaller infarct volumes than controls (Vehicle) (**b**). Distal middle cerebral artery occlusion (dMCAO) resulted in consistent cortical infarction (**c**). 4 weeks after transplantation, the majority of human HuNu+ cells are found in the ischemic boundary zone (**d, e**). Most of the cells remain in a undifferentiated state (Nestin+; **f**), while smaller portions show astrocytic (GFAP; **g**) or neuronal (TuJ1; **h**) differentiation. Confocal immunofluorescence photomicrographs of Nestin+ (**i**), GFAP+ (**j**) and TuJ1+ (**k**) cells co-localizing with HuNu in the peri-infarct area. hNPC-grafted animals (hNPC) demonstrate improved functional recovery as compared to sham-operated controls (Vehicle) using the vibrissae-elicited forelimb placing test (**l**), the elevated body swing test (**m**), the postural reflex test (**n**), and the cylinder test (**o**). Scale bars: **e**: 50 μm, **i**–**k**: 10 μm. Mean ± SEM; $*p < 0.05$, $**p < 0.01$. *Data partially published in Andres et al. Brain 2011 Jun;134(Pt 6):1777-89. doi: 10.1093/brain/awr094, with kind permission from Oxford University Press*

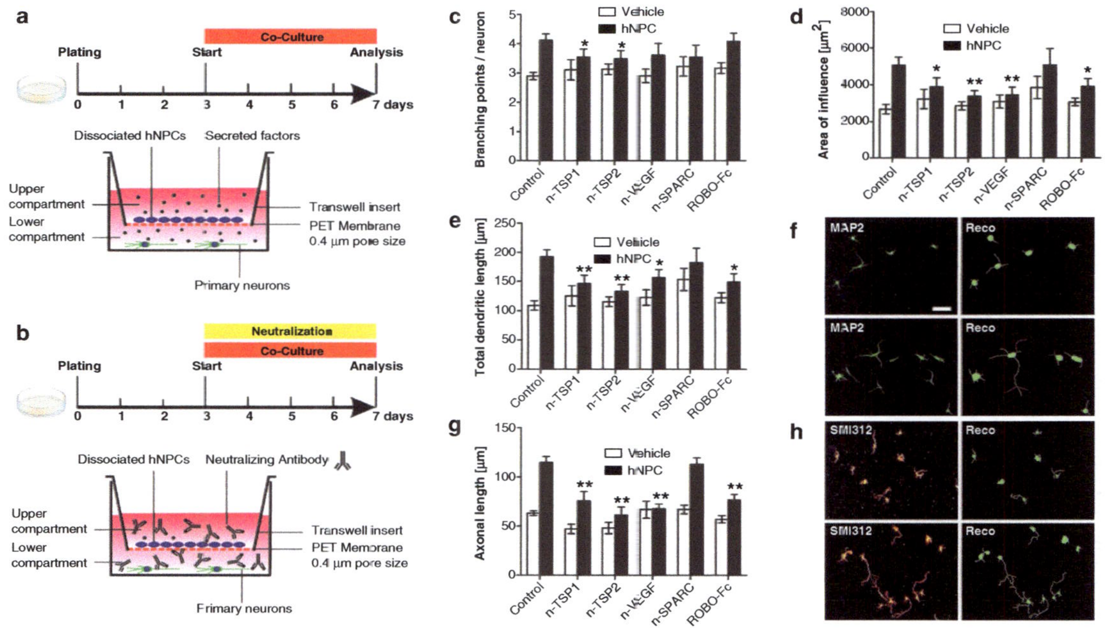

Fig. 7.2 Identification of human NPC (hNPC)-secreted factors mediating the effects on dendritic plasticity and axonal outgrowth *in vitro*. Experimental setup of co-culture experiments (**a**) and immunodepletion studies for identification of specific molecules (**b**). hNPCs were indirectly co-cultured with rat E14 primary cortical neurons for 7 days. Dendritic complexicity and axonal outgrowth were quantified using MAP2 (dendritic marker) and SMI312 (axonal marker) immunofluorescence and automated high-throughput analysis. As compared to untreated controls, the presence of hNPCs significantly promoted dendritic branching (**c**), total dendritic length (**d**), axonal outgrowth (**e**), and area of influence per individual neuron (**f**). Neutralization of factors significantly reduced the effects of hNPCs on dendritic branching (**c**; TSP1, TSP2), total dendritic length (**d**; TSP1, TSP2, VEGF, ROBO-Fc), axonal length (**e**; TSP1, TSP2, VEGF, ROBO-Fc), and area of influence (**f**; TSP1, TSP2, VEGF, ROBO-Fc). Representative photomicrographs and digital reconstructions (Reco) of neurons co-cultured with (hNPC) and without (Vehicle) hNPCs stained for MAP2 (**g**) and SMI312 (**h**). Scale bar: 50 µm, Mean ± SEM; $^*p < 0.05$, $^{**}p < 0.01$. *Data partially published in Andres et al. Brain 2011 Jun;134(Pt 6):1777-89. doi: 10.1093/brain/awr094, with kind permission from Oxford University Press*

like BMSCs, UCBCs, ATSCs, and ESCs as well, might have distinct action profiles (for review see *Andres et al.* [110]).

In another study, we investigated the effects of grafted murine NPCs on microglial activation, proliferation and phagocytosis [111]. VEGF secreted by NPCs was identified to mediate potent effects after grafting NPCs in mice. Thus, neural precursor cells are not only influenced by surrounding microglia, but also regulate microglia functions and activity. This might also play an important role in stroke and needs to be addressed in further studies.

Intravascular, i.e. intraarterial or intravenous administration of NPCs, is another feasible approach for cell transplantation in stroke. The advantage of this technique is the widespread distribution of the grafted cells in larger ischemic areas, which is not accomplishable by means of focal stereotactic transplantation. NPCs are recruited across the BBB by mechanisms similar to transendothelial homing of immune cells, including endothelial attachment and rolling along the endothelial surface. This process is facilitated by the integrin very large antigen-4 (VLA-4) expressed on immune cells as well as NPCs, which supports tethering and rolling on flow on vascular cell adhesion molecule 1 (VCAM-1). Expression of VCAM-1 is upregulated on the endothelial surface after stroke, leading to transendothelial recruitment of immune cells from the systemic circulation into the ischemic brain parenchyma. We have previously demonstrated that enrichment of NPCs by fluorescence activated cell sorting for VLA-4 and intracarotid delivery promoted cell homing to the area of stroke in mice and improved behavioral recovery [112]. In a second step, the interaction between CC-chemokine ligand-2 (CCL2) in the perivascular space and its receptor CC-chemokine ligand receptor-2 (CCR2) expressed on the plasma surface of NPCs was identified to be critical for targeted homing of intravascularly delivered NPCs [113]. Blocking CCR2 or using NPCs derived from CCR2 knock out animals led to a significant reduction of transendothelial migration as shown by bioluminescence and immunohistochemical studies. On the other hand, increasing the expression of chemokine receptors on the cellular surface by chemical pretreatment of the cells with brain derived neurotrophic factor (BDNF) augmented the transendothelial migration [114]. According to the temporal profile of adhesion molecule upregulation and chemokine expression after stroke the are ideal therapeutic windows for intravascular cell delivery. In an ischemic reperfusion rat stroke model we have demonstrated that intraarterial cell injection was most efficient 48 h after the ictus [114].

Recent clinical studies demonstrated the feasibility of transplantation of modified BMSCs (SB623) with improvement in clinical outcome 1 year following stable, chronic stroke [115].

4 Cell-Free Treatment Strategies

The demand of therapeutic options complementing thrombolysis and the better understanding of the endogenous repair mechanisms after stroke have favored the development of interventions based on cell transplantation. The observation that neurovascular plasticity is crucial in brain recovery has brought different cell types directly involved in angiogenesis and neurogenesis in the focus. In the regenerative scenarios following the ischemic insult, in addition to NSCs and NPCs, endothelial progenitor cells (EPC) are of particular importance being both targets and effectors. This is not surprising in view of the functional and anatomical coupling of vascular and neuronal cells (which is particularly evident in the neurogenic niche). Indeed, preclinical studies have confirmed that NPCs transplanted in stroked rodents are capable to generate neurons and promote angiogenesis. Similarly, EPC not only integrate in the brain vasculature, but also display neuroprotective actions. Another feature that makes EPCs suitable for therapy is their capacity to be readily recruited to the ischemic site.

Despite the evidence that intravenously transplanted EPCs engraft in the brain capillaries of stroked rats [116], it is now clear that soluble factors released by transplanted cells promote cell viability by providing trophic support, modulating local immunity and activating tissue remodeling processes, including angiogenesis. Thus, the paracrine actions of transplanted cells are not considered any more as 'bystander effects', but present with major tissue regenerative activities that precede the eventual differentiation and replacement of injured cells (if occurring at all). This concept and the presence of limitations inherent to cell transplantation such as the poor efficacy due to extensive death of grafted cells, microemboli, and tumor formation. have inspired a new type of therapeutic strategies designed on the administration of cell-secreted factors also referred as secretome. Indeed, using a rat hindlimb ischemia model, we have demonstrated that the EPC secretome in the form of conditioned medium (EPC-CM) has the same or even superior therapeutic capacity than transplanted cells [117]. These observations have been extended in a rat model of stroke [118]. In addition, we could demonstrate that intraventricular infusion of EPC-CM significantly increased the number of DCX-positive neuronal precursors in the SVZ of adult naïve rats (Fig. 7.3).

For the translation into clinical practice, the effectors of the diverse secretomes and their mechanisms of action still need to be fully elucidated [119]. We and others have recently reported that the EPC secretome-induced effect on viability of rat brain microvascular ECs under standard culture conditions is mediated by PI3K/AKT and MAPK/ERK activation [120, 121]. Moreover, EPC-CM significantly protected rat brain microvascular ECs from an ischemic insult induced by an oxygen-glucose deprivation. Importantly, these effects seem to be mediated not only by growth factors such as BDNF [121], but also involve not yet identified lipidic factors [122].

5 Pharmacological Treatment

Small molecules, e.g. the neuroprotective and differentiation-inducing ergogenic amino compound creatine [123], and many others, can be systemically administered in order to promote endogenous repair processes after stroke or to improve the fate of grafted NPCs. However, this is usually limited by low penetration of the BBB, resulting in poor CNS bioavailability, and systemic side effects.

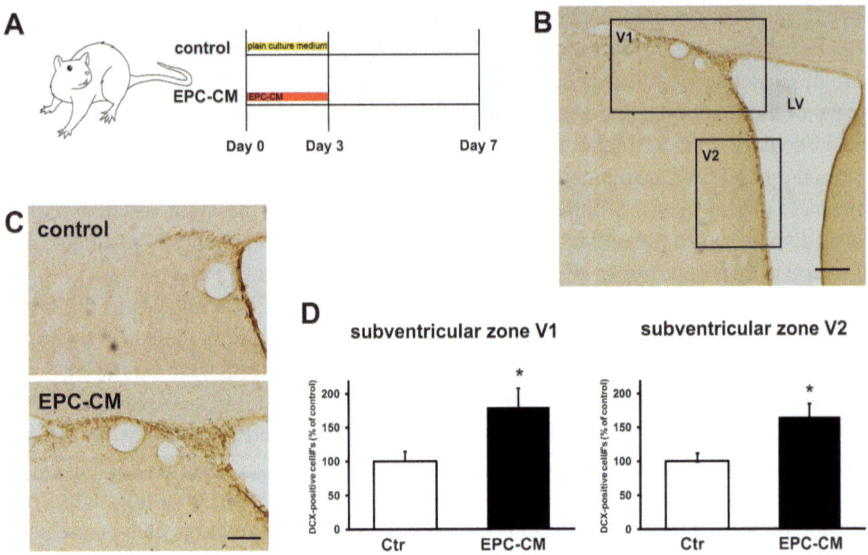

Fig. 7.3 Effects of endothelial progenitor cells-derived conditioned medium (EPC-CM) on number of doublecortin (DCX) positive neuronal precursors in the subventricular zone of adult rats. EPC-CM or control medium was infused into the right lateral ventricle (LV) by means of an Alzet pump for a period of 3 days. After 7 days, brains were removed and processed for immunohistological analyses (**a**). DCX-positive cells were examined in two regions of the subventricular zone (V1, V2) as indicated with *boxes* (**b**). The digitalized representative photomicrographs demonstrate a higher number of DCX-positive cells in V1 after EPC-CM treatment as compared to controls (**c**). Scale bars: 200 μm (**b**), 100 μm (**c**). Significantly higher DCX-positive cell numbers were observed in the EPC-CM groups (*filled bars*) as compared to controls (Ctr, *open bars*) in both the V1 and V2 regions (**d**). Values are expressed as percentage of controls and are given as mean ± SEM. *$p < 0.05$ vs. corresponding controls

Inhibition of tonic (extrasynaptic) gamma aminobutyric acid (GABA) signaling during the repair phase in the post-ischemic brain has been shown to promote functional recovery in mice, suggesting that GABA plays an important role in modulating brain repair [124]. Administration of N,N-Dimethyl-2-(6-methyl-2-p-tolylimidazo[1,2-a]pyridin-3-yl)acetamide (Zolpidem), a GABA agonist specific to the α-1 receptor subtype, was demonstrated to improve behavioral recovery [124].

Zinc (Zn) homeostasis, which is integral to normal CNS functioning, might also be involved in regenerative processes after ischemia and other CNS disorders [125]. Zn ions (Zn^{2+}) have been shown to play a crucial role in the modulation of synaptic transmission as well as in cortical plasticity [126]. Zn is required for the mammalian brain development and physiology. Under normal circumstances, Zn is tightly bound to many proteins in the CNS, whereas ionic Zn is a major etiological factor in CNS damage or diseases due to its toxicity [125, 127–129]. As thus, intracellular Zn^{2+}-concentrations are tightly regulated, as proper homeostasis is critical in the maintenance of cellular processing [130, 131]. Excessive exposure to extracellular Zn^{2+} on the other hand damages neurons of the CNS. Namely, transient forebrain ischemia in rats leads to an accumulation of chelatable, ionic Zn in degenerating CA1 neurons of the hippocampus, as well as in the cerebral cortex, thalamus, striatum, and the amygdala [127]. Interestingly, this accumulation precedes neurodegenerative processes, which could be prevented by the intraventricular injection of a Zn chelating agent, wherefore the early occurring toxic release of Zn^{2+} may be a key mechanism underlying selective neuronal cell death after ischemia [127]. As thus, administration of Zn^{2+}-chelating agents such as N,N,N′,N′-tetrakis-(2-pyridylmethyl) ethylenediamine (TPEN) could represent a potential cell-targeted therapy. TPEN significantly suppressed cell death, apoptosis, and neuronal glutamate release in primary cultured neurons undergoing a hypoxic-ischemic insult [132]. Moreover, there is a striking feature of a delayed rise in intracellular free Zn^{2+} in CA1 neurons just before the onset of histologically detectable cell death. Intrahippocampal injection of 1-naphthyl acetyl spermine, a selective channel blocker of GluR2-lacking alpha-amino-3-hydroxy-5-methyl-4-isoxazolepropionic acid (AMPA)-type glutamate receptors at *Schaffer* collateral to CA1 synapses 9–40 h after transient ischemia greatly reduced the late rise in intracellular free Zn^{2+} in postischemic CA1 neurons and afforded partial protection against ischemia-induced cell death. This receptor subtype appears to be an important therapeutic target for prevention of ischemia-induced neuronal death in humans [133].

Beside Zn^{2+}, nitric oxide (NO) is implicated in the pathogenesis of post-ischemic neuronal damage. The addition of exogenous NO or N-methyl-D-aspartate (NMDA) in order to increase endogenous NO leads to peroxynitrite (ONOO-) formation and consecutive Zn^{2+} release from intracellular stores in cerebrocortical neurons. Free Zn^{2+} in turn induces respiratory block, mitochondrial permeability transition, cytochrome c release, generation of reactive oxygen species, and p38 MAP kinase activation. This crosstalk between NO and Zn^{2+} dependent apoptotic signal transduction pathways may contribute to the delayed loss of neurons after ischemia [134].

Furthermore, we recently showed that Zn^{2+}-dyshomeostasis represents a major suppressor of axonal regeneration in the CNS, with Zn^{2+}-chelation (i.e. with TPEN) leading to persistent survival of many damaged neurons [135]. Thus, synaptic Zn^{2+} represents a previously unknown, critical suppressor of regeneration that might become a crucial player in neuroprotective and plasticity-enhancing strategies after stroke.

A greater understanding of the role of Zn^{2+} for cellular processes following CNS injuries where aberrant metal homeostasis is implicated in disease pathogenesis may therefore allow for the development of new potentially promising therapeutic approaches.

6 Conclusions

Ischemic stroke is the leading cause of severe long-term disability in the Western population, with very few therapeutic options. After an ischemic lesion, the neurovascular units are niches for NSCs and NPCs, supporting the regeneration potential, and the brain is being reshaped by means of neuronal sprouting or by unmasking the existing, but functionally silent connections. *Kreisel et al.* proposed in 2006 a timeline classification of recovery processes after stroke, differentiating between five distinct stages: (1) hyperacute phase from the stroke event up to 6 h after; (2) acute phase lasting up to the fourth day after stroke characterized by secondary events; (3) subacute phase from the second day up to 2–3 weeks characterized by brain remapping and functional plasticity; (4) consolidation period lasting up to several months and being characterized by functional alteration; (5) chronic phase characterized with the tendency of the events to become static [136, 137]. Despite intense research efforts during the last decade, effective therapeutic agents that promote the repair phase of recovery are still missing. There is a high heterogeneity in the preclinical data, making it unable to be synthetized and translated to the next clinical level. Elucidating underlying mechanisms of endogenous repair processes and plasticity of the brain is critical for the development of new therapeutic strategies for stroke in humans.

Acknowledgements This research was funded by the Swiss National Science Foundation (Grants No. 31-064975.01, 31-050824, 31-102075/1, 3100A0-112529, 31003A-135565, 406340-128124, 146632, PBBEB-117034, PASMP3-123221/1, PBBEB-146099 and PBBEB-155299), the Swiss Parkinson Foundation, the HANELA Foundation, a Research Grant from the University Hospital Berne, Inselspital (RGI-84800855), the Evelyn L. Neizer Fund, and by the Department of Clinical Research at the University of Berne.

References

1. von Constantin M. (1853-1930), neurobiologic philosopher. JAMA. 1970;211(6):1003–4.
2. Pribram KH, Spinelli DN, Reitz SL. The effects of radical disconnexion of occipital and temporal cortex on visual behaviour of monkeys. Brain. 1969;92(2):301–12.
3. Seshadri S, Wolf PA. Lifetime risk of stroke and dementia: current concepts, and estimates from the Framingham Study. Lancet Neurol. 2007;6(12):1106–14.
4. Payne BR, Lomber SG. Reconstructing functional systems after lesions of cerebral cortex. Nat Rev Neurosci. 2001;2(12):911–9.
5. Carmichael ST. Plasticity of cortical projections after stroke. Neuroscientist. 2003;9(1):64–75.
6. Carmichael ST, Wei L, Rovainen CM, Woolsey TA. New patterns of intracortical projections after focal cortical stroke. Neurobiol Dis. 2001;8(5):910–22.
7. Rossini PM, Calautti C, Pauri F, Baron JC. Post-stroke plastic reorganisation in the adult brain. Lancet Neurol. 2003;2(8):493–502.
8. Jin K, Minami M, Lan JQ, Mao XO, Batteur S, Simon RP, et al. Neurogenesis in dentate subgranular zone and rostral subventricular zone after focal cerebral ischemia in the rat. Proc Natl Acad Sci U S A. 2001;98(8):4710–5.
9. Roger VL, Go AS, Lloyd-Jones DM, Benjamin EJ, Berry JD, Borden WB, et al. Heart disease and stroke statistics--2012 update: a report from the American Heart Association. Circulation. 2012;125(1):e2–e220.
10. Lo EHA. new penumbra: transitioning from injury into repair after stroke. Nat Med. 2008;14(5):497–500.
11. Busch HJ, Buschmann IR, Mies G, Bode C, Hossmann KA. Arteriogenesis in hypoperfused rat brain. J Cereb Blood Flow Metab. 2003;23(5):621–8.

12. Cramer SC, Chopp M. Recovery recapitulates ontogeny. Trends Neurosci. 2000;23(6):265–71.
13. Berkhemer OA, Fransen PS, Beumer D, van den Berg LA, Lingsma HF, Yoo AJ, et al. A randomized trial of intraarterial treatment for acute ischemic stroke. N Engl J Med. 2015;372(1):11–20.
14. Balami JS, Sutherland BA, Edmunds LD, Grunwald IQ, Neuhaus AA, Hadley G, et al. A systematic review and meta-analysis of randomized controlled trials of endovascular thrombectomy compared with best medical treatment for acute ischemic stroke. Int J Stroke. 2015;10(8):1168–78.
15. Ginsberg MD, Pulsinelli WA. The ischemic penumbra, injury thresholds, and the therapeutic window for acute stroke. Ann Neurol. 1994;36(4):553–4.
16. Wahlgren NG, Ahmed N. Neuroprotection in cerebral ischaemia: facts and fancies--the need for new approaches. Cerebrovasc Dis. 2004;17(Suppl 1):153–66.
17. Repici M, Mariani J, Borsello T. Neuronal death and neuroprotection: a review. Methods Mol Biol. 2007;399:1–14.
18. Ferrer I. Apoptosis: future targets for neuroprotective strategies. Cerebrovasc Dis. 2006;21(Suppl 2):9–20.
19. Gill R, Andine P, Hillered L, Persson L, Hagberg H. The effect of MK-801 on cortical spreading depression in the penumbral zone following focal ischaemia in the rat. J Cereb Blood Flow Metab. 1992;12(3):371–9.
20. Iijima T, Mies G, Hossmann KA. Repeated negative DC deflections in rat cortex following middle cerebral artery occlusion are abolished by MK-801: effect on volume of ischemic injury. J Cereb Blood Flow Metab. 1992;12(5):727–33.
21. Griesdale DE, Honey CR. Aquaporins and brain edema. Surg Neurol. 2004;61(5):418–21.
22. Hirt L, Ternon B, Price M, Mastour N, Brunet JF, Badaut J. Protective role of early aquaporin 4 induction against postischemic edema formation. J Cereb Blood Flow Metab. 2009;29(2):423–33.
23. Endres M, Engelhardt B, Koistinaho J, Lindvall O, Meairs S, Mohr JP, et al. Improving outcome after stroke: overcoming the translational roadblock. Cerebrovasc Dis. 2008;25(3):268–78.
24. Stroemer RP, Rothwell NJ. Cortical protection by localized striatal injection of IL-1ra following cerebral ischemia in the rat. J Cereb Blood Flow Metab. 1997;17(6):597–604.
25. Savitz SI, Fisher M. NXY-059 for the treatment of stroke. N Engl J Med. 2007;357(21):2198. author reply -9.
26. Savitz SI. A critical appraisal of the NXY-059 neuroprotection studies for acute stroke: a need for more rigorous testing of neuroprotective agents in animal models of stroke. Exp Neurol. 2007;205(1):20–5.
27. Gould E, Reeves AJ, Graziano MS, Gross CG. Neurogenesis in the neocortex of adult primates. Science. 1999;286(5439):548–52.
28. Gould E, Beylin A, Tanapat P, Reeves A, Shors TJ. Learning enhances adult neurogenesis in the hippocampal formation. Nat Neurosci. 1999;2(3):260–5.
29. Krupinski J, Kaluza J, Kumar P, Kumar S, Wang JM. Role of angiogenesis in patients with cerebral ischemic stroke. Stroke. 1994;25(9):1794–8.
30. Stroemer RP, Kent TA, Hulsebosch CE. Neocortical neural sprouting, synaptogenesis, and behavioral recovery after neocortical infarction in rats. Stroke. 1995;26(11):2135–44.
31. Lois C, Alvarez-Buylla A. Long-distance neuronal migration in the adult mammalian brain. Science. 1994;264(5162):1145–8.
32. Clark SG, Chiu C. C. elegans ZAG-1, a Zn-finger-homeodomain protein, regulates axonal development and neuronal differentiation. Development. 2003;130(16):3781–94.
33. Pleasure SJ, Collins AE, Lowenstein DH. Unique expression patterns of cell fate molecules delineate sequential stages of dentate gyrus development. J Neurosci. 2000;20(16):6095–105.
34. Gage R. How Old Brains Got New Neurons. Cell. 2016;167(4):875–9.
35. Gage FH, Kempermann G, Palmer TD, Peterson DA, Ray J. Multipotent progenitor cells in the adult dentate gyrus. J Neurobiol. 1998;36(2):249–66.
36. Alvarez-Buylla A, Garcia-Verdugo JM. Neurogenesis in adult subventricular zone. J Neurosci. 2002;22(3):629–34.

37. Gritti A, Bonfanti L, Doetsch F, Caille I, Alvarez-Buylla A, Lim DA, et al. Multipotent neural stem cells reside into the rostral extension and olfactory bulb of adult rodents. J Neurosci. 2002;22(2):437–45.
38. Yan YP, Sailor KA, Lang BT, Park SW, Vemuganti R, Dempsey RJ. Monocyte chemoattractant protein-1 plays a critical role in neuroblast migration after focal cerebral ischemia. J Cereb Blood Flow Metab. 2007;27(6):1213–24.
39. Zhang R, Zhang Z, Wang L, Wang Y, Gousev A, Zhang L, et al. Activated neural stem cells contribute to stroke-induced neurogenesis and neuroblast migration toward the infarct boundary in adult rats. J Cereb Blood Flow Metab. 2004;24(4):441–8.
40. Zhang L, Zhang ZG, Zhang RL, Lu M, Adams J, Elliott PJ, et al. Postischemic (6-Hour) treatment with recombinant human tissue plasminogen activator and proteasome inhibitor PS-519 reduces infarction in a rat model of embolic focal cerebral ischemia. Stroke. 2001;32(12):2926–31.
41. Dempsey RJ, Sailor KA, Bowen KK, Tureyen K, Vemuganti R. Stroke-induced progenitor cell proliferation in adult spontaneously hypertensive rat brain: effect of exogenous IGF-1 and GDNF. J Neurochem. 2003;87(3):586–97.
42. Zhu Y, Sun Y, Xie L, Jin K, Sheibani N, Greenberg DA. Hypoxic induction of endoglin via mitogen-activated protein kinases in mouse brain microvascular endothelial cells. Stroke. 2003;34(10):2483–8.
43. Iwai M, Sato K, Omori N, Nagano I, Manabe Y, Shoji M, et al. Three steps of neural stem cells development in gerbil dentate gyrus after transient ischemia. J Cereb Blood Flow Metab. 2002;22(4):411–9.
44. Tsai TH, CH L, Wallace CG, Chang WN, Chen SF, Huang CR, et al. Erythropoietin improves long-term neurological outcome in acute ischemic stroke patients: a randomized, prospective, placebo-controlled clinical trial. Crit Care. 2015;19:49.
45. Tsai PT, Ohab JJ, Kertesz N, Groszer M, Matter C, Gao J, et al. A critical role of erythropoietin receptor in neurogenesis and post-stroke recovery. J Neurosci. 2006;26(4):1269–74.
46. Wang Y, Jin K, Mao XO, Xie L, Banwait S, Marti HH, et al. VEGF-overexpressing transgenic mice show enhanced post-ischemic neurogenesis and neuromigration. J Neurosci Res. 2007;85(4):740–7.
47. Thored P, Arvidsson A, Cacci E, Ahlenius H, Kallur T, Darsalia V, et al. Persistent production of neurons from adult brain stem cells during recovery after stroke. Stem Cells. 2006;24(3):739–47.
48. Arvidsson A, Collin T, Kirik D, Kokaia Z, Lindvall O. Neuronal replacement from endogenous precursors in the adult brain after stroke. Nat Med. 2002;8(9):963–70.
49. Zhao BQ, Wang S, Kim HY, Storrie H, Rosen BR, Mooney DJ, et al. Role of matrix metalloproteinases in delayed cortical responses after stroke. Nat Med. 2006;12(4):441–5.
50. Wang L, Zhang ZG, Zhang RL, Gregg SR, Hozeska-Solgot A, LeTourneau Y, et al. Matrix metalloproteinase 2 (MMP2) and MMP9 secreted by erythropoietin-activated endothelial cells promote neural progenitor cell migration. J Neurosci. 2006;26(22):5996–6003.
51. Carmeliet P. Angiogenesis in health and disease. Nat Med. 2003;9(6):653–60.
52. Pugh CW, Ratcliffe PJ. Regulation of angiogenesis by hypoxia: role of the HIF system. Nat Med. 2003;9(6):677–84.
53. Hermann DM, Zechariah A. Implications of vascular endothelial growth factor for postischemic neurovascular remodeling. J Cereb Blood Flow Metab. 2009;29(10):1620–43.
54. Carmeliet P. Blood vessels and nerves: common signals, pathways and diseases. Nat Rev Genet. 2003;4(9):710–20.
55. Lambrechts D, Storkebaum E, Morimoto M, Del-Favero J, Desmet F, Marklund SL, et al. VEGF is a modifier of amyotrophic lateral sclerosis in mice and humans and protects motoneurons against ischemic death. Nat Genet. 2003;34(4):383–94.
56. Simons M, Gordon E, Claesson-Welsh L. Mechanisms and regulation of endothelial VEGF receptor signalling. Nat Rev Mol Cell Biol. 2016;17(10):611–25.
57. Seylaz J, Charbonne R, Nanri K, Von Euw D, Borredon J, Kacem K, et al. Dynamic in vivo measurement of erythrocyte velocity and flow in capillaries and of microvessel diameter in the rat brain by confocal laser microscopy. J Cereb Blood Flow Metab. 1999;19(8):863–70.

58. Marti HJ, Bernaudin M, Bellail A, Schoch H, Euler M, Petit E, et al. Hypoxia-induced vascular endothelial growth factor expression precedes neovascularization after cerebral ischemia. Am J Pathol. 2000;156(3):965–76.
59. Zhang ZG, Zhang L, Jiang Q, Zhang R, Davies K, Powers C, et al. VEGF enhances angiogenesis and promotes blood-brain barrier leakage in the ischemic brain. J Clin Invest. 2000;106(7):829–38.
60. Stanfield BB, O'Leary DD, Fricks C. Selective collateral elimination in early postnatal development restricts cortical distribution of rat pyramidal tract neurones. Nature. 1982;298(5872):371–3.
61. Stanfield BB, O'Leary DD. Fetal occipital cortical neurones transplanted to the rostral cortex can extend and maintain a pyramidal tract axon. Nature. 1985;313(5998):135–7.
62. Stanfield BB. The development of the corticospinal projection. Prog Neurobiol. 1992;38(2):169–202.
63. Carmichael ST. Cellular and molecular mechanisms of neural repair after stroke: making waves. Ann Neurol. 2006;59(5):735–42.
64. Mohajerani MH, Aminoltejari K, Murphy TH. Targeted mini-strokes produce changes in interhemispheric sensory signal processing that are indicative of disinhibition within minutes. Proc Natl Acad Sci U S A. 2011;108(22):E183–91.
65. Biernaskie J, Chernenko G, Corbett D. Efficacy of rehabilitative experience declines with time after focal ischemic brain injury. J Neurosci. 2004;24(5):1245–54.
66. Papadopoulos CM, Tsai SY, Alsbiei T, O'Brien TE, Schwab ME, Kartje GL. Functional recovery and neuroanatomical plasticity following middle cerebral artery occlusion and IN-1 antibody treatment in the adult rat. Ann Neurol. 2002;51(4):433–41.
67. Seymour AB, Andrews EM, Tsai SY, Markus TM, Bollnow MR, Brenneman MM, et al. Delayed treatment with monoclonal antibody IN-1 1 week after stroke results in recovery of function and corticorubral plasticity in adult rats. J Cereb Blood Flow Metab. 2005;25(10):1366–75.
68. Buchli AD, Schwab ME. Inhibition of Nogo: a key strategy to increase regeneration, plasticity and functional recovery of the lesioned central nervous system. Ann Med. 2005;37(8):556–67.
69. Ding G, Jiang Q, Li L, Zhang L, Zhang ZG, Ledbetter KA, et al. Angiogenesis detected after embolic stroke in rat brain using magnetic resonance T2*WI. Stroke. 2008;39(5):1563–8.
70. Wiessner C, Bareyre FM, Allegrini PR, Mir AK, Frentzel S, Zurini M, et al. Anti-Nogo-A antibody infusion 24 hours after experimental stroke improved behavioral outcome and corticospinal plasticity in normotensive and spontaneously hypertensive rats. J Cereb Blood Flow Metab. 2003;23(2):154–65.
71. Greenberg DA, Jin K. From angiogenesis to neuropathology. Nature. 2005;438(7070):954–9.
72. Grutzendler J, Kasthuri N, Gan WB. Long-term dendritic spine stability in the adult cortex. Nature. 2002;420(6917):812–6.
73. Corbett D, Giles T, Evans S, McLean J, Biernaskie J. Dynamic changes in CA1 dendritic spines associated with ischemic tolerance. Exp Neurol. 2006;202(1):133–8.
74. Brown CE, Li P, Boyd JD, Delaney KR, Murphy TH. Extensive turnover of dendritic spines and vascular remodeling in cortical tissues recovering from stroke. J Neurosci. 2007;27(15):4101–9.
75. Dijkhuizen RM, Ren J, Mandeville JB, Wu O, Ozdag FM, Moskowitz MA, et al. Functional magnetic resonance imaging of reorganization in rat brain after stroke. Proc Natl Acad Sci U S A. 2001;98(22):12766–71.
76. Shanina EV, Schallert T, Witte OW, Redecker C. Behavioral recovery from unilateral photothrombotic infarcts of the forelimb sensorimotor cortex in rats: role of the contralateral cortex. Neuroscience. 2006;139(4):1495–506.
77. Dancause N, Barbay S, Frost SB, Plautz EJ, Chen D, Zoubina EV, et al. Extensive cortical rewiring after brain injury. J Neurosci. 2005;25(44):10167–79.
78. Brown CE, Boyd JD, Murphy TH. Longitudinal in vivo imaging reveals balanced and branch-specific remodeling of mature cortical pyramidal dendritic arbors after stroke. J Cereb Blood Flow Metab. 2010;30(4):783–91.

79. Brown CE, Aminoltejari K, Erb H, Winship IR, Murphy TH. In vivo voltage-sensitive dye imaging in adult mice reveals that somatosensory maps lost to stroke are replaced over weeks by new structural and functional circuits with prolonged modes of activation within both the peri-infarct zone and distant sites. J Neurosci. 2009;29(6):1719–34.
80. Gonzalez CL, Kolb B. A comparison of different models of stroke on behaviour and brain morphology. Eur J Neurosci. 2003;18(7):1950–62.
81. Rowntree S, Kolb B. Blockade of basic fibroblast growth factor retards recovery from motor cortex injury in rats. Eur J Neurosci. 1997;9(11):2432–41.
82. Mostany R, Portera-Cailliau C. Absence of large-scale dendritic plasticity of layer 5 pyramidal neurons in peri-infarct cortex. J Neurosci. 2011;31(5):1734–8.
83. David S, Aguayo AJ. Axonal elongation into peripheral nervous system "bridges" after central nervous system injury in adult rats. Science. 1981;214(4523):931–3.
84. Schwab ME, Caroni P. Oligodendrocytes and CNS myelin are nonpermissive substrates for neurite growth and fibroblast spreading in vitro. J Neurosci. 1988;8(7):2381–93.
85. Huttenlocher PR, de Courten C. The development of synapses in striate cortex of man. Hum Neurobiol. 1987;6(1):1–9.
86. Huttenlocher PR. Morphometric study of human cerebral cortex development. Neuropsychologia. 1990;28(6):517–27.
87. Huttenlocher PR, Dabholkar AS. Regional differences in synaptogenesis in human cerebral cortex. J Comp Neurol. 1997;387(2):167–78.
88. Abbott LF, Nelson SB. Synaptic plasticity: taming the beast. Nat Neurosci. 2000;3(Suppl):1178–83.
89. Amantea D, Russo R, Gliozzi M, Fratto V, Berliocchi L, Bagetta G, et al. Early upregulation of matrix metalloproteinases following reperfusion triggers neuroinflammatory mediators in brain ischemia in rat. Int Rev Neurobiol. 2007;82:149–69.
90. Rosell A, Lo EH. Multiphasic roles for matrix metalloproteinases after stroke. Curr Opin Pharmacol. 2008;8(1):82–9.
91. Lindvall O, Kokaia Z. Stem cells in human neurodegenerative disorders--time for clinical translation? J Clin Invest. 2010;120(1):29–40.
92. Bliss TM, Andres RH, Steinberg GK. Optimizing the success of cell transplantation therapy for stroke. Neurobiol Dis. 2010;37(2):275–83.
93. Hermann DM, Peruzzotti-Jametti L, Schlechter J, Bernstock JD, Doeppner TR, Pluchino S. Neural precursor cells in the ischemic brain - integration, cellular crosstalk, and consequences for stroke recovery. Front Cell Neurosci. 2014;8:291.
94. Martino G, Pluchino S. The therapeutic potential of neural stem cells. Nat Rev Neurosci. 2006;7(5):395–406.
95. Martino G, Bacigaluppi M, Peruzzotti-Jametti L. Therapeutic stem cell plasticity orchestrates tissue plasticity. Brain. 2011;134(Pt 6):1585–7.
96. Bacigaluppi M, Pluchino S, Peruzzotti-Jametti L, Kilic E, Kilic U, Salani G, et al. Delayed post-ischaemic neuroprotection following systemic neural stem cell transplantation involves multiple mechanisms. Brain. 2009;132(Pt 8):2239–51.
97. Ourednik J, Ourednik V, Lynch WP, Schachner M, Snyder EY. Neural stem cells display an inherent mechanism for rescuing dysfunctional neurons. Nat Biotechnol. 2002;20(11):1103–10.
98. Einstein O, Karussis D, Grigoriadis N, Mizrachi-Kol R, Reinhartz E, Abramsky O, et al. Intraventricular transplantation of neural precursor cell spheres attenuates acute experimental allergic encephalomyelitis. Mol Cell Neurosci. 2003;24(4):1074–82.
99. Hayase M, Kitada M, Wakao S, Itokazu Y, Nozaki K, Hashimoto N, et al. Committed neural progenitor cells derived from genetically modified bone marrow stromal cells ameliorate deficits in a rat model of stroke. J Cereb Blood Flow Metab. 2009;29(8):1409–20.
100. Liao JK. Statins and ischemic stroke. Atheroscler Suppl. 2002;3(1):21–5.
101. Prockop DJ, Olson SD. Clinical trials with adult stem/progenitor cells for tissue repair: let's not overlook some essential precautions. Blood. 2007;109(8):3147–51.

102. Uccelli A, Moretta L, Pistoia V. Mesenchymal stem cells in health and disease. Nat Rev Immunol. 2008;8(9):726–36.
103. Andres RH, Horie N, Slikker W, Keren-Gill H, Zhan K, Sun G, et al. Human neural stem cells enhance structural plasticity and axonal transport in the ischaemic brain. Brain. 2011;134(Pt 6):1777–89.
104. Nudo RJ. Mechanisms for recovery of motor function following cortical damage. Curr Opin Neurobiol. 2006;16(6):638–44.
105. Carmichael ST, Archibeque I, Luke L, Nolan T, Momiy J, Li S. Growth-associated gene expression after stroke: evidence for a growth-promoting region in peri-infarct cortex. Exp Neurol. 2005;193(2):291–311.
106. Murphy TH, Corbett D. Plasticity during stroke recovery: from synapse to behaviour. Nat Rev Neurosci. 2009;10(12):861–72.
107. Southwell DG, Froemke RC, Alvarez-Buylla A, Stryker MP, Gandhi SP. Cortical plasticity induced by inhibitory neuron transplantation. Science. 2010;327(5969):1145–8.
108. Reitmeir R, Kilic E, Kilic U, Bacigaluppi M, ElAli A, Salani G, et al. Post-acute delivery of erythropoietin induces stroke recovery by promoting perilesional tissue remodelling and contralesional pyramidal tract plasticity. Brain. 2011;134(Pt 1):84–99.
109. Reitmeir R, Kilic E, Reinboth BS, Guo Z, ElAli A, Zechariah A, et al. Vascular endothelial growth factor induces contralesional corticobulbar plasticity and functional neurological recovery in the ischemic brain. Acta Neuropathol. 2012;123(2):273–84.
110. Andres RH, Choi R, Steinberg GK, Guzman R. Potential of adult neural stem cells in stroke therapy. Regen Med. 2008;3(6):893–905.
111. Mosher KI, Andres RH, Fukuhara T, Bieri G, Hasegawa-Moriyama M, He Y, et al. Neural progenitor cells regulate microglia functions and activity. Nat Neurosci. 2012;15(11):1485–7.
112. Guzman R, De Los Angeles A, Cheshier S, Choi R, Hoang S, Liauw J, et al. Intracarotid injection of fluorescence activated cell-sorted CD49d-positive neural stem cells improves targeted cell delivery and behavior after stroke in a mouse stroke model. Stroke. 2008;39(4):1300–6.
113. Andres RH, Choi R, Pendharkar AV, Gaeta X, Wang N, Nathan JK, et al. The CCR2/CCL2 interaction mediates the transendothelial recruitment of intravascularly delivered neural stem cells to the ischemic brain. Stroke. 2011;42(10):2923–31.
114. Rosenblum S, Smith TN, Wang N, Chua JY, Westbroek E, Wang K, et al. BDNF pretreatment of human embryonic-derived neural stem cells improves cell survival and functional recovery after transplantation in hypoxic-ischemic stroke. Cell Transplant. 2015;24(12):2449–61.
115. Steinberg GK, Kondziolka D, Wechsler LR, Lunsford LD, Coburn ML, Billigen JB, et al. Clinical outcomes of transplanted modified bone marrow-derived mesenchymal stem cells in stroke: a phase 1/2a study. Stroke. 2016;47(7):1817–24.
116. Garbuzova-Davis S, Haller E, Lin R, Borlongan CV. Intravenously transplanted human bone marrow endothelial progenitor cells engraft within brain capillaries, preserve mitochondrial morphology, and display pinocytotic activity toward blood-brain barrier repair in ischemic stroke rats. Stem Cells. 2017;35(5):1246–58.
117. Di Santo S, Yang Z, Wyler von Ballmoos M, Voelzmann J, Diehm N, Baumgartner I, et al. Novel cell-free strategy for therapeutic angiogenesis: in vitro generated conditioned medium can replace progenitor cell transplantation. PLoS One. 2009;4(5):e5643.
118. Rosell A, Morancho A, Navarro-Sobrino M, Martinez-Saez E, Hernandez-Guillamon M, Lope-Piedrafita S, et al. Factors secreted by endothelial progenitor cells enhance neurorepair responses after cerebral ischemia in mice. PLoS One. 2013;8(9):e73244.
119. Di Santo S, Widmer HR. Paracrine factors for neurodegenerative disorders: special emphasis on Parkinson's disease. Neural Regen Res. 2016;11(4):570–1.
120. Di Santo S, Seiler S, Fuchs AL, Staudigl J, Widmer HR. The secretome of endothelial progenitor cells promotes brain endothelial cell activity through PI3-kinase and MAP-kinase. PLoS One. 2014;9(4):e95731.
121. Wang J, Chen Y, Yang Y, Xiao X, Chen S, Zhang C, et al. Endothelial progenitor cells and neural progenitor cells synergistically protect cerebral endothelial cells from Hypoxia/reoxygenation-induced injury via activating the PI3K/Akt pathway. Mol Brain. 2016;9:12.

122. Di Santo S, Fuchs AL, Periasamy R, Seiler S, Widmer HR. The cytoprotective effects of human endothelial progenitor cell-conditioned medium against an ischemic insult are not dependent on VEGF and IL-8. Cell Transplant. 2016;25(4):735–47.
123. Andres RH, Ducray AD, Andereggen L, Hohl T, Schlattner U, Wallimann T, et al. The effects of creatine supplementation on striatal neural progenitor cells depend on developmental stage. Amino Acids. 2016;48(8):1913–27.
124. Hiu T, Farzampour Z, Paz JT, Wang EH, Badgely C, Olson A, et al. Enhanced phasic GABA inhibition during the repair phase of stroke: a novel therapeutic target. Brain. 2016;139(Pt 2):468–80.
125. Frederickson CJ, Koh JY, Bush AI. The neurobiology of zinc in health and disease. Nat Rev Neurosci. 2005;6(6):449–62.
126. Bitanihirwe BK, Cunningham MG. Zinc: the brain's dark horse. Synapse. 2009;63(11):1029–49.
127. Koh JY, Suh SW, Gwag BJ, He YY, Hsu CY, Choi DW. The role of zinc in selective neuronal death after transient global cerebral ischemia. Science. 1996;272(5264):1013–6.
128. Land PW, Aizenman E. Zinc accumulation after target loss: an early event in retrograde degeneration of thalamic neurons. Eur J Neurosci. 2005;21(3):647–57.
129. Sensi SL, Paoletti P, Bush AI, Sekler I. Zinc in the physiology and pathology of the CNS. Nat Rev Neurosci. 2009;10(11):780–91.
130. Maret W. Analyzing free zinc(II) ion concentrations in cell biology with fluorescent chelating molecules. Metallomics: Integrated Biometal. Science. 2015;7(2):202–11.
131. Pan E, Zhang XA, Huang Z, Krezel A, Zhao M, Tinberg CE, et al. Vesicular zinc promotes presynaptic and inhibits postsynaptic long-term potentiation of mossy fiber-CA3 synapse. Neuron. 2011;71(6):1116–26.
132. Zhang F, Ma XL, Wang YX, He CC, Tian K, Wang HG, et al. TPEN, a specific Zn2+ chelator, inhibits sodium dithionite and glucose deprivation (SDGD)-induced neuronal death by modulating apoptosis, glutamate signaling, and voltage-gated K+ and Na+ channels. Cell Mol Neurobiol. 2017;37(2):235–50.
133. Noh KM, Yokota H, Mashiko T, Castillo PE, Zukin RS, Bennett MV. Blockade of calcium-permeable AMPA receptors protects hippocampal neurons against global ischemia-induced death. Proc Natl Acad Sci U S A. 2005;102(34):12230–5.
134. Bossy-Wetzel E, Talantova MV, Lee WD, Scholzke MN, Harrop A, Mathews E, et al. Crosstalk between nitric oxide and zinc pathways to neuronal cell death involving mitochondrial dysfunction and p38-activated K+ channels. Neuron. 2004;41(3):351–65.
135. Li Y, Andereggen L, Yuki K, Omura K, Yin Y, Gilbert HY, et al. Mobile zinc increases rapidly in the retina after optic nerve injury and regulates ganglion cell survival and optic nerve regeneration. Proc Natl Acad Sci U S A. 2017;114(2):E209–E18.
136. Kreisel SH, Hennerici MG, Bazner H. Pathophysiology of stroke rehabilitation: the natural course of clinical recovery, use-dependent plasticity and rehabilitative outcome. Cerebrovasc Dis. 2007;23(4):243–55.
137. Kreisel SH, Bazner H, Hennerici MG. Pathophysiology of stroke rehabilitation: temporal aspects of neuro-functional recovery. Cerebrovasc Dis. 2006;21(1-2):6–17.

Chapter 8
Stem Cell-Paved Biobridge: A Merger of Exogenous and Endogenous Stem Cells Toward Regenerative Medicine in Stroke

Hung Nguyen, M. Grant Liska, Marci G. Crowley, and Cesario V. Borlongan

Abstract Stroke is a significant unmet clinical need with therapeutic options limited to tissue-type plasminogen activator (tPA), which has a small therapeutic window and risk for hemorrhagic transformation. Stroke is a multiphasic disease with a complex pathology. After the initial insult, a cascade of events occur causing secondary cell death and the expansion of the penumbra. The major contributing factors to this secondary cell death are depletion of growth factors, neuroinflammation, and disruption of the neurovascular unit. There is a need for more innovative and effective therapies that can target the diverse pathological consequences of stroke. To this end, stem cell therapy is a promising approach for stroke. Pre-clinical studies have demonstrated the potential of stem cells for treating neurological disorders, including stroke. Here, we discuss diverse stem cell types which have generated encouraging results for advancing to the clinic. Then, we examine the mechanisms of action of stem cells—cell replacement, by stander effect, and a novel biobridge concept advanced by our laboratory. These mechanisms work in concert to afford the neuroprotection and neuroregeneration after stroke. We envision that an in-depth understanding of the benefits and drawbacks of various stem cells and their mechanisms of action will guide the translational entry of stem cell therapy from the laboratory into the clinical setting.

Keywords Adult-derived stem cells • Ischemia pathology • Stem cell mechanisms • Stem cell migration • Neuroregeneration • Neuroprotection • Extracellular matrix remodeling • Stem cell therapies • Translational research

---

H. Nguyen • M.G. Liska • M.G. Crowley • C.V. Borlongan, Ph.D. (✉)
Center of Excellence for Aging and Brain Repair, Department of Neurosurgery and Brain Repair, University of South Florida Morsani College of Medicine, 12901 Bruce B. Downs Blvd, Tampa, FL 33612, USA
e-mail: cborlong@health.usf.edu

P.A. Lapchak, J.H. Zhang (eds.), *Cellular and Molecular Approaches to Regeneration and Repair*, Springer Series in Translational Stroke Research,
https://doi.org/10.1007/978-3-319-66679-2_8

Abbreviations

BDNF	Brain-derived neurotrophic factor
BM-MSC	Bone marrow-derived mesenchymal stem cells
CCI	Controlled cortical impact
ECM	Extracellular matrix
EGF	Epithelial growth factor
FGF	Fibroblast growth factor
GDNF	Glial cell line-derived neurotrophic factor
hBMSCs	Human bone marrow stromal cells
IA	Intra-arterial
IC	Intracranial
IL-1β	Interleukin-1-beta
iPSCs	Induced pluripotent stem cells
IV	Intravenous
MCAO	Middle cerebral artery occlusion
MMP	Metallomatrix protein
NGF	Nerve growth factor
NSCs	Neural stem cells
SCF	Stem cell factor
SDF-1	Stromal-derived factor 1
SGZ	Subgranular zone
SVZ	Subventricular zone
TBI	Traumatic brain injury
TNF-α	Tumor necrosis-alpha
tPA	Tissue-type plasminogen activator
VEGF	Vascular endothelial growth factor

1 Introduction

Stroke continues to be a leading cause of death and disability in America, with approximately 800,000 people being affected annually [1]. Responsible for roughly 5% of American deaths—the fifth leading cause [2]—long term consequences for stroke survivors can range from mild functional impairments to severe disability [1]. Accounting for healthcare costs and loss of productivity, an economic burden of $33.6 billion is attributed to stroke, with this figure projected to increase in the future [1]. In fact, the economic burden of stroke has increased notably in recent years, largely due to improved treatment protocols and a resulting decreased mortality rate [1]. Despite posing such a prevalent medical and economic burden, therapeutic options for stroke have been limited to tissue-type plasminogen activator (tPA) and physical therapies to alleviate symptoms. Unfortunately, the clinical benefits of tPA are minimized by its narrow therapeutic window, with the risk of hemorrhagic transformation rising sharply and its efficacy decreasing significantly over the initial 1–6 h timeframe [3–5]. As a result, the search for innovative and effective therapies which maintain their therapeutic value over the acute, sub-acute, and chronic pathological stages of stroke continues.

Stem cell therapies have been explored as a possible treatment to this unmet clinical need, having demonstrated both neuroprotective effects in the acute stage, as well as regenerative capacity in later stages of stroke [6–10]. Furthermore, stem cell therapies offer unique advantages over traditional pharmaceuticals by providing a dynamic and adaptive therapeutic profile—a likely requisite for any intervention capable of providing substantial functional recovery from the complex neurodegenerative pathology of stroke [11–15]. Apparent from the completed clinical trials of stem cell transplantation is their relative safety via both intracerebral and intravenous administration [16] (NCT01501773, NCT00535197, NCT00859014, NCT01716481). Unfortunately, clearly demonstrating their efficacy has proven more difficult due to a number of practical difficulties in outcome measurements, patient enrollment numbers, and trial design [17, 18]. As a result, basic and translational laboratories have engaged in a concerted effort to better understand the mechanisms by which stem cells offer their therapeutic effects in the hopes of inspiring more successful clinical trials. Following the recent *in vitro* and *in vivo* studies of our laboratory, we have proposed a third mechanism by which stem cells convey therapeutic effects, the *bio bridge*, which works cooperatively with the two well-established mechanisms of cell replacement and bystander effects (secretion of trophic factors, cytokines, and anti-inflammatory molecules, among others) [19–21]. This novel mechanism, whereby transplanted stem cells assist the migration of endogenous stem cells from neurogenic niches in the subventricular zone (SVZ) and the subgranular zone (SGZ) to the region of damaged tissue via extracellular matrix remodeling, was demonstrated in a controlled cortical impact (CCI) model of traumatic brain injury (TBI) [21]. Here, we expand this concept by revealing preliminary data which indicate the formation of a similar structure in the middle cerebral artery occlusion (MCAO) model of ischemic stroke. When contemplating the clinical feasibility of cell-based therapies for the treatment of stroke, the biobridge concept advances the notion that transplanted stem cells can work in synchrony with endogenous stem cell repair mechanisms. This provides a clearer understanding of the mechanisms by which stem cells confer their therapeutic benefits, and also supports their safety by demonstrating that long-term effects generated by cell therapy may not require transplanted stem cell survival per se, but rather endogenous stem cells can subsequently continue the regenerative process despite non-survival fates of the grafted cells.

2 The Many Facets of Stroke Pathology

Stroke is defined as a pathological state whereby a reduction in blood flow effects one or more regions of the brain, which may be caused by an obstructed vessel resulting in ischemic stroke or a ruptured blood vessel, leading to hemorrhagic stroke [1]. Ischemic stroke is more common and has a lower mortality rate [1]. The cells that directly lose their supply of glucose and oxygen die quickly, as neurons are exceedingly sensitive to metabolic stress. This ischemic tissue region comprises the infarct core; these cells are vulnerable to primary cell death processes and are less amenable to therapeutic intervention [22, 23]. Oxygen and nutrient deprivation causes mitochondrial damage and an increase in reactive oxygen species, both of which contribute to cell death cascades [22]. Additionally, without proper energy supply, the cell membrane is no longer able to uphold ionic homeostasis, which drives improper calcium ion concentrations within the cell, further contributing to cell death pathways [24]. The acute damage to these cells ultimately leads to cell death, with little opportunity for intervention.

Despite stroke being an acute event, the resulting pathophysiology of this event persists chronically, a product of a phenomenon known as secondary cell death [23]. The necrotic cells within the infarct core leave in their wake a toxic microenvironment. Leaked substances from these cells have the capacity to reach adjacent healthy cells and cause harm [23]. For example, following stroke, high levels of glutamate are released into the microenvironment and reach concentrations that lead to excitotoxicity in neighboring cells [23]. This region of cells susceptible to secondary cell death is referred to as the penumbra. Researchers often focus on this region of cells due to a higher likelihood of restoration and a wider therapeutic window. Unlike the infarct core, the penumbra is not fixed—this region of secondary cell death may continue to expand over weeks, months and even years [25, 26].

There are many components contributing to secondary cells death after stroke including depleted growth factors, neuroinflammation, and blood-brain barrier (BBB) breakdown [27–30]. Appropriate growth factor levels within the microenvironment must be sustained for cell survival, with loss of these factors resulting in apoptosis. Several types of growth factors contribute to neuron homeostasis including, but not limited to, glial cell line-derived neurotrophic factor (GDNF), brain-derived neurotrophic factor (BDNF), nerve growth factor (NGF), fibroblast growth factor (FGF), vascular endothelial growth factor (VEGF), stromal-derived factor 1 (SDF-1), epithelial growth factor (EGF), and stem cell factor (SCF). Highlighting their importance, preclinical studies have displayed neuroprotective benefits using GDNF, BDNF VEGF, SDF-1, and SCF treatments following cerebrovascular injury [30].

The neuroinflammatory response after stroke is a double-edged sword. While inflammation plays an important neuroprotective role in the acute phase, chronic inflammation perpetuates secondary cell death [22]. The neuroinflammatory process is triggered by damage-associated molecular patterns (DAMPs) propagated by dying and dead cells. Some of the DAMPs are high mobility group box-1 (HMGB1), heat shock proteins, and hyaluronan [31]. Once the inflammation process is initiated, the vulnerable cells within the penumbra are activated and secrete pro-inflammatory cytokines including tumor necrosis-alpha (TNF-α), interleukin-1-beta (IL-1β), and interleukin-6 (IL-6), among others [30]. This stroke-induced inflammatory response further exacerbates cell death and BBB breakdown. The BBB is a part of the dynamic neurovascular unit which is composed of vascular cells (endothelial cells, pericytes, and smooth muscle cells), supporting glial cells (astrocytes, microglia, and oligodendrocytes), neurons, and extracellular matrix [32]. Aberrant neuroinflammation dramatically disrupts the interactions between components of the neurovascular unit. Pro-inflammatory cytokines interfere with the tight connections between the astrocytic end-feet, pericytes, and endothelial cells, causing a leaky BBB [33]. The damaged BBB permits the entry of circulating cells and substances which are typically excluded or tightly-regulated from the brain, inducing further inflammation and upsetting the homeostatic solute balance, which results in intracranial edema [33]. Also, a number of molecular factors which are upregulated following injury—such as Notch, HMBG1, and SPARC—prompt microglia toward an M1-like phenotype, favoring mobility, phagocytosis, and the section of additional pro-inflammatory cytokines [34–36]. Finally, the inflammatory cytokines promote the upregulation of adhesion molecules (i.e. ICAM1, E-selectin, P-selection) on the endothelial cells and attract peripheral immune cells to adhere and enter the brain [32, 35–38]. Altogether, this inflammation contributes to a hostile environment which, if prolonged, can cause further damage to neural cells.

The pathology of cerebrovascular diseases is not isolated within the central nervous system (Fig. 8.1). Peripheral body systems have received increasing recognition for their role in cerebrovascular disease progression. Inflammatory signals that result post-injury travel through the circulatory system and impact systemic inflammation which may propagate cerebral inflammation. This brain-to-periphery interplay is both permitted and heightened by BBB breakdown, as peripheral lymphocytes and monocytes easily pass through compromised vessels, migrating toward the inflammatory signals originating from the site of injury. Treatment options will be

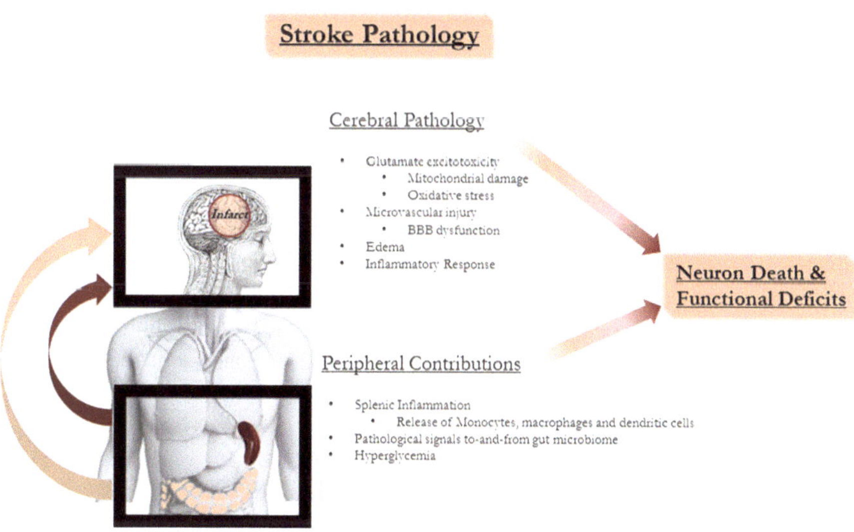

Fig. 8.1 A diagram of stroke pathology which includes both cerebral damage and peripheral contributions. Importantly, the loss of BBB fidelity permits the transfer of pathologically relevant molecules to and from the periphery. Changes in the peripheral organs—especially the spleen and gut—have been shown to accompany and contribute to worsening outcomes

rendered most effective if they consider peripheral body systems, due to their capacity to exacerbate brain injury. For example, preclinical studies suggest that mitigation of peripheral inflammation—particularly in the spleen—may be a primary mechanism of intravenous stem cell injection after stroke [9]. Indeed, the spleen is a significant contributor to systemic inflammation as a consequence of cerebral insult [9, 39, 40]. Following stroke, the physical size and function of the spleen alter, impacting brain health [41–43]. Under pathological conditions, the spleen will release splenocytes into circulation causing further neurodegeneration. Animals that receive splenectomies prior to cerebrovascular insults display improved cognitive function and decreased lesion volumes [44]. While this method is not practical for clinical use, this knowledge of the spleen-brain inflammatory axis highlights the critical role of the spleen in neuropathology. In addition to the spleen, research has also revealed that the gut microbiome plays a vital role in stroke pathology. Depletion of the proper intestinal flora leads to poorer outcomes in animal models of stroke, testifying to the significance of the microbiome to global health [45]. Our understanding of stroke as a global disease-state gives insight on how to properly assess and develop effective treatments [46, 47]. As we will discuss in great depth later, intravenous transplantation of stem cells has the unique ability to utilize trophic mechanisms to abrogate central and peripheral inflammation, in addition to forming the biobridge structure which helps facilitate repair by way of endogenous stem cell optimization, all working to reduce the pathological consequence of stroke.

Stroke pathology is complex and multiphasic. The initial metabolic restriction and glutamate toxicity are not the only factors that cause damage to the neurovascular unit. In fact, the subsequent secondary cell death in the form of growth factors deficiency, neuroinflammation, and BBB breakdown can further exacerbate the injury for an extensive period of time. Current stroke treatment is limited to restoration of the blood flow through tPA or mechanical means which is only effective when targeting the supracute stage of pathology [3]. While the body has a small capacity to repair and regenerate neural damage, these efforts are insufficient to overcome the overwhelming damage of secondary cell death. Therefore, there is a tremendous need for novel therapeutic strategies that can address this multifaceted pathology of stroke. The complexity of stroke pathology necessitates a therapy that has as an equally complex and diverse array of therapeutic mechanisms. To this end, we and others have proposed stem cell therapy as a promising therapeutic strategy. Briefly, stem cells exert their therapeutic benefits through replacing loss or damaged cells, providing trophic factors and anti-inflammatory cytokines, and via the novel concept of the stem cell-paved biobridge. These mechanisms will be expanded upon in later sections.

3 The Evolution of Stem Cell Research

Stem cells are a small population of cells which possess specific characteristics, including the ability to self-replicate, to differentiate into various cell lineages, and to express specific cell markers [48]. Self-replication gives stem cells the ability to preserve their characteristics and maintain a reservoir population of stem cells within several niches of the body. The capability of stem cells to differentiate into different cell types is vital to their role in preserving homeostasis and to the maintenance of various body systems [49]. For example, the body continually regenerates red blood cells to replace the old by using stem cells within the bone marrow. Collectively, the capacity of stem cells to self-replicate and differentiate into various lineages is referred to as the property of *stemness* [50]. Each type of stem cell has a characteristic level of stemness which is an important factor to be considered when contemplating any potential therapeutic treatment.

There are several ways to classify stem cells. The most common type of classification is based on the origin of the harvested stem cells. For example, umbilical cord stem cells and adipose stem cells are harvested from the umbilical cord and adipose tissue, respectively. Depending on a stem cell's potency, defined as the number of cell types a stem cell can differentiate into, a stem cell can be classified as totipotent, pluripotent, or multipotent [51]. Totipotent stem cells can become all cell types including extraembryonic cells, whereas, pluripotent stem cells can develop into all cell types except for extraembryonic and placental cells. Multipotent stem cells can give rise to various cell types, yet much more limited than totipotent and pluripotent stem cells [51]. In general, the earlier the cell is harvested within the developmental process, the higher the stem cells' potency (i.e. embryonic). Additionally, stem cells can be classified molecularly based on their profile of expressed cell markers. Bone marrow-derived mesenchymal stem cells, for example, are positive for CD29, CD44, CD105, CD73, CD90, CD106, and CD166 markers, while negative for CD14, CD34 and CD45 [52, 53]. In this section, we will discuss the unique properties and pros/cons of specific stem cell types which have shown promising preclinical results, with an emphasis on the relevance and feasibility for clinical translation.

3.1 The Early Era of Stem Cell Research and Initial Cell Sources

When the stem cell research field first developed, stem cells were primarily isolated from fetal tissues. Fetal stem cells have been shown to afford therapeutic benefits in preclinical models of many neurological disorders, including stroke, and were the cornerstone of early stem cell research in the 1970s and 1980s [54, 55]. These benefits include neuroprotective and neuroregenerative effects through secreting anti-inflammation molecules, releasing growth factors and differentiating into neuronal cells [56]. Furthermore, fetal stem cells demonstrate greater graft survival and ability to hone in on sites of injury when compared to adult stem cells [57, 58]. Unfortunately, fetal stem cells have been plagued by notions of immorality since their discovery, with opponents citing a lack of respect for human life and a possible justification for abortion as grounds for restricting research efforts [55]. From 1987 to 1992, these ethical concerns manifested as a moratorium—a legislative suspension of all funds related to fetal stem cell research—which pushed scientists to search for non-fetal stem cell sources [55].

In an attempt to avoid these ethical concerns, varying methods have been used to develop and harvest alternative stem cell sources which produce potent therapies in lieu of fetal tissue. One such effort in the neurological field involved creating neuron-like hNT by exposing NT2-N embryonic carcinoma-derived stem cells to retinoic acid. These cells terminally differentiate into post-mitotic neurons, and were shown to survive and integrate into host neural networks [59]. Despite promising preclinical data [60], this line of cells was beset by concerns of tumorigenicity [61]. In a Phase I clinical trial, 12 patients—9 male and 3 female—with an age range from 44 through 74 years old, were transplanted with hNT cells developed by Layton Bioscience Inc. [62]. The study concluded that the transplantation of the hNT cells was safe and feasible, however consensus on the efficacy could not be reached due to small sample size [62]. The first postmortem analysis of a participant was reported 27 months after implantation [63]. The analysis showed that the hNT cells survived at 27 months after implantation with no evidence of tumor, additional infarcts, or neurodegenerative diseases [63]. However, this patient did not show motor recovery after transplantation [63]. While Phase I and Phase II clinical trials ultimately revealed the safety of these cells—with no adverse cell-related serological effects [60, 63], and moderate functional improvements—the inadequate patient sample size and ongoing concerns over their cancerous origin and high proliferative capacity would severely cripple investigations into this cell line. In light of cell lines such as hNT, the genetic modification of stem cells emerged as a potential solution to a number of issues which dampened progression into the clinic, such as artificially reducing proliferation/tumorigenicity, improving graft survival, and heightening anti-inflammatory effects [64].

Cell lines such as the conditionally immortalized human neural stem cell, CTX0E03 or CTX, developed by ReNeuron aimed to maintain all facets of stem cell therapeutic efficacy while eliminating tumorigenic risks [65]. ReNeuron utilized c-mycER(TAM) technology in human first trimester fetal cortical cells to develop conditional growth control with a fusion protein containing the growth promoting gene, c-myc, and a hormone receptor regulated by the synthetic drug, 4-hydroxytamoxifen (4-OHT) in producing the CTX-DP immortalized cell line [66]. This allowed the cells to be cultured to large quantities *in vitro* with 4-OHT-containing media, yet have their growth cycle arrested upon transplantation in the absence of 4-OHT [66]. With the support of promising preclinical data, CTX cells entered a phase 1 clinical trial named PISCES in 2010 (NCT01151124) and were shown to improve primary outcome measurements in male stroke patients [67]. A narrow patient pool of 11 males aged 60+, and the open-label, single-arm study design calls into question the extent to which reliable conclusions can be made regarding the efficacy of CTX cell implantation on functional recovery (NCT01151124). Arguably, the modified nature of these CTX cells may have negative effects on their stemness and therapeutic characteristics. In particular, with the lineage commitment of the cells artificially restricted to neuronal phenotype, the ability of these neuronal-like cells to migrate is likely reduced, thus compromising their efficacy. This underscores the important balance which must be found when genetically modifying stem cells; being that stem cells are such complex biologics, scientists must be mindful not to unintentionally diminish major therapeutic mechanisms of stem cells by modifying dynamic and far-reaching pathways. When compared to unmanipulated or minimally-manipulated cell types, CTX cells (as well as SB623, which will be discussed shortly) took significantly longer to gain clinical approval, largely due to additional regulatory obstacles including long term *in vivo* preclinical studies and safety mechanism demonstrations which were required for all modified cell types [68]. The complications and dangers of genetic modification were first made evident in clinical trials of viral vector gene therapy which displayed the risk of fatal side-effects in some patients, producing an atmosphere of fear and apprehension surrounding all forms of genetically modified therapies [69]. This had the result of dampening and greatly delaying the clinical entry of genetically engineered stem cell types, such as CTX, which faced the skepticism of a wary Food and Drug Administration with the tragic loss of life fresh in their memory, and a negative public perception of all things genetically modified [70]. These unfavorable attitudes severely crippled the clinical progress of genetically modified stem cells—which objectively possess unique therapeutic potential.

Turning to embryonic stem cells, being from an early stage of development, these cells are considered the gold standard for stemness, with intrinsically high potency and high proliferative rates. In fact, only embryonic stem cells from the first few cell divisions after fertilization have true totipotent characteristics and are free from replicative senescence. These qualities make embryonic stem cells diverse in their applications. The use of embryonic stem cells arose from scientific efforts to steer clear of fetal-derived and cancer-derived cell lines, seeing as both were fraught with public image issues. Formed from *in vitro* blastocysts fertilization [71], embryonic stem cells evaded a portion of the moral issues surrounding fetal stem cells, yet fell short of acquiring complete public acceptance. While preclinical evidence has repeatedly demonstrated the efficacy of embryonic stem cells in neurological disorders [72], their wide-scale use has been similarly hindered by ethical, moral, and tumorigenic concerns.

Pressure from politicians and public opinion concerning embryonic and fetal stem cells, as well as the failed clinical trials of gene therapy which negatively affected the view of genetically engineered stem cells, have pushed scientists in the field of adult stem cells to look for alternative sources. For the past few decades, scientists have been able to identify and isolate adult-derived stem cells from various sources. These stem cells circumvent the ethical issues faced with embryonic stem cells, however, they pose challenges of their own. Some of the adult-derived stem cells which will be discussed are bone marrow-derived mesenchymal stem cells (BM-MSCs), extraembryonic stem cells, and induced pluripotent stem cells (iPSCs). While there are many other stem cells, these cell types currently hold the most potential to advance to the clinic.

3.2 Transitioning to Adult-Sourced Stem Cells

3.2.1 Bone-Marrow Mesenchymal Stem Cells

Mesenchymal stem cells (MSCs) are a class of multipotent stem cells that can be harvested from many adult mesenchymal tissues such as bone marrow, adipose tissue, and placenta. Of these, BM-MSCs are the most common and also the most studied adult stem cells, with multiple pre-clinical studies showing their therapeutic benefits in various neurological disorders such as TBI, amyotrophic lateral sclerosis, and particularly in stroke [9, 73–75]. One of the major advantages of BM-MSCs is the ability for autologous transplantation. BM-MSCs can be harvested and amplified from a patient's own tissues, thereby eliminating the concern of post-transplant immunologic rejection. However, an argument can be made against using autologous stem cells because the patients' BM-MSCs might be less potent than healthy donors'. Recent studies have linked stroke neurological deficits with changes in the peripheral systems such as inflammation in the spleen or alterations in the gut microbiome [9, 76]. These peripheral alterations could negatively affect the health and therapeutic efficacy of the patient's BM-MSCs. In terms of tumorigenicity, some studies have reported that BM-MSCs may induce tumor formation [77–79], however, BM-MSCs have been deemed safe in both pre-clinical and clinical studies [70, 75, 79–82]. These advantages of BM-MSCs are particularly relevant in attempting to transition BM-MSCs into the clinic. Importantly, BM-MSCs can still exert their beneficial effects despite short survival time and lack of neuronal differentiation [9, 16, 81, 83]. While appearing paradoxical, BM-MSCs' mechanisms of action rely more heavily on immunomodulation, modifying the microenvironment, and secreting trophic factors rather than differentiating and integrating into neural networks [53]. This is a distinct advantage of these cells, as clinically it is more feasible to give a stem cell "booster shot" to compensate for the low survival rate rather than attempting to control the formation of tumors inherent in other cell types. However, BM-MSCs also pose challenges that must be considered. BM-MSCs may behave differently depending on their location, method of extraction, isolation, and culture [79]. Therefore, it could be difficult to have a consistent and homogenous pool of BM-MSCs in mass production. Another limitation of BM-MSCs is that it requires time to collect, isolate and amplify the autologous BM-MSCs before they can be transplanted back into the patient, limiting the accessibility of these stem cells and their availability to the population at large.

Homogenous subpopulations of BM-MSCs may offer distinctive benefits. One such cell type, multilineage-differentiating stress-enduring (Muse) stem cells can be found within bone marrow (in addition to all connective tissues), and have displayed characteristics which make them highly appealing for therapeutic exploration [84]. These pluripotent cells have shown a unique ability to remain viable within highly stressful microenvironments [84]. Furthermore, the asymmetrical divisions and low telomerase activity of Muse cells mean low tumorigenicity and minimal risk of teratoma formation. Another subpopulation of MSCs found within bone marrow is the

Very small embryonic-like (VSEL) stem cell [85]. These cells are roughly half the size of hematopoietic stem cells, and maintain the pluripotent ability to differentiate into cells from all three germ layers [86]. Additionally, these cells have been shown to form small clusters that resemble embryonic bodies *in vitro*, which could have implications in the efficiency these cells can be cultured, as well as preserving their highly potent characteristics [86].

BM-MSCs have experienced success within the clinic. A GDNF-releasing, Notch-induced human bone marrow-derived mesenchymal stem cell line—SB623—was employed in a Phase I clinical trial beginning in 2011 (NCT01287936), after displaying significant amelioration of stroke symptomology in preclinical animal models. GDNF—glial cell line-derived neurotrophic factor—confers potent pro-survival effects. SB623 stem cells undergo *ex vivo* gene delivery to heighten their neurotrophic properties via enhanced GDNF secretion. As of now, the SB623 clinical trial has demonstrated relative safety, and preliminary reports of efficacy in chronic stroke patients [80]. Similar to the CTX cells described previously, consideration must be given to the genetically modified nature of these stem cells, and how these modifications may inadvertently affect the stemness and therapeutic properties of the cells.

3.2.2 Extraembryonic Stem Cells

Extraembryonic stem cells are a collective term for the adult-derived stem cells found in the placenta, the umbilical cord, the amnion, and Wharton's jelly [83, 87–89]. Placenta-derived MSCs, umbilical cord blood-derived MSCs (UCB-MSCs), and amnion-derived MSCs are the focus of many current investigations. Considering these stem cells' common origins, they share many therapeutic properties with BM-MSCs, such as modulating neuroinflammation, stimulating endogenous neurogenesis, releasing trophic factors, and promoting functional recovery in pre-clinical animal models of stroke [87, 90–92]. However, extraembryonic stem cells can differentiate into more cell types than BM-MSCs [89]. Recent studies have demonstrated that UCB-MSCs and placenta-derived MSCs can differentiate into neuronal cells that express markers such as Nestin or β-tubulin III—important markers of neuronal identity and function [93]. Similarly, recent studies have also reported that Wharton's jelly-derived MSCs can differentiate into various cell types such as glial, neuronal, and endothelial cell [94]. There are several advantages of extraembryonic stem cells compared to embryonic, fetal, and bone marrow-derived stem cells. These tissues are currently considered waste products, thus posing no health risk to the mother or baby, and circumventing any ethical issue related to the extraction of these extraembryonic stem cells. In the case of amnion-derived MSCs, the stem cells can be collected during amniocentesis—a safe, routine procedure during pregnancy [95]. These stem cells can then be expanded and ready to treat any disease associated with childbirth, such as hypoxia, or cryogenically preserved for future catastrophic events such as stroke or TBI [95]. However, there are also downfalls associated with extraembryonic stem cells. These extraembryonic tissues contain a variety of cells, making it difficult to isolate a homogeneous population of stem cells. Moreover, the amount of stem cells in these tissues is limited, especially in amnion fluid, requiring more time to amplify these stem cells prior to transplantation. In addition, it is expensive and unrealistic to maintain all extraembryonic tissues for every baby. Only a small portion of the population can afford to cryogenically preserve these tissues for an extended time.

3.3 Induced Pluripotent Stem Cells: A New Horizon for Stem Cell Research

Contrary to previous dogma, recent studies have demonstrated that differentiated adult cells can be reverted back to earlier stem cell states. Through molecular manipulation, these cells can regain their stemness, especially their proliferative property [89]. These cells are termed induced pluripotent stem cells (iPSCs). One of the challenges of adult stem cells is the limited number of passages before the cells stop proliferating. iPSCs are molecularly enhanced to increase the stemness (both proliferative and differentiating capacities) of the cells and can be scaled to large quantities. Furthermore, iPSCs also bypass the ethical issues associated with harvesting embryonic or fetal stem cells. In pre-clinical studies of stroke, iPSCs have shown promising results for improving neurological deficits, decreasing neuroinflammation, promoting neurogenesis and increasing angiogenesis [96–99]. Another major advantage of iPSCs is their ability to be redirected to differentiate into various cell lineages. For example, iPSCs can be induced into neural cells such as neurons, astrocytes, microglia, and vascular endothelial cells. While iPSCs have many advantages, tumorigenesis is a major concern when using iPSCs. In most cases, cancerous genes are used to induce the iPSCs, therefore it is important to control the tumorigenic property before iPSCs can advance further into the clinic.

3.4 Challenges in Translating Stem Cell Therapies to the Clinic

Finally, it is worth noting that there are many other logistic challenges that must be considered before any of the stem cells discussed can successfully advance into, through, and beyond clinical trials. These challenges include reaching a consensus on ideal cell type, dosage, number of transplants, timing, and route of administration. Indeed, the current clinical trials mentioned above (NCT01151124, NCT01287936) are being carefully analyzed and scrutinized for sub-optimal design and small patient pools.

The ideal timing and route of the administration depend on the intended purpose of the stem cell transplantation. Within the context of stroke, the distinct acute and chronic pathological phases must be considered. Intracranial (IC) transplantation is preferable in the acute and subacute phase of stroke. In these time frames, the presence of stem cells at the penumbra dampens the hostile environment and reduces the spread of the infarct core. Conversely, in the chronic phase, the inflammation both in the brain and the periphery is the main concern. Therefore intravenous (IV) or intra-arterial (IA) injection of stem cells may pose as better alternatives. In addition, if there is a need for multiple transplantations or injections, IV and IA are much more desirable choices. Of note, during the IV and IA injection, the majority of stem cells are trapped in the peripheral organs such as lung and spleen. However, the route of administration does not have to be mutually exclusive; an appealing option may be first transplanting via IC injection followed by IV booster shots for maximizing effectiveness.

The growing number of unique stem cell types begs the question of which is the best candidate stem cell type for clinical application. As discussed previously, each of the various stem cell types has their specific strengths and weaknesses. A well-designed preclinical research effort geared toward evaluating the safety, efficacy and mechanism of action of each stem cell type may reveal the optimal transplantation regimen of cell therapy for clinical trials. In particular, determining the appropriate stem cell dosage, timing, and route of delivery in animals with direct human application will be critical in advancing cell therapy to the clinic.

Stem cell therapies for stroke are at a pivotal point currently. Preclinical evidence has continued to accumulate for the past four decades which indicates that transplantation of stem cells offers significant amelioration of stroke-induced deficits, both when delivered acutely as well as chronically. Furthermore, IV and IC administration have displayed unique benefits and practical advantages which broaden the applicability of stem cell transplantation and heighten their far-reaching potential. The issues described above, however, have crippled the advancement of this therapy, resulting in limited clinical trials with inconsistent measures of efficacy. Careful evaluation of the six most recent clinical trials of BM-derived stem cell therapies in stroke—four within the subacute phase of stroke (NCT01716481, NCT00859014, NCT01501773, NCT00535197), and two within the chronic phase (NCT01151124, NCT01287936)—confirms the disconnect between lab and clinic, and reveals the gaps which still exist in our knowledge of stem cell therapies. As additional clinical trials proceed with enlisting larger cohorts of patients, pursuing long-term follow-up, and thoroughly assessing the status of the transplanted cells, we will be able to further evaluate the safety, efficacy, and mechanisms of action of stem cell therapy for stroke. Indeed, the mechanisms of action by which stem cell confer their therapeutic benefits in stroke are yet to be fully understood. How stem cells achieve this regenerative process stands as the primary challenge for stem cell researchers within the field, and is a vital step in designing more successful clinical trials [68, 100, 101]. The following section will discuss the canonical mechanisms of action for stem cells, as well as explore the concept of the biobridge and how it advances our understanding of the host-transplant interactions which mediate stem cells' therapeutic effects.

4 Stem Cell Therapy: Moving Beyond the Cell Replacement Paradigm

Given the multifaceted pathology of stroke, therapies targeting only a single pathology are unlikely to resurrect the motor and cognitive deficits caused by stroke, particularly at the chronic stages. Stem cell therapy is unique in its potential to be beneficial over a wide therapeutic window and its capacity to mitigate the diverse pathological processes observed after stroke [102]. The two known and widely-accepted mechanisms by which stem cells elicit neuroprotective and neuroregenerative effects after stroke are cell replacement and bystander effects [103, 104].

Initially, it was proposed that transplanted stem cells would serve the same function as they do within the body—generating new cells and replacing dead or damaged tissue. Transplanted stem cells were predicted to differentiate and directly replace loss cells, however, studies have demonstrated that within the injured brain, this notion is at best partially correct due to various factors [105, 106]. First, the majority of transplanted stem cells do not survive even when immunogenicity is accounted for through autologous transplant or Immunosuppressants [107]. Second, while many stem cells have demonstrated that they can differentiate into neuronal cells *in vitro* under highly-controlled conditions, they failed to do so in large numbers within *in vivo* model [108, 109]. One explanation for both issues is that transplanted stem cells enter a hostile microenvironment which is not conducive to long-term survival, differentiation and maturation. Thus, merely increasing the number of transplanted cells would not solve the problem. Furthermore, even with the small number of differentiated and living cells, there is little evidence to support that these cells integrate into neural networks to a significant extent, hence cell replacement is not considered a primary mechanism of action of stem cells.

Instead, evidence supports that the therapeutic capacity of stem cells lies largely within its bystander effects in which the stem cells secrete trophic factors and anti-inflammatory cytokines [110]. Stem cells secrete a cocktail of vital growth factors and, as mentioned previously, a reduction in growth factors is a key player in secondary cell death [111, 112]. For example, in animal studies, BM-MSCs secrete a variety of trophic factors which stimulate the neuroregeneration process [113]. Some of the notable trophic factors are VEGF, BDNF, NGF, insulin growth factor-1, and hepatocyte growth factor [113]. Similarly, several growth factors such as VEGF and BDNF were elevated after the transplantation of UBC-MSCs or placenta derived-MSCs [114]. In addition to growth factors, stem cells secrete anti-inflammatory molecules that mitigate neuroinflammation [115]. Stem cells secrete microvesicles and exosomes known to contain growth factors, proteins, anti-inflammatory cytokines such as IL-10 and IL-4 [9, 74, 116, 117], microRNA and lncRNA such as nuclear enriched abundant transcript 1 (NEAT1) and metastasis associated adenocarcinoma transcript 1 (MALAT1) which play key roles in inflammation, gene expression, and cell survival [74]. When transplanted after stroke, not only do stem cells have the capacity to sequester inflammation at the ischemic source, but also throughout the periphery. Intravenous administration of human bone marrow stromal cells (hBMSCs) in rats following stroke resulted in the preferential migration of stem cells to the spleen compared to the brain [9]. Treated animals presented with lower infarct volumes, and reduced cerebral and splenic inflammation [9]. Interestingly, this study reported that a greater number of hBMSCs observed in the spleen correlated to decreased infarct and peri-infarct volume, as well as lower TNF-α density in the spleen [9]. Viewed holistically, these results indicate that peripheral implantation of stem cells may afford neuroprotection indirectly by moderating the overactive and global inflammatory response following stroke by similar anti-inflammatory mechanisms as observed in IC injection.

Mounting evidence has shifted the consensus respecting the primary mechanism of action from cell replacement paradigm toward bystander effect [118]. Indeed, stem cells are now well known for their therapeutic trophic mechanisms that contribute to neuroprotection. However, even combined, both mechanisms do not fully explain the endogenous recovery effect observed after transplantation. While the trophic factors can stimulate endogenous stem cells to proliferate and differentiate, it is unclear how these endogenous stem cells can then migrate to the injured brain regions [119]. Migration is a challenging and complex process, especially in a mature adult brain. Without external support and guidance, inflammatory cytokines are not enough to attract the endogenous stem cells over long distances. To this end, we propose a third mechanism of action for stem cell transplants that our lab has revealed—the formation of a stem cell-paved biobridge—which furthered our understanding of how endogenous stem cells achieve migration from deep neurogenic niches to distal injured regions of the brain.

5 The Biobridge: Exogenous Stem Cells Guide Endogenous NSCs Towards Repair

For the past five decades, the scientific community has been aware of the neurogenic capacity of the adult mammalian brain [120], however, the precise role and regulation of neural stem cells (NSCs) remains an active area of research. Evidence contradicts the original assumption that the primary role of endogenous NSCs is to regenerate damaged tissue after brain injury. Instead, NSCs take part in brain plasticity by both direct and indirect mechanisms which are crucial for certain types of hippocampal and/or olfactory bulb-dependent learning and memory [121]. Unfortunately, NSCs' capacity for tissue regeneration after brain injury is extremely limited despite an increase in activation following such injurious events. Poor cell survival and proliferation, lack of commitment to neuronal lineage, and limited migration are all challenges that prevent these endogenous NSCs from facilitating significant regeneration after brain injury [21].

Much like peripheral inflammatory cells, transplanted stem cells are drawn towards molecular signals from the peri-infarct area. Extracellular matrix (ECM) remodeling allows these cells to move through the brain parenchyma. Interestingly, the process of migrating exogenous stem cells benefits endogenous neural stem cells (NSCs) as well [21]. NSCs are not ubiquitous throughout the brain, but are instead restricted to neurogenic niches in two brain regions—the SVZ of the lateral ventricles and the SGZ of the dentate gyrus of the hippocampus (although quiescent NSCs have been identified in other brain regions) [121]. When brain injury occurs at sites distal to these locations, the potential for robust repair or neuroprotection afforded by endogenous NSCs is diminished due to their limited capacity for migration. In a previous study, our lab discovered that this shortcoming of NSCs may be compensated for by additional mechanisms of transplanted stem cells [21]. In this investigation, a controlled cortical impact was delivered to the frontal cortex of Sprague-Dawley (SD) rats, a common model for TBI. Intracerebral injection of Notch-induced hBMSCs (referred to as SB623, supplied by SanBio Inc.—see Sect. 3.2.1) [8, 122] was performed 7 days post TBI. Locomotor and neurological tests were completed prior to TBI, pre-transplantation, and monthly following transplantation for up to 3 months. As expected, at 1, 2 and 3 months post TBI, treated animals displayed significant improvements in motor and neurological tasks. Histological analysis at both 1 and 3 month time-points also showed reduced lesion size and improved cell survival in the peri-impact area. Notably, the engraftment rate for the transplantation was minimal, at only 0.60% at 1 month post-transplantation and 0.16% at 3 months.

While these findings were similar to other reports of stem cell transplantation after TBI, immunohistochemistry and laser capture revealed a previously unreported phenomenon in which exogenous stem cells form a cellular bridge between the neurogenic SVZ and the lesion within cortex. With the formation of this biobridge came successful endogenous stem cell migration; a pathway was observed alongside the same trajectory of the migrating injected stem cells. The pattern of endogenous stem cell migration was remarkably different between treated and untreated animals. In vehicle injected animals, endogenous cells were sparse throughout peri-impact cortical regions and newly formed neural cells within the SVZ were nearly absent. Additionally, cell proliferation and neural differentiation was stunted in non-treated animals. By contrast, in animals that received the SB623 cells, at 1 month post-transplantation, robust endogenous cell proliferation (Ki67) and immature neural differentiation (nestin) was observed in peri-impact cortical regions and the SVZ, with migrating cells (DCX) along the corpus callosum. Immunohistochemistry revealed hBMSCs localized within the impacted region, down into the cortex, across the corpus callosum and along the ventricles to the location of neurogenic niche. At 3 months post transplantation, $DCX^+/HuNu^+$ (human nuclei) cells were identified alongside the hBMSCs transplanted cells indicating that non-transplanted cells were able to navigate through the ECM that was likely recently remodeled by the migrating hBMSCs. It is important to note that the transplanted stem cells survival was largely diminished by 3 months, suggesting that even though these cells did not persist, endogenous cells were still able to utilize the same route through the ECM where they could continue to migrate through and thrive, sustaining endogenous recovery efforts despite the absence of transplanted stem cells.

To better understand the mechanism of biobridge formation, we explored metallomatrix protein (MMP) expression, specifically, MMP-9. Molecular analysis via laser-capture revealed increased MMP-9 expression along the migratory pathway [21]. Notably, TBI-vehicle animals also displayed an increased MMP-9 expression following stroke, however, this upregulation reverted to levels comparable to sham animals at 3-months post-injury. In SB623 transplant animals, MMP-9 expression doubled that of TBI-vehicle animals at 1 month post-TBI and expression increased ninefold by month 3. This data suggest the importance of this neurovascular proteinase in the long-term neural regenerative efforts of transplanted stem cells. While these results indicate that endogenous cells alone increase MMP-9 expression after brain injury, stem cell transplantation promotes a more robust mechanism for ECM remodeling than unaided endogenous stem cells by leaving a direct pathway for the endogenous stem cells to utilize.

Complementary to these *in vivo* results, an *in vitro* study presented SB623-promoted cell migration via an ECM-mediated mechanism [21]. Primary rat cortical cells were grown by themselves or co-cultured with SB623 cells in two different conditions—with or without Cyclosporin-A, an MMP-9 inhibitor. Co-culture of SB623 cells without the presence of MMP-9 inhibitor significantly enhanced the migration of primary cortical rat cells. The migration of primary cortical rat cells into the chamber containing the SB623 cells was significantly reduced when treated with Cyclosporin-A, with no significant difference compared to the cultures without stem cells. This study further supports that stem cells, particularly SB623 cells, promote cell migration mediated largely via MMP-induced ECM remodeling.

Moreover, it is believed that migratory trophic factors released by the exogenous stem cells such as cysteine-x-cysteine motif chemokine ligand 14 (CXCL14) and monocyte chemoattractant protein 1 (MCP1) further promote endogenous stem cell migration from the neurogenic niche. It is important to note that the transplanted MSC's long-term survival was not necessary for functional improvements in this study. Instead, the therapeutic benefit was attributed to their ability to manipulate the microenvironment and stimulate endogenous stem cell migration, proliferation, and differentiation. These findings positively address some of the tumorigenic concerns mentioned in previous sections, as eventual death of transplanted stem cells and loss of stemness characteristics are increasingly regarded as important in preventing tumorigenesis.

To further investigate this novel stem cell mechanism of action, we designed a pilot study to investigate if a similar biobridge formation occurs after stem cell transplantation in the MCAO stroke model. We would like to share our promising ongoing study. Normal male SD rats (n = 10, average weight = 200 g) were subjected to MCAO surgery. Three days post stroke, the animals were split into two groups that received a one-time transplantation of human BM-MSCs by either IC (n = 5) with 1.0×10^6 cells or IV (n = 5) with 4.0×10^6 cells. The animals were sacrificed and processed for immunohistological staining at day 7 post-stroke. Similar to previous reports, we observed MSCs in the cortex (Fig. 8.2) and striatum in both IC and IV groups (Fig. 8.3), showing that the MSCs can infiltrate the brain either through IC or IV transplantation. Interestingly, the transplanted MSCs from the IC injection group mainly traveled along the corpus callosum, while the MSCs from the IV group disperse throughout the striatum and cortex. DCX^+ stain-

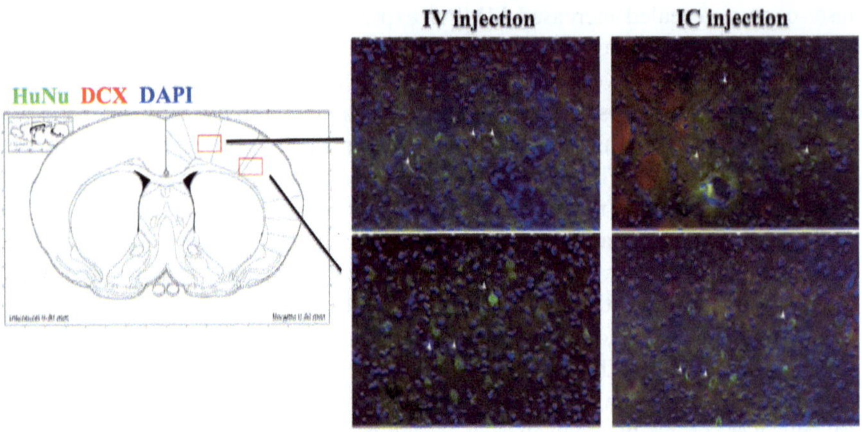

Fig. 8.2 Distribution of human BM-MSCs (HuNu⁺) in the cortex. After transplantation, BM-MSCs (HuNu⁺) succesfully infiltrated the ischemic brains in both IC and IV route of administration. HuNu⁺ cells were detected, however DCX⁺ cells were not found in the cortex

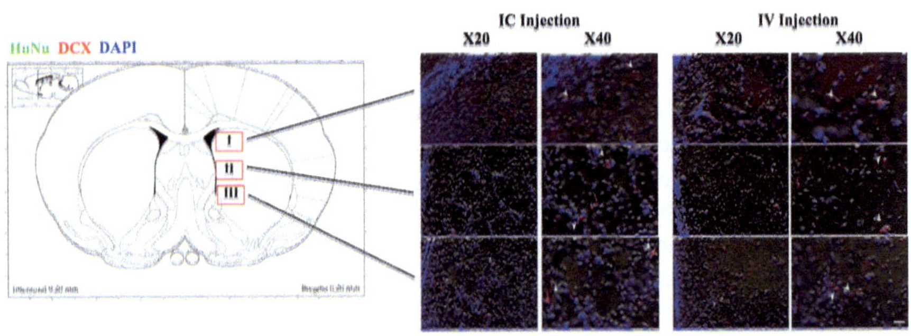

Fig. 8.3 Distribution of human BM-MSCs (HuNu⁺) and immature neurons (DCX⁺) in the striatum. HuNu⁺ cells from the IC group traveled along the corpus callosum, whereas HuNu⁺ cells from the IV group dispersed throughout. Transplanted human BM-MSCs were found around DCX⁺ cells, paving the way for these immature neurons to migrate toward the penumbra

ing, a marker for cell migration, revealed that DCX⁺ cells from the IV group traveled along the corpus callosum and the ventricle wall (Fig. 8.3). The DCX⁺ cells found in the IC group traveled more laterally. However, it is worth noting that the DCX⁺ cells from the IV group traveled further into the striatum compared to the IC group. The migration pattern of the immature neurons is summarized in Fig. 8.4. To further validate our findings, we performed another set of staining for proliferating neuronal cells (Ki67⁺/Nestin⁺). Similar to the DCX staining results, Ki67⁺/Nestin⁺ cells were found along the corpus callosum and ventricle wall. In addition, fewer Ki67⁺/Nestin⁺ cells were found further into the striatum (Fig. 8.5). In conclusion, we have demonstrated in this pilot study that a similar phenomenon

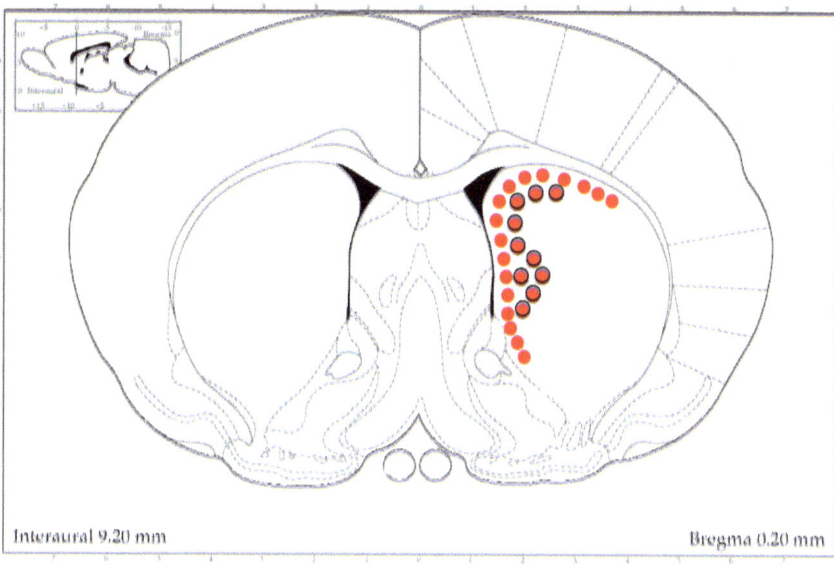

Fig. 8.4 Schematic of the distribution of DCX⁺ cells between the IV and IC group. DCX⁺ cells from the IV group travel along the corpus callosum and the ventricle wall. DCX⁺ cells from the IC group travel more horizontally compared to IV group. However, more DCX⁺ cells from IV group travel further into the striatum compared to IC group

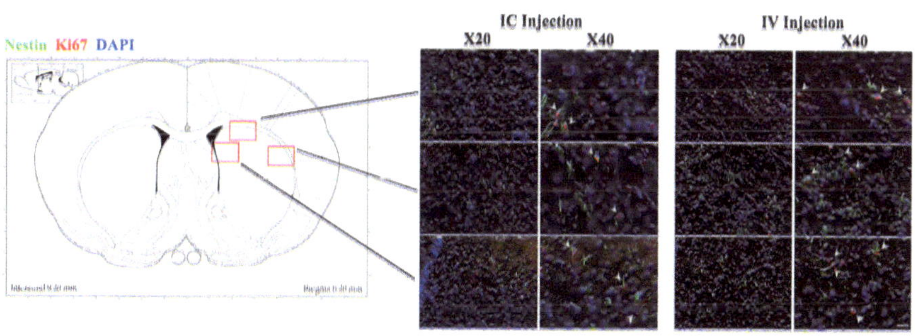

Fig. 8.5 Distribution of proliferating neuronal cells (Nestin⁺Ki67⁺) in the striatum. Nestin⁺Ki67⁺ cells have similar migration pattern compared to the DCX⁺ immature neurons. In both IC and IV groups, Nestin⁺Ki67⁺ cells were found along the wall of the ventricle and the corpus callosum

reported in TBI, whereby transplanted BM-MSCs modify the environment to facilitate the migration of proliferating and immature neurons, also occurs in stroke. Interestingly, with the help of transplanted stem cells, both DCX⁺ and Nestin⁺Ki67⁺ cells utilized the corpus callosum as a highway to travel further into the penumbra.

6 Future Directions for Advancing the Biobridge Concept

Although the biobridge concept has now been demonstrated in TBI and preliminarily in stroke, barriers still exist to translating these findings into being clinically relevant. A more complete understanding of the cellular and molecular processes which define the biobridge formation and how the assisted migration of endogenous stem cells can be optimized by exogenous transplantation must be unveiled before patients can benefit from these findings. Future studies should aim to more fully characterize the underlying molecular changes that produce the biobridge. Our group has revealed the role of MMP-9 in extracellular matrix remodeling in vitro, yet this single protein is unlikely to account for the totality of the extensive remodeling seen within the biobridge region. Conditional MMP-9 knock-out animals could be valuable in further illustrating the role this protein has in vivo [123–125]. Moreover, transplanted stem cells modified to overexpress MMP-9 and other remodeling factors may reveal a target for heightening the graft-host cell interactions, providing an avenue by which this new mechanism could be utilized to improve clinical outcomes. Importantly, data on the global effects of MMP-9 after stroke are inconclusive, and thus exploring the biobridge formation in MMP-9 knockout mice could help characterize the complex roles which MMP-9 has after brain injury, perhaps playing protective and detrimental roles in different capacities.

Future research efforts should investigate the molecular interactions and cross-talk of the transplant and host stem cells. Here, we describe the remodeling processes observed in brain regions where host stem cells overlap with transplant stem cells. Importantly, transplanted MSCs have been shown to secrete factors which not only promote the survival of host neurons, but are also likely to promote survival of the endogenous stem cells which they come into close contact with. The vast pro-survival secretion profile of transplanted hMSCs, such as wnt3a, VEGF, and BDNF, among others [110], could mean that endogenous stem cells are both guided, and nurtured, by transplanted cells, thereby heightening their regenerative capacity upon arrival to the peri-injured regions. Additionally, factors such as wtn3a and VEGF have been shown to inhibit the quiescent state of host stem cells, wherein their migratory and regenerative properties are stagnated [121]. Beyond the ECM remodeling discussed extensively above, exploring how transplanted MSCs enhance the therapeutic capabilities of host stem cells through cell-to-cell interactions will further enhance our understanding of the robust benefits offered by stem cell transplantation.

The chronological characteristics of the biobridge also deserve additional evaluation—both with regards to its structure and composition over time, as well as how its development varies with different transplant time points. To date, our group has only investigated the progress of the biobridge formation out to 3 days in stroke, making it imperative for additional studies which investigate the biobridge structure and formation through the sub-acute and chronic phases. Understanding how ongoing molecular changes encourage the migration of endogenous stem cells could provide indications for the effects which acute biobridge formation, and sub-acute progression, have in ameliorating chronic deficits. Discrepancies in the ideal time point for stem cell transplantation post-injury already exist, so careful consideration must be given in determining the transplant time which not only augments the biobridge formation but also gives equal consideration to the various other therapeutic mechanisms occurring concurrently.

Finally, the prevalence of the biobridge concept in other neurological disorders should also be explored. That this process has been demonstrated in two different disease models indicates that this graft-host cell interaction is a more general mechanism of stem cell therapies, and not specific to the pathologies of a single disease. Indeed, this therapeutic mechanism may have far-reaching implications in other neurological diseases amenable to cell transplantation, although the intricacies of its formation may vary greatly between diseases with and without focal damage. This was partly demonstrated in our TBI versus stroke comparison, with TBI brains showing a more unidirectional biobridge and stroke brains displaying a three-dimensional, multi-directional biobridge. How this biobridge concept manifests in neurological disease without focal lesions—i.e. amyotrophic lateral sclerosis, multiple sclerosis, Alzheimer's disease, transient global ischemic events, and neonatal hypoxic—will need to be further explored.

7 Conclusion

Tailoring the use of stem cell therapies in stroke, TBI, and other neurological disorders is an ongoing scientific effort. The unique pathology of neurodegenerative diseases poses a challenge seemingly too large for traditional pharmaceuticals to compensate for, and thus alternative therapeutic options—namely stem cells and regenerative medicine—have received increased attention. Stroke, in particular, has received significant attention as a possible beneficiary of stem cell transplantations. The various pathological processes which accompany stroke appear highly compatible with the dynamic therapeutic profile of stem cells. Transplanted stem cells' ability to secrete anti-inflammatory factors, pro-survival/anti-apoptotic molecules, and to integrate into the host parenchyma contribute to the benefits which they confer. The therapeutic capacity of stem cells in stroke has been demonstrated repeatedly in pre-clinical investigations, yet translating this promise into widely-available clinical treatment options has been slow. This is in no small part to the inherent complications which accompany non-traditional pharmaceuticals, including issues of dose, timing, route of administration, and stem cell source.

The shortcomings of clinical trials of cell transplantation have resulted in a renewed effort to explore the basic science mechanisms of stem cell therapies. The path to successful clinical trials will likely be paved by basic science discoveries concerning the complex therapeutic mechanisms of stem cells. Here, we described a novel therapeutic mechanism of stem cells, the biobridge, which works in conjunction with the established mechanisms to produce the functional improvements observed following stroke, as well as TBI. The discovery of this mechanism has both basic science, as well as translational, Implications; that exogenous stem cells interact with and encourage the movement of endogenous host stem cells to regions of damage aids in explaining the seemingly paradoxically-robust functional recovery seen in stem cell transplantations despite minimal graft survival rates. Moreover, understanding the extracellular matrix remodeling capacity of transplanted stem cells provides a novel bioengineering target for genetically enhancing stem cells. These findings, in the context of the larger scientific effort to better understand the details of stem cell therapeutic modalities, assist in providing the preclinical basis for more effective clinical trials, bringing stem cell therapies closer to positively impacting stroke patient recovery.

References

1. Mozaffarian D, Benjamin EJ, Go AS, Arnett DK, Blaha MJ, Cushman M, et al. Heart disease and stroke statistics—2015 update: a report from the American Heart Association. Circulation. 2015;131(4):e29–322.
2. Writing Group Members, Mozaffarian D, Benjamin EJ, Go AS, Arnett DK, Blaha MJ, et al. Heart disease and stroke statistics—2016 update: a report from the American Heart Association. Circulation. 2016;133(4):e38–60.
3. Goldstein LB. Modern medical management of acute ischemic stroke. Methodist Debakey Cardiovasc J. 2014;10(2):99–104.
4. Gumbinger C, Reuter B, Stock C, Sauer T, Wietholter H, Bruder I, et al. Time to treatment with recombinant tissue plasminogen activator and outcome of stroke in clinical practice: retrospective analysis of hospital quality assurance data with comparison with results from randomised clinical trials. BMJ. 2014;348:g3429.
5. Kanazawa M, Takahashi T, Nishizawa M, Shimohata T. Therapeutic strategies to attenuate hemorrhagic transformation after tissue plasminogen activator treatment for acute ischemic stroke. J Atheroscler Thromb. 2017;24(3):240–53.
6. van Velthoven CT, Dzietko M, Wendland MF, Derugin N, Faustino J, Heijnen CJ, et al. Mesenchymal stem cells attenuate MRI-identifiable injury, protect white matter, and improve long-term functional outcomes after neonatal focal stroke in rats. J Neurosci Res. 2017;95(5):1225–36.
7. Zhang H, Sun F, Wang J, Xie L, Yang C, Pan M, et al. Combining injectable plasma scaffold with mesenchymal stem/stromal cells for repairing infarct cavity after ischemic stroke. Aging Dis. 2017;8(2):203–14.
8. Yasuhara T, Matsukawa N, Hara K, Maki M, Ali MM, Yu SJ, et al. Notch-induced rat and human bone marrow stromal cell grafts reduce ischemic cell loss and ameliorate behavioral deficits in chronic stroke animals. Stem Cells Dev. 2009;18(10):1501–14.
9. Acosta SA, Tajiri N, Hoover J, Kaneko Y, Borlongan CV. Intravenous bone marrow stem cell grafts preferentially migrate to spleen and abrogate chronic inflammation in stroke. Stroke. 2015;46(9):2616–27.
10. Tajiri N, Acosta S, Portillo-Gonzales GS, Aguirre D, Reyes S, Lozano D, et al. Therapeutic outcomes of transplantation of amniotic fluid-derived stem cells in experimental ischemic stroke. Front Cell Neurosci. 2014;8:227.
11. Xing C, Arai K, Lo EH, Hommel M. Pathophysiologic cascades in ischemic stroke. Int J Stroke. 2012;7(5):378–85.

12. Markus HS, Bevan S. Mechanisms and treatment of ischaemic stroke—insights from genetic associations. Nat Rev Neurol. 2014;10(12):723–30.
13. Lambertsen KL, Biber K, Finsen B. Inflammatory cytokines in experimental and human stroke. J Cereb Blood Flow Metab. 2012;32(9):1677–98.
14. Tajiri N, Lau T, Glover LE, Shinozuka K, Kaneko Y, van Loveren H, et al. Cerebral aneurysm as an exacerbating factor in stroke pathology and a therapeutic target for neuroprotection. Curr Pharm Des. 2012;18(25):3663–9.
15. Shinozuka K, Dailey T, Tajiri N, Ishikawa H, Kim DW, Pabon M, et al. Stem cells for neurovascular repair in stroke. J Stem Cell Res Ther. 2013;4(4):12912.
16. Hess DC, Borlongan CV. Cell-based therapy in ischemic stroke. Expert Rev Neurother. 2008;8(8):1193–201.
17. Savitz SI, Chopp M, Deans R, Carmichael T, Phinney D, Wechsler L, et al. Stem cell therapy as an emerging paradigm for stroke (STEPS) II. Stroke. 2011;42(3):825–9.
18. Liska MG, Crowley MG, Borlongan CV. Regulated and unregulated clinical trials of stem cell therapies for stroke. Transl Stroke Res. 2017;8(2):93–103.
19. Drago D, Cossetti C, Iraci N, Gaude E, Musco G, Bachi A, et al. The stem cell secretome and its role in brain repair. Biochimie. 2013;95(12):2271–85.
20. Haas S, Weidner N, Winkler J. Adult stem cell therapy in stroke. Curr Opin Neurol. 2005;18(1):59–64.
21. Tajiri N, Kaneko Y, Shinozuka K, Ishikawa H, Yankee E, McGrogan M, et al. Stem cell recruitment of newly formed host cells via a successful seduction? Filling the gap between neurogenic niche and injured brain site. PLoS One. 2013;8(9):e74857.
22. Broughton BR, Reutens DC, Sobey CG. Apoptotic mechanisms after cerebral ischemia. Stroke. 2009;40(5):e331–9.
23. Moskowitz MA, Lo EH, Iadecola C. The science of stroke: mechanisms in search of treatments. Neuron. 2010;67(2):181–98.
24. Mergenthaler P, Dirnagl U, Meisel A. Pathophysiology of stroke: lessons from animal models. Metab Brain Dis. 2004;19(3–4):151–67.
25. Xiong XY, Liu L, Yang QW. Functions and mechanisms of microglia/macrophages in neuroinflammation and neurogenesis after stroke. Prog Neurobiol. 2016;142:23–44.
26. Lozano D, Gonzales-Portillo GS, Acosta S, de la Pena I, Tajiri N, Kaneko Y, et al. Neuroinflammatory responses to traumatic brain injury: etiology, clinical consequences, and therapeutic opportunities. Neuropsychiatr Dis Treat. 2015;11:97–106.
27. Yang Y, Rosenberg GA. Blood-brain barrier breakdown in acute and chronic cerebrovascular disease. Stroke. 2011;42(11):3323–8.
28. Doll DN, Hu H, Sun J, Lewis SE, Simpkins JW, Ren X. Mitochondrial crisis in cerebrovascular endothelial cells opens the blood-brain barrier. Stroke. 2015;46(6):1681–9.
29. Libby P. Inflammation in atherosclerosis. Arterioscler Thromb Vasc Biol. 2012;32(9):2045–51.
30. Nguyen H, Aum D, Mashkouri S, Rao G, Vega Gonzales-Portillo JD, Reyes S, et al. Growth factor therapy sequesters inflammation in affording neuroprotection in cerebrovascular diseases. Expert Rev Neurother. 2016;16(8):915–26.
31. Chen GY, Nunez G. Sterile inflammation: sensing and reacting to damage. Nat Rev Immunol. 2010;10(12):826–37.
32. Maki T, Hayakawa K, Pham LD, Xing C, Lo EH, Arai K. Biphasic mechanisms of neurovascular unit injury and protection in CNS diseases. CNS Neurol Disord Drug Targets. 2013;12(3):302–15.
33. Iadecola C, Anrather J. The immunology of stroke: from mechanisms to translation. Nat Med. 2011;17(7):796–808.
34. Taylor RA, Sansing LH. Microglial responses after ischemic stroke and intracerebral hemorrhage. Clin Dev Immunol. 2013;2013:746068.
35. Morganti JM, Riparip LK, Rosi S. Call off the dog(ma): M1/M2 polarization is concurrent following traumatic brain injury. PLoS One. 2016;11(1):e0148001.

36. Morganti JM, Riparip LK, Chou A, Liu S, Gupta N, Rosi S. Age exacerbates the CCR2/5-mediated neuroinflammatory response to traumatic brain injury. J Neuroinflammation. 2016;13(1):80.
37. Morganti JM, Jopson TD, Liu S, Riparip LK, Guandique CK, Gupta N, et al. CCR2 antagonism alters brain macrophage polarization and ameliorates cognitive dysfunction induced by traumatic brain injury. J Neurosci. 2015;35(2):748–60.
38. Moore KJ, Sheedy FJ, Fisher EA. Macrophages in atherosclerosis: a dynamic balance. Nat Rev Immunol. 2013;13(10):709–21.
39. Seifert HA, Leonardo CC, Hall AA, Rowe DD, Collier LA, Benkovic SA, et al. The spleen contributes to stroke induced neurodegeneration through interferon gamma signaling. Metab Brain Dis. 2012;27(2):131–41.
40. Ajmo CT Jr, Vernon DO, Collier L, Hall AA, Garbuzova-Davis S, Willing A, et al. The spleen contributes to stroke-induced neurodegeneration. J Neurosci Res. 2008;86(10):2227–34.
41. Sahota P, Vahidy F, Nguyen C, Bui TT, Yang B, Parsha K, et al. Changes in spleen size in patients with acute ischemic stroke: a pilot observational study. Int J Stroke. 2013;8(2):60–7.
42. Chiu NL, Kaiser B, Nguyen YV, Welbourne S, Lall C, Cramer SC. The volume of the spleen and its correlates after acute stroke. J Stroke Cerebrovasc Dis. 2016;25(12):2958–61.
43. Seifert HA, Hall AA, Chapman CB, Collier LA, Willing AE, Pennypacker KR. A transient decrease in spleen size following stroke corresponds to splenocyte release into systemic circulation. J Neuroimmune Pharmacol. 2012;7(4):1017–24.
44. Zhang BJ, Men XJ, Lu ZQ, Li HY, Qiu W, Hu XQ. Splenectomy protects experimental rats from cerebral damage after stroke due to anti-inflammatory effects. Chin Med J. 2013;126(12):2354–60.
45. Winek K, Engel O, Koduah P, Heimesaat MM, Fischer A, Bereswill S, et al. Depletion of cultivatable gut microbiota by broad-spectrum antibiotic pretreatment worsens outcome after murine stroke. Stroke. 2016;47(5):1354–63.
46. Gonzales-Portillo C, Ishikawa H, Shinozuka K, Tajiri N, Kaneko Y, Borlongan CV. Stroke and cardiac cell death: two peas in a pod. Clin Neurol Neurosurg. 2016;142:145–7.
47. Acosta SA, Mashkouri S, Nwokoye D, Lee JY, Borlongan CV. Chronic inflammation and apoptosis propagate in ischemic cerebellum and heart of non-human primates. Oncotarget. 2017. 10.18632/oncotarget.18330.
48. Potten CS, Loeffler M. Stem cells: attributes, cycles, spirals, pitfalls and uncertainties. Lessons for and from the crypt. Development. 1990;110(4):1001–20.
49. Alison MR, Islam S. Attributes of adult stem cells. J Pathol. 2009;217(2):144–60.
50. Cai J, Weiss ML, Rao MS. In search of "stemness". Exp Hematol. 2004;32(7):585–98.
51. de Kretser D. Totipotent, pluripotent or unipotent stem cells: a complex regulatory enigma and fascinating biology. J Law Med. 2007;15(2):212–8.
52. Mafi R, Hindocha S, Mafi P, Griffin M, Khan WS. Sources of adult mesenchymal stem cells applicable for musculoskeletal applications - a systematic review of the literature. Open Orthop J. 2011;5(Suppl 2):242–8.
53. Li G, Yu F, Lei T, Gao H, Li P, Sun Y, et al. Bone marrow mesenchymal stem cell therapy in ischemic stroke: mechanisms of action and treatment optimization strategies. Neural Regen Res. 2016;11(6):1015–24.
54. Liu XY, Wang CP, Liu M, Ji G, Guo JC. Transplantation of human embryonic neural stem cells protects rats against cerebral ischemic injury. Sheng Li Xue Bao. 2014;66(6):691–701.
55. Ishii T, Eto K. Fetal stem cell transplantation: past, present, and future. World J Stem Cells. 2014;6(4):404–20.
56. Maya-Espinosa G, Collazo-Navarrete O, Millan-Aldaco D, Palomero-Rivero M, Guerrero-Flores G, Drucker-Colin R, et al. Mouse embryonic stem cell-derived cells reveal niches that support neuronal differentiation in the adult rat brain. Stem Cells. 2015;33(2):491–502.
57. O'Donoghue K, Chan J. Human fetal mesenchymal stem cells. Curr Stem Cell Res Ther. 2006;1(3):371–86.

58. O'Donoghue K, Fisk NM. Fetal stem cells. Best Pract Res Clin Obstet Gynaecol. 2004;18(6):853–75.
59. Hartley RS, Trojanowski JQ, Lee VM. Differential effects of spinal cord gray and white matter on process outgrowth from grafted human NTERA2 neurons (NT2N, hNT). J Comp Neurol. 1999;415(3):404–18.
60. Hara K, Yasuhara T, Maki M, Matsukawa N, Masuda T, Yu SJ, et al. Neural progenitor NT2N cell lines from teratocarcinoma for transplantation therapy in stroke. Prog Neurobiol. 2008;85(3):318–34.
61. Newman MB, Misiuta I, Willing AE, Zigova T, Karl RC, Borlongan CV, et al. Tumorigenicity issues of embryonic carcinoma-derived stem cells: relevance to surgical trials using NT2 and hNT neural cells. Stem Cells Dev. 2005;14(1):29–43.
62. Kondziolka D, Wechsler L, Goldstein S, Meltzer C, Thulborn KR, Gebel J, et al. Transplantation of cultured human neuronal cells for patients with stroke. Neurology. 2000;55(4):565–9.
63. Nelson PT, Kondziolka D, Wechsler L, Goldstein S, Gebel J, DeCesare S, et al. Clonal human (hNT) neuron grafts for stroke therapy: neuropathology in a patient 27 months after implantation. Am J Pathol. 2002;160(4):1201–6.
64. Phillips MI, Tang YL. Genetic modification of stem cells for transplantation. Adv Drug Deliv Rev. 2008;60(2):160–72.
65. Sinden JD, Hicks C, Stroemer P, Vishnubhatla I, Corteling R. Human neural stem cell therapy for chronic ischemic stroke: charting progress from laboratory to patients. Stem Cells Dev. 2017;26(13):933–47.
66. Pollock K, Stroemer P, Patel S, Stevanato L, Hope A, Miljan E, et al. A conditionally immortal clonal stem cell line from human cortical neuroepithelium for the treatment of ischemic stroke. Exp Neurol. 2006;199(1):143–55.
67. Kalladka D, Sinden J, Pollock K, Haig C, McLean J, Smith W, et al. Human neural stem cells in patients with chronic ischaemic stroke (PISCES): a phase 1, first-in-man study. Lancet (London, England). 2016;388(10046):787–96.
68. Napoli E, Borlongan CV. Stem cell recipes of bone marrow and fish: just what the stroke doctors ordered. Stem Cell Rev. 2017;13(2):192–7.
69. Sibbald B. Death but one unintended consequence of gene-therapy trial. CMAJ. 2001;164(11):1612.
70. Borlongan CV. Preliminary reports of stereotaxic stem cell transplants in chronic stroke patients. Mol Ther. 2016;24(10):1710–1.
71. Wobus AM, Boheler KR. Embryonic stem cells: prospects for developmental biology and cell therapy. Physiol Rev. 2005;85(2):635–78.
72. Buhnemann C, Scholz A, Bernreuther C, Malik CY, Braun H, Schachner M, et al. Neuronal differentiation of transplanted embryonic stem cell-derived precursors in stroke lesions of adult rats. Brain. 2006;129(Pt 12):3238–48.
73. Lee JY, Kim E, Choi SM, Kim DW, Kim KP, Lee I, et al. Microvesicles from brain-extract-treated mesenchymal stem cells improve neurological functions in a rat model of ischemic stroke. Sci Rep. 2016;6:33038.
74. Tajiri N, Acosta SA, Shahaduzzaman M, Ishikawa H, Shinozuka K, Pabon M, et al. Intravenous transplants of human adipose-derived stem cell protect the brain from traumatic brain injury-induced neurodegeneration and motor and cognitive impairments: cell graft biodistribution and soluble factors in young and aged rats. J Neurosci. 2014;34(1):313–26.
75. Garbuzova-Davis S, Kurien C, Thomson A, Falco D, Ahmad S, Staffetti J, et al. Endothelial and astrocytic support by human bone marrow stem cell grafts into symptomatic ALS mice towards blood-spinal cord barrier repair. Sci Rep. 2017;7(1):884.
76. Singh V, Roth S, Llovera G, Sadler R, Garzetti D, Stecher B, et al. Microbiota dysbiosis controls the neuroinflammatory response after stroke. J Neurosci. 2016;36(28):7428–40.
77. Tolar J, Nauta AJ, Osborn MJ, Panoskaltsis Mortari A, McElmurry RT, Bell S, et al. Sarcoma derived from cultured mesenchymal stem cells. Stem Cells. 2007;25(2):371–9.

78. McAndrews KM, McGrail DJ, Ravikumar N, Dawson MR. Mesenchymal stem cells induce directional migration of invasive breast cancer cells through TGF-beta. Sci Rep. 2015;5:16941.
79. Subramanian A, Shu-Uin G, Kae-Siang N, Gauthaman K, Biswas A, Choolani M, et al. Human umbilical cord Wharton's jelly mesenchymal stem cells do not transform to tumor-associated fibroblasts in the presence of breast and ovarian cancer cells unlike bone marrow mesenchymal stem cells. J Cell Biochem. 2012;113(6):1886–95.
80. Steinberg GK, Kondziolka D, Wechsler LR, Lunsford LD, Coburn ML, Billigen JB, et al. Clinical outcomes of transplanted modified bone marrow-derived mesenchymal stem cells in stroke: a phase 1/2a study. Stroke. 2016;47(7):1817–24.
81. Hsieh J-Y, Wang H-W, Chang S-J, Liao K-H, Lee I-H, Lin W-S, et al. Mesenchymal stem cells from human umbilical cord express preferentially secreted factors related to neuroprotection, neurogenesis, and angiogenesis. PLoS One. 2013;8(8):e72604.
82. Liu Z, Li Y, Zhang RL, Cui Y, Chopp M. Bone marrow stromal cells promote skilled motor recovery and enhance contralesional axonal connections after ischemic stroke in adult mice. Stroke. 2011;42(3):740–4.
83. Tajiri N, Acosta S, Glover LE, Bickford PC, Jacotte Simancas A, Yasuhara T, et al. Intravenous grafts of amniotic fluid-derived stem cells induce endogenous cell proliferation and attenuate behavioral deficits in ischemic stroke rats. PLoS One. 2012;7(8):e43779.
84. Simerman AA, Phan JD, Dumesic DA, Chazenbalk GD. Muse cells: nontumorigenic pluripotent stem cells present in adult tissues-a paradigm shift in tissue regeneration and evolution. Stem Cells Int. 2016;2016:1463258.
85. Ratajczak MZ, Zuba-Surma EK, Wysoczynski M, Ratajczak J, Kucia M. Very small embryonic-like stem cells: characterization, developmental origin, and biological significance. Exp Hematol. 2008;36(6):742–51.
86. Zuba-Surma EK, Kucia M, Ratajczak J, Ratajczak MZ. "Small stem cells" in adult tissues: very small embryonic-like stem cells stand up! Cytometry A. 2009;75(1):4–13.
87. Park DH, Borlongan CV, Willing AE, Eve DJ, Cruz LE, Sanberg CD, et al. Human umbilical cord blood cell grafts for brain ischemia. Cell Transplant. 2009;18(9):985–98.
88. Elias M, Hoover J, Nguyen H, Reyes S, Lawton C, Borlongan CV. Stroke therapy: the potential of amniotic fluid-derived stem cells. Future Neurol. 2015;10(4):321–6.
89. Dailey T, Metcalf C, Mosley YI, Sullivan R, Shinozuka K, Tajiri N, et al. An update on translating stem cell therapy for stroke from bench to bedside. J Clin Med. 2013;2(4):220–41.
90. Ou Y, Yu S, Kaneko Y, Tajiri N, Bae EC, Chheda SH, et al. Intravenous infusion of GDNF gene-modified human umbilical cord blood CD34+ cells protects against cerebral ischemic injury in spontaneously hypertensive rats. Brain Res. 2010;1366:217–25.
91. Chen J, Shehadah A, Pal A, Zacharek A, Cui X, Cui Y, et al. Neuroprotective effect of human placenta-derived cell treatment of stroke in rats. Cell Transplant. 2013;22(5):871–9.
92. Iskander A, Knight RA, Zhang ZG, Ewing JR, Shankar A, Varma NR, et al. Intravenous administration of human umbilical cord blood-derived AC133+ endothelial progenitor cells in rat stroke model reduces infarct volume: magnetic resonance imaging and histological findings. Stem Cells Transl Med. 2013;2(9):703–14.
93. Jin W, Xu YP, Yang AH, Xing YQ. In vitro induction and differentiation of umbilical cord mesenchymal stem cells into neuron-like cells by all-trans retinoic acid. Int J Ophthalmol. 2015;8(2):250–6.
94. Dalous J, Larghero J, Baud O. Transplantation of umbilical cord-derived mesenchymal stem cells as a novel strategy to protect the central nervous system: technical aspects, preclinical studies, and clinical perspectives. Pediatr Res. 2012;71(4 Pt 2):482–90.
95. Rennie K, Haukenfrers J, Ribecco-Lutkiewicz M, Ly D, Jezierski A, Smith B, et al. Therapeutic potential of amniotic fluid-derived cells for treating the injured nervous system. Biochem Cell Biol. 2013;91(5):271–86.
96. Jiang Q, Ding S, Wu J, Liu X, Wu Z. Norepinephrine stimulates mobilization of endothelial progenitor cells after limb ischemia. PLoS One. 2014;9(7):e101774.

97. Lee IH, Huang SS, Chuang CY, Liao KH, Chang LH, Chuang CC, et al. Delayed epidural transplantation of human induced pluripotent stem cell-derived neural progenitors enhances functional recovery after stroke. Sci Rep. 2017;7(1):1943.
98. Yuan T, Liao W, Feng NH, Lou YL, Niu X, Zhang AJ, et al. Human induced pluripotent stem cell-derived neural stem cells survive, migrate, differentiate, and improve neurologic function in a rat model of middle cerebral artery occlusion. Stem Cell Res Ther. 2013;4(3):73.
99. Eckert A, Huang L, Gonzalez R, Kim HS, Hamblin MH, Lee JP. Bystander effect fuels human induced pluripotent stem cell-derived neural stem cells to quickly attenuate early stage neurological deficits after stroke. Stem Cells Transl Med. 2015;4(7):841–51.
100. Pabon MM, Acosta S, Guedes VA, Tajiri N, Kaneko Y, Borlongan CV. Brain region-specific histopathological effects of varying trajectories of controlled cortical impact injury model of traumatic brain injury. CNS Neurosci Ther. 2016;22(3):200–11.
101. Borlongan CV. Age of PISCES: stem-cell clinical trials in stroke. Lancet (London, England). 2016;388(10046):736–8.
102. Tang YH, Ma YY, Zhang ZJ, Wang YT, Yang GY. Opportunities and challenges: stem cell-based therapy for the treatment of ischemic stroke. CNS Neurosci Ther. 2015;21(4):337–47.
103. Tate CC, Fonck C, McGrogan M, Case CC. Human mesenchymal stromal cells and their derivative, SB623 cells, rescue neural cells via trophic support following in vitro ischemia. Cell Transplant. 2010;19(8):973–84.
104. Woodbury D, Schwarz EJ, Prockop DJ, Black IB. Adult rat and human bone marrow stromal cells differentiate into neurons. J Neurosci Res. 2000;61(4):364–70.
105. Yasuhara T, Matsukawa N, Hara K, Yu G, Xu L, Maki M, et al. Transplantation of human neural stem cells exerts neuroprotection in a rat model of Parkinson's disease. J Neurosci. 2006;26(48):12497–511.
106. Freed CR. Will embryonic stem cells be a useful source of dopamine neurons for transplant into patients with Parkinson's disease? Proc Natl Acad Sci U S A. 2002;99(4):1755–7.
107. English K, Wood KJ. Mesenchymal stromal cells in transplantation rejection and tolerance. Cold Spring Harb Perspect Med. 2013;3(5):a015560.
108. Quinn C, Flake AW. In vivo differentiation potential of mesenchymal stem cells: prenatal and postnatal model systems. Transfus Med Hemother. 2008;35(3):239–47.
109. Bae JS, Han HS, Youn DH, Carter JE, Modo M, Schuchman EH, et al. Bone marrow-derived mesenchymal stem cells promote neuronal networks with functional synaptic transmission after transplantation into mice with neurodegeneration. Stem Cells. 2007;25(5):1307–16.
110. Madrigal M, Rao KS, Riordan NH. A review of therapeutic effects of mesenchymal stem cell secretions and induction of secretory modification by different culture methods. J Transl Med. 2014;12:260.
111. Horie N, Pereira MP, Niizuma K, Sun G, Keren-Gill H, Encarnacion A, et al. Transplanted stem cell-secreted vascular endothelial growth factor effects poststroke recovery, inflammation, and vascular repair. Stem Cells. 2011;29(2):274–85.
112. Uzun G, Subhani D, Amor S. Trophic factors and stem cells for promoting recovery in stroke. J Vasc Interv Neurol. 2010;3(1):3–12.
113. Kim DH, Yoo KH, Choi KS, Choi J, Choi SY, Yang SE, et al. Gene expression profile of cytokine and growth factor during differentiation of bone marrow-derived mesenchymal stem cell. Cytokine. 2005;31(2):119–26.
114. Borlongan CV, Hadman M, Sanberg CD, Sanberg PR. Central nervous system entry of peripherally injected umbilical cord blood cells is not required for neuroprotection in stroke. Stroke. 2004;35(10):2385–9.
115. Zhang R, Liu Y, Yan K, Chen L, Chen XR, Li P, et al. Anti-inflammatory and immunomodulatory mechanisms of mesenchymal stem cell transplantation in experimental traumatic brain injury. J Neuroinflammation. 2013;10:106.
116. Egashira Y, Sugitani S, Suzuki Y, Mishiro K, Tsuruma K, Shimazawa M, et al. The conditioned medium of murine and human adipose-derived stem cells exerts neuroprotective effects against experimental stroke model. Brain Res. 2012;1461:87–95.

117. Ribeiro CA, Fraga JS, Graos M, Neves NM, Reis RL, Gimble JM, et al. The secretome of stem cells isolated from the adipose tissue and Wharton jelly acts differently on central nervous system derived cell populations. Stem Cell Res Ther. 2012;3(3):18.
118. Misra V, Ritchie MM, Stone LL, Low WC, Janardhan V. Stem cell therapy in ischemic stroke: role of IV and intra-arterial therapy. Neurology. 2012;79(13 Suppl 1):S207–12.
119. Kan I, Barhum Y, Melamed E, Offen D. Mesenchymal stem cells stimulate endogenous neurogenesis in the subventricular zone of adult mice. Stem Cell Rev. 2011;7(2):404–12.
120. Altman J, Das GD. Autoradiographic and histological evidence of postnatal hippocampal neurogenesis in rats. J Comp Neurol. 1965;124(3):319–35.
121. Bond AM, Ming GL, Song H. Adult mammalian neural stem cells and neurogenesis: five decades later. Cell Stem Cell. 2015;17(4):385–95.
122. Dezawa M, Kanno H, Hoshino M, Cho H, Matsumoto N, Itokazu Y, et al. Specific induction of neuronal cells from bone marrow stromal cells and application for autologous transplantation. J Clin Invest. 2004;113(12):1701–10.
123. Mori T, Wang X, Aoki T, Lo EH. Downregulation of matrix metalloproteinase-9 and attenuation of edema via inhibition of ERK mitogen activated protein kinase in traumatic brain injury. J Neurotrauma. 2002;19(11):1411–9.
124. Asahi M, Asahi K, Jung JC, del Zoppo GJ, Fini ME, Lo EH. Role for matrix metalloproteinase 9 after focal cerebral ischemia: effects of gene knockout and enzyme inhibition with BB-94. J Cereb Blood Flow Metab. 2000;20(12):1681–9.
125. Wang X, Mori T, Jung JC, Fini ME, Lo EH. Secretion of matrix metalloproteinase-2 and -9 after mechanical trauma injury in rat cortical cultures and involvement of MAP kinase. J Neurotrauma. 2002;19(5):615–25.

Chapter 9
Bone-Marrow-Derived Cell Therapies in Stroke: Immunomodulatory Effects

Laith Maali and David C. Hess

Abstract Cell therapies have attracted significant attention in treating multiple neurological disorders including stroke. The preclinical studies have paved the road in understanding the potential clinical applications of cell therapies in stroke recovery. Cells can be obtained from multiple sources and transplanted through different routes. Animal and human studies suggest that cell therapies exert their effect via paracrine and immunomodulatory effects rather than physically replacing the damaged cells. Clinical studies are still in the early phases but show safety and feasibility and some hints at efficacy.

Keywords Cell therapies • Stem cells • Bone marrow derived • Mesenchymal stem cells • Multipotent adult progenitor cells • Stroke • Phase 2 clinical trial • Immune modulation • Immunomodulation • Neuroprotection • MultiStem

Abbreviations

ACTH	Adrenocorticotropic hormone
APCs	Antigen presenting cells
DAMPs	Danger-associated molecular pattern molecules
GABA	gamma-Aminobutyric acid
IA	Intra-arterial
IL	Interleukin
INF	Interferon
IP	IFN-γ-inducible protein
IV	Intravenous
MAPC	Multipotent adult progenitor cells

L. Maali • D.C. Hess, M.D. (✉)
Department of Neurology, Medical College of Georgia, Augusta University, 1120 15th street, BI 3076, Augusta, GA 30912, USA
e-mail: laith.maali@gmail.com; DHESS@augusta.edu

P.A. Lapchak, J.H. Zhang (eds.), *Cellular and Molecular Approaches to Regeneration and Repair*, Springer Series in Translational Stroke Research,
https://doi.org/10.1007/978-3-319-66679-2_9

MCP	Monocyte chemoattractant protein
MHC	Major histocompatibility complex
MIP	Macrophage inflammatory protein
miRNA	microRNA
MMP	Matrix metalloproteinases
MNC	Mononuclear cells
MSC	Mesenchymal stem cells
NIHSS	National Institute of Health Stroke Scale
RANTES	Regulated on activation normal T cell expressed and secreted
RNA	Ribonucleic acid
r-tPA	Recombinant tissue plasminogen activator
SCID	Severe combined immunodeficiency
TLRs	Toll-like receptors
TNF	Tumor necrosis factor

1 What Are Cell Therapies?

Stroke is the second cause of death and the most common cause of adult neurological disability worldwide [1]. The main goal of current stroke management is to achieve a rapid recanalization to limit brain tissue damage. Currently, the only approved drug to be used in such a treatment is recombinant tissue plasminogen activator (r-tPA) [2]. On the other hand, this drug can only be used in the first 4.5 h of symptoms onset and only about 5% of acute ischemic stroke patients receive this therapy and despite that about half of them will end up with significant long term disability [3]. Even with the improved functional recovery with endovascular thrombectomy, the therapeutic window for such a treatment remains under 6 h according to current guidelines, although this time window will likely expand [3].

Cellular therapies emerge as a promising additional treatment to help limit the brain damage and improve functional outcome after acute stroke beyond the limited therapeutic window of r-tPA and endovascular thrombectomy. Cell therapy refers to cellular material that is able to exert a desired biological effect. In the 1990s, cell therapies were for the first time shown to be a potential treatment in patients with Parkinson's disease, as studies showed a positive effect of fetal striatal graft in increasing GABA release and reorganizing GABA receptors that led to physical and intellectual benefits [4]. Since that time, cell therapies evolved and other cell types were identified to reduce neurological injury, including positive studies in animal models of spinal cord injury [5] and traumatic brain injury [6] and stroke [7–11].

It is now believed that cell therapies do not mainly act by replacing injured cells but rather they exert their benefit by the activation of injured tissue to remodel and prevent further injury [12]. Many types of cell therapies release growth factors and cytokines that lead to immunomodulation, brain repair and cell survival [13].

2 Cell Therapy Types

There are multiple cell types involved in cell therapies, but they can be classified according to their original source: embryonic, induced pluripotent stem cells and adult cells. Adult derived cells are usually taken either from the bone marrow, adipose tissue or dental pulp cells. Another source is from the umbilical cord, the amniotic fluid or the placenta at the time of birth. We will focus in this chapter on the adult bone marrow derived cell as it is type furthest along in the translation from the bench to the bedside.

The bone marrow contains multiple types of cells, and based on the culturing techniques, we can generate mononuclear cells (MNC), mesenchymal stem cells (MSC) or multipotent adult progenitor cells (MAPC). MNC contains mature and immature cells of lymphoid, myeloid and erythroid origin. They do not require culture and can be quickly isolated which make them easy to use in acute to subacute phases of stroke as an autologous transplant. MNC have be shown to improve neurological outcomes in animal model, they reduced stroke lesion size, promoted angiogenesis and suppressed lymphocytic infiltration [7–9, 14]. Based on this, human MNC therapies were initiated and they are still in the early phases, but so far showing feasibility and safety with different routes of administration [15–17].

MSC are plastic adherent cells and require longer times to expand in culture than MNC. They can be easily derived from multiple sources including the bone marrow, adipose tissue and umbilical cord, but the majority of animal studies of MSC focused on the bone marrow derived cells. Bone marrow-derived MSC reside in the stromal part of bone marrow and sometimes termed marrow stromal cells [10], they have low immunogenicity and have strong immunomodulation capabilities. Autologous transplantation usually requires weeks to months to culture which limit its use in the acute phase of stroke [18], but their low immunogenicity suggest their safety and benefit for allogeneic and autologous transplantation in human [18–20] and animal studies [21].

MAPC are more primitive cells compared to MSC they have a broad differentiation capability. As MSC they have low immunogenicity and strong immunomodulation capabilities. They require long term culture so their use in acute stroke is limited unless as an allogeneic transplant [22]. Compared to MSC they have distinct phenotype protein and gene expression pattern [23]. They showed benefit in a stroke animal model [24] and safety in human studies [25].

3 Route of Administration

There are three main routes for cell administration: Intravenous (IV), intra-arterial and intracerebral. Intravenous route is the preferred method given the ease of use and especially if the main goal from the cells is to exert their effect through a systemic process. With IV administration the majority of MSC, for example, get entrapped passively in the lung vasculature given their relative large size, but some cells are able to migrate to the ischemic area and the peri-infarct zone [26–28]. Based on that, the possibility that MSC benefit patients with stroke by becoming brain cells is unlikely [12]. The first reported IV MSC transplantation in human was performed by Bang et al. [29]; in their study they showed significant improvement in the modified Rankin score and Barthel index up to 6 months after IV MSC transplantation compared to control group.

Intra-arterial (IA) route provides some benefits with selective delivery of cells to the area of injury, deliver a higher number of cells and bypass the peripheral systemic entrapment. On the other hand, IA route carries some clear risks, like microvascular plugging with large sized cell delivery [30–32].

Intracerebral route using stereotactic injection places the cells in a specific area, helps control the dose and leads to a better survival for stem cells [33], and is the preferred route if migration outside the brain is not wanted. This route showed benefit when neural stem cells were used in an animal model [34], also was feasible and was tolerated well in humans without significant serious adverse events [33].

The best route of cell delivery is not yet clear and depends on the cell type. Multiple animal model studies showed clear benefit with IV route and when compared to IA and intracerebral routes the benefits were similar or greater [31, 35]. In general, bone marrow derived cells may be optimized for an intravenous route while neural stem cells are more optimal if delivered intracerebrally.

4 Stroke Affects the Immune System

Stroke is a multiorgan systemic disease and not solely a brain lesion. Stroke induces immune changes and systemic inflammation in both animal and human models [36–38]. The inflammatory process in stroke begins immediately after the vascular occlusion with activation of complement, platelets and endothelial cells [39, 40].

Shortly after ischemia, a variety of signals are released from dying cells and cells under stress, called danger-associated molecular pattern molecules (DAMPs) [41]. These signals which include purines, cytokines, and chemokines activate the innate immune system via the activation of the toll-like receptors (TLRs) and scavenger receptors on microglia, astrocytes, endothelial cells and perivascular macrophages [42, 43]. The activated microglia initiate the inflammatory response by the release of inflammatory cytokines (like IL-1β, TNF-α), chemokines to recruit macrophages and leukocytes infiltration [44]. The activated perivascular macrophages also release pro-inflammatory cytokines and chemokines [45]. Adhesion molecules are up-regulated shortly after stroke, they appear to be involved in the migration of neutrophils into the stroke area [46]. Neutrophils release pro-inflammatory molecules upon activation, these molecules, including MMP-9, contribute to blood brain barrier breakdown a leukocytes activation and infiltration [47].

The adaptive immune system requires longer time to get activated. B lymphocytes secret antibodies against the brain-derived antigens as part of the humoral immune response [48]. Antigen presenting cells (APCs) including dendritic cells, increase in number in the ischemic brain while decrease in the periphery [49, 50]. These APCs process brain antigens and present them with MHC molecule that get recognized by T lymphocyte receptors and that leads to their recruitment into the ischemic brain [41, 50].

Multiple studies favor the cellular adaptive immunity as a mechanism of damage in stroke [49–52]; this cell mediated damage was found to be non-antigen or cell receptor driven [53]. In animal model studies, focal ischemia led to rapid splenic activation of T lymphocytes and the release of local cytokines in the acute phase which was accompanied by a reduction in the number of immune cells in the peripheral lymphoid organs (spleen and thymus) and eventually decreased levels of tumor necrotic factor-α (TNF-α) and interferon-γ (INF-γ) that contributed to immunosuppression and infections after stroke [37, 54].

Offner et al. studied the effect of the loss of T and B lymphocytes in animal stroke model, based on their previous work [54] which indicated a detrimental effect of the immune system in stroke. They used mice with severe combined immunodeficiency (SCID) and found that in the absence of T and B cells, post-stroke inflammatory mediators were largely suppressed in the acute phase both in the brain and spleen of SCID mice, with improvement in early post-ischemic histological damage. Also, the post-stoke intra-splenic cytokines and chemokines expression were reduced as well as the loss of splenocytes compared to their control mice. Yilmaz et al. also showed smaller infarct volumes in lymphocyte deficient mice [55]. The role of T regulatory cells in stoke is controversial; Liesz et al. showed that T-regulatory cell depletion increased brain damage and deteriorated functional outcome [52], while Kleinschnitz et al. showed that T-regulatory cell depletion dramatically reduced stroke size [56].

Early after stroke, ischemic cortex shows a significant increase in the expression of inflammatory cytokines (TNF-α, IL-1β, IL-6) and chemokines (RANTES, IP-10, MIP-2) [57], plasma levels of IL-6 also increases [57, 58]. In the spleen, inflammatory factors (TNF-α, INF-γ, IL-6, MCP-1, IL-2) levels significantly increase along with anti-inflammatory factor IL-10 [57]. These studies confirm that stroke is not a local brain disease but it also affects the spleen and the immune system response.

Another study showing the instrumental role of spleen in mediating immune response after stroke; performing splenectomy in their animal model resulted in reduction in the neurodegeneration and the immune response in the splenectomized animal after middle cerebral artery occlusion [59].

Infection after stroke is another major problem, and is seen more often with major strokes [60]. There is an accumulating clinical evidence suggesting that acute stroke induces immunological changes that facilitate the appearance of infections. Acute stroke leads to the release of inflammatory cytokines, which in turn induce the pituitary gland to secrete ACTH which induces glucocorticoids secretion from the adrenal cortex [61], which has a strong antiproliferative and apoptotic properties on the immune system and at the same time suppress further cytokine synthesis [62].

5 How Do Cell Therapies Work?

Cell therapies are not believed to work through replacing injured cells, but rather through the release of biological factors that promote recovery and suppress further damage [10, 12, 13]. Different cell types have different ways to exert their beneficial effects. These mechanisms include: neuroprotection, promoting angiogenesis, promoting neurogenesis, glial scar prevention, local and systemic anti-inflammatory effects [10, 12, 13, 63–65]. For example, bone marrow derived hematopoietic stem were able to counter-regulate the up-regulation of proinflammatory cytokines and chemokines receptor gene transcripts in the spleen, thus preventing the activation of the immune gene transcripts in the splenocytes which result in a reduction of the number of cells entering the blood circulation and brain tissue [66]. Neural stem cells were found to produce a broad spectrum of trophic factors (Nerve growth factor, brain-derived neurotrophic factor, glia-derived neurotrophic factor) [67] which play an important role in neuroprotection by promoting cell survival [68], they also promote angiogenesis and restoration of blood-brain barrier integrity via the production of endothelial growth factors [69]. MSC was found to induce cell proliferation in the subventricular and subgranular zones which suggest their neurogenesis benefits [65, 70, 71], also they release bioactive substances that promote the proliferation of glial cells [72, 73]. Cell therapies modulate the immune system rather than suppressing it, they down-regulate pro-inflammatory, pro-apoptotic cytokines like TNF-α and INF-γ [74], secrete biological factors that suppress immune cell migration and infiltration in to the brain [75], and overall suppress subtypes of T-cells [76–79] and modulate B-cells function [80].

The effect of cell therapies on the immune system, as mentioned, is selective. They were shown to improve mortality, reduce inflammation and enhance bacterial clearance in sepsis animal model [81]. The selective suppression effect and inflammation reduction is potentially the reason that cell therapies might reduce infection rates seen in stroke. In the MASTERS trail, there was a signal for lower infection rates in the therapy group compared to placebo [25].

The mechanisms that cell therapies interact and communicate with the injured brain cells are still not fully understood, but there is an increasing evidence to support the paracrine effect by stem cells via the release of extracellular vesicles able to exert a biological activity [82–85]. An important type of these vesicles are exosomes which can be released by different types of cells [86] and especially produced in large amounts from MSCs [87]. Most exosomes contain several proteins, cell surface receptors, trophic factors, RNAs and Micro-RNAs (miRNAs) [84, 86, 88–91]. The content and the amount of exosomes released can be altered by cellular damage and stress like hypoxia [92–94]. miRNAs are a small non-coding sequences of RNA that have the capability to regulate genes, pathways and biological activities within cells [95–97]. Multiple studies have shown that MSC exosomes modify recipient cell characteristics and regulate their protein expression through miRNA transfer [98–100] and thus participating in stroke recovery [101]. Xin et al. were able to demonstrate for the first time a systemic treatment of cell-free exosomes derived from MSCs and showed an increase in neural plasticity and recovery after stroke in their animal model [82, 102].

6 Clinical Trials in Cell Therapies

Clinical trials in cell therapies still for the most part in early phases, mainly focusing on bone marrow derived cells with autologous transplantation but recently allogeneic transplantation.

Based on the animal studies, Savitz et al. showed the safety of IV autologous MNC infusion given 24–72 h after stroke [15]. Friedrich et al. also showed the safety of intra-arterial infusion of autologous MNC 3–7 days after stroke [103]. Other studies also have shown the good safety profile of MNC infusion in humans [104].

Most of MSC human studies were done on patients with chronic stroke, given the relative longer culture time required to obtain these cells. The first reported MSC transplantation in human was performed by Bang et al., they used IV route with autologous cells infused 4–9 weeks after stroke symptoms, they reported improvement in both modified Rankin score and Barthel index up to 6 months, also no adverse events were reported up to 1 year later [29], a 5-year follow study showed sustained improvement in functional outcome without significant side effects or a change in mortality compared to the control group [105]. Other studies also reported the safety of IV MSC transplantation in humans [18, 20, 104].

MAPC have been shown to enhance recovery in animal model through modulating the immune response and targeting the spleen inflammatory response [11, 106] (see Fig. 9.1). MultiStem® a proprietary cell therapy from Athersys is a MAPC derived from the bone marrow, and has shown promising potential to treat other conditions other than stroke. Based on the promising results of the preclinical studies, Hess et al. conducted a phase II, multicenter, double-blinded, randomized, placebo-control study; in the MultiStem in Acute Stroke Treatment to Enhance Recovery Study (MASTERS), they aimed to establish the highest safe, well tolerated IV dose of MultiStem® while assessing for any efficacy on stroke recovery. They enrolled patients aged 18–83 years with moderate to severe acute ischemic stroke and a National Institute of Health Stroke Scale (NIHSS) score of 8–20 within 24–48 h after stroke onset. This was a dose escalation trial where patients were first randomized in Group 1 to a receive a low dose (400 million cells) or placebo and once this was determined to be safe to be randomized to a high dose (1200 million cells) or placebo in Group 2. The high dose was determined to be without safety issues, so patients were randomized in Group 3 to 1200 million cells or placebo. Groups 2 and 3 were combined for the primary safety and efficacy analyses. There were no dose limiting toxicity, difference in adverse events or difference in stroke recovery at 90 days [25]. On the other hand, an exploratory analysis of their data suggested an increase in excellent outcome (mRS $\leq$ 1, NIHSS $\leq$ 1 and Barthel index $\geq$95) at 1 year follow up. Multiple biomarkers were also measured at 2, 7 and 30 days post treatment and their analysis showed that MultiStem® reduced T lymphocytes (CD3+ and FoxP3+ T-cells) at 2 days and inflammatory cytokines (TNF-α, IL-1β, IL-6) at 7 days. This is the first data to show that cell therapies modulate the immune system after acute stroke in humans [25].

Neural stem cells were studied in the human neural stem cells in patients with chronic ischemic stroke (PISCES), a phase I, single arm, non randomized, dose

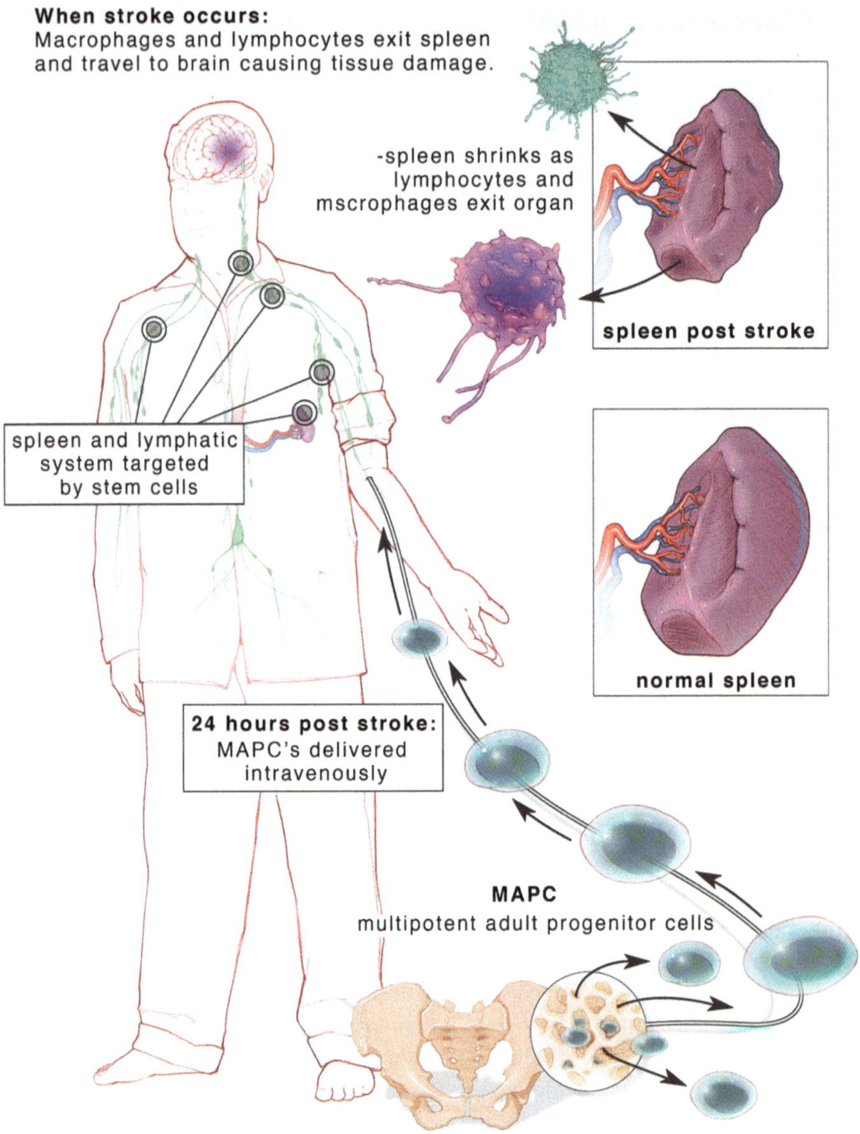

Fig. 9.1 Cell therapies—modulating the immune system. After stroke, proinflammatory signals leads to splenic lymphocytes activation and release. These lymphocytes target the ischemic brain and cause further damage. Cell therapies, such as MAPC, act by targeting the spleen and preventing immune system activation and later immunodepression

escalation clinical trial of 11 subjects to assess neural stem cell transplantation safety and feasibility via intracerebral stereotactic implantation 6–60 months after ischemic stroke. There were no significant adverse events related to the cells and there was improvement in neurological outcome at 12 months follow up [33].

Currently, PISCES II, a phase II trial is ongoing. Steinberg et al. reported an interim results from their 2 year phase 1/2a study single arm, non randomized trial of stereotactic administration of modified bone marrow-derived MSC in patients with chronic stroke; their analysis showed safety and feasibility, along with improvement in neurological function at 12 months [107]. However, the lack of a control group makes activity of the therapy difficult to evaluate.

7 Cell Therapies Direction and Development

The Stem Cell Therapies as an Emerging Paradigm for Stroke (STEPS) committee has published a series of recommendation for both preclinical and clinical research in stem cell therapy to help guide stem cell research. STEPS I publication focused on: the validation of animal studies, build robust safety studies in humans, and selecting the right time and patients for cell therapies [108]. STEPS II focused on providing guidelines for human safety trails and the need to obtain a signal for efficacy [109]. STEPS III focused on the design of clinical trials in phase IIb and III, and covered topics on patient selection, timing of therapy, desired endpoints, assessing cell therapies activity through biomarkers, concomitant rehabilitation therapy and their potential application in chronic stroke [110].

It is still too early to know whether cell therapy in stroke will be successful. To date, cell therapy trials have shown safety and feasibility. This is crucial as it is important to avoid the mistakes of gene therapy trials. The MASTERS trial suggests that an earlier time window of 18–36 h may be optimal and planned clinical trials in Japan, North American and Europe will enroll patients in the 18–36 h time window.

References

1. Johnston SC, Mendis S, Mathers CD. Global variation in stroke burden and mortality: estimates from monitoring, surveillance, and modelling. Lancet Neurol. 2009;8(4):345–54.
2. Ansari S, Rahman M, Waters MF, Hoh BL, Mocco J. Recanalization therapy for acute ischemic stroke, part 1: surgical embolectomy and chemical thrombolysis. Neurosurg Rev. 2011;34(1):1–9.
3. Goyal M, Menon BK, Van Zwam WH, Dippel DW, Mitchell PJ, Demchuk AM, et al. Endovascular thrombectomy after large-vessel ischaemic stroke: a meta-analysis of individual patient data from five randomised trials. Lancet. 2016;387(10029):1723–31.
4. Freed CR, Breeze RE, Rosenberg NL, Schneck SA, Kriek E, Qi J, et al. Survival of implanted fetal dopamine cells and neurologic improvement 12 to 46 months after transplantation for Parkinson's disease. N Engl J Med. 1992;327(22):1549–55.
5. Chopp M, Zhang XH, Li Y, Wang L, Chen J, Lu D, et al. Spinal cord injury in rat: treatment with bone marrow stromal cell transplantation. Neuroreport. 2000;11(13):3001–5.
6. Mahmood A, Lu D, Li Y, Chen JL, Chopp M. Intracranial bone marrow transplantation after traumatic brain injury improving functional outcome in adult rats. J Neurosurg. 2001;94(4):589–95.

7. Wang J, Yu L, Jiang C, Chen M, Ou C, Wang J. Bone marrow mononuclear cells exert long-term neuroprotection in a rat model of ischemic stroke by promoting arteriogenesis and angiogenesis. Brain Behav Immun. 2013;34:56–66.
8. Franco EC, Cardoso MM, Gouvêia A, Pereira A, Gomes-Leal W. Modulation of microglial activation enhances neuroprotection and functional recovery derived from bone marrow mononuclear cell transplantation after cortical ischemia. Neurosci Res. 2012;73(2):122–32.
9. Brenneman M, Sharma S, Harting M, Strong R, Cox CS Jr, Aronowski J, et al. Autologous bone marrow mononuclear cells enhance recovery after acute ischemic stroke in young and middle-aged rats. J Cereb Blood Flow Metab. 2010;30(1):140–9.
10. Eckert MA, Vu Q, Xie K, Yu J, Liao W, Cramer SC, et al. Evidence for high translational potential of mesenchymal stromal cell therapy to improve recovery from ischemic stroke. J Cereb Blood Flow Metab. 2013;33(9):1322–34.
11. Yang B, Hamilton J, Strong R, Xi X, Mays R, Savitz S. Human multipotential bone marrow stem cells exert immunomodulatory effects, prevent splenic contraction, and enhance functional recovery in a rodent model of ischemic stroke. In: Stroke. Philadelphia, PA: Lippincott Williams & Wilkins; 2011. p. E67.
12. Chopp M, Li Y. Treatment of neural injury with marrow stromal cells. Lancet Neurol. 2002;1(2):92–100.
13. Crigler L, Robey RC, Asawachaicharn A, Gaupp D, Phinney DG. Human mesenchymal stem cell subpopulations express a variety of neuro-regulatory molecules and promote neuronal cell survival and neuritogenesis. Exp Neurol. 2006;198(1):54–64.
14. Baker AH, Sica V, Work LM, Williams-Ignarro S, De Nigris F, Lerman LO, et al. Brain protection using autologous bone marrow cell, metalloproteinase inhibitors, and metabolic treatment in cerebral ischemia. Proc Natl Acad Sci U S A. 2007;104(9):3597–602.
15. Savitz SI, Misra V, Kasam M, Juneja H, Cox CS, Alderman S, et al. Intravenous autologous bone marrow mononuclear cells for ischemic stroke. Ann Neurol. 2011;70(1):59–69.
16. Moniche F, Gonzalez A, Gonzalez-Marcos J-R, Carmona M, Piñero P, Espigado I, et al. Intra-arterial bone marrow mononuclear cells in ischemic stroke. Stroke. 2012; 43(8):2242–4.
17. Prasad K, Mohanty S, Bhatia R, Srivastava MVP, Garg A, Srivastava A, et al. Autologous intravenous bone marrow mononuclear cell therapy for patients with subacute ischaemic stroke: a pilot study. Indian J Med Res. 2012;136(2):221.
18. Honmou O, Houkin K, Matsunaga T, Niitsu Y, Ishiai S, Onodera R, et al. Intravenous administration of auto serum-expanded autologous mesenchymal stem cells in stroke. Brain. 2011;134(Pt 6):1790–807.
19. William TT, Pendleton JD, Beyer WM, Egalka MC, Guinan EC. Suppression of allogeneic T-cell proliferation by human marrow stromal cells: implications in transplantation. Transplantation. 2003;75(3):389–97.
20. Bhasin A, Srivastava MP, Mohanty S, Bhatia R, Kumaran SS, Bose S. Stem cell therapy: a clinical trial of stroke. Clin Neurol Neurosurg. 2013;115(7):1003–8.
21. Li Y, McIntosh K, Chen J, Zhang C, Gao Q, Borneman J, et al. Allogeneic bone marrow stromal cells promote glial–axonal remodeling without immunologic sensitization after stroke in rats. Exp Neurol. 2006;198(2):313–25.
22. Kovacsovics-Bankowski M, Streeter PR, Mauch KA, Frey MR, Raber A, van't Hof W, et al. Clinical scale expanded adult pluripotent stem cells prevent graft-versus-host disease. Cell Immunol. 2009;255(1):55–60.
23. Boozer S, Lehman N, Lakshmipathy U, Love B, Raber A, Maitra A, et al. Global characterization and genomic stability of human multiStem, a multipotent adult progenitor cell. J Stem Cells. 2008;4(1):17–28.
24. Mays RW, Borlongan CV, Yasuhara T, Hara K, Maki M, Carroll JE, et al. Development of an allogeneic adherent stem cell therapy for treatment of ischemic stroke. J Exp Stroke Transl Med. 2010;3(1):34–46.
25. Hess DC, Wechsler LR, Clark WM, Savitz SI, Ford GA, Chiu D, et al. Safety and efficacy of multipotent adult progenitor cells in acute ischaemic stroke (MASTERS): a randomised, double-blind, placebo-controlled, phase 2 trial. Lancet Neurol. 2017;16(5):360–8.

26. Eglitis MA, Dawson D, Park K-W, Mouradian MM. Targeting of marrow-derived astrocytes to the ischemic brain. Neuroreport. 1999;10(6):1289–92.
27. Hillyer C. Turning blood into brain: cells bearing neuronal antigens generated in vivo from bone marrow: E. Mezey, KJ Chandross, G. Garta, et al. Science 290:1779–1782, 2000. Transfus Med Rev. 2001;15(3):248–9.
28. Li Y, Chopp M. Marrow stromal cell transplantation in stroke and traumatic brain injury. Neurosci Lett. 2009;456(3):120–3.
29. Bang OY, Lee JS, Lee PH, Lee G. Autologous mesenchymal stem cell transplantation in stroke patients. Ann Neurol. 2005;57(6):874–82.
30. Grieve SM, Bhindi R, Seow J, Doyle A, Turner AJ, Tomka J, et al. Microvascular obstruction by intracoronary delivery of mesenchymal stem cells and quantification of resulting myocardial infarction by cardiac magnetic resonance. Circ Heart Fail. 2010;3(3):e5–6.
31. Willing AE, Lixian J, Milliken M, Poulos S, Zigova T, Song S, et al. Intravenous versus intrastriatal cord blood administration in a rodent model of stroke. J Neurosci Res. 2003;73(3):296–307.
32. Janowski M, Lyczek A, Engels C, Xu J, Lukomska B, Bulte JW, et al. Cell size and velocity of injection are major determinants of the safety of intracarotid stem cell transplantation. J Cereb Blood Flow Metab. 2013;33(6):921–7.
33. Kalladka D, Sinden J, Pollock K, Haig C, McLean J, Smith W, et al. Human neural stem cells in patients with chronic ischaemic stroke (PISCES): a phase 1, first-in-man study. Lancet. 2016;388(10046):787–96.
34. Stroemer P, Patel S, Hope A, Oliveira C, Pollock K, Sinden J. The neural stem cell line CTX0E03 promotes behavioral recovery and endogenous neurogenesis after experimental stroke in a dose-dependent fashion. Neurorehabil Neural Repair. 2009;23(9):895–909.
35. Yang B, Migliati E, Parsha K, Schaar K, Xi X, Aronowski J, et al. Intra-arterial delivery is not superior to intravenous delivery of autologous bone marrow mononuclear cells in acute ischemic stroke. Stroke. 2013;44(12):3463–72.
36. Audebert HJ, Rott MM, Eck T, Haberl RL. Systemic inflammatory response depends on initial stroke severity but is attenuated by successful thrombolysis. Stroke. 2004;35(9):2128–33.
37. Prass K, Meisel C, Höflich C, Braun J, Halle E, Wolf T, et al. Stroke-induced immunodeficiency promotes spontaneous bacterial infections and is mediated by sympathetic activation reversal by poststroke T helper cell type 1-like immunostimulation. J Exp Med. 2003;198(5):725–36.
38. Meisel C, Schwab JM, Prass K, Meisel A, Dirnagl U. Central nervous system injury-induced immune deficiency syndrome. Nat Rev Neurosci. 2005;6(10):775–86.
39. Peerschke EI, Yin W, Ghebrehiwet B. Complement activation on platelets: implications for vascular inflammation and thrombosis. Mol Immunol. 2010;47(13):2170–5.
40. Pinsky DJ, Naka Y, Liao H, Oz MC, Wagner DD, Mayadas TN, et al. Hypoxia-induced exocytosis of endothelial cell Weibel-Palade bodies. A mechanism for rapid neutrophil recruitment after cardiac preservation. J Clin Invest. 1996;97(2):493.
41. Kono H, Rock KL. How dying cells alert the immune system to danger. Nat Rev Immunol. 2008;8(4):279.
42. Benakis C, Garcia-Bonilla L, Iadecola C, Anrather J. The role of microglia and myeloid immune cells in acute cerebral ischemia. Front Cell Neurosci [Internet]. 2014;8. Available from: https://www.ncbi.nlm.nih.gov/pmc/articles/PMC4294142/
43. Marsh BJ, Williams-Karnesky RL, Stenzel-Poore MP. Toll-like receptor signaling in endogenous neuroprotection and stroke. Neuroscience. 2009;158(3):1007–20.
44. Benakis C, Garcia-Bonilla L, Iadecola C, Anrather J. The role of microglia and myeloid immune cells in acute cerebral ischemia. Front Cell Neurosci. 2014;8:461.
45. Konsman JP, Drukarch B, Van Dam A-M. (Peri)vascular production and action of pro-inflammatory cytokines in brain pathology. Clin Sci. 2007;112(1):1–25.
46. Yilmaz G, Granger DN. Cell adhesion molecules and ischemic stroke. Neurol Res. 2008;30(8):783–93.
47. Justicia C, Panés J, Solé S, Cervera Á, Deulofeu R, Chamorro Á, et al. Neutrophil infiltration increases matrix metalloproteinase-9 in the ischemic brain after occlusion/reperfusion of the middle cerebral artery in rats. J Cereb Blood Flow Metab. 2003;23(12):1430–40.

48. Ak A. Basic immunology updated edition: functions and disorders of the immune system. Philadelphia, PA: Saunders; 2010.
49. Yilmaz A, Fuchs T, Dietel B, Altendorf R, Cicha I, Stumpf C, et al. Transient decrease in circulating dendritic cell precursors after acute stroke: potential recruitment into the brain. Clin Sci. 2009;118(2):147–57.
50. Gelderblom M, Leypoldt F, Steinbach K, Behrens D, Choe C-U, Siler DA, et al. Temporal and spatial dynamics of cerebral immune cell accumulation in stroke. Stroke. 2009;40(5):1849–57.
51. Bornstein NM, Aronovich B, Korczyn AD, Shavit S, Michaelson DM, Chapman J. Antibodies to brain antigens following stroke. Neurology. 2001;56(4):529–30.
52. Liesz A, Suri-Payer E, Veltkamp C, Doerr H, Sommer C, Rivest S, et al. Regulatory T cells are key cerebroprotective immunomodulators in acute experimental stroke. Nat Med. 2009;15(2):192–9.
53. Kleinschnitz C, Schwab N, Kraft P, Hagedorn I, Dreykluft A, Schwarz T, et al. Early detrimental T-cell effects in experimental cerebral ischemia are neither related to adaptive immunity nor thrombus formation. Blood. 2010;115(18):3835–42.
54. Offner H, Subramanian S, Parker SM, Wang C, Afentoulis ME, Lewis A, et al. Splenic atrophy in experimental stroke is accompanied by increased regulatory T cells and circulating macrophages. J Immunol. 2006;176(11):6523–31.
55. Yilmaz G, Arumugam TV, Stokes KY, Granger DN. Role of T lymphocytes and interferon-γ in ischemic stroke. Circulation. 2006;113(17):2105–12.
56. Kleinschnitz C, Kraft P, Dreykluft A, Hagedorn I, Göbel K, Schuhmann MK, et al. Regulatory T cells are strong promoters of acute ischemic stroke in mice by inducing dysfunction of the cerebral microvasculature. Blood. 2013;121(4):679–91.
57. Offner H, Subramanian S, Parker SM, Afentoulis ME, Vandenbark AA, Hurn PD. Experimental stroke induces massive, rapid activation of the peripheral immune system. J Cereb Blood Flow Metab. 2006;26(5):654–65.
58. Emsley HC, Smith CJ, Gavin CM, Georgiou RF, Vail A, Barberan EM, et al. An early and sustained peripheral inflammatory response in acute ischaemic stroke: relationships with infection and atherosclerosis. J Neuroimmunol. 2003;139(1):93–101.
59. Ajmo CT, Vernon DO, Collier L, Hall AA, Garbuzova-Davis S, Willing A, et al. The spleen contributes to stroke-induced neurodegeneration. J Neurosci Res. 2008;86(10):2227–34.
60. Hamidon BB, Raymond AA, Norlinah MI, Jefferelli SB. The predictors of early infection after an acute ischaemic stroke. Singap Med J. 2003;44(7):344–6.
61. Turnbull AV, Rivier CL. Regulation of the hypothalamic-pituitary-adrenal axis by cytokines: actions and mechanisms of action. Physiol Rev. 1999;79(1):1–71.
62. Tuosto L, Cundari E, Montani MSG, Piccolella E. Analysis of susceptibility of mature human T lymphocytes to dexamethasone-induced apoptosis. Eur J Immunol. 1994;24(5):1061–5.
63. Ohtaki H, Ylostalo JH, Foraker JE, Robinson AP, Reger RL, Shioda S, et al. Stem/progenitor cells from bone marrow decrease neuronal death in global ischemia by modulation of inflammatory/immune responses. Proc Natl Acad Sci U S A. 2008;105(38):14638–43.
64. Toyama K, Honmou O, Harada K, Suzuki J, Houkin K, Hamada H, et al. Therapeutic benefits of angiogenetic gene-modified human mesenchymal stem cells after cerebral ischemia. Exp Neurol. 2009;216(1):47–55.
65. Bao X, Feng M, Wei J, Han Q, Zhao H, Li G, et al. Transplantation of Flk-1+ human bone marrow-derived mesenchymal stem cells promotes angiogenesis and neurogenesis after cerebral ischemia in rats. Eur J Neurosci. 2011;34(1):87–98.
66. Schwarting S, Litwak S, Hao W, Bähr M, Weise J, Neumann H. Hematopoietic stem cells reduce postischemic inflammation and ameliorate ischemic brain injury. Stroke. 2008;39(10):2867–75.
67. Lu P, Jones LL, Snyder EY, Tuszynski MH. Neural stem cells constitutively secrete neurotrophic factors and promote extensive host axonal growth after spinal cord injury. Exp Neurol. 2003;181(2):115–29.

68. Nguyen N, Lee SB, Lee YS, Lee K-H, Ahn J-Y. Neuroprotection by NGF and BDNF against neurotoxin-exerted apoptotic death in neural stem cells are mediated through Trk receptors, activating PI3-kinase and MAPK pathways. Neurochem Res. 2009;34(5):942–51.
69. Roitbak T, Li L, Cunningham LA. Neural stem/progenitor cells promote endothelial cell morphogenesis and protect endothelial cells against ischemia via HIF-1α-regulated VEGF signaling. J Cereb Blood Flow Metab. 2008;28(9):1530–42.
70. Chen J, Li Y, Katakowski M, Chen X, Wang L, Lu D, et al. Intravenous bone marrow stromal cell therapy reduces apoptosis and promotes endogenous cell proliferation after stroke in female rat. J Neurosci Res. 2003;73(6):778–86.
71. Steffenhagen C, Dechant F-X, Oberbauer E, Furtner T, Weidner N, Küry P, et al. Mesenchymal stem cells prime proliferating adult neural progenitors toward an oligodendrocyte fate. Stem Cells Dev. 2011;21(11):1838–51.
72. Mauri M, Lentini D, Gravati M, Foudah D, Biella G, Costa B, et al. Mesenchymal stem cells enhance GABAergic transmission in co-cultured hippocampal neurons. Mol Cell Neurosci. 2012;49(4):395–405.
73. Ribeiro CA, Salgado AJ, Fraga JS, Silva NA, Reis RL, Sousa N. The secretome of bone marrow mesenchymal stem cells-conditioned media varies with time and drives a distinct effect on mature neurons and glial cells (primary cultures). J Tissue Eng Regen Med. 2011;5(8):668–72.
74. Bacigaluppi M, Pluchino S, Jametti LP, Kilic E, Kilic Ü, Salani G, et al. Delayed post-ischaemic neuroprotection following systemic neural stem cell transplantation involves multiple mechanisms. Brain. 2009;awp174.
75. Yoo S-W, Chang D-Y, Lee H-S, Kim G-H, Park J-S, Ryu B-Y, et al. Immune following suppression mesenchymal stem cell transplantation in the ischemic brain is mediated by TGF-β. Neurobiol Dis. 2013;58:249–57.
76. Di Nicola M, Carlo-Stella C, Magni M, Milanesi M, Longoni PD, Matteucci P, et al. Human bone marrow stromal cells suppress T-lymphocyte proliferation induced by cellular or nonspecific mitogenic stimuli. Blood. 2002;99(10):3838–43.
77. Nasef A, Mathieu N, Chapel A, Frick J, François S, Mazurier C, et al. Immunosuppressive effects of mesenchymal stem cells: involvement of HLA-G. Transplantation. 2007;84(2):231–7.
78. Meisel R, Zibert A, Laryea M, Göbel U, Däubener W, Dilloo D. Human bone marrow stromal cells inhibit allogeneic T-cell responses by indoleamine 2,3-dioxygenase-mediated tryptophan degradation. Blood. 2004;103(12):4619–21.
79. Aggarwal S, Pittenger MF. Human mesenchymal stem cells modulate allogeneic immune cell responses. Blood. 2005;105(4):1815–22.
80. Corcione A, Benvenuto F, Ferretti E, Giunti D, Cappiello V, Cazzanti F, et al. Human mesenchymal stem cells modulate B-cell functions. Blood. 2006;107(1):367–72.
81. Mei SH, Haitsma JJ, Dos Santos CC, Deng Y, Lai PF, Slutsky AS, et al. Mesenchymal stem cells reduce inflammation while enhancing bacterial clearance and improving survival in sepsis. Am J Respir Crit Care Med. 2010;182(8):1047–57.
82. Xin H, Li Y, Cui Y, Yang JJ, Zhang ZG, Chopp M. Systemic administration of exosomes released from mesenchymal stromal cells promote functional recovery and neurovascular plasticity after stroke in rats. J Cereb Blood Flow Metab. 2013;33(11):1711–5.
83. Camussi G, Deregibus MC, Cantaluppi V. Role of stem-cell-derived microvesicles in the paracrine action of stem cells. Biochem Soc Trans. 2013;41(1):283–7.
84. Lai RC, Chen TS, Lim SK. Mesenchymal stem cell exosome: a novel stem cell-based therapy for cardiovascular disease. Regen Med. 2011;6(4):481–92.
85. Collino F, Deregibus MC, Bruno S, Sterpone L, Aghemo G, Viltono L, et al. Microvesicles derived from adult human bone marrow and tissue specific mesenchymal stem cells shuttle selected pattern of miRNAs. PLoS One. 2010;5(7):e11803.
86. Xin H, Li Y, Chopp M. Exosomes/miRNAs as mediating cell-based therapy of stroke. Front Cell Neurosci. 2014;8:377.
87. Yeo RWY, Lai RC, Zhang B, Tan SS, Yin Y, Teh BJ, et al. Mesenchymal stem cell: an efficient mass producer of exosomes for drug delivery. Adv Drug Deliv Rev. 2013;65(3):336–41.

88. Shen B, Wu N, Yang J-M, Gould SJ. Protein targeting to exosomes/microvesicles by plasma membrane anchors. J Biol Chem. 2011;286(16):14383–95.
89. Yang J-M, Gould SJ. The *cis*-acting signals that target proteins to exosomes and microvesicles. Biochem Soc Trans. 2013;41(1):277–82.
90. Koh W, Sheng CT, Tan B, Lee QY, Kuznetsov V, Kiang LS, et al. Analysis of deep sequencing microRNA expression profile from human embryonic stem cells derived mesenchymal stem cells reveals possible role of let-7 microRNA family in downstream targeting of hepatic nuclear factor 4 alpha. BMC Genomics. 2010;11(1):S6.
91. Tomasoni S, Longaretti L, Rota C, Morigi M, Conti S, Gotti E, et al. Transfer of growth factor receptor mRNA via exosomes unravels the regenerative effect of mesenchymal stem cells. Stem Cells Dev. 2012;22(5):772–80.
92. Greenwalt TJ. The how and why of exocytic vesicles. Transfusion (Paris). 2006;46(1):143–52.
93. Horstman LL, Jy W, Minagar A, Bidot CJ, Jimenez JJ, Alexander JS, et al. Cell-derived microparticles and exosomes in neuroinflammatory disorders. Int Rev Neurobiol. 2007;79:227–68.
94. Sano S, Izumi Y, Yamaguchi T, Yamazaki T, Tanaka M, Shiota M, et al. Lipid synthesis is promoted by hypoxic adipocyte-derived exosomes in 3T3-L1 cells. Biochem Biophys Res Commun. 2014;445(2):327–33.
95. Chen J, Ning R, Zacharek A, Cui C, Cui X, Yan T, et al. MiR-126 contributes to human umbilical cord blood cell-induced neurorestorative effects after stroke in Type-2 diabetic mice. Stem Cells. 2016;34(1):102–13.
96. Fiore R, Siegel G, Schratt G. MicroRNA function in neuronal development, plasticity and disease. Biochim Biophys Acta BBA Gene Regul Mech. 2008;1779(8):471–8.
97. Zhang B, Wang Q, Pan X. MicroRNAs and their regulatory roles in animals and plants. J Cell Physiol. 2007;210(2):279–89.
98. Xin H, Li Y, Buller B, Katakowski M, Zhang Y, Wang X, et al. Exosome-mediated transfer of miR-133b from multipotent mesenchymal stromal cells to neural cells contributes to neurite outgrowth. Stem Cells. 2012;30(7):1556–64.
99. Lee J-K, Park S-R, Jung B-K, Jeon Y-K, Lee Y-S, Kim M-K, et al. Exosomes derived from mesenchymal stem cells suppress angiogenesis by down-regulating VEGF expression in breast cancer cells. PLoS One. 2013;8(12):e84256.
100. Lusardi TA, Murphy SJ, Phillips JI, Chen Y, Davis CM, Young JM, et al. MicroRNA responses to focal cerebral ischemia in male and female mouse brain. Front Mol Neurosci. 2014;7:11.
101. Liu FJ, Lim KY, Kaur P, Sepramaniam S, Armugam A, Wong PTH, et al. microRNAs involved in regulating spontaneous recovery in embolic stroke model. PLoS One. 2013;8(6):e66393.
102. Xin H, Katakowski M, Wang F, Qian J-Y, Liu XS, Ali MM, et al. MicroRNA cluster miR-17-92 cluster in exosomes enhance neuroplasticity and functional recovery after stroke in rats. Stroke. 2017;48(3):747–53.
103. Friedrich MA, Martins MP, Araújo MD, Klamt C, Vedolin L, Garicochea B, et al. Intra-arterial infusion of autologous bone marrow mononuclear cells in patients with moderate to severe middle cerebral artery acute ischemic stroke. Cell Transplant. 2012;21(1):S13–21.
104. Bhasin A, Srivastava MV, Bhatia R, Mohanty S, Kumaran SS, Bose S. Autologous intravenous mononuclear stem cell therapy in chronic ischemic stroke. J Stem Cells Regen Med. 2012;8(3):181.
105. Lee JS, Hong JM, Moon GJ, Lee PH, Ahn YH, Bang OY. A long-term follow-up study of intravenous autologous mesenchymal stem cell transplantation in patients with ischemic stroke. Stem Cells. 2010;28(6):1099–106.
106. Yang B, Schaar K, Hamilton J, Xi X, Mays R, Savitz SI. Abstract 198: The spleen is a pivotal target of functional recovery after treatment with multistem for acute ischemic stroke. Stroke. 2012;43(Suppl 1):A198.
107. Steinberg GK, Kondziolka D, Wechsler LR, Lunsford LD, Coburn ML, Billigen JB, et al. Clinical outcomes of transplanted modified bone marrow-derived mesenchymal stem cells in stroke: a phase 1/2a study. Stroke. 2016;47(7):1817–24.

108. Reynolds BA, et al. Stem Cell Therapies as an Emerging Paradigm in Stroke (STEPS): bridging basic and clinical science for cellular and neurogenic factor therapy in treating stroke. Stroke. 2009;40(2):510–5.
109. Savitz SI, Chopp M, Deans R, Carmichael ST, Phinney D, Wechsler L, et al. Stem Cell Therapy as an Emerging Paradigm for Stroke (STEPS) II. Stroke. 2011;42(3):825–9.
110. Savitz SI, Cramer SC, Wechsler L. Stem cells as an emerging paradigm in stroke 3. Stroke. 2014;45(2):634–9.

Chapter 10
The Combination of Stem Cell Factor (SCF) and Granulocyte-Colony Stimulating Factor (G-CSF) in Repairing the Brain Post-acute Stroke

Li-Ru Zhao, Suning Ping, and Fei Hao

Abstract Stroke represents the leading cause of long-term disability in adults worldwide. Most stroke survivors suffer from lifelong neurological deficits. Developing a pharmaceutical approach to enhance brain repair and improve functional outcomes post-acute stroke is a very important but less investigated area in stroke research. Stem cell factor (SCF) and granulocyte-colony stimulating factor (G-CSF) are the well-characterized vital hematopoietic growth factors for regulating hematopoiesis. Increasing evidence supports that SCF and G-CSF also play roles in the nervous system. Over the past decade, preclinical studies have demonstrated that SCF in combination with G-CSF synergistically enhances stroke recovery in the subacute or chronic phase. In this chapter, we have reviewed the biological function of SCF and G-CSF in hematopoiesis, neural plasticity, and neurogenesis, and summarized the preclinical studies illustrating the neurorestorative effects of SCF and G-CSF post-acute stroke.

Keywords Stem cell factor • Granulocyte-colony stimulating factor • Brain repair • Stroke • Subacute phase • Chronic phase

Abbreviations

AD	Alzheimer's disease
BBB	Blood-brain barrier
BDA	Biotinylated dextran amine

L.-R. Zhao, M.D., Ph.D. (✉) • S. Ping • F. Hao
Department of Neurosurgery, State University of New York Upstate Medical University, 750 E. Adams Street, Syracuse, NY 13210, USA
e-mail: ZHAOL@upstate.edu

P.A. Lapchak, J.H. Zhang (eds.), *Cellular and Molecular Approaches to Regeneration and Repair*, Springer Series in Translational Stroke Research,
https://doi.org/10.1007/978-3-319-66679-2_10

BDNF	Brain-derived neurotrophic factor
CADASIL	Cerebral autosomal dominant arteriopathy with subcortical infarcts and leucoencephalopathy
CNS	Central nervous system
CSF	Colony stimulating factor
CXCR4	C-X-C chemokine receptor type 4
DRGs	Dorsal root ganglia neurons
ECs	Endothelial cells
G-CSF	Granulocyte-colony stimulating factor
GCSFR	G-CSF receptor
GM-CSF	Granulocyte macrophage-colony stimulating factor
HPCs	Hematopoietic progenitor cells
HSCs	Hematopoietic stem cells
LTP	Long-term potential
NSCs/NPCs	Neural stem/progenitor cells
PPF	Paired-pulse facilitation
SCF	Stem cell factor
SDF-1	Stromal cell-derived factor 1
SGZ	Subgranular zone
SHRs	Spontaneous hypertensive rats
Sl	Steel gene
SVZ	Subventricular zone
tPA	Tissue plasminogen activator
U-type spines	Uncertain type spines
W	White-spotting gene
YFP	Yellow fluorescent protein

1 Introduction

Stroke remains the leading cause of long-term disability in adults worldwide [1, 2]. Stroke not only represents a serious medical condition but it also causes huge medical and financial burdens throughout the world [1–3].

A stroke has three clinical phases: the acute phase, subacute phase, and chronic phase. The exact time frame of these three phases varies among individuals as the duration of the three phases is dependent upon the size and location of the infarcts, the responsive capacity of cerebrovascular collateral circulation, the metabolic state of brain tissue, and patient's age and medical comorbidities. In general, the acute phase is the first 48 h after stroke symptom onset, the subacute phase represents the period from 48 h up to 3 or 6 months post-stroke, whereas the chronic phase starts 3 or 6 months after stroke [4–10].

Currently, there are only two therapeutic approaches available for stroke patients. The *first one* is the thrombolytic/thrombectomy treatment for ischemic stroke patients in the acute phase. The therapeutic time windows for thrombolytic/thrombectomy approach are limited up to 4.5 h post-stroke for thrombolysis by tissue plasminogen activator (tPA) [11, 12] and within 6–8 h post-stroke onset for the thrombectomy [13–15]. The *other* treatment for stroke patients is physical therapy. The therapeutic window for physical therapy is restricted to the first 6 months post-stroke [16–18]. Due to the narrow time window and intracerebral hemorrhage risk of the thrombolytic/thrombectomy treatment [11, 13, 19], the majority of stroke patients are not able to receive this treatment in the acute phase [20]. In addition, many stroke survivors do not receive or complete the physical therapy post-acute stroke because of financial or family-related issues. Developing new therapeutic strategies, therefore, is highly important to reduce stroke-induced disability and enhance stroke recovery.

Over the past two decades, the vast majority of stroke research has targeted the neuroprotection in the acute phase, and little attention has been paid to enhancing stroke recovery in the subacute or chronic phase of stroke. In fact, the neuroprotective agents have all failed in clinical trials [21]. Searching for therapeutic approaches to improve stroke recovery post-acute phase becomes highly recognized in the stroke research field today.

Stem cell factor (SCF) and granulocyte-colony stimulating factor (G-CSF) are the essential hematopoietic growth factors that critically regulate hematopoiesis [22–24]. Beside their roles in the hematopoietic systems, a large body of evidence shows that SCF and G-CSF also play roles in the nervous system. SCF and G-CSF do not only promote neural plasticity [25–30] and neurogenesis [31–33], but they can also enhance brain repair in both the subacute [34] and chronic phases [35–41].

In this chapter, we have reviewed the discovery of SCF and G-CSF in the hematopoietic system and current knowledge concerning the biological function of SCF and G-CSF in hematopoiesis, highlighted the studies demonstrating the effects of SCF and G-CSF in promoting neural plasticity and neurogenesis, and summarized up-to-date research progress regarding the effective and mechanistic determinations of SCF and G-CSF on brain repair in the subacute and chronic phases of experimental stroke.

2 The Discovery and Essential Role of SCF and G-CSF in the Hematopoietic System

SCF and G-CSF are the hematopoietic growth factors that are critically involved in regulation of blood cell production and mobilization of bone marrow stem cells. Since the discovery of SCF and G-CSF, great effort has been made to elucidate their biological function. Over the past six decades, there have been many breakthroughs in understanding the mechanisms underlying SCF- and G-CSF-regulated hematopoiesis and in developing potential therapies for using SCF and G-CSF in clinical trials. In this section, we have summarized the current understanding of the essential role of SCF and G-CSF in the hematopoietic system.

The discovery of SCF (also known as kit ligand, steel factors and mast cell growth factor) and its receptor c-kit took place in 1990s [42]. Observations of white spots on a few mice among thousands of laboratory mice lead to the identification for the loci of steel (Sl) and white-spotting (W), which encode SCF and c-kit receptor, respectively [43]. Mutations at either of these two loci result in similar phenotypes with coat color alterations, anemia, and lack of mast cells in the tissue and neonatal mortality [44]. These findings offer critical information concerning the *in vivo* function of SCF and c-kit, and highlighting its important roles in hematopoiesis, melanogenesis and fertility. In addition, it has been demonstrated that the W mutation-induced c-kit dysfunction affects hematopoietic stem cells and hematopoietic progenitor cells (HSCs/HPCs), while the Sl mutation impairs stromal cell function. These findings are in accordance with the *in vitro* study that was reported in 1977 [45]. In this *in vitro* study, Dexter and Moore demonstrated the stromal-dependent hematopoietic cells culture, and proposed that Sl and W encoded a ligand-receptor pair.

Many studies have revealed that there are two forms of natural SCF due to alternative splicing of the DNA transcripts. A shorter form consists of 220 amino acids and produces a membrane-bound form of SCF. The soluble SCF, which consists of 165 amino acids, is derived from a full length 248 amino acids cleaved in the extracellular domain [46]. Both the soluble and membrane-bound forms of SCF are biologically active. However, the two forms of SCF have distinct but overlapping roles [47]. Membrane-bound SCF is expressed on stromal cells, endothelial cells (ECs) and fibroblasts in the bone marrow and induces more persistent tyrosine kinase activation than soluble SCF [48]. In 1991, Brannan and colleagues reported that Steel-Dickie mice exhibited anemia, pigmentation and germ cell defects as these mice only produced soluble SCF due to genome deletion affecting the transmembrane and cytoplasmic domain [49]. These research findings suggest that membrane-bound SCF plays a unique biological role in the stromal cells, ECs and fibroblasts in the bone marrow. C-kit is expressed on normal hematopoietic cells and several other cell types, including mast cells [50], melanocytes [51] and a wide range of non-hematopoietic cell types as ECs [52], interstitial cells [53] and astrocytes [43]. Interaction between SCF and c-kit is the initial and key step for triggering the downstream signaling. It has been shown that the SCF/c-kit system has an important function not only in mouse but also in humans and other primates due to its pleiotropic effects on hematopoietic cell survival, proliferation, differentiation and mobilization [54]. SCF acts directly on HSCs/HPCs, promotes HSC/HPC entry to the cell cycle, and facilitates HSC/HPC proliferation [55]. SCF enhances the primitive HSC survival by suppression of apoptosis [56, 57]. In 1992, Valent and colleagues reported that SCF induced mast cell development from immature hematopoietic cells in human bone marrow [58], suggesting the effect of SCF on cell differentiation. This study was also confirmed by Irani and colleagues, who observed similar results in human fetal liver [59]. In addition, SCF is also a potent agent for mobilization of murine and human HSCs from bone marrow to peripheral blood [43], which shares a similar role as G-CSF.

The discovery of G-CSF occurred much earlier than SCF. The identification of G-CSF happened in the 1960s by *in vitro* assays measuring the ability of the growth factors to stimulate colony formation of bone marrow cells (see review by [60]). During the 1970s, a detailed category of colony stimulating factor (CSF) had been published, and G-CSF had been defined as a stimulator specific for colonies containing predominantly neutrophils [61]. G-CSF is produced by a variety of cells, of which, monocyte/macrophage lineage cells are the most prominent source [60]. G-CSF is also produced by normal mesothelial cells [62], fibroblasts [63] and ECs [64]. G-CSF, like other growth factors, exerts its biological functions by binding to the G-CSF specific receptor, G-CSFR. G-CSFR expression has been found on a variety of hematopoietic cells, including myeloid progenitors, mature neutrophils, monocytes, myeloid cells, lymphoid leukemia cells, and normal B and T cells [65].

Generally, G-CSF is known to have multiple functions in regulation of HSC/HPC proliferation, differentiation and mobilization, neutrophil production and mobilization from the bone marrow, neutrophil progenitor cell proliferation and differentiation, and the state of functional activation of neutrophils. In 1987, Tamura and colleagues reported that G-CSF mobilized large numbers of hematopoietic cells from the bone marrow into the circulation [66]. In addition, increased progenitor cells of all lineages were detected in the spleen of G-CSF-treated mice. These results were further confirmed by Duhrsen and colleagues in cancer patients with G-CSF treatment [67]. The administration of G-CSF to the patients showed significant increases of circulating HSCs, followed by a slight reduction in the frequency of bone marrow progenitor cells. However, the absolute number of the progenitor cells in the bone marrow was still increased. Together, all these data support the efficacy of G-CSF on HSC/HPC mobilization and proliferation. As peripheral blood is one of the important sources for stem cell transplantation, the biological effects of G-CSF in HSC/HPC mobilization and proliferation therefore allow autologous and allogeneic HSC transplantation in the clinical setting [68]. G-CSF is also a strong stimulator for neutrophil activation. Masja and colleagues reported that G-CSF increased the release of inflammatory granules [69]. In addition, G-CSF stimulates the survival and primitive proliferation of progenitor cells *in vitro* by combination with other factors. McNiece and colleagues observed more numerous and larger colonies of progenitor cells after combination treatment of G-CSF and GM-CSF as compared to either single factor [70].

G-CSF also shows a synergistic effect with SCF in regulating many important biological responses. As stated earlier, both SCF and G-CSF have effects on regulating survival, proliferation, differentiation, and mobilization of HSCs/HPCs and hematopoietic lineage; the combination of SCF and G-CSF shows an enhanced effect. SCF in combination with G-CSF increases more progenitor cell mobilization in peripheral blood than SCF or G-CSF alone [71]. Many clinical trials have further confirmed this finding. Combined treatment of SCF and G-CSF show beneficial effects on peripheral blood progenitor cell mobilization with an increased number of CD34+ cells/kg in circulating system in patients who have received high dose chemotherapy for lymphoma [72, 73], breast cancer [74] and multiple myeloma

[75]. The synergistic effect of SCF + G-CSF on HSC/HPC proliferation has also been illustrated. The synergistic effect of combined treatment of SCF and G-CSF in HSC/HPC proliferation is not only because of the enhanced ligand/receptor interaction [76], but it also due to the marked shortening of the duration of G0/G1 phase [76]. A direct effect of SCF and G-CSF on cell cycle distribution has been identified, and this effect is mainly induced by the regulation of cyclin-dependent kinase inhibitor p27kip1 [77]. Besides, the combination of SCF and G-CSF also shows a synergistic enhancement of STAT3 and MAPK signaling [76], which is involved in promoting the cell proliferation.

3 The Role of SCF and G-CSF in the Central Nervous System: Neural Plasticity and Neurogenesis

In addition to the effects of SCF and G-CSF in the hematopoietic system, increasing evidence shows that SCF and G-CSF also play a role in the central nervous system (CNS). Receptors for SCF and G-CSF have been found to express in the brain [78, 79], particularly in the neural stem cells/neural progenitor cells (NSCs/NPCs) [31–33, 80], and in cerebral neurons [32, 80] of adult mice and rats. It has been demonstrated that both SCF and G-CSF can pass through the blood-brain barrier [32, 81]. These findings suggest that hematopoietic growth factors, SCF and G-CSF, may have biological function in the CNS.

Numerous *in vitro* and *in vivo* studies have examined the contribution of SCF and G-CSF in the neuronal plasticity. SCF and G-CSF have been shown to play a key role in regulation of the neural plasticity in both the developing and adult brains. *In vitro* studies have shown that SCF supports the survival of c-kit-positive dorsal root ganglia neurons (DRGs) and promotes the neurite outgrowth of mouse embryonic DRGs through the c-kit receptor tyrosine kinase activity [25]. In addition, SCF has also shown to increase the neurite outgrowth of cultured cortical neurons [30]. In cultured brain slices, SCF selectively promotes outgrowth of commissural axons, which highly express SCF receptor [82]. *In vivo* studies, commissural axons fail to exit the floor plate in SCF and c-kit mutant mice [82]. In addition, c-kit conditional knockout mice show delayed extension of callosal fibers within the contralateral cortex and fail to innervate their target area [83]. At the functional level, SCF mutant mice exhibit a reduction of baseline synaptic transmission between dentate gyrus and hippocampal CA3 pathway and show deficits in spatial learning and memory [27]. C-kit mutant rats and mice both display impairments of paired-pulse facilitation (PPF) and long-term potential (LTP) in the hippocampal mossy fiber-CA3 pathway and a deficit in performance in Morris water maze task [28, 84]. In mouse brain slices, SCF binding to c-kit receptor activates PI3K/PLA_2 intracellular pathway, modulates PPF and LTP, and regulates synaptic transmission in the hippocampus [84]. These studies suggest that SCF/c-kit signaling is involved in the structural and functional regulation of synaptic plasticity. In addition to SCF, G-CSF also participates in neural plasticity.

G-CSF knockout mice show impaired LTP, reduced densities of NMDA receptors and dendritic complexity of hippocampal neurons in the dentate gyrus and the CA1 region, and deficiency in spatial learning and memory [29]. G-CSF treatment restores impaired long-term depression (LTD) in a mouse model of Alzheimer's disease (AD) [85]. In addition, subcutaneous injection of G-CSF increases the dendritic length and complexity of pyramidal neurons in the peri-infarct cortex in the cerebral ischemia rats [86]. Furthermore, it has been demonstrated that G-CSF in combination with SCF synergistically promotes neurite outgrowth and network formation of cultured cortical neurons through the PI3K/AKT/NF-kB/BDNF pathway [30]. Collectively, these research data suggest that SCF and G-CSF, the two hematopoietic growth factors, act as neurotrophic factors to regulate the neural plasticity during development and maturity.

The role of SCF and G-CSF in promoting neurogenesis and directing NSCs/NPCs to give rise to neurons has been illustrated in both *in vitro* and *in vivo* studies. There are two neurogenic regions in the adult mammalian brain, including the subventricular zone (SVZ) surrounding the anterior part of lateral ventricles and subgranular zone (SGZ) of the hippocampal dentate gyrus. NSCs/NPCs in these regions have regenerative potential, which has been postulated as a likely source for neural repair. Infusing SCF into the cerebrolateral ventricle has been shown to increase the number of newborn neurons in the SVZ [31]. Injection of anti-c-kit antibody into the cisterna magnum increases the number of cell death and results in thinning of the cerebral cortex, suggesting essential role of SCF/c-kit for cortical progenitor cell survival [87]. In cultured NSCs/NPCs, G-CSF is shown to promote the differentiation of NSCs/NPCs into neurons in a dose dependent manner [32, 88]. In G-CSF knockout mice, hippocampal neurogenesis is strongly diminished, and the mice show deficits in behavioral plasticity [29]. Peripheral or intraventricular administration of G-CSF has been demonstrated to increase the neurogenesis and promote the proliferation and differentiation of NSCs/NPCs, not only in the intact mice and rats [32, 88], but also in the animal models of neurological disorders, including the cerebral ischemia, perinatal hypoxia, irradiation-induced brain injury, traumatic brain injury, AD, cerebral autosomal dominant arteriopathy with subcortical infarcts and leukoencephalopathy (CADASIL) and bacterial meningitis [32, 34, 89–98]. G-CSF-induced neurogenesis is probably associated with G-CSF receptor-mediated phosphorylation of transcription factor STAT3/5 [88]. Besides, combination of SCF and G-CSF has been reported to have a synergistic effect in facilitating the proliferation of intrinsic NSCs/NPCs in a mouse model of cerebral ischemia [34]. When adding SCF and G-CSF into the culture medium during the proliferating stage of NSCs/NPCs, SCF in combination with G-CSF (SCF + G-CSF) shows a dual function in directing cell cycle arrest and promoting neuronal fate commitment through the regulation of neurogenin 1 [33]. Together, these studies reveal that SCF and G-CSF are involved in the regulation of NSC/NPC proliferation and neurogenesis.

In addition to promoting the proliferation and differentiation of intrinsic NSCs/NPCs, the combination of SCF and G-CSF also mobilizes bone marrow-derived cells, causing them to migrate into the brain and differentiate into various types of cells, including neurogenesis. The fate of bone marrow-derived cells in the brain is dependent upon the microenvironment of the brain. In the subacute and chronic stroke brain, SCF + G-CSF treatment augments bone marrow-derived endothelial cells and neurons [34, 36]. In the brains of CADASIL mice, SCF + G-CSF selectively directs bone marrow-derived cells toward neuronal fate commitment [99]. In the APP/PS1 transgenic mice, bone marrow-derived microglial cells are significantly increased in the brain following SCF + G-CSF treatment, suggesting that SCF + G-CSF treatment leads to an enhancement in microglial fate commitment of bone marrow-derived cells in the brain with β-amyloid deposits [92, 100, 101]. G-CSF treatment has also been shown to mobilize bone marrow-derived mesenchymal stem cells, promote the migration and differentiation of mesenchymal stem cells into the neurons, and contribute to neurogenesis in the brains of AD mice [102]. The C-X-C chemokine receptor type 4 (CXCR4)/stromal cell-derived factor 1 (SDF-1) has been shown to be a key mediator in G-CSF-based recruitment of bone marrow-derived cells [102]. Together, these studies suggest that although bone marrow-derived cells possess different phenotypes in various brain conditions, these cells may participate in the neurogenesis and brain repair.

4 SCF and G-CSF Combination in Brain Repair Post-acute Stroke: Effective and Mechanistic Determinations

As stated in the previous section, substantial evidence has revealed the capacity of SCF and G-CSF in promoting neural plasticity and neurogenesis, and the permeability of the BBB to SCF and G-CSF. These discoveries provide a scientific base for seeking to determine the therapeutic effectiveness of SCF and G-CSF in enhancing brain repair and stroke recovery in the subacute phase and/or the chronic phase of stroke.

Here we highlight the preclinical studies demonstrating the efficacy and possible mechanisms of SCF and G-CSF in brain repair during subacute or chronic phases of experimental stroke.

4.1 The Effects of SCF and G-CSF on Brain Repair in the Subacute Phase of Stroke

There are a few preclinical studies demonstrating the therapeutic efficacy of SCF and G-CSF in the subacute phase of stroke. Using a transient focal ischemia model in Sprague-Dawley rats, Lee and co-workers [103] reported that intraperitoneal injections of G-CSF for 3 days beginning at 4 or 7 days post-ischemia led to motor function improvement, infarction size reduction, and increased angiogenesis. Kawada and colleagues [34] injected SCF and G-CSF subcutaneously during the period of 11–20 days after induction of focal cerebral ischemia in C57BL mice, and observed that the SCF and G-CSF treatment increased the number of bone marrow-derived neuronal cells in the ipsilesional hemisphere and promoted the proliferation of intrinsic NSCs/NPCs in the SVZ. In addition, they also found that the SCF + G-CSF synergistically enhanced NSC/NPC proliferation in the SVZ when compared with treatment of SCF or G-CSF alone [34]. How SCF + G-CSF optimally repairs the brain in the subacute phase of stroke has not been clarified. Using the same treatment paradigm as reported by Kawada and colleagues [34], SCF + G-CSF treatment was found to upregulate IL-10, an anti-inflammatory cytokine, and to reduce infiltration of microglial/macrophages in the infarcted brain [104]. Although inhibiting inflammation by SCF + G-CSF may provide a favorable microenvironment for neurogenesis in the subacute phase of stroke, the causal link among the SCF + G-CSF-induced neurogenesis, anti-inflammation, and motor function improvement remains to be elucidated.

4.2 *The Effects of SCF and G-CSF on Brain Repair in the Chronic Phase of Stroke*

Most stroke patients still carry different degrees of disability when they enter into the chronic phase of stroke although many of them have received thrombolytic therapy in the acute phase [19] and physical therapy during the subacute phase [16]. However, in the chronic phase, there has been no therapy available for enhancing stroke recovery as it has been believed that the opportunity for obtaining recovery is largely ended by the time stroke patients enter the chronic phase [17, 105].

Brain plasticity is an intrinsic ability of the brain to reorganize its function and modify its structure in response to stimuli and injuries from both internal and external sources. Accumulating evidence supports that brain plasticity exists throughout a person's lifespan [106–111]. Accordingly, there is a possibility that a stroke-damaged brain may still be reparable during the chronic phase.

Over the past decade, our research team has demonstrated the safety, efficacy and possible mechanisms of SCF and G-CSF on stroke recovery in the chronic phase of stroke using rat and mouse models of cerebral cortical ischemia.

4.2.1 The Efficacy, Safety and Effective Dosage of SCF and G-CSF on Stroke Recovery in the Chronic Phase of Stroke

Systemic administration of SCF and G-CSF during the period of 3.5–6 months after cerebral cortical ischemia has been demonstrated and validated to be effective for brain repair in spontaneously hypertensive rats (SHRs), C57BL mice, or transgenic mice with C57BL genetic background [35–39, 41]. SHRs are used for making a stroke model because hypertension is the most important risk factor for stroke in humans [112]. Chronic hypertension leads to extensive pathological changes in the cerebrovasculature [113, 114]. Numerous studies have illustrated that the cerebral cortical ischemia model in SHRs shows a more consistent and larger infarction in the cortex than in normotensive rats due to poor collateral circulation [35, 80, 114–121]. This model also causes permanent deficits in somatosensorimotor function that last up to the chronic phase of stroke [35, 80, 118–122]. In addition to the cortical infarct model in SHRs, we also use C57BL mice or transgenic mice with C57BL genetic background to make the cerebral cortical infarct model for exploring the mechanisms behind the SCF and G-CSF-enhanced brain repair in chronic stroke.

First of all, the therapeutic efficacy of SCF and G-CSF on stroke recovery in the chronic phase has been examined using the cerebral cortical ischemia model in SHRs. SCF (200 μg/kg), G-CSF (50 μg/kg), or SCF + G-CSF was subcutaneously injected daily for 7 days beginning at 3.5 months post-ischemic stroke. Among the treatment groups, only the SCF + G-CSF treatment led to a stable and long-term (17 weeks) improvement in somatosensory motor function. SCF alone treatment improved functional outcomes but the improvement did not present as stable as the SCF + G-CSF combination treatment. G-CSF alone treatment, however, did not result in functional benefits. The research data of field-evoked potentials lent further support to the neurobehavioral findings and revealed a reestablished normal pattern of somatosensory pathways by SCF + G-CSF treatment [35]. These findings provide first evidence that SCF + G-CSF combination treatment in the chronic phase of stroke can enhance stroke recovery.

Given the fact that stroke has the highest incidence in the elderly [2], the safety, efficacy, and optimal dosage of SCF + G-CSF combination treatment on chronic stroke recovery have been assessed in experimental stroke using aged SHRs and C57BL mice [37]. Six dosages of SCF + G-CSF ranging from 5 μg/kg (SCF) and 2.5 μg/kg (G-CSF) to 200 μg/kg (SCF) and 50 μg/kg (G-CSF) have been examined [37]. The treatment was initiated at 3–4 months post-experimental stroke. All the tested dosages did not show either acute or chronic toxicity to the livers and kidneys, demonstrating the safety of SCF + G-CSF treatment for chronic stroke in the aged population. The higher dosages (SCF/G-CSF: 200/50,100/25, and 50/25 μg/kg) showed the most effective outcomes in mobilizing circulating stem cells and in stably improving functional recovery. The intermediate dose of SCF + G-CSF (20/10 μg/kg) displayed a short-term improvement, whereas the dosages less than 20/10 μg/kg did not lead to functional improvement in chronic stroke in aged SHRs. These findings demonstrate that SCF + G-CSF treatment for chronic stroke recovery is a safe and effective therapeutic approach for the aged population and acts in a dose dependent manner.

4.2.2 The Possible Mechanisms Underlying the SCF + G-CSF-Enhanced Recovery in the Chronic Phase of Stroke

We have employed the approaches of bone marrow-derived cell tracking, molecular manipulation, live brain imaging, whole brain imaging, axon tracking, immunohistochemistry, confocal imaging, and neurobehavioral testing to determine how SCF + G-CSF repairs a stroke-damaged brain in the chronic phase.

By tracking bone marrow-derived cells through bone marrow transplantation, our study has revealed that increased bone marrow-derived endothelial cells and bone marrow-derived neurons are involved in SCF + G-CSF-enhanced angiogenesis and neurogenesis in the brain of chronic stroke [36].

Previous studies have shown that the receptors for SCF and G-CSF are expressed in cerebral neurons [32, 80] and cerebral endothelial cells [81] of adult mice and rats, and that both the SCF and G-CSF can pass through the blood-brain barrier of the adult rodent brain [32, 81]. Can SCF + G-CSF treatment in chronic stroke remodel the neural networks in an aged brain? To address this question, we used 2-photon microscopy to scan the brain area adjacent to the infarct cavity before and after SCF + G-CSF treatment in aged Thy-1-YFPH mice (C57BL background) [38]. In the brains of Thy-1-YFPH mice, the yellow fluorescent protein (YFP) is exclusively expressed in the layer V pyramidal neurons [123]. The mushroom spines with large heads on the dendrites are unique spines forming functioning synapses [124, 125]. Before SCF + G-CSF treatment, the mushroom spines of layer V pyramidal neurons were decreased, and the uncertain type (U-type) spines, which cannot build synapses with other neurons, were increased in the chronic stroke brain. This observation indicates that reduced synaptic circuits occur in the peri-infarct cavity cortex in the chronic stroke brain. However, 6 weeks after treatment, increased mushroom spines with decreased U-type spines were found in the brains of SCF + G-CSF-treated stroke mice. In addition, the densities of dendrites and PSD-95 were also increased in the ipsilesional cortex by SCF + G-CSF treatment. These findings demonstrate that SCF + G-CSF intervention in the chronic phase of stroke enhances synaptic network regeneration in the ipsilesional cortex of aged brains.

To distinguish whether SCF + G-CSF can directly modulate neural network formation, we carried out an *in vitro* study by determining neurite outgrowth of primary cortical neurons [30]. We found the expression of SCF and G-CSF receptors on the neurite growth cones. SCF + G-CSF showed synergistic effects in promoting neurite extension, activating NF-kB, and upregulating brain-derived neurotrophic factor (BDNF). Blockage of NF-kB activation eliminated the SCF + G-CSF-increased neurite outgrowth and BDNF production [30]. These data demonstrate the direct and synergistic efficacy of SCF + G-CSF in promoting neurite outgrowth, which is the initial step for generating neural networks. SCF + G-CSF enhances neurite extension through the NF-kB signaling.

Based on the *in vitro* findings, we then sought to use NF-kB inhibitor for blocking SCF + G-CSF-promoted neural network regeneration and to elucidate whether there is a dependent link between the SCF + G-CSF-enhanced neural network remodeling in the ipsilesional cortex and the SCF + G-CSF-improved motor function in chronic stroke. In an *in vivo* study [39], the NF-kB inhibitor was infused into the lateral ventricle through an osmotic pump for 7 days beginning at 1 h before a 7 day treatment (s.c.) of SCF + G-CSF, which was initiated 4 months after cortical ischemia. To track axons projecting from the contralesional hemisphere, an anterograde neuronal tracer, biotinylated dextran amine (BDA), was injected into the somatosensorimotor cortex in the contralesional hemisphere. After motor function testing 2 and 6 weeks after treatment, mice were sacrificed at 10 weeks post-treatment. Our findings have revealed that SCF + G-CSF-increased BDA-labeled axons, PSD-95 accumulation, and blood vessel density in the peri-infarct cavity is eliminated by NF-kB inhibitor. In addition, the SCF + G-CSF-induced motor functional improvement is also prevented by NF-kB inhibitor. These data suggest that the SCF + G-CSF-improved functional outcome in chronic stroke may depend on the regeneration of neural networks and vasculature in the peri-infarct cavity cortex. However, this terminal determination study is limited to clarify the dynamically causal link between the SCF + G-CSF-promoted neural network rewiring and functional improvement in chronic stroke.

To over come this limitation, we conducted a unique study combining live brain imaging and motor function evaluation to simultaneously examine the dependent relationship between the SCF + G-CSF-enhanced synaptic network remodeling and motor function improvement in the chronic phase of experimental stroke [40]. To prevent the influence of behavioral testing-induced neural network remodeling, the following two sets of experiments were carried out simultaneously: (1) Thy1-YFPH mice with cortical infarction for live brain imaging at 2 and 6 weeks post-SCF + G-CSF treatment, and (2) Thy1-YFPH mice with cortical infarction for motor function assessment at 2 and 6 weeks post-SCF + G-CSF treatment. The SCF + G-CSF treatment was initiated at 6 months post-experimental stroke. We observed that once the SCF + G-CSF-increased mushroom spines in the ipsilesional motor cortex were eliminated by NF-kB inhibitor, the SCF + G-CSF-improved motor function was simultaneously prevented. This observation provides solid evidence validating a clear causal link between SCF + G-CSF-promoted neural network remodeling and motor functional improvement and strongly supporting that the SCF + G-CSF-enhanced neural network rewiring in the ipsilesional motor cortex is required for SCF + G-CSF-improved motor function in the chronic stroke.

It is worth noting that the SCF + G-CSF-increased dendritic spine head size, PSD-95 accumulation, and blood vessel density in the peri-infarct cortex are much greater than in the contralesional hemisphere of chronic stroke brain and in the intact control mouse brain [39, 40]. Blocking the SCF + G-CSF-induced "over growth" of synaptic networks and vasculature in the peri-infarct cortex by NF-kB inhibitor leads to abolition of the SCF + G-CSF-improved motor function in chronic stroke [39, 40]. These findings reveal that SCF + G-CSF-strengthened synaptic function in the peri-infarct motor cortex plays a vital role in motor functional improvement in chronic stroke.

5 Concluding Remarks

SCF and G-CSF were initially discovered as critical hematopoietic growth factors to regulate hematopoiesis. SCF in combination with G-CSF has been demonstrated to have synergistic effects in promoting the proliferation, differentiation and survival of HSCs/HPCs, and in mobilization of HSCs/HPCs into the blood.

Numerous studies have demonstrated that both SCF and G-CSF are crucially involved in neural plasticity and neurogenesis. These findings significantly advance our knowledge of these two hematopoietic growth factors: the biological function of SCF and G-CSF is not only limited in the hematopoietic system but it also acts in the CNS. In addition to the synergistic effects of SCF + G-CSF in the hematopoietic system, our research team has, for the first time, illustrated that the combination of SCF and G-CSF also synergistically promote neurite outgrowth of primary cortical neurons.

Over the past decade, the contribution of SCF and G-CSF in brain repair post-acute stroke has been determined. Importantly, it has been demonstrated that SCF in combination with G-CSF synergistically enhances brain repair in the subacute phase (by Kawada's group) and chronic phase (by our research team) of experimental stroke. These findings extend current understanding concerning the neurorestorative efficacy of SCF + G-CSF in brain repair post-acute stroke and provide a new approach for enhancing stroke recovery.

Acknowledgments The authors would like to thank Sandy McGillis for her assistance in editing the manuscript. This work was partially supported by the National Institutes of Health, National Institute of Neurological Disorders and Stroke (R01 NS060911), National Institute on Aging (1R01AG051674), and Department of Veterans Affairs (I01 RX002125) in the United States.

References

1. Feigin VL, Forouzanfar MH, Krishnamurthi R, Mensah GA, Connor M, Bennett DA, et al. Global and regional burden of stroke during 1990–2010: findings from the Global Burden of Disease Study 2010. Lancet. 2014;383(9913):245–55.
2. Mozaffarian D, Benjamin EJ, Go AS, Arnett DK, Blaha MJ, Cushman M, et al. Heart Disease and Stroke Statistics-2016 update: a report from the American Heart Association. Circulation. 2016;133(4):e38–e360.
3. Skolarus LE, Freedman VA, Feng C, Wing JJ, Burke JF. Care received by elderly US stroke survivors may be underestimated. Stroke. 2016;47(8):2090–5.
4. Bernheisel CR, Schlaudecker JD, Leopold K. Subacute management of ischemic stroke. Am Fam Physician. 2011;84(12):1383–8.
5. Donnan GA, Fisher M, Macleod M, Davis SM. Stroke. Lancet. 2008;371(9624):1612–23.
6. Kang DW, Latour LL, Chalela JA, Dambrosia JA, Warach S. Early and late recurrence of ischemic lesion on MRI: evidence for a prolonged stroke-prone state? Neurology. 2004;63(12):2261–5.
7. Maraka S, Jiang Q, Jafari-Khouzani K, Li L, Malik S, Hamidian H, et al. Degree of corticospinal tract damage correlates with motor function after stroke. Ann Clin Transl Neurol. 2014;1(11):891–9.
8. Parsons MW, Li T, Barber PA, Yang Q, Darby DG, Desmond PM, et al. Combined (1)H MR spectroscopy and diffusion-weighted MRI improves the prediction of stroke outcome. Neurology. 2000;55(4):498–505.
9. Poh T. Time course of ischemic stroke on non-enhanced CT. Brain Stories. 2013:1–8.
10. van Delden AE, Peper CE, Beek PJ, Kwakkel G. Unilateral versus bilateral upper limb exercise therapy after stroke: a systematic review. J Rehabil Med. 2012;44(2):106–17.
11. National Institute of Neurological Disorders and Stroke rt-PA Stroke Study Group. Tissue plasminogen activator for acute ischemic stroke. N Engl J Med. 1995;333(24):1581–7.

12. Hacke W, Kaste M, Bluhmki E, Brozman M, Davalos A, Guidetti D, et al. Thrombolysis with alteplase 3 to 4.5 hours after acute ischemic stroke. N Engl J Med. 2008;359(13):1317–29.
13. Penumbra Pivotal Stroke Trial I. The penumbra pivotal stroke trial: safety and effectiveness of a new generation of mechanical devices for clot removal in intracranial large vessel occlusive disease. Stroke. 2009;40(8):2761–8.
14. Castano C, Dorado L, Guerrero C, Millan M, Gomis M, Perez de la Ossa N, et al. Mechanical thrombectomy with the Solitaire AB device in large artery occlusions of the anterior circulation: a pilot study. Stroke. 2010;41(8):1836–40.
15. Saver JL, Goyal M, Bonafe A, Diener HC, Levy EI, Pereira VM, et al. Stent-retriever thrombectomy after intravenous t-PA vs. t-PA alone in stroke. N Engl J Med. 2015;372(24):2285–95.
16. Kwakkel G, Kollen BJ, van der Grond J, Prevo AJ. Probability of regaining dexterity in the flaccid upper limb: impact of severity of paresis and time since onset in acute stroke. Stroke. 2003;34(9):2181–6.
17. Hendricks HT, van Limbeek J, Geurts AC, Zwarts MJ. Motor recovery after stroke: a systematic review of the literature. Arch Phys Med Rehabil. 2002;83(11):1629–37.
18. Schaechter JD. Motor rehabilitation and brain plasticity after hemiparetic stroke. Prog Neurobiol. 2004;73(1):61–72.
19. Emberson J, Lees KR, Lyden P, Blackwell L, Albers G, Bluhmki E, et al. Effect of treatment delay, age, and stroke severity on the effects of intravenous thrombolysis with alteplase for acute ischaemic stroke: a meta-analysis of individual patient data from randomised trials. Lancet. 2014;384(9958):1929–35.
20. Warlow C, Sudlow C, Dennis M, Wardlaw J, Sandercock P. Stroke. Lancet. 2003;362(9391):1211–24.
21. Moskowitz MA, Lo EH, Iadecola C. The science of stroke: mechanisms in search of treatments. Neuron. 2010;67(2):181–98.
22. Welte K, Platzer E, Lu L, Gabrilove JL, Levi E, Mertelsmann R, et al. Purification and biochemical characterization of human pluripotent hematopoietic colony-stimulating factor. Proc Natl Acad Sci U S A. 1985;82(5):1526–30.
23. Zsebo KM, Wypych J, McNiece IK, Lu HS, Smith KA, Karkare SB, et al. Identification, purification, and biological characterization of hematopoietic stem cell factor from buffalo rat liver—conditioned medium. Cell. 1990;63(1):195–201.
24. Zsebo KM, Williams DA, Geissler EN, Broudy VC, Martin FH, Atkins HL, et al. Stem cell factor is encoded at the Sl locus of the mouse and is the ligand for the c-kit tyrosine kinase receptor. Cell. 1990;63(1):213–24.
25. Hirata T, Morii E, Morimoto M, Kasugai T, Tsujimura T, Hirota S, et al. Stem cell factor induces outgrowth of c-kit-positive neurites and supports the survival of c-kit-positive neurons in dorsal root ganglia of mouse embryos. Development. 1993;119(1):49–56.
26. Hirata T, Kasugai T, Morii E, Hirota S, Nomura S, Fujisawa H, et al. Characterization of c-kit-positive neurons in the dorsal root ganglion of mouse. Brain Res Dev Brain Res. 1995;85(2):201–11.
27. Motro B, Wojtowicz JM, Bernstein A, van der Kooy D. Steel mutant mice are deficient in hippocampal learning but not long-term potentiation. Proc Natl Acad Sci U S A. 1996;93(5):1808–13.
28. Katafuchi T, Li AJ, Hirota S, Kitamura Y, Hori T. Impairment of spatial learning and hippocampal synaptic potentiation in c-kit mutant rats. Learn Mem. 2000;7(6):383–92.
29. Diederich K, Sevimli S, Dorr H, Kosters E, Hoppen M, Lewejohann L, et al. The role of granulocyte-colony stimulating factor (G-CSF) in the healthy brain: a characterization of G-CSF-deficient mice. J Neurosci. 2009;29(37):11572–81.
30. Su Y, Cui L, Piao C, Li B, Zhao LR. The effects of hematopoietic growth factors on neurite outgrowth. PLoS One. 2013;8(10):e75562.
31. Jin K, Mao XO, Sun Y, Xie L, Greenberg DA. Stem cell factor stimulates neurogenesis in vitro and in vivo. J Clin Invest. 2002;110(3):311–9.
32. Schneider A, Kruger C, Steigleder T, Weber D, Pitzer C, Laage R, et al. The hematopoietic factor G-CSF is a neuronal ligand that counteracts programmed cell death and drives neurogenesis. J Clin Invest. 2005;115(8):2083–98.

33. Piao CS, Li B, Zhang LJ, Zhao LR. Stem cell factor and granulocyte colony-stimulating factor promote neuronal lineage commitment of neural stem cells. Differentiation. 2012;83(1):17–25.
34. Kawada H, Takizawa S, Takanashi T, Morita Y, Fujita J, Fukuda K, et al. Administration of hematopoietic cytokines in the subacute phase after cerebral infarction is effective for functional recovery facilitating proliferation of intrinsic neural stem/progenitor cells and transition of bone marrow-derived neuronal cells. Circulation. 2006;113(5):701–10.
35. Zhao LR, Berra HH, Duan WM, Singhal S, Mehta J, Apkarian AV, et al. Beneficial effects of hematopoietic growth factor therapy in chronic ischemic stroke in rats. Stroke. 2007;38(10):2804–11.
36. Piao CS, Gonzalez-Toledo ME, Xue YQ, Duan WM, Terao S, Granger DN, et al. The role of stem cell factor and granulocyte-colony stimulating factor in brain repair during chronic stroke. J Cereb Blood Flow Metab. 2009;29(4):759–70.
37. Piao CS, Gonzalez-Toledo ME, Gu X, Zhao LR. The combination of stem cell factor and granulocyte-colony stimulating factor for chronic stroke treatment in aged animals. Exp Transl Stroke Med. 2012;4(1):25.
38. Cui L, Murikinati SR, Wang D, Zhang X, Duan WM, Zhao LR. Reestablishing neuronal networks in the aged brain by stem cell factor and granulocyte-colony stimulating factor in a mouse model of chronic stroke. PLoS One. 2013;8(6):e64684.
39. Cui L, Duchamp NS, Boston DJ, Ren X, Zhang X, Hu H, et al. NF-kappaB is involved in brain repair by stem cell factor and granulocyte-colony stimulating factor in chronic stroke. Exp Neurol. 2015;263:17–27.
40. Cui L, Wang D, McGillis S, Kyle M, Zhao LR. Repairing the brain by SCF+G-CSF treatment at 6 months postexperimental stroke: mechanistic determination of the causal link between neurovascular regeneration and motor functional recovery. ASN Neuro. 2016;8(4). https://doi.org/10.1177/1759091416655010.
41. Liu Y, Popescu M, Longo S, Gao M, Wang D, McGillis S, et al. Fibrinogen reduction and motor function improvement by hematopoietic growth factor treatment in chronic stroke in aged mice: a treatment frequency study. Cell Transplant. 2016;25(4):729–34.
42. Galli SJ, Tsai M, Wershil BK. The c-kit receptor, stem cell factor, and mast cells. What each is teaching us about the others. Am J Pathol. 1993;142(4):965–74.
43. Ashman LK. The biology of stem cell factor and its receptor C-kit. Int J Biochem Cell Biol. 1999;31(10):1037–51.
44. Russell ES. Hereditary anemias of the mouse: a review for geneticists. Adv Genet. 1979;20:357–459.
45. Dexter TM, Moore MA. In vitro duplication and "cure" of haemopoietic defects in genetically anaemic mice. Nature. 1977;269(5627):412–4.
46. Lennartsson J, Ronnstrand L. Stem cell factor receptor/c-Kit: from basic science to clinical implications. Physiol Rev. 2012;92(4):1619–49.
47. Smith MA, Court EL, Smith JG. Stem cell factor: laboratory and clinical aspects. Blood Rev. 2001;15(4):191–7.
48. Miyazawa K, Williams DA, Gotoh A, Nishimaki J, Broxmeyer HE, Toyama K. Membrane-bound Steel factor induces more persistent tyrosine kinase activation and longer life span of c-kit gene-encoded protein than its soluble form. Blood. 1995;85(3):641–9.
49. Brannan CI, Lyman SD, Williams DE, Eisenman J, Anderson DM, Cosman D, et al. Steel-Dickie mutation encodes a c-kit ligand lacking transmembrane and cytoplasmic domains. Proc Natl Acad Sci U S A. 1991;88(11):4671–4.
50. Mayrhofer G, Gadd SJ, Spargo LD, Ashman LK. Specificity of a mouse monoclonal antibody raised against acute myeloid leukaemia cells for mast cells in human mucosal and connective tissues. Immunol Cell Biol. 1987;65(Pt 3):241–50.
51. Nocka K, Majumder S, Chabot B, Ray P, Cervone M, Bernstein A, et al. Expression of c-kit gene products in known cellular targets of W mutations in normal and W mutant mice—evidence for an impaired c-kit kinase in mutant mice. Genes Dev. 1989;3(6):816–26.
52. Broudy VC, Kovach NL, Bennett LG, Lin N, Jacobsen FW, Kidd PG. Human umbilical vein endothelial cells display high-affinity c-kit receptors and produce a soluble form of the c-kit receptor. Blood. 1994;83(8):2145–52.

53. Torihashi S, Ward SM, Nishikawa S, Nishi K, Kobayashi S, Sanders KM. c-kit-dependent development of interstitial cells and electrical activity in the murine gastrointestinal tract. Cell Tissue Res. 1995;280(1):97–111.
54. Broudy VC. Stem cell factor and hematopoiesis. Blood. 1997;90(4):1345–64.
55. Leary AG, Zeng HQ, Clark SC, Ogawa M. Growth factor requirements for survival in G0 and entry into the cell cycle of primitive human hemopoietic progenitors. Proc Natl Acad Sci U S A. 1992;89(9):4013–7.
56. Carson WE, Haldar S, Baiocchi RA, Croce CM, Caligiuri MA. The c-kit ligand suppresses apoptosis of human natural killer cells through the upregulation of bcl-2. Proc Natl Acad Sci U S A. 1994;91(16):7553–7.
57. Li CL, Johnson GR. Stem cell factor enhances the survival but not the self-renewal of murine hematopoietic long-term repopulating cells. Blood. 1994;84(2):408–14.
58. Valent P, Spanblochl E, Sperr WR, Sillaber C, Zsebo KM, Agis H, et al. Induction of differentiation of human mast cells from bone marrow and peripheral blood mononuclear cells by recombinant human stem cell factor/kit-ligand in long-term culture. Blood. 1992;80(9):2237–45.
59. Irani AA, Nilsson G, Ashman LK, Schwartz LB. Dexamethasone inhibits the development of mast cells from dispersed human fetal liver cells cultured in the presence of recombinant human stem cell factor. Immunology. 1995;84(1):72–8.
60. Demetri GD, Griffin JD. Granulocyte colony-stimulating factor and its receptor. Blood. 1991;78(11):2791–808.
61. Stanley ER, Hansen G, Woodcock J, Metcalf D. Colony stimulating factor and the regulation of granulopoiesis and macrophage production. Fed Proc. 1975;34(13):2272–8.
62. Demetri GD, Zenzie BW, Rheinwald JG, Griffin JD. Expression of colony-stimulating factor genes by normal human mesothelial cells and human malignant mesothelioma cells lines in vitro. Blood. 1989;74(3):940–6.
63. Koeffler HP, Gasson J, Ranyard J, Souza L, Shepard M, Munker R. Recombinant human TNF alpha stimulates production of granulocyte colony-stimulating factor. Blood. 1987;70(1):55–9.
64. Zsebo KM, Yuschenkoff VN, Schiffer S, Chang D, McCall E, Dinarello CA, et al. Vascular endothelial cells and granulopoiesis: interleukin-1 stimulates release of G-CSF and GM-CSF. Blood. 1988;71(1):99–103.
65. van de Geijn GJ, Aarts LH, Erkeland SJ, Prasher JM, Touw IP. Granulocyte colony-stimulating factor and its receptor in normal hematopoietic cell development and myeloid disease. Rev Physiol Biochem Pharmacol. 2003;149:53–71.
66. Tamura M, Hattori K, Nomura H, Oheda M, Kubota N, Imazeki I, et al. Induction of neutrophilic granulocytosis in mice by administration of purified human native granulocyte colony-stimulating factor (G-CSF). Biochem Biophys Res Commun. 1987;142(2):454–60.
67. Duhrsen U, Villeval JL, Boyd J, Kannourakis G, Morstyn G, Metcalf D. Effects of recombinant human granulocyte colony-stimulating factor on hematopoietic progenitor cells in cancer patients. Blood. 1988;72(6):2074–81.
68. Martino M, Laszlo D, Lanza F. Long-active granulocyte colony-stimulating factor for peripheral blood hematopoietic progenitor cell mobilization. Expert Opin Biol Ther. 2014;14(6):757–72.
69. de Haas M, Kerst JM, van der Schoot CE, Calafat J, Hack CE, Nuijens JH, et al. Granulocyte colony-stimulating factor administration to healthy volunteers: analysis of the immediate activating effects on circulating neutrophils. Blood. 1994;84(11):3885–94.
70. McNiece I, Andrews R, Stewart M, Clark S, Boone T, Quesenberry P. Action of interleukin-3, G-CSF, and GM-CSF on highly enriched human hematopoietic progenitor cells: synergistic interaction of GM-CSF plus G-CSF. Blood. 1989;74(1):110–4.
71. McNiece IK, Briddell RA, Hartley CA, Smith KA, Andrews RG. Stem cell factor enhances in vivo effects of granulocyte colony stimulating factor for stimulating mobilization of peripheral blood progenitor cells. Stem cells (Dayton, Ohio). 1993;11(Suppl 2):36–41.

72. Moskowitz CH, Stiff P, Gordon MS, McNiece I, Ho AD, Costa JJ, et al. Recombinant methionyl human stem cell factor and filgrastim for peripheral blood progenitor cell mobilization and transplantation in non-Hodgkin's lymphoma patients—results of a phase I/II trial. Blood. 1997;89(9):3136–47.
73. Stiff P, Gingrich R, Luger S, Wyres M, Brown R, LeMaistre C, et al. A randomized phase 2 study of PBPC mobilization by stem cell factor and filgrastim in heavily pretreated patients with Hodgkin's disease or non-Hodgkin's lymphoma. Bone Marrow Transplant. 2000;26(5):471.
74. Shpall EJ, Wheeler CA, Turner SA, Yanovich S, Brown RA, Pecora AL, et al. A randomized phase 3 study of peripheral blood progenitor cell mobilization with stem cell factor and filgrastim in high-risk breast cancer patients. Blood. 1999;93(8):2491–501.
75. Facon T, Harousseau JL, Maloisel F, Attal M, Odriozola J, Alegre A, et al. Stem cell factor in combination with filgrastim after chemotherapy improves peripheral blood progenitor cell yield and reduces apheresis requirements in multiple myeloma patients: a randomized, controlled trial. Blood. 1999;94(4):1218–25.
76. Duarte RF, Franf DA. The synergy between stem cell factor (SCF) and granulocyte colony-stimulating factor (G-CSF): molecular basis and clinical relevance. Leuk Lymphoma. 2002;43(6):1179–87.
77. Kaplan MH, Daniel C, Schindler U, Grusby MJ. Stat proteins control lymphocyte proliferation by regulating p27Kip1 expression. Mol Cell Biol. 1998;18(4):1996–2003.
78. Hirota S, Ito A, Morii E, Wanaka A, Tohyama M, Kitamura Y, et al. Localization of mRNA for c-kit receptor and its ligand in the brain of adult rats: an analysis using in situ hybridization histochemistry. Brain Res Mol Brain Res. 1992;15(1–2):47–54.
79. Manova K, Bachvarova RF, Huang EJ, Sanchez S, Pronovost SM, Velazquez E, et al. c-kit receptor and ligand expression in postnatal development of the mouse cerebellum suggests a function for c-kit in inhibitory interneurons. J Neurosci. 1992;12(12):4663–76.
80. Zhao LR, Singhal S, Duan WM, Mehta J, Kessler JA. Brain repair by hematopoietic growth factors in a rat model of stroke. Stroke. 2007;38(9):2584–91.
81. Zhao LR, Navalitloha Y, Singhal S, Mehta J, Piao CS, Guo WP, et al. Hematopoietic growth factors pass through the blood-brain barrier in intact rats. Exp Neurol. 2007;204(2):569–73.
82. Gore BB, Wong KG, Tessier-Lavigne M. Stem cell factor functions as an outgrowth-promoting factor to enable axon exit from the midline intermediate target. Neuron. 2008;57(4):501–10.
83. Guijarro P, Wang Y, Ying Y, Yao Y, Jieyi X, Yuan X. In vivo knockdown of cKit impairs neuronal migration and axonal extension in the cerebral cortex. Dev Neurobiol. 2013;73(12):871–87.
84. Kondo T, Katafuchi T, Hori T. Stem cell factor modulates paired-pulse facilitation and long-term potentiation in the hippocampal mossy fiber-CA3 pathway in mice. Brain Res. 2002;946(2):179–90.
85. Song S, Wang X, Sava V, Weeber EJ, Sanchez-Ramos J. In vivo administration of granulocyte colony-stimulating factor restores long-term depression in hippocampal slices prepared from transgenic APP/PS1 mice. J Neurosci Res. 2014;92(8):975–80.
86. Frauenknecht K, Diederich K, Leukel P, Bauer H, Schabitz WR, Sommer CJ, et al. Functional improvement after photothrombotic stroke in rats is associated with different patterns of dendritic plasticity after G-CSF treatment and G-CSF treatment combined with concomitant or sequential constraint-induced movement therapy. PLoS One. 2016;11(1):e0146679.
87. Mashayekhi F, Gholizadeh L. Administration of anti-c-kit antibody into the cerebrospinal fluid leads to increased cell death in the developing cerebral cortex. Saudi J Biol Sci. 2011;18(3):261–6.
88. Jung KH, Chu K, Lee ST, Kim SJ, Sinn DI, Kim SU, et al. Granulocyte colony-stimulating factor stimulates neurogenesis via vascular endothelial growth factor with STAT activation. Brain Res. 2006;1073–1074:190–201.
89. Chen WF, Hsu JH, Lin CS, Jong YJ, Yang CH, Huang LT, et al. Granulocyte-colony stimulating factor alleviates perinatal hypoxia-induced decreases in hippocampal synaptic efficacy and neurogenesis in the neonatal rat brain. Pediatr Res. 2011;70(6):589–95.

90. Kim JS, Yang M, Jang H, Oui H, Kim SH, Shin T, et al. Granulocyte-colony stimulating factor ameliorates irradiation-induced suppression of hippocampal neurogenesis in adult mice. Neurosci Lett. 2010;486(1):43–6.
91. Prakash A, Medhi B, Chopra K. Granulocyte colony stimulating factor (GCSF) improves memory and neurobehavior in an amyloid-beta induced experimental model of Alzheimer's disease. Pharmacol Biochem Behav. 2013;110:46–57.
92. Sanchez-Ramos J, Song S, Sava V, Catlow B, Lin X, Mori T, et al. Granulocyte colony stimulating factor decreases brain amyloid burden and reverses cognitive impairment in Alzheimer's mice. Neuroscience. 2009;163(1):55–72.
93. Schmidt AK, Reich A, Falkenburger B, Schulz JB, Brandenburg LO, Ribes S, et al. Adjuvant granulocyte colony-stimulating factor therapy results in improved spatial learning and stimulates hippocampal neurogenesis in a mouse model of pneumococcal meningitis. J Neuropathol Exp Neurol. 2015;74(1):85–94.
94. Sehara Y, Hayashi T, Deguchi K, Zhang H, Tsuchiya A, Yamashita T, et al. Potentiation of neurogenesis and angiogenesis by G-CSF after focal cerebral ischemia in rats. Brain Res. 2007;1151:142–9.
95. Sehara Y, Hayashi T, Deguchi K, Zhang H, Tsuchiya A, Yamashita T, et al. G-CSF enhances stem cell proliferation in rat hippocampus after transient middle cerebral artery occlusion. Neurosci Lett. 2007;418(3):248–52.
96. Tsai KJ, Tsai YC, Shen CK. G-CSF rescues the memory impairment of animal models of Alzheimer's disease. J Exp Med. 2007;204(6):1273–80.
97. Yang DY, Chen YJ, Wang MF, Pan HC, Chen SY, Cheng FC. Granulocyte colony-stimulating factor enhances cellular proliferation and motor function recovery on rats subjected to traumatic brain injury. Neurol Res. 2010;32(10):1041–9.
98. Yang YN, Lin CS, Yang CH, Lai YH, Wu PL, Yang SN. Neurogenesis recovery induced by granulocyte-colony stimulating factor in neonatal rat brain after perinatal hypoxia. Pediatr Neonatol. 2013;54(6):380–8.
99. Liu XY, Gonzalez-Toledo ME, Fagan A, Duan WM, Liu Y, Zhang S, et al. Stem cell factor and granulocyte colony-stimulating factor exhibit therapeutic effects in a mouse model of CADASIL. Neurobiol Dis. 2015;73:189–203.
100. Li B, Gonzalez-Toledo ME, Piao CS, Gu A, Kelley RE, Zhao LR. Stem cell factor and granulocyte colony-stimulating factor reduce beta-amyloid deposits in the brains of APP/PS1 transgenic mice. Alzheimers Res Ther. 2011;3(2):8.
101. Simard AR, Soulet D, Gowing G, Julien JP, Rivest S. Bone marrow-derived microglia play a critical role in restricting senile plaque formation in Alzheimer's disease. Neuron. 2006;49(4):489–502.
102. Wu CC, Wang IF, Chiang PM, Wang LC, Shen CJ, Tsai KJ. G-CSF-mobilized bone marrow mesenchymal stem cells replenish neural lineages in Alzheimer's disease mice via CXCR4/SDF-1 chemotaxis. Mol Neurobiol. 2016; PMID:27709493, DOI:10.1007/s12035-016-0122-x
103. Lee ST, Chu K, Jung KH, Ko SY, Kim EH, Sinn DI, et al. Granulocyte colony-stimulating factor enhances angiogenesis after focal cerebral ischemia. Brain Res. 2005;1058(1–2):120–8.
104. Morita Y, Takizawa S, Kamiguchi H, Uesugi T, Kawada H, Takagi S. Administration of hematopoietic cytokines increases the expression of anti-inflammatory cytokine (IL-10) mRNA in the subacute phase after stroke. Neurosci Res. 2007;58(4):356–60.
105. Fama ME, Turkeltaub PE. Treatment of poststroke aphasia: current practice and new directions. Semin Neurol. 2014;34(5):504–13.
106. Johansson BB. Brain plasticity and stroke rehabilitation. The Willis lecture. Stroke. 2000;31(1):223–30.
107. Nudo R. Adaptive plasticity in motor cortex: implications for rehabilitation after brain injury. J Rehabil Med. 2003;41:7–10.
108. Burke SN, Barnes CA. Neural plasticity in the ageing brain. Nat Rev Neurosci. 2006;7(1):30–40.
109. Anderson V, Spencer-Smith M, Wood A. Do children really recover better? Neurobehavioural plasticity after early brain insult. Brain. 2011;134(Pt 8):2197–221.

110. Bavelier D, Levi DM, Li RW, Dan Y, Hensch TK. Removing brakes on adult brain plasticity: from molecular to behavioral interventions. J Neurosci. 2010;30(45):14964–71.
111. Cramer SC, Sur M, Dobkin BH, O'Brien C, Sanger TD, Trojanowski JQ, et al. Harnessing neuroplasticity for clinical applications. Brain. 2011;134(6):1591–609.
112. Johansson BB, Auer LM, Sayama I. Reaction of pial arteries and veins to hypercapnia in hypertensive and normotensive rats. Stroke. 1985;16(2):320–3.
113. Johansson BB. Cerebral vascular bed in hypertension and consequences for the brain. Hypertension. 1984;6(6 Pt 2):III81–6.
114. Barone FC, Price WJ, White RF, Willette RN, Feuerstein GZ. Genetic hypertension and increased susceptibility to cerebral ischemia. Neurosci Biobehav Rev. 1992;16(2):219–33.
115. Coyle P. Different susceptibilities to cerebral infarction in spontaneously hypertensive (SHR) and normotensive Sprague-Dawley rats. Stroke. 1986;17(3):520–5.
116. Duverger D, MacKenzie ET. The quantification of cerebral infarction following focal ischemia in the rat: influence of strain, arterial pressure, blood glucose concentration, and age. J Cereb Blood Flow Metab. 1988;8(4):449–61.
117. Grabowski M, Nordborg C, Brundin P, Johansson BB. Middle cerebral artery occlusion in the hypertensive and normotensive rat: a study of histopathology and behaviour. J Hypertens. 1988;6(5):405–11.
118. Stroemer RP, Kent TA, Hulsebosch CE. Neocortical neural sprouting, synaptogenesis, and behavioral recovery after neocortical infarction in rats. Stroke. 1995;26(11):2135–44.
119. Zhao LR, Mattsson B, Johansson BB. Environmental influence on brain-derived neurotrophic factor messenger RNA expression after middle cerebral artery occlusion in spontaneously hypertensive rats. Neuroscience. 2000;97(1):177–84.
120. Zhao LR, Risedal A, Wojcik A, Hejzlar J, Johansson BB, Kokaia Z. Enriched environment influences brain-derived neurotrophic factor levels in rat forebrain after focal stroke. Neurosci Lett. 2001;305(3):169–72.
121. Zhao LR, Duan WM, Reyes M, Keene CD, Verfaillie CM, Low WC. Human bone marrow stem cells exhibit neural phenotypes and ameliorate neurological deficits after grafting into the ischemic brain of rats. Exp Neurol. 2002;174(1):11–20.
122. Ohlsson AL, Johansson BB. Environment influences functional outcome of cerebral infarction in rats. Stroke. 1995;26(4):644–9.
123. Feng G, Mellor RH, Bernstein M, Keller-Peck C, Nguyen QT, Wallace M, et al. Imaging neuronal subsets in transgenic mice expressing multiple spectral variants of GFP. Neuron. 2000;28(1):41–51.
124. Kasai H, Fukuda M, Watanabe S, Hayashi-Takagi A, Noguchi J. Structural dynamics of dendritic spines in memory and cognition. Trends Neurosci. 2010;33(3):121–9.
125. Noguchi J, Nagaoka A, Watanabe S, Ellis-Davies GC, Kitamura K, Kano M, et al. In vivo two-photon uncaging of glutamate revealing the structure-function relationships of dendritic spines in the neocortex of adult mice. J Physiol. 2011;589(Pt 10):2447–57.

Chapter 11
Mesenchymal Stromal Cell Therapy of Stroke

Yi Shen, Poornima Venkat, Michael Chopp, and Jieli Chen

Abstract Stroke is a major cause of high mortality, morbidity and long-term disability worldwide. Development of neuroprotective and neurorestorative therapies for stroke has been a target of intense research. Accumulating preclinical literature has identified that bone marrow mesenchymal stromal cell (MSC) treatment of stroke improves neurological functional outcome after stroke. This chapter focuses on the therapeutic effects and molecular mechanisms underlying MSC treatment of stroke, such as angiogenesis, arteriogenesis, neurogenesis and white matter remodeling, as well as a discussion on the interaction/coupling among these restorative events. In addition, the role of microRNAs (miRNAs) and MSC secreted exosomes in mediating intercellular communication between MSCs and parenchymal cells of the brain, and their effects on the regulation of neurovascular remodeling and white matter remodeling after stroke are discussed.

Keywords Mesenchymal stromal cell • Stroke • Angiogenesis • Neurogenesis • White matter remodeling • MicroRNA • Exosome • Neurorestoration

Y. Shen, M.D. • J. Chen, M.D. (✉)
Department of Neurology, Henry Ford Hospital,
2799 W Grand Blvd, Detroit, MI 48202, USA

Gerontology & Neurological Institute, Department of Neurology,
Tianjin Medical University General Hospital, Tianjin 300052, China
e-mail: jchen4@hfhs.org

P. Venkat, Ph.D.
Department of Neurology, Henry Ford Hospital,
2799 W Grand Blvd, Detroit, MI 48202, USA

M. Chopp, Ph.D. (✉)
Department of Neurology, Henry Ford Hospital,
2799 W Grand Blvd, Detroit, MI 48202, USA

Department of Physics, Oakland University, Rochester, MI 48309, USA
e-mail: mchopp1@hfhs.org

P.A. Lapchak, J.H. Zhang (eds.), *Cellular and Molecular Approaches to Regeneration and Repair*, Springer Series in Translational Stroke Research,
https://doi.org/10.1007/978-3-319-66679-2_11

Abbreviations

BBB	Brain-blood barrier
BDNF	Brain-derived neurotrophic factor
BMMNC	Autologous bone marrow mononuclear cell
CRT	Corticorubral tract
CST	Corticospinal tract
DM	Diabetes mellitus
FGF-2	Fibroblast growth factor-2
GDNF	Glial cell line-derived neurotrophic factor
HGF	Hepatocyte growth factor
HLA-DR	Human leukocyte antigen-antigen D related
IA	Intra-arterial
IC	Intracerebral
ICH	Intracerebral hemorrhage
ICV	Intracerebro ventricular
IGF	Insulin-like growth factor
IN	Intranasal
IV	Intravenous
miRNA	MicroRNA
MSC	Mesenchymal stromal cell
MSC-Exo	Exosome derived from MSCs
mTOR	Mammalian target of rapamycin
NGF	Nerve growth factor
NPCs	Neural progenitor cells
NSCs	Neural stem cells
OPC	Oligodendrocyte progenitor cell
PGF	Placental growth factor
PTEN	Phosphatase and tensin homologue
SHRSP	Spontaneously hypertensive stroke prone
T1DM	Type 1 diabetic
tPA	Tissue plasminogen activator
VEGF	Vascular endothelial growth factor

1 Introduction

Stroke remains one of most common causes of death and major cause of disability all over the world [1]. Tissue plasminogen activator (tPA) remains the only pharmacological agent approved by the US Food and Drug Administration (FDA) for the treatment of ischemic stroke. There is a compelling need to develop novel therapies specifically designed to reduce neurological functional deficits after stroke. Decades of preclinical research and the failure of several clinical trials have drawn attention to the limitations of acute neuroprotection [2, 3]. Therapeutic and repair potential of cell-based therapies, including bone marrow mesenchymal stromal cells (MSC) have been investigated in stroke and multiple neurodegenerative disease models [4]. MSCs are being evaluated in human clinical trials for efficacy in treating genetic diseases of bone, and in reducing the severity of graft versus host disease [5–7]. The therapeutic effects of MSCs have been measured in myocardial, limb and brain ischemia [8–12]. In this chapter, we discuss the therapeutic effects and mechanisms of MSC treatment of stroke.

1.1 Mesenchymal Stromal Cells

Mesenchymal stromal cells were first isolated from the bone marrow of adult guinea-pigs in the 1960s [13, 14]. They were described as cells which could rapidly adhere to the plastic of culture flasks and were characterized by their fibroblast-like morphology [13–15]. Now MSCs are recognized as a heterogeneous subset of multipotent precursors residing in the stromal fraction of many adult tissues [16]. To discriminate MSCs from other stem cells, the International Society for Cell Therapy has provided three criteria: (a) adherence to plastic of culture flasks; (b) specific surface antigen expression; (c) multipotent differentiation potential [17].

MSCs can be isolated from a variety of adult rodents tissue such as the bone marrow, adipose tissue, cord blood cells, skin, lung, muscle, etc. [14, 16, 18–20]. The bone marrow is a major source of MSCs and has been widely studied as a treatment option for stroke [17–19, 21]. MSCs have great potential as therapeutic agents, since they are easy to harvest and can be expanded from patient's own bone marrow without serious ethical or technical problems [22]. MSCs also represent a promising source for autologous cell transplantation therapies [22].

MSCs have been distinguished from other hematopoietic stem cells according to the expression of surface molecules CD73, CD90, and CD105 but in the absence of CD34, CD45, Human Leukocyte Antigen-antigen D Related (HLA-DR), CD14 or CD11b, CD79a, or CD19 [14, 17, 23, 24]. MSCs can differentiate into osteoblasts, adipocytes and chondroblasts [16, 25]. MSCs can also differentiate into cerebral parenchymal cells such as neurons and astrocytes in animals subject to experimental ischemic stroke model [26–28]. Autologous MSC treatment has been employed in patients with cerebral infarction and spinal cord injury and has the advantage of a lower risk of graft-versus-host disease [4].These advantages make MSCs therapy a new avenue for stroke treatment.

1.2 MSC Treatment Improves Stroke Outcome in Preclinical Studies

During the past two decades, numerous studies have employed MSC treatment for ischemic stroke in rodents and demonstrated that MSC administration in the acute phase (1–3 day) of ischemic stroke improves sensorimotor function, and in the ischemic penumbra promotes arteriogenesis, angiogenesis, reduces apoptosis, promotes endogenous cell proliferation, augments synaptogenesis, increases nerve regeneration and regulates inflammatory and immune responses in the ischemic brain [2, 24, 29–39]. MSC therapy initiated at a delayed time point can also reduce neurological functional deficits when administered intravenously at 7 days after stroke and even at 1 month after stroke by promoting neurorestorative effects in the (IBZ) [29, 36, 40].

Apart from ischemic stroke, several pre-clinical and clinical studies have demonstrated the safety and efficacy of MSC treatment for hemorrhagic stroke [41, 42]. In intracerebral hemorrhage (ICH) and subarachnoid hemorrhage, MSC treatment attenuates neurological deficits, decreases brain-blood barrier (BBB) leakage, reduces apoptosis, activates axonal remodeling, enhances endogenous neurogenesis and promotes anti-inflammatory effects [43–48]. While there are only a few studies that investigated the effects of MSC treatment in the sub-acute and chronic phase of ICH; MSC treatment has been reported to enhance neural progenitor cells (NPCs) proliferation from 48 h to 14 days after ICH [45]. Rehabilitation of patients is a vital component of ICH treatment and further studies are required to investigate neurorestorative therapy and long term functional outcome using MSCs for ICH.

Poor clinical translation of successful experimental therapies for stroke in part has in part been attributed to the widespread use of healthy, male, young animals in pre-clinical experiments [49, 50]. In reality, stroke patients suffer from one or more co-morbidities such as diabetes mellitus (DM), old age or hypertension [51, 52]. Co-morbidities aggravate stroke outcome and result in poor long term recovery after ischemic stroke. For example, diabetes mellitus is associated with microvascular and macrovascular disease and is a predisposing risk factor for stroke [53–56]. Neurovascular alterations induced by DM also influence the safety and efficacy of stroke therapies [54, 56, 57]. Stroke in type 1 diabetic rats (T1DM) increases mortality and BBB leakage resulting in reduced functional recovery compared to non-diabetic stroke rats [58]. MSC therapy administered in the acute phase (1 day after stroke) of stroke in T1DM rats induced worse functional outcome and adverse effects such as increased mortality, aggravated BBB permeability, higher risk of brain hemorrhage, and more macrophage infiltration [59]. However, MSC therapy initiated at a delayed time point (3 day after stroke) has been demonstrated to reduce vascular damage, promote endogenous neurorestorative effects and improve the functional outcomes post stroke in type two diabetes rats [57, 60, 61]. Therefore, MSC treatment initiation time point may be crucial in determining safety and efficacy of MSC therapy in the diabetic stroke population. Advancing age and hypertension are other major risk factors for stroke [62, 63]. In aged rats subject to stroke, MSC treatment decreased neurological impairment, reduced ischemic lesion volume, and promoted brain remodeling processes including angiogenesis, neurogenesis, and synaptogenesis [62].

Hypertension leads to worse functional outcome and increases ischemic lesion burden after stroke [46, 64]. Treatment of stroke with MSCs in spontaneously hypertensive stroke prone (SHRSP) rats decreased neurological deficits and improved BBB integrity [46]. Overall, preclinical findings indicate that MSC treatment improves recovery both in the acute and chronic periods of the stroke. Furthermore, when administered with specific protocol, MSC induced therapeutic effects persist in the setting of risk factors of stroke.

1.3 Gene Modification of MSCs to Enhance Therapeutic Benefits

To amplify the efficacy of MSC therapy for stroke, several gene modifications have been investigated. MSCs can be easily transfected via vectors such as lentivirus, adenovirus or adeno-associated virus carrying target genes [4, 65, 66]. A wide variety of genes have been incorporated into the MSCs and reported to induce greater functional recovery including microRNA-145 [61], nerve growth factor (NGF) or Noggin (an antagonist of bone morphogenetic protein) [65, 67], microRNA-133 [68], hepatocyte growth factor (HGF) [66], fibroblast growth factor-2 [69], brain-derived neurotrophic factor (BDNF) [70], vascular endothelial growth factor (VEGF) [71] in ischemic stroke; and glial cell line-derived neurotrophic factor (GDNF) in hemorrhagic stroke [72]. Collectively, these studies indicate that gene modification of the MSCs can enhance its therapeutic effects, but additional pre-clinical and clinical trials are needed to investigate the safety, efficacy, risk of geno-toxicity and other potential adverse effects of such modification.

1.4 Comparison of Administration Routes of MSCs for Stroke Treatment

MSC administration after stroke via different routes such as intracerebral (IC), intracerebro ventricular (ICV), intravenous (IV), intra-arterial (IA) and intranasal (IN) have been reported to promote functional outcome after stroke [4, 24, 30, 73–76]. Among these cell delivery routes, which is the optimal delivery mode for MSCs? Currently, there is no definitive answer and in this section we discuss the pros and cons of various delivery routes.

1.4.1 Intracerebral (IC) and Intracerebro Ventricular (ICV) Injections

IC injection is an invasive and clinically challenging method although it can directly deliver cells to the ischemic brain tissue. IC administration facilitates the trafficking of transplanted cells to the ischemic region, but can also lead to secondary brain trauma and poor distribution of cells within the target lesion [76]. Many experimental stroke studies have shown that IC injection is a reliable route for delivery of MSCs and may provide strong therapeutic effects for stroke recovery [11, 77, 78]. For instance, IC delivery of ~4×10^5 MSCs cultured with NGF at 24 h after stroke in rats significantly improved neurological functional outcome [11]. Steinberg GK et al. found that IC injection of MSCs in 18 patients with chronic stroke is safe and associated with improvement in clinical neurological outcome end points at 12 months [79]. However, several clinical trials report that IC injections may lead to several adverse events such as seizures, syncope, asymptomatic subdural hematoma and so on [80].Therefore, IC route is limited for use in animal experiments. ICV injections are less invasive, can deliver more cells, and may produce more extensive cell seeding than IC route [81]. In a clinical trial involving ten chronic stroke patients (seven ischemic and three hemorrhagic), researchers investigating ICV delivery route found that some patients developed fever and meningeal signs 48 h after delivery of cells via ICV injection [76, 82].

1.4.2 Intravascular Injections

A majority of studies employ IV or IA routes for delivery of stem cells. Intravascular administrations are less invasive and MSCs systemically infused into ischemic rats have been observed to migrate to the injured brain tissue [83, 84]. Mounting evidence indicates that IV injections of MSCs can successfully traffic cells to the ischemic zone which plays a vital role in recovery and stroke outcome in preclinical and clinical trials [29, 61, 65, 85, 86]. Chen et al. was the first to demonstrate that IV administration of MSCs in rats subject to stroke resulted in significant improvement in neurological outcome [29]. IV administration of autologous MSCs has already been demonstrated to be safe and efficacious in small clinical trials [87–90]. However, intravascular modes of cell therapy also face some challenges. For example, a large fraction of IV transplanted MSCs home to peripheral organs such as lungs, liver, and spleen, thereby limiting potential engraftment in the ischemic lesion in the brain [91, 92]. In 2001, Li et al. first demonstrated that intra-artery injection of MSCs significantly increases MSC migration into brain and improves functional outcome in stroke rats [93]. Selective IA administration of human MSCs have substantially increased migration of transplanted hBM-MSCs in the target brain than IV administration [94]. Although cells IA administered pass the BBB and traffic to ischemic brain, IV and IA achieve similar structural and functional outcomes after stroke at low and high doses of autologous bone marrow mononuclear cell (BMMNC) treatment [95]. One preclinical study comparing IV and IA BMMNC delivery found significant reduction in infarct volume, greater cell engraftment, and improved motor function with IV administration than with IA delivery [96].

1.4.3 Intranasal (IN) Delivery

In recent years, IN delivery has emerged as a novel method to transplant MSCs. IN administered MSCs pass through the BBB, and facilitates cell migration from the nasal mucosa to the central nervous system through the cribriform plate and movea into the brain parenchyma along the olfactory neural pathways, corpus callosum, and blood vessels [4, 97, 98]. IN delivery is non-invasive, and it decreases any adverse effects associated with intravascular administration. Transplantation of hypoxic preconditioned MSCs reached the injured cortex and were deposited outside of blood vessels 1.5 h after administration, and decreased ischemic lesion volume, and improved motor function in experimental stroke [73, 99]. These therapeutic effects were also observed in hemorrhagic stroke and ICH [100].

1.5 Stroke Clinical Trial for MSCs Therapy for Stroke

In a recent meta-analysis of stem cell therapies for patients with brain ischemia, MSC therapy was found to significantly enhance neurological function and quality of life, but additional investigations may be required to further support the safety and efficacy of stem cell transplantation in stroke patients. A number of clinical trials have investigated the optimal time point, dose and delivery mode for cell therapy for stroke, and are listed in Table 11.1.

2 Mechanisms of the Action of MSCs Treatment for Stroke

The mechanisms underlying neuroprotective and neurorestorative effects of MSCs are still unclear. Currently, it is widely accepted that MSCs can induce brain protection and remodeling primarily by secreting factors that enhance brain tissue repair by promoting vascular remodeling and white matter remodeling and neurogenesis [2–4, 24, 45, 86, 101]. Recently, researchers have found that exosomes derived from MSCs and the exosome cargo microRNA may mediate MSCs induced restorative effects after stroke [16, 102].

2.1 MSC Induced Therapeutic Effects via Cell Replacement

Originally, the mechanism of action of MSC induced therapeutics in stroke was thought to be cell replacement. When progenitor and stem cells are placed in the injured brain, they can differentiate into brain cells [103]. MSCs transplanted after stroke can migrate to the ischemic brain, and since MSCs are multipotent [104–108], they can differentiate into various tissue lineages, such as neurons, astrocytes, and endothelial cells [67, 109, 110]. However, preclinical studies have also shown that upon intravascular delivery, only a small fraction (1–10%) of MSCs migrate to the ischemic lesion [11, 12, 29, 93, 111, 112], and only 2–20% MSCs possibly differentiate into brain cells [11, 29, 111], and approximately only 2% of transplanted MSCs differentiate into endothelial cells in the ischemic brain, 14 days after

Table 11.1 Summary of recent and ongoing clinical trials for stroke treatment using MSCs

Study phase	Mode of delivery	Cell type	Time from onset	Primary outcome measure	Secondary outcome measure	Clinical identifier
I	IV; IT	Umbilical cord, mesenchymal	IV, 7–14 days; IT, 1 week after IV	Function outcome	Function outcome; MRI changes	NCT01389453
I/IIa	IV	Autologous bone marrow mononuclear	24–72 h	Safety; feasibility	Function outcome	NCT00859014
I/IIa	IV	Autologous mesenchymal stem	Within 6 weeks	Safety; feasibility	Function outcome	NCT00875654
I/II	IA	Autologous bone marrow CD 34+ stem	7 days	Safety	Function outcome	NCT00535197
I	IV	Autologous mesenchymal stem	7–30 days	Safety; feasibility	Function outcome	NCT01501773

administration [12]. This shows that at most, only a minor subpopulation of MSCs assume a parenchymal brain cell phenotype, and probably do not contribute to functional recovery [12].

2.2 MSCs Induced Therapeutic Effects via Secretion of Trophic Factors

MSCs secrete large amounts of angiogenic, anti-apoptotic, and mitogenic factors [113]. MSCs do not incorporate into the adult growing vasculature; but may function as supporting parenchymal cells and microglia [114, 115]. MSCs express mRNAs for a wide spectrum of angiogenic/arteriogenic cytokines including VEGF, BFGF2, Angiopoietin-1 (Ang1) and placental growth factor (PGF), insulin-like growth factor (IGF) and SDF [116–120].These growth factors play a crucial role in maintaining and augmenting brain plasticity process such as neurovascular remodeling and white matter remodeling and synaptic plasticity. BMSCs also stimulate brain parenchymal cell production of trophic factors [24, 121, 122]. These cytokines and trophic factors have both paracrine and autocrine activities [121]. MSCs behave as small biochemical and molecular "factories", producing many cytokines [123–125] and trophic factors that may affect the compromised brain tissue neurovascular and white matter remodeling in the ischemic border zone [24].

2.3 *MSCs Treatment Promotes Endogenous Brain Plasticity*

2.3.1 Vascular Remodeling

The stimulation of vascular remodeling is an important therapeutic target for recovery after stroke [126–128]. Stroke patients with a higher cerebral blood vessel density appear to make better progress and survive longer than patients with lower vascular density [129]. Vascular remodeling includes angiogenesis and arteriogenesis. Arteriolar collateral growth (Arteriogenesis) and new capillaries (Angiogenesis) support restored perfusion in the ischemic border after stroke, and increase long-term neurological functional recovery [130]. Numerous angiogenic factors, growth factors and cytokines have been discovered in the MSC secretome, all of which alter endothelial cell behavior and promote capillary tube formation in vitro and increase angiogenesis in vivo [131]. Increasing VEGF/Flk1 and Ang1/Tie2 signaling pathways plays an important role in MSC treatment induced angiogenesis and vascular stabilization [132]. Very early transplantation of human MSCs (1 h after MCAO) produced increased neurological recovery and decreased infarction volume as well as promoted angiogenesis [133]. Therefore, MSCs have the ability to effectively recruit and participate in vascular remodeling [11, 12, 24, 29, 33, 36, 38, 134].

2.3.2 White Matter Remodeling

Promoting functional outcome after stroke is not only related with vascular remodeling, but also involves white matter remodeling and rewiring of neuronal circuits [3, 135]. White matter remodeling includes axonal remodeling, oligodendrogenesis and remyelination. Oligodendrogenesis involves oligodendrocyte progenitor cell (OPC) proliferation and differentiation into mature oligodendrocytes (OLs) which form myelin sheath around axons. Since injured OLs do not form new myelin and mature OLs do not proliferate, oligodendrogenesis is essential to form myelin sheaths around the new sprouting axons after stroke [136]. MSC treatment stimulates oligodendrogenesis identified by increasing OPC proliferation and differentiation into OL as well as promotes myelination and axonal outgrowth post stroke [3, 61, 99, 110, 137, 138]. Functional recovery post stroke is positively correlated with enhanced white matter integrity in the ischemic brain [138].

In addition, the corticospinal tract (CST) neuronal pathway is also required for motor functional recovery. The severity of motor impairment after stroke is correlated with the extent of ischemic injury to the CST in all phases of stroke [139, 140]. After stroke, bilateral innervations occur through axonal sprouting of the uninjured corticorubral tract (CRT) and CST [141]. CST axonal remodeling in the spinal cord and pyramidal neuronal reorganization in the bilateral cortices promotes neuronal communication and partially contributes to spontaneous functional recovery in the chronic phase of stroke [142]. MSC treatment significantly increases synaptic proteins in the denervated motoneurons and increases axonal restructuring on the de-afferented red nucleus and the denervated spinal motoneurons [141]. MSCs significantly promote neuronal remodeling of the CST originating from the contralesional cortex in mice subjected to unilateral pyramidotomy after ischemic stroke which may contribute to motor recovery after stroke [143]. Axonal sprouting and rewiring is highly correlated with improved functional outcome after stroke [141, 144].

2.3.3 Neurogenesis

Neurogenesis is defined as the proliferation and differentiation of neural stem cells (NSCs) and neural progenitor cells (NPCs) into neurons. In response to stroke or brain injury, NSCs increase in the subventricular zone (SVZ) and subsequently increase neuroblasts [145]. The newly generated neuroblasts can migrate from the SVZ to peri-infarct regions [145–147]. Therapies that increase post stroke neurogenesis have been associated with improved neurological recovery in rodents [2, 101, 148]. MSC treatment increases NPC proliferation and migration from the SVZ to the ischemic boundary zone where they differentiate into parenchymal cells [149–151]. IV or IC delivery of MSCs in adult or neonatal stroke rats promotes neurogenesis and synaptogenesis, which are associated with functional recovery after stroke [12, 36, 152]. Enhanced neurogenesis and axonal remodeling likely underlie the improved functional outcome following MSC treatment after ischemic brain injury [153].

2.4 Exosome and microRNA Mediates MSC Induced Therapeutic Effects

Stem-like cells, such as MSCs, have been shown to directly secrete paracrine factors, but also membrane vesicles including exosomes [154]. Exosomes are a subpopulation of cellular secreted vesicles (also referred to as extracellular vesicles) that range ~30–100 nm in diameter. Compared to other cell types, the MSC is the most prolific exosome producer [155]. Most exosomes contain conserved proteins such as CD81, CD63, and CD9, Alix and Tsg101, as well as the unique tissue/cell type specific proteins that reflect their cellular source [156]. The exosome membranes are enriched with cholesterol, sphingomyelin, and ceramide [157]. Exosomes contain a variety of biologically active molecules, such as proteins, messenger RNAs (mRNAs), and microRNAs, and these bioactive molecules can mediate exosomal intercellular communication. Using proteomics analysis of exosome derived from MSCs (MSC-Exo), Otero-Ortega et al. identified more than 2000 proteins in MSC-Exo that could be implicated in brain repair [158]. MSCs induce neurological recovery post stroke and neural injury primarily by paracrine via exosomes produced by these cells (MSC-Exo), which mediate restorative actions of MSCs [68, 159–161]. Administration of MSC-Exo significantly improves stroke outcome by promoting neurovascular remodeling [162, 163]. Similarly, systemically injected human MSC-Exo improved long-term neuroprotection, promoted neuroregeneration, enhanced neurological recovery, and modulated peripheral post-stroke immune responses in mice [164]. MSC-Exo treatment significantly decreases lesion size, increases white matter remodeling identified by fiber tract integrity, axonal sprouting and white matter repair as well as improves functional recovery compared with the control group at 28 days after ICH [158]. Treatment of stroke with MSC-Exo improves functional outcome, as well as enhances angiogenesis, neurogenesis and neurite remodeling [160, 165, 166]. Improved neurological outcome after MSC-Exo therapy for stroke is comparable to the therapeutic effect observed with MSC therapy, suggesting that MSC-Exo-mediated cell-cell communication may contribute to the therapeutic effect of the MSC therapy.

Exosomes play an important role in intercellular communication by transferring exosomal protein and RNA cargo between source and target cells in the brain [167–169]. Many studies have identified that various miRNAs are present in MSC-Exos, and MSCs release functional small RNAs via exosomes that seem to convey the essential features of MSCs [68, 159–161]. MicroRNAs are short sequences of non-coding RNA (containing about 22 nucleotides) found in plants, animals and some viruses, that function in RNA silencing and post-transcriptional regulation of gene expression [170]. Among its myriad of functional properties, microRNAs regulate angiogenesis, neurogenesis, inflammation and stem cell biology [171]. Cell-based therapy, by IV injection release exosomes that contain enriched microRNAs that also stimulate endogenous brain cells to subsequently release microRNAs or genes, ultimately promoting brain plasticity after stroke [172]. MSCs treated with ischemic brain extract produced exosomes with neuroprotective effects in a stroke model in rats, inducing functional recovery mediated by transfer of microRNA-133b in exosomes [68]. MSC-Exo increase neuron and astrocyte miR-133b expression and MSCs communicate with brain parenchymal cells and regulate neurite outgrowth by transfer of miR-133b to neural cells via exosomes [166]. Using primary cortical neuron culture, Zhang et al. found that MSC-released exosomes promoted axonal growth, and inhibition of argonaut 2 protein (a primary microRNA machinery protein) significantly abolished MSC-exosomes induced axonal growth [165]. In addition, MSCs inhibit macrophage activation by shedding miRNA-containing exosomes [173]. Our previous study also shows that diabetic MSCs (DM-MSCs) and exosomes derived from DM-MSCs have decreased microRNA-145 expression [174]. DM-MSC treatment of stroke in DM-stroke improved functional outcome and vascular and white matter remodeling, as well as decreased serum microRNA-145 expression compared with vehicle treatment [174]. These data demonstrate that MSCs mediate their functional benefit in stroke at least partially by the transfer of exosomes with active microRNAs to parenchymal cells.

MSCs-Exo can be manipulated to deliver microRNAs to enhance recovery of injured tissues [167, 175]. Treatment of stroke with tailored exosomes enriched with the miR-17-92 cluster significantly increases neural plasticity and functional recovery after stroke [176]. Tailored exosomes derived from MSC further enhance neurite growth via the Phosphatase and tensin homologue/Mammalian target of rapamycin (PTEN/mTOR) signals by increasing the microRNA-17-92 cluster

[177]. Exosomes derived from microRNA-133b-overexpressed MSCs also improved neurite remodeling/brain plasticity and neurological functional recovery after stroke compared to MSC-Exo treatment group [178]. Therapeutic modulation of microRNAs generated by MSCs, and either mimicking or antagonizing microRNA actions, may enhance MSC induced therapeutic efficacy. Hence, the use of microRNAs as novel regulators and therapeutic modulators of individual microRNAs of MSCs have been proposed to improve therapeutic efficiency [179].

3 Conclusions

Early or delayed MSCs treatment of stroke (ischemic or hemorrhage stroke) significantly improves functional outcome in neonatal, adult and aging populations as well as in diabetic and hypertension stroke populations. The mechanisms underlying neuroprotective and neurorestorative effects of MSCs are still unclear. However, MSCs secrete trophic factors, exosomes with their cargo microRNAs, that may modulate brain tissue repair and restorative mechanisms such as neurovascular remodeling and white matter remodeling. We thus have major insight into the restorative mechanism by which MSCs promote neurological recovery. Preclinical studies in many models of neurological injury and disease have demonstrated therapeutic efficacy of MSC treatments, and there have been clinical trials that support the therapeutic approach of MSC therapy for stroke and other forms of neural injury. Thus, in concert, this compendium of studies and data, support the further development and clinical application of this promising therapy for the treatment of stroke, and other forms of neural injury.

References

1. Adeoye O, Hornung R, Khatri P, Kleindorfer D. Recombinant tissue-type plasminogen activator use for ischemic stroke in the United States: a doubling of treatment rates over the course of 5 years. Stroke. 2011;42(7):1952–5.
2. Chen J, Chopp M. Neurorestorative treatment of stroke: cell and pharmacological approaches. NeuroRx. 2006;3(4):466–73.
3. Chen J, Venkat P, Zacharek A, Chopp M. Neurorestorative therapy for stroke. Front Hum Neurosci. 2014;8:382.
4. Liu X, Ye R, Yan T, Yu SP, Wei L, Xu G, et al. Cell based therapies for ischemic stroke: from basic science to bedside. Prog Neurobiol. 2014;115:92–115.
5. Mazzini L, Mareschi K, Ferrero I, Miglioretti M, Stecco A, Servo S, et al. Mesenchymal stromal cell transplantation in amyotrophic lateral sclerosis: a long-term safety study. Cytotherapy. 2012;14(1):56–60.
6. Kim SJ, Moon GJ, Chang WH, Kim YH, Bang OY. Intravenous transplantation of mesenchymal stem cells preconditioned with early phase stroke serum: current evidence and study protocol for a randomized trial. Trials. 2013;14:317.
7. Ringden O, Uzunel M, Rasmusson I, Remberger M, Sundberg B, Lonnies H, et al. Mesenchymal stem cells for treatment of therapy-resistant graft-versus-host disease. Transplantation. 2006;81(10):1390–7.

8. Krause DS. Plasticity of marrow-derived stem cells. Gene Ther. 2002;9(11):754–8.
9. Menasche P. Cell transplantation for the treatment of heart failure. Semin Thorac Cardiovasc Surg. 2002;14(2):157–66.
10. Wang JS, Shum-Tim D, Galipeau J, Chedrawy E, Eliopoulos N, Chiu RC. Marrow stromal cells for cellular cardiomyoplasty: feasibility and potential clinical advantages. J Thorac Cardiovasc Surg. 2000;120(5):999–1005.
11. Chen J, Li Y, Wang L, Lu M, Zhang X, Chopp M. Therapeutic benefit of intracerebral transplantation of bone marrow stromal cells after cerebral ischemia in rats. J Neurol Sci. 2001;189(1–2):49–57.
12. Li Y, Chen J, Chen XG, Wang L, Gautam SC, Xu YX, et al. Human marrow stromal cell therapy for stroke in rat: neurotrophins and functional recovery. Neurology. 2002;59(4):514–23.
13. Friedenstein AJ, Chailakhjan RK, Lalykina KS. The development of fibroblast colonies in monolayer cultures of guinea-pig bone marrow and spleen cells. Cell Tissue Kinet. 1970;3(4):393–403.
14. Si YL, Zhao YL, Hao HJ, Fu XB, Han WD. MSCs: biological characteristics, clinical applications and their outstanding concerns. Ageing Res Rev. 2011;10(1):93–103.
15. Ye X, Hu J, Cui G. Therapy effects of bone marrow stromal cells on ischemic stroke. Oxidative Med Cell Longev. 2016;2016:7682960.
16. Nombela-Arrieta C, Ritz J, Silberstein LE. The elusive nature and function of mesenchymal stem cells. Nat Rev Mol Cell Biol. 2011;12(2):126–31.
17. Dominici M, Le Blanc K, Mueller I, Slaper-Cortenbach I, Marini F, Krause D, et al. Minimal criteria for defining multipotent mesenchymal stromal cells. The International Society for Cellular Therapy position statement. Cytotherapy. 2006;8(4):315–7.
18. Bunnell BA, Flaat M, Gagliardi C, Patel B, Ripoll C. Adipose-derived stem cells: isolation, expansion and differentiation. Methods. 2008;45(2):115–20.
19. Williams AR, Hare JM. Mesenchymal stem cells: biology, pathophysiology, translational findings, and therapeutic implications for cardiac disease. Circ Res. 2011;109(8):923–40.
20. Futami I, Ishijima M, Kaneko H, Tsuji K, Ichikawa-Tomikawa N, Sadatsuki R, et al. Isolation and characterization of multipotential mesenchymal cells from the mouse synovium. PLoS One. 2012;7(9):e45517.
21. da Silva Meirelles L, Chagastelles PC, Nardi NB. Mesenchymal stem cells reside in virtually all post-natal organs and tissues. J Cell Sci. 2006;119(Pt 11):2204–13.
22. Dezawa M, Ishikawa H, Hoshino M, Itokazu Y, Nabeshima Y. Potential of bone marrow stromal cells in applications for neuro-degenerative, neuro-traumatic and muscle degenerative diseases. Curr Neuropharmacol. 2005;3(4):257–66.
23. Peister A, Mellad JA, Larson BL, Hall BM, Gibson LF, Prockop DJ. Adult stem cells from bone marrow (MSCs) isolated from different strains of inbred mice vary in surface epitopes, rates of proliferation, and differentiation potential. Blood. 2004;103(5):1662–8.
24. Chen J, Zhang ZG, Li Y, Wang L, Xu YX, Gautam SC, et al. Intravenous administration of human bone marrow stromal cells induces angiogenesis in the ischemic boundary zone after stroke in rats. Circ Res. 2003;92(6):692–9.
25. Weissman IL. Stem cells: units of development, units of regeneration, and units in evolution. Cell. 2000;100(1):157–68.
26. Li Y, Chopp M, Chen J, Wang L, Gautam SC, Xu YX, et al. Intrastriatal transplantation of bone marrow nonhematopoietic cells improves functional recovery after stroke in adult mice. J Cereb Blood Flow Metab. 2000;20(9):1311–9.
27. Sanchez-Ramos J, Song S, Cardozo-Pelaez F, Hazzi C, Stedeford T, Willing A, et al. Adult bone marrow stromal cells differentiate into neural cells in vitro. Exp Neurol. 2000;164(2):247–56.
28. Kopen GC, Prockop DJ, Phinney DG. Marrow stromal cells migrate throughout forebrain and cerebellum, and they differentiate into astrocytes after injection into neonatal mouse brains. Proc Natl Acad Sci U S A. 1999;96(19):10711–6.
29. Chen J, Li Y, Wang L, Zhang Z, Lu D, Lu M, et al. Therapeutic benefit of intravenous administration of bone marrow stromal cells after cerebral ischemia in rats. Stroke. 2001;32(4):1005–11.

30. Zhao LR, Duan WM, Reyes M, Keene CD, Verfaillie CM, Low WC. Human bone marrow stem cells exhibit neural phenotypes and ameliorate neurological deficits after grafting into the ischemic brain of rats. Exp Neurol. 2002;174(1):11–20.
31. Weng JS, Liu N, Du HW, Chen RH, Zhang YX, Wang JH, et al. Effects of bone marrow-derived mesenchymal stem cells transplantation on recovery of neurological functions and expression of synaptophysin in focal cerebral infarction in rats. Xi Bao Yu Fen Zi Mian Yi Xue Za Zhi. 2008;24(1):34–7.
32. Huang W, Mo X, Qin C, Zheng J, Liang Z, Zhang C. Transplantation of differentiated bone marrow stromal cells promotes motor functional recovery in rats with stroke. Neurol Res. 2013;35(3):320–8.
33. Pavlichenko N, Sokolova I, Vijde S, Shvedova E, Alexandrov G, Krouglyakov P, et al. Mesenchymal stem cells transplantation could be beneficial for treatment of experimental ischemic stroke in rats. Brain Res. 2008;1233:203–13.
34. Yoo KH, Jang IK, Lee MW, Kim HE, Yang MS, Eom Y, et al. Comparison of immunomodulatory properties of mesenchymal stem cells derived from adult human tissues. Cell Immunol. 2009;259(2):150–6.
35. Scheibe F, Ladhoff J, Huck J, Grohmann M, Blazej K, Oersal A, et al. Immune effects of mesenchymal stromal cells in experimental stroke. J Cereb Blood Flow Metab. 2012;32(8):1578–88.
36. Chen J, Li Y, Katakowski M, Chen X, Wang L, Lu D, et al. Intravenous bone marrow stromal cell therapy reduces apoptosis and promotes endogenous cell proliferation after stroke in female rat. J Neurosci Res. 2003;73(6):778–86.
37. Tohill M, Mantovani C, Wiberg M, Terenghi G. Rat bone marrow mesenchymal stem cells express glial markers and stimulate nerve regeneration. Neurosci Lett. 2004;362(3):200–3.
38. Al-Khaldi A, Eliopoulos N, Martineau D, Lejeune L, Lachapelle K, Galipeau J. Postnatal bone marrow stromal cells elicit a potent VEGF-dependent neoangiogenic response in vivo. Gene Ther. 2003;10(8):621–9.
39. Cui X, Chopp M, Zacharek A, Roberts C, Lu M, Savant-Bhonsale S, et al. Chemokine, vascular and therapeutic effects of combination Simvastatin and BMSC treatment of stroke. Neurobiol Dis. 2009;36(1):35–41.
40. Shen LH, Li Y, Chen J, Zacharek A, Gao Q, Kapke A, et al. Therapeutic benefit of bone marrow stromal cells administered 1 month after stroke. J Cereb Blood Flow Metab. 2007;27(1):6–13.
41. Park WS, Sung SI, Ahn SY, Sung DK, Im GH, Yoo HS, et al. Optimal timing of mesenchymal stem cell therapy for neonatal intraventricular hemorrhage. Cell Transplant. 2016;25(6):1131–44.
42. Rosado-de-Castro PH, de Carvalho FG, de Freitas GR, Mendez-Otero R, Pimentel-Coelho PM. Review of preclinical and clinical studies of bone marrow-derived cell therapies for intracerebral hemorrhage. Stem Cells Int. 2016;2016:4617983.
43. Cui J, Cui C, Cui Y, Li R, Sheng H, Jiang X, et al. Bone marrow mesenchymal stem cell transplantation increases GAP-43 expression via ERK1/2 and PI3K/Akt pathways in intracerebral hemorrhage. Cell Physiol Biochem. 2017;42(1):137–44.
44. Ding R, Lin C, Wei S, Zhang N, Tang L, Lin Y, et al. Therapeutic benefits of mesenchymal stromal cells in a rat model of hemoglobin-induced hypertensive intracerebral hemorrhage. Mol Cells. 2017;40(2):133–42.
45. Otero L, Zurita M, Bonilla C, Aguayo C, Rico MA, Rodriguez A, et al. Allogeneic bone marrow stromal cell transplantation after cerebral hemorrhage achieves cell transdifferentiation and modulates endogenous neurogenesis. Cytotherapy. 2012;14(1):34–44.
46. Wang C, Fei Y, Xu C, Zhao Y, Pan Y. Bone marrow mesenchymal stem cells ameliorate neurological deficits and blood-brain barrier dysfunction after intracerebral hemorrhage in spontaneously hypertensive rats. Int J Clin Exp Pathol. 2015;8(5):4715–24.
47. Wang SP, Wang ZH, Peng DY, Li SM, Wang H, Wang XH. Therapeutic effect of mesenchymal stem cells in rats with intracerebral hemorrhage: reduced apoptosis and enhanced neuroprotection. Mol Med Rep. 2012;6(4):848–54.

48. Khalili MA, Sadeghian-Nodoushan F, Fesahat F, Mir-Esmaeili SM, Anvari M, Hekmati-Moghadam SH. Mesenchymal stem cells improved the ultrastructural morphology of cerebral tissues after subarachnoid hemorrhage in rats. Exp Neurobiol. 2014;23(1):77–85.
49. Jolkkonen J, Kwakkel G. Translational hurdles in stroke recovery studies. Transl Stroke Res. 2016;7(4):331–42.
50. Pennypacker KR, Bix G, Fraser JF. Correcting the trajectory of stroke therapeutic research. Transl Stroke Res. 2017;8(1):65–6.
51. Lackland DT, Roccella EJ, Deutsch AF, Fornage M, George MG, Howard G, et al. Factors influencing the decline in stroke mortality: a statement from the American Heart Association/American Stroke Association. Stroke. 2014;45(1):315–53.
52. Moskowitz MA, Lo EH, Iadecola C. The science of stroke: mechanisms in search of treatments. Neuron. 2010;67(2):181–98.
53. Ghosh P, Sahoo R, Vaidya A, Chorev M, Halperin JA. Role of complement and complement regulatory proteins in the complications of diabetes. Endocr Rev. 2015;36(3):272–88.
54. Vija L, Farge D, Gautier JF, Vexiau P, Dumitrache C, Bourgarit A, et al. Mesenchymal stem cells: stem cell therapy perspectives for type 1 diabetes. Diabetes Metab. 2009;35(2):85–93.
55. Ye X, Chopp M, Cui X, Zacharek A, Cui Y, Yan T, et al. Niaspan enhances vascular remodeling after stroke in type 1 diabetic rats. Exp Neurol. 2011;232(2):299–308.
56. Zang L, Hao H, Liu J, Li Y, Han W, Mu Y. Mesenchymal stem cell therapy in type 2 diabetes mellitus. Diabetol Metab Syndr. 2017;9:36.
57. Ding G, Chen J, Chopp M, Li L, Yan T, Li Q, et al. Cell treatment for stroke in type two diabetic rats improves vascular permeability measured by MRI. PLoS One. 2016;11(2):e0149147.
58. Yan T, Ye X, Chopp M, Zacharek A, Ning R, Venkat P, et al. Niaspan attenuates the adverse effects of bone marrow stromal cell treatment of stroke in type one diabetic rats. PLoS One. 2013;8(11):e81199.
59. Chen J, Ye X, Yan T, Zhang C, Yang XP, Cui X, et al. Adverse effects of bone marrow stromal cell treatment of stroke in diabetic rats. Stroke. 2011;42(12):3551–8.
60. Yan T, Venkat P, Chopp M, Zacharek A, Ning R, Roberts C, et al. Neurorestorative responses to delayed human mesenchymal stromal cells treatment of stroke in type 2 diabetic rats. Stroke. 2016;47(11):2850–8.
61. Cui C, Ye X, Chopp M, Venkat P, Zacharek A, Yan T, et al. miR-145 regulates diabetes-bone marrow stromal cell-induced neurorestorative effects in diabetes stroke rats. Stem Cells Transl Med. 2016;5(12):1656–67.
62. Popa-Wagner A, Buga A-M, Doeppner TR, Hermann DM. Stem cell therapies in preclinical models of stroke associated with aging. Front Cell Neurosci. 2014;8:347.
63. Johansson BB. Hypertension mechanisms causing stroke. Clin Exp Pharmacol Physiol. 1999;26(7):563–5.
64. Calio ML, Marinho DS, Ko GM, Ribeiro RR, Carbonel AF, Oyama LM, et al. Transplantation of bone marrow mesenchymal stem cells decreases oxidative stress, apoptosis, and hippocampal damage in brain of a spontaneous stroke model. Free Radic Biol Med. 2014;70:141–54.
65. Chen C, Cheng Y, Chen J. Transfection of Noggin in bone marrow stromal cells (BMSCs) enhances BMSC-induced functional outcome after stroke in rats. J Neurosci Res. 2011;89(8):1194–202.
66. Zhao MZ, Nonoguchi N, Ikeda N, Watanabe T, Furutama D, Miyazawa D, et al. Novel therapeutic strategy for stroke in rats by bone marrow stromal cells and ex vivo HGF gene transfer with HSV-1 vector. J Cereb Blood Flow Metab. 2006;26(9):1176–88.
67. Ding J, Cheng Y, Gao S, Chen J. Effects of nerve growth factor and Noggin-modified bone marrow stromal cells on stroke in rats. J Neurosci Res. 2011;89(2):222–30.
68. Xin H, Li Y, Liu Z, Wang X, Shang X, Cui Y, et al. MiR-133b promotes neural plasticity and functional recovery after treatment of stroke with multipotent mesenchymal stromal cells in rats via transfer of exosome-enriched extracellular particles. Stem Cells. 2013;31(12):2737–46.

69. Ikeda N, Nonoguchi N, Zhao MZ, Watanabe T, Kajimoto Y, Furutama D, et al. Bone marrow stromal cells that enhanced fibroblast growth factor-2 secretion by herpes simplex virus vector improve neurological outcome after transient focal cerebral ischemia in rats. Stroke. 2005;36(12):2725–30.
70. Kurozumi K, Nakamura K, Tamiya T, Kawano Y, Kobune M, Hirai S, et al. BDNF gene-modified mesenchymal stem cells promote functional recovery and reduce infarct size in the rat middle cerebral artery occlusion model. Mol Ther. 2004;9(2):189–97.
71. Zhang ZG, Zhang L, Jiang Q, Zhang R, Davies K, Powers C, et al. VEGF enhances angiogenesis and promotes blood-brain barrier leakage in the ischemic brain. J Clin Invest. 2000;106(7):829–38.
72. Yang C, Zhou L, Gao X, Chen B, Tu J, Sun H, et al. Neuroprotective effects of bone marrow stem cells overexpressing glial cell line-derived neurotrophic factor on rats with intracerebral hemorrhage and neurons exposed to hypoxia/reoxygenation. Neurosurgery. 2011;68(3):691–704.
73. Wei N, Yu SP, Gu X, Taylor TM, Song D, Liu XF, et al. Delayed intranasal delivery of hypoxic-preconditioned bone marrow mesenchymal stem cells enhanced cell homing and therapeutic benefits after ischemic stroke in mice. Cell Transplant. 2013;22(6):977–91.
74. Lim JY, Jeong CH, Jun JA, Kim SM, Ryu CH, Hou Y, et al. Therapeutic effects of human umbilical cord blood-derived mesenchymal stem cells after intrathecal administration by lumbar puncture in a rat model of cerebral ischemia. Stem Cell Res Ther. 2011;2(5):38.
75. Wang L, Lin Z, Shao B, Zhuge Q, Jin K. Therapeutic applications of bone marrow-derived stem cells in ischemic stroke. Neurol Res. 2013;35(5):470–8.
76. Misra V, Ritchie MM, Stone LL, Low WC, Janardhan V. Stem cell therapy in ischemic stroke: role of IV and intra-arterial therapy. Neurology. 2012;79(13 Suppl 1):S207–12.
77. Yoo SW, Kim SS, Lee SY, Lee HS, Kim HS, Lee YD, et al. Mesenchymal stem cells promote proliferation of endogenous neural stem cells and survival of newborn cells in a rat stroke model. Exp Mol Med. 2008;40(4):387–97.
78. Irons H, Lind JG, Wakade CG, Yu G, Hadman M, Carroll J, et al. Intracerebral xenotransplantation of GFP mouse bone marrow stromal cells in intact and stroke rat brain: graft survival and immunologic response. Cell Transplant. 2004;13(3):283–94.
79. Steinberg GK, Kondziolka D, Wechsler LR, Lunsford LD, Coburn ML, Billigen JB, et al. Clinical outcomes of transplanted modified bone marrow-derived mesenchymal stem cells in stroke: a phase 1/2a study. Stroke. 2016;47(7):1817–24.
80. Kondziolka D, Steinberg GK, Wechsler L, Meltzer CC, Elder E, Gebel J, et al. Neurotransplantation for patients with subcortical motor stroke: a phase 2 randomized trial. J Neurosurg. 2005;103(1):38–45.
81. Jin K, Sun Y, Xie L, Mao XO, Childs J, Peel A, et al. Comparison of ischemia-directed migration of neural precursor cells after intrastriatal, intraventricular, or intravenous transplantation in the rat. Neurobiol Dis. 2005;18(2):366–74.
82. Rabinovich SS, Seledtsov VI, Banul NV, Poveshchenko OV, Senyukov VV, Astrakov SV, et al. Cell therapy of brain stroke. Bull Exp Biol Med. 2005;139(1):126–8.
83. Jiang Q, Zhang ZG, Ding GL, Zhang L, Ewing JR, Wang L, et al. Investigation of neural progenitor cell induced angiogenesis after embolic stroke in rat using MRI. NeuroImage. 2005;28(3):698–707.
84. Modo M, Stroemer RP, Tang E, Patel S, Hodges H. Effects of implantation site of stem cell grafts on behavioral recovery from stroke damage. Stroke. 2002;33(9):2270–8.
85. Chen JJ, Zhou SH. Mesenchymal stem cells overexpressing MiR-126 enhance ischemic angiogenesis via the AKT/ERK-related pathway. Cardiol J. 2011;18(6):675–81.
86. Guo F, Lv S, Lou Y, Tu W, Liao W, Wang Y, et al. Bone marrow stromal cells enhance the angiogenesis in ischaemic cortex after stroke: involvement of notch signalling. Cell Biol Int. 2012;36(11):997–1004.
87. Suarez-Monteagudo C, Hernandez-Ramirez P, Alvarez-Gonzalez L, Garcia-Maeso I, de la Cuetara-Bernal K, Castillo-Diaz L, et al. Autologous bone marrow stem cell neurotransplantation in stroke patients. An open study. Restor Neurol Neurosci. 2009;27(3):151–61.

88. Wise AF, Williams TM, Kiewiet MB, Payne NL, Siatskas C, Samuel CS, et al. Human mesenchymal stem cells alter macrophage phenotype and promote regeneration via homing to the kidney following ischemia-reperfusion injury. Am J Physiol Renal Physiol. 2014;306(10):F1222–35.
89. Bhasin A, Srivastava MV, Kumaran SS, Mohanty S, Bhatia R, Bose S, et al. Autologous mesenchymal stem cells in chronic stroke. Cerebrovasc Dis Extra. 2011;1(1):93–104.
90. Lee JS, Hong JM, Moon GJ, Lee PH, Ahn YH, Bang OY. A long-term follow-up study of intravenous autologous mesenchymal stem cell transplantation in patients with ischemic stroke. Stem Cells. 2010;28(6):1099–106.
91. Steiner B, Roch M, Holtkamp N, Kurtz A. Systemically administered human bone marrow-derived mesenchymal stem home into peripheral organs but do not induce neuroprotective effects in the MCAo-mouse model for cerebral ischemia. Neurosci Lett. 2012;513(1):25–30.
92. Kraitchman DL, Tatsumi M, Gilson WD, Ishimori T, Kedziorek D, Walczak P, et al. Dynamic imaging of allogeneic mesenchymal stem cells trafficking to myocardial infarction. Circulation. 2005;112(10):1451–61.
93. Li Y, Chen J, Wang L, Lu M, Chopp M. Treatment of stroke in rat with intracarotid administration of marrow stromal cells. Neurology. 2001;56(12):1666–72.
94. Byun JS, Kwak BK, Kim JK, Jung J, Ha BC, Park S. Engraftment of human mesenchymal stem cells in a rat photothrombotic cerebral infarction model: comparison of intra-arterial and intravenous infusion using MRI and histological analysis. J Korean Neurosurg Soc. 2013;54(6):467–76.
95. Yang B, Migliati E, Parsha K, Schaar K, Xi X, Aronowski J, et al. Intra-arterial delivery is not superior to intravenous delivery of autologous bone marrow mononuclear cells in acute ischemic stroke. Stroke. 2013;44(12):3463–72.
96. Kamiya N, Ueda M, Igarashi H, Nishiyama Y, Suda S, Inaba T, et al. Intra-arterial transplantation of bone marrow mononuclear cells immediately after reperfusion decreases brain injury after focal ischemia in rats. Life Sci. 2008;83(11–12):433–7.
97. Dhuria SV, Hanson LR, Frey WH II. Intranasal delivery to the central nervous system: mechanisms and experimental considerations. J Pharm Sci. 2010;99(4):1654–73.
98. Jiang Y, Zhu J, Xu G, Liu X. Intranasal delivery of stem cells to the brain. Expert Opin Drug Deliv. 2011;8(5):623–32.
99. van Velthoven CT, Sheldon RA, Kavelaars A, Derugin N, Vexler ZS, Willemen HL, et al. Mesenchymal stem cell transplantation attenuates brain injury after neonatal stroke. Stroke. 2013;44(5):1426–32.
100. Sun J, Wei ZZ, Gu X, Zhang JY, Zhang Y, Li J, et al. Intranasal delivery of hypoxia-preconditioned bone marrow-derived mesenchymal stem cells enhanced regenerative effects after intracerebral hemorrhagic stroke in mice. Exp Neurol. 2015;272:78–87.
101. Zhang R, Wang Y, Zhang L, Zhang Z, Tsang W, Lu M, et al. Sildenafil (Viagra) induces neurogenesis and promotes functional recovery after stroke in rats. Stroke. 2002;33(11):2675–80.
102. Pistoia V, Raffaghello L. Mesenchymal stromal cells and autoimmunity. Int Immunol. 2017;29(2):49–58.
103. Riess P, Zhang C, Saatman KE, Laurer HL, Longhi LG, Raghupathi R, et al. Transplanted neural stem cells survive, differentiate, and improve neurological motor function after experimental traumatic brain injury. Neurosurgery. 2002;51(4):1043–52. discussion 52–4
104. Van Damme A, Vanden Driessche T, Collen D, Chuah MK. Bone marrow stromal cells as targets for gene therapy. Curr Gene Ther. 2002;2(2):195–209.
105. Hanada K, Dennis JE, Caplan AI. Stimulatory effects of basic fibroblast growth factor and bone morphogenetic protein-2 on osteogenic differentiation of rat bone marrow-derived mesenchymal stem cells. J Bone Miner Res. 1997;12(10):1606–14.
106. Martin DR, Cox NR, Hathcock TL, Niemeyer GP, Baker HJ. Isolation and characterization of multipotential mesenchymal stem cells from feline bone marrow. Exp Hematol. 2002;30(8):879–86.

107. Alberti-Amador E, Garcia-Miniet R. Bone marrow stromal cells. An alternative source of restorative therapy in degenerative diseases of the central nervous system. Rev Neurol. 2003;37(8):752–8.
108. Prockop DJ, Gregory CA, Spees JL. One strategy for cell and gene therapy: harnessing the power of adult stem cells to repair tissues. Proc Natl Acad Sci U S A. 2003;100(Suppl 1):11917–23.
109. Shichinohe H, Kuroda S, Maruichi K, Osanai T, Sugiyama T, Chiba Y, et al. Bone marrow stromal cells and bone marrow-derived mononuclear cells: which are suitable as cell source of transplantation for mice infarct brain? Neuropathology. 2010;30(2):113–22.
110. Zhang J, Li Y, Zhang ZG, Lu M, Borneman J, Buller B, et al. Bone marrow stromal cells increase oligodendrogenesis after stroke. J Cereb Blood Flow Metab. 2009;29(6):1166–74.
111. Shen LH, Li Y, Chen J, Cui Y, Zhang C, Kapke A, et al. One-year follow-up after bone marrow stromal cell treatment in middle-aged female rats with stroke. Stroke. 2007;38(7):2150–6.
112. Chen J, Shehadah A, Pal A, Zacharek A, Cui X, Cui Y, et al. Neuroprotective effect of human placenta-derived cell treatment of stroke in rats. Cell Transplant. 2012;22(5):871–9.
113. Nagaya N, Kangawa K, Itoh T, Iwase T, Murakami S, Miyahara Y, et al. Transplantation of mesenchymal stem cells improves cardiac function in a rat model of dilated cardiomyopathy. Circulation. 2005;112(8):1128–35.
114. Ziegelhoeffer T, Fernandez B, Kostin S, Heil M, Voswinckel R, Helisch A, et al. Bone marrow-derived cells do not incorporate into the adult growing vasculature. Circ Res. 2004;94(2):230–8.
115. Hess DC, Abe T, Hill WD, Studdard AM, Carothers J, Masuya M, et al. Hematopoietic origin of microglial and perivascular cells in brain. Exp Neurol. 2004;186(2):134–44.
116. Tang YL, Zhao Q, Qin X, Shen L, Cheng L, Ge J, et al. Paracrine action enhances the effects of autologous mesenchymal stem cell transplantation on vascular regeneration in rat model of myocardial infarction. Ann Thorac Surg. 2005;80(1):229–36. discussion 36–7
117. Kinnaird T, Stabile E, Burnett MS, Epstein SE. Bone marrow-derived cells for enhancing collateral development: mechanisms, animal data, and initial clinical experiences. Circ Res. 2004;95(4):354–63.
118. Ponte AL, Marais E, Gallay N, Langonne A, Delorme B, Herault O, et al. The in vitro migration capacity of human bone marrow mesenchymal stem cells: comparison of chemokine and growth factor chemotactic activities. Stem Cells. 2007;25(7):1737–45.
119. Wu Y, Chen L, Scott PG, Tredget EE. Mesenchymal stem cells enhance wound healing through differentiation and angiogenesis. Stem Cells. 2007;25(10):2648–59.
120. Kinnaird T, Stabile E, Burnett MS, Lee CW, Barr S, Fuchs S, et al. Marrow-derived stromal cells express genes encoding a broad spectrum of arteriogenic cytokines and promote in vitro and in vivo arteriogenesis through paracrine mechanisms. Circ Res. 2004;94(5):678–85.
121. Matsuda-Hashii Y, Takai K, Ohta H, Fujisaki H, Tokimasa S, Osugi Y, et al. Hepatocyte growth factor plays roles in the induction and autocrine maintenance of bone marrow stromal cell IL-11, SDF-1 alpha, and stem cell factor. Exp Hematol. 2004;32(10):955–61.
122. Annabi B, Lee YT, Turcotte S, Naud E, Desrosiers RR, Champagne M, et al. Hypoxia promotes murine bone-marrow-derived stromal cell migration and tube formation. Stem Cells. 2003;21(3):337–47.
123. Eaves CJ, Cashman JD, Kay RJ, Dougherty GJ, Otsuka T, Gaboury LA, et al. Mechanisms that regulate the cell cycle status of very primitive hematopoietic cells in long-term human marrow cultures. II. Analysis of positive and negative regulators produced by stromal cells within the adherent layer. Blood. 1991;78(1):110–7.
124. Majumdar MK, Thiede MA, Mosca JD, Moorman M, Gerson SL. Phenotypic and functional comparison of cultures of marrow-derived mesenchymal stem cells (MSCs) and stromal cells. J Cell Physiol. 1998;176(1):57–66.
125. Seshi B, Kumar S, Sellers D. Human bone marrow stromal cell: coexpression of markers specific for multiple mesenchymal cell lineages. Blood Cells Mol Dis. 2000;26(3):234–46.
126. Plate KH. Mechanisms of angiogenesis in the brain. J Neuropathol Exp Neurol. 1999;58(4):313–20.

127. Renner O, Tsimpas A, Kostin S, Valable S, Petit E, Schaper W, et al. Time- and cell type-specific induction of platelet-derived growth factor receptor-beta during cerebral ischemia. Brain Res Mol Brain Res. 2003;113(1–2):44–51.
128. Slevin M, Kumar P, Gaffney J, Kumar S, Krupinski J. Can angiogenesis be exploited to improve stroke outcome? Mechanisms and therapeutic potential. Clin Sci (Lond). 2006;111(3):171–83.
129. Krupinski J, Kaluza J, Kumar P, Kumar S, Wang JM. Role of angiogenesis in patients with cerebral ischemic stroke. Stroke. 1994;25(9):1794–8.
130. Wei L, Erinjeri JP, Rovainen CM, Woolsey TA. Collateral growth and angiogenesis around cortical stroke. Stroke. 2001;32(9):2179–84.
131. Bronckaers A, Hilkens P, Martens W, Gervois P, Ratajczak J, Struys T, et al. Mesenchymal stem/stromal cells as a pharmacological and therapeutic approach to accelerate angiogenesis. Pharmacol Ther. 2014;143(2):181–96.
132. Zacharek A, Chen J, Cui X, Li A, Li Y, Roberts C, et al. Angiopoietin1/Tie2 and VEGF/Flk1 induced by MSC treatment amplifies angiogenesis and vascular stabilization after stroke. J Cereb Blood Flow Metab. 2007;27(10):1684–91.
133. Nam HS, Kwon I, Lee BH, Kim H, Kim J, An S, et al. Effects of mesenchymal stem cell treatment on the expression of matrix metalloproteinases and angiogenesis during ischemic stroke recovery. PLoS One. 2015;10(12):e0144218.
134. Doeppner TR, Hermann DM. Mesenchymal stem cells in the treatment of ischemic stroke: progress and possibilities. Stem Cells Cloning. 2010;3:157–63.
135. Jiang Q, Zhang ZG, Ding GL, Silver B, Zhang L, Meng H, et al. MRI detects white matter reorganization after neural progenitor cell treatment of stroke. NeuroImage. 2006;32(3):1080–9.
136. Zhang R, Chopp M, Zhang ZG. Oligodendrogenesis after cerebral ischemia. Front Cell Neurosci. 2013;7:201.
137. Ye X, Yan T, Chopp M, Zacharek A, Ning R, Venkat P, et al. Combination BMSC and Niaspan treatment of stroke enhances white matter remodeling and synaptic protein expression in diabetic rats. Int J Mol Sci. 2013;14(11):22221–32.
138. van Velthoven CT, Dzietko M, Wendland MF, Derugin N, Faustino J, Heijnen CJ, et al. Mesenchymal stem cells attenuate MRI-identifiable injury, protect white matter, and improve long-term functional outcomes after neonatal focal stroke in rats. J Neurosci Res. 2017;95(5):1225–36.
139. Maraka S, Jiang Q, Jafari-Khouzani K, Li L, Malik S, Hamidian H, et al. Degree of corticospinal tract damage correlates with motor function after stroke. Ann Clin Transl Neurol. 2014;1(11):891–9.
140. Zhu LL, Lindenberg R, Alexander MP, Schlaug G. Lesion load of the corticospinal tract predicts motor impairment in chronic stroke. Stroke. 2010;41(5):910–5.
141. Liu Z, Li Y, Qu R, Shen L, Gao Q, Zhang X, et al. Axonal sprouting into the denervated spinal cord and synaptic and postsynaptic protein expression in the spinal cord after transplantation of bone marrow stromal cell in stroke rats. Brain Res. 2007;1149:172–80.
142. Liu Z, Zhang RL, Li Y, Cui Y, Chopp M. Remodeling of the corticospinal innervation and spontaneous behavioral recovery after ischemic stroke in adult mice. Stroke. 2009;40(7):2546–51.
143. Liu Z, Li Y, Zhang RL, Cui Y, Chopp M. Bone marrow stromal cells promote skilled motor recovery and enhance contralesional axonal connections after ischemic stroke in adult mice. Stroke. 2011;42(3):740–4.
144. Liu Z, Li Y, Zhang X, Savant-Bhonsale S, Chopp M. Contralesional axonal remodeling of the corticospinal system in adult rats after stroke and bone marrow stromal cell treatment. Stroke. 2008;39(9):2571–7.
145. Zhang R, Zhang Z, Wang L, Wang Y, Gousev A, Zhang L, et al. Activated neural stem cells contribute to stroke-induced neurogenesis and neuroblast migration toward the infarct boundary in adult rats. J Cereb Blood Flow Metab. 2004;24(4):441–8.
146. Xiong Y, Mahmood A, Chopp M. Angiogenesis, neurogenesis and brain recovery of function following injury. Curr Opin Investig Drugs. 2010;11(3):298–308.

147. Kojima T, Hirota Y, Ema M, Takahashi S, Miyoshi I, Okano H, et al. Subventricular zone-derived neural progenitor cells migrate along a blood vessel scaffold toward the post-stroke striatum. Stem Cells. 2010;28(3):545–54.
148. Zhang RL, Zhang Z, Zhang L, Wang Y, Zhang C, Chopp M. Delayed treatment with sildenafil enhances neurogenesis and improves functional recovery in aged rats after focal cerebral ischemia. J Neurosci Res. 2006;83(7):1213–9.
149. Bao X, Wei J, Feng M, Lu S, Li G, Dou W, et al. Transplantation of human bone marrow-derived mesenchymal stem cells promotes behavioral recovery and endogenous neurogenesis after cerebral ischemia in rats. Brain Res. 2011;1367:103–13.
150. Chen J, Li Y, Zhang R, Katakowski M, Gautam SC, Xu Y, et al. Combination therapy of stroke in rats with a nitric oxide donor and human bone marrow stromal cells enhances angiogenesis and neurogenesis. Brain Res. 2004;1005(1–2):21–8.
151. Zhang J, Li Y, Chen J, Yang M, Katakowski M, Lu M, et al. Expression of insulin-like growth factor 1 and receptor in ischemic rats treated with human marrow stromal cells. Brain Res. 2004;1030(1):19–27.
152. Gutierrez-Fernandez M, Rodriguez-Frutos B, Ramos-Cejudo J, Teresa Vallejo-Cremades M, Fuentes B, Cerdan S, et al. Effects of intravenous administration of allogenic bone marrow- and adipose tissue-derived mesenchymal stem cells on functional recovery and brain repair markers in experimental ischemic stroke. Stem Cell Res Ther. 2013;4(1):11.
153. van Velthoven CT, Kavelaars A, Heijnen CJ. Mesenchymal stem cells as a treatment for neonatal ischemic brain damage. Pediatr Res. 2012;71(4 Pt 2):474–81.
154. Thery C, Ostrowski M, Segura E. Membrane vesicles as conveyors of immune responses. Nat Rev Immunol. 2009;9(8):581–93.
155. Yeo RW, Lai RC, Zhang B, Tan SS, Yin Y, Teh BJ, et al. Mesenchymal stem cell: an efficient mass producer of exosomes for drug delivery. Adv Drug Deliv Rev. 2013;65(3):336–41.
156. Xin H, Li Y, Chopp M. Exosomes/miRNAs as mediating cell-based therapy of stroke. Front Cell Neurosci. 2014;8:377.
157. Thery C, Amigorena S, Raposo G, Clayton A. Isolation and characterization of exosomes from cell culture supernatants and biological fluids. Curr Protoc Cell Biol. 2006;Chapter 3:Unit 3.22.
158. Otero-Ortega L, Gomez de Frutos MC, Laso-Garcia F, Rodriguez-Frutos B, Medina-Gutierrez E, Lopez JA, et al. Exosomes promote restoration after an experimental animal model of intracerebral hemorrhage. J Cereb Blood Flow Metab. 2017:271678x17708917.
159. Xu JF, Yang GH, Pan XH, Zhang SJ, Zhao C, Qiu BS, et al. Altered microRNA expression profile in exosomes during osteogenic differentiation of human bone marrow-derived mesenchymal stem cells. PLoS One. 2014;9(12):e114627.
160. Xin H, Li Y, Cui Y, Yang JJ, Zhang ZG, Chopp M. Systemic administration of exosomes released from mesenchymal stromal cells promote functional recovery and neurovascular plasticity after stroke in rats. J Cereb Blood Flow Metab. 2013;33(11):1711–5.
161. Baglio SR, Rooijers K, Koppers-Lalic D, Verweij FJ, Perez Lanzon M, Zini N, et al. Human bone marrow- and adipose-mesenchymal stem cells secrete exosomes enriched in distinctive miRNA and tRNA species. Stem Cell Res Ther. 2015;6:127.
162. Xiong Y, Mahmood A, Chopp M. Emerging potential of exosomes for treatment of traumatic brain injury. Neural Regen Res. 2017;12(1):19–22.
163. Xiong Y, Zhang Y, Mahmood A, Chopp M. Investigational agents for treatment of traumatic brain injury. Expert Opin Investig Drugs. 2015;24(6):743–60.
164. Doeppner TR, Herz J, Gorgens A, Schlechter J, Ludwig AK, Radtke S, et al. Extracellular vesicles improve post-stroke neuroregeneration and prevent postischemic immunosuppression. Stem Cells Transl Med. 2015;4(10):1131–43.
165. Zhang Y, Chopp M, Liu XS, Katakowski M, Wang X, Tian X, et al. Exosomes derived from mesenchymal stromal cells promote axonal growth of cortical neurons. Mol Neurobiol. 2017;54(4):2659–73.

166. Xin H, Li Y, Buller B, Katakowski M, Zhang Y, Wang X, et al. Exosome-mediated transfer of miR-133b from multipotent mesenchymal stromal cells to neural cells contributes to neurite outgrowth. Stem Cells. 2012;30(7):1556–64.
167. Zhang ZG, Chopp M. Exosomes in stroke pathogenesis and therapy. J Clin Invest. 2016;126(4):1190–7.
168. Chopp M, Zhang ZG. Emerging potential of exosomes and noncoding microRNAs for the treatment of neurological injury/diseases. Expert Opin Emerg Drugs. 2015;20(4):523–6.
169. Zhang ZG, Chopp M. Promoting brain remodeling to aid in stroke recovery. Trends Mol Med. 2015;21(9):543–8.
170. Ambros V. The functions of animal microRNAs. Nature. 2004;431(7006):350–5.
171. Sen CK. MicroRNAs as new maestro conducting the expanding symphony orchestra of regenerative and reparative medicine. Physiol Genomics. 2011;43(10):517–20.
172. Juranek JK, Geddis MS, Song F, Zhang J, Garcia J, Rosario R, et al. RAGE deficiency improves postinjury sciatic nerve regeneration in type 1 diabetic mice. Diabetes. 2013;62(3):931–43.
173. Phinney DG, Di Giuseppe M, Njah J, Sala E, Shiva S, St Croix CM, et al. Mesenchymal stem cells use extracellular vesicles to outsource mitophagy and shuttle microRNAs. Nat Commun. 2015;6:8472.
174. Cui C, Ye X, Chopp M, Venkat P, Zacharek A, Yan T, et al. miR-145 regulates diabetes-bone marrow stromal cell-induced neurorestorative effects in diabetes stroke rats. Stem Cells Transl Med. 2016;
175. Wang XQ, Zhu XJ, Zou P. Research progress of mesenchymal stem cell-derived microvesicle. Zhongguo Shi Yan Xue Ye Xue Za Zhi. 2013;21(1):227–30.
176. Xin H, Katakowski M, Wang F, Qian JY, Liu XS, Ali MM, et al. MicroRNA cluster miR-17-92 cluster in exosomes enhance neuroplasticity and functional recovery after stroke in rats. Stroke. 2017;48(3):747–53.
177. Zhang Y, Ueno Y, Liu XS, Buller B, Wang X, Chopp M, et al. The MicroRNA-17-92 cluster enhances axonal outgrowth in embryonic cortical neurons. J Neurosci. 2013;33(16):6885–94.
178. Xin H, Wang F, Li Y, Lu QE, Cheung WL, Zhang Y, et al. Secondary release of exosomes from astrocytes contributes to the increase in neural plasticity and improvement of functional recovery after stroke in rats treated with exosomes harvested from MicroRNA 133b-overexpressing multipotent mesenchymal stromal cells. Cell Transplant. 2017;26(2):243–57.
179. Wen Z, Zheng S, Zhou C, Yuan W, Wang J, Wang T. Bone marrow mesenchymal stem cells for post-myocardial infarction cardiac repair: microRNAs as novel regulators. J Cell Mol Med. 2012;16(4):657–71.

Chapter 12
Combination Treatment of Mesenchymal Stem Cells (MSCs) and Angelica sinensis' Active Ingredients for Ischemic Stroke

Qian Zhang and Yonghua Zhao

Abstract At present, mesenchymal stem cells (MSCs) are regarded as a candidate for neovascularization and tissue regeneration after ischemic stroke. Numerous studies reported that Angelica (also called Dong quai, a well-known Chinese herbal medicine) extracts and its active ingredients such as ligustilide, n-Butylphthalide and sodium ferulate had significant effects of anti-inflammatory, anti-activation of oxygen free radicals, angiogenesis, anti-platelet aggregation, neuroprotection and so on. Angelica' active compositions facilitated MSCs to migrate into infarcted zone and differentiation. Moreover, MSCs combined with angelica' active components improved neurological function and decreased infarcted volume, advanced neovascularization and neurogenesis, regulated astrocytes characteristics, enhanced regional cerebral blood flow and glucose metabolism, as well as reduced brain-blood barrier permeability in infarction. Consequently, the structure and function of neurovascular unit in infarct region partly obtained recovery. Therefore, the combination treatment was a valuable therapy aimed at improving post-stroke restoration.

Keywords Angelica sinensis • Combination treatment • Ischemic stroke • Mesenchymal stem cells

Abbreviations

BBB	Blood-brain barrier
BDNF	Brain-derived neurotrophic factor
BMP	Bone morphogenetic proteins
BP	n-Butylidenephthalide

Q. Zhang • Y. Zhao (✉)
State Key Laboratory of Quality Research in Chinese Medicine, Faculty of Chinese Medicine, Macau University of Science and Technology,
Avenida Wai Long, Taipa, Macau 999078, China
e-mail: yhzhao@must.edu.mo

P.A. Lapchak, J.H. Zhang (eds.), *Cellular and Molecular Approaches to Regeneration and Repair*, Springer Series in Translational Stroke Research,
https://doi.org/10.1007/978-3-319-66679-2_12

BrdU	5-Bromo-2′-deoxyuridine
CBF	Cerebral blood flow
CXCR4	Chemokine (CXC motif) receptor-4
DCX	Doublecortin
DG	Dentate gyrus
EC	Endothelial cell
ERK	Extracellular signal-regulated kinases
FA	Ferulic acid
FDA	Food and Drug Administration
FDG	^{18}F-2-deoxy-glucose
GDNF	Glial cell line-derived neurotrophic factor
GFAP	Glial fibrillary acidic protein
HIF	Hypoxia-inducible factors
HUVEC	Human umbilical vein endothelial cell
MCAo	Middle cerebral artery occlusion
MRI	Magnetic resonance imaging
MSC	Mesenchymal stem cell
mTOR	Mammalian target of rapamycin
NBP	n-Butylphthalide
PET/CT	Positron emission tomography–computed tomography
PWI	Perfusion-weighted imaging
SDF-1	Stromal cell-derived factor-1
SF	Sodium ferulate
STEPS	Stem Cells as an Emerging Paradigm in Stroke
SVZ	Subventricular zone
TTC	2,3,5-Triphenyltetrazolium chloride
Tuj-1	Neuron-specific class III beta-tubulin
VEGF	Vascular endothelial growth factor
vWF	Von Willebrand factor

1 Introduction

Ischemic stroke is the most common cerebralvascular disease. Due to blood flow blockage by arterial thrombus, amounts of neurons in ischemic central and penumbral regions occur to necrosis and apoptosis, which resulted in attenuation of neurological function. Evidence indicated that stroke was the second leading cause of death and the major cause of disability globally, especially in developing countries [1]. There are 15 million individuals suffer from stroke in every year worldwide, and in the United States, among 800,000 stroke patients, 75% of them have never experienced stroke before, and 25% undergo recurrent attack [2]. Moreover, mortality of ischemic stroke is predicted to nearly double by 2032 [3]. Although stroke has been major threat to life expectancy and quality, there are relatively few treatment options available to ameliorate neurological function due to complicated etiological and pathophysiological evolutions after ischemic stroke [4]. At present, recombinant tissue plasminogen activator (rt-PA) is still approved by United States Food and Drug Administration (FDA) for dissolution of thrombus and improvement of cerebral flow, but narrow therapeutic time window (3–6 h) and multiplicative individual exclusion criteria limit it to be widely applied [5, 6].

In recent years, cell therapy is regarded as a promising approach. Bone marrow-derived mesenchymal stem cells (MSCs) have been demonstrated to be able to differentiate into neuronal cells and replace injured neurons after cerebral ischemia, as well as activate endogenous restorative responses (e.g. neurogenesis, angiogenesis and synaptogenesis) against injured brain, so autologous MSCs were transplanted into stroke patients in 2005 [7–11]. Evidence indicated that five stroke patients accepted stereotactically transplanted MSCs treatment, and the therapeutic results showed that their neurological functions were improved and no complications happened after 1 year's observation [12]. In 2007, the National Institutes of Health and FDA issued consensus-based guidelines on the development of cell therapies for stroke, entitled "Stem Cells as an Emerging Paradigm in Stroke" (STEPS). Current STEPS 3 had discussed how to successful complete translation from animal models to patients and optimize clinical trial designs for acute and chronic stroke [13].

Angelica (Angelica sinensis (Oliv.) Diels), commonly called Dong quai in Chinese, is a dried root derived from an herb in the family Apiaceae, which is used over thousands of years as a well-known Chinese medicine. Since 1980s, angelica extract and its active ingredients began to be used to treat ischemic cerebrovascular disease. Liu and colleagues observed 1404 patients of acute cerebral infarction, and 692 of them treated with angelica injection, 390 of them used Danshen (Salvia miltiorrhiza) injection, 322 of them treated with low molecular dextran. Consequently, the total effective rate was 78.7%, 63.6% and 59.3% respectively after treatment, suggesting neurological functional recovery in angelica injection group was better than those in other two groups [14]. Study indicated that Z-Ligustilide, a main component of volatile oil in angelica, could reduce infarct volume and cerebral edema in a dose-dependent way, and ameliorate injured neurological function after 2 h in middle cerebral artery occlusion (MCAo) rats, suggesting it had obvious neuroprotective effect [15]. n-Butylphthalide (NBP) derived from phthalides compounds in angelica has been identified as a new drug for the treatment of ischemic cerebrovascular disease by China FDA. Previous study indicated that it could improve cognitive deficits in rats with chronic cerebral ischemia [16]. Moreover, in a randomized, double-blind and multi-center research on 535 stroke patients, it showed that patients' neurological functional scores in NBP group were obviously higher than those in ozagrel group [17]. The extract and active ingredients in angelica exert multi-efficacy in the treatment of ischemic stroke.

It has been demonstrated that the effect of combining MSCs with pharmacological agents on stroke is superior to MSCs treatment alone or pharmacological agents. For example, combined treatment of BMSCs with simvastatin could further facilitate MSCs' migration and differentiation, as well as enhance arteriogenesis and angiogenesis and reduce infarction volume, which contributed to the amelioration of functional outcome after cerebral ischemia [18–20]. Our previous study also evidenced that simvastatin combined with MSCs could obviously activate astrocytes and increase astrocyte-derived stromal cell-derived factor-1α (SDF-1α), vascular endothelial growth factor (VEGF), brain-derived neurotrophic factor (BDNF) and glial cell line-derived neurotrophic factor (GDNF) expressions post-stroke, as well as up-regulate Akt/mammalian target of rapamycin (mTOR) signaling pathway in oxygen glucose deprived astrocytes [21]. Recently, the optimization of pharmacological agents in combination treatment began to transfer to Chinese medicines. Astragaloside IV, Naomai Yihao capsules, Buyang huanwu tang and Tongxinluo combined with MSCs had been demonstrated to notably promote angiogenesis and attenuate ischemic injury [22–25]. Based on angelica extract and active ingredients' multiple efficacy on ischemic stroke, our research group devotes the investigation of combination treatment of angelica and MSCs and finds some synergic functions and mechanisms.

2 Amelioration of Neurological Outcome and Reduction of Infarcted Volume

According to the evaluation of Garcia JH neurological score which included six sections: (1) evaluating animals' spontaneous activity; (2) symmetry of four limbs' movements when rat was held suspended by the tail and symmetric forepaws were assessed, (3) climbing wall of wire cage; (4) body proprioception; (5) reaction to touch on either side of the trunk; (6) response to vibrissa touch [26], we observed that derived from angelica's active ingredients, sodium ferulate (SF), as well as SF and n-Butylidenephthalide (BP) combined with MSCs began to improve neurological functional outcomes from day 3, and the increased trend always kept to day 7 after ischemia, suggesting the amelioration of neurological outcome was obviously superior to MSCs alone [27, 28]. Additionally, Bederson scale was administrated for neurological assessment following stroke, which included forelimb flexion, resistance to lateral push and circling behavior [29]. The scores are as below when SF (60 mg/kg) and BP (10 mg/kg) combined with MSCs (2×10^6 cells/ml, intravenous injection) were applied for the treatment of MCAo model in rats.

The scoring scale indicated that ischemic animals would have more significant neurological deficits than non-ischemic animals, resulting in a higher score. It indicated that neurological functional scores in SF + BP + MSC group began to reduce at day 3, showing neurological deficit had been attenuated (Table 12.1).

Magnetic resonance imaging (MRI) scanning and 2,3,5-triphenyltetrazolium chloride (TTC) staining showed whatever SF or SF and BP combined MSCs, both of therapeutic methods notably reduce infarcted volume post-stroke [30, 31].

Table 12.1 Bederson scale for neurological functional assessment (n = 20, means ± SD)

Group	3 h	1 day	3 days	7 days
Sham	0.02 ± 0.01	0.03 ± 0.03	0.02 ± 0.02	0.05 ± 0.02
MCAo	2.50 ± 0.21	2.56 ± 0.02	2.26 ± 0.14	2.08 ± 0.13
MSC	2.50 ± 0.12	2.64 ± 0.05	2.14 ± 0.12	1.45 ± 0.14*
SF + BP + MSC	2.50 ± 0.21	2.43 ± 0.17	1.35 ± 0.15*	1.09 ± 0.02*

*P < 0.05, vs. MCAo group

3 Acceleration of MSCs Differentiation and Migration

As a main organic acid in angelica, ferulic acid (FA) had been demonstrated that it could decrease infarction size and improve neurological function in MCAo rats through anti-oxidative and anti-inflammatory actions [32]. SF is the sodium salt of FA, which been used as an important agent for cardiovascular and cerebrovascular diseases. In 2005, Wang and colleagues firstly found that SF could induce Human MSCs to express neural proteins, such as nestin, neuron specific enolase and glial fibrillary acidic protein (GFAP), as well as advance MSCs to differentiate into neural-like cells in vitro [33]. Our previous study also indicated that SF could enhance 5-bromo-2′-deoxyuridine (BrdU)-labeled bone-derived MSCs to express nestin, GFAP and neuron-specific class III beta-tubulin (Tuj-1) in ischemic rat stroke model, suggesting it facilitated the differentiations of MSCs into astrocytic- and neuronal-like cells. Another experiment showed that adipose-derived MSCs were incubated with SF (5 μg/ml) and BP (0.75 μg/ml) for 3 and 7 days. Cultured MSCs in SF + BP group obviously presented neuronal morphology at day 3, and some cells even possessed long neuronal-like synapses under light microscopic observations. At day 7, fluorescence staining results also showed that combination of SF and BP could noticeable advance MSCs to express Tuj-1 (Fig. 12.1).

In order to illustrate the differentiated mechanisms, we investigated bone morphogenetic proteins (BMP) 2/4 and Notch-1 signaling pathways. As a member in the transforming growth factor beta superfamily, BMPs and their signaling systems play important roles in regulation of neural activity and rescue of injured neurons, and they could selectively and dose-dependently increase multipotent progenitor cells in murine embryonic subventricular zone (SVZ) to differentiate into astroglial lineage [34, 35]. It is reported that BMP2 and BMP7 levels began to enhance from 1 to 4 weeks in mice ischemic brain, which were associated with astrogliosis [36]. In our study, we found that SF combined with MSCs up-regulated BMP2/4 pathway post-stroke, which might be related to the differentiation of MSCs into astrocytic-like cells [30]. The mammalian family of Notch receptors consists of four members, Notch-1 through Notch-4, all of which are single pass transmembrane proteins. Study evidenced that activated Notch receptor promoted the survival and numbers of murine somatic and human embryonic stem cells by induced the expression of the specific target genes hairy and enhancer of split 3 (Hes3) and Sonic hedgehog, resulting in amelioration of motor skills after ischemic injury [37]. Not only Notch signaling plays a role in keeping the progenitors from differentiating into neurons,

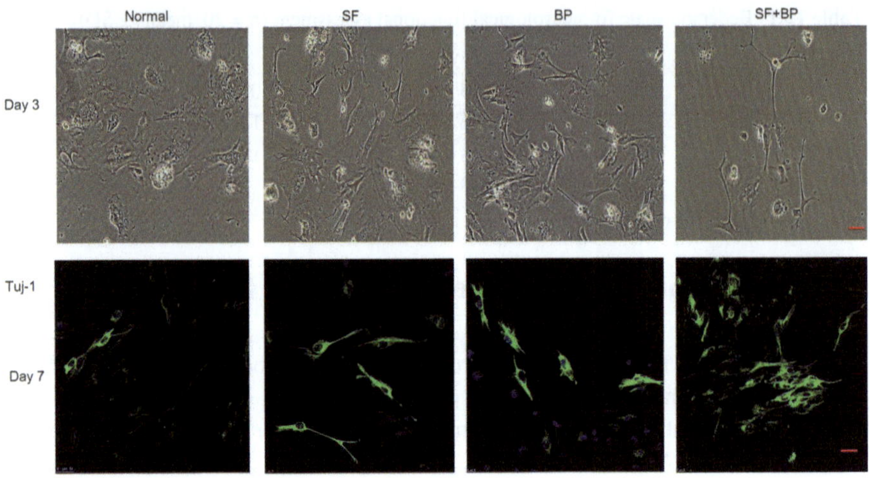

Fig. 12.1 Morphology and differentiation of cultured MSCs incubated with SF and BP. *First panel*: morphology of cultured adipose-derived MSCs incubated with SF and BP for 3 days under optical microscope; *Second panel*: immunofluorescence staining of Tuj-1 in MSCs after 7 days (scale bar: 20 μm)

but also down-regulated Notch1 signaling accelerated striatal astrocytes to carry a latent neurogenic program after stroke [38]. Moreover, there was cross-talk between Notch and BMP signaling pathways, which embodied that BMP2 enhanced Notch-induced transcriptional activation of Hes-5 and Hes-1 in mouse neuroepithelial cells [39]. In our study, it showed that the expressions of Notch-1, Hes1 and Hes5 in combination treatment of SF and MSCs group decreased, which might contribute to the differentiation of MSCs into neural-like cells [30].

Due to low migration efficiency of the transplanted BMSCs into the lesion area, MSCs treatment is limited. Wang and colleagues reported that SDF-1α and chemokine (CXC motif) receptor-4 (CXCR4) could systemically regulate transplanted MSCs towards ischemic zone in the MCAo rat model [40]. In the bone marrow, CXCR4 on endothelial cells and MSCs recruited peripheral blood SDF-1 to translocate into bone marrow, subsequently resulting in the homing of transplanted human $CD34^+$ hematopoietic progenitors to the bone marrow, and the effect was crucial related to SDF-1 gradient [41, 42]. Evidence indicated that up-regulated SDF-1/CXCR4 axis increased SVZ neuroblast cell migration after stroke [43]. Therefore, improving SDF-1α gradient in cerebral damaged tissue might contribute to MSCs recruitment into ischemic zone. Through Western blot and RT-PCR assay, it suggested that SF combined MSCs significantly up-regulate SDF-1/CXCR4 axis, which was beneficial to recruit more stem/progenitor cells to migrate into infarcted lesion [27]. Additionally, luciferase labeled adipose-derived MSCs was injected into the margin of laser illuminated area in photochemically induced stroke model, and it showed that SF and BP could obviously promote MSCs' abilities of survival and migration (Fig. 12.2).

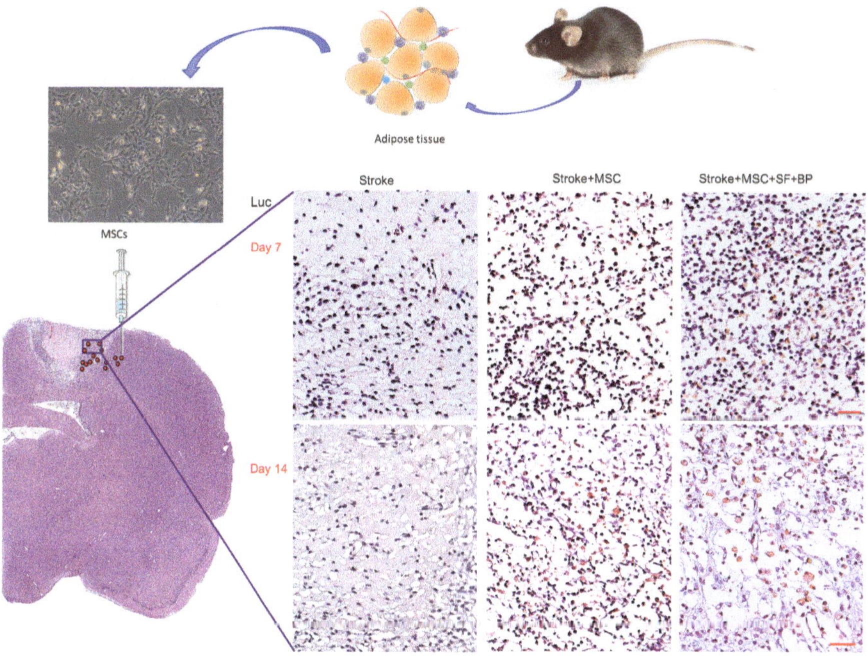

Fig. 12.2 Migration of Luciferase labeled adipose-derived MSCs at day 7 and 14 post-stroke. Luciferase immunohistochemistry staining images suggested that SF and BP could advance migration of adipose-derived MSCs into infarcted zone (scale bar: 50 μm)

4 Enhancement of Angiogenesis and Neurogenesis

Being a potential therapeutic candidate in the treatment of ischemic stroke, MSCs are capable of promoting angiogenesis and neurogenesis after cerebral ischemia [11, 44, 45]. Exogenous transplanted MSCs not only directly differentiated vascular endothelial cells (ECs), but also induced endogenous angiogenic responses to amplify angiogenesis and vascular stabilization after stroke [11, 46]. In addition, angiogenetic gene-modified MSCs, e.g. MSCs transfected with the angiopoietin-1, placental growth factor, VEGF and Flk-1 gene showed the greatest structural-functional recovery and notably angiogenesis and neurogenesis post-stroke [47–49]. Komatsu and colleagues thought that angiogenesis accounted for the main therapeutic effects, although there were several hypotheses in the treatment of MSCs [50]. Our previous review summarized the actions of new vessel formation after ischemic stroke, which included the improvement of cerebral blood flow (CBF) and metabolism in infarction lesion, removal of necrotic debris, enhancement of neurotrophic components for neuronal remodeling and endogenous stem/progenitor cells migration [51]. Based on the efficacy of angiogenesis, it has been recognized to be the basis and prerequisite for neurogenesis. However, not all angiogenesis is advantageous for stroke by MSCs transplantation. In Type 1 diabetic MCAo rats, angiogenesis by grafted MSCs deteriorated internal carotid artery neointimal formation and blood-brain barrier (BBB) leakage, which possibly was due to increased expression of angiogenin. The adverse effects promoted mortality and the risk of brain hemorrhage [52]. Thereby, the homeostasis of angiogenesis should be taken into account after cerebral ischemic stroke.

More and more evidences indicated that the occurrences of angiogenesis and neurogenesis were coupled processes rather than separate after stroke [53, 54]. It suggested that coculture of neural progenitor cells from SVZ with cerebral ECs from the stroke boundary notably increased neural progenitor cell proliferation and neuronal differentiation, and inhibition of VEGF receptor 2 decreased these beneficial effects on neurogenesis and angiogenesis [53]. Sun and colleagues also demonstrated that VEGF exerted primary role in neuroprotection, survival of new neurons and angiogenesis after cerebral ischemia [55]. In addition to promoting synaptic and axonal plasticity and advancing neurogenesis of MSCs, BDNF also involved in angiogenesis after stroke [56–58]. Phosphorylated AKT could activate mTOR in sequent up-regulate Hypoxia-inducible factors (HIF)-1α expression, consequently improve VEGF expression which contributed to angiogenesis post-stroke, another hand AKT/mTOR had been demonstrated as a crucial target to regulate new neuron development and was essential to maintain endogenous neuronal progenitor pool [59, 60]. Additionally, reports also showed that BDNF could bind to Tropomyosin receptor kinase B (TrkB) receptor and then activate AKT/mTOR signaling resulted in neuroprotective actions in stroke [61, 62]. AKT/mTOR pathway is a central regulated approach of angiogenesis.

Evidence showed that angelica extract and active ingredients were able to improve angiogenesis. Lam and colleagues demonstrated that human umbilical vein endothelial cells (HUVECs) and zebra fish intestine capillaries incubated with angelica extract presented obvious angiogenic abilities, whose mechanism was mainly related to p38 and Jun N-terminal protein kinase 1/2 phosphorylation [63]. FA could advance HUVECs to secret VEGF, platelet-derived growth factor and HIF-1α, consequently, promote the ability of angiogenesis via activation of PI3K signaling pathway [64]. As a new drug for cerebralvascular disease approved by China FDA, NBP was capable of increasing brain microvessels density against stroke, whose mechanisms were associated with enhanced expressions of VEGF, VEGFR and HIF-1α as well as activation of extracellular signal-regulated kinases (ERK)1/2 and PI3K/Akt-endothelial nitric oxide synthase (eNOS) signal pathways [65–68].

It has been observed that angiogenesis and neurogenesis simultaneously taken place in the penumbra, and newly born, immature neurons derived from neural stem cells (NSCs) in SVZ and dentate gyrus (DG) closely associate with the remodeling vasculature in this neurovascular niche [69]. In rat permanent bilateral common carotid artery occlusion model, angelica extract that contained the component Z-ligustilide improved neurogenesis in the hippocampus and cognitive decline due to hypoperfusion though enhanced expressions of BDNF and phosphorylated cyclic adenosine monophosphate-responsive element binding protein and γ-aminobutyric acid [70]. In vitro, FA promoted proliferated ability of cultured neural stem/progenitor cells derived from embryonic telencephalon and the number and size of secondary formed neurospheres; in vivo, it increased the number of newly generated cells in the hippocampal DG of corticosterone-treated mice [71]. Previous study suggest that BP, a kind of alkylphthalide derived from the volatile oil of angelica, had ECs protective, vasorelaxing, antiplatelet and antianginal effects, as well as maintained stem cells pluripotency [72–76]. Based on the characteristics of SF and BP, we chose SF and BP as representative constituents of angelica and discuss the effects and mechanisms combined with BMSC on angiogenesis and neurogenesis after ischemic stroke. In our study, we think that SF and BP was a "Trigger point" which embodied that they advanced MSCs to synthesize VEGF and BDNF, subsequently AKT/mTOR cascade was activated in cerebral parenchymal cells, consequently the combination treatment improved angiogenesis and neurogenesis. In order to define the role of astrocytes on angiogenesis in the treatment of MSCs combined with SF and BP, we investigated astrocyte-derived neurovascular trophic factors and found that combined treatment could obviously increase the expressions of astrocyte-derived VEGF and BDNF via activation of astrocytic AKT/mTOR signaling, resulting in migration and tube formation of HUVECs [28, 31]. Additionally, Immunohistochemistry staining images indicated that adipose-derived MSCs combined SF and BP promoted Von Willebrand factor (vWF)$^+$ capillary density compared with SF + BP group and MSC group and Western blotting showed combination treatment notably enhance Doublecortin (DCX) expression in ischemic boundary zone, suggesting the enhancements of angiogenesis and neurogenesis (Fig. 12.3).

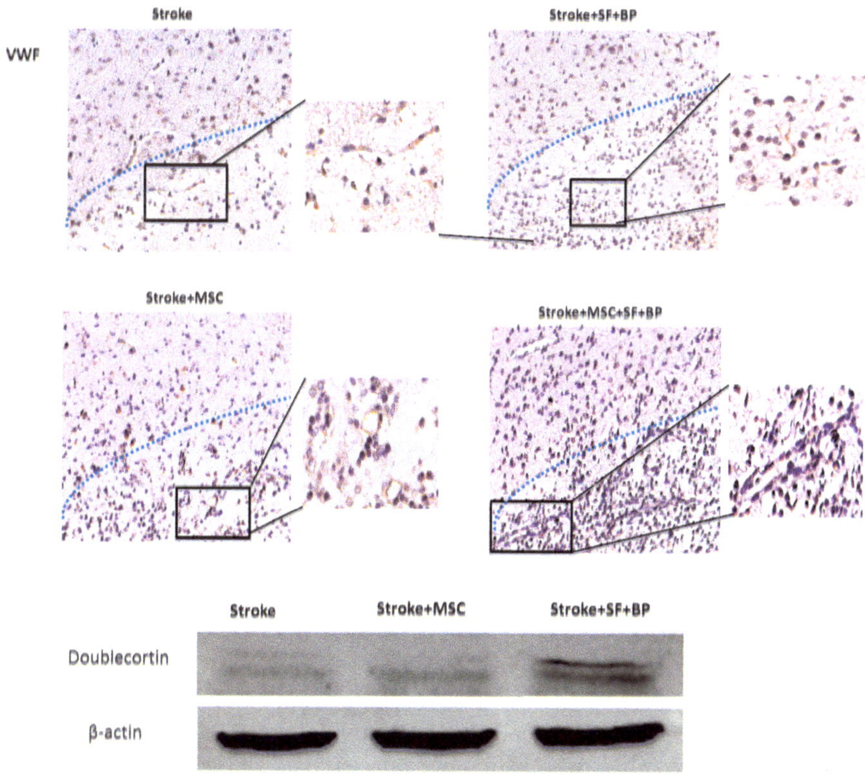

Fig. 12.3 vWF positive capillary and DCX expression in ischemic boundary zone after ischemia. Immunohistochemistry staining images and western blotting showed SF and BP combined with MSCs significantly improve angiogenesis and neurogenesis

5 Regulation of Astrocytes, Activation or Inhibition?

Evidences indicate reactive astrocytes play an important neuroprotective role through enhanced number of mitochondria and antioxidant enzyme activity, reabsorption of glutamate, antitoxic action of free radicals and anti-apoptosis, and regulating immunological response against ischemic brain injury [77]. In the study of cerebral energy metabolism, Kajihara and colleagues demonstrated that astrocytes increased cytoplasmic storage capacity of glycogen granules in the ischemic penumbra after ischemic stroke, and protoplasmic astrocytes gradually became into fibrous astrocytes as ischemic time went by [78]. Previous evidence showed that MSCs treatment could reduce thickness of glial scar formation by reactive astrocytes, consequently decrease inhibition of axonal and synaptic growth, as well as neuronal functional regeneration in the later stage of cerebral ischemia [79, 80]. However, the newest report suggested that scar formation by regulation of astrocyte had advantageous action for axonal regeneration in severe spinal cord injury [81]. On the other hand, in the maintenance of BBB integrity, reactive astrocytes are traditionally thought as detrimental actions which present promoted endothelial permeability and VEGF secretion, as well as decreased occludin and claudin-5 proteins expressions [82–85]. Therefore, it should be compromised evaluation between the beneficial and adverse effects of reactive astrocytes post-stroke.

Present study showed that NBP reduced GFAP-positive astrocytes induced by chronic cerebral ischemia, and inhibited the amyloid β (Aβ)-induced astrocyte activation and pro-inflammatory molecules, which contributed to against ischemic stroke and Alzheimer's disease [16, 86]. Our previous study indicated that SF and bone-derived MSCs respectively activate astrocytes at day 3, and SF combined with MSCs more significantly promoted GFAP expression in ischemic penumbra. Interestingly, when SF and BP combined with adipose-derived MSCs were used to treat photothrombotic stroke, we found SF and BP notably inhibited activation of astrocytes in ischemic boundary zone, but adipose-derived MSCs activated astrocytes at day 7 after ischemia. Simultaneously, the combination treatment also suppressed GFAP expression to a certain extent (Fig. 12.4). The results suggested that different composition of angelica might exert differential actions on astrocytes post-stroke.

The detrimental effect of reactive astrocytes on BBB permeability mainly attributed to its VEGF secretion. Study indicated that knockdown of VEGF no longer damaged endothelial barrier, so astrocyte-derived VEGF has been described as a key mechanism in BBB breakdown [84]. In addition to reactive astrocytes, adult human dental pulp stem cells also were found to secret VEGF-A resulted in enhancement of permeability across an in vitro model of BBB [87]. Shimotake and colleagues demonstrated that VEGF receptor-2 inhibition advanced ischemic injury and reduced endothelial cell proliferation in neonatal rats, whereas it attenuated BBB permeability in diabetic mice after stroke, whose action was associated with enhanced endothelial transcytosis rather than tight junctions [88, 89]. Recent report evidenced that astrocyte-derived Pentraxin 3 bound to VEGF, subsequently notably decreased VEGF-induced endothelial permeability in vitro, and astrocytes existed at least two subclasses by different migratory abilities after

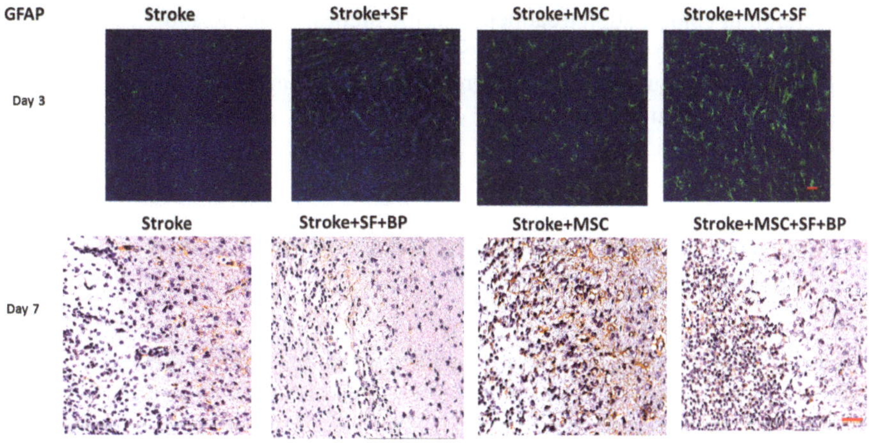

Fig. 12.4 Angelica' active ingredients combined with MSCs regulated GFAP expression in penumbra post-stroke. *First panel*: Immunofluorescence staining indicated that combining SF and bone-derived MSCs could activate astrocyte after 3 days in MCAo model; *Second panel*: SF and BP combined with adipose-derived MSCs inhibited reactive astrocytes after 7 days in photothrombotic stroke model (scale bar: 50 μm)

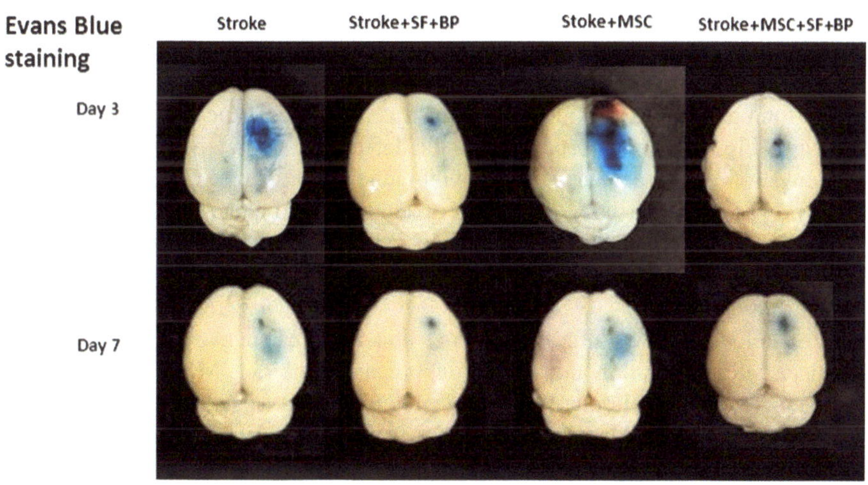

Fig. 12.5 Evaluation of blood-brain-barrier integrity after photothrombotic stroke. Representative images of brain EB staining in rat at day 3 and 7 were presented

cerebral stoke [90, 91]. Therefore, the results reflect the multifaceted actions of VEGF and astrocytes on BBB integrity and endothelial function. In our study, Evans blue staining showed that SF + BP group distinctly reduced BBB leakage compared with other three groups whatever at day 3 or 7 after stroke, which might attribute to inhibitions of astrocytes activation, and SF and BP combined with MSCs presented the effect of maintenance of BBB integrity which was notably superior to MSC alone treatment, suggesting that SF and BP could be against side effect of increased BBB leakage of adipose-derived MSCs (Fig. 12.5). But VEGF expression in combination treatment was the highest (Data no show), whether the therapy influenced subtypes of astrocytes, astrocyte-derived Pentraxin 3, endothelial function or regulation of VEGF receptors, exact mechanisms need to be further illustrated in future experiment.

6 Improvement of Cerebral Blood Flow and Glucose Metabolism

Neurovascular coupling is responsible for controlling regional CBF by neurons directly or astrocytes indirectly secreting vasodilator factors targeting on vascular cells under physiological condition [92]. Moreover, it reported that reactive astrocytes exerted key effect on rCBF regulation, and they could modulate rCBF longer than neurons under pathological conditions [93]. We examined perfusion-weighted imaging (PWI) by MRI to evaluate CBF after photothrombotic stroke. It showed that SF and BP combined with adipose-derived MSCs could significantly ameliorate CBF in the infarction zone (Fig. 12.6a). Previous study had demonstrated that the combination treatment could significantly enhance angiogenesis and neurogenesis, as well as regulate astrocytes, so we thought the interactions of newly neuron, astrocyte and neovascularization contributed to enhancement of CBF in infarction. Additionally, ^{18}F-2-deoxy-glucose (FDG)-positron emission tomography–computed tomography (PET/CT) was administrated to assess glucose metabolism. As shown in Fig. 12.6b, the cortical metabolic defect partially recovered with the time prolongation, and it seemed to more obvious amelioration in MSC + SF + BP groups at day 14, suggesting that combination treatment could enhance glucose metabolism. We also observed that combing SF and MSCs promoted Glucose transporter 1 expression in ischemic boundary zone [30], which might be a mechanism of glucose metabolism.

In summary, these studies uncovered that angelica' active ingredients could advance MSCs migration and differentiation, and combining MSCs with angelica' active ingredients ameliorated neurological function and reduced infarction volume, improved neovascularization and neurogenesis, regulate astrocyte characteristics, enhanced CBF and glucose metabolism, and decrease BBB permeability, which restitute the structure and function of neurovascular unit in infarction zone after stroke. Therefore, the combination treatment should be a more effective therapy due to synergic functions of supplementary approaches.

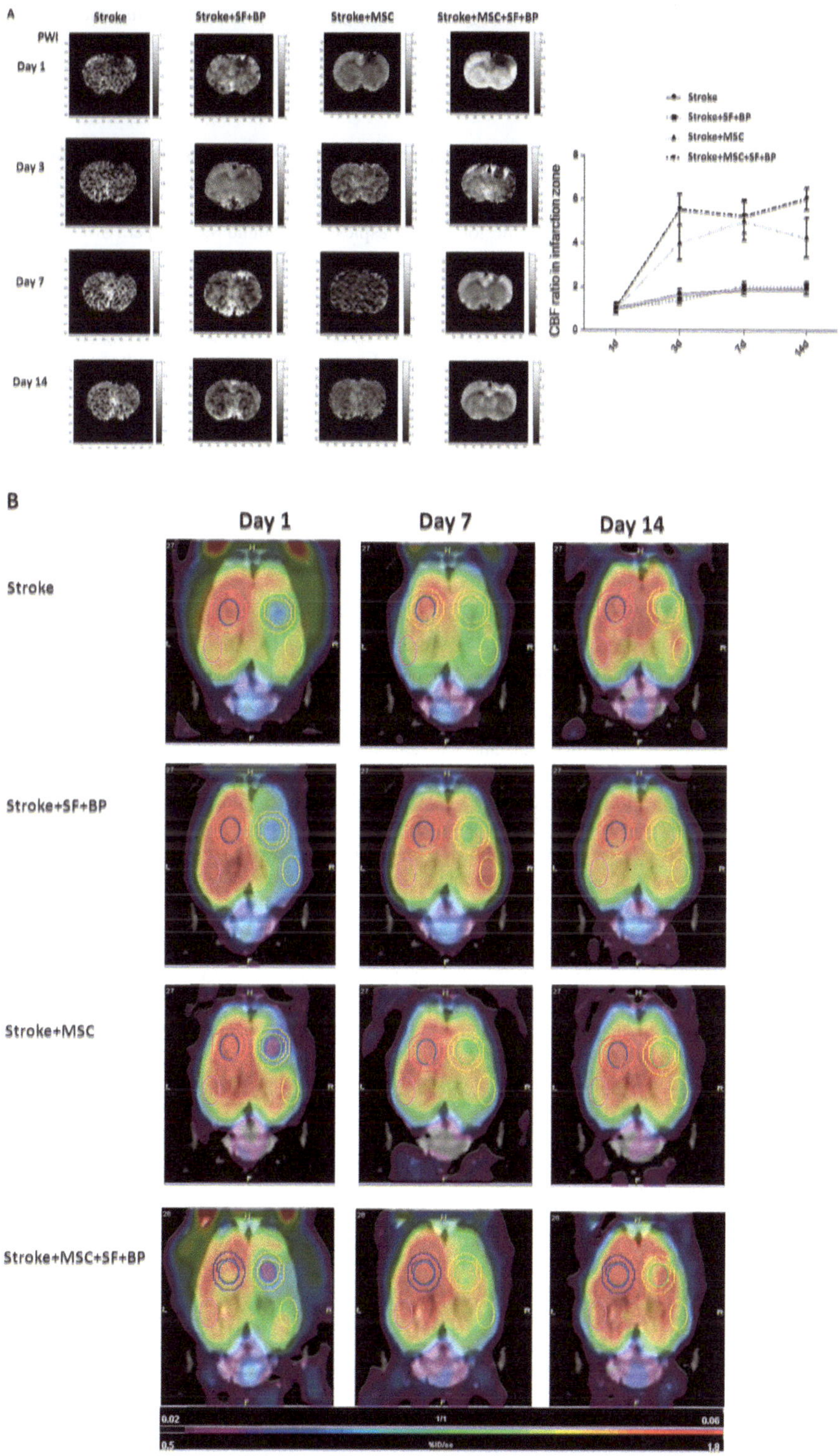

Fig. 12.6 Evolution of cerebral blood flow and glucose metabolism in infarct lesion. (**a**) Images of PWI at day 1, 3, 7 and 14 after ischemia were presented and quantitative analysis of CBF. (**b**) ^{18}F-FDG PET imaging at day 1, 7 and 14 was showed

References

1. Macrez R, Ali C, Toutirais O, Le Mauff B, Defer G, Dirnagl U, Vivien D. Stroke and the immune system: from pathophysiology to new therapeutic strategies. Lancet Neurol. 2011;10:471–80.
2. Roger VL, Go AS, Lloyd-Jones DM, Adams RJ, Berry JD, Brown TM, Carnethon MR, Dai S, de Simone G, Ford ES, Fox CS, Fullerton HJ, Gillespie C, Greenlund KJ, Hailpern SM, Heit JA, Ho PM, Howard VJ, Kissela BM, Kittner SJ, Lackland DT, Lichtman JH, Lisabeth LD, Makuc DM, Marcus GM, Marelli A, Matchar DB, McDermott MM, Meigs JB, Moy CS, Mozaffarian D, Mussolino ME, Nichol G, Paynter NP, Rosamond WD, Sorlie PD, Stafford RS, Turan TN, Turner MB, Wong ND, Wylie-Rosett J, American Heart Association Statistics Committee and Stroke Statistics Subcommittee. American Heart Association Statistics Committee and Stroke Statistics Subcommittee. Heart disease and stroke statistics–2011 update: a report from the American Heart Association. Circulation. 2011;123:e18–e209.
3. Elkins JS, Johnston CC. Thirty-year projections for deaths from ischemic stroke in the United States. Stroke. 2003;34:2109–13.
4. Woodruff TM, Thundyil J, Tang SC, Sobey CG, Taylor SM, Arumugam TV. Pathophysiology, treatment, and animal and cellular models of human ischemic stroke. Mol Neurodegener. 2011;6:11.
5. Adams HP Jr, Brott TG, Furlan AJ, Gomez CR, Grotta J, Helgason CM, Kwiatkowski T, Lyden PD, Marler JR, Torner J, Feinberg W, Mayberg M, Thies W. Guidelines for thrombolytic therapy for acute stroke: a supplement to the guidelines for the management of patients with acute ischemic stroke: a statement for healthcare professionals from a Special Writing Group of the Stroke. Circulation. 1996;94:1167–74.
6. Demaerschalk BM, Kleindorfer DO, Adeoye OM, Demchuk AM, Fugate JE, Grotta JC, Khalessi AA, Levy EI, Palesch YY, Prabhakaran S, Saposnik G, Saver JL, Smith EE, American Heart Association Stroke Council and Council on Epidemiology and Prevention. Scientific rationale for the inclusion and exclusion criteria for intravenous alteplase in acute ischemic stroke: a statement for healthcare professionals from the American Heart Association/American Stroke Association. Stroke. 2016;47:581–641.
7. Dezawa M, Kanno H, Hoshino M, Cho H, Matsumoto N, Itokazu Y, et al. Specific induction of neuronal cells from bone marrow stromal cells and application for autologous transplantation. J Clin Invest. 2004;113:1701–10.
8. Bang OY, Lee JS, Lee PH, Lee G. Autologous mesenchymal stem cell transplantation in stroke patients. Ann Neurol. 2005;57:874–82.
9. Chopp M, Li Y. Treatment of neural injury with marrow stromal cells. Lancet Neurol. 2002;1:92–100.
10. Lee J, Kuroda S, Shichinohe H, Ikeda J, Seki T, Hida K, Tada M, Sawada K, Iwasaki Y. Migration and differentiation of nuclear fluorescence-labeled bone marrow stromal cells after transplantation into cerebral infarct and spinal cord injury in mice. Neuropathology. 2003;23:169–80.
11. Wei L, Fraser JL, Lu ZY, Hu X, Yu SP. Transplantation of hypoxia preconditioned bone marrow mesenchymal stem cells enhances angiogenesis and neurogenesis after cerebral ischemia in rats. Neurobiol Dis. 2012;46:635–45.
12. Suárez-Monteagudo C, Hernández-Ramírez P, Alvarez-González L, García-Maeso I, de la Cuétara-Bernal K, Castillo-Díaz L, Bringas-Vega ML, Martínez-Aching G, Morales-Chacón LM, Báez-Martín MM, Sánchez-Catasús C, Carballo-Barreda M, Rodríguez-Rojas R, Gómez-Fernández L, Alberti-Amador E, Macías-Abraham C, Balea ED, Rosales LC, Del Valle Pérez L, Ferrer BB, González RM, Bergado JA. Autologous bone marrow stem cell neurotransplantation in stroke patients. An open study. Restor Neurol Neurosci. 2009;27:151–61.
13. Savitz SI, Cramer SC, Wechsler L, STEPS 3 Consortium. Stem cells as an emerging paradigm in stroke 3: enhancing the development of clinical trials. Stroke. 2014;45:634–9.
14. Liu YM, Zhang JJ, Jiang J. Observation on clinical effect of Angelica injection in treating acute cerebral infarction. Zhongguo Zhong Xi Yi Jie He Za Zhi. 2004;24:205–8.

15. Peng HY, Du JR, Zhang GY, Kuang X, Liu YX, Qian ZM, Wang CY. Neuroprotective effect of Z-ligustilide against permanent focal ischemic damage in rats. Biol Pharm Bull. 2007;30:309–12.
16. Xu J, Wang Y, Li N, Xu L, Yang H, Yang Z. L-3-n-butylphthalide improves cognitive deficits in rats with chronic cerebral ischemia. Neuropharmacology. 2012;62:2424–9.
17. Cui LY, Zhu YC, Gao S, Wang JM, Peng B, Ni J, Zhou LX, He J, Ma XQ. Ninety-day administration of dl-3-n-butylphthalide for acute ischemic stroke: a randomized, double-blind trial. Chin Med J. 2013;126:3405–10.
18. Cui X, Chopp M, Zacharek A, Roberts C, Lu M, Savant-Bhonsale S, Chen J. Chemokine, vascular and therapeutic effects of combination Simvastatin and BMSC treatment of stroke. Neurobiol Dis. 2009;36:35–41.
19. Xu J, Liu X, Chen J, Zacharek A, Cui X, Savant-Bhonsale S, Liu Z, Chopp M. Simvastatin enhances bone marrow stromal cell differentiation into endothelial cells via notch signaling pathway. Am J Physiol Cell Physiol. 2009;296:C535–43.
20. Pirzad Jahromi G, Seidi S, Sadr SS, Shabanzadeh AP, Keshavarz M, Kaka GR, Hosseini SK, Sohanaki H, Charish J. Therapeutic effects of a combinatorial treatment of simvastatin and bone marrow stromal cells on experimental embolic stroke. Basic Clin Pharmacol Toxicol. 2012;110:487–93.
21. Zhao Y, Zhang Q, Chen Z, Liu N, Ke C, Xu Y, Wu W. Simvastatin combined with bone marrow stromal cells treatment activates astrocytes to ameliorate neurological function after ischemic stroke in rats. Turk J Biol. 2016;40:519–28.
22. Wang Y, Li WY, Li MQ, Guan YQ, Zhang XB, Lv WZ. Effects of astragaloside IV combined with bone mesenchymal stem cell transplantation on angiogenesis after cerebral ischemia-reperfusion in rats. Anat Res. 2011;33:323–6.
23. Guo JW, Chen C, Huang Y, Li B. Combinatorial effects of Naomai Yihao capsules and vascular endothelial growth factor gene-transfected bone marrow mesenchymal stem cells on angiogenesis in cerebral ischemic tissues in rats. J Tradit Chin Med. 2012;32:87–92.
24. Hu XY, Wang WX, Yu MJ, Liu XB, Wu RR, Gao F, Huang X, Cao J, Xie XJ, Wang JA. Tongxinluo promotes mesenchymal stem cell tube formation in vitro. J Zhejiang Univ Sci B. 2011;12:644–51.
25. Zhang YK, Han XY, Che ZY. Effects of buyang huanwu tang combined with bone marrow mesenchymal stem cell transplantation on the expression of VEGF and Ki-67 in the brain tissue of the cerebral ischemia-reperfusion model rat. J Tradit Chin Med. 2010;30:278–82.
26. Garcia JH, Wagner S, Liu KF, Hu XJ. Neurological deficit and extent of neuronal necrosis attributable to middle cerebral artery occlusion in rats. Statistical validation. Stroke. 1995;26:627–34.
27. Zhao Y, Guan Y, Xu Y, Li Y, Wu W. Sodium Ferulate combined with bone marrow stromal cell treatment ameliorating rat brain ischemic injury after stroke. Brain Res. 2012;1450:157–65.
28. Zhang Q, Chen ZW, Zhao YH, Liu BW, Liu NW, Ke CC, Tan HM. Bone marrow stromal cells combined with sodium ferulate and n-butylidenephthalide promote the effect of therapeutic angiogenesis via advancing astrocyte-derived trophic factors after ischemic stroke. Cell Transplant. 2017;26:229–42.
29. Bederson JB, Pitts LH, Tsuji M, Nishimura MC, Davis RL, Bartowski H. Rat middle cerebral artery occlusion: evaluation of the model and development of a neurological examination. Stroke. 1986;17:472–6.
30. Zhao Y, Lai W, Xu Y, Li L, Chen Z, Wu W. Exogenous and endogenous therapeutic effects of combination Sodium Ferulate and bone marrow stromal cells (BMSCs) treatment enhance neurogenesis after rat focal cerebral ischemia. Metab Brain Dis. 2013;28:655–66.
31. Zhang Q, Zhao Y, Xu Y, Chen Z, Liu N, Ke C, Liu B, Wu W. Sodium ferulate and n-butylidenephthalate combined with bone marrow stromal cells (BMSCs) improve the therapeutic effects of angiogenesis and neurogenesis after rat focal cerebral ischemia. J Transl Med. 2016;14:223.

32. Cheng CY, Ho TY, Lee EJ, Su SY, Tang NY, Hsieh CL. Ferulic acid reduces cerebral infarct through its antioxidative and anti-inflammatory effects following transient focal cerebral ischemia in rats. Am J Chin Med. 2008;36:1105–19.
33. Wang Y, Deng Z, Lai X, Tu W. Differentiation of human bone marrow stromal cells into neural-like cells induced by Sodium Ferulate in vitro. Cell Mol Immunol. 2005;2:225–9.
34. Ebendal T, Bengtsson H, Soderstrom S. Bone morphogenetic proteins and their receptors: potential functions in the brain. J Neurosci Res. 1998;51:139–46.
35. Gross RE, Mehler MF, Mabie PC, Zang Z, Santschi L, Kessler JA. Bone morphogenetic proteins promote astroglial lineage commitment by mammalian subventricular zone progenitor cells. Neuron. 1996;17:595–606.
36. Shin JA, Kang JL, Lee KE, Park EM. Different temporal patterns in the expressions of bone morphogenetic proteins and noggin during astroglial scar formation after ischemic stroke. Cell Mol Neurobiol. 2012;32:587–97.
37. Androutsellis-Theotokis A, Leker RR, Soldner F, Hoeppner DJ, Ravin R, Poser SW, Rueger MA, Bae SK, Kittappa R, McKay RD. Notch signalling regulates stem cell numbers in vitro and in vivo. Nature. 2006;442:823–6.
38. Magnusson JP, Göritz C, Tatarishvili J, Dias DO, Smith EM, Lindvall O, Kokaia Z, Frisén J. A latent neurogenic program in astrocytes regulated by Notch signaling in the mouse. Science. 2014;346:237–41.
39. Takizawa T, Ochiai W, Nakashima K, Taga T. Enhanced gene activation by Notch and BMP signaling cross-talk. Nucleic Acids Res. 2003;31:5723–31.
40. Wang Y, Deng YB, Zhou GQ. SDF-1alpha/CXCR4-mediated migration of systemically transplanted bone marrow stromal cells towards ischemic brain lesion in a rat model. Brain Res. 2008;1195:104–12.
41. Dar A, Goichberg P, Shinder V, Kalinkovich A, Kollet O, Netzer N, Margalit R, Zsak M, Nagler A, Hardan I, Resnick I, Rot A, Lapidot T. Chemokine receptor CXCR4-dependent internalization and resecretion of functional chemokine SDF-1 by bone marrow endothelial and stromal cells. Nat Immunol. 2005;6:1038–46.
42. Avigdor A, Goichberg P, Shivtiel S, Dar A, Peled A, Samira S, Kollet O, Hershkoviz R, Alon R, Hardan I, Ben-Hur H, Naor D, Nagler A, Lapidot T. CD44 and hyaluronic acid cooperate with SDF-1 in the trafficking of human $CD34^+$ stem/progenitor cells to bone marrow. Blood. 2004;103:2981–9.
43. Cui X, Chen J, Zacharek A, Roberts C, Yang Y, Chopp M. Nitric oxide donor up-regulation of SDF1/CXCR4 and Ang1/Tie2 promotes neuroblast cell migration after stroke. J Neurosci Res. 2009;87:86–95.
44. Chen J, Zhang ZG, Li Y, Wang L, Xu YX, Gautam SC, Lu M, Zhu Z, Chopp M. Intravenous administration of human bone marrow stromal cells induces angiogenesis in the ischemic boundary zone after stroke in rats. Circ Res. 2003;92:692–9.
45. Chen J, Li Y, Zhang R, Katakowski M, Gautam SC, Xu Y, Lu M, Zhang Z, Chopp M. Combination therapy of stroke in rats with a nitric oxide donor and human bone marrow stromal cells enhances angiogenesis and neurogenesis. Brain Res. 2004;1005:21–8.
46. Zacharek A, Chen J, Cui X, Li A, Li Y, Roberts C, Feng Y, Gao Q, Chopp M. Angiopoietin1/Tie2 and VEGF/Flk1 induced by MSC treatment amplifies angiogenesis and vascular stabilization after stroke. J Cereb Blood Flow Metab. 2007;27:1684–91.
47. Toyama K, Honmou O, Harada K, Suzuki J, Houkin K, Hamada H, Kocsis JD. Therapeutic benefits of angiogenetic gene-modified human mesenchymal stem cells after cerebral ischemia. Exp Neurol. 2009;216:47–55.
48. Liu H, Honmou O, Harada K, Nakamura K, Houkin K, Hamada H, Kocsis JD. Neuroprotection by PlGF gene-modified human mesenchymal stem cells after cerebral ischaemia. Brain. 2006;129:2734–45.
49. Bao X, Feng M, Wei J, Han Q, Zhao H, Li G, Zhu Z, Xing H, An Y, Qin C, Zhao RC, Wang R. Transplantation of Flk-1^+human bone marrow-derived mesenchymal stem cells promotes angiogenesis and neurogenesis after cerebral ischemia in rats. Eur J Neurosci. 2011;34:87–98.

50. Komatsu K, Honmou O, Suzuki J, Houkin K, Hamada H, Kocsis JD. Therapeutic time window of mesenchymal stem cells derived from bone marrow after cerebral ischemia. Brain Res. 2010;1334:84–92.
51. Zhang Q, Zhao YH. Therapeutic angiogenesis after ischemic stroke: Chinese medicines, bone marrow stromal cells (BMSCs) and their combinational treatment. Am J Chin Med. 2014;42:61–77.
52. Chen J, Ye X, Yan T, Zhang C, Yang XP, Cui X, Cui Y, Zacharek A, Roberts C, Liu X, Dai X, Lu M, Chopp M. Adverse effects of bone marrow stromal cell treatment of stroke in diabetic rats. Stroke. 2011;42:3551–8.
53. Teng H, Zhang ZG, Wang L, Zhang RL, Zhang L, Morris D, Gregg SR, Wu Z, Jiang A, Lu M, Zlokovic BV, Chopp M. Coupling of angiogenesis and neurogenesis in cultured endothelial cells and neural progenitor cells after stroke. J Cereb Blood Flow Metab. 2008;28(4):764–71.
54. Ruan L, Wang B, ZhuGe Q, Jin K. Coupling of neurogenesis and angiogenesis after ischemic stroke. Brain Res. 1623;2015:166–73.
55. Sun Y, Jin K, Xie L, Childs J, Mao XO, Logvinova A, Greenberg DA. VEGF-induced neuroprotection, neurogenesis, and angiogenesis after focal cerebral ischemia. J Clin Invest. 2003;111:1843–51.
56. Schäbitz WR, Steigleder T, Cooper-Kuhn CM, Schwab S, Sommer C, Schneider A, Kuhn HG. Intravenous brain-derived neurotrophic factor enhances poststroke sensorimotor recovery and stimulates neurogenesis. Stroke. 2007;38:2165–72.
57. Lim JY, Park SI, Oh JH, Kim SM, Jeong CH, Jun JA, Lee KS, Oh W, Lee JK, Jeun SS. Brain-derived neurotrophic factor stimulates the neural differentiation of human umbilical cord blood-derived mesenchymal stem cells and survival of differentiated cells through MAPK/ERK and PI3K/Akt-dependent signaling pathways. J Neurosci Res. 2008;86:2168–78.
58. Fouda AY, Alhusban A, Ishrat T, Pillai B, Eldahshan W, Waller JL, Ergul A, Fagan SC. Brain-derived neurotrophic factor knockdown blocks the angiogenic and protective effects of angiotensin modulation after experimental stroke. Mol Neurobiol. 2017;54:661–70.
59. Chong ZZ, Yao Q, Li HH. The rationale of targeting mammalian target of rapamycin for ischemic stroke. Cell Signal. 2013;25:1598–607.
60. Kim JY, Duan X, Liu CY, Jang MH, Guo JU, Pow-anpongkul N, Kang E, Song H, Ming GL. DISC1 regulates new neuron development in the adult brain via modulation of AKT-mTOR signaling through KIAA1212. Neuron. 2009;63:761–73.
61. Qi D, Ouyang C, Wang Y, Zhang S, Ma X, Song Y, Yu H, Tang J, Fu W, Sheng L, Yang L, Wang M, Zhang W, Miao L, Li T, Huang X, Dong H. HO-1 attenuates hippocampal neurons injury via the activation of BDNF-TrkB-PI3K/Akt signaling pathway in stroke. Brain Res. 2014;1577:69–76.
62. Guo S, Som AT, Waeber C, Lo EH. Vascular neuroprotection via TrkB- and Akt-dependent cell survival signaling. J Neurochem. 2012;123(Suppl 2):58–64.
63. Lam HW, Lin HC, Lao SC, Gao JL, Hong SJ, Leong CW, Yue PY, Kwan YW, Leung AY, Wang YT, Lee SM. The angiogenic effects of Angelica sinensis extract on HUVEC in vitro and zebrafish in vivo. J Cell Biochem. 2008;103:195–211.
64. Lin CM, Chiu JH, Wu IH, Wang BW, Pan CM, Chen YH. Ferulic acid augments angiogenesis via VEGF, PDGF and HIF-1 alpha. J Nutr Biochem. 2010;21:627–33.
65. Liu CL, Liao SJ, Zeng JS, Lin JW, Li CX, Xie LC, Shi XG, Huang RX. dl-3n-butylphthalide prevents stroke via improvement of cerebral microvessels in RHRSP. J Neurol Sci. 2007;260:106–13.
66. Zhang L, Lü L, Chan WM, Huang Y, Wai MS, Yew DT. Effects of DL-3-n-butylphthalide on vascular dementia and angiogenesis. Neurochem Res. 2012;37:911–9.
67. Liao SJ, Lin JW, Pei Z, Liu CL, Zeng JS, Huang RX. Enhanced angiogenesis with dl-3n-butylphthalide treatment after focal cerebral ischemia in RHRSP. Brain Res. 2009;1289:69–78.
68. Lu XL, Luo D, Yao XL, Wang GL, Liu ZY, Li ZX, Li W, Chang FJ, Wen L, Lee SM, Zhang ZJ, Li L, Zeng JS, Huang RX, Pei Z, Ou JS. dl-3n-Butylphthalide promotes angiogenesis via the

extracellular signal-regulated kinase 1/2 and phosphatidylinositol 3-kinase/Akt-endothelial nitric oxide synthase signaling pathways. J Cardiovasc Pharmacol. 2012;59:352–62.
69. Zhang K, Zhu L, Fan M. Oxygen, a key factor regulating cell behavior during neurogenesis and cerebral diseases. Front Mol Neurosci. 2011;4:5.
70. Xin J, Zhang J, Yang Y, Deng M, Xie X. Radix Angelica sinensis that contains the component Z-ligustilide promotes adult neurogenesis to mediate recovery from cognitive impairment. Curr Neurovasc Res. 2013;10:304–15.
71. Yabe T, Hirahara H, Harada N, Ito N, Nagai T, Sanagi T, Yamada H. Ferulic acid induces neural progenitor cell proliferation in vitro and in vivo. Neuroscience. 2010;165:515–24.
72. Yang J, Yang S, Yuan YJ. Integrated investigation of lipidome and related signaling pathways uncovers molecular mechanisms of tetramethylpyrazine and butylidenephthalide protecting endothelial cells under oxidative stress. Mol BioSyst. 2012;8:1789–97.
73. Chan SS, Choi AO, Jones RL, Lin G. Mechanisms underlying the vasorelaxing effects of butylidenephthalide, an active constituent of Ligusticumchuanxiong, in rat isolated aorta. Eur J Pharmacol. 2006;537:111–7.
74. Ko WC, Liao CC, Shih CH, Lei CB, Chen CM. Relaxant effects of butylidenephthalide in isolated dog blood vessels. Planta Med. 2002;68:1004–9.
75. Teng CM, Chen WY, Ko WC, Ouyang CH. Antiplatelet effect of butylidenephthalide. Biochim Biophys Acta. 1987;924:375–82.
76. Liu SP, Harn HJ, Chien YJ, Chang CH, Hsu CY, Fu RH, Huang YC, Chen SY, Shyu WC, Lin SZ. n-butylidenephthalide (BP) maintains stem cell pluripotency by activating Jak2/Stat3 pathway and increases the efficiency of iPS cells generation. PLoS One. 2012;7:e44024.
77. Gabryel B, Trzeciak HI. Role of astrocytes in pathogenesis of ischemic brain injury. Neurotox Res. 2001;3:205–21.
78. Kajihara H, Tsutsumi E, Kinoshita A, Nakano J, Takagi K, Takeo S. Activated astrocytes with glycogen accumulation in ischemic penumbra during the early stage of brain infarction: immunohistochemical and electron microscopic studies. Brain Res. 2001;909:92–101.
79. Shen LH, Li Y, Gao Q, Savant-Bhonsale S, Chopp M. Down-regulation of neurocan expression in reactive astrocytes promotes axonal regeneration and facilitates the neurorestorative effects of bone marrow stromal cells in the ischemic rat brain. Glia. 2008;56:1747–54.
80. Pekny M, Pekna M. Astrocyte reactivity and reactive astrogliosis: costs and benefits. Physiol Rev. 2014;94:1077–98.
81. Anderson MA, Burda JE, Ren Y, Ao Y, O'Shea TM, Kawaguchi R, Coppola G, Khakh BS, Deming TJ, Sofroniew MV. Astrocyte scar formation aids central nervous system axon regeneration. Nature. 2016;532:195–200.
82. Bush TG, Puvanachandra N, Horner CH, Polito A, Ostenfeld T, Svendsen CN, Mucke L, Johnson MH, Sofroniew MV. Leukocyte infiltration, neuronal degeneration, and neurite outgrowth after ablation of scar-forming, reactive astrocytes in adult transgenic mice. Neuron. 1999;23:297–308.
83. Argaw AT, Gurfein BT, Zhang Y, Zameer A, John GR. VEGF-mediated disruption of endothelial CLN-5 promotes blood-brain barrier breakdown. Proc Natl Acad Sci U S A. 2009;106:1977–82.
84. Argaw AT, Asp L, Zhang J, Navrazhina K, Pham T, Mariani JN, Mahase S, Dutta DJ, Seto J, Kramer EG, Ferrara N, Sofroniew MV, John GR. Astrocyte-derived VEGF-A drives blood-brain barrier disruption in CNS inflammatory disease. J Clin Invest. 2012;122:2454–68.
85. Li YN, Pan R, Qin XJ, Yang WL, Qi Z, Liu W, Liu KJ. Ischemic neurons activate astrocytes to disrupt endothelial barrier via increasing VEGF expression. J Neurochem. 2014;129:120–9.
86. Wang HM, Zhang T, Huang JK, Sun XJ. 3-N-butylphthalide (NBP) attenuates the amyloid-β-induced inflammatory responses in cultured astrocytes via the nuclear factor-κB signaling pathway. Cell Physiol Biochem. 2013;32:235–42.
87. Winderlich JN, Kremer KL, Koblar SA. Adult human dental pulp stem cells promote blood-brain barrier permeability through vascular endothelial growth factor-a expression. J Cereb Blood Flow Metab. 2016;36:1087–97.

88. Shimotake J, Derugin N, Wendland M, Vexler ZS, Ferriero DM. Vascular endothelial growth factor receptor-2 inhibition promotes cell death and limits endothelial cell proliferation in a neonatal rodent model of stroke. Stroke. 2010;41:343–9.
89. Reeson P, Tennant KA, Gerrow K, Wang J, Weiser Novak S, Thompson K, Lockhart KL, Holmes A, Nahirney PC, Brown CE. Delayed inhibition of VEGF signaling after stroke attenuates blood-brain barrier breakdown and improves functional recovery in a comorbidity-dependent manner. J Neurosci. 2015;35:128–43.
90. Shindo A, Maki T, Mandeville ET, Liang AC, Egawa N, Itoh K, Itoh N, Borlongan M, Holder JC, Chuang TT, McNeish JD, Tomimoto H, Lok J, Lo EH, Arai K. Astrocyte-derived pentraxin 3 supports blood-brain barrier integrity under acute phase of stroke. Stroke. 2016;47:1094–100.
91. Wagner DC, Scheibe J, Glocke I, Weise G, Deten A, Boltze J, Kranz A. Object-based analysis of astroglial reaction and astrocyte subtype morphology after ischemic brain injury. Acta Neurobiol Exp (Wars). 2013;73:79–87.
92. Girouard H, Iadecola C. Neurovascular coupling in the normal brain and in hypertension, stroke, and Alzheimer disease. J Appl Physiol (1985). 2006;100:328–35.
93. Kowiański P, Lietzau G, Steliga A, Waśkow M, Moryś J. The astrocytic contribution to neurovascular coupling—still more questions than answers? Neurosci Res. 2013;75:171–83.

Chapter 13
Gene Therapy for Cognitive Recovering After Ischemic Stroke

Johanna Gutierrez-Vargas, Rafael Posada-Duque, and Gloria Patricia Cardona-Gómez

Abstract Cerebrovascular accident (CVA) is the second leading cause of death in the world and the first cause of disability in adults, being a 34% of affected people younger than 65 years old. Which is an important consequence by sedentary lifestyle and a high intake of fats and sugars. One of the major shortcomings of current therapeutical approach is the lack of comorbidity studies, intervention time (less than 4.5 h) and the short time of protection or follow-up study, which unprotect for long-term sequelae in the patients. Gene therapy has been shown to be a very useful tool for the treatment of neurodegenerative diseases; specifically in cerebral ischemia there are few experimental studies, which are mentioned in this chapter. The most of them have a pretreatment approach, which does not facilitate the clinical translation, therefore, a major challenge of gene therapy is that it to be implemented as post-injury therapy, which is supported by our results using shRNAmiR carried out in adeno associated viral vector, preventing and reversing neurodegeneration, neurovascular unit uncoupling and cognitive impairment, which could be relevant in the field of translational medicine.

Keywords Stroke • Neurodegeneration • Cognitive impairment • Dementia • Gene therapy • RNA interference • Translational medicine

J. Gutierrez-Vargas • R. Posada-Duque • G.P. Cardona-Gómez, Ph.D. (✉)
Cellular and Molecular Neurobiology Area, Group of Neuroscience, Faculty of Medicine,
Sede de Investigación Universitaria (SIU), University of Antioquia,
Calle 70 No. 52-21, Medellín, Colombia
e-mail: patricia.cardonag@udea.edu.co

P.A. Lapchak, J.H. Zhang (eds.), *Cellular and Molecular Approaches to Regeneration and Repair*, Springer Series in Translational Stroke Research,
https://doi.org/10.1007/978-3-319-66679-2_13

Abbreviations

AAV	Adeno associated viral vectors
AD	Alzheimer's disease
ALS	Amyotrophic lateral sclerosis
BDNF	Brain-derived neurotrophic factor
CDK5	Cyclin-dependent kinase 5
CVA	Cerebrovascular accident
Glu	Glutamate
HMG-CoA	3-Hydroxy-3-methylglutaryl coenzyme A
LTD	Long term depression
LTP	Long term potentiation
MCAO	Middle cerebral artery occlusion
NFTs	Neurofibrillary tangles
NMDA	N-Methyl-D-aspartic acid
PD	Parkinson's disease
RNAi	RNA interference
Rosc	Roscovitine
rt-PA	Recombinant tissue plasminogen activator
shRNAs	Short hairpin RNAs
TEER	Transendothelial resistance
WHO	World Health Organization

1 Background and Significance of Cerebral Ischemia

Cerebrovascular accident (CVA) is the second leading cause of death in the world and the first cause of disability in adults [1] According to the World Health Organization (WHO), 15 million people suffer a stroke each year; of them 5.5 million die (10% of all deaths) and another 5 million are permanently disabled [2]. Being a 34% of affected people younger than 65 years old [2]. In terms of disability, CVA are among the top five contributors, which entails huge economic burdens on health systems because of the resources needed to care for acute patients and the long-term care of survivors [1, 2].

The two main mechanisms that cause brain damage in stroke are ischemia and hemorrhage [3]. Ischemic stroke accounts 87% of strokes [2], because a thrombus or plunger lead to the cerebral artery occlusion producing anoxia in the affected cerebral parenchyma [4, 5], which causes a decrease of brain metabolism due to the non-availability of the necessary substrates. The ischemia effects are rapid since the brain does not store the necessary glucose as the main energy substrate, making it unable to perform anaerobic metabolism [6]. Most severe degree of blood hypoperfusion progresses rapidly towards irreversible damage due to necrotic death, which represents the *ischemic nucleus* (Fig. 13.1). This zone has low cerebral blood flow

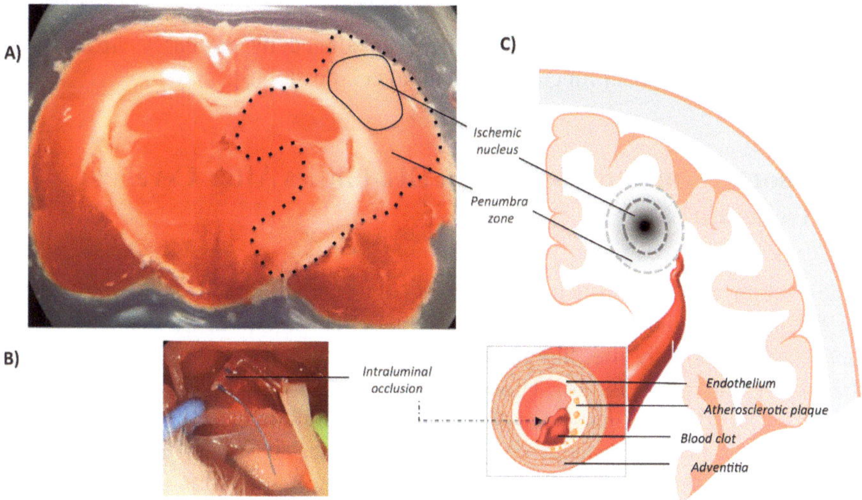

Fig. 13.1 Stroke after focal ischemia. This is mainly divided into the ischemic nucleus and the penumbra zone (**a, c**). (**a**) Infarct area produced by (**b**) intraluminal occlusion of middle cerebral artery during 60 minutes using a 4.0 nylon monofilament recovered with poly-L-lisine and 24 hours of survival in a tMCAO in vivo model in rats. (**c**) Stroke scheme, which usually occurs due to the formation of an atherosclerotic plaque or a blood clot at the level of the middle cerebral artery in humans. -.-.- analogue situation

(<10% of baseline) and irreversible damage of the energy metabolism. The remaining hypoperfused tissue surrounding the ischemic nucleus has an imbalance in the mechanisms of self-regulation of blood flow and is known as the *penumbra zone* [7] (Fig. 13.1). In this region, neurovascular unit show dysfunction, retaining a minimal metabolic activity and altered structural integrity following a pattern of cellular uncoupling and apoptotic death [8–10]. The penumbra is potentially recoverable and represents a key zone for therapeutic intervention in cerebral ischemia. This critical period of viability that this volume of brain tissue is at risk, is known as the "window of opportunity", since the deficit created by ischemia can be partially or completely reversed within a critical period of 2–4.5 h. However, unless perfusion improves, cells in the penumbra zone are at risk of dying within a few hours by necrosis [11].

At the molecular level, the ischemic neuronal injury development is largely due to an increase in the release of excitatory neurotransmitters into the synaptic space, mainly glutamate. This process called "excitotoxicity" is a canonical response as a consequence of ion pump failure and energy deficit, as well as the neurotransmitter uptake mechanism insufficiency by astrocytic cells [12]. So, the glutamate overload leads to prolonged stimulation of ionotropic glutamate receptors of AMP (α-amino-3-hydroxy-5-methyl-4-isoxazolepropionic acid) and NMDA (N-methyl-D-aspartic acid) receptors, increasing drastically the calcium (Ca^{2+}), sodium (Na^{+}), potassium

(K^+) and water influx in neurons. Excessive ion accumulation and simultaneous deregulation of several signaling pathways mediated by proteases, lipases, and nucleases, alter neuronal function and lead to cell death [13].

2 Short-Term and Long-Term Disabilities After Cerebral Ischemia

The total physical disability in the survivors of a cerebral infarction is simultaneous or secondary to the disease, temporarily or permanently, causes damages in the activities of the daily life, like feeding, dressing, care of personal hygiene, use of transport, among others [14]. However, cognitive impairment following ischemic event is almost always ignored in patients, due precisely to the severity of the physical disability that most of them present. These patients have a degree of repercussion that can be divided into mild cognitive impairment and dementia [15]. According to Nys et al., a high survivors proportion of cerebral infarction have cognitive impairment after 3 months and sometimes dementia is only detected after several years of injury [16]. The prevalence of cognitive impairment ranges from 20 to 80% is defined according to each country's case series, diagnostic criteria and post-injury time [15] although post-ischemia cognitive impairment is underestimated in survivors yet. According to WHO, latinamerican people with dementia will increase by 368% in 2050, higher than USA and Europe [17–19]; because sedentary lifestyle and a high fats and sugars intake.

There is evidence to suggest that hippocampal deterioration is associated with post-ischemia dementia. A study by Szabo et al. suggests that lesion in the hippocampus could lead to persistent memory deterioration, considered to be the usual consequence of occlusion of the posterior cerebral artery [20]. In recent studies, Gemmell et al., when analyzing hippocampal volume in postmortem samples of patients with dementia, suggested that altered brain volume was between 10 and 20% in the CA1 and CA2 regions and 20% in the CA3 and CA4 regions of the hippocampus [21, 22]. However, the mechanism of post-ischemia hippocampal injury related to cognitive impairment is still uncertain. Some studies, including ours, have shown that Tau hyperphosphorylation, neurodegeneration hallmark, is closely correlated with cognitive disorder and dementia after cerebral ischemia [23–25].

So far there is no effective treatment to treat cognitive impairment post-ischemia. Some drugs used for the treatment of dementia in Alzheimer's disease have shown some positive effects on cognitive impairment post-ischemia [26]: for example, anti-inflammatory agents, blood-brain barrier modulators, endothelin antagonist, flavonoids, immunosuppressive agents, antidepressants, neurotrophic agents, among others. However, although none of them has been established as an effective treatment, somes strategies have been defined for prevention of cognitive impairment, which focus on acute treatment (to limit damage from the onset of injury and prevent early recurrence) and the prevention of long-term recurrence, but does not exist a standard gold for post-ischemia treatment yet.

3 Neuroprotective Therapies in Cerebral Ischemia: Translational Medicine Perspective

Over the past two decades, neuroprotective agents designed to block cell death have been investigated in animal models of cerebral ischemia. Numerous drugs have been found that reduce the size of infarction in ischemia models in rodents, rabbits and primates. However, neuroprotection as a strategy for the treatment of stroke, which have been effective in experimental models have failed in phase III clinical trials [27]. Currently, the only approved therapy for cerebral infarction is limited to the treatment with recombinant tissue plasminogen activator (rt-PA), which although improving functional prognosis, is used in less than 3% of patients [28].

Several difficulties have been identified in establishing effective therapy that limits the devastating consequences of reduced blood flow in the penumbra zone and prevents secondary insults induced by reperfusion [8]. For example, the focus on a single event of the ischemic cascade may not be sufficient to decrease the consequences of a multifactorial condition such as cerebral ischemia [29]. In this context, a treatment for one specific target over time under a controlled physiological condition may be ineffective. However, if we used stem events in the ischemia/reperfusion phenomenon, such as glutamate excitoxicity and avoided its propagation wave by local neural circuits, we could block several associated downstream pathophysiological phenomena, such as the activation of calpains, caspases, enzymatic activity, substrate dysregulation, activation of death pathways, depolymerization of the microtubule and actin cytoskeleton, among others.

For example, the use of statins has shown beneficial effects following an ischemic event [9, 30]. Statins, competitive inhibitors of 3-hydroxy-3-methylglutaryl coenzyme A (HMG-CoA) reductase, are cholesterol-lowering drugs and with multiple pleiotropic effects in cerebral ischemia, such as enhancement of endothelial function, have potential anti-oxidant, anti-thrombotic and anti-inflammatory, among others [6]. Our studies have shown that post-ischemic statin therapy recovers from neurological deficit, memory deficit and learning, as well as active survival pathways and synaptic plasticity associated with neuronal recovery [31]. These drugs also regulated NMDA-like glutamate receptors, which play a crucial role in glutamate excitotoxicity, as a pathogenic primary event. Also, another alternative has been proposed, where the combination of two or more potential neuroprotective agents, each one against different targets respect to the neurodegenerative cascade, achieve synergistic protective action on ischemic injury, such as statins and rt-PA [32]. However, minimum adverse reactions should be demonstrated in combinated therapies [27].

Another, large discrepancy between the results obtained in the laboratory and the clinical trials makes difficult to translate many therapies, because is not considered the morbidity condition of the patients [27]. For example, in stroke models, researchers often choose young and healthy animals; however, patients with cerebral infarction are often elderly and suffer multiple chronic diseases, such as arteriosclerosis, hypertension, diabetes and hyperlipidemia [33]. In addition, the most studies in ischemia models have been performed at short term, based on the fact that treatments established in the first few hours of symptom onset are more likely to become effective therapeutics. However, many agents considered effective in acute phase are not effective in the sub-acute or chronic phase. Therefore, long-term protection studies and to follow-up the health progress of the treated patients for longer time is completely necessary for post-ischemic stroke patient care protocols and to propose a post-ischemia gold standard treatment.

4 Gene Therapy as Treatment for Neurodegeneration

Gene therapy is an alternative treatment to pharmacological, surgical and conventional approaches that are being developed both experimentally and clinically. Basically, it consists of the specific modification of a gene to prevent or remedy a pathological condition for the organism. This technique is based on interfering, correcting or replacing the defective gene within cells expressing the pathology by the normal gene and its correct functional protein to slow down, stabilize or reverse the course of the disease [34].

Depending on the type of disorder, gene therapy is performed through one of the following approaches [35]: Gene addition: consists of introducing a correct copy of the functional gene so that the corresponding protein in the tissue is produced in adequate quantities to be treated. It is the most used procedure. Gene Correction: we seek to correct the altered gene by replacing it with the correct gene by homologous recombination, although for the moment this method is inefficient. Instead, it is possible to exchange a specific mutated nucleotide, which is useful for treating diseases with point mutations. Gene suppression: consists in reducing the expression of a particular gene. This is achieved using RNA interference (RNAi), including short hairpin RNAs (shRNAs), that induce the degradation or silencing of target messenger RNA (mRNA) [36].

In the framework of therapy based on gene interference, shRNA administration has been tested in experimental models and clinical trials using various vehicles including viral vectors, liposomes and nanoparticles [37]. Viral vectors have a greater efficiency in DNA-RNA transfer, but are more likely to generate an undesired immune response in the patient. Integration of the gene of interest into the vector requires removal of the genes conferring its virulence by incorporating in its place the desired sequence. Among the viral vectors, the most used are retroviruses, adenoviruses, lentiviruses, herpes simplex virus and adeno-associated viruses [38, 39]. The latter are those with less inflammatory response. Non-viral vectors are less effective in transference, but safer and simpler to elaborate, characterize and manipulate [36].

Neurodegenerative diseases have been one of most evaluated in gene therapy, after cancer and genetic and metabolic diseases. Over the past decade, scientists have made great strides in evaluating shRNA-based therapies for diseases affecting the central nervous system (CNS) [37].

For a number of neurodegenerative disorders (e.g., Alzheimer's disease (AD), Parkinson's disease (PD) and post-stroke infarction) abnormal accumulation of proteins seems to play a central role in the onset of disease and/or in progression [40]. Thus, a modest reduction in neurotoxic protein levels is expected to provide significant therapeutic relief. Using a shRNA-based approach, researchers have successfully inhibited the expression of disease-causing proteins in animal models, where in most cases this inhibition correlates with improved neuropathological and behavioral phenotypes. For example, AD is the leading cause of dementia worldwide and is characterized by the presence of amyloid plaques and neurofibrillary tangles (NFTs) in the brain. In our previous research the silencing of CKD5, an enzyme responsible of tau hyperphosphorylation in dementia, reduced the NFTs production in a triple transgenic AD mice model [41]. Also, in PD the second most common neurodegenerative disease and histopathologically characterized by the formation of Lewy bodies in the brain, composed mainly of alpha-synuclein (α-syn) protein [42]. A single mutation in α-syn as well as genetic duplication or triplication of the gene are linked to hereditary parkinsonism, which was blocked using shRNA in PD model in rats [43].

4.1 Preclinical Studies of Gene Therapy in Cerebral Infarction

shRNA is shown as a potential therapeutic tool in animal models of acute cerebral injuries [44] with satisfactory results (Table 13.1). Initially, researchers focused on neuroprotective strategies to limit the spread of apoptosis. For example, shRNAs against Beclin1, a protein responsible for cellular autophagy and apoptosis, against caspase-3 and for kinase regulator of apoptotic kinase 1 (ASK1) showed decreased infarct volume and improved neurological outcomes respectively [45–47]. In an effort to target vascular compartments was reduced protease-1 activated receptor (PAR-1) in a mouse t-MCAO model. PAR-1 is involved in blood clotting and serves as another viable option for thrombolytic strategy along with rt-PA. shRNA against PAR-1 7 days before the injury resulted in a decrease in infarct volume and lower neurological deficit at 24 and 72 h post-injury [48]. However, it is important to emphasize in those studies, shRNA administration occurred before the ischemic event, which limits clinical translational approaches. But, the hypoxia-inducible factor 1 (HIF-1), implicated in neurovascular unit dysfunction, was decreased by shRNA one hour after t-MCAO in rats, avoiding blood-brain barrier disruption and improving behavior skills, in addition to a lower expression of the p53 protein and less activity of caspase-3 [49]. This study brings high hopes in the field of translational therapy, because shRNA was administered after ischemic injury and the results were satisfactory.

Another important mechanism of injury is neuroinflammation. Previous studies have shown the G protein coupled to receptor 17 (GPR17) is involving in neuroinflammation; particularly, GPR17 inhibited by an antisense oligonucleotide showed beneficial effects after ischemic stroke [50]. Zhao et al. specifically examined the

Table 13.1 *In vivo* studies using shRNA in Cerebral ischemia Models

Therapeutic target	Patho-physiological event	Ischemia model	shRNA administration method	Intervention time	Findings
Beclin 1 [45]	Apoptosis and autophagy	t-MCAO in rats	Intraventricular injection	7 days before	Reduction of infarct volume and improvement in functional tests
Caspase 3 [46]	Apoptosis	Entotelin-1 injection in rats	Cortical injection	24 h before 24 h after	Reduced behavioral deficit (significantly 24 h before ischemia)
Ask1 [47]	Apoptosis	t-MCAO in mice	Intraventricular osmotic mini-pump	3 days	Reduction of infarct volume
PARK1 [48]	Coagulation cascade	t-MCAO in mice	Intraventricular injection	7 days before	Reduction of infarct volume and neurological deficit
HIF1 α [49]	Hypoxia	t-MCAO in rats	Intraparenchymal injection	<1 h after	Reduction in mortality, neurological deficit, blood-brain barrier injury, and cell death
HMGB1 [50]	Neuro-inflammation	t-MCAO in rats	Intranasal	1 h before	Reduction of infarct volume and behavioral improvement
GPR17 [51]	Microgliosis neuro-inflammation	t-MCAO in rats	Intraventricular injection	2 days before until 7 days after	Reduction in neuronal loss, neurological deficit, and infarct volume

effect of shRNA against GPR17 on microglial activation in the acute and chronic stages, finding a microgliosis reduction after 14 days, but not 24 h after injury [51]. Another important protein during the post-ischemic neuroinflammatory phase is the high mobility group Box 1 protein (HMGB1), secreted by necrotic cells for the recruitment of pro-inflammatory cells [52]. An intracortical injection of shRNA against HMGB1 was found to have a neuroprotective effect after ischemic stroke through reduction of microglial activation and neuronal apoptosis [53]. In subsequent experiments, intranasal administration of this shRNA in a rat model of t-MCAO resulted in a significant reduction of HMGB1 in various regions of the brain but not in the liver, lung, kidney, or heart. In addition, this effective reduction correlated with improvements in behavioral tests [54]. It is important to highlight that intranasal administration is one of the most promising tools for a clinical translational perspective, not only for the delivery of shRNA in the brain, but also for the delivery of other types of drugs [55, 56].

4.2 CDK5 as a Potential Therapeutic Target for Cerebral Ischemia

Cyclin-5 kinase (CDK5) is a serine/threonine kinase, a member of the cyclin-dependent kinase family. Similar to other members of this kinase group, CDK5 binds to p35 and p39 regulatory subunits to be activated, similar to cyclins [57]. However, CDK5 is not directly involved in cell cycle progression and its activity predominates in postmitotic neurons [58], since their p35 and p39 activators are expressed mainly in the central nervous system (CNS) [59]. However, although CDK5 activity is vital for the proper development of the CNS, and synaptic plasticity [60], but its deregulation has been shown to play a critical role in the chronic neurodegenerative events of several diseases, including Alzheimer's disease (AD), Parkinson's disease (PD) and Amyotrophic lateral sclerosis (ALS).

4.2.1 CDK5 at the Synapse: Functions and Dysfunctions

CDK5 plays an important role in a variety of physiological and pathological processes. This multifunctionality includes participation in the development of the nervous system, synaptic plasticity and neurodegeneration [57]. CNS requires programmed migration, differentiation and neurons connection to form functional circuits capable of expressing synaptic plasticity. Studies in several lines of mutant mice have shown that CDK5 is critical for all these stages of CNS development [60]. Phenotypically null mutant mice for CDK5, as well as double mutants for p35/p39 exhibit an alteration of cortical laminar architecture, as well as cytoarchitecture disorders in the cerebellum, brainstem and hippocampus [61].

Molecular mechanisms of synaptic plasticity occur at the pre and post-synaptic levels and involve the regulation of vesicle release, changes in the conductance of ion channels and modulation of kinase and phosphatase activities [62]. Various studies have shown that CDK5 has an important role in synaptic plasticity directing phosphorylation of key substrates for synaptic plasticity, e.g. ionotropic glutamate receptors, cell adhesion proteins, and cytoskeletal proteins [57, 63]. Within the cytoskeleton proteins that are a direct substrate of CDK5, is the Tau protein, which participates in the assembly and disassembly of microtubules through the dynamic incorporation of tubulin monomers to form neuronal axons; which contribute to the maintenance of cellular form and transport. Tau also establishes links between microtubules and other cytoskeletal elements, such as neurofilaments, spectrin and actin filaments. In the normal brain, the balance between phosphorylation as well as the dephosphorylation of tau by CDK5, other kinases and phosphatases, leads to structural and conformational changes that regulate cytoskeletal stability and axonal morphology [64]. Another of the great functions of CDK5 is in learning and memory [60]. A positive role of CDK5 was identified in $p35^{-/-}$ mice, which show depotentiation, LTP reduction and defective induction of long-term depression (LTD), electrophysiological mechanisms crucial for memory formation [65]. Similarly, $Cdk5^{flox/flox}$ T29 mice in which there is no CDK5 expression in hippocampal CA1 pyramidal neurons, revealed memory impairment and synaptic plasticity [66]. However, Hawasli et al. showed that the initial suppression of 50% of CDK5 protein in a mouse knockout conditional model enhances LTP and NMDAR-mediated synaptic plasticity and improves learning and memory skills [67].

In spite of the functions that CDK5 fulfills in plasticity, neuronal development and memory, its over-activation triggers neurodegenerative events ranging from cell death, microtubule destabilization, alteration of the actin cytoskeleton, loss of cellular adhesion and alterations of memory and learning. Deregulation of CDK5 activity begins with the cleavage of its p35 and p39 activators through the calpain, which are a group of cytosolic proteases activated by intracellular calcium [68]. After an ischemic event, the overactivation of NMDA and AMPA-type glutamate receptors leads to an increase of intracellular calcium, activating the calpain, which cleaves the p35 and p39 activators, generating p25 and p29 respectively (Fig. 13.2). The half-life of p25 and p29 is significantly longer (about threefold greater than p35 and p39), as well as the binding to the kinase of these activators is much stronger than p35 and p39 binding [69, 70], which results in higher p25/CDK5 (or p29) activity, compared to p35/CDK5 (or p39). On another hand, p25 and p29 lack an amino-terminal myristoylation site, a necessary modification to maintain the protein at the plasma membrane [71]. Thus, the interaction of p25 with CDK5 not only leads to an activation of the kinase in a sustained manner, but also modifies its cellular distribution, concentrating on the cytoplasm and nucleus, altering its substrate specificity and triggering cell death [71–73]. Also, CDK5 expression as well as p35 cleavage has been found to increase in penumbra region after middle cerebral artery occlusion (MCAO) in rats [74] and an increase in expression of CDK5 and cleaved fragment p25 has been associated with neuronal damage in brains of patients affected by cerebral ischemia [75].

Additionally, *in vivo* models indicate that the sustained increase of CDK5 activity in Alzheimer's disease correlates with aberrant tau hyperphosphorylation, forming aggregates of this protein in the cell body [41] and consequently induce cellular death and learning and memory decline [63]. Mice overexpressing long-term p25 levels have been shown to undergo LTP impairment in the hippocampus and memory deficits along with significant neuronal loss. However, when p25 is expressed transiently, LTP is enhanced, the number of dendritic spines and synapses increases, without observing neurodegeneration [76].

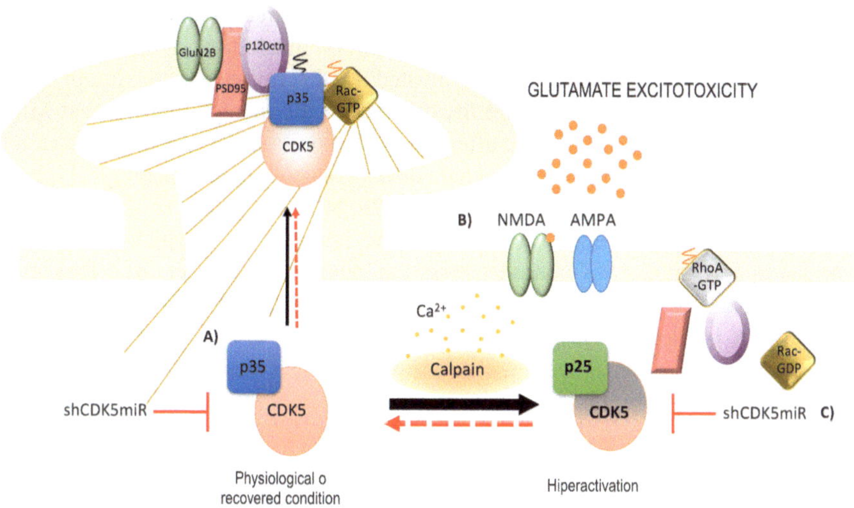

Fig. 13.2 CDK5 regulation in mature neurons. (**a**) CDK5 alone is an inactive catalytic subunit. This kinase is activated by p35 and then recruited to the membrane through myristoylation of its N-terminal region. p35 is a protein with a short life span and is degraded by the proteosome. (**b**) When a glutamate excitotoxicity condition occurs as in cerebral ischemia, calpain is activated by increasing intracellular calcium concentrations and cuts to p35 on a C-terminal fragment generating p25 fragment. p25 has a longer half-life and dissociates from membranes, where it is able to sustained CDK5 hyperactivation. (**c**) However, when is reduced the CDK5 overactivation by shCDK5miR in adult brain or mature neurons, p35 may to be associated to p120ctn, PSD95 and NR2B subunit in a Rac1-active dependent mode in the cell membrane and promote or strengthen the synapses

4.2.2 Strategies to Control the Over-Activation of CDK5

Because of the importance of CDK5 in neurodegeneration, it is presumed that this kinase is a good therapeutic target to prevent or even stop pathologies associated with Tau hyperphosphorylation, as we demonstrated in an aged triple transgenic Alzheimer's disease mice model already [25, 41]. In fact, various *in vitro* and *in vivo* studies have shown that blocking over-activity of CDK5 may have a beneficial effect and generates neuroprotection. Two main strategies have been used for these purposes: direct inhibition with the use of CDK5 inhibitors and indirect action by preventing the excessive generation of the associated p25 activator through the use of calpain inhibitors [57]. So, roscovitine, a pharmacological inhibitor of CDK5, exerts a neuroprotective effect *in vivo*, following systemic pre and post-ischemia administration in experimental models of stroke [77]. Roscovitine acts on different cell types (neurons and glial cells) and through various mechanisms: anti-apoptotic, anti-excitotoxicity and possibly anti-inflammatory pathways. Thus, even when roscovitine is administered 2 h post-ischemia reduces the volume of infarction in rats subjected to transient focal ischemia [57]. On another hand, the best-known calpain inhibitor, MDL 28170, causes reduction in infarct volume when was gave 30 min after occlusion of the middle cerebral artery. In addition, in this same study, several therapeutic windows were evaluated to determine the maximum delay between the onset of ischemia and the efficacy of the therapy, finding that MDL 28170 reduced the infarct volume after 0.5, 3, 4 and 6 h onset of ischemia, but not after 8-h delayed [78]. However, the main difficulties of the pharmacological inhibitors is the lack of specificity or off-targeting [79, 80].

On another side, a peptide was designed to inhibit CDK5, which protected in various *in vivo* models of neurodegeneration, such as Alzheimer [81], Parkinson [82] and cerebral infarction [83]. This peptide of 24 amino acids, can cross the blood-brain barrier, *in vitro* inhibit the activity of p35/CDK5 and p25/CDK5, whereas in rodent cortical neurons, inhibits p25/CDK5 without affecting the endogenous p35/CDK5 activity. In a model of cerebral ischemia, a p5-TAT peptide does not alter the levels of p35, p39, although it reduces the phosphorylation of Tau through the inhibition of p25/CDK5 activity. In addition, p5-TAT reduces levels of caspase-3 and the cerebral infarction volume; even when this was administered up to 24 h after ischemic injury promote long-term functional recovery [83]. Also, in our investigations, using a CDK5 shRNAmiR carried on adeno associated viral vectors (AAV) injected into the hippocampus of a triple transgenic mice for Alzheimer's disease, reduced tau hyperphosphorylation and amyloid beta plaques, improved learning and spatial memory to short and long-term of treatment [25, 41, 84], analog effects were obtained in post-ischemic treatment in rats [85, 86], whose studies will be detailed below.

5 Silencing of CDK5 as Post-ischemia Therapy

The above background shows that the reduction of CDK5 may be a key tool in reversing damage following cerebral ischemia. This is based on the fact that blocking CDK5 can prevent several pathological events since first hours to even months post-injury. For example, neuronal death after cerebral ischemia occurs as an early event of pathophysiological cascade, which is primarily associated with p25-mediated hyperactivation of CDK5. Because, the silencing of CDK5 may be playing a critical role at early times regulating the rate of degradation and internalization of NMDA receptors, which are essential for the excitotoxic pathway that occurs after ischemic damage, avoiding the consequent increase of calcium influx in cell, calpain and cell death pathways activation, because GluN2B (subunit of the NMDA receptor), calcium ionotropic receptor, is activated by CDK5, and its phosphorylation rate determines the dynamics of receptor internalization [87].

However, we have suggest that the silencing of CDK5 to long term post-ischemia represents an advantage to the control of CDK5 over-activation because of its high specificity and the long-term expression in the tissue preventing pathophysiological events after several post-ischemia months.

The CDK5 silencing strategy is based on the use of an endogenous microRNA (miR30) that carries CDK5 (exogenous) shRNA, which is transduced into the cells using an AAV 2.5 adeno-associated viral vector. This strategy has several advantage points because the miR-30 skeleton confers low toxicity, high processing and expression in the cell compared to the use of only the CDK5 shRNA [36]. In addition, the use of AAV, specifically serotype 2.5, confers: (1) tropism to neural cells, astrocytes and neurons; (2) decreased inflammatory response or interferon response, (3) low insertion of exogenous DNA into the host cell DNA, which decreases the probability of a tumor focus, (4) stable expression of CDK5 RNAi up to 1 year in mice and (5) ease of expression monitoring with the enhanced green fluorescent protein reporter [25, 41, 63].

Those is based in our recent research which show that the gene therapy directed to silence CDK5 in the hippocampus of ischemic rats by the middle cerebral artery occlusion (tMCAO), did not generate changes in the physiological parameters (blood pressure, pH, pO_2, pCO_2), decreased CDK5 protein levels resulting in a neurological and motor improvement during the first week after the ischemia. It also CDK5 shRNAmiR prevented dysfunctions in learning, memory and reversal learning at 1 month [85]. This effect was maintained at 4 months, preventing neuronal loss, tauopathy, microglial hyperreactivity and generating branched astrocytes [86]. CDK5 silencing increased the expression of brain derived neurotrophic factor (BDNF) and activated the TRKB/CREB/CaMKII pathway in the hippocampus, involving calcium modulation in the spines and induction of LTP in a TRKB receptor-dependent mode in neurons [86, 88]. Also, CDK5 RNAi protected against glutamate-mediated excitotoxicity (major excitatory neurotransmitter and uncontrollably upregulated in a cerebral stroke) in primary neuronal cultures. This protection was dependent on a concomitant increase of p35, since it was blocked by the use of an RNAi against p35, which affected downstream activity of the RhoGTPases (proteins responsible of cytoskeletal remodeling and dendritic spine formation). In addition, overexpression of p35 and constitutively active Rac1 mimicked the neuroprotection exerted by the silencing of CDK5 [63]. Also, another synaptic proteins also participated in neuroprotection and synaptic plasticity induced by the CDK5 shRNAmiR, such as: PSD95, NR2B, p120ctn, N-cadherin and β-catenin. Inhibition or knockdown of CDK5 resulted in the increase of p120ctn and the neuroprotection induced on depended the expression of p120ctn in a model of glutamate-induced excitotoxicity. Thus the p35/p120ctn/PSD95/NR2B complex is involved in the synaptic recovery induced by the silencing of CDK5 [89] (Fig. 13.2).

On another hand, astrocytes perform metabolic and structural support functions in the brain and contribute to the integrity of the blood-brain barrier. Astrocytes influence neuronal survival and prevent gliotoxicity by uptaking glutamate (Glu), reactive oxygen species and nutrients. CDK5 may have a double effect on the endothelium and astrocytes as it is involved in migration, senescence and angiogenesis, and its hyperactivity is associated with dysfunction of glutamate recapture and hypoxia. Therefore, treating a possible deregulation of CDK5 with RNAi is other strategy that we have proposed as a treatment for neurodegeneration. In the C6 cell line astroglioma and primary astrocytes, CDK5 RNAi prevented the activation of cells generated by glutamate-induced gliotoxicity, and this finding was corroborated by CDK5 pharmacological inhibition with roscovitine (Rosc). This effect was associated with the occurrence of lamellipodia, and ramifications, the activation of Rac1, the release of BDNF by astrocytes, which correlated with the protection of neurons exposed to glutamate excitotoxicity. Interestingly, inhibition of Rac1 in astrocytes blocked the release of BDNF and astrocyte-mediated neuroprotection [90]. Thus, the remodeling of the actin cytoskeleton in the hyperreactive astrocytes suggests a functional phenotype for the release of BDNF, which promotes neuroprotection.

Complementarily, in the bEnd.3 cell line from mice cerebral microvasculature, Rosc was found to recover endothelial adhesion altered by glutamate toxicity; this effect was corroborated by the increase of TEER (transendothelial resistance), p120ctn levels in membrane and decreased intercellular gaps. In addition, CDK5 RNAi increased the primary processes in bEnd.3 cells. Also, endothelial cells that were co-cultured with CDK5-knock down (KD) astrocytes showed a recovery in cell viability, an increased adhesion proteins (p120ctn and PECAM-1) and BDNF release. These findings suggest that inhibition of CDK5 or its silencing in astrocytes protects the endothelium, which promotes the BDNF release and endothelial adhesion (Posada-Duque et al. unpublished data). Taking together these findings suggests CDK5-knockdown astrocytes as a cell-gene therapy to favor the protection of neurovascular unit. Therefore, we transplanted astrocytes-CDK5 KD in the somatosensory cortex after cerebral ischemia in rats. In these experiments we obtained a general prevention of neuronal loss by transplantation of CDK5-KD astrocytes, which induced a significant stimulation in the arborization of endogenous astrocytes, involving blood vessels, accompanied by increased immunoreactivity of PECAM-1 in the motor and somatosensory areas, as well as an increased Ki67 immunostaining (proliferation marker) in the lateral ventricles, partially associated with BDNF production, which suggest us also morphological and physiological protection benefits by cell therapy base in CDK5-KD astrocyte [91], stable to 4 months post-ischemia with side effects (Becerra-Calixto et al. unpublished data).

Overall, silencing of CDK5 protects neurons, astrocytes and endothelium from adult brain suggesting neurovascular unit has a critical role in the functional and morphological recovery after a cerebral ischemia and [10, 92]. In addition, our results suggest that CDK5-KD astrocytes are a paracrine source of BDNF production, which generates neuronal and endovascular protection that would be a novel strategy to protect BBB integrity after stroke.

Therefore, CDK5 silencing could be an ideal gene therapy strategy because block stem events by the ischemia/reperfusion phenomenon, as glutamate excitoxicity associated to downstream pathophysiological phenomena, decreased calpain, sustained enzymatic activation, prevented the tau hyperphosphorylation and cell death (Fig. 13.3), generating protection of synapses, short and long term morphological and physiological recovery in the ischemic brain.

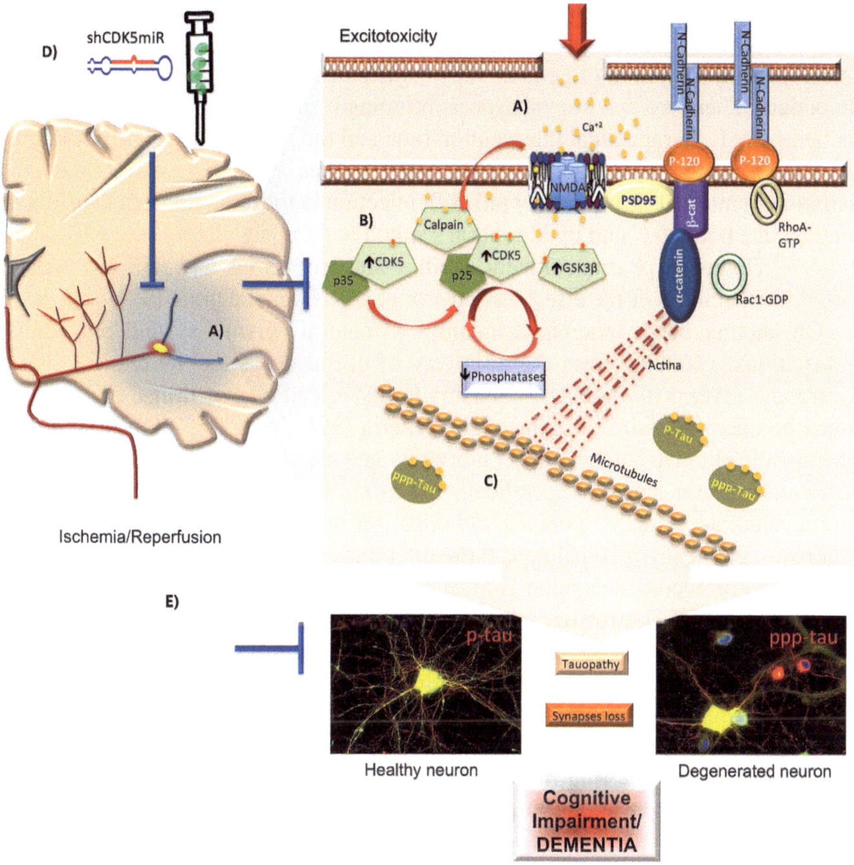

Fig. 13.3 Hypothetic model of post-ischemia gene therapy based on the silencing of CDK5 blocking excitotoxicity spreading and preventing cognitive impairment. (**a**) Anoxia / Reperfusion, and excitoxicity spreading, (**b**) Activation of CDK5, disbalance kinases / phosphatases, tau hyperphosphorylation, (**c**) Actin and microtubule cytoskeleton disassembly, loss of synapses, cognitive impairment and dementia. (**d**) Gene therapy using CDK5 RNAi blocks **a–c** events. (**e**) Recovering synapses and cognitive function. pTau: Tau phosphorylation, pppTau: tau hyperphosphorylation

6 Current and Future Challenges of Gene Therapy in Ischemic Stroke for Translational Medicine

Therefore, despite identifying potential molecular targets, including our own studies, there are considerable limitations inherent in the nervous system, such as crossing the blood-brain barrier and the difficulty of targeting specific neuronal populations. The presence of these obstacles has led to the search for new strategies for the treatment of neurodegenerative diseases, the use of viral vectors, the design of nanoparticles to improve the distribution, and future approaches for systemic

administration or functionalised nutrition, including safety and effectiveness studies, which together represent a current challenge.
An important limitation of gene therapy studies in cerebral infarction is translation to patient. There are two crucial aspects bringing results from preclinical studies to patient: one is therapeutical intervention time and the second one is shRNA administration route to the infarcted area [44]. Some studies report diverse system administration as internal carotid artery [45, 47], injection in the tail with wide distribution in the mice body [93] and by intranasal method without significant side effects [54, 94–97]. However, pharmacokinetic studies are necessaries and will available to provide assurances of the effective delivery of shRNA in the brain [98].

On another hand, inherent difficulties of central nervous system procedures, translational medicine from the discovery of useful drugs for the clinic needs to overcome several limitations as sensitivity and specificity of drugs, and patients must be classified with clear inclusion criteria [99]. Promoting the translation of therapeutic alternatives that impact morbidity and mortality is an urgent need worldwide and in Latin America. Specifically in Colombia, we need to take a step forward in translational medicine policies, although our developing countries need to first improve emergency medical care for acute injuries [100]. We also need to update norms and protocols to develop clinical trials with original candidates or strategies proposed by our Latin American countries, and gather efforts of scientists, medical specialists, pharmaceutical industry and government, accompanied by economic and social support. We must offer experimental treatments, without false expectations to the patient/family, based on the rigor of preclinical scientific evidence, in direct dialogue with the medical team, peer monitoring of phases I, II and III, and strict long-term monitoring of potential unwanted side effects [101], which could progressively strengthen the contribution to solving mental health problems [19].

7 Conclusion and Perspective

Cerebral ischemia has become a health problem worldwide due to its high mortality rate and disability in patients suffering from it. Many therapeutic strategies have been evaluated both in experimental models and in clinical trials without obtaining good results. One of the major shortcomings of these therapeutic investigations is the limitation in intervention time (less than 4.5 h) and the short time of protection or follow-up of therapy, which generates long-term sequelae in patients.

Gene therapy has been shown to be a very useful tool for the treatment of neurodegenerative diseases. However, specifically in cerebral ischemia, there are few studies at the experimental level that show neuroprotective effect, reduction in infarct volume and functional improvement, since reference is made to a pretreatment which does not facilitate an approach to the clinical problem. Therefore, a major challenge of gene therapy is that it to be implemented as post-injury therapy, which is supported by our results, as this situation recreates the problem at the clinical level and the experimental results are more relevant in the field of translational medicine.

The silencing of CDK5 has become a highly important therapeutic strategy to reduce cell death in several neurodegenerative diseases, mainly in those where the tauopathy as main hallmark of cognitive deterioration and dementia, as in cerebral ischemia (Fig. 13.3). However, although pharmacological inhibitors have been tried for this kinase, it is necessary to use methodologies that allow the directed and efficient reduction of CDK5, avoiding side effects generated by nonspecific pharmacological blocks towards other proteins and signaling pathways. It is here that gene therapy becomes very important, being a tool aimed at controlling the kinase overactivation, not only in times immediate to the ischemic lesion where cell death responses are given, but, to the long-term, where events such as the loss of spines and dendrites are key pieces in the alteration of function and appearance of delayed post-ischemia sequelae, which can be avoided. However, it is necessary to align the scientific interests of the government and the company to support the development of therapies proposed from our own country and that would help to avoid or improve the quality of life of patients with physical and mental disability post-cerebral stroke.

Acknowledgement We are grateful to the Cellular and Molecular Neurobiology Area, Antioquia Neuroscience Group (GNA), University of Antioquia for constructive academic discussion. This research was supported by projects of Colciencias (code # 111551928905, # 111554531400) (GPC-G) and CODI # 2014-970 of the University of Antioquia, Republic Bank # 3494.

Statement of Conflict of Interest: The authors declare no conflict of interest.

References

1. WHO. 2017. http://www.who.int/mediacentre/factsheets/fs310/en/
2. CDC. 2017. https://www.cdc.gov/stroke/facts.htm
3. Assarzadegan F, Tabesh H, Shoghli A, Ghafoori Yazdi M, Tabesh H, Daneshpajooh P, Yaseri M. Relation of stroke risk factors with specific stroke subtypes and territories. Iran J Public Health. 2015;44:1387–94.
4. Durukan A, Tatlisumak T. Acute ischemic stroke: overview of major experimental rodent models, pathophysiology, and therapy of focal cerebral ischemia. Pharmacol Biochem Behav. 2007;87:179–97.
5. Amlie-Lefond C, Chan AK, Kirton A, DeVeber G, Hovinga CA, Ichord R, Stephens D, Zaidat OO. Thrombolysis in acute childhood stroke: design and challenges of the thrombolysis in pediatric stroke clinical trial. Neuroepidemiology. 2009;32:279–86.
6. Goldstein LB, Bushnell CD, Adams RJ, Appel LJ, Braun LT, Chaturvedi S, Creager MA, Culebras A, Eckel RH, Hart RG, Hinchey JA, Howard VJ, Jauch EC, Levine SR, Meschia JF, Moore WS, Nixon JV, Pearson TA, American Heart Association Stroke Council, Council on Cardiovascular Nursing, Council on Epidemiology and Prevention, Council for High Blood Pressure Research, Council on Peripheral Vascular Disease, and Interdisciplinary Council on Quality of Care and Outcomes Research. Guidelines for the primary prevention of stroke: a guideline for healthcare professionals from the American Heart Association/American Stroke Association. Stroke. 2011;42:517–84.
7. Moustafa RR, Baron JC. Pathophysiology of ischaemic stroke: insights from imaging, and implications for therapy and drug discovery. Br J Pharmacol. 2008;153(Suppl 1):S44–54.
8. Moskowitz MA, Lo EH, Iadecola C. The science of stroke: mechanisms in search of treatments. Neuron. 2010;67:181–98.

9. Gutiérrez-Vargas JA, Cespedes-Rubio A, Cardona-Gómez GP. Perspective of synaptic protection after post-infarction treatment with statins. J Transl Med. 2015;13:118. https://doi.org/10.1186/s12967-015-0472-6.
10. Posada-Duque RA, Barreto GE, Cardona-Gomez GP. Protection after stroke: cellular effectors of neurovascular unit integrity. Front Cell Neurosci. 2014;8:231.
11. Lo EH. A new penumbra: transitioning from injury into repair after stroke. Nat Med. 2008;14:497–500.
12. Choi DW, Rothman SM. The role of glutamate neurotoxicity in hypoxic-ischemic neuronal death. Annu Rev Neurosci. 1990;13:171–82.
13. Won SJ, Kim DY, Gwag BJ. Cellular and molecular pathways of ischemic neuronal death. J Biochem Mol Biol. 2002;35:67–86.
14. Dobkin BH, Dorsch A. New evidence for therapies in stroke rehabilitation. Curr Atheroscler Rep. 2013;15:331.
15. Sun JH, Tan L, Yu JT. Post-stroke cognitive impairment: epidemiology, mechanisms and management. Ann Transl Med. 2014;2:80.
16. Nys GM, van Zandvoort MJ, de Kort PL, Jansen BP, Kappelle LJ, de Haan EH. Restrictions of the Mini-Mental State Examination in acute stroke. Arch Clin Neuropsychol. 2005;20:623–9.
17. WHO. 2015. http://www.who.int/mediacentre/factsheets/fs362/en/
18. ADI/BUPA Inform. La demencia en América: EL costo y la prevalencia del Alzheimer y otros tipos de demencia. 2013.
19. Cardona-Gómez GP, Lopera F. Dementia, preclinical studies and its potential for translational medicine in South America. Front Aging Neurosci. 2016;8:304. https://doi.org/10.3389/fnagi.2016.00304.
20. Szabo K, Szabo K, Förster A, Jäger T, Kern R, Griebe M, Hennerici MG, Gass A. Hippocampal lesion patterns in acute posterior cerebral artery stroke: clinical and MRI findings. Stroke. 2009;40:2042–5.
21. Gemmell E, Bosomworth H, Allan L, Hall R, Khundakar A, Oakley AE, Deramecourt V, Polvikoski TM, O'Brien JT, Kalaria RN. Hippocampal neuronal atrophy and cognitive function in delayed poststroke and aging-related dementias. Stroke. 2012;43:808–14.
22. Gemmell E, Tam E, Allan L, Hall R, Khundakar A, Oakley AE, Thomas A, Deramecourt V, Kalaria RN. Neuron volumes in hippocampal subfields in delayed poststroke and aging-related dementias. J Neuropathol Exp Neurol. 2014;73:305–11.
23. Pluta R, Jolkkonen J, Cuzzocrea S, Pedata F, Cechetto D, Popa-Wagner A. Cognitive impairment with vascular impairment and degeneration. Curr Neurovasc Res. 2011;8:342–50.
24. Castro-Alvarez JF, Gutierrez-Vargas J, Darnaudéry M, Cardona-Gómez GP. ROCK inhibition prevents tau hyperphosphorylation and p25/CDK5 increase after global cerebral ischemia. Behav Neurosci. 2011;125:465–72.
25. Castro-Alvarez JF, Uribe-Arias SA, Kosik KS, Cardona-Gómez GP. Long- and short-term CDK5 knockdown prevents spatial memory dysfunction and tau pathology of triple transgenic Alzheimer's mice. Front Aging Neurosci. 2014;6:243.
26. Brainin M, Tuomilehto J, Heiss WD, Bornstein NM, Bath PM, Teuschl Y, Richard E, Guekht A, Quinn T. Post-stroke cognitive decline: an update and perspectives for clinical research. Eur J Neurol. 2015;22:229–38.
27. Cheng YD, Al-Khoury L, Zivin JA. Neuroprotection for ischemic stroke: two decades of success and failure. NeuroRx. 2004;1:36–45.
28. Auriel E, Bornstein NM. Neuroprotection in acute ischemic stroke—current status. J Cell Mol Med. 2010;14:2200–2.
29. Chavez JC, Hurko O, Barone FC, Feuerstein GZ. Pharmacologic interventions for stroke: looking beyond the thrombolysis time window into the penumbra with biomarkers, not a stopwatch. Stroke. 2009;40:e558–63.
30. Cespedes-Rubio A, Céspedes-Rubio A, Jurado FW, Cardona-Gómez GP. p120 catenin/αN-catenin are molecular targets in the neuroprotection and neuronal plasticity mediated by atorvastatin after focal cerebral ischemia. J Neurosci Res. 2010;88(16):3621–34. https://doi.org/10.1002/jnr.22511.

31. Gutierrez-Vargas JA, Muñoz-Manco JI, Garcia-Segura LM, Cardona-Gómez GP. GluN2B N-methyl-D-aspartic acid receptor subunit mediates atorvastatin-induced neuroprotection after focal cerebral ischemia. J Neurosci Res. 2014;92(11):1529–48. https://doi.org/10.1002/jnr.23426.
32. Zhang L, Zhang ZG, Chopp M. The neurovascular unit and combination treatment strategies for stroke. Trends Pharmacol Sci. 2012;33:415–22.
33. Demchuk AM, Buchan AM. Predictors of stroke outcome. Neurol Clin. 2000;18:455–73.
34. Grimm D, Kay MA. RNAi and gene therapy: a mutual attraction. Hematology Am Soc Hematol Educ Program. 2007:473–81.
35. Fillat C. Perspectivas actuales de la terapia génica. BSCP Can Ped. 2004;28:203–7.
36. Boudreau RL, Rodriguez-Lebron E, Davidson BL. RNAi medicine for the brain: progresses and challenges. Hum Mol Genet. 2011;20:R21–7.
37. Boudreau RL, Davidson BL. RNAi therapeutics for CNS disorders. Brain Res. 2010;1338:112–21.
38. Kay MA, Glorioso JC, Naldini L. Viral vectors for gene therapy: the art of turning infectious agents into vehicles of therapeutics. Nat Med. 2001;7:33–40.
39. CLINIGENE. European Network for the Advancement of Clinical Gene Transfer and Therapy. Available from: http://www.clinigene.eu/.
40. Aguzzi A, O'Connor T. Protein aggregation diseases: pathogenicity and therapeutic perspectives. Nat Rev Drug Discov. 2010;9:237–48.
41. Piedrahita D, Hernández I, López-Tobón A, Fedorov D, Obara B, Manjunath BS, Boudreau RL, Davidson B, Laferla F, Gallego-Gómez JC, Kosik KS, Cardona-Gómez GP. Silencing of CDK5 reduces neurofibrillary tangles in transgenic alzheimer's mice. J Neurosci. 2010;30:13966–76.
42. Halliday GM, McCann H. The progression of pathology in Parkinson's disease. Ann N Y Acad Sci. 2010;1184:188–95.
43. Sapru MK, Yates JW, Hogan S, Jiang L, Halter J, Bohn MC. Silencing of human alpha-synuclein in vitro and in rat brain using lentiviral-mediated RNAi. Exp Neurol. 2006;198:382–90.
44. Fukuda AM, Badaut J. siRNA Treatment: "A Sword-in-the-Stone" for Acute Brain Injuries. Genes (Basel). 2013;4:435–56.
45. Zheng YQ, Liu JX, Li XZ, Xu L, Xu YG. RNA interference-mediated downregulation of Beclin1 attenuates cerebral ischemic injury in rats. Acta Pharmacol Sin. 2009;30:919–27.
46. Al-Jamal KT, Gherardini L, Bardi G, Nunes A, Guo C, Bussy C, Herrero MA, Bianco A, Prato M, Kostarelos K, Pizzorusso T. Functional motor recovery from brain ischemic insult by carbon nanotube-mediated siRNA silencing. Proc Natl Acad Sci U S A. 2011;108:10952–7.
47. Kim HW, Cho KJ, Lee SK, Kim GW. Apoptosis signal-regulating kinase 1 (Ask1) targeted small interfering RNA on ischemic neuronal cell death. Brain Res. 2011;1412:73–8.
48. Price M, Badaut J, Thevenet J, Hirt L. Activation of c-Jun in the nuclei of neurons of the CA-1 in thrombin preconditioning occurs via PAR-1. J Neurosci Res. 2010;88:1338–47.
49. Chen C, Hu Q, Yan J, Yang X, Shi X, Lei J, Chen L, Huang H, Han J, Zhang JH, Zhou C. Early inhibition of HIF-1alpha with small interfering RNA reduces ischemic-reperfused brain injury in rats. Neurobiol Dis. 2009;33:509–17.
50. Lecca D, Trincavelli ML, Gelosa P, Sironi L, Ciana P, Fumagalli M, Villa G, Verderio C, Grumelli C, Guerrini U, Tremoli E, Rosa P, Cuboni S, Martini C, Buffo A, Cimino M, Abbracchio MP. The recently identified P2Y-like receptor GPR17 is a sensor of brain damage and a new target for brain repair. PLoS One. 2008;3:e3579.
51. Zhao B, Zhao CZ, Zhang XY, Huang XQ, Shi WZ, Fang SH, YB L, Zhang WP, Xia Q. The new P2Y-like receptor G protein-coupled receptor 17 mediates acute neuronal injury and late microgliosis after focal cerebral ischemia in rats. Neuroscience. 2012;202:42–57.
52. Bianchi ME, Manfredi AA. High-mobility group box 1 (HMGB1) protein at the crossroads between innate and adaptive immunity. Immunol Rev. 2007;220:35–46.
53. Kim ID, Lim CM, Kim JB, Nam HY, Nam K, Kim SW, Park JS, Lee JK. Neuroprotection by biodegradable PAMAM ester (e-PAM-R)-mediated HMGB1 siRNA delivery in primary cortical cultures and in the postischemic brain. J Control Release. 2010;142:422–30.

54. Kim ID, Shin JH, Kim SW, Choi S, Ahn J, Han PL, Park JS, Lee JK. Intranasal delivery of HMGB1 siRNA confers target gene knockdown and robust neuroprotection in the postischemic brain. Mol Ther. 2012;20:829–39.
55. Thorne RG, Hanson LR, Ross TM, Tung D, Frey WH 2nd. Delivery of interferon-beta to the monkey nervous system following intranasal administration. Neuroscience. 2008;152:785–97.
56. Thorne RG, Pronk GJ, Padmanabhan V, Frey WH 2nd. Delivery of insulin-like growth factor-I to the rat brain and spinal cord along olfactory and trigeminal pathways following intranasal administration. Neuroscience. 2004;127:481–96.
57. Lopes JP, Agostinho P. Cdk5: multitasking between physiological and pathological conditions. Prog Neurobiol. 2011;94:49–63.
58. Tsai LH, Takahashi T, Caviness VS Jr, Harlow E. Activity and expression pattern of cyclin-dependent kinase 5 in the embryonic mouse nervous system. Development. 1993; 119: 1029-1040.
59. Zheng M, Leung CL, Liem RK. Region-specific expression of cyclin-dependent kinase 5 (cdk5) and its activators, p35 and p39, in the developing and adult rat central nervous system. J Neurobiol. 1998;35:141–59.
60. Angelo M, Plattner F, Giese KP. Cyclin-dependent kinase 5 in synaptic plasticity, learning and memory. J Neurochem. 2006;99:353–70.
61. Ko J, Humbert S, Bronson RT, Takahashi S, Kulkarni AB, Li E, Tsai LH. p35 and p39 are essential for cyclin- dependent kinase 5 function during neurodevelopment. J Neurosci. 2001; 21: 6758-6771.
62. Vautrin J, Barker JL. Presynaptic quantal plasticity: Katz's original hypothesis revisited. Synapse. 2003;47:184–99.
63. Posada-Duque RA, López-Tobón A, Piedrahita D, González-Billault C, Cardona-Gomez GP. p35 and Rac1 underlie the neuroprotection and cognitive improvement induced by CDK5 silencing. J Neurochem. 2015;134:354–70.
64. Kimura T, Ishiguro K, Hisanaga S. Physiological and pathological phosphorylation of tau by Cdk5. Front Mol Neurosci. 2014;7:65.
65. Ohshima T, Ogura H, Tomizawa K, Hayashi K, Suzuki H, Saito T, Kamei H, Nishi A, Bibb JA, Hisanaga S, Matsui H, Mikoshiba K. Impairment of hippocampal long-term depression and defective spatial learning and memory in p35 mice. J Neurochem. 2005;94:917–25.
66. Guan JS, SC S, Gao J, Joseph N, Xie Z, Zhou Y, Durak O, Zhang L, Zhu JJ, Clauser KR, Carr SA, Tsai LH. Cdk5 is required for memory function and hippocampal plasticity via the cAMP signaling pathway. PLoS One. 2011;6:e25735.
67. Hawasli AH, Benavides DR, Nguyen C, Kansy JW, Hayashi K, Chambon P, Greengard P, Powell CM, Cooper DC, Bibb JA. Cyclin-dependent kinase 5 governs learning and synaptic plasticity via control of NMDAR degradation. Nat Neurosci. 2007;10:880–6.
68. Camins A, Verdaguer E, Folch J, Pallàs M. Involvement of calpain activation in neurodegenerative processes. CNS drug reviews. 2006;12:135–48.
69. Patrick GN, Zhou P, Kwon YT, Howley PM, Tsai LH. p35, the neuronal-specific activator of cyclin-dependent kinase 5 (Cdk5) is degraded by the ubiquitin-proteasome pathway. The Journal of biological chemistry. 1998; 273: 24057-22464.
70. Amin ND, Albers W, Pant HC. Cyclin-dependent kinase 5 (cdk5) activation requires interaction with three domains of p35. J Neurosci Res. 2002;67:354–62.
71. Asada A, Yamamoto N, Gohda M, Saito T, Hayashi N, Hisanaga S. Myristoylation of p39 and p35 is a determinant of cytoplasmic or nuclear localization of active cyclin-dependent kinase 5 complexes. J Neurochem. 2008;106:1325–36.
72. Kusakawa G, Saito T, Onuki R, Ishiguro K, Kishimoto T, Hisanaga S. Calpain-dependent proteolytic cleavage of the p35 cyclin-dependent kinase 5 activator to p25. J Biol Chem. 2000;275:17166–72.
73. Patrick GN, Zukerberg L, Nikolic M, De la Monte S, Dikkes P, Tsai LH. Conversion of p35 to p25 deregulates Cdk5 activity and promotes neurodegeneration. Nature. 1999;402:615–22.

74. Wen Y, Yang SH, Liu R, Perez EJ, Brun-Zinkernagel AM, Koulen P, Simpkins JW. Cdk5 is involved in NFT-like tauopathy induced by transient cerebral ischemia in female rats. Biochimica et biophysica acta. 2007;1772:473–83.
75. Mitsios N, Pennucci R, Krupinski J, Sanfeliu C, Gaffney J, Kumar P, Kumar S, Juan-Babot O, Slevin M. Expression of cyclin-dependent kinase 5 mRNA and protein in the human brain following acute ischemic stroke. Brain pathology. 2007;17:11–23.
76. Fischer A, Sananbenesi F, Pang PT, Lu B, Tsai LH. Opposing roles of transient and prolonged expression of p25 in synaptic plasticity and hippocampus-dependent memory. Neuron. 2005;48:825–38.
77. Menn B, Bach S, Blevins TL, Campbell M, Meijer L, Timsit S. Delayed treatment with systemic (S)-roscovitine provides neuroprotection and inhibits in vivo CDK5 activity increase in animal stroke models. PLoS One. 2010;5:e12117.
78. Markgraf CG, Velayo NL, Johnson MP, McCarty DR, Medhi S, Koehl JR, Chmielewski PA. Linnik, M.D. Six- hour window of opportunity for calpain inhibition in focal cerebral ischemia in rats. Stroke. 1998;29:152–8.
79. Fischer PM. Recent advances and new directions in the discovery and development of cyclin-dependent kinase inhibitors. Curr Opin Drug Discov Devel. 2001;4:623–34.
80. Glicksman MA, Cuny GD, Liu M, Dobson B, Auerbach K, Stein RL, Kosik KS. New approaches to the discovery of cdk5 inhibitors. Curr Alzheimer Res. 2007;4:547–9.
81. Binukumar BK, Shukla V, Amin ND, Bhaskar M, Skuntz S, Steiner J, Winkler D, Pelech SL, Pant HC. Analysis of the Inhibitory Elements in the p5 Peptide Fragment of the CDK5 Activator, p35, CDKR1 Protein. J Alzheimers Dis. 2015;48:1009–17.
82. Binukumar BK, Shukla V, Amin ND, Grant P, Bhaskar M, Skuntz S, Steiner J, Pant HC. Peptide (TFP5/TP5), derived from Cdk5 activator P35, provides neuroprotection in the MPTP model of Parkinson's disease. Mol Biol Cell. 2015;26:4478–91.
83. Tan X, Chen Y, Li J, Li X, Miao Z, Xin N, Zhu J, Ge W, Feng Y, Xu X. The inhibition of Cdk5 activity after hypoxia/ischemia injury reduces infarct size and promotes functional recovery in neonatal rats. Neuroscience. 2015;290:552–60.
84. Castro-Alvarez JF, Uribe-Arias SA, Cardona-Gomez GP. Cyclin-Dependent kinase 5 targeting prevents beta- Amyloid aggregation involving glycogen synthase kinase 3beta and phosphatases. J Neurosci Res. 2015;93:1258–66.
85. Gutiérrez-Vargas JA, Múnera A, Cardona-Gómez GP. CDK5 knockdown prevents hippocampal degeneration and cognitive dysfunction produced by cerebral ischemia. J Cereb Blood Flow Metab. 2015;35:1937–49.
86. Gutiérrez-Vargas JA, Moreno H, Cardona-Gómez GP. Targeting CDK5 post-stroke provides long-term neuroprotection and rescues synaptic plasticity. J Cereb Blood Flow Metab. 2017;37(6):2208–23. https://doi.org/10.1177/0271678X16662476.
87. Zhang S, Edelmann L, Liu J, Crandall JE, Morabito MA. Cdk5 regulates the phosphorylation of tyrosine 1472 NR2B and the surface expression of NMDA receptors. J Neurosci. 2008;28:415–24.
88. Posada-Duque RA, Ramirez O, Härtel S, Inestrosa NC, Bodaleo F, González-Billault C, Kirkwood A, Cardona-Gómez GP. CDK5 downregulation enhances synaptic plasticity. Cell Mol Life Sci. 2017;74:153–72. https://doi.org/10.1007/s00018-016-2333-8.
89. Uribe-Arias A, Posada-Duque RA, González-Billault C, Villegas A, Lopera F, Cardona-Gómez GP. p120- catenin is necessary for neuroprotection induced by CDK5 silencing in models of Alzheimer's disease. J Neurochem. 2016;138:624–39.
90. Posada-Duque RA, Palacio-Castañeda V, Cardona-Gómez GP. CDK5 knockdown in astrocytes provide neuroprotection as a trophic source via Rac1. Mol Cell Neurosci. 2015;68:151–66. https://doi.org/10.1016/j.mcn.2015.07.001.
91. Becerra-Calixto A, Cardona-Gómez GP. Neuroprotection Induced by Transplanted CDK5 Knockdown Astrocytes in Global Cerebral Ischemic Rats. Mol Neurobiol. 2016. https://doi.org/10.1007/s12035-016-0162-2.

92. Becerra-Calixto A, Cardona-Gómez GP. The Role of Astrocytes in Neuroprotection after Brain Stroke: Potential in Cell Therapy. Front Mol Neurosci. 2017;10:88. https://doi.org/10.3389/fnmol.2017.00088.
93. Campbell M, Hanrahan F, Gobbo OL, Kelly ME, Kiang AS, Humphries MM, Nguyen AT, Ozaki E, Keaney J, Blau CW, Kerskens CM, Cahalan SD, Callanan JJ, Wallace E, Grant GA, Doherty CP, Humphries P. Targeted suppression of claudin-5 decreases cerebral oedema and improves cognitive outcome following traumatic brain injury. Nat Commun. 2012;3:849.
94. Danielyan L, Klein R, Hanson LR, Buadze M, Schwab M, Gleiter CH, Frey WH. Protective effects of intranasal losartan in the APP/PS1 transgenic mouse model of Alzheimer disease. Rejuvenation Res. 2010;13:195–201.
95. Gomez D, Martinez JA, Hanson LR, Frey WH 2nd, Toth CC. Intranasal treatment of neurodegenerative diseases and stroke. Front Biosci (Schol Ed). 2012;4:74–89.
96. Renner DB, Frey WH 2nd, Hanson LR. Intranasal delivery of siRNA to the olfactory bulbs of mice via the olfactory nerve pathway. Neurosci Lett. 2012;513:193–7.
97. Bortolozzi A, Castañé A, Semakova J, Santana N, Alvarado G, Cortés R, Ferrés-Coy A, Fernández G, Carmona MC, Toth M, Perales JC, Montefeltro A, Artigas F. Selective siRNA-mediated suppression of 5-HT1A autoreceptors evokes strong anti-depressant-like effects. Mol Psychiatry. 2012;17:612–23.
98. Thorne RG, Frey WH 2nd. Delivery of neurotrophic factors to the central nervous system: pharmacokinetic considerations. Clin Pharmacokinet. 2001;40:907–46.
99. Drucker E, Krapfenbauer K. Pitfalls and limitations in translation from biomarker discovery to clinical utility in predictive and personalised medicine. The EPMA journal. 2013;4:1–10.
100. Razzak J, Kellermann A. Emergency medical care in developing countries is it worthwhile?. Bulltetin of the World Health. Organization. 2002;80:900–5.
101. Main H, Munsie M, O'Connor MD. Managing the potential and pitfalls during clinical translation of emerging stem cell therapies. Clin Transl Med. 2014;3:10. https://doi.org/10.1186/2001-1326-3-10.

Chapter 14
SB623 Preclinical and Clinical Trial Experience

Eric S. Sussman and Gary K. Steinberg

Abstract Stroke affects more than 15 million individuals each year, and is the second leading cause of mortality worldwide [1]. In the United States alone, there are nearly 800,000 strokes annually, and over seven million individuals (approximately 3% of the adult population) live with the sequelae of a prior stroke [2]. More than half of these individuals suffer from long-term limitation of functional mobility [3]. From an economic perspective, the direct and indirect costs related to stroke are as high as $65 billion per year in the United States [4].

Keywords Stroke • Neuroregeneration • Cell-based therapy • Stem cells • Mesenchymal stem cells • SB623 cells

Abbreviations

AE	Adverse event
BDNF	Brain-derived neurotrophic factor
bFGF	Basic fibroblast growth factor
DWI	Diffusion-weighted imaging
ECM	Extracellular matrix
EGF	Epidermal growth factor
ESC	Embryonic stem cells
ESS	European Stroke Scale
FDA	Food and Drug Administration
FMA	Fugl-Meyer Assessment
GDNF	Glial cell line-derived neurotrophic factor
IGF-1	Insulin-like growth factor 1
MRI	Magnetic resonance imaging

E.S. Sussman, M.D. (✉) • G.K. Steinberg, M.D., Ph.D.
Department of Neurosurgery and Stanford Stroke Center, Stanford University School of Medicine and Stanford Health Care, Stanford University, 300 Pasteur Drive, MC 5327, Stanford, CA 94305, USA
e-mail: esussman@stanford.edu

P.A. Lapchak, J.H. Zhang (eds.), *Cellular and Molecular Approaches to Regeneration and Repair*, Springer Series in Translational Stroke Research,
https://doi.org/10.1007/978-3-319-66679-2_14

mRS	Modified Rankin Scale
MSC	Mesenchymal stem cells
NIHSS	National Institute of Health Stroke Scale
NSC	Neural stem cells
RCT	Randomized controlled trial MCA, middle cerebral artery
T2 FLAIR	T2-weighted-fluid-attenuated inversion recovery
TEAE	Treatment emergent adverse event
VEGF	Vascular endothelial growth factor

1 Introduction

Stroke affects more than 15 million individuals each year, and is the second leading cause of mortality worldwide [1]. In the United States alone, there are nearly 800,000 strokes annually, and over seven million individuals (approximately 3% of the adult population) live with the sequelae of a prior stroke [2]. More than half of these individuals suffer from long-term limitation of functional mobility [3]. From an economic perspective, the direct and indirect costs related to stroke are as high as $65 billion per year in the United States [4].

Current management of acute stroke is focused on early recognition and attempted rapid restoration of cerebral perfusion. This has traditionally been accomplished with intravenous or intra-arterial thrombolytic therapy [5]. Recently, endovascular thrombectomy has become a new standard of care in a carefully selected subgroup of patients [6, 7]. Despite these recent advances, the vast majority of stroke patients are not eligible for acute stroke interventions due to factors such as delayed presentation to a healthcare setting, lack of access, or presence of exclusion criteria. Beyond the acute period, physical rehabilitation is the cornerstone of stroke management however the benefits of physical rehabilitation taper over time, and existing functional deficits typically plateau by 6 months after stroke onset [8, 9]. At the present time, there are no FDA-approved neuro-restorative treatments for chronic stroke.

Cell-based therapy is a promising neuro-restorative option for chronic neurologic disability associated with stroke. Preclinical and early clinical studies have reported improvements in functional outcome in subjects treated with experimental stem cell-based therapies.

2 Cell-Based Therapies for Stroke

Broadly speaking, cell-based therapies involve the implantation of cellular material into a recipient to effect beneficial changes in patient sequelae. Multiple stem cell types have been studied as cell-based therapies for ischemic stroke these are broadly classified into neural and non-neural stem cell subgroups. Neural stem cells (NSCs) have the inherent capacity to differentiate into neuronal and glial cells, and different NSC preparations give rise to varying proportions of neurons and glia. The specific types of glial cells that arise (astrocytes, oligodendrocytes and ependymal cells) also vary between NSC preparations [10]. NSCs have the theoretical advantage over non-NSCs, in that they can directly replace damaged neural and glial tissue, and several preclinical studies have demonstrated differentiation and engraftment of transplanted NSC-derived neurons and glia in stroke-damaged brain parenchyma. Kelly and colleagues demonstrated in 2004 that human fetal CNS-derived cells implanted into the ischemic rat brain survive and differentiate into site-appropriate neurons and glial cells, and that these implanted cells are capable of migrating long distances towards an ischemic lesion [11]. Similarly, Bühnemann and colleagues showed in 2006 that embryonic stem cell (ESC)-derived neural precursors implanted into infarcted territory in an experimental rat model of stroke can differentiate into mature glial cells, as well as diverse neuronal cell populations that form functional synaptic connectivity with host neurons [12]. In 2009, Daadi and colleagues further utilized MRI, bioluminescence, electrophysiology and immuno-electron microscopy to evaluate the survival and integration of ESC-derived human neural stem cells following experimental stroke in rats they demonstrated differentiation into functionally mature neurons (that expressed synaptic proteins, made synapses with endogenous neurons, and generated voltage dependent responses as well as spontaneous EPSCs), oligodendrocytes and astrocytes with prolonged survival of up to 2 months following implantation [13].

Multiple studies have also demonstrated improved neurologic function in experimental stroke models implanted with human NSCs [14, 15]. Interestingly, the studies have not consistently demonstrated an association between NSC engraftment/survival and the degree of functional recovery. For instance, Bliss and colleagues demonstrated in 2006 the robust survival and neurite extension from human neural precursor cells in animal models of stroke. This did not, however, correspond with an improvement in functional outcome [16]. This has led to the hypothesis that alternative mechanisms underlie functional improvements with cell-based therapies beyond the *direct* replacement of injured neural and glial tissue by the transplanted cells.

A variety of alternative mechanisms of therapeutic benefits derived from stem cell-based therapies have been reported. Intracarotid injection of bone marrow stromal cells (containing both mesenchymal stem and progenitor cells) in experimental stroke leads to axonal remodeling and remyelination in infarcted tissue [17]. Mesenchymal stem cell (MSC) treatments have also been shown to increase the levels of various neurotrophic factors (EGF, VEGF, bFGF, IGF-1, BDNF, GDNF) in animal stroke model brains [18, 19], and elevated levels of neurotrophic factors have been associated with increased cell proliferation and decreased apoptosis in the subventricular and subgranular zones [20]. Interestingly, axonal sprouting of neural tissue cultures in response to stem cell-based therapies can be halted by various neurotrophic factor-specific inhibitors [21]. These results suggest that enhanced neural plasticity, neurotrophic support, and endogenous neurogenesis all may underlie the therapeutic benefits observed with stem cell-based therapies. Numerous studies have also identified increased vascularity and neovascularization in and around the ischemic territory in stem cell-treated experimental stroke [17, 22, 23]. Stem cells have been shown to elicit an immunosuppressive effect on host tissue, thereby enhancing blood-brain barrier integrity and attenuating the host immune response to xenotransplants [24–26]. Thus the functional benefits of stem cell-based therapies in stroke are likely multi-modal and potentially involve changes in neural plasticity, neovascularization, neurotrophic support, endogenous neurogenesis, and immunomodulation the precise interplay of these mechanisms has not yet been fully elucidated. Nonetheless, these purported mechanisms provide the rationale for non-NSC transplantation.

A wide range of non-NSCs have been studied in experimental stroke including cells derived from bone marrow, umbilical cord blood, peripheral blood, and mesenchymal tissue. Bone marrow-derived MSCs are multipotent stem cells that are particularly suitable as cell-based therapies for stroke due to their: (1) ease of isolation, (2) immunomodulatory properties [27–29], (3) ability to selectively target infarcted brain parenchyma [30, 31], (4) well-documented safety profile in the clinical setting [32], and (5) avoidance of the ethical concerns associated with using embryonic and fetal stem cells. In a meta-analysis of experimental stroke studies, MSC therapy was consistently associated with improvements in multiple measures of behavioral function [33].

Mesenchymal stem cells have been evaluated in several clinical trials. A 2005 Phase 1/2 randomized controlled trial (RCT) of intravenously-administered autologous MSCs in subacute middle cerebral artery (MCA) infarcts revealed a statistically significant improvement in Barthel Index at 3 and 6 (but not at 12) months in addition there was a non-significant trend towards improved modified Rankin Scale (mRS) scores that decreased in magnitude at each successive time point. No adverse effects occurred in the MSC-treated cohort [34]. This same research group also evaluated the long-term (5-year follow up) safety and efficacy of intravenously-administered MSCs in a larger cohort of patients. They reported sustained safety, as well as a statistically significant improvement in mRS and a non-significant trend towards reduced mortality in the MSC-treated cohort compared with controls [35]. Another Phase 1/2 non-randomized trial of intra-arterially administered autologous MSCs in subacute stroke patients revealed a modest but non-significant trend towards improved mRS scores in cell-treated subjects when compared with controls. Notably, two of ten MSC-treated patients (20%) in this study experienced isolated simple partial seizures at 3 months post-transplantation [36]. Another group administered autologous MSCs to chronic stroke patients in two separate non-randomized observational trials. One of these trials noted a significant improvement in Barthel Index at 2 and 6 month follow ups, however the remainder of clinical outcome assessments were not significantly different between MSC-treated patients and controls during the course of the 6 month follow up. Of note, MSC-treated patients in both studies were reported to have increased activation of primary and supplementary motor cortex on post-transplantation functional MR imaging. No treatment-related adverse events occurred [37, 38]. In 2014, Aldagen Inc. announced the results of the small Phase 2 RECOVER-Stroke Trial (not published), in which a proprietary MSC (ALD-401) was administered intra-arterially to subacute stroke patients. There were no serious adverse events associated with this cell therapy, however the trial failed to demonstrate any significant change in the primary endpoint (mean mRS), nor any secondary endpoints, at 90-day follow up. Another Phase 2 RCT of intravenously-administered autologous MSCs in subacute stroke also failed to identify an improvement in any outcome measure including Barthel Index, mRS, NIHSS, or change in infarct volume at 180 days, however there were again, no adverse events associated with cell therapy in this trial [39]. In contrast, a 2015 Phase 1/2a non-randomized trial by another group also administered autologous MSCs intravenously to subacute stroke patients and noted statistically significant improvements in mRS, and a trend towards improved NIHSS and Barthel Index at discharge, as compared with a cohort of historical controls. Notably, this benefit was more pronounced in the cohort of patients treated with high-dose cell infusions, as compared with those treated with low-dose infusions. No adverse events occurred in MSC-treated patients [40]. The recently published MASTERS trial was a Phase 2 RCT in which allogeneic bone marrow-derived multipotent adult progenitor cells were intravenously-administered within 24–48 h of stroke onset. There was no difference in the primary safety endpoint (dose-limiting toxic events at 7 days after administration) or in the primary efficacy endpoint (global stroke recovery) between patients treated with 1.2×10^9 cells versus those treated with placebo. Notably, a post-hoc analysis identified improvements in multiple outcome measures among the subset of patients treated within 24–36 h [41]. A Phase 3 RCT is now being planned with a 24–36 h treatment window.

In summary, clinical trials of MSCs in ischemic stroke have consistently verified the safety and feasibility of MSC therapy in ischemic stroke however efficacy has not yet been reliably demonstrated [42]. It is important to note that each of the trials discussed here are early phase clinical trials designed to assess primarily safety and feasibility and not efficacy, which has typically been evaluated as a secondary outcome measure.

3 Optimizing Cell-Based Therapies for Stroke

Despite promising preclinical and early stage clinical trial data, numerous questions remain unanswered regarding the optimization of cell-based therapies for stroke patients. Existing preclinical and clinical investigations of cell-based stroke therapies have varied extensively with regard to cell type, cell dose, and timing and route of cell administration, and the ideal combination of these variables remains largely unknown. The variety of NSCs and non-NSCs that have been investigated in experimental and clinical stroke was described above. There have been no comparative studies to suggest superiority of any particular cell type to date.

When designing any cell based stroke therapy, a critical component to optimize is the cell dose itself. A recent preclinical study noted significantly reduced efficacy with low dose transplantation (5×10^5 cells), and a significantly higher incidence of adverse outcomes and mortality with high dose transplantation (5×10^7 cells). Based on this, an intermediate dose of 5×10^6 cells was selected to optimize efficacy while minimizing cell-related complications [43]. Another preclinical study identified a positive correlation between cell dose and efficacy, with improved behavioral outcomes noted with doses of 1×10^6–10^7 cells, but not with 1×10^4–10^5 cells [44]. In contrast, a meta-analysis of experimental stroke studies noted an inverse correlation between MSC dose and behavioral outcome [33]. As noted previously, a Phase 1/2a clinical trial identified a dose-dependent effect of intravenously-administered autologous MSCs in patients with subacute ischemic stroke [40] however, this was a small non-randomized open-label study that used a matched cohort of historical controls.

Timing of cell dose administration is also an important variable for optimizing cell-based therapies for stroke. Several preclinical studies have noted superior results on a range of outcome assessments when cell therapy is administered early after stroke onset. For instance in one preclinical study, cell administration 48 h after stroke onset was associated with a significant improvement in cell survival compared with treatments administered at 6 weeks. Notably, there was no significant difference in the extent of neuronal differentiation or distance of grafted cell migration noted between the two therapeutic windows [45]. Another preclinical study identified marked reductions in infarct volume and behavioral deficits when MSCs were administered 3 h after stroke onset, and this remained statistically significant when cells were administered 24 h post-stroke. These benefits were not observed when MSCs were administered 7 days after stroke onset [44]. A meta-analysis of preclinical studies identified a larger effect size on behavioral outcomes when MSCs were administered early (i.e. 0–8 h) after stroke onset [33]. Importantly however, no head-to-head clinical studies have been conducted that directly compare the range of possible therapeutic time windows. It is important to note that cell therapy at different time points following stroke (e.g. acute vs. subacute vs. chronic) may have fundamentally different therapeutic benefits. It is possible that administration during the acute period may serve to minimize primary and secondary injury from stroke, for example, whereas subacute or chronic delivery of cell-based therapeutics is more likely to provide a neuro-restorative benefit.

In the clinical stroke setting, the most commonly studied routes of cell-based therapy delivery are intravenous, intra-arterial, and intracerebral [26]. Direct intracerebral administration implants cells directly into the desired target, and thereby theoretically maximizes the therapeutic benefit while potentially minimizing untoward systemic effects. There is conflicting data as to whether intracerebral administration offers additional therapeutic benefit over intravenous or intra-arterial systemic delivery [46, 47]. Notably, a meta-analysis of preclinical studies demonstrated the greatest effect size with intracerebral implantation of cell-based therapies [33]. It should be noted, however, that significant improvements in behavioral outcomes were noted with all routes of delivery, and that intracerebral administration techniques are inherently invasive.

4 Using Modified Mesenchymal Stem Cells for Stroke: The SB623 Experience

SB623 cells originate from bone marrow-derived MSCs developed by SanBio, Inc. (Mountain View, CA, USA) as an allogeneic cell therapy for stroke. These cells are modified by transfection with a plasmid vector containing the human *Notch1*-intracellular domain, followed by the administration of specific trophic factors. This process induces MSCs to differentiate in a highly specific manner into post-mitotic neurons [48]. *In vitro* studies have demonstrated distinct differences of SB623 cells over unmodified MSCs, which may have important implications for SB623 cells as a cell-based therapy for stroke. For instance, embryonic rat brain neurons exhibit a substantial increase in metabolic activity when grown on SB623-cell derived extracellular matrix (ECM), as compared with cells grown on unmodified MSC-derived ECM [49]. SB623 cells have also been shown to induce more potent expression of neural and glial markers than parental MSCs [50]. *In vitro* assays of angiogenesis demonstrate enhanced vasculogenesis and a significant increase in vessel branching in the presence of SB623-derived conditioned medium, as compared with unmodified MSC-derived medium [51]. In addition, *in vitro* studies have demonstrated that SB623 cells have immunosuppressive and neurotrophic supportive properties that are equal or superior to unmodified MSCs [52, 53]. In a preclinical study, implantation of SB623 cells into the striatum of experimental stroke rats one month post-injury resulted in significant improvements in locomotor and neurologic function, and significant reductions in peri-infarct cell loss [54].

Steinberg and colleagues [55] recently published the 12-month interim data from a 2-year, open-label single-arm study (ClinicalTrials.gov Identifier: NCT01287936) designed to evaluate the safety, feasibility, and clinical outcomes of the stereotactic intracerebral implantation of SB623 cells in patients with stable, chronic stroke-related motor deficits. These results demonstrated that SB623 cell therapy is generally safe and well-tolerated, and associated with significant improvements in chronic neurologic deficits at both 6 months and 1 year. With regard to the primary outcome measure (European Stroke Scale, ESS), treated patients in this study demonstrated a statistically significant improvement from baseline at 6 months ($P < 0.01$), and this improvement remained significant at 12-month follow up ($P < 0.001$). Treated patients also had improvements in various secondary outcome measures, including NIHSS score at 12 months ($P < 0.001$) and Fugl-Meyer Assessment (FMA) FMA motor score at 12 months ($P < 0.001$). Notably, mRS was not significantly improved at any follow up time point. On radiographic follow up, 72% of patients demonstrated new T2-weighted-fluid-attenuated inversion recovery (T2 FLAIR) hyperintensity on MRI 1 week after cell transplantation, which was primarily in or adjacent to the premotor cortex and along the cannula tract. The size of this hyperintense signal ranged from 0.5–9.2 cm^2 to 0.6–3.5 cm^2 in maximal diameter, and was notably diffusion-weighted imaging (DWI)-negative in all patients. This FLAIR signal was not present on the day 1 post-transplant MRI, and resolved in all patients by the month 1 or 2 MRI. There were significant Pearson correlations between the size of the FLAIR signal hyperintensity and the change from baseline on all clinical outcome measures (ESS +0.82, $P < 0.001$ NIHSS −0.69, $P < 0.01$ FMA total score +0.71, $P < 0.01$ FMA motor score +0.67, $P < 0.01$).

With regard to safety, all patients experienced at least one treatment-emergent adverse event (TEAE), however these were typically only mild to moderate in severity the vast majority were determined by study personnel to be related to the surgical transplantation of the cell-based treatment, rather than to the actual SB623 cells. There was no significant correlation between SB623 cell dose and the incidence of any individual TEAE. Post-operative headache was the most frequent procedure-associated TEAE, occurring in 78% of patients. There were six serious AEs (defined as requiring additional hospitalization) in six patients. Specifically, one patient was noted to have an asymptomatic subdural fluid collection during the post-procedure period, which was effectively treated by burr-hole drainage. One patient had a seizure on day 70, which was determined to be unrelated to the cell treatment, but probably related to the procedure. One patient underwent stenting of asymptomatic carotid artery stenosis on day 291, which was unrelated to either the cell treatment or the procedure. One patient suffered a transient ischemic attack on day 334, which was also determined to be unrelated to either the cell treatment or the surgical procedure on the basis that it occurred 11 months after treatment administration. Finally, one patient developed a urinary tract infection and another developed pneumonia during the peri-operative period. All serious AEs were determined to be recovered or resolved by the time of the 12-month interim analysis. There were no clinically relevant changes in an extensive range of hematologic, biochemical or metabolic laboratory parameters that were monitored throughout the 12-month follow up course, nor were there any clinically significant changes in vital signs during this period. No humoral sensitization to SB623 cells was identified in any patient. It is important to recognize that this was a small, open-label, single arm Phase I/IIa trial with several limitations these included a small sample size, non-blinded non-randomized design, and highly selective eligibility criteria. Thus, the results of this trial should be interpreted in context, and as such are not necessarily applicable to the general stroke population. Nonetheless, the safety and feasibility of intracerebral implantation of SB623 cells in this trial, in combination with the neurologic improvement seen on multiple well-validated stroke outcome scales, have laid the foundation for a larger multicenter, randomized, controlled Phase 2b clinical trial of SB623 cell therapy for stroke patients that is currently underway.

5 Future Directions

Tremendous progress has been made over the past decade with regard to designing effective cell-based therapies for stroke. At this time, the results of early-stage clinical trials are just becoming available. The results to date have consistently demonstrated the safety and feasibility of utilizing stem cells in stroke patients, and preliminary data suggests that such cell-based therapies could be most promising for improving neurologic function in this patient population. Nevertheless, there is still much to be learned from further preclinical *in vivo* and *in vitro* investigation, as well as further early stage clinical evaluation of cell-based therapies for stroke. Future studies should be aimed at defining the mechanisms of therapeutic benefit of cell-based therapies for stroke. Defining the optimal combination of cell type, dose, timing and route of delivery is also critical to provide maximum therapeutic benefit. These data will be essential for informing the design of later stage clinical trials.

Acknowledgments We thank Christine Plant for assistance with the manuscript.

Funding: This work was supported in part by funding from Bernard and Ronni Lacroute, the William Randolph Hearst Foundation and Marc Paskin.

References

1. Lozano R, Naghavi M, Foreman K, Lim S, Shibuya K, Aboyans V, et al. Global and regional mortality from 235 causes of death for 20 age groups in 1990 and 2010: a systematic analysis for the Global Burden of Disease Study 2010. Lancet. 2012 Dec;380(9859):2095–128.
2. Roger VL, Go AS, Lloyd-Jones DM, Benjamin EJ, Berry JD, Borden WB, et al. Executive summary: heart disease and stroke statistics—2012 update: a report from the American Heart Association. Circulation. 2012;3125(1):188–97.
3. Writing Group Members, Mozaffarian D, Benjamin EJ, Go AS, Arnett DK, Blaha MJ, et al. Heart disease and stroke statistics-2016 update: a report from the American Heart Association. Circulation. 2016;26133(4):e38–60.
4. Ma VY, Chan L, Carruthers KJ. Incidence, prevalence, costs, and impact on disability of common conditions requiring rehabilitation in the United States: stroke, spinal cord injury, traumatic brain injury, multiple sclerosis, osteoarthritis, rheumatoid arthritis, limb loss, and back pain. Arch Phys Med Rehabil. 2014 May;95(5):986–995.e1.
5. National Institute of Neurological Disorders and Stroke rt-PA Stroke Study Group. Tissue plasminogen activator for acute ischemic stroke. N Engl J Med. 1995;14333(24):1581–7.
6. Jayaraman MV, Hussain MS, Abruzzo T, Albani B, Albuquerque FC, Alexander MJ, et al. Embolectomy for stroke with emergent large vessel occlusion (ELVO): report of the Standards and Guidelines Committee of the Society of NeuroInterventional Surgery: Table 1. J Neurointerv Surg. 2015;137(5):316–21.
7. Powers WJ, Derdeyn CP, Biller J, Coffey CS, Hoh BL, Jauch EC, et al. 2015 American Heart Association/American Stroke Association focused update of the 2013 guidelines for the early management of patients with acute ischemic stroke regarding endovascular treatment. Stroke. 2015;2846(10):3020–35.
8. Pollock A, Baer G, Campbell P, Choo PL, Forster A, Morris J, et al. Physical rehabilitation approaches for the recovery of function and mobility following stroke. Chichester, UK: Wiley; 1996.
9. Kwakkel G, van Peppen R, Wagenaar RC, Wood Dauphinee S, Richards C, Ashburn A, et al. Effects of augmented exercise therapy time after stroke: a meta-analysis. Stroke. 2004;2835(11):2529–39.
10. Bliss T, Guzman R, Daadi M, Steinberg GK. Cell transplantation therapy for stroke. Stroke. 2007 Feb;38(2 Suppl):817–26.
11. Kelly S, Bliss TM, Shah AK, Sun GH, Ma M, Foo WC, et al. Transplanted human fetal neural stem cells survive, migrate, and differentiate in ischemic rat cerebral cortex. Proc Natl Acad Sci U S A. 2004;10101(32):11839–44.

12. Bühnemann C, Scholz A, Bernreuther C, Malik CY, Braun H, Schachner M, et al. Neuronal differentiation of transplanted embryonic stem cell-derived precursors in stroke lesions of adult rats. Brain. 2006 Dec;129(Pt 12):3238–48.
13. Daadi MM, Li Z, Arac A, Grueter BA, Sofilos M, Malenka RC, et al. Molecular and magnetic resonance imaging of human embryonic stem cell-derived neural stem cell grafts in ischemic rat brain. Mol Ther. 2009 Jul;17(7):1282–91.
14. Ishibashi S, Sakaguchi M, Kuroiwa T, Yamasaki M, Kanemura Y, Shizuko I, et al. Human neural stem/progenitor cells, expanded in long-term neurosphere culture, promote functional recovery after focal ischemia in Mongolian gerbils. J Neurosci Res. 2004;78(2):215–23.
15. Chu K, Kim M, Park K-I, Jeong S-W, Park H-K, Jung K-H, et al. Human neural stem cells improve sensorimotor deficits in the adult rat brain with experimental focal ischemia. Brain Res. 2004 Aug;1016(2):145–53.
16. Bliss TM, Kelly S, Shah AK, Foo WC, Kohli P, Stokes C, et al. Transplantation of hNT neurons into the ischemic cortex: cell survival and effect on sensorimotor behavior. J Neurosci Res. 2006;183(6):1004–14.
17. Shen LH, Li Y, Chen J, Zhang J, Vanguri P, Borneman J, et al. Intracarotid transplantation of bone marrow stromal cells increases axon-myelin remodeling after stroke. Neuroscience. 2006;137(2):393–9.
18. Wakabayashi K, Nagai A, Sheikh AM, Shiota Y, Narantuya D, Watanabe T, et al. Transplantation of human mesenchymal stem cells promotes functional improvement and increased expression of neurotrophic factors in a rat focal cerebral ischemia model. J Neurosci Res. 2009;88(5):1017–25.
19. Kurozumi K, Nakamura K, Tamiya T, Kawano Y, Ishii K, Kobune M, et al. Mesenchymal stem cells that produce neurotrophic factors reduce ischemic damage in the rat middle cerebral artery occlusion model. Mol Ther. 2005 Jan;11(1):96–104.
20. Bao X, Wei J, Feng M, Lu S, Li G, Dou W, et al. Transplantation of human bone marrow-derived mesenchymal stem cells promotes behavioral recovery and endogenous neurogenesis after cerebral ischemia in rats. Brain Res. 2011 Jan;1367:103–13.
21. Lladó J, Haenggeli C, Maragakis NJ, Snyder EY, Rothstein JD. Neural stem cells protect against glutamate-induced excitotoxicity and promote survival of injured motor neurons through the secretion of neurotrophic factors. Mol Cell Neurosci. 2004 Nov;27(3):322–31.
22. Zhang ZG, Zhang L, Jiang Q, Chopp M. Bone marrow-derived endothelial progenitor cells participate in cerebral neovascularization after focal cerebral ischemia in the adult mouse. Circ Res. 2002;2290(3):284–8.
23. Jiang Q, Zhang ZG, Ding GL, Zhang L, Ewing JR, Wang L, et al. Investigation of neural progenitor cell induced angiogenesis after embolic stroke in rat using MRI. NeuroImage. 2005;1528(3):698–707.
24. Horie N, Pereira MP, Niizuma K, Sun G, Keren-Gill H, Encarnacion A, et al. Transplanted stem cell-secreted vascular endothelial growth factor effects poststroke recovery, inflammation, and vascular repair. Stem Cells. 2011 Feb;29(2):274–85.
25. Modo M, Rezaie P, Heuschling P, Patel S, Male DK, Hodges H. Transplantation of neural stem cells in a rat model of stroke: assessment of short-term graft survival and acute host immunological response. Brain Res. 2002 Dec;958(1):70–82.
26. Bliss T, Guzman R, Daadi M, Steinberg GK. Cell transplantation therapy for stroke. Stroke. 2007;2938(2):817–26.
27. Niemeyer P, Krause U, Kasten P, Kreuz P, Henle P, Sudkamp N, et al. Mesenchymal stem cell-based HLA-independent cell therapy for tissue engineering of bone and cartilage. Curr Stem Cell Res Ther. 2006;11(1):21–7.
28. Jones BJ, McTaggart SJ. Immunosuppression by mesenchymal stromal cells: from culture to clinic. Exp Hematol. 2008 Jun;36(6):733–41.
29. Ankrum JA, Ong JF, Karp JM. Mesenchymal stem cells: immune evasive, not immune privileged. Nat Biotechnol. 2014 Mar;32(3):252–60.
30. Ji JF, He BP, Dheen ST, Tay SSW. Interactions of chemokines and chemokine receptors mediate the migration of mesenchymal stem cells to the impaired site in the brain after hypoglossal nerve injury. Stem Cells. 2004 May;22(3):415–27.

31. Satake K, Lou J, Lenke LG. Migration of mesenchymal stem cells through cerebrospinal fluid into injured spinal cord tissue. Spine. 2004;29(18):1971–9.
32. Lalu MM, McIntyre L, Pugliese C, Fergusson D, Winston BW, Marshall JC, et al. Safety of cell therapy with mesenchymal stromal cells (SafeCell): a systematic review and meta-analysis of clinical trials. PLoS One. 2012 Oct;257(10):e47559.
33. Vu Q, Xie K, Eckert M, Zhao W, Cramer SC. Meta-analysis of preclinical studies of mesenchymal stromal cells for ischemic stroke. Neurology. 2014;882(14):1277–86.
34. Bang OY, Lee JS, Lee PH, Lee G. Autologous mesenchymal stem cell transplantation in stroke patients. Ann Neurol. 2005 Jun;57(6):874–82.
35. Lee JS, Hong JM, Moon GJ, Lee PH, Ahn YH, Bang OY. A long-term follow-up study of intravenous autologous mesenchymal stem cell transplantation in patients with ischemic stroke. Stem Cells. 2010;128(6):1099–106.
36. Moniche F, Gonzalez A, Gonzalez-Marcos JR, Carmona M, Pinero P, Espigado I, et al. Intra-arterial bone marrow mononuclear cells in ischemic stroke: a pilot clinical trial. Stroke. 2012;2343(8):2242–4.
37. Bhasin A, Srivastava MVP, Kumaran SS, Mohanty S, Bhatia R, Bose S, et al. Autologous mesenchymal stem cells in chronic stroke. Cerebrovasc Dis Extra. 2011;1(1):93–104.
38. Bhasin A, Srivastava M, Bhatia R, Mohanty S, Kumaran S, Bose S. Autologous intravenous mononuclear stem cell therapy in chronic ischemic stroke. J Stem Cells Regen Med. 2012;8(3):181–9.
39. Prasad K, Sharma A, Garg A, Mohanty S, Bhatnagar S, Johri S, et al. Intravenous autologous bone marrow mononuclear stem cell therapy for ischemic stroke: a multicentric, randomized trial. Stroke. 2014 Dec;45(12):3618–24.
40. Taguchi A, Sakai C, Soma T, Kasahara Y, Stern DM, Kajimoto K, et al. Intravenous autologous bone marrow mononuclear cell transplantation for stroke: phase 1/2a clinical trial in a homogeneous group of stroke patients. Stem Cells Dev. 2015 Oct;24(19):2207–18.
41. Hess DC, Wechsler LR, Clark WM, Savitz SI, Ford GA, Chiu D, et al. Safety and efficacy of multipotent adult progenitor cells in acute ischaemic stroke (MASTERS): a randomised, double-blind, placebo-controlled, phase 2 trial. Lancet Neurol. 2017 May;16(5):360–8.
42. Wang Q, Duan F, Wang M-X, Wang X-D, Liu P, Ma L-Z. Effect of stem cell-based therapy for ischemic stroke treatment: a meta-analysis. Clin Neurol Neurosurg. 2016 Jul;146:1–11.
43. Yang Z, Cai X, Xu A, Xu F, Liang Q. Bone marrow stromal cell transplantation through tail vein injection promotes angiogenesis and vascular endothelial growth factor expression in cerebral infarct area in rats. Cytotherapy. 2015 Sep;17(9):1200–12.
44. Wang L-Q, Lin Z-Z, Zhang H-X, Shao B, Xiao L, Jiang H-G, et al. Timing and dose regimens of marrow mesenchymal stem cell transplantation affect the outcomes and neuroinflammatory response after ischemic stroke. CNS Neurosci Ther. 2014;720(4):317–26.
45. Darsalia V, Allison SJ, Cusulin C, Monni E, Kuzdas D, Kallur T, et al. Cell number and timing of transplantation determine survival of human neural stem cell grafts in stroke-damaged rat brain. J Cereb Blood Flow Metab. 2011 Jan;31(1):235 42.
46. Modo M, Stroemer RP, Tang E, Patel S, Hodges H. Effects of implantation site of stem cell grafts on behavioral recovery from stroke damage. Stroke. 2002 Sep;33(9):2270–8.
47. Jin K, Sun Y, Xie L, Mao XO, Childs J, Peel A, et al. Comparison of ischemia-directed migration of neural precursor cells after intrastriatal, intraventricular, or intravenous transplantation in the rat. Neurobiol Dis. 2005 Mar;18(2):366–74.
48. Dezawa M, Kanno H, Hoshino M, Cho H, Matsumoto N, Itokazu Y, et al. Specific induction of neuronal cells from bone marrow stromal cells and application for autologous transplantation. J Clin Investig. 2004;15113(12):1701–10.
49. Aizman I, Tate CC, McGrogan M, Case CC. Extracellular matrix produced by bone marrow stromal cells and by their derivative, SB623 cells, supports neural cell growth. J Neurosci Res. 2009;187(14):3198–206.
50. Aizman I, Tirumalashetty BJ, McGrogan M, Case CC. Comparison of the neuropoietic activity of gene-modified versus parental mesenchymal stromal cells and the identification of soluble and extracellular matrix-related neuropoietic mediators. Stem Cell Res Ther. 2014;265(1):29.

51. Dao M, Tate CC, McGrogan M, Case CC. Comparing the angiogenic potency of naïve marrow stromal cells and Notch-transfected marrow stromal cells. J Transl Med. 2013;2711(1):81.
52. Tate CC, Fonck C, McGrogan M, Case CC. Human mesenchymal stromal cells and their derivative, SB623 cells, rescue neural cells via trophic support following in vitro ischemia. Cell Transplant. 2010;19(8):973–84.
53. Dao MA, Tate CC, Aizman I, McGrogan M, Case CC. Comparing the immunosuppressive potency of naïve marrow stromal cells and Notch-transfected marrow stromal cells. J Neuroinflammation. 2011;78(1):133.
54. Yasuhara T, Matsukawa N, Hara K, Maki M, Ali MM, Yu SJ, et al. Notch-induced rat and human bone marrow stromal cell grafts reduce ischemic cell loss and ameliorate behavioral deficits in chronic stroke animals. Stem Cells Dev. 2009 Dec;18(10):1501–14.
55. Steinberg GK, Kondziolka D, Wechsler LR, Lunsford LD, Coburn ML, Billigen JB, et al. Clinical outcomes of transplanted modified bone marrow-derived mesenchymal stem cells in stroke: a phase 1/2a study. Stroke. 2016 Jul;47(7):1817–24.

Chapter 15
Preparing for Future Stem Cell Clinical Trials

Keith W. Muir

Abstract Clinical trials in stem cell therapy for stroke have predominantly been small, single-centre and safety focused studies, few with blinding or concurrent control groups, and typically with wide treatment time windows and clinical entry criteria. Only recently have trials begun to consider the evolving preclinical evidence base and strategies that might translate this successfully into clinical use. The next few years will witness clinical trials that are likely to establish whether or not there is worthwhile therapeutic potential.

The accumulated experimental evidence has led to two distinct paradigms for cell therapy in stroke. In the first, systemically administered cells are delivered in the acute or early subacute phase, with a mechanism of action that is likely to be predominantly reliant on anti-inflammatory and trophic effects. With intravascular delivery, cells do not enter the central nervous system (CNS) in any significant numbers, if at all, and neither CNS nor systemic engraftment has been established. This approach reflects the great majority of experimental studies. Its likely translational route replicates established acute stroke trial paradigms. Trial designs in this area have had the advantage of evolution since the 1990s such that inclusion and exclusion criteria are well understood, as are trial endpoints.

The second paradigm is of later stage cell delivery to enhance recovery in subacute or chronic stroke. The experimental support is thinner, there being few animal models of this scenario, and there are fewer clinical trials in this time frame from which to draw designs. On the other hand, this represents a huge area of unmet clinical need lacking any very effective intervention.

Both paradigms are being addressed by currently planned or ongoing clinical trials of cell therapy. This chapter will review the main issues that require to be considered.

Keywords Clinical trials • Stem cells • Trial design • Placebo

K.W. Muir, M.D., F.R.C.P (✉)
Institute of Neuroscience and Psychology, University of Glasgow, Glasgow, G12 8QQ, UK

Queen Elizabeth University Hospital, Glasgow, G51 4TF, UK
e-mail: Keith.muir@glasgow.ac.uk

P.A. Lapchak, J.H. Zhang (eds.), *Cellular and Molecular Approaches to Regeneration and Repair*, Springer Series in Translational Stroke Research,
https://doi.org/10.1007/978-3-319-66679-2_15

Abbreviations

ARAT	Action research arm test
BI	Barthel index
BMMCs	Bone marrow-derived mononuclear cells
CNS	Central nervous system
MCAO	Middle cerebral artery occlusion
mRS	Modified Rankin Scale
NIHSS	National Institutes of Health Stroke Scale
NINDS	National Institutes of Neurological Disorders
PISCES-1	Preliminary investigation of stem cell effects in stroke
STEPS	Stem cells as an emerging paradigm in stroke
TMS	Transcortical magnetic stimulation
VISTA	Virtual international stroke trials archive

1 Introduction

Clinical trial design for stem cell therapy evaluation has been considered in a series of workshops involving academia, industry and regulators to produce recommendations for the translation of this potential treatment modality, the Stem Cells as an Emerging Paradigm in Stroke (STEPS) meetings [1–3]. The content and concerns of the four STEPS meetings (three reported and published, the report of a fourth being prepared at the time of writing) reflect the evolution of the field from a focus on animal models through to early phase and now later phase clinical research, and the developing experimental data in the field. Issues relating to clinical trial design and progress in the cell therapy field have additionally exercised the thinking of many investigators [4–8]. In particular, thinking has evolved from the initial concept of cell or tissue replacement as the dominant biological effect mediator, to an appreciation that much (or indeed all) of the therapeutic benefits from cell therapy might be mediated by indirect effects [6], and most recently that only certain components of cells might be necessary for a therapeutic action [9].

2 Acute Cell Therapy

It is recognised that cells delivered within the first hours after induction of cerebral ischaemia in animal models limit the extent of brain injury, and are associated with improved functional outcome [10–12]. Cells of various types and delivered by various routes have been reported to be effective, but some concerns regarding methodological quality and reporting bias have been identified [10]. More specific focus on studies of intravenous delivery of bone marrow-derived cells in rodent middle cerebral artery occlusion (MCAO) has yielded a more consistent evidence base [12]. Typical time to administration has been prior to induction of ischaemia or within a few hours of ischaemia.

Biodistribution studies in mouse hypoxia and focal ischaemia indicated limited distribution of mouse neural stem cells to the brain [13]: intravenously delivered cells were cleared almost entirely within 1 week, while intra-arterial delivery produced higher concentrations in brain, but almost complete clearance within 2 weeks. In rat MCAO, intravenous bone marrow stromal cells were distributed predominantly to the lungs, with transient distribution to the brain, and negligible cell persistence even by 8 days [14]. In another rat MCAO study, no preferential brain uptake in MCAO rats compared to controls was seen, and again distribution was predominantly to lung [15]. In human studies, intravenously delivered bone marrow-derived mononuclear cells (BMMCs) were distributed predominantly to lungs and spleen by 2 and 24 h after injection [16]. It therefore seems unlikely that there will be significant distribution of cells to the CNS after intravascular administration, and any engraftment is likely to be very limited. With respect to outcomes, differences between intravenous and intra-arterial routes have not been observed consistently in animal studies [17]. Therapeutic effects have nonetheless been seen with most cell types in animal models of stroke, therefore it is assumed that "bystander" effects reliant on release of trophic factors and immune modulation underpin the action of stem cells delivered in this manner, and there is experimental evidence of multiple systemic effects of potential relevance. Systemic engraftment of some cell types in the lung may allow more sustained action.

Within the first hours after ischaemic stroke, intravascular delivery of a therapeutic agent is the most feasible option: intravenous delivery is straightforward, and intra-arterial delivery is increasingly an option with the wider use of endovascular thrombectomy. Intravenous delivery is clearly more practical since almost universally available, and possible at any stage after stroke.

Intra-arterial delivery requires appropriately skilled personnel and facilities, and while thrombectomy is becoming more widely available, endovascular treatment is indicated and possible in only a minority of patients, and speed of intervention is critical. It is logistically challenging to add intra-arterial cell delivery to an emergency procedure such as thrombectomy, and it would be entirely impractical to infuse autologous cells due to the requirements for cell harvest and preparation. In addition, animal studies of intra-arterial cell delivery have been complicated by cell clumping and downstream arterial occlusion [18] that produced significant complication rates potentially outweighing any advantage of improved CNS retention of cells, and requiring modification of injection rates to overcome [19, 20]. The practicality of intra-arterial delivery in the later subacute period is also questionable—the logistics and safety of intra-arterial procedures potentially requiring anaesthesia outside the acute time window for thrombectomy are challenging, and targeted delivery may be impossible due to persistent occlusion of the relevant artery. Very slow recruitment was evident in one pilot clinical study of subacute intra-arterial cell delivery, with only 6% of screened patients proving to be suitable and many exclusions due to medical instability and major arterial occlusion precluding vascular access [21]. Whether such an approach is advantageous is unclear: while the evidence above suggests that this may allow some degree of cell delivery to the brain, it is uncertain whether this leads to long term engraftment, or whether this is necessary for a therapeutic effect [22].

Several small clinical studies using intravenously delivered cells have reported findings, but with broad patient inclusion criteria and a wide time window [23–27]. The acute use of autologous cells is extremely challenging since invasive harvest procedures are likely to be more hazardous in the face of recent thrombolytic and antithrombotic drug treatments, and neurological impairments will impair cooperation by many patients with significant stroke deficits. Two moderately large controlled clinical trials have reported early subacute use, one using autologous bone marrow cells and the other allogeneic cells [28, 29].

The phase 2 Indian multi-centre trial of Prasad and colleagues [28] had a time window of 7–30 days, treating patients with autologous bone marrow mononuclear cells infused intravenously. Randomisation to cell infusion or control was undertaken, and assessors for the outcome measures were blinded to treatment allocation. The control arm patients underwent no invasive procedures—neither cell harvest from bone marrow, nor intravenous infusion.

The MASTERS trial [29] of the Athersys "multistem" allogeneic bone marrow-derived multipotent adult progenitor cells aimed to treat patients 24–36 h after stroke onset, but requirements for cell processing in specialist facilities caused the trial to expand the time window to 48 h for practical reasons. Eligible patients were randomised to either cell infusion (400 or 1200 million cells) or placebo. The global results showed no difference in outcomes compared with control, although subgroup analysis of those treated within 36 h of onset was interpreted as supporting this earlier time window.

No safety issues were identified in these studies, and similar paradigms of intravenous infusion of cells have been deployed in other therapeutic areas including myocardial ischaemia and multiple sclerosis.

As outcome measures, these trials used well-characterised clinical scales that are familiar from other therapeutic modalities in the acute stroke setting: general measures of neurological function (the National Institutes of Health Stroke Scale [NIHSS]) [30, 31], of activities of daily living (the Barthel Index [BI]) [32] and of disability (the modified Rankin Scale [mRS]) [33, 34]. These scales are advantageous in being well understood, widely used in routine practice as well as trial settings, acceptable to regulators, and being applicable to all potential patients irrespective of stroke mechanism, location or size. For an acutely delivered therapy, especially within the first 36 h, detailed patient selection is not practical. An inclusive approach to trial design also improves generalisability of results and would allow widespread adoption of treatments if efficacy is established. The properties of the scales are well understood, and specific training is available for NIHSS and mRS to minimise inter-observer variability. Standardised structured interviews and centralised video interpretation can be used to further reduce mRS variability, with the additional possibility of ensuring blinded independent outcome scoring.

Dosing can be reasonably controlled for allogeneic cell therapies, but is highly variable in autologous cell delivery [25, 27, 28]. The population of cells present in bone marrow mononuclear cell aspirates is also mixed, with a range of cell types including haematopoietic progenitor cells and mesenchymal stem cells, and unless additional ex-vivo culture expansion is undertaken (with the additional time incurred in this step delaying any potential therapeutic use), it is extremely difficult to characterise the cells that are actually delivered.

3 Chronic Stroke Cell Therapy

Early investigation of teratocarcinoma-derived neural cells [35–37] or porcine xenografts [37] established the feasibility and basic methodological approach for delivery of cells by direct cerebral implantation. Subsequent trials have adapted these methods for cell delivery [38]. Chronic stroke—arbitrarily proposed to be 6 months or more after the ictus—is a huge therapeutic need, several million people worldwide living with long-term neurological disability as a consequence of stroke and having high costs of social and medical care [39]. Current medical interventions are limited to secondary prevention of further events, prevention or treatment of physical complications such as spasticity or mechanical joint disruption, and behavioural adaptations to deficits. The challenge is uncertainty about whether there is useful plasticity in the injured brain at late stages after stroke [40], with very limited animal model data of uncertain relevance to this time point in human stroke.

Chronic, stable patients allow trials to adopt more careful selection, planning of procedures, and targeting of cell delivery. Intracerebral implantation of cells ensures that cells can be delivered in a defined dose to a specific location. While the mechanism of cell action in chronic stroke is not fully characterised, it more plausibly includes engraftment and integration than does intravascular administration, in addition to local anti-inflammatory, immunomodulatory and trophic effects, including stimulation of endogenous neurogenesis and angiogenesis [41–43]. What proportion of cells survive and engraft long-term is unknown, as is the differentiation fate of these cells.

Two trials using human stem cells have reported recently, the Preliminary Investigation of Stem Cell Effects in Stroke (PISCES-1) trial, and the SanBio phase 1 trial. Each included small numbers of patients with no control group, in order to establish safety and tolerability of different doses.

PISCES-1 [44] used ReNeuron's human foetal cortical neural stem cell line genetically modified with the c-myc$^{ER\text{-}TAM}$ transgene to allow large-scale cell production for allogeneic cell therapy, CTX0E03 [45–47]. The study followed a similar paradigm to that studied in rodent MCAO models of "chronic" stroke, with intraputaminal implantation of doses of up to 20 million cells. In PISCES-1, 11 patients received doses of between 2 and 20 million cells, and no cell-related safety issues were identified over a 2 year follow-up period.

The SanBio trial [48] included 18 subjects administered up to ten million cells placed stereotactically in small deposits around the margin of the infarct, using genetically modified human bone marrow-derived mesenchymal stem cells. Both studies included patients 6–60 months after ischaemic stroke, with a median time to treatment of around 2.5 years. Some improvements in neurological and functional scales over the first 3 months after implantation were described, with static function thereafter. The time scale for neurological change was unexpected and not clearly consistent with cell differentiation and replacement as the major mode of action.

For phase 2 studies, trials of these agents have opted to target patients with motor deficits specifically, and employed more specific neurological scales focusing on motor function as their primary end-points. The PISCES-2 study completed recruitment in summer 2016 and reported preliminary favourable early functional improvements sufficient to justify further clinical trial development (http://4965zs3ha21125fk78zkozo3.wpengine.netdna-cdn.com/wp-content/uploads/ReNeuron-PISCES-II-data.pdf, accessed 7 July 2017). PISCES-2 selected a population of subacute stroke patients 3–12 months after onset, with major upper limb dysfunction, and its primary endpoint was recovery of useful upper limb function defined on the Action Research Arm Test (ARAT) [49].

4 Trial Design

Rigorous standards of trial design and interpretation are key to the credibility and ultimate adoption into practice of clinical trial results.

4.1 Controls

Clinical investigations of stem cells have hitherto justified small studies without concurrent controls or blinding on grounds of practicality and patient acceptability, but controlled trials will be necessary to advance clinical practice. Randomisation, placebo controls, and blinding are critical to the integrity of trials but the invasive nature of cell harvest for autologous cell preparation and of several delivery approaches means that some compromise may be required. As noted above in relation to the experience of acute intravenous cell delivery trials, optimal design is feasible for an allogeneic cell approach such as that of the Athersys multipotent adult progenitor cell studies, but blinding becomes problematic when autologous cell harvest is required such as in the Prasad trials. For more invasive intracerebral approaches, ethical issues become significant. The acceptability of placebo surgical procedures is debated. While investigators and regulators encourage this approach in recognition of the importance of placebo effects, [50, 51] patient groups have questioned it [52]. Opinions derived from patient experience in other diseases may not be applicable in stroke: in a neurodegenerative process such as Parkinson's Disease, deferred cell implantation has been considered to be an acceptable offer for those trial participants randomised to the control arm (assuming that test implantation is established to be effective). In stroke, however, the injury is acute and the mechanisms by which cell therapy might be effective are heavily weighted towards the early subacute period, so deferred treatment cannot be reasonably assumed to have the same potential value as early treatment.

Placebo surgery is deemed to be acceptable and is preferred by some regulators. What constitutes an appropriate balance between placebo procedures and reasonable risk is a grey area. The phase 2 SanBio trial uses a partial thickness Burr hole under local anaesthesia, without dural incision or any intraparenchymal injection, allowing a similar procedural duration and blinding the participant and trial team (outside the operating theatre environment). Other trials are likely to adopt the same approach. This has potential advantages, but two alternatives merit consideration. First, the potential harmful effect of surgery cannot be assessed by this approach. Surgery may have negative consequences, including local infection, pain or bleeding, adverse effects of sedative medication for the procedure, and of temporary cessation of preventative antithrombotic medication, as well as many more if procedures are undertaken under general anaesthesia (as was the case in the PISCES trials, for example). An alternative design would be randomisation to surgical implantation or to a non-surgical control group, allowing evaluation of the net effect (both benefit and potential harm) of the procedure, but with the disadvantage of lacking control for a placebo effect.

A second alternative would be a more invasive approach, where the control group undergoes identical intraparenchymal injection of vehicle solution. This would offer additional blinding (except possibly for the surgeon), would also control for potential non-specific (positive) effects of injecting a volume of fluid to introduce a focal lesion in the brain, and allows investigators to distinguish specifically cell-related adverse effects from those of the procedure: for example, do the T2 hyperintensities seen around needle tracts in both the SanBio and PISCES trials represent a specific tissue reaction to cell implantation, or a non-specific reaction to vehicle, or some other aspect of the process? This more invasive approach has scientific merit but is likely to be deemed to expose the control group to unacceptable risk.

It is important to frame discussions with patients in terms that recognise the experimental nature of stem cell administration, since there is a widespread assumption that benefit is expected, or indeed inevitable. Uncritical reporting of early phase clinical trial findings (for example http://www.dailymail.co.uk/health/article-3622589/Major-breakthrough-doctors-REVERSE-symptoms-stroke-Patients-walk-talk-live-normal-life-stem-cell-treatment-3-YEARS-later.html, accessed 7 July 2017) and the widespread unregulated online advertising of supposed "stem cell therapy" clinics contribute to this environment. The potential for harm (including very limited long-term safety data for most cell types) is one component of a complex discussion [53].

4.2 Sample Size Estimation and Endpoints

In the acute setting, clinical trial design is informed by a large body of experience that will allow realistic estimates of credible effect sizes using the same general outcome scales as have been deployed in other trials. The mRS benefits from development over many years to minimise subjective inter-observer variability in scoring, including rater training, independent video assessment [54] and structured interviews [55–58]. Large databases such as the Virtual International Stroke Trials Archive (VISTA) can be interrogated to model the impact of inclusion and exclusion criteria on expected outcomes in the control population [59]. Statistical methods have been developed to maximise study power (or reduce sample size), particularly through analysis of the entire distribution of the mRS rather than arbitrary dichotomous outcomes [60, 61]. The use of a broad disability scale as a primary endpoint has the advantage of applicability to all types of stroke deficit: inclusion and exclusion criteria can therefore be less restrictive, with advantages for recruitment rates and generalisability.

Sample size is likely to be moderate or large for a credible range of treatment effects, particularly with subacute interventions. While there is an analogy with acute trials in terms of design, a less appealing analogy is the failure of all clinical trials in acute stroke other than those involving reperfusion, a source of much soul-searching in the neuroprotectant field, among other notable translational failures [62, 63]. Many of the deficiencies that were postulated to underpin the failure of neuroprotectant drug trials might also apply to cell therapies: overestimates of effect size leading to trials that were too small, insufficient phenotypic detail to select relevant patients with biological targets, and a tendency to seek "responder populations" by tortured analysis of small phase 2 trials and their subgroups with consequent restrictive and (with hindsight) misdirected patient selection criteria in repeat phase 2b/3 trials. Nonetheless, there is now a track record of success for these general approaches in acute stroke through both thrombolytic drugs and thrombectomy, and the design features are recognised by clinicians and regulators.

Trials in chronic stroke face potentially greater challenges as the methodology is less standardised, largely a reflection of the absence of positive clinical trials in the rehabilitation and regeneration field [64]. Trials have elected to target what is effectively a human model system, motor deficits of upper, or both upper and lower, limb. Similar approaches have been taken in trials of rehabilitation strategies such as constraint-induced motor therapy [65]. While there is logic in the approach, since motor deficits are common after stroke and scales for motor function assessment are available, potential difficulties are illustrated by previous motor rehabilitation studies, which have been characterised by slow recruitment through the need to target patient populations with very specific deficits, and high drop-out rates. Despite enthusiasm for specific motor function scales such as the Fugl-Meyer scale among specialists in rehabilitation [66, 67], the scale is less widely understood among stroke physicians and its acceptability to regulators is unclear. Familiarity with more specialised scales such as the ARAT is less still, and specific equipment and training are required to perform the assessment. There is also uncertainty over the extent of change on ARAT that is meaningful to patients [68].

Motor function change represents only one component of a complex multi-dimensional deficit. Whilst emphasis has been placed on inclusion of adjunctive physical therapy to minimise between-site variation in multicentre trials, defining the minimum necessary duration and intensity of therapy input has proved to be challenging [69], and the content of therapy programmes may vary widely. Physical therapy represents only one of several therapy inputs that a patient is likely to receive, and even description of rehabilitative inputs has proved challenging, let alone quantification.

Combination of several different outcome scales has been proposed in order to better capture the multi-dimensional nature of stroke recovery [70], and has been an effective strategy in some acute trials such as the National Institutes of Neurological Disorders and Stroke (NINDS) trial of thrombolysis [71]. This statistical approach may strengthen findings when there is a common direction of effect, but may weaken study power if effects diverge.

Sample size ultimately depends on the variance of the outcome measure and the magnitude of the effect of the intervention. An uncomfortable reality for the field is that the costs, invasive nature of the interventions, and complexity of cell supply, are all likely to place practical constraints on trial size in cell therapies. The magnitude of the effect is difficult to estimate, and cannot be extrapolated reliably from animal studies where cell delivery has typically been much earlier than has been attempted (or is likely achievable) in the clinic. The schematic representation of different processes contributing to brain injury after stroke as a series of waves of differing size, latency and duration [72] is a useful conceptual framework, which recognises the dominant effect of very early cell necrosis due to severe ischaemia. Imaging studies support the very short time window during which the greatest part of an ischaemic lesion becomes damaged irreversibly. Thereafter there may be smaller contributions to the final infarct from late processes such as inflammation and apoptosis, but these appear to be minor contributors to the physical extent of an infarct. Clinically it has been difficult to demonstrate either delayed infarct growth (other than oedema), or any measurable neurological consequence of this, although there may be a relationship [73]. Interventions delivered at 36 h after stroke onset and likely to have a predominantly anti-inflammatory action may thus be anticipated to have a small treatment effect and to require very large sample sizes for convincing demonstration of any therapeutic action. The failure of other anti-inflammatory treatment strategies [74–78] may reflect ineffective drugs, late delivery, or trials that were too small, but equally it may signify that this mechanism has insufficient impact on tissue fate at late stage to be detected with anything other than an extremely large trial.

4.3 Patient Selection

Sample size is greatly inflated if trials include patients with no relevant biological target for a therapeutic intervention [79]. Recent acute trials have highlighted that modest sample sizes can produce highly significant and persuasive results when the relevant target population is selected—for example with selection of MCA occlusion and appropriate volumes of viable tissue on perfusion imaging, for both intravenous thrombolysis and for thrombectomy [80–82]. The relevant clinical phenotype has not been clearly defined for acute systemic cell therapy, but demonstration of a worthwhile clinical effect is likely to follow similar principles to other acute stroke trials. In chronic stroke trials based on motor recovery paradigms, variable outcomes are well recognised and are not reliably predictable from clinical scales alone [83], especially for the more severe deficits that are likely to be over-represented in any clinical trial of an invasive therapy. Selection of patients with potential motor response might be possible using combinations of clinical scales, brain imaging (for example to define the integrity of the corticospinal tract) and excitability of the motor system using transcortical magnetic stimulation (TMS) or similar [84–87]. Refining practical methods for patient selection that could be applied consistently across multiple clinical centres in a trial is a challenge: very complex and time-consuming imaging analysis may be difficult and may not be feasible for all scanners [88], and methods such as TMS are poorly standardised and not widely available. The adverse consequences of insufficiently informed patient selection are, however, clear, and can be documented both by head-counts of neutral and failed trials, and by the huge accompanying cost to the academic and pharmaceutical industry communities [89]. In addition, insufficient levels of phenotypic detail will impede any attempt to better define a potential "responder" population for future studies. Inclusion of imaging at least sufficient to offer an opportunity to enhance mechanistic understanding would be advantageous both for individual trials and for the field as a whole. In an analogous setting where clinical scales offered limited insight, multiple sclerosis, the identification of an imaging biomarker was the key that unlocked the door to disease modifying therapy development, by enabling clinical trials of reasonable size to be undertaken. Stem cell therapies would benefit from the same approach.

4.4 *Time Windows*

The nature of the intervention places constraints on cell supply for trials, as noted above, and this introduces additional limits on time windows. Experimental models of stroke indicate a complex and dynamic environment after ischaemia, and biological targets change both over time and anatomically. Underlying disease states may modify cell therapy effects and are rarely investigated in model systems. Patients undergo complex rehabilitation interventions, receive a variety of secondary preventative medications, and are prone to complications that reflect physical disabilities as well as systemic effects of stroke such as immunodepression, all of which might influence recovery patterns. To deviate far from the experimental evidence underpinning a particular cell therapy adds another confounding factor, and it would appear logical to limit at least this element that can be under the control of the trial investigators. Acute intervention in animal models has been overwhelmingly delivered in early acute stages; "chronic" intervention at most a few weeks after stroke in rodents. Thrombolytic therapy for stroke could not show benefit when delivered an average of 4.5 h after onset [90]. It was only when the NINDS trial adhered rigidly to the narrower time window of 3 h based on animal model data (and insisted on even earlier treatment in 50% of participants by design) that benefit was eventually evident [91]. We may ignore the lessons of animal model time windows at our peril.

5 Summary and Conclusions

Two distinct paradigms for cell therapy in stroke are under investigation.

Acute systemic administration of cells follows a pathway reasonably well informed by other acute trials, but has to consider what might be a credible effect size and an appropriate time window for delivery. Allogeneic cells appear to have an advantage over autologous cells in this setting. Only limited patient selection is feasible, but generalisability of results is straightforward.

Chronic stroke allows for more targeted intracerebral administration of cells, but the invasive nature of the procedure places constraints on trial design and size, and the human model of motor system recovery is less standardised in the absence of successful clinical trial interventions. Patient selection, employing complex imaging and other advanced methods, and development of biomarkers will likely be critical to success.

References

1. Stem Cell Therapies as an Emerging Paradigm in Stroke P. Stem cell therapies as an emerging paradigm in stroke (STEPS): bridging basic and clinical science for cellular and neurogenic factor therapy in treating stroke. Stroke. 2009;40(2):510–5.
2. Savitz SI, Chopp M, Deans R, Carmichael T, Phinney D, Wechsler L, et al. Stem cell therapy as an emerging paradigm for stroke (STEPS) II. Stroke. 2011;42(3):825–9.
3. Savitz SI, Cramer SC, Wechsler L, Aronowski J, Boltze J, Borlongan C, et al. Stem cells as an emerging paradigm in stroke 3 enhancing the development of clinical trials. Stroke. 2014;45(2):634–9.
4. Bliss TM, Andres RH, Steinberg GK. Optimizing the success of cell transplantation therapy for stroke. Neurobiol Dis. 2010;37(2):275–83.
5. Savitz SI. Developing cellular therapies for stroke. Stroke. 2015;46(7):2026–31.
6. Janowski M, Wagner DC, Boltze J. Stem cell-based tissue replacement after stroke: factual necessity or notorious fiction? Stroke. 2015;46(8):2354–63.
7. Dobkin BH, Carmichael ST. The specific requirements of neural repair trials for stroke. Neurorehabil Neural Repair. 2015;30(5):470–8.
8. Muir KW. Clinical trial design for stem cell therapies in stroke: what have we learned? Neurochem Int. 2017;106:108–13.
9. Zhang Y, Chopp M, Zhang ZG, Katakowski M, Xin H, Qu C, et al. Systemic administration of cell-free exosomes generated by human bone marrow derived mesenchymal stem cells cultured under 2D and 3D conditions improves functional recovery in rats after traumatic brain injury. Neurochem Int. 2016;pii:S0197-0186(16)30251-0.

10. Lees JS, Sena ES, Egan KJ, Antonic A, Koblar SA, Howells DW, et al. Stem cell-based therapy for experimental stroke: a systematic review and meta-analysis. Int J Stroke. 2012;7(7):582–8.
11. Vu Q, Xie K, Eckert M, Zhao W, Cramer SC. Meta-analysis of preclinical studies of mesenchymal stromal cells for ischemic stroke. Neurology. 2014;82(14):1277–86.
12. Vahidy FS, Rahbar MH, Zhu H, Rowan PJ, Bambhroliya AB, Savitz SI. Systematic review and meta-analysis of bone marrow-derived mononuclear cells in animal models of ischemic stroke. Stroke. 2016;47(6):1632–9.
13. Pendharkar AV, Chua JY, Andres RH, Wang N, Gaeta X, Wang H, et al. Biodistribution of neural stem cells after intravascular therapy for hypoxic-ischemia. Stroke. 2010;41(9):2064–70.
14. Goldmacher GV, Nasser R, Lee DY, Yigit S, Rosenwasser R, Iacovitti L. Tracking transplanted bone marrow stem cells and their effects in the rat MCAO stroke model. PLoS One. 2013;8(3):e60049.
15. Detante O, Moisan A, Dimastromatteo J, Richard MJ, Riou L, Grillon E, et al. Intravenous administration of 99mTc-HMPAO-labeled human mesenchymal stem cells after stroke: in vivo imaging and biodistribution. Cell Transplant. 2009;18(12):1369–79.
16. Rosado-de-Castro PH, Schmidt Fda R, Battistella V, Lopes de Souza SA, Gutfilen B, Goldenberg RC, et al. Biodistribution of bone marrow mononuclear cells after intra-arterial or intravenous transplantation in subacute stroke patients. Regen Med. 2013;8(2):145–55.
17. Yang B, Migliati E, Parsha K, Schaar K, Xi X, Aronowski J, et al. Intra-arterial delivery is not superior to intravenous delivery of autologous bone marrow mononuclear cells in acute ischemic stroke. Stroke. 2013;44(12):3463–72.
18. Li L, Jiang Q, Ding GL, Zhang L, Zhang ZG, Li QJ, et al. Effects of administration route on migration and distribution of neural progenitor cells transplanted into rats with focal cerebral ischemia, an MRI study. J Cereb Blood Flow Metab. 2010;30(3):653–62.
19. Janowski M, Lyczek A, Engels C, Xu J, Lukomska B, Bulte JW, et al. Cell size and velocity of injection are major determinants of the safety of intracarotid stem cell transplantation. J Cereb Blood Flow Metab. 2013;33(6):921–7.
20. Cui LL, Kerkela E, Bakreen A, Nitzsche F, Andrzejewska A, Nowakowski A, et al. The cerebral embolism evoked by intra-arterial delivery of allogeneic bone marrow mesenchymal stem cells in rats is related to cell dose and infusion velocity. Stem Cell Res Ther. 2015;6:11.
21. Banerjee S, Bentley P, Hamady M, Marley S, Davis J, Shlebak A, et al. Intra-arterial immunoselected CD34+ stem cells for acute ischemic stroke. Stem Cells Transl Med. 2014;3(11):1322–30.
22. Rodriguez-Frutos B, Otero-Ortega L, Gutierrez-Fernandez M, Fuentes B, Ramos-Cejudo J, Diez-Tejedor E. Stem cell therapy and administration routes after stroke. Transl Stroke Res. 2016;7(5):378–87.
23. Bang OY, Lee JS, Lee PH, Lee G. Autologous mesenchymal stem cell transplantation in stroke patients. Ann Neurol. 2005;57(6):874–82.
24. Lee JS, Hong JM, Moon GJ, Lee PH, Ahn YH, Bang OY, et al. A long-term follow-up study of intravenous autologous mesenchymal stem cell transplantation in patients with ischemic stroke. Stem Cells. 2010;28(6):1099–106.
25. Honmou O, Houkin K, Matsunaga T, Niitsu Y, Ishiai S, Onodera R, et al. Intravenous administration of auto serum-expanded autologous mesenchymal stem cells in stroke. Brain. 2011;134(Pt 6):1790–807.
26. Savitz SI, Misra V, Kasam M, Juneja H, Cox CS Jr, Alderman S, et al. Intravenous autologous bone marrow mononuclear cells for ischemic stroke. Ann Neurol. 2011;70(1):59–69.
27. Prasad K, Mohanty S, Bhatia R, Srivastava MV, Garg A, Srivastava A, et al. Autologous intravenous bone marrow mononuclear cell therapy for patients with subacute ischaemic stroke: a pilot study. Indian J Med Res. 2012;136(2):221–8.
28. Prasad K, Sharma A, Garg A, Mohanty S, Bhatnagar S, Johri S, et al. Intravenous autologous bone marrow mononuclear stem cell therapy for ischemic stroke: a multicentric, randomized trial. Stroke. 2014;45(12):3618–24.

29. Hess DC, Wechsler LR, Clark WM, Savitz SI, Ford GA, Chiu D, et al. Safety and efficacy of multipotent adult progenitor cells in acute ischaemic stroke (MASTERS): a randomised, double-blind, placebo-controlled, phase 2 trial. Lancet Neurol. 2017;16(5):360–8.
30. Brott T, Adams HP, Olinger CP, Marler JR, Barsan WG, Biller J, et al. Measurements of acute cerebral infarction: a clinical examination scale. Stroke. 1989;20(7):864–70.
31. Goldstein LB, Bertels C, Davis J. Interrater reliability of the NIH stroke scale. Arch Neurol. 1989;46:660–2.
32. Mahoney FI, Barthel DW. Functional evaluation: the Barthel index. Md State Med J. 1965;14:61–5.
33. Rankin J. Cerebral vascular accidents in patients over the age of 60. 2: Prognosis. Scott Med J. 1957;2:200–15.
34. van Swieten JC, Koudstaal PJ, Visser MC, Schouten HJ, van Gijn J. Interobserver agreement for the assessment of handicap in stroke patients. Stroke. 1988;19:604–7.
35. Kondziolka D, Wechsler L, Goldstein S, Meltzer C, Thulborn KR, Gebel J, et al. Transplantation of cultured human neuronal cells for patients with stroke. Neurology. 2000;55(4):565–9.
36. Kondziolka D, Steinberg GK, Wechsler L, Meltzer CC, Elder E, Gebel J, et al. Neurotransplantation for patients with subcortical motor stroke: a phase 2 randomized trial. J Neurosurg. 2005;103(1):38–45.
37. Savitz SI, Dinsmore J, Wu J, Henderson GV, Stieg P, Caplan LR. Neurotransplantation of fetal porcine cells in patients with basal ganglia infarcts: a preliminary safety and feasibility study. Cerebrovasc Dis. 2005;20(2):101–7.
38. Muir KW, Sinden J, Miljan E, Dunn L. Intracranial delivery of stem cells. Transl Stroke Res. 2011;2(3):266–71.
39. Feigin VL, Forouzanfar MH, Krishnamurthi R, Mensah GA, Connor M, Bennett DA, et al. Global and regional burden of stroke during 1990–2010: findings from the Global Burden of Disease Study 2010. Lancet. 2014;383(9913):245–54.
40. Carmichael ST. Emergent properties of neural repair: elemental biology to therapeutic concepts. Ann Neurol. 2016;79(6):895–906.
41. Hassani Z, O'Reilly J, Pearse Y, Stroemer P, Tang E, Sinden J, et al. Human neural progenitor cell engraftment increases neurogenesis and microglial recruitment in the brain of rats with stroke. PLoS One. 2012;7(11):e50444.
42. Hicks C, Stevanato L, Stroemer RP, Tang E, Richardson S, Sinden JD. In vivo and in vitro characterization of the angiogenic effect of CTX0E03 human neural stem cells. Cell Transplant. 2013;22(9):1541–52.
43. Katare R, Stroemer P, Hicks C, Stevanato L, Patel S, Corteling R, et al. Clinical-grade human neural stem cells promote reparative neovascularization in mouse models of hindlimb ischemia. Arterioscler Thromb Vasc Biol. 2014;34(2):408–18.
44. Kalladka D, Sinden J, Pollock K, Haig C, McLean J, Smith W, et al. Human neural stem cells in patients with chronic ischaemic stroke (PISCES): a phase 1, first-in-man study. Lancet. 2016;388(10046):787–96.
45. Stroemer P, Patel S, Hope A, Oliveira C, Pollock K, Sinden J. The neural stem cell line CTX0E03 promotes behavioral recovery and endogenous neurogenesis after experimental stroke in a dose-dependent fashion. Neurorehabil Neural Repair. 2009;23(9):895–909.
46. Thomas RJ, Hope AD, Hourd P, Baradez M, Miljan EA, Sinden JD, et al. Automated, serum-free production of CTX0E03: a therapeutic clinical grade human neural stem cell line. Biotechnol Lett. 2009;31(8):1167–72.
47. Sinden JD, Hicks C, Stroemer P, Vishnubhatla I, Corteling RL. Human neural stem cell therapy for chronic ischemic stroke: charting progress from laboratory to patients. Stem Cells Dev. 2017;26(13):933–47.
48. Steinberg GK, Kondziolka D, Wechsler LR, Lunsford LD, Coburn ML, Billigen JB, et al. Clinical outcomes of transplanted modified bone marrow-derived mesenchymal stem cells in stroke: a phase 1/2a study. Stroke. 2016;47(7):1817–24.

49. Lang CE, Wagner JM, Dromerick AW, Edwards DF. Measurement of upper-extremity function early after stroke: properties of the action research arm test. Arch Phys Med Rehabil. 2006;87(12):1605–10.
50. Diederich NJ, Goetz CG. The placebo treatments in neurosciences: new insights from clinical and neuroimaging studies. Neurology. 2008;71(9):677–84.
51. George AJT, Collett C, Carr AJ, Holm S, Bale C, Burton S, et al. When should placebo surgery as a control in clinical trials be carried out? Bull R College Surg Engl. 2016;98(2):75–9.
52. Cohen PD, Isaacs T, Willocks P, Herman L, Stamford J, Riggare S, et al. Sham neurosurgical procedures: the patients' perspective. Lancet Neurol. 2012;11(12):1022.
53. Misra V, Hicks WJ, Vahidy F, Alderman S, Savitz SI. Recruiting patients with stroke into cell therapy trials: a review. JAMA Neurol. 2016;73(9):1141–4.
54. McArthur KS, Johnson PC, Quinn TJ, Higgins P, Langhorne P, Walters MR, et al. Improving the efficiency of stroke trials: feasibility and efficacy of group adjudication of functional end points. Stroke. 2013;44(12):3422–8.
55. Wilson J, Hareendran A, Grant M, Baird T, Schulz U, Muir K, et al. Improving the assessment of outcomes in stroke – use of a structured interview to assign grades on the modified Rankin Scale. Stroke. 2002;33(9):2243–6.
56. Wilson J, Hareendran A, Hendry A, Potter J, Bone I, Muir K. Reliability of the modified rankin scale across multiple raters – benefits of a structured interview. Stroke. 2005;36(4):777–81.
57. Quinn TJ, McArthur K, Dawson J, Walters MR, Lees KR. Reliability of structured modified rankin scale assessment. Stroke. 2010;41(12):e602. Author reply e3
58. Saver JL, Filip B, Hamilton S, Yanes A, Craig S, Cho M, et al. Improving the reliability of stroke disability grading in clinical trials and clinical practice: the rankin focused assessment (RFA). Stroke. 2010;41(5):992–5.
59. Abdul-Rahim AH, Fulton RL, Sucharew H, Kleindorfer D, Khatri P, Broderick JP, et al. National institutes of health stroke scale item profiles as predictor of patient outcome: external validation on independent trial data. Stroke. 2015;46(2):395–400.
60. Bath PMW, Lees KR, Schellinger PD, Altman H, Bland M, Hogg C, et al. Statistical analysis of the primary outcome in acute stroke trials. Stroke. 2012;43(4):1171–8.
61. Howard G, Waller JL, Voeks JH, Howard VJ, Jauch EC, Lees KR, et al. A simple, assumption-free, and clinically interpretable approach for analysis of modified Rankin outcomes. Stroke. 2012;43(3):664–9.
62. Grotta J. Why do all drugs work in animals but none in stroke patients? 2. Neuroprotective therapy. J Intern Med. 1995;237:89–94.
63. Muir K, Teal P. Why have neuroprotectants failed? Lessons learned from stroke trials. J Neurol. 2005;252(9):1011–20.
64. Langhorne P, Coupar F, Pollock A. Motor recovery after stroke: a systematic review. Lancet Neurol. 2009;8(8):741–54.
65. Wolf SL, Winstein CJ, Miller JP, Taub E, Uswatte G, Morris D, et al. Effect of constraint-induced movement therapy on upper extremity function 3 to 9 months after stroke: the EXCITE randomized clinical trial. J Am Med Assoc. 2006;296(17):2095–104.
66. Bushnell C, Bettger JP, Cockroft KM, Cramer SC, Edelen MO, Hanley D, et al. Chronic stroke outcome measures for motor function intervention trials: expert panel recommendations. Circ Cardiovasc Qual Outcomes. 2015;8(6 Suppl 3):S163–9.
67. Cramer SC, Wolf SL, Adams HP, Chen D, Dromerick AW, Dunning K, et al. Stroke recovery and rehabilitation research: issues, opportunities, and the National Institutes of Health StrokeNet. Stroke. 2017;48(3):813–9.
68. Lang CE, Edwards DF, Birkenmeier RL, Dromerick AW. Estimating minimal clinically important differences of upper-extremity measures early after stroke. Arch Phys Med Rehabil. 2008;89(9):1693–700.
69. Winstein CJ, Wolf SL, Dromerick AW, Lane CJ, Nelsen MA, Lewthwaite R, et al. Effect of a task-oriented rehabilitation program on upper extremity recovery following motor stroke: the ICARE randomized clinical trial. JAMA. 2016;315(6):571–81.

70. Hommel M, Detante O, Favre I, Touze E, Jaillard A. How to measure recovery? Revisiting concepts and methods for stroke studies. Transl Stroke Res. 2016;7(5):388–94.
71. Tilley BC, Marler J, Geller NL, Lu M, Legler J, Brott T, et al. Use of a global test for multiple outcomes in stroke trials with application to the National Institute of Neurological Disorders and Stroke t-PA stroke trial. Stroke. 1996;27:2136–42.
72. Dirnagl U, Iadecola C, Moskowitz MA. Pathobiology of ischaemic stroke: an integrated view. Trends Neurosci. 1999;22(9):391–7.
73. Alawneh JA, Jones PS, Mikkelsen IK, Cho TH, Siemonsen S, Mouridsen K, et al. Infarction of 'non-core-non-penumbral' tissue after stroke: multivariate modelling of clinical impact. Brain. 2011;134(Pt 6):1765–76.
74. Enlimomab Acute Stroke Trial Investigators. Use of anti-ICAM-1 therapy in ischemic stroke: results of the enlimomab acute stroke trial. Neurology. 2001;57(8):1428–34.
75. Krams M, Lees KR, Hacke W, Grieve AP, Orgogozo JM, Ford GA, et al. Acute stroke therapy by inhibition of neutrophils (ASTIN): an adaptive dose-response study of UK-279,276 in acute ischemic stroke. Stroke. 2003;34(11):2543–8.
76. Lees KR, Zivin JA, Ashwood T, Davalos A, Davis SM, Diener HC, et al. NXY-059 for acute ischemic stroke. N Engl J Med. 2006;354(6):588–600.
77. Shuaib A, Lees KR, Lyden P, Grotta J, Davalos A, Davis SM, et al. NXY-059 for the treatment of acute ischemic stroke. N Engl J Med. 2007;357(6):562–71.
78. Diener HC, Lees KR, Lyden P, Grotta J, Davalos A, Davis SM, et al. NXY-059 for the treatment of acute stroke: pooled analysis of the SAINT I and II Trials. Stroke. 2008;39(6):1751–8.
79. Muir K. Heterogeneity of stroke pathophysiology and neuroprotective clinical trial design. Stroke. 2002;33(6):1545–50.
80. Parsons MW, Spratt N, Bivard A, Campbell B, Chung K, Miteff F, et al. A randomised trial of tenecteplase versus alteplase for acute ischaemic stroke. N Engl J Med. 2012;366:1099–107.
81. Campbell BC, Mitchell PJ, Kleinig TJ, Dewey HM, Churilov L, Yassi N, et al. Endovascular therapy for ischemic stroke with perfusion-imaging selection. N Engl J Med. 2015;372(11):1009–18.
82. Bivard A, Huang X, McElduff P, Levi CR, Campbell BCV, Cheripelli BK, et al. Impact of computed tomography perfusion imaging on the response to tenecteplase in ischemic stroke analysis of 2 randomized controlled trials. Circulation. 2017;135(5):440.
83. Patel AT, Duncan PW, Lai SM, Studenski S. The relation between impairments and functional outcomes poststroke. Arch Phys Med Rehabil. 2000;81(10):1357–63.
84. Stinear C. Prediction of recovery of motor function after stroke. Lancet Neurol. 2010;9(12):1228–32.
85. Stinear CM, Barber PA, Petoe M, Anwar S, Byblow WD. The PREP algorithm predicts potential for upper limb recovery after stroke. Brain. 2012;135(Pt 8):2527–35.
86. Byblow WD, Stinear CM, Barber PA, Petoe MA, Ackerley SJ. Proportional recovery after stroke depends on corticomotor integrity. Ann Neurol. 2015;78(6):848–59.
87. Stinear CM, Byblow WD, Ackerley SJ, Smith MC, Borges VM, Barber PA. Proportional motor recovery after stroke: implications for trial design. Stroke. 2017;48(3):795–8.
88. Baron JC, Cohen LG, Cramer SC, Dobkin BH, Johansen-Berg H, Loubinoux I, et al. Neuroimaging in stroke recovery: a position paper from the first international workshop on neuroimaging and stroke recovery. Cerebrovasc Dis. 2004;18(3):260–7.
89. Kidwell CS, Liebeskind DS, Starkman S, Saver JL. Trends in acute ischemic stroke trials through the 20th century. Stroke. 2001;32(6):1349–59.
90. Hacke W, Kaste M, Fieschi C, Toni D, Lesaffre E, von Kummer R, et al. Intravenous thrombolysis with recombinant tissue plasminogen activator for acute hemispheric stroke. The European cooperative acute stroke study (ECASS). JAMA. 1995;274(13):1017–25.
91. The National Institute of Neurological DaSrSSG. Tissue plasminogen activator for acute ischemic stroke. N Engl J Med. 1995;333:1581–7.

Part II
Exosomes-miRNA

Chapter 16
Extra-Cellular Vesicles: A Promising Approach for Translating Cell-Based Therapy

Benjamin Buller, Michael Chopp, and Zheng Gang Zhang

Abstract As cell-based therapies have demonstrated efficacy in the treatment of experimental and clinical stroke, their mechanisms of action warrant intense investigation and are being investigated in greater depth. It is becoming increasingly clear that one of the main ways that cell therapies based on mesenchymal stem cells (MSC) and other cells impart functional benefits to animals is through release of exosomes and other extracellular vesicles *in vivo*. Mounting evidence shows that MSCs release exosomes, and that these exosomes induce predictable and impactful changes in recipient cells. These exosome-induced cellular changes are likely mediated through the content of the exosomes, which comprise mRNA, miRNA, proteins, and other macromolecules. Many studies that have been published in the last several years have shown that treatment of animals with exosomes, harvested from MSCs and other cells, after stroke and traumatic brain injury (TBI) recapitulate the effect of the parent cells. Exosomes lack the safety and manufacturability issues that plague cell therapy, and they therefore may represent the next generation of cell-free therapies. Their biology and potential use as therapies for CNS injuries are discussed herein.

Keywords MSC • Stroke • TBI • Exosome • Microvesicle • Extracellular vesicle • Neuroresoration

Abbreviations

Ago2	Argonaute-2
Ang1	Angiopoietin-1
BBB	Blood-brain barrier

B. Buller • Z.G. Zhang (✉)
Department of Neurology, Henry Ford Health System, Detroit, MI, USA
e-mail: zhazh@neuro.hfh.edu

M. Chopp
Department of Neurology, Henry Ford Health System, Detroit, MI, USA

Department of Physics, Oakland University, Rochester, MI, USA

P.A. Lapchak, J.H. Zhang (eds.), *Cellular and Molecular Approaches to Regeneration and Repair*, Springer Series in Translational Stroke Research,
https://doi.org/10.1007/978-3-319-66679-2_16

CSF	Cerebral-spinal fluid
dll4	Delta-like 4
EV	Extracellular vesicle
lncRNA	Long non-coding RNA
MCAo	Middle cerebral artery occlusion
miRNA	MicroRNA
MSC	Mesenchymal stem/stromal cell
MVB	Multivesicular body
RISC	RNA-induced silencing complex
TBI	Traumatic brain injury

1 Introduction

Cell-based therapy may be the most promising approach to achieving a clinically viable restorative therapy for stroke and other neurologic diseases. In a series of preclinical studies first published by our laboratory beginning in 2000, we showed that administration of bone marrow mesenchymal stromal cells (MSC), leads to enhanced neurological recovery in rats that are subjected to middle cerebral artery occlusion (MCAo) [1–4]. This work has been repeated and reproduced by many laboratories around the world [5–8], and other cell types have also been shown to be effective in aiding recovery from stroke [9–12]. Despite the acceptance of cell therapy as an effective treatment for animal models of stroke, the mechanisms by which it imparts functional recovery have remained elusive.

It was first hypothesized that MSC transdifferentiate into neural cells and thereby regrow brain tissue. This hypothesis was abandoned relatively quickly in favor of a paracrine hypothesis—that the cells secrete factors that stimulate growth [13]. This hypothesis too evoked skepticism, because treating animals with any number of secreted growth factors from cells has never reproduced the effect of treating with cells directly.

Extracellular vesicles (EV), including exosomes and microvesicles, are small, membrane bound spheroids of approximately 30–200 nm in diameter. They contain macromolecular cargo that includes receptors, ligands, and nucleic acids. The nucleic acid cargo of EVs comprises a mixture of mRNA, tRNA, vault RNA, microRNAs (miRNA), and long non-coding RNA (lncRNA). While the biologic functions of the majority of the contents of EVs remain unclear, their miRNA contents have been shown to be functional *in vivo*, which is particularly important given that many roles have been described for miRNAs in both the pathogenesis of, and recovery from, stroke (for review of the miRNA and stroke, see e.g. [14]). The role of exosomes and other EVs in mediating cell therapy repair, with particular attention paid to MSCs, and the putative mechanism by which they operate are discussed below.

2 EV Biogenesis Pathways

EVs comprise several subtypes, and they are secreted by virtually all cells [15]. They exist in all body fluids, including blood, cerebrospinal fluid (CSF), urine, ascites, and saliva [16–20]. The two most well studied types of EVs are exosomes and microvesicles. Each of these two types has unique attributes and a distinctive biogenesis pathway. Exosomes are spheroids that have an approximate size of 30–100 nm [21] that are generated by the endosomal pathway [22]. In this pathway, the cell membrane invaginates to form an endosome, and then successive invaginations of the endosome create a multivesicular body (MVB). Fusion of the MVB with the cell membrane releases exosomes to the extracellular space where they may dock locally or distally with other cells, or perhaps be taken up by endocytosis or macropinocytosis [23]. By contrast, microvesicles are thought to bud directly from the membrane; they have more amorphous shapes than exosomes, and they have a much larger average diameter, perhaps any size up to 2000 nm [24]. Figure 16.1 describes these separate biogenesis pathways.

Exosomes and microvesicles can be distinguished by their surface markers. Exosomes, uniform lipid bilayer spheroids, are generally marked by tetraspanin proteins including CD63, CD81, and CD9, as well as flotillin and Alix [25, 26]. Microvesicles, in contrast, generally lack tetraspanins and are of varying size, shape, and density [27]. Separating exosomes from microvesicles is difficult practically, as

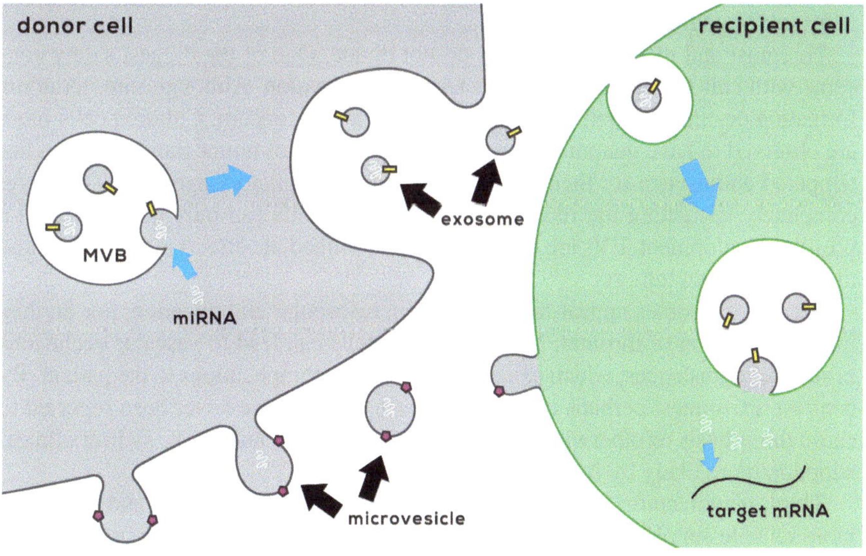

Fig. 16.1 Extracellular vesicles are shed by two primary mechanisms. In the endosomal pathway, exosomes are created by successive invaginations of the plasma membrane to create an MVB. The MVB then fuses with the plasma membrane to release the exosomes into the extracellular space. Microvesicles are created by direct budding of the membrane. EVs can be taken up by the recipient cell by direct membrane fusion or by endocytosis/macropinocytosis

their densities and sizes have a significant amount of overlap. However, separation of subpopulations of EVs may be more important from a technical scientific standpoint than a medical one. It is unclear which EV subtype, if any, contributes more to their therapeutic effects. Some studies have implicated exosomes [28], while others have implicated microvesicles [29] as the more important vesicle type in mediating the effects of parent cells. This debate may not be settled soon; however, exosomes are generally thought to be the more biologically relevant EV subtype, and the majority of the literature therefore focuses on exosomes, although it should be noted that much of the early work in EVs was clouded by a lack of consistent nomenclature to distinguish between the various subtypes of EVs.

3 Potential Therapeutic Applications

EVs have been shown to be therapeutic for many of the same neurologic diseases and injuries as have been demonstrated for their parent cells, including traumatic brain injuries and stroke [30, 31], and they are the only cell product that has been shown to recapitulate the therapeutic effect of the parent cells. EVs have only begun to be deployed clinically, so it is not entirely clear what their eventual impact may be; however, we can speculate that their adoption will be swift should they be shown to be as efficacious as parent cells in treatment of injury and disease. They have several inherent advantages over cells that make them ideal replacements for or adjuvants to cell-based therapy.

Foremost and most obvious, EVs do not divide. One of the biggest safety concerns with cell therapy is the risk of teratoma formation. Although rare, teratoma formation or other uncontrolled cell division is a real concern. Cultured cells often are observed to have genomic aberrations, and the risk of tumor transformation has tempered enthusiasm for their use, especially from regulatory agencies. Therefore, completely mitigating the risks posed by dividing cells can only be counted as a positive development. EVs are not cells, so their immediate effects are transient, and they cannot form tumors.

Second, exosomes appear to not cause microvascular embolization, nor do they induce formation of thrombi. MSC and other cells can lead to vascular occlusions in some circumstances, which can cause significant complications to the patient. By contrast, exosomes, perhaps owing to their small size, have never been reported to cause thrombosis or otherwise occlude vessels. This fact may make swifter clinical adoption more likely by further reducing risk.

Third, manufacture and delivery of EVs may prove to be simpler and produce a more reliable supply chain than cells. Cell therapy requires that cells be grown for each patient, and that they then be delivered intact and sterilely at the point of care. This requires the infrastructure to thaw and formulate the cells on site, or else a manufacturing facility in very close proximity. By contrast, EVs are stable at 4 °C for relatively long periods of time, with little detectable difference in the cargo of exosomes that were collected fresh or stored for several weeks [32]. This remarkable property makes central manufacturing and formulation much nearer to reality than could ever be possible with parent cells. For example, EVs could potentially be loaded into premade IV bags that could be stored on site at hospitals that serve stroke and TBI patients. Because of their relative stability, the product could be on site for immediate use, with a time buffer of potentially many weeks. Pre-formulation of a hypothetical exosome product obviates the need for experienced technicians to prepare treatments on site on a patient-by-patient basis, and allows for central quality control in a way that cell therapy does not.

The most pressing barrier to quick clinical adoption of EVs for treatment of stroke is their relatively short history compared to cell therapy. Some of the earliest investigations of EVs for treatment of any neurologic disease were published by our lab in 2013 [30]. These first reports demonstrated that MSC derived EVs could impart therapeutic benefits to rats after stroke when delivered at 24 h after MCAo, and that the functional recovery of these animals is caused by enhanced white matter remodeling, including new axon growth and myelination, as well as angiogenesis. MSCs have long been known to cause remodeling of neurites and angiogenesis [33–36], further evidence that MSCs enhance functional outcomes via release of exosomes and other EVs. This finding that MSC exosomes promote recovery after stroke has been reproduced and verified by several independent labs in rats and in mice [37, 38]. Furthermore, using human cells to generate exosomes does not impact their ability to enhance neurologic recovery in rats subjected to TBI [39]. The above renders it likely that exosome-induced functional recovery after neurologic injury is generalizable across multiple species, and thus also likely applicable to human disease.

To date, the therapeutic potential of EVs derived from MSCs has been investigated most extensively preclinically. MSCs are a robust source of exosomes and other EVs, producing an abundance of them compared to other cell types [40]. However, the majority of cell types produce exosomes and microvesicles, and several of these cell types have been explored as potential sources of therapeutic EVs. For example, exosomal miR-126 is pro-angiogenic [41], and may underlie human cord blood cell mediated recovery from stroke in diabetic animals [42]; endothelial cell derived exosomes have been used to treat hindlimb ischemia [43]; and dendritic and other immune cell derived exosomes are being explored extensively as a therapy for cancer [44–48].

Most of the clinical work focused on exosomes has been dedicated to their potential as biomarkers (for review see e.g., [49, 50]). Despite their short history as a therapeutic agent, exosomes have begun to appear in clinical trials. Table 16.1 is a list of all current registered trials on clinicaltrials.gov for which 'exosome' is a keyword and that are targeting therapy and not biomarkers. The range of diseases is diverse, and only one so far uses MSC as a source. However, this is likely to change rapidly in the coming years.

Table 16.1 List of trials using exosomes as a therapeutic

Identifier	Institution	Disease	Source	Phase
NCT02565264	Kumamoto University	Cutaneous wound healing/ulcers	Plasma	I
NCT02138331	General Committee of Teaching Hospitals and Institutes, Egypt	Type I diabetes mellitus	MSC	II/III
NCT01159288	Gustave Roussy, Cancer Campus, Grand Paris	Non-small cell lung cancer	Dendritic cells	II
NCT01668849	James Graham Brown Cancer Center, University of Louisville	Chemoradiation-induced oral mucositis	Grape	I
NCT01294072	James Graham Brown Cancer Center, University of Louisville	Colon cancer	Curcumin-loaded plant	I

3.1 *Mechanism*

It is apparent that among the most important cargo that exosomes carry are miRNAs. miRNAs are often highly conserved across disparate organisms, and although they are frequently gained during evolution, they are rarely lost [51]. The number of miRNA that a species possesses correlates well with morphologic complexity [52], and any given miRNA may target many genes in a single gene network, thereby possessing the ability to efficiently shut down redundant systems [53]. More than 700 miRNAs can be detected in exosomes and other EVs [54], and they are mostly bound to Argonaute 2 (Ago2) [55, 56], a major constituent of the RNA-induced silencing complex (RISC). Silencing of targets by miRNA is RISC-dependent, so the fact that miRNA in exosomes are bound to Ago2 suggests that they are destined to bind to mRNAs in recipient cells (i.e. be functional). All this points to exosomes being a potent system to pass "information" from cell to cell in a manner that other macromolecules cannot.

It has been shown that miRNA expressed in one cell can suppress protein expression in another cell through innate mechanisms [57]. Although more than one pathway for targeted inhibition of translation from one cell to another may exist, exosomes represent a major mechanism by which this information transfer occurs. It has been shown in many studies across multiple independent labs that specific proteins can be suppressed in cells in a predictable way when the cells are incubated with exosomes containing targeting miRNA [57–60]. Therefore, the likeliest way that exosomes function is to release their miRNA contents into target cells upon being internalized, thereby affecting gene networks in the recipient cells. This hypothesis is supported by a number of studies in which the miRNA cargo of exosomes was altered to target specific genes. Xin et al. showed that over-expressing miR-133b in MSC exosomes enhances functional recovery after MCAo to an even greater extent that naïve exosomes [61, 62]. Additionally, miR-17-92 cluster expression can target neurons and promote axonal growth via suppression of PTEN [63], and exosomes enriched in miR-17~92 constituents promote functional recovery and axonal growth more efficiently than naïve exosomes [64, 65]. It is likely that in the future, better methods of expression and more predictive targeting algorithms will allow for even more refined tuning of the therapeutic properties of exosomes.

3.2 *Neurovascular Niche*

Recovery from stroke is dependent on remodeling of the neurovascular niche [66, 67]. Exosomes have been shown to affect multiple aspects of the neurovascular unit during recovery from stroke and brain injury, and in *in vitro* injury models. For example, when exposed to MSC exosomes, astrocytes are stimulated to release exosomes of their own, which in turn induce downstream remodeling of axons [68]. Indeed, exosome treatment after stroke is associated with improved axonal growth and myelination [30, 64, 68]. In an apparent feedback loop, neuronal exosomes also contain biomolecules that target astrocytes, including PTEN and miR-124, which limit astrocyte proliferation and increase expression of the amino acid transporter GLT-1, respectively [69]. Furthermore, oligodendrocytes secrete exosomes that impact neuronal behavior, helping to coordinate myelination [70, 71] and supplying protective molecules in stress conditions [72].

As the name suggests, the other half of the neurovascular unit comprises cerebral blood vessels, whose function after stroke is coupled to recovery of brain parenchyma, and exosomes from MSCs promote angiogenesis [30, 37, 73]. Endothelial cells communicate with each other via exosome secretion. For example, endothelial cell exosomes contain miR-214 and miR-126, both of which are pro-angiogenic miRNAs, and they also contain angiopoietin-1 (Ang1), the primary ligand of Tie2 receptor and a potent inducer of angiogenesis [43, 74]. Endothelial exosomes also contain delta-like 4 (dll4), a notch ligand that maintains endothelial stasis [75]. The exosome system may represent a way for a large and distributed tissue such as the endothelium to maintain homeostasis over a large surface area and long distances. Treatment of hindlimb ischemia with endothelial derived exosomes significantly improves recovery of function and angiogenic sprouting [76], suggesting that supporting the natural cell communication system in the endothelium could be a strategy for treating cardiovascular disease. Dysfunction of the natural endothelial cell exosome axis may lead to pathologic conditions that are prevalent in stroke and other cardiovascular disease, such as atherosclerosis. Endothelial cells from sclerotic vessels secrete exosomes with cargo that is distinct from healthy cells [77], which may trigger damage and recruitment of inflammatory cells.

The other relevant question in cell therapy with respect to the cerebral endothelium is whether MSC exosomes can cross the blood-brain barrier (BBB). Several lines of evidence suggest that they can. First, brain tumor exosomes can readily be detected in blood, which suggests crossage of the BBB [78]. Second, *in vitro* evidence shows that endothelial cells can actively transport exosomes across the BBB [79]. Although circumstantial, these reports provide clear evidence of instances in which exosomes can in fact cross the BBB intact. Due to their heterogeneous nature as aggregates of biomolecules that can be disaggregated *in vivo*, directly observing exosomes *in vivo* after injection is difficult, but it seems likely that therapeutic exosomes can enter the brain [80].

3.3 *Inflammatory System*

The inflammatory system, both in the brain and in the periphery, may mediate cell therapy after stroke [81]. When introduced IV, exosomes encounter macrophages and other immune cells of the periphery almost immediately, and macrophage depleted animals clear exosomes much slower than wild-type animals [82]. The half-life of injected exosomes in wild type rats may be as little as 2 min, with total clearance happening by 4 h [83]. The exact role of the peripheral immune system in mediating cell therapy has not been fully described, but some evidence suggests that its presence is necessary for enhancing recovery [84, 85]. Additionally, the majority of injected exosomes lodge in the peripheral organs, including the lungs, liver, and spleen [82, 83, 86], although these studies do not agree as to which organ is the primary point of exosome uptake, which may be cell source dependent. An open question for scientists who are developing exosome therapies is whether the interaction of exosomes with peripheral organs contributes to or inhibits their effectiveness as therapeutic agents. It may be that one of the ways in which exosomes from MSCs and other cells impart functional benefits is by "reprogramming" the immune system to behave in a way that supports recovery. Secreted vesicles from MSCs suppress secretion of pro-inflammatory cytokines from stimulated microglia *in vitro* [87]. In turn, secreted inflammatory factors from microglia, such as IL-1α and TNFα, stimulate astrocytes to suppress synapse formation [88], which may have serious deleterious effects on recovery. Indeed, microglia help coordinate tissue remodeling after injury, and can encourage oligodendrocyte differentiation and myelination during recovery [89–91]. Conversely, neuronal secreted exosomes can recruit microglia to prune synapses [92], which may be an innate mechanism for normal function, but also could potentially contribute to dysfunction in degenerative disease states, as aberrant synapse pruning is a hallmark of early Alzheimer's disease and other forms of dementia, for example [93, 94]. Therefore, the potential of MSC exosomes to reprogram microglia to adopt a pro-recovery phenotype is perhaps one of their greatest assets.

4 Summary and Conclusion

The use of cell therapies for recovery from stroke has gained prominence and traction in recent years due to their effectiveness in treating animal models of brain injury. Their use in clinical trials, of which hundreds now are registered, is therefore warranted, as no other regenerative or restorative treatment is available to patients.

However, exosomes are only now beginning to be investigated as a potential next generation replacement for or adjuvant to cell therapy, but awareness of them is rising quickly. Several clinical trials have been registered to investigate the use of exosomes for diseases such as cancer, wound healing, and diabetes. Should they prove to be safe and effective, exosomes will become one of regenerative medicine's best hopes for treating patients with stroke and other debilitating CNS diseases. In animal models of stroke, TBI, cognitive decline, and other CNS diseases, they have been shown to have a great impact at lessening the disease severity and quickening and deepening recovery.

The therapeutic impact of exosomes is multifactorial, but is certainly dependent on three identifiable factors: (a) the surface proteins that determine the targeted cell type; (b) the miRNA cargo that determines their function in target cells; and (c) secondary release of exosomes and paracrine factors from target cells. Deeper understanding of each of these factors will doubtlessly affect our ability to design custom treatments for stroke and other CNS diseases that are currently untreatable. Exosomes represent a unique opportunity to advance cell therapy to a place of safety, efficacy, and manufacturability that currently does not exist.

References

1. Li Y, Chopp M, Chen J, Wang L, Gautam SC, Xu YX, et al. Intrastriatal transplantation of bone marrow nonhematopoietic cells improves functional recovery after stroke in adult mice. J Cereb Blood Flow Metab. 2000;20(9):1311–9.
2. Chen J, Li Y, Wang L, Lu M, Zhang X, Chopp M. Therapeutic benefit of intracerebral transplantation of bone marrow stromal cells after cerebral ischemia in rats. J Neurol Sci. 2001;189(1-2):49–57.
3. Li Y, Chen J, Wang L, Lu M, Chopp M. Treatment of stroke in rat with intracarotid administration of marrow stromal cells. Neurology. 2001;56(12):1666–72.
4. Chen J, Li Y, Katakowski M, Chen X, Wang L, Lu D, et al. Intravenous bone marrow stromal cell therapy reduces apoptosis and promotes endogenous cell proliferation after stroke in female rat. J Neurosci Res. 2003;73(6):778–86.
5. Zhao LR, Duan WM, Reyes M, Keene CD, Verfaillie CM, Low WC. Human bone marrow stem cells exhibit neural phenotypes and ameliorate neurological deficits after grafting into the ischemic brain of rats. Exp Neurol. 2002;174(1):11–20.
6. Kurozumi K, Nakamura K, Tamiya T, Kawano Y, Kobune M, Hirai S, et al. BDNF gene-modified mesenchymal stem cells promote functional recovery and reduce infarct size in the rat middle cerebral artery occlusion model. Mol Ther. 2004;9(2):189–97.
7. Kurozumi K, Nakamura K, Tamiya T, Kawano Y, Ishii K, Kobune M, et al. Mesenchymal stem cells that produce neurotrophic factors reduce ischemic damage in the rat middle cerebral artery occlusion model. Mol Ther. 2005;11(1):96–104.
8. Hanabusa K, Nagaya N, Iwase T, Itoh T, Murakami S, Shimizu Y, et al. Adrenomedullin enhances therapeutic potency of mesenchymal stem cells after experimental stroke in rats. Stroke. 2005;36(4):853–8.
9. Liu YP, Seckin H, Izci Y, Du ZW, Yan YP, Baskaya MK. Neuroprotective effects of mesenchymal stem cells derived from human embryonic stem cells in transient focal cerebral ischemia in rats. J Cereb Blood Flow Metab. 2009;29(4):780–91.
10. Lin YC, Ko TL, Shih YH, Lin MY, Fu TW, Hsiao HS, et al. Human umbilical mesenchymal stem cells promote recovery after ischemic stroke. Stroke. 2011;42(7):2045–53.

11. Mohamad O, Drury-Stewart D, Song M, Faulkner B, Chen D, Yu SP, et al. Vector-free and transgene-free human iPS cells differentiate into functional neurons and enhance functional recovery after ischemic stroke in mice. PLoS One. 2013;8(5):e64160.
12. Yuan T, Liao W, Feng NH, Lou YL, Niu X, Zhang AJ, et al. Human induced pluripotent stem cell-derived neural stem cells survive, migrate, differentiate, and improve neurologic function in a rat model of middle cerebral artery occlusion. Stem Cell Res Ther. 2013;4(3):73.
13. Chopp M, Li Y. Treatment of neural injury with marrow stromal cells. Lancet Neurol. 2002;1(2):92–100.
14. Bhalala OG. The emerging impact of microRNAs in neurotrauma pathophysiology and therapy. In: Kobeissy FH, editor. Brain neurotrauma: molecular, neuropsychological, and rehabilitation aspects. Boca Raton, FL: Frontiers in Neuroengineering; 2015.
15. Maas SL, Breakefield XO, Weaver AM. Extracellular vesicles: unique intercellular delivery vehicles. Trends Cell Biol. 2017;27(3):172–88.
16. Zocco D, Ferruzzi P, Cappello F, Kuo WP, Fais S. Extracellular vesicles as shuttles of tumor biomarkers and anti-tumor drugs. Front Oncol. 2014;4:267.
17. Cheow ES, Cheng WC, Lee CN, de Kleijn D, Sorokin V, Sze SK. Plasma-derived extracellular vesicles contain predictive biomarkers and potential therapeutic targets for myocardial ischemic (MI) injury. Mol Cell Proteomics. 2016;15(8):2628–40.
18. Foster BP, Balassa T, Benen TD, Dominovic M, Elmadjian GK, Florova V, et al. Extracellular vesicles in blood, milk and body fluids of the female and male urogenital tract and with special regard to reproduction. Crit Rev Clin Lab Sci. 2016;53(6):379–95.
19. Keller S, Ridinger J, Rupp AK, Janssen JW, Altevogt P. Body fluid derived exosomes as a novel template for clinical diagnostics. J Transl Med. 2011;9:86.
20. Choi DS, Park JO, Jang SC, Yoon YJ, Jung JW, Choi DY, et al. Proteomic analysis of microvesicles derived from human colorectal cancer ascites. Proteomics. 2011;11(13):2745–51.
21. Rak J. Extracellular vesicles - biomarkers and effectors of the cellular interactome in cancer. Front Pharmacol. 2013;4:21.
22. Trajkovic K, Hsu C, Chiantia S, Rajendran L, Wenzel D, Wieland F, et al. Ceramide triggers budding of exosome vesicles into multivesicular endosomes. Science. 2008;319(5867):1244–7.
23. Tian T, Zhu YL, Zhou YY, Liang GF, Wang YY, Hu FH, et al. Exosome uptake through clathrin-mediated endocytosis and macropinocytosis and mediating miR-21 delivery. J Biol Chem. 2014;289(32):22258–67.
24. Vader P, Breakefield XO, Wood MJ. Extracellular vesicles: emerging targets for cancer therapy. Trends Mol Med. 2014;20(7):385–93.
25. Park JE, Tan HS, Datta A, Lai RC, Zhang H, Meng W, et al. Hypoxic tumor cell modulates its microenvironment to enhance angiogenic and metastatic potential by secretion of proteins and exosomes. Mol Cell Proteomics. 2010;9(6):1085–99.
26. Kowal J, Arras G, Colombo M, Jouve M, Morath JP, Primdal-Bengtson B, et al. Proteomic comparison defines novel markers to characterize heterogeneous populations of extracellular vesicle subtypes. Proc Natl Acad Sci U S A. 2016;113(8):E968–77.
27. Mrvar-Brecko A, Sustar V, Jansa V, Stukelj R, Jansa R, Mujagic E, et al. Isolated microvesicles from peripheral blood and body fluids as observed by scanning electron microscope. Blood Cells Mol Dis. 2010;44(4):307–12.
28. Bruno S, Tapparo M, Collino F, Chiabotto G, Deregibus MC, Soares Lindoso R, et al. Renal regenerative potential of different extra-cellular vesicle populations derived from bone marrow mesenchymal stromal cells. Tissue Eng Part A. 2017;PMID:28471327.
29. Akyurekli C, Le Y, Richardson RB, Fergusson D, Tay J, Allan DS. A systematic review of preclinical studies on the therapeutic potential of mesenchymal stromal cell-derived microvesicles. Stem Cell Rev. 2015;11(1):150–60.
30. Xin H, Li Y, Cui Y, Yang JJ, Zhang ZG, Chopp M. Systemic administration of exosomes released from mesenchymal stromal cells promote functional recovery and neurovascular plasticity after stroke in rats. J Cereb Blood Flow Metab. 2013;33(11):1711–5.

31. Zhang Y, Chopp M, Meng Y, Katakowski M, Xin H, Mahmood A, et al. Effect of exosomes derived from multipluripotent mesenchymal stromal cells on functional recovery and neurovascular plasticity in rats after traumatic brain injury. J Neurosurg. 2015;122(4):856–67.
32. Ge Q, Zhou Y, Lu J, Bai Y, Xie X, Lu Z. miRNA in plasma exosome is stable under different storage conditions. Molecules. 2014;19(2):1568–75.
33. Zacharek A, Chen J, Cui X, Li A, Li Y, Roberts C, et al. Angiopoietin1/Tie2 and VEGF/Flk1 induced by MSC treatment amplifies angiogenesis and vascular stabilization after stroke. J Cereb Blood Flow Metab. 2007;27(10):1684–91.
34. Xin H, Li Y, Shen LH, Liu X, Wang X, Zhang J, et al. Increasing tPA activity in astrocytes induced by multipotent mesenchymal stromal cells facilitate neurite outgrowth after stroke in the mouse. PLoS One. 2010;5(2):e9027.
35. van Velthoven CT, Kavelaars A, Heijnen CJ. Mesenchymal stem cells as a treatment for neonatal ischemic brain damage. Pediatr Res. 2012;71(4 Pt 2):474–81.
36. Guo F, Lv S, Lou Y, Tu W, Liao W, Wang Y, et al. Bone marrow stromal cells enhance the angiogenesis in ischaemic cortex after stroke: involvement of notch signalling. Cell Biol Int. 2012;36(11):997–1004.
37. Doeppner TR, Herz J, Gorgens A, Schlechter J, Ludwig AK, Radtke S, et al. Extracellular vesicles improve post-stroke neuroregeneration and prevent postischemic immunosuppression. Stem Cells Transl Med. 2015;4(10):1131–43.
38. Otero-Ortega L, Gomez de Frutos MC, Laso-Garcia F, Rodriguez-Frutos B, Medina-Gutierrez E, Lopez JA, et al. Exosomes promote restoration after an experimental animal model of intracerebral hemorrhage. J Cereb Blood Flow Metab. 2017;1:271678X17708917.
39. Zhang Y, Chopp M, Zhang ZG, Katakowski M, Xin H, Qu C, et al. Systemic administration of cell-free exosomes generated by human bone marrow derived mesenchymal stem cells cultured under 2D and 3D conditions improves functional recovery in rats after traumatic brain injury. Neurochem Int. 2016;pii:S0197-0186(16)30251-0.
40. Yeo RW, Lai RC, Zhang B, Tan SS, Yin Y, Teh BJ, et al. Mesenchymal stem cell: an efficient mass producer of exosomes for drug delivery. Adv Drug Deliv Rev. 2013;65(3):336–41.
41. Du W, Zhang K, Zhang S, Wang R, Nie Y, Tao H, et al. Enhanced proangiogenic potential of mesenchymal stem cell-derived exosomes stimulated by a nitric oxide releasing polymer. Biomaterials. 2017;133:70–81.
42. Chen J, Ning R, Zacharek A, Cui C, Cui X, Yan T, et al. MiR-126 contributes to human umbilical cord blood cell-induced neurorestorative effects after stroke in type-2 diabetic mice. Stem Cells. 2016;34(1):102–13.
43. Mathiyalagan P, Liang Y, Kim D, Misener S, Thorne T, Kamide CE, et al. Angiogenic mechanisms of human CD34+ stem cell exosomes in the repair of ischemic hindlimb. Circ Res. 2017;120(9):1466–76.
44. Besse B, Charrier M, Lapierre V, Dansin E, Lantz O, Planchard D, et al. Dendritic cell-derived exosomes as maintenance immunotherapy after first line chemotherapy in NSCLC. Oncoimmunology. 2016;5(4):e1071008.
45. Que RS, Lin C, Ding GP, Wu ZR, Cao LP. Increasing the immune activity of exosomes: the effect of miRNA-depleted exosome proteins on activating dendritic cell/cytokine-induced killer cells against pancreatic cancer. J Zhejiang Univ Sci B. 2016;17(5):352–60.
46. Lu Z, Zuo B, Jing R, Gao X, Rao Q, Liu Z, et al. Dendritic cell-derived exosomes elicit tumor regression in autochthonous hepatocellular carcinoma mouse models. J Hepatol. 2017;pii:S0168-8278(17)32055-X.
47. Bu N, Wu H, Zhang G, Zhan S, Zhang R, Sun H, et al. Exosomes from dendritic cells loaded with chaperone-rich cell lysates elicit a potent T cell immune response against intracranial glioma in mice. J Mol Neurosci. 2015;56(3):631–43.
48. Romagnoli GG, Zelante BB, Toniolo PA, Migliori IK, Barbuto JA. Dendritic cell-derived exosomes may be a tool for cancer immunotherapy by converting tumor cells into immunogenic targets. Front Immunol. 2014;5:692.

49. Kawikova I, Askenase PW. Diagnostic and therapeutic potentials of exosomes in CNS diseases. Brain Res. 2015;1617:63–71.
50. Rao P, Benito E, Fischer A. MicroRNAs as biomarkers for CNS disease. Front Mol Neurosci. 2013;6:39.
51. Hertel J, Lindemeyer M, Missal K, Fried C, Tanzer A, Flamm C, et al. The expansion of the metazoan microRNA repertoire. BMC Genomics. 2006;7:25.
52. Berezikov E. Evolution of microRNA diversity and regulation in animals. Nat Rev Genet. 2011;12(12):846–60.
53. Gurtan AM, Sharp PA. The role of miRNAs in regulating gene expression networks. J Mol Biol. 2013;425(19):3582–600.
54. Zhang X, Yuan X, Shi H, Wu L, Qian H, Xu W. Exosomes in cancer: small particle, big player. J Hematol Oncol. 2015;8:83.
55. Melo SA, Sugimoto H, O'Connell JT, Kato N, Villanueva A, Vidal A, et al. Cancer exosomes perform cell-independent microRNA biogenesis and promote tumorigenesis. Cancer Cell. 2014;26(5):707–21.
56. McKenzie AJ, Hoshino D, Hong NH, Cha DJ, Franklin JL, Coffey RJ, et al. KRAS-MEK signaling controls Ago2 sorting into exosomes. Cell Rep. 2016;15(5):978–87.
57. Katakowski M, Buller B, Wang X, Rogers T, Chopp M. Functional microRNA is transferred between glioma cells. Cancer Res. 2010;70(21):8259–63.
58. Katakowski M, Buller B, Zheng X, Lu Y, Rogers T, Osobamiro O, et al. Exosomes from marrow stromal cells expressing miR-146b inhibit glioma growth. Cancer Lett. 2013;335(1):201–4.
59. Bovy N, Blomme B, Freres P, Dederen S, Nivelles O, Lion M, et al. Endothelial exosomes contribute to the antitumor response during breast cancer neoadjuvant chemotherapy via microRNA transfer. Oncotarget. 2015;6(12):10253–66.
60. Wei Y, Lai X, Yu S, Chen S, Ma Y, Zhang Y, et al. Exosomal miR-221/222 enhances tamoxifen resistance in recipient ER-positive breast cancer cells. Breast Cancer Res Treat. 2014;147(2):423–31.
61. Xin H, Li Y, Buller B, Katakowski M, Zhang Y, Wang X, et al. Exosome-mediated transfer of miR-133b from multipotent mesenchymal stromal cells to neural cells contributes to neurite outgrowth. Stem Cells. 2012;30(7):1556–64.
62. Xin H, Li Y, Liu Z, Wang X, Shang X, Cui Y, et al. MiR-133b promotes neural plasticity and functional recovery after treatment of stroke with multipotent mesenchymal stromal cells in rats via transfer of exosome-enriched extracellular particles. Stem Cells. 2013;31(12):2737–46.
63. Zhang Y, Ueno Y, Liu XS, Buller B, Wang X, Chopp M, et al. The MicroRNA-17-92 cluster enhances axonal outgrowth in embryonic cortical neurons. J Neurosci. 2013;33(16):6885–94.
64. Xin H, Katakowski M, Wang F, Qian JY, Liu XS, Ali MM, et al. MicroRNA cluster miR-17-92 cluster in exosomes enhance neuroplasticity and functional recovery after stroke in rats. Stroke. 2017;48(3):747–53.
65. Zhang Y, Chopp M, Liu XS, Katakowski M, Wang X, Tian X, et al. Exosomes derived from mesenchymal stromal cells promote axonal growth of cortical neurons. Mol Neurobiol. 2017;54(4):2659–73.
66. Lo EH, Dalkara T, Moskowitz MA. Mechanisms, challenges and opportunities in stroke. Nat Rev Neurosci. 2003;4(5):399–415.
67. Zhang ZG, Chopp M. Neurorestorative therapies for stroke: underlying mechanisms and translation to the clinic. Lancet Neurol. 2009;8(5):491–500.
68. Xin H, Wang F, Li Y, Lu QE, Cheung WL, Zhang Y, et al. Secondary release of exosomes from astrocytes contributes to the increase in neural plasticity and improvement of functional recovery after stroke in rats treated with exosomes harvested from microRNA 133b-overexpressing multipotent mesenchymal stromal cells. Cell Transplant. 2017;26(2):243–57.
69. Morel L, Regan M, Higashimori H, Ng SK, Esau C, Vidensky S, et al. Neuronal exosomal miRNA-dependent translational regulation of astroglial glutamate transporter GLT1. J Biol Chem. 2013;288(10):7105–16.

70. Bakhti M, Winter C, Simons M. Inhibition of myelin membrane sheath formation by oligodendrocyte-derived exosome-like vesicles. J Biol Chem. 2011;286(1):787–96.
71. Fruhbeis C, Frohlich D, Kuo WP, Kramer-Albers EM. Extracellular vesicles as mediators of neuron-glia communication. Front Cell Neurosci. 2013;7:182.
72. Fruhbeis C, Frohlich D, Kuo WP, Amphornrat J, Thilemann S, Saab AS, et al. Neurotransmitter-triggered transfer of exosomes mediates oligodendrocyte-neuron communication. PLoS Biol. 2013;11(7):e1001604.
73. Bian S, Zhang L, Duan L, Wang X, Min Y, Yu H. Extracellular vesicles derived from human bone marrow mesenchymal stem cells promote angiogenesis in a rat myocardial infarction model. J Mol Med (Berl). 2014;92(4):387–97.
74. van Balkom BW, de Jong OG, Smits M, Brummelman J, den Ouden K, de Bree PM, et al. Endothelial cells require miR-214 to secrete exosomes that suppress senescence and induce angiogenesis in human and mouse endothelial cells. Blood. 2013;121(19):3997–4006. S1–15
75. Schulz GB, Wieland E, Wustehube-Lausch J, Boulday G, Moll I, Tournier-Lasserve E, et al. Cerebral cavernous malformation-1 protein controls DLL4-notch3 signaling between the endothelium and pericytes. Stroke. 2015;46(5):1337–43.
76. Hu GW, Li Q, Niu X, Hu B, Liu J, Zhou SM, et al. Exosomes secreted by human-induced pluripotent stem cell-derived mesenchymal stem cells attenuate limb ischemia by promoting angiogenesis in mice. Stem Cell Res Ther. 2015;6:10.
77. Goetzl EJ, Schwartz JB, Mustapic M, Lobach IV, Daneman R, Abner EL, et al. Altered cargo proteins of human plasma endothelial cell-derived exosomes in atherosclerotic cerebrovascular disease. FASEB J. 2017;31(8):3689–94.
78. Graner MW, Alzate O, Dechkovskaia AM, Keene JD, Sampson JH, Mitchell DA, et al. Proteomic and immunologic analyses of brain tumor exosomes. FASEB J. 2009;23(5):1541–57.
79. Chen CC, Liu L, Ma F, Wong CW, Guo XE, Chacko JV, et al. Elucidation of exosome migration across the blood-brain barrier model in vitro. Cell Mol Bioeng. 2016;9(4):509–29.
80. Zhuang X, Xiang X, Grizzle W, Sun D, Zhang S, Axtell RC, et al. Treatment of brain inflammatory diseases by delivering exosome encapsulated anti-inflammatory drugs from the nasal region to the brain. Mol Ther. 2011;19(10):1769–79.
81. Satani N, Savitz SI. Is immunomodulation a principal mechanism underlying how cell-based therapies enhance stroke recovery? Neurotherapeutics. 2016;13(4):775–82.
82. Imai T, Takahashi Y, Nishikawa M, Kato K, Morishita M, Yamashita T, et al. Macrophage-dependent clearance of systemically administered B16BL6-derived exosomes from the blood circulation in mice. J Extracellular Vesicles. 2015;4:26238.
83. Takahashi Y, Nishikawa M, Shinotsuka H, Matsui Y, Ohara S, Imai T, et al. Visualization and in vivo tracking of the exosomes of murine melanoma B16-BL6 cells in mice after intravenous injection. J Biotechnol. 2013;165(2):77–84.
84. Vendrame M, Gemma C, Pennypacker KR, Bickford PC, Davis Sanberg C, Sanberg PR, et al. Cord blood rescues stroke-induced changes in splenocyte phenotype and function. Exp Neurol. 2006;199(1):191–200.
85. Walker PA, Shah SK, Jimenez F, Aroom KR, Harting MT, Cox CS Jr. Bone marrow-derived stromal cell therapy for traumatic brain injury is neuroprotective via stimulation of non-neurologic organ systems. Surgery. 2012;152(5):790–3.
86. Sakamoto W, Masuno T, Yokota H, Takizawa T. Expression profiles and circulation dynamics of rat mesenteric lymph microRNAs. Mol Med Rep. 2017;15(4):1989–96.
87. Jaimes Y, Naaldijk Y, Wenk K, Leovsky C, Emmrich F. Mesenchymal stem cell-derived microvesicles modulate lipopolysaccharides-induced inflammatory responses to microglia cells. Stem Cells. 2017;35(3):812–23.
88. Liddelow SA, Guttenplan KA, Clarke LE, Bennett FC, Bohlen CJ, Schirmer L, et al. Neurotoxic reactive astrocytes are induced by activated microglia. Nature. 2017;541(7638):481–7.
89. Doring A, Sloka S, Lau L, Mishra M, van Minnen J, Zhang X, et al. Stimulation of monocytes, macrophages, and microglia by amphotericin B and macrophage colony-stimulating factor promotes remyelination. J Neurosci. 2015;35(3):1136–48.

90. Fu PC, Tang RH, Wan Y, Xie MJ, Wang W, Luo X, et al. ROCK inhibition with fasudil promotes early functional recovery of spinal cord injury in rats by enhancing microglia phagocytosis. J Huazhong Univ Sci Technolog Med Sci. 2016;36(1):31–6.
91. Lin X, Zhao T, Walker M, Ding A, Lin S, Cao Y, et al. Transplantation of pro-oligodendroblasts, preconditioned by LPS-stimulated microglia, promotes recovery after acute contusive spinal cord injury. Cell Transplant. 2016;25(12):2111–28.
92. Bahrini I, Song JH, Diez D, Hanayama R. Neuronal exosomes facilitate synaptic pruning by up-regulating complement factors in microglia. Sci Rep. 2015;5:7989.
93. Lui H, Zhang J, Makinson SR, Cahill MK, Kelley KW, Huang HY, et al. Progranulin deficiency promotes circuit-specific synaptic pruning by microglia via complement activation. Cell. 2016;165(4):921–35.
94. Southam KA, Vincent AJ, Small DH. Do microglia default on network maintenance in alzheimer's disease? J Alzheimers Dis. 2016;51(3):657–69.

Chapter 17
Exosome and MiRNA in Stroke

Ji Bihl, Jinju Wang, Xiaotang Ma, Yi Yang, Bin Zhao, and Yanfang Chen

Abstract Stroke is one of the leading causes of death and disability worldwide. Various types of stem cells have been applied to treat stroke and have been shown promising potential. The principal mechanism of therapeutic action has been partially ascribed to their strong paracrine capacity. Exosomes are small vesicles released from all kinds of cells and mediate intercellular communication by transferring exosomal protein and microRNA (miRNA) cargoes between cells in the brain. Among these cargoes, miRNAs play a key role in mediating biological function due to their prominent roles in gene regulation. Emerging data suggest that stem cell-released exosomes have advantages over stem cells to treat stroke, because exosomes could cross the blood bran barrier and easily to be modified and handled. Here, we first review the biogenesis, cargoes, and detection of exosomes. Then, we discussed the role of miRNAs in stroke. At last, we highlight the use of stem cell-released exosomes as biomarkers and therapeutic avenues in stroke. Perspectives on the developing role of stem cell-released exosomes mediated transfer of miRNAs as a therapeutic approach will also be discussed.

Keywords Stroke • Exosomes • miRNAs • Brain microenvironment • Biomarker • Therapy

---

J. Bihl • J. Wang • Y. Chen (✉)
Department of Pharmacology and Toxicology, Wright State University,
Dayton, OH 45435, USA
e-mail: yanfang.chen@wright.edu

X. Ma • B. Zhao
Guangdong Key Laboratory of Age-Related Cardiac and Cerebral Diseases,
Institute of Neurology, Guangdong Medical University, Zhanjiang 524001, China

Y. Yang
College of Health Science, Wuhan Sports University, Wuhan, Hubei 430000, China

P.A. Lapchak, J.H. Zhang (eds.), *Cellular and Molecular Approaches to Regeneration and Repair*, Springer Series in Translational Stroke Research,
https://doi.org/10.1007/978-3-319-66679-2_17

Abbreviations

Ago2	Argonaute 2
AMPA	Alpha-amino-3-hydroxy-5-methyl-4-isoxazolepropionic acid
BBB	Blood-brain barrier
BDNF	Brain-derived neurotrophic factor
CD	Cluster of differentiation
CNS	Central nervous system
CSF	Cerebrospinal fluid
CSPGs	Chondroitin sulfate proteoglycans
CTGF	Connective tissue growth factor
DCs	Dendritic cells
Dll4	Delta-like 4
ECs	Endothelial cells
EpCAM	Epithelial cell adhesion molecule
EPCs	Endothelial progenitor cells
EPC-EXs	EPC-released exosomes
ESCART	Endosomal-sorting complex responsible for transport
EVs	Extracellular vesicles
EXs	Exosomes
FGF2	Fibroblast growth factor 2
GFAP	Glial fibrillary acidic protein
GFP	Green fluorescence protein
GluR2/3	Glutamate receptor AMPA R2/3
H/R	Hypoxia/reoxygenation
HMGA2	High mobility group AT-hook 2
HSCs	Hematopoietic stem cells
IFN-γ	The interferon gamma
IGF	The insulin-like growth factor
L1CAM	Neuronal-specific protein L1 cell adhesion molecule
Lamp-2	Lysosomal-associated membrane protein 2
MAP 1b	Microtubule associated protein 1b
MCAO	Middle cerebral artery occlusion
miR-126-EPC-EXs	Exosomes released from miR-126 primed EPCs
miRNA	MicroRNA
MOR	Opioid receptor mu
mRNA	Messenger RNA
MSCs	Mesenchymal stromal cells
MVB	Multivesicular bodies
MVs	Microvesicles
NPCs	Neural progenitor cells
NPC-EXs	NPCs-released exosomes
NTA	Nanoparticle tracking analysis

PEG	Polyethyleneglycol
PTEN	Phosphatase and tensin homolog
Rab5	Ras-related protein
RARβ	Retinoic acid receptor β2
RhoA	ras Homolog family member A
RISC	RNA-induced silencing complex
RVG	Rabies virus glycoprotein
SGZ	Subgranular zone
Shh	Sonic hedgehog
STAT1	Signal transducer and activator of transcription 1
SVZ	Subventricular zone
TNFα	Tumor necrosis factor-α
VEGF	Vascular endothelial growth factor
VEGFR2	Vascular endothelial growth factor receptor 2
VPS4	Vacuolar protein sorting 4

1 Introduction of Exosomes

Exosomes are extracellular nanozided (30–100 nm) vesicles generated from endosomal membranes. Exosomes are generated inside cells from multivesicular bodies (MVB) and released into extracellular space via exocytosis. MVB are endosomal organelles that are characterized by internal membrane-enveloped vesicles, which were described for the first time, in neurons, by Palay and Palade in 1955 [1]. Originally, these vesicles were regarded to be pre-lysosomal structures involved in protein degradation [2]; however, recent evidence indicates that MVB mediate diverse intra- and intercellular trafficking of molecules [3].

Nearly three decades after the discovery of intracellular MVB, Trams et al. analyzed cell-free supernatants collected from human neuronal neoplastic cell lines. The supernatants were span at the speed of 100,000 × *g* for 90 min, and the pellets of these supernatants were then studied by electron microscope [4], which resulted in the discovery of extracellular membrane-enveloped vesicles ranging in size from 40 to 1000 nm. The supernatant pellets possessed enzymatic activities, including 5-nucleotidase and ATPase, suggesting that the vesicles could play some physiological roles rather than just be cellular waste products [4]. Figure 17.1 shows the process of isolating microvesicles and exosomes from cell-free supernatants.

The extracellular vesicles (EVs) found by these researchers that are today known as exosomes (<100 nm), as well as larger ectosomes (also known as microvesicles or microparticles) that typically range in the size 0.1–1.0 μm (Table 17.1). While exosomes have been shown to be derived from MVB, ectosomes shad from cellular membranes [5–9] and differ in their content, implying that exosomes and ectosomes may have distinct functions [10, 11]. In this chapter, we focus on exosomes.

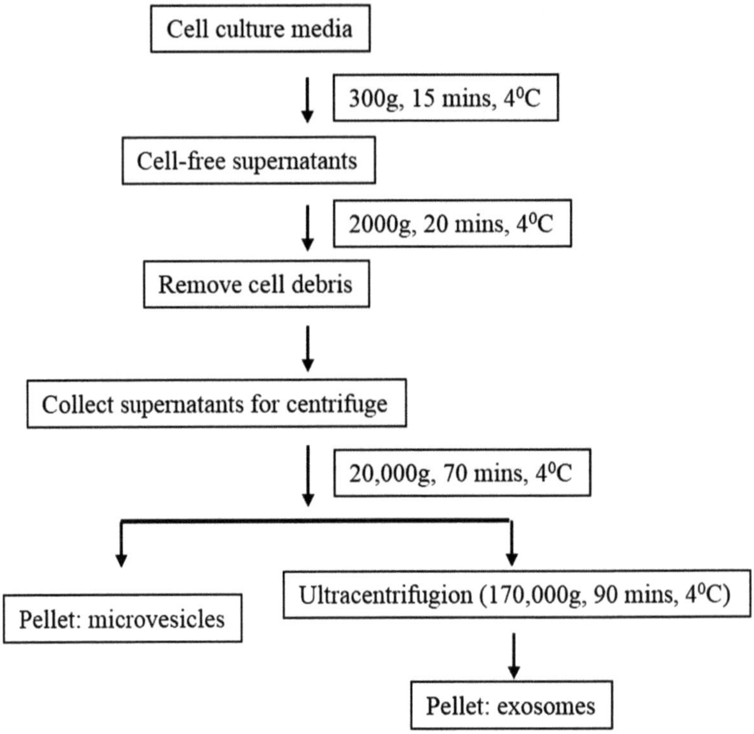

Fig. 17.1 Exosome isolation from cell culture media. The cell culture medium was collected and centrifuged at 300 × *g* for 15 min, followed by centrifugation at 2000 × *g* for 20 min to remove cells and cell debris. The cell-free culture medium was centrifuged at 20,000 × *g* for 70 min and then ultracentrifuged at 170,000 × *g* for 90 min to pellet exosomes

Table 17.1 Key features of microvesicles and exosomes

	Microvesicles	Exosomes
Size range	0.1–1 μm in diameter	30–100 nm in diameter
Biogenesis	Budding from plasma membrane	Exocytosis of MVBs
Markers	Annexin V binding, tissue factor and cell-specific markers	CD63, CD81, CD9, and Tsg101

2 Biogenesis of Exosomes

Exosomes are formed by inward budding. Therefore, the orientation of proteins and lipids of the exosome membrane is equivalent to the plasma membrane of the parental cell [12]. MVBs are late endosomes, which are refined by early endosome maturation. This maturation results in gradual changes in content and composition of the membrane [13]. It is suggested that phosphatidylinositol-3 kinase activity is required for both the formation of MVBs and the formation of exosomes [12, 14]. A previous study showed that the inhibition of phosphatidylinositol-3 kinase results in swelling of several endocytic compartments and inhibition of MVB biogenesis [14].

The machinery involved in the formation of MVBs is relevant to exosome production and function. There are two possible fates for the MVBs: fuse with the plasma membrane, or fuse with a lysosome followed by digestion of the cargo [15]. It was reported that cholesterol rich MVBs are prone to release exosome and cholesterol-poor MVBs are targeted to lysosomal digestion [16]. Another study showed that incorporation of membrane proteins such as growth factor receptors designates MVB to lysosomal degradation [12]. The molecular machinery involved in the biogenesis of exosomes can be dependent on or independent from endosomal-sorting complex responsible for transport (ESCART). The ESCART-dependent system is associated with accessory proteins such as programmed cell death 6 interacting protein and vacuolar protein sorting 4 (VPS4), which are used as exosome markers in many studies [15]. Alternatively, in the presence of sphingomyelinase, the exosome biogenesis might be independent from ESCART. Trajkovic et al. showed that inhibition of sphingomyelinase significantly decreases the release of exosomes [17]. Simons et al. showed that MVB formation can even be independent from ESCART and sphingomyelinase. By their results, tetraspanin proteins enriched in MVBs can play a major role in the formation of exosomes [18].

MVBs designated to exocytosis release the exosomes into the extracellular space by fusion with the plasma membrane. The release of exosomes can be constitutive or inducible depending on the cell type and the state of cell activation [19, 20]. For instance, immature dendritic cells (DCs) and epithelial cells release exosomes in a constitutive manner [21]. Another study has indicated that members of Rab family are involved in classical intracellular trafficking and in fusion of cellular compartment. It was also observed that a subset of this family, such as Rab27a and Rab27b, is involved in the secretion of exosomes [22].

3 Isolation and Characterization of Exosomes

Exosomes can be isolated from cell culture supernatant and plasma using several different techniques. The most common isolation technique is differential centrifugation, whereby large particles and cell debris in the culture medium/plasma are separated using centrifugal force between 200–100,000 × g and the exosomes are separated from supernatant by centrifuging at 100,000 × g [12, 23, 24]. Exosome purity can be improved by centrifuging the samples using flotation density gradient centrifugation with sucrose or Optiprep, which results in highly purified and enriched exosomes [25, 26]. Another common technique for exosome isolation is the monoclonal antibody based method. Antibodies against exosome-associated antigens—such as cluster of differentiation (CD) molecules CD63, CD81, CD82, CD9, epithelial cell adhesion molecule (EpCAM), and Ras-related protein (Rab5)—are used for separation [25, 26]. The antibodies can be immobilized in different media conditions and combined with magnetic beads, chromatographic matrix, plates, and microfluidic devices for separation [26]. A draw back to this technique is that non-exosomes vesicles that carry the antigens also bind to the antibody, reducing the purity of the extracted exosomes [25]. Ultrafiltration is another way to isolate exosomes based on their size differences. This method is less time consuming compared to ultracentrifugation and does not require special equipment [27]. One alternative method for isolating exosomes based on size is using high performance liquid chromatography, which provides highly purified exosomes [25, 27].

Exosomes can be characterized based on their size, protein content, and lipid content. Several methods have been used to characterize exosomes, including flow cytometry, nanoparticle tracking analysis, dynamic light scattering, western blot, mass spectrometry, and microscopy techniques [28]. Exosomes can also be characterized and marked based on their protein compositions. There are a number of proposed exosome reference markers such as lysosomal-associated membrane protein 2 (Lamp-2) and Rab5B, which is a member of the RAS oncogene family [27]. Also, proteins such as Pdcd6ip, TSG101, tetraspanin proteins resulting from exosome formation in MVBs (CD9, CD63, CD81 and CD82) and proteins enabling intracellular membrane fusion and transportation [21] are often used as markers to detect and identify exosomes [29]. There are kits to detect exosomes in vivo such as Exosomal Cyto-Tracer, the lentivector based Cyto-Tracer (Biocat GmbH) expresses the tetraspanin CD63, CD9 or CD8, which are fused to GFP or RFP. In this way, exosomes are marked and long-term and in-depth experimentation is enabled. However, these proposed markers are just general markers for total exosomes or even microvesicles and apoptotic bodies [30].

In future studies, a method to isolate/detect a pure specific population of exosomes is needed and it is important for exosome research. Related, the authors recently published a protocol to combine microbeads and Qdot to detect endothelial cells (ECs) and endothelial progenitor cells (EPCs) specific exosomes [31, 32]. In this protocol (Fig. 17.2), we used ECs or EPCs specific makers: CD105/CD144 or CD34/vascular endothelial growth factor receptor 2 (VEGFR2) to detect the EC-exosomes and EPC-exosomes. After selected with microbeads conjugated first antibodies (CD105 or CD34), the concentration of EC-exosomes and EPC-exosomes from cell culture and plasm were recognized by Qdot-labelled second antibodies (CD144 or VEGFR2) and measured by Nanoparticle Tracking Analysis system (NTA) (NS300, Malvern Instruments). Compared to the transitional flow cytometry method, the NTA has better sensitivity and specificity to detect the total exosomes, EC-exosomes and EPC-exosomes.

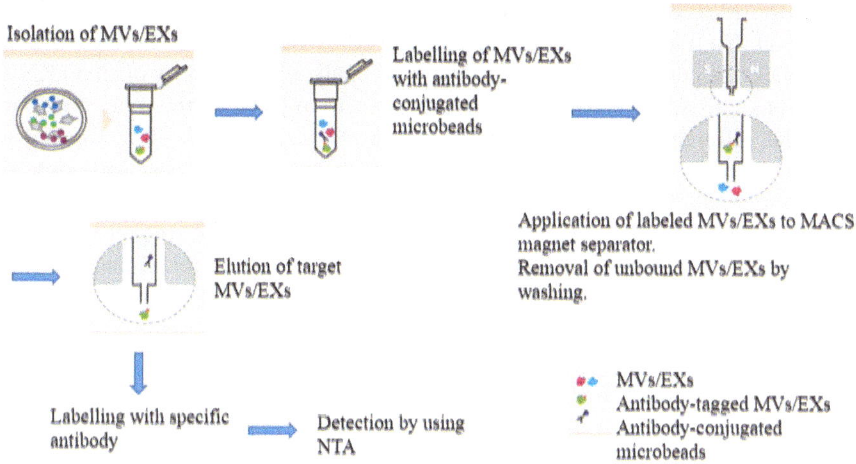

Fig. 17.2 Application of microbeads and Qdot for detection of specific MVs and EXs. The pelleted EXs were incubated with 10 μL of Biotin-conjugated specific antibodies (Miltenyi Biotec), such as anti-CD105 for ECs or anti-CD34 for EPCs, in a 100 μL reaction volume for 2 h, followed by adding 10 μL of anti-Biotin microbeads (Miltenyi Biotec). Then, a magnet module was applied to separate microbeads-labeled EXs from the total EXs. After an overnight separation, the fluid was gently removed from the magnet. The microbeads bound EXs were resuspended with 100 μL filtered PBS and added with 10 μL of multisort release reagent (Miltenyi Biotec) for 10 min to cleave off the microbeads from EXs. Afterwards, the second antibodies, such as anti-CD144 for ECs or anti-VEGFR2 for EPCs, were added to the selected EXs. After overnight labeling, the fluid was collected and enumerated by using the NTA NS300 system (Malvern Instruments). *MVs* microvesicles, *EXs* exosomes, *MACS* magnetic-activated cell sorting, *NTA* nanosight tracking analysis

4 Cargoes of Exosomes

At the beginning, it was assumed that the content of exosomes is random due to the engulfed part of the cytoplasm packaged by the membrane blebbing [33]. Later, it was observed that the content of exosomes released by mesenchymal stromal cells (MSCs) differs from their parental cells, probably caused by selective packaging. Although the sorting mechanisms of nucleic acids and proteins are poorly understood, there are some suggestions for sorting proteins inside exosomes, such as sorting via ESCART, via lipid and/or protein affinity or via sorting by protein incorporation into detergent-resistant protein complexes [19]. It is mentioned that exosomes contain some common and also cell-type-specific proteins [13].

Thery et al. reported that exosomes do not contain proteins that originated from the nucleus, mitochondrion, endoplasmic reticulum or Golgi apparatus. Instead, proteins identified in exosomes were observed in the cytosol and plasma membrane [21]. In contrast, a subsequent study by Record et al. reported that exosomal proteins could also originate from the endocytotic compartment, Golgi and nucleus, but rarely from the endoplasmic reticulum or mitochondria [19]. RNAs detected in exosomes of MSCs consist of mainly messenger RNA (mRNA) and microRNA (miRNA). No track of 18S or 28S ribosomal RNA was detected. Indeed, in these exosomes, the portion of mRNAs is relatively small, while majority of small RNAs as well as miRNAs were observed in precursor form [34]. Furthermore, there are essential differences between mRNA transcripts in parental cells compared to mRNA transcripts detected in exosomes [35]. Such a selection was also observed for a number of detected miRNAs, which are assumed to be exclusively packed into exosomes [36]. Baglio et al. have observed that some miRNAs are present in both exosomes and parental cells [37]. Our research group further found that the miR-126, miR-210 and miR-18a are present in both exosomes and parental cells, with higher level in exosomes.

This observation supports the existence of control mechanisms for selective packaging of miRNAs for at least MSCs [36]. Likewise, in AZ-P7a cells (a metastatic gastric cancer cell line) an enrichment of let-7 miRNA family members in exosomes was observed. The exosomal release of let-7 miRNAs into the extracellular environment maintained the oncogenesis and invasiveness of AZ-P7a cells by at least partially neutralizing the inhibitory effects of let-7 miRNAs on their targeting oncogenes such as RAS and high mobility group AT-hook 2 (HMGA2) [35, 38, 39]. In another study, it was shown that the enrichment of miRNAs in the exosomes derived from DCs is selective as some miRNAs are detected in parental cells but not in the exosomes and vice versa [40].

As a matter of fact, miRNAs are delivered to distant cells by either miRNA containing exosomes or via free miRNA molecules bound to Argonaute 2 (Ago2). It was shown that Ago2 is not only bound to free extracellular miRNAs but also to miRNAs within exosomes and other membrane-derived vesicles [38]. In most cases, miRNA was found single stranded in exosomes, yet precursor hairpin miRNAs were also detected [34, 38]. It is not clear if Ago2 is necessary for miRNA export [38]. The regulatory functions of miRNAs are accomplished through the RNA-induced silencing complex (RISC) [41]. It is believed that the mature miRNA not connected to RISC is not functional. Pre-miRNA can be loaded with RISC followed by cleavage into functionally mature RISC-loaded miRNA. The favored secretion of pre-miRNA into exosomes suggests an important physiological role after being taken up by target cells and also supports the conception that the content of exosomes is not just random [34]. Notwithstanding, miRNAs isolated from plasma exosomes, which are mostly platelet derived, has a significant different composition compared to platelets and peripheral blood mononuclear cells. As a consequence, many miRNAs are uniquely present in exosomes isolated from plasma [42].

The number and content of exosomes consistently vary based on the microenvironmental conditions of the cells and, particularly, if cells are subjected to stress factors [36]. It is shown that the miRNA content of exosomes in the plasma is different between normal and tumor-induced tissues [42]. Breast cancer cells produce exosomes with a changed pattern of miRNA [43]. Another example is that the stimulation of increased intracellular Ca^{2+} in neutrophils affect the content of exosomes released from that neutrophils [44, 45]. We also demonstrated that the miR-126 and caspase 3 levels of microvesicles released from EPCs under tumor necrosis factor-α (TNFα) stimulation are different from the microvesicles released from EPCs under serum free condition [46]. It is important to mention that apoptotic bodies, microvesicles and exosomes contain fundamentally different RNA profiles. For instance, microvesicles isolated from cell culture often do not contain a considerable amount of miRNA. Ribosomal RNA is primarily found in apoptotic bodies [30].

Finally, the miRNA content of exosomes also depends on the maturation state of the parental cell. The miRNA content of exosomes derived from mature DCs, which promote immunity, and the miRNA content of exosomes derived from immature DCs, which downregulate T cell responses, were observed to be significantly different. However, there is no significant difference in the amount of miRNA between exosomes derived from mature DC and immature DC [40]. A selective loading is not only observed for protein and RNA but also for other types of molecules packed into exosomes. Regarding DNA as a content of exosomes, there exists a disagreement. One study announced that exosomes contain no DNA [35]; later, it was observed that astrocyte-derived exosomes might contain mitochondrial DNA [47, 48]. This discrepancy in the reported data demonstrates the need of further investigations in this field.

5 Exosomes as an Important Player in the Brain Microenvironment

The brain is composed of different cell types, including neurons, astrocytes, oligodendrocytes and microglia. Specific cell-cell contacts and particular extracellular cues originated both locally and distantly. Cells communicate reciprocally with other cells by (1) intercellular contacts, and (2) secreted molecules, such as growth factors, cytokines, hormones, etc. (paracrine or endocrine communication). A novel way of cell-to-cell communication mediated by exosomes, which carry a specific cargo of proteins, lipids and nucleic acids and are currently considered one of the most complex and physiologically relevant messengers in brain microenvironment. These extracellular vesicles are secreted by neural cells under both normal and pathological conditions and have been isolated not just from the cerebrospinal fluid [49] but also from adult human brain [50]. The suggested roles of exosomes in brain microenvironment as messengers for communication between neural and vascular cells. They are important by secreting and transporting multilevel information, including signaling, factors, and regulatory molecules. Initially thought to have a function merely in waste disposal, the involvement of exosomes in neuronal and vascular development, maintenance, and regeneration through its paracrine and endocrine signaling functions has drawn particular attention in recent years. The physiological function of brain is supported by different cellular and molecular components in the brain, balancing quiescence with proliferation, and regulating cell differentiation.

5.1 Exosomes in Neuron–Neuron Communication

Classical inter-neuronal communication involves synaptic transmission, a dynamic and plastic process that is tightly regulated by neuronal activity [51]. Exosome-mediated communication between pre and post synaptic cells participates in synaptic plasticity, as it has been shown in the Drosophila neuromuscular junction [52]. Using cultures of mixed hippocampal cells with exosomes derived from the neuroblastoma cell line N2a and labeled with green fluorescence protein (GFP)-CD63 and GFP-TTC, it was found that they interact either with neurons, astrocytes or oligodendrocytes. On the other hand, exosomes released by cortical neurons upon synaptic activation interact with neurons but not with glial fibrillary acidic protein (GFAP) + astrocytes. Furthermore, some exosomes co-localize with synaptophysin indicating that they bind to pre-synaptic sites [53].

5.2 Exosomes in Neuron–Glia Communication

The communication between neurons and glia is important for brain physiology during both development and adulthood. The different glial cell types help to maintain neuronal activity. Oligodendrocytes protect axons with the myelins heath and also provide trophic support to neurons [54]. To maintain these functions over time there is a constant communication between neurons and oligodendrocytes, but the mechanisms underlying this phenomenon are not well understood. Frühbeis et al. demonstrated that upon glutamate stimulation, oligodendrocytes secrete exosomes, which are endocytosed by neurons. Furthermore, exosomal cargoes improve neuronal metabolism and viability in situations of nutrient deprivation or oxidative stress exposure [55]. It is also noteworthy that this work demonstrated that the internalization of exosomes by neurons occurs through aclathrin and dynamin-dependent mechanism, shedding light on the mechanisms that may be involved in exosome internalization. On the other hand, the selective elimination of synaptic connections comprises the engulfment of neurites. In a recent study, it was shown that neuron-derived exosomes stimulate microglial phagocytosis of neurites via upregulation of complement factors [56].

5.3 Exosomes in Glia–Glia Communication

The communication between glial cells through exosomes has been studied to a lesser extent. Exosomes secreted by oligodendrocytes are selectively internalized through macropinocytosis by microglia, both in vitro and in vivo [57]. Remarkably, only those microglial cells that do not show antigen-presenting capacity endocytose exosomes, thus supporting the idea that different types of microglial cells co-exist and are differentially involved in immune functions.

5.4 Exosomes in Neuronal-Vascular Cell Communication

Cross talk among different classes of brain cells is also essential to generate the blood-brain barrier (BBB), which then maintains the brain internal milieu, by controlling trafficking of molecules and ions between the brain and the blood [58]. For example, in a transwell coculture system, containing both rat cortical astrocytes and neurons, brain capillary ECs were found to form over time a functional barrier layer, even in the absence of cell-to-cell contacts [59, 60]. EVs released from astrocytes/neurons have been found in ECs. Astrocytes, neurons, and ECs, metabolically labeled with 35S-methionine, as well as from unlabeled ECs, incubated for 24 h with microvesicles shed from labeled astrocytes or neurons. The results indicate the presence of metabolic protein in both labeled neurons and ECs incubated with EVs from neurons. This provides evidence to support the possibility that central nervous system (CNS) derived exosomes may potentially interact with ECs within the brain, and that they may potentially find their way to the bloodstream, where they could interact with ECs and with cells of the immune system.

5.5 Exosomes as Novel Regulators of Adult Neurogenic Niches

Niches are defined by their ability to anatomically house stem cells and functionally control their development in vivo [61]. In the mammalian brain, there are defined regions termed neurogenic niches, areas with the proper environment that are able to support and modulate neurogenesis during adulthood [62]. The first validated and most studied neurogenic niches of the brain are the subventricular zone (SVZ) of the lateral ventricles and the subgranular zone (SGZ) of the dentate gyrus in the hippocampus [63]. Nevertheless, other brain regions have been proposed as having putative neurogenic niches (e.g., substantia nigra, cerebellum, and amygdala) though the extent at which this happens in vivo and in humans remains controversial for some of them. Neurogenesis has been shown to occur in the brain after stroke [64], and recent studies have shown that adult neurogenesis is active in the hippocampus [65] and in the striatum [66, 67].

An efficient and well-regulated communication between cells is vital to ensure brain homeostasis and plasticity throughout life, particularly in the adult neurogenic niches. Thus, exosomes can serve as physiological or pathological messengers between cellular components of the neurogenic niche and coordinate the function of the adult neurogenic niches. First, neural stem/progenitor cells and most of the cell types present in the CNS, including those cells that constitute and regulate the neurogenic niche, secrete and/or are target of exosomes. Furthermore, some of the biomolecules expressed (and secreted) by niche cells have been reported to be present in exosomes under physiological or pathological conditions. Interestingly, several examples highlight the role of exosomes as (1) messengers between neural cells (neurons and glial cells) either locally or distantly [via cerebrospinal fluid (CSF) or volume transmission]; (2) blood-CNS communicators (including their potential as therapeutic vehicles); and (3) modulators of several stem cell niches [68].

5.6 Exosomes in the CSF as Volume Transmission Vehicles

It has been proposed that the CSF compartment plays an essential role in volume transmission within the central CNS; thus, molecules or messengers secreted in one brain region may reach the CSF and exert their function in sites located far from its secretion site [69, 70]. Given the close contact between the CSF and the interstitial fluid of several brain areas, including the SVZ, it is conceivable that exosomes originated in the brain parenchyma can be found in the CSF and vice versa. Actually, isolation of membrane vesicle-enriched fractions and further proteomic studies have demonstrated the presence of exosomes in the human CSF [49, 71, 72]. Furthermore, the exosome content of the CSF is supposed to reflect ongoing brain processes, and especially those related to plasticity, disease or repair. Proteins related to the onset or progression of some CNS diseases such as Alzheimer's disease, prion disease, and Parkinson's disease, among others, have been found in the exosomal fraction of CSF-samples [73]. Exosomes in the CSF decrease with age while those derived from the embryonic CSF positively act on the stem cell niche [49, 74], revealing their influence on recipient cells. A clear demonstration of exosomal secretion into the CSF has been recently obtained in epithelial cells. Using cell-culture assays, human CSF analyses and in vivo tracing experiments, the authors describe a novel pathway of exosome-mediated folate delivery into the CSF and subsequently, into the brain parenchyma [72].

6 MiRNAs in Stroke

Stroke is one of the leading causes of death and disability worldwide. Stroke is classified into ischemic, which accounts for around 85% of cases, and hemorrhagic, which accounts for the remaining 15% [75]. Ischemic stroke occurs as a result of an obstruction within a blood vessel supplying blood to the brain, while the hemorrhagic stroke is the rupture of a blood vessel in the brain.

Ischemic stroke is a sudden loss of neurologic function resulting from a focal disturbance of cerebral blood flow due to the occluding of blood vessels. The goal of acute stroke therapies is to normalize perfusion and to preserve the maximal amount of penumbral tissue ischemia. Because it is a multifactorial disease with a short therapeutic window many clinical stroke trials have failed and the only currently approved therapy is thrombolysis.

Primary hemorrhagic stroke is caused by small arteriole ruptures, which stem from vascular pathological changes [76, 77]. The vascular pathologies include intimal thickening, plaque formation, vascular remodeling, atherosclerosis and aneurysms, which are vulnerable to vessel ruptures. After vessel rupture, the hemoglobin enters brain tissue and induces a series of consequential pathological changes, such as edema, cell damage/death and inflammation, which ultimately cause neurological deficit.

MiRNAs are a novel and abundant class of 19- to 22-nucleotide (nt) noncoding RNAs that control gene expression at the post-transcriptional level. They play a role in the regulation of gene expression at the post-transcriptional level, via degradation or translational inhibition of their target mRNAs. miRNAs are especially important candidates for stroke therapeutics because of their ability to simultaneously regulate many target genes and since to date targeting single genes for therapeutic intervention has not yet succeeded in the clinic.

6.1 *The Role of miRNAs in Ischemic Stroke*

Several studies have demonstrated alterations in the cerebral "miRNA-ome" following ischemia/reperfusion [78–80] suggesting that miRNA may be an important factor in modulating the gene expression cascade that occurs in response to isch emia/reperfusion. Changes in miRNAs with ischemic brain injury have been identified using miRNA profiling techniques in a rat middle cerebral artery occlusion (MCAO) model [78, 79, 81] and in forebrain ischemia [82] as well as in stroke patients [80]. An acute alteration of the miRNA profile following cerebral ischemia would suggest that miRNAs play a role in the early stress response to ischemia in the brain, as either a negative or positive regulator of cell survival. Dharap et al. [81] demonstrated in a rat model of MCAO that while the expression of several miRNAs was altered up to 3 days post-ischemia/reperfusion, a progressive increase in miR-140, miR-145 and miR-331 was observed as early as 3 h following reperfusion.

Recently a few studies have evaluated the significance of individual miRNAs in ischemic brain damage [82–85]. miR-15a has been shown to contribute to the pathogenesis of ischemic vascular injury. Gain or loss of miR-15a significantly reduced or increased oxygen–glucose deprivation induced cerebral vascular EC death, respectively [84]. In vivo repression of miR-497 using antagomirs was found to effectively lower miR-497 levels, reduce MCAO induced infarct, and improve neurological deficits [85]. Transfection of miR-200 b and miR-200 c into Neuro-2a cells increased neural cell survival when subjected to oxygen glucose deprivation [86]. Additionally, miR-210 is positively correlated with better prognosis in stroke patients [87]. A key finding demonstrated by Buller et al. was regional expression of miR-121, primarily in the ischemic penumbra [88]. Such regional specificity of expression adds another layer of complexity to miRNA expression profiles, and may explain seemingly conflicting results in the literature. For example, in a rat model of global cerebral ischemia, Yuan et al. [82] reported that hippocampal miR-181a was upregulated following 30 min of reperfusion, however no change in brain miR-181 was observed either following permanent focal ischemia [79] or following transient focal ischemia [81]. Ouyang et al. demonstrated [83] in a mouse model of MCAO that regional expression of miR-181 differed according to the distribution of blood flow and that anti-miR-181 can protect the brain from ischemia.

6.2 The Role of miRNAs in Hemorrhagic Stroke

The miRNA expression of a hemorrhagic insult to the brain have been investigated [80]. Total of 381 miRNAs were screened with TaqMan miRNA arrays 24 h following intraventricular injection of fresh blood, lysed blood or thrombin. Different patterns of miRNA expression were observed in both blood and brain following hemorrhage. Two miRNAs (miR-498 and miR-200b) were upregulated and one (miR-155) downregulated in both brain and blood in at least two paradigms. However, none of these were significantly different from controls in both brain and blood in at least two paradigms. Following injection of lysed blood, 17 miRNAs were upregulated and 12 downregulated more than twofold in brain, while 21 miRNAs were upregulated and 20 downregulated more than twofold in blood. A report has examined circulating miRNA expression profile in patients with intracerebral hemorrhage, focusing on discriminating intracerebral hemorrhage with hematoma enlargement from intracerebral hemorrhage without hematoma enlargement [89]. In total, 866 miRNAs were screened with microarray analysis, 30 of which were differentially expressed between intracerebral hemorrhage patients with hematoma enlargement and intracerebral hemorrhage patients without hematoma enlargement (Table 17.2). At present, there are no human studies investigating in detail the differential expression of circulating miRNAs in intracerebral hemorrhage compared to normal controls. The ability to predict in the clinic which patients is likely to develop hemorrhage would be more helpful. Furthermore, the authors found that the biological processes implicated by the differentially expressed miRNAs include apoptosis, inflammation, coagulation and collagen biosynthesis [89].

6.3 The Potential Role of miRNAs in Stroke Therapy

There are two major approaches to develop miRNA-based therapeutics: mimics to increase effective levels of a miRNA, and inhibitors or antagomirs to reduce them. miRNA mimics are small, chemically modified, double stranded RNA molecules that load the active strand into the RISC which then binds the target mRNA to induce translational silencing. miRNA mimics can be used to restore a loss of function of beneficial miRNAs. miRNA inhibitors and antagomirs (which differ in their chemical modifications and intended use in vivo) are modified single stranded antisense oligonucleotides harboring the full or partial complementary sequence to the mature miRNA, to reduce endogenous levels of the miRNA and increase expression of its mRNA targets. miRNA inhibitors/antagomirs can inhibit endogenous

Table 17.2 Microarray profiling of microRNAs after stroke

MicroRNAs	Types of stroke	Expression in stroke	Models	References
miR-7, -9, -27a, -29 (b and c), -30e, -92, -98, -101a, -137, -148b, -152a, -204, -218, -301, -338, -335, -369-5p, -376 (b and b), and -424	Ischemia/reperfusion	Down-regulation	Rat-brain	[78]
miR-134, -138, -145, 206, -210, -214, -215, -223, -290, -292-5p, -298, -324-3p, -327, -422b, -451, -494, and -497	Ischemia/reperfusion	Up-regulation	Rat-brain	[78]
miR-155, -362-3p, -223, -210	Ischemia	Down-regulation	Rat-brain	[79]
miR-10a, -182, -200b, -298	Ischemia	Up-regulation	Rat-brain	[79]
hsa-let-7f, miR-126, -1259, -142-3p, -15b, -186, -519e, -768-5p	Ischemia	Down-regulation	Human -blood	[80]
hsa-let-7e, miR-1184, -1246, -1261, -1275, -1285, -1290, -181a, -25, -513a-5p, -550, -602, -665, -891a, -933, -939, -923	Ischemia	Up-regulation	Human-blood	[80]
miR-376b-5p, -153, -29c, -98, -204, -26b, -29b, -338, -301, -341, -377, -664	Transient MCAO	Down-regulation	Rat-brain	[81]
miR-140, -145, -331, -290, -214, -324-5p, -324-3p, -344-3p	Transient MCAO	Up-regulation	Rat-brain	[81]
let-7e, miR-98, -125a-5p, -139-5p, -150, -204, -323, -329, -352, -384-5p, -539, -7a, -7b, -92a, -338 and -92b	Global cerebral ischemia/reperfusion	Down-regulation	Rat-brain	[82]
miR-143, -16, -181a and -495	Global cerebral ischemia/reperfusion	Up-regulation	Rat-brain	[82]
miR-155, -20b-3p, -200a	Hemorrhage	Down-regulation	Rat-brain	[79]
miR-298,a -200b,a - 205, a-345, -423-5p, -298,a-423-5p,a -10a, -345-5p, -674	Hemorrhage	Up-regulation	Rat-brain	[79]

miRNAs and could be applied to reduce miRNAs with pathogenic function in stressed cells or diseased tissues. Endogenous circulating miRNAs have been found to be stable because of their packaging and secretion into the blood within exosomes [33]. Increasing evidence supports the key role of miRNAs in the regulation of exosome function [35]. Our previous studies have reported the expression of miR-126 in EPC-released microvesicles. Using miR-126 mimics or inhibitors could up-regulate and down-regulate the level of miR-126 in both EPCs and EPC-released microvesicles [90]. Moreover, we found that over-expressing of miR-126 could enhance the function of EPCs and EPC-released microvesicles. More recently, our pilot study show the similar results on the EPC-released exosomes (data not published yet). Taken together, exosomes might play an important role in stroke by transferring their carried miRNAs.

7 Exosomes and miRNAs in Stroke Pathogenesis and Pathophysiological Processes

All brain cells release exosomes [55]. Emerging data suggest that exosome-mediated intercellular communication contributes to brain remodeling by transferring cargo from source cells to target cells. Exosomes can be isolated from biofluids such as CSF and from the supernatant of cells cultured in exosome-free medium by centrifugation and other methods [55]. Exosomes are generally enriched with tetraspanin proteins (CD63, CD81), the regulator of endosomal trafficking Alix, and the chaperone protein HSP70, although the content of exosomes varies with cell origin and physiological and pathological conditions [55, 91]. Tetraspanins, Alix, and HSP70 have been used as exosomal markers [91]. Proteomic and RNA analyses have demonstrated that exosomes carry cargoes of lipids, proteins, and RNAs, including mRNAs and miRNAs [55, 91]; however, it is unclear how biological materials are loaded into individual exosomes.

More recently, miRNAs have been reported to mediate the function of exosomes and play an important role in strokes [35, 92]. For example, downregulation of miR-15a in cerebral vessels in a mouse model of focal cerebral ischemia promotes stroke-induced angiogenesis in the peri-infarct region by increasing fibroblast growth factor 2 (FGF2) and vascular endothelial growth factor (VEGF) levels [93]. VEGF released by angiogenic endothelial cells also interacts with its receptor VEGFR2 in neural progenitor cells to promote their proliferation and neuronal differentiation [94]. Moreover, cerebral endothelial cells in white matter are involved in regeneration of myelinating oligodendrocytes through brain-derived neurotrophic factor (BDNF) and FGF2 in injured brain [95]. Stroke-induced limited axonal sprouting and remyelination in the periinfarct region are also regulated by miRNAs. In vitro and in vivo studies showed that stroke-induced downregulation of miR-9 and miR-200b expression in white matter mediates remyelination [96]. Chondroitin sulfate proteoglycans (CSPGs) produced by reactive astrocytes inhibit axonal regrowth [97]. Overexpression of the miR-17–92 cluster or miR-27a in cultured cortical neurons activates neuronal intrinsic growth signals by suppressing phosphatase and tensin homolog (PTEN) and ras homolog family member A (RhoA) signals, thereby overcoming the CSPG inhibitory effect [98, 99]. In vivo studies of spinal cord injury in adult animals have shown that suppression of the PTEN signaling pathway within neurons enhances axonal sprouting even in the presence of CSPGs [98, 100]. Thus, the miRNA and mRNA networks play a pivotal role in mediating brain-repair processes [101].

8 Exosomes and miRNAs in Regeneration and Repair After Stroke

Angiogenesis is a vital component of tissue repair processes after stroke. Besides angiogenesis, neurogenesis is another potential target for treatment of stroke. There is evidence showing that angiogenesis accompanies neuroprotection and neurogenesis during the recovery process of ischemic stroke [102, 103]. Blockade of angiogenesis attenuates the process of neuron generation [103]. Furthermore, neuroblasts migrate along these regenerated vessels to achieve neurogenesis in peri-infarct areas [103–105].

8.1 Exosomes and miRNAs in Cerebral Angiogenesis

In vitro and in vivo experiments have shown that exosomes from circulating EPCs transfer cargo mRNAs associated with the PI3K/Akt signaling pathway and proangiogenic miRNAs, such as miR-126 and miR-296, into recipient ECs [106, 107]. Within the recipient ECs, these miRNAs activate the PI3K/Akt signaling pathway, leading to angiogenesis [106, 107]. In the brain, exosomes from cultured glioblastoma cells induce angiogenesis by delivering their contents of proangiogenic proteins, mRNAs, and miRNAs into cerebral ECs [108]. Additionally, immortalized human brain microvascular ECs secrete exosomes [109]. Proteomic analysis has demonstrated that exosomes released by human cerebral ECs contain 1179 proteins, including several receptors that carry macromolecules across the BBB, such as transferrin receptor and insulin receptor [109]. The role of these exosomal proteins has not been investigated, but interactions between cerebral endothelial exosomes and pericytes have been studied [110]. Exosomes secreted by immortalized mouse cerebral ECs stimulated by lipopolysaccharide and cytokines transferred cargo miRNAs and increased VEGF-B mRNA and protein levels in recipient cerebral vascular pericytes [110].

In addition, activation of the Notch signaling pathway between cerebral ECs and pericytes is required for cerebral angiogenesis and BBB integrity [111]. For example, Delta-like 4 (Dll4), a membrane-bound Notch ligand expressed by cerebral ECs, stimulates Notch3 receptors on pericytes to keep the cerebral vascular structure quiescent [112]. Exosomes released from human microvascular ECs and human umbilical vein ECs contain Dll4 proteins and have been shown to regulate development of angiogenesis [113, 114]. The Notch signaling pathway interacts with the VEGF signaling pathway [115, 116]. Together, these data suggest that cerebral endothelial exosomes could communicate with pericytes to mediate angiogenesis and to maintain BBB integrity through the VEGF and Notch signaling pathways.

8.2 Exosomes and miRNAs in Neurogenesis

Neural stem cells in the SVZ exist in a unique niche where they contact blood vessels, neighboring cells, and CSF, constantly exchanging molecular signals [117]. There is evidence that exosomes in CSF and neural stem cells mediate neural stem cell function and immune system function, respectively, by regulating intercellular pathways [74, 118]. Exosomes isolated from embryonic CSF of rats and humans contain protein and miRNA components of the insulin-like growth factor (IGF) signaling pathway [74]. CSF-exosome cargoes, including both proteins and miRNAs, are highly conserved between rodent and human [74]. Incubation of embryonic neural stem cells with CSF exosomes activate the IGF/mTORC1 pathway in the neural stem cells and promoted stem cell proliferation [74]. Exposure of neural stem cells derived from SVZ of adult mouse to proinflammatory cytokines leads to release of exosomes enriched with mRNAs encoding components of the interferon gamma (IFN-γ) signaling pathway [118]. These exosomes activate signal transducer and activator of transcription 1 (STAT1) signaling in recipient cells through exosome-associated IFN-γ and its receptor IFNGR1 [118]. Stroke activates innate and adaptive immune responses [119]; thus, exosomes released by neural stem cells may also communicate with the immune system after stroke.

Moreover, exosomes released by cultured primary cerebral ECs and neural progenitor cells (NPCs) harvested from nonischemic and ischemic animals. Proteomic and miRNA array analyses revealed that stroke substantially changed exosomal cargo proteins and miRNAs compared with exosomes from those nonischemic cells, indicating that stroke alters exosomal contents from cerebral ECs and NPCs. Exosomes derived from ischemic NPCs promoted primary EC migration and capillary tube formation, whereas exosomes from ischemic cerebral ECs enhanced NPC proliferation and neuronal differentiation. These data suggest that exosomes secreted by cerebral ECs and NPCs contribute to the observed coupling of neurogenesis and angiogenesis during brain repair processes after stroke. In addition, cerebral endothelial exosomes could also actively engage in brain remodeling by communicating with brain cells, including neurons and glia, and with remote cells in other organs during stroke recovery [120].

8.3 Exosomes and miRNAs in Neuronal Plasticity

Neurons and glia actively communicate with each other to coordinate axonal growth and myelination. Emerging data suggest that exosomes released by neurons and glia contribute to these processes [55, 121, 122]. Exosomes released by cultured cortical neurons carried the neuronal-specific protein L1 cell adhesion molecule (L1CAM) and the glutamate receptor AMPA R2/3 (GluR2/3) subunits, but not the NR1 subunits, of glutamate receptors [123, 124]. Increasing cytosolic calcium in neurons and neuronal depolarization augmented the secretion of exosomes [123–125].

Notably, exosomes released by neurons contain alpha-amino-3-hydroxy-5-methyl-4-isoxazolepropionic acid (AMPA) receptors, while exosomes secreted from neurites of depolarized neurons are enriched with microtubule associated protein 1b (MAP1b) and miRNAs that target genes involved in neurite plasticity [124, 125]. Exosomes from neurons treated with a retinoic acid receptor β2 (RARβ) agonist had a dual effect on neurons and astrocytes to inactivate PTEN signaling, leading to enhancement of neurite outgrowth [126]. RARβ agonist inactivated cortical neuron PTEN signaling by releasing exosomes enriched with PTEN. These PTEN-enriched neuronal exosomes transferred PTEN proteins into astrocytes to suppress astrocyte proliferation [126]. In addition to neurons, exosomes released by cortical neurons transferred miR-124 to astrocytes and increased the expression of the excitatory amino acid transporter GLT-1 in astrocytes, suggesting a role for neuronal exosomes in regulating astrocyte function [127]. AMPA receptors and MAP1b are key regulators of synaptic and dendritic plasticity and axonal spouting [124, 125, 128]. The astrocyte glutamate transporter GLT-1 in rodent regulates extracellular glutamate levels and modulates synaptic activation [127]. Activation of AMPA receptors contributes to motor function recovery after stroke [129]. Suppression of neuronal PTEN signals and reduction of an astrocyte scar promote axonal sprouting in adult CNS after spinal cord injury and stroke [98, 100, 130]. Together, these data suggest that neuronal exosomes mediate synaptic and axonal plasticity by synaptic transfer of their cargo between neurons and by communication with astrocytes [124, 131], which potentially mediate axonal and synaptic remodeling in the ischemic brain.

9 Exosomes and miRNAs as New Noninvasive Biomarkers or Diagnostic Tools of Stroke

Bodily fluids are also promising sources of molecular biomarkers, which can be divided into three categories: molecules, proteins, and mRNAs/miRNAs [132]. The advantage of biomarkers obtained from bodily fluids (i.e., CSF, blood, plasma, serum, saliva, and urine) is the possibility of searching for a large number of molecules at once, for example, by the use of proteomics or genomics, at earlier stages than those could be detected by imaging. Having a biomarker that can be accessed from a body fluid has the added advantage of not relying on expensive technology (such as imaging equipment) and, in some instances (e.g., saliva and urine), of avoiding invasive method. The main disadvantage of obtaining biomarkers from body fluids is the low levels of molecules and the heterogeneity of these, as such samples arise from a wide number of tissues. Thus, to circumvent this difficulty and to improve the specificity of the biomarker, exosomes have the most potential. Exosomes can be isolated from all bodily fluids, and they carry a complex cargo consisting of miRNAs, proteins, lipids, and DNA that in part depends on the tissue of origin and its "health or disease" state [133, 134]. Catalytically active enzymes like PTEN in neurons, via can be transferred by exosomes to astrocytes to prevent glial scar formation and induce spinal cord regeneration [126]. Exosomes carry a set of common proteins considered as "exosome markers," most of them related to their biogenesis [25, 26], such as CD36, CD9, and etc. In addition, they carry molecules that derived from their parent cells, for example, membrane and intraluminal proteins. In the case of transmembrane proteins, they can be used to immunoisolate exosomes of a specific cellular origin, separating them from other exosomes and thus improving the sensitivity of exosomes as biomarkers [31, 32]. Of all the molecules carried by exosomes, miRNAs are the ones that have gathered the most interest in the last years. One of the aspects that make miRNAs as promising biomarkers is that they can be found in body fluids that are easily accessible, such as plasma, where they appear to be transported by lipoproteins and exosomes [135]. Exosomes provide an enriched source of miRNAs for biomarker profiling [136], and miRNAs present in blood-derived exosomes have been linked to strokes [31, 137].

There are increasing reports on the application of exosome-derived miRNAs as biomarkers for vascular diseases [138]. The possibility of extracting high quality miRNAs and profiling them using well-established methods has also contributed to making them a favored area of study in the search for biomarkers [139]. Researchers have attempted to profile the miRNA identity from body fluids of patients with ischemic stroke, hemorrhagic stroke, ischemic preconditioning and hypoxic injury, however, there are important difficulties that need to be considered if some of these molecules are going to be proposed as reliable biomarkers. The methodologies used for these profiles are not consistent between laboratories, and the sample sizes are usually small; thus, validating the miRNAs associated with neurological disorders has proven difficult [140]. It is still an ongoing process with a considerable degree of variability and efforts are constantly made to increase specificity and sensitivity.

It is well known that different kind of cells produces exosomes with a specific parental molecular signature [141]. For example, B cell receptor is selectively expressed on B cell derived exosomes, as CD11c, a specific marker of DC, is present on DC-derived exosomes [142]. Similarly oligodendrocyte derived exosomes contain the myelin associated proteins PLP/DM20 [143]. As we reported, the exosomes derived from ECs contain EC specific markers, such as CD144 and CD105. Similarly, exosomes-released from EPCs show positive to the EPC specific markers, VEGFR2 and CD34 [31]. Coherently, exosome cargo depends on the physiological/pathophysiological state of the cell when produces it [144]; for instance, inflammatory and hypoxic stimuli change the protein and RNA content of EC-derived exosomes [145]. In the acute ischemic stroke, exosomal miR-223 expression in stroke patients with poor outcomes was higher than those with good outcomes. Increased exosomal miR-223 was associated with acute ischemic stroke occurrence, stroke severity, and short-term outcomes [137]. Thus, increased circulating exosomal miRNA-223 could be a novel biomarker for ischemic stroke diagnosis. Similarly, another study has found that serum exosomal miR-9 and miR-124 are promising biomarkers for diagnosing acute ischemic stroke and evaluating the degree of damage caused by ischemic injury [146]. Larger sample are needed for further verification. Moreover, so far, there is no report on the biomarker role of exosomes for the hemorrhagic stroke.

Thus, exosomes have a great potential as noninvasive diagnostic tools for stroke. More important, considering that pathologies are related to alterations of many different cells, it is intriguing to know whether changes in the molecular signature of specific cell population are induced in response to strokes. A good challenge to solve this question is to identify a specific transmembrane protein that might be used as a marker to capture peripherally the exosomes derived from different cell types. For examples, we have identified EC-exosomes and EPC-exosomes by using the microbeads and Qdot methods following by the NTA analysis. We found that the circulating EC-exosomes and EPC-exosomes are increased after stroke on day 1, 3 and 5 in a time-dependent manner [32].

10 A Novel Approach of Stem Cell-Released Exosomes in Treating Stroke

For the treatment of ischemic stroke, thrombolytic therapy is limited by a 4 h therapeutic time window, and interventional therapies, such as angioplasty and stenting, having a high rate (20%) of re-stroke within the first year [147]. At this time, there are no effective neuroprotective drugs. Unlike the ischemic infarct, which often has an acute onset, hemorrhagic stroke usually has a progressive onset. Current treatment for hemorrhagic stroke is based on the prevention of secondary brain injury, including rebleeding and secondary brain ischemia.

The aim of stroke therapy is to restore the lost neural tissue or stimulate brain plasticity to improve the functional outcome, which subsequently improves the quality of life of patients with permanent disabilities. Stem cell based therapies have been shown to be a promising approach in achieving such results [148]. Different types of stem cells also have the potential to induce or accelerate functional recovery in animal models of intracerebral hemorrhagic stroke and subarachnoid hemorrhage [149–151]. For examples, bone marrow-derived MSCs have been shown to improve neurological outcome after stroke [105, 152], and MSC therapy is in clinical trials for stroke [153, 154]. Preclinical studies have demonstrated that MSCs promote angiogenesis, neurogenesis, and white matter remodeling in the injured brain by secreting factors to trigger the signaling pathways that are involved in brain repair [105]. Intravenous infusion of MSCs inhibits intracranial hemorrhage after recombinant tissue plasminogen activator therapy for transient MCAO in rats [155]. We have reported that bone marrow-derived EPCs have also been shown to protect the brain from acute ischemic injury via cerebrovascular protection and promote neurological recovery via increasing angiogenesis and neurogenesis [64, 156]. NPCs have also been shown to offer neuroprotection by modulating the BBB and microglial functions [157], and stimulating post-ischemic angiogenesis [158].

However, one of the key limits for a noninvasive systemic therapy of strokes is the fact that several substances are not able to cross the BBB, which is a multicellular interface composed of pericytes, astrocytes, and epithelium that becomes paracellularly impermeable to certain molecules (e.g., drugs) and to most of the cells of the blood stream [159]. As a novel cell–cell communicator, exosomes released from stem cell have several advantages over stem cells: (1) exosomes could cross the BBB; (2) exosomes would not cause tumorigenesis because of lacking self-proliferation; (3) ex vivo preparation of stem cell-released exosomes can avoid these limitations that the function of stem cells could be impaired by the risk factors; (4) In addition, exosomes are relatively easy to be modified, stored and administrated. Therefore, stem cell-released exosomes hold the most potential for stroke therapy.

10.1 Exosomes as Drug Delivery Vehicles to Cross the BBB

Extensive research has been done using exosomes as vehicles for therapeutic drug delivery. One study involved the use of exosomes to deliver curcumin and treat an inflammatory disease [160]. Exosomes are employed to form a complex with curcumin for the purpose of enhancing curcumin's effectiveness. Clinical trials have also shown its efficacy and safety for cancer patients [161]. Besides enhancing the properties of drugs, exosomes are also employed to carry small molecular drugs across the BBB. Indeed, 98% of potent central nervous system drugs cannot cross the BBB and their conceptual efficacy shown in labs have not been successful in clinical trials [162]. Many Nano-formulations have been employed to solve the problems associated with the permeability of drugs across the BBB. However, other problems, such as nano-toxicity and rapid drug clearance by the mononuclear phagocyte system, have also been observed [163]. To compensate for these complications, polyethyleneglycol (PEG) has been introduced to decrease mononuclear phagocyte system drug uptake. However, this resulted in reduced interaction between target cells, consequently decreasing drug distribution in the brain [164, 165]. In this case, exosomes, an product of the body's own cells, can cross the BBB, thus improving drug transport to the brain by decreasing mononuclear phagocyte system drug clearance. Meanwhile, research has been carried out encapsulating anticancer drugs such as paclitaxel and doxorubicinin to exosomes, showing the potential of exosomes for brain delivery across the BBB and explaining their transport mechanisms, using zebrafish as an animal model [166]. In that study, exosomes were isolated from various cell lines, including gliobastoma astrocytoma U-87 MG, endothelial bEND.3, neuroecto dermal tumor PFSK-1, and glioblastoma A-172, using Invitrogen® total exosome RNA and a protein isolation kit. Exosomes were loaded with rhodamine123 and paclitaxelor doxorubicin through mixing and incubation. Experiments were then performed to characterize the isolated exosomes, the cellular uptake of the exosomes containing rhodamine123, and the cytotoxicity of the delivered anticancer drugs within the exosomes of U-87 MG and bEND.3 cells. In vivo, the ability of exosomes to deliver drugs across the BBB was examined by injecting bEND.3-derived exosomes loaded with rhodamine123 and doxorubicin, orpaclitaxel, in zebrafish embryos. At the end of the experiments, brain tissue was examined for the presence of rhodamine123 fluorescence. The results undeniably showed the drug's distribution in the brain region of the zebrafish embryos, suggesting the ability of exosomes to deliver drugs across the BBB. In subsequent experiments, a primary brain cancer model was developed using zebrafish and anticancer drugs loaded within and without exosomes, which were compared. The data showed significant therapeutic efficacy in the zebrafish brain model treated with exosomes loaded doxorubicin compared to doxorubicin alone. Overall, the results obtained from the study showed the potential of exosomes to deliver small molecule drugs across the BBB to treat both brain cancers and neurological disorders. Our pilot study showed that intravenously infused exosomes could across BBB in the stroke brain. By tagging exosomes with red fluorescence PKH26, we observed that EPC-released exosomes could merge with ECs, astrocytes and neurons dominantly in the peri-infarct area after intravenous administration.

10.2 Exosomes as Drug Delivery Vehicles for miRNAs

Exosomes are known to naturally carry miRNA and, hence, it is logical to use exosomes as a therapeutic vehicle to deliver miRNA to targeted cells. For example, let-7a was introduced into GE11 exosomes by the lipofection method. Then, GE11 exosomes containing let-7a were intravenously injected into tumor-bearing mice. Since previous studies showed let-7a inhibited tumor growth by reducing the expression level of RAS and HMGA2, the expressions of these genes were examined in injected tumor-bearing mice, using real-time reverse transcription-PCR analysis, immunoblotting, and immunostaining. The results showed that let-7a delivered by GE11-exosomes strongly inhibited the expression of HMGA2 in cancer cells.

Moreover, intranasal administration of Odyssey 800 dye-labeled exosomes derived from a glioblastoma cell line led to distribution of fluorescent particles throughout the brain, mainly in the olfactory bulb in mice [167]. Using the Cre-loxP system, studies have demonstrated that intrahippocampal injection of Cre-recombinase mRNA containing exosomes into mice with a ROSA26-lacZ reporter activated a lacZ reporter in the hippocampal neurons [168], indicating that Cre-recombinase mRNA within the exosomes activates the reporter gene in recipient neurons. Furthermore, intravenous injection of exosomes expressing a fusion protein consisting of the neuronspecific rabies virus glycoprotein (RVG) peptide with the exosomal membrane protein LAMP2B demonstrated targeting of neurons, microglia, and oligodendrocytes in the brain [169]. Intravenous administration of the RVG peptide-expressing exosomes carrying siRNA against opioid receptor mu (MOR) enhanced the movement of exosomes across the BBB and inhibited MOR expression in the brain [170]. This data suggests that exosomes not only cross the BBB, but also deliver functional cargo to trigger gene expression in specific recipient cell types in the brain.

Our study found that EPC-released exosomes (EPC-EXs) and NPCs-released exosomes (NPC-EXs) could transfer their carried miRNAs to the vascular and brain cells. For examples, EPC-EXs transfer their carried miR-126 to the ECs, neurons and astrocytes after co-incubation. Up-regulating the exogenous miRNAs, such as miR-210, miR-126 and miR-18a, in EPCs and NPCs could increase the expression levels of miR-210, miR-126 and miR-18a in EPC-EXs and/or NPC-EXs; and the miR-126 or miR-210 enriched EPC-EXs and/or NPC-EXs could deliver those miRNAs to the ECs, neurons and astrocytes after co-incubation. This indicates that exosomes do successfully deliver their cargo to the target cells, showing promising characteristics for drug delivery [171].

10.3 MSC Derived Exosomes and Therapies

Cultured MSCs secrete a large quantity of exosomes [172]. Emerging data from independent laboratories indicate that exosomes released from MSCs provide therapeutic benefits in stroke by modulating the brain microenvironment [173–178]. Intravenous administration of MSC-derived exosomes to rats subjected to focal cerebral ischemia substantially improved neurological function by promoting neurovascular remodeling during stroke and TBI recovery [176–178]. Subsequently, the therapeutic effect of MSC-derived exosomes has also been demonstrated by independent laboratories in the mouse subjected to stroke [173, 174]. Systemic administration of MSC-derived extracellular vesicles to ischemic mice markedly reduced motor coordination deficits and enhanced angiogenesis and neurogenesis, while treatment of TBI mice with human MSC derived extracellular vesicles substantially preserved spatial leaning ability [173, 174]. Improved neurological outcomes from these MSC-derived exosome studies are comparable to the therapeutic effect observed with MSC therapy, suggesting that MSC-derived exosome–mediated cell-cell communication may contribute to the therapeutic effect of the MSC therapy.

Exosomes transfer their cargo miRNAs to recipient cells [179, 180]. The effect of engineered MSC-derived exosomes that carry elevated miRNAs on brain remodeling after stroke has been investigated in vitro and in vivo [175, 177]. Treatment of stroke models with MSCs abolished stroke-induced downregulation of miR-133b in the ischemic brain [175]. When MSCs were cultured with extracts harvested from ischemic brain tissues, they released exosomes enriched with miR-133b. Tailored MSC-derived exosomes with elevated or reduced miR-133b were harvested from the supernatant of MSCs transfected with lentiviral vectors carrying pre–miR-133b or anti-miR-133b, respectively [175, 177]. Intravenous administration of tailored MSC-derived exosomes with increased or decreased miR-133b to rats with stroke led to enhancement or exacerbation, respectively, of axonal remodeling and neurological function compared with naturally occurring MSC-derived exosomes [177]. Connective tissue growth factor (CTGF) and RhoA are putative targets of miR-133b and are known to suppress neurite growth [175]. In vitro, incubation of cortical neurons with miR-133b-elevated exosomes downregulated RhoA and enhanced neurite outgrowth, whereas treatment of astrocytes with miR-133b-elevated exosomes suppressed CTGF, which is mainly expressed by astrocytes [177, 181].

Collectively, this data indicates that MSC derived exosomes may be used as vehicles to transport miRNAs that modulate genes in the recipient neurons and astrocytes.

10.4 Hematopoietic Stem Cells (HSCs)-Derived Exosomes and Therapies

Studies from myocardial ischemia have shown that engineered exosomes with elevated sonic hedgehog (Shh) derived from CD34+ HSCs transferred functional Shh and activated the Shh signaling pathway in recipient cells, enhancing angiogenesis in the border zone of infarction and preserving cardiac function [182]. Shh [183–185] plays an important role in the regulation of adult neurogenesis under physiological and pathological conditions [186–189]. These experiments suggest that exosomes can deliver functional proteins to modulate cellular function of recipient cells and that treatment of stroke with tailored Shh-exosomes could facilitate brain remodeling.

10.5 EPC-Derived Exosomes and Therapies

EPCs, defined as bone marrow-derived immature cells with the ability to differentiate into mature ECs, are known to participate in vascular homeostasis and angiogenesis [190–192]. EPCs have been shown to have therapeutic effects on ischemic stroke by promoting angiogenesis, neuroprotection and neurogenesis during the recovery process of ischemic stroke [102, 103]. The mechanism has been partially ascribed to EPC released growth factors and EVs, including microvesicles and exosomes. Recent studies suggest that EPC-released microvesicles could promote EC survival and proliferation, and enhance EC function [107, 193]. Our in vitro studies have shown that EPC-released microvesicles protect ECs from hypoxia/reoxygenation (H/R)-induced injury [46] and protect cardiomyocytes from angiotensin II-induced injury [194]. Additionally, another study showed that EPC-released microvesicles enhance angiogenesis of human pancreatic islets [106]. Moreover, an in vivo study demonstrated that intravenous administration of EPC-released microvesicles protects kidneys from ischemia/reperfusion injury [195] and improve neovascularization in a murine model of hindlimb ischemia [196].

Functions of EVs in physiological and pathological processes depend on the ability of EVs to interact with recipient cells to deliver their contents (proteins, RNAs and miRs) [35]. EPC-released microvesicles have been shown to enhance angiogenesis of human pancreatic islets via carried miR-126 and miR-296 [106] and protect kidneys from ischemia/reperfusion injury through transferring pro-angiogenic miR-126 and miR-296 to target cells [195]. Our recent work demonstrated the expression of miR-126 in EPC-released microvesicles and containing high amounts of miR-126 is associated with the protective effects of EPC-released microvesicles on ECs [46]. As we know, here are two types of EVs: microvesicles and exosomes [197]. Recent studies found that most of the miRs in the circulation are carried by exosomes [35, 198]. Of not, our recent study found that EPC-exosomes have protective effects on H/R-induced EC and neuron injury.

On top of this, gene-modified exosomes that carry specific miRs have been shown to exert better efficacy than null exosomes for treating stroke. Our previous work has shown that miR-126 primed EPC-released microvesicles have better protective effects than EPC-released microvesicles on protecting EPCs from high glucose-induced injury [90]. More important, we recently found that exosomes released from miR-126 primed EPCs (miR-126-EPC-EXs) exhibited better efficacy than EPC-EXs in persevering EC and neuron functions under H/R condition (data not published). These evidence suggests that EPC-EXs or gene-modified EPC-EXs would offer a novel approach for treating ischemic stroke by protecting brain from acute injury and promoting angiogenesis and neurogenesis for the long-term recovery.

10.6 NPC-Derived Exosomes and Therapies

As one type of stem cells in the brain, NPCs residing in the subventricular zone contact the blood vessels and directly juxtapose to ECs [199]. The two types of cells could interact with each other through direct physical contact or through paracrine mechanisms with potentially different biological effects. As showed in our previous study [200], NPCs can decrease H/R-induced ROS overproduction on ECs. NPCs could also offer neuroprotection by modulating the BBB and microglial functions [157]. Besides neuronal protection, transplanted NPCs can stimulate post-ischemic angiogenesis.

Brain functions rely on intercellular communication between neural cells, astrocytes and vascular cells. The potential role of stem cell exosomes in brain "microenvironment" manifests in several aspects such as protection of brain cells (neurons, glial cells, ECs and NPCs), maintenance of BBB hemostasis, and tissue repair mechanisms (angiogenesis and neurogenesis). Our recent data revealed that NPC-EXs have protective effects on H/R-induced neuron and astrocyte injury.

10.6.1 Perspectives

Although many studies dealing with the role of exosomes in stem cell differentiation and the use of exosomes isolated from stem cells for treatment of several diseases have been published, the involved mechanisms remain largely unknown. Further understanding of these mechanisms, which include the involved cellular pathways, may improve the use of exosomes in diagnostic and treatment methods, especially for those involving stem cells.

One of the promising avenues of exosomes-based therapy for strokes might be the combination of different stem cell-released exosomes, which could provide synergistic effects for treating strokes. We have recently discovered that EPCs and NPCs act synergistically to protect cerebral ECs from H/R-induced injury through VEGF and BDNF pathways [200]. Moreover, we found that co-transplantation of EPCs and NPCs provides synergistic effects of treating ischemic stroke by alleviating infarct volume and promoting angiogenesis and neurogenesis (data not published). Altogether, these demonstrate a new conception of angioneurogenesis coupling [157, 201]. Of note, we found that EPC-EXs and NPC-EXs synergistically protect ECs and neurons from H/R-induced apoptosis. Theoretically, combination of EPC-EXs and NPC-EXs could provide a therapeutic strategy protecting neurovascular unit by targeting both the acute and later neurological recovery phases.

Overall, an exosome-based delivery system has particular benefits such as (1) specificity, as the exosomes deliver their cargo to a specific target; (2) safety, as self-derived exosomes promote no undesired immunogenicity; and (3) stability, not only the exosomes itself as nanostructures circulating in the blood were reported stable but also the content of exosomes are protected from RNases and proteases and, therefore, can be delivered in an intact form to the target cell [29]. Despite these benefits, there are some, until now, unsolved problems such as identifying and purifying a single subpopulation of endogenous cell-specific exosomes.

References

1. SL P, GE P. The fine structure of neurons. J Biophys Biochem Cytol. 1955;1(1):69–88.
2. Piper RC, Katzmann DJ. Biogenesis and function of multivesicular bodies. Annu Rev Cell Dev Biol. 2007;23:519–47.
3. Von Bartheld CS, Altick AL. Multivesicular bodies in neurons: distribution, protein content, and trafficking functions. Prog Neurobiol. 2011;93(3):313–40.
4. Trams EG, Lauter CJ, Salem N Jr, Heine U. Exfoliation of membrane ecto-enzymes in the form of micro-vesicles. Biochim Biophys Acta. 1981;645(1):63–70.
5. Fleury A, Martinez MC, Le LS. Extracellular vesicles as therapeutic tools in cardiovascular diseases. Front Immunol. 2014;5:370.
6. Harding C, Heuser J, Stahl P. Endocytosis and intracellular processing of transferrin and colloidal gold-transferrin in rat reticulocytes: demonstration of a pathway for receptor shedding. Eur J Cell Biol. 1984;35(2):256–63.
7. Heijnen HF, Debili N, Vainchencker W, Breton-Gorius J, Geuze HJ, Sixma JJ. Multivesicular bodies are an intermediate stage in the formation of platelet alpha-granules. Blood. 1998;91(7):2313–25.
8. Heijnen HF, Schiel AE, Fijnheer R, Geuze HJ, Sixma JJ. Activated platelets release two types of membrane vesicles: microvesicles by surface shedding and exosomes derived from exocytosis of multivesicular bodies and alpha-granules. Blood. 1999;94(11):3791–9.
9. Thery M, Piel M. Adhesive micropatterns for cells: a microcontact printing protocol. Cold Spring Harb Protoc. 2009;2009(7):db.
10. Lazaro-Ibanez E, Sanz-Garcia A, Visakorpi T, Escobedo-Lucea C, Siljander P, Ayuso-Sacido A, et al. Different gDNA content in the subpopulations of prostate cancer extracellular vesicles: apoptotic bodies, microvesicles, and exosomes. Prostate. 2014;74(14):1379–90.

11. Revenfeld AL, Baek R, Nielsen MH, Stensballe A, Varming K, Jorgensen M. Diagnostic and prognostic potential of extracellular vesicles in peripheral blood. Clin Ther. 2014;36(6):830–46.
12. Denzer K, Kleijmeer MJ, Heijnen HF, Stoorvogel W, Geuze HJ. Exosome: from internal vesicle of the multivesicular body to intercellular signaling device. J Cell Sci. 2000;113(Pt 19):3365–74.
13. Stoorvogel W, Kleijmeer MJ, Geuze HJ, Raposo G. The biogenesis and functions of exosomes. Traffic. 2002;3(5):321–30.
14. Fernandez-Borja M, Wubbolts R, Calafat J, Janssen H, Divecha N, Dusseljee S, et al. Multivesicular body morphogenesis requires phosphatidyl-inositol 3-kinase activity. Curr Biol. 1999;9(1):55–8.
15. Raposo G, Stoorvogel W. Extracellular vesicles: exosomes, microvesicles, and friends. J Cell Biol. 2013;200(4):373–83.
16. Mobius W, Ohno-Iwashita Y, van Donselaar EG, Oorschot VM, Shimada Y, Fujimoto T, et al. Immunoelectron microscopic localization of cholesterol using biotinylated and non-cytolytic perfringolysin O. J Histochem Cytochem. 2002;50(1):43–55.
17. Trajkovic K, Hsu C, Chiantia S, Rajendran L, Wenzel D, Wieland F, et al. Ceramide triggers budding of exosome vesicles into multivesicular endosomes. Science. 2008;319(5867):1244–7.
18. Simons M, Raposo G. Exosomes--vesicular carriers for intercellular communication. Curr Opin Cell Biol. 2009;21(4):575–81.
19. Record M, Subra C, Silvente-Poirot S, Poirot M. Exosomes as intercellular signalosomes and pharmacological effectors. Biochem Pharmacol. 2011;81(10):1171–82.
20. Thery C, Ostrowski M, Segura E. Membrane vesicles as conveyors of immune responses. Nat Rev Immunol. 2009;9(8):581–93.
21. Thery C, Zitvogel L, Amigorena S. Exosomes: composition, biogenesis and function. Nat Rev Immunol. 2002;2(8):569–79.
22. Ostrowski M, Carmo NB, Krumeich S, Fanget I, Raposo G, Savina A, et al. Rab27a and Rab27b control different steps of the exosome secretion pathway. Nat Cell Biol. 2010;12(1):19–30.
23. Lamparski HG, Metha-Damani A, Yao JY, Patel S, Hsu DH, Ruegg C, et al. Production and characterization of clinical grade exosomes derived from dendritic cells. J Immunol Methods. 2002;270(2):211–26.
24. Zitvogel L, Regnault A, Lozier A, Wolfers J, Flament C, Tenza D, et al. Eradication of established murine tumors using a novel cell-free vaccine: dendritic cell-derived exosomes. Nat Med. 1998;4(5):594–600.
25. Lai RC, Yeo RW, Tan KH, Lim SK. Exosomes for drug delivery – a novel application for the mesenchymal stem cell. Biotechnol Adv. 2013;31(5):543–51.
26. Vlassov AV, Magdaleno S, Setterquist R, Conrad R. Exosomes: current knowledge of their composition, biological functions, and diagnostic and therapeutic potentials. Biochim Biophys Acta. 2012;1820(7):940–8.
27. Qin J, Xu Q. Functions and application of exosomes. Acta Pol Pharm. 2014;71(4):537–43.
28. Ha D, Yang N, Nadithe V. Exosomes as therapeutic drug carriers and delivery vehicles across biological membranes: current perspectives and future challenges. Acta Pharm Sin B. 2016;6(4):287–96.
29. Kooijmans SA, Vader P, van Dommelen SM, van Solinge WW, Schiffelers RM. Exosome mimetics: a novel class of drug delivery systems. Int J Nanomedicine. 2012;7:1525–41.
30. Crescitelli R, Lasser C, Szabo TG, Kittel A, Eldh M, Dianzani I, et al. Distinct RNA profiles in subpopulations of extracellular vesicles: apoptotic bodies, microvesicles and exosomes. J Extracell Vesicles. 2013;2:eCollection 2013.
31. Wang J, Guo R, Yang Y, Jacobs B, Chen S, Iwuchukwu I, et al. The novel methods for analysis of exosomes released from endothelial cells and endothelial progenitor cells. Stem Cells Int. 2016;2016:2639728.
32. Wang J, Zhong Y, Ma X, Xiao X, Cheng C, Chen Y, et al. Analyses of endothelial cells and endothelial progenitor cells released microvesicles by using microbead and Q-dot based nanoparticle tracking analysis. Sci Rep. 2016;6:24679.

33. Hunter MP, Ismail N, Zhang X, Aguda BD, Lee EJ, Yu L, et al. Detection of microRNA expression in human peripheral blood microvesicles. PLoS One. 2008;3(11):e3694.
34. Chen TS, Lai RC, Lee MM, Choo AB, Lee CN, Lim SK. Mesenchymal stem cell secretes microparticles enriched in pre-microRNAs. Nucleic Acids Res. 2010;38(1):215–24.
35. Valadi H, Ekstrom K, Bossios A, Sjostrand M, Lee JJ, Lotvall JO. Exosome-mediated transfer of mRNAs and microRNAs is a novel mechanism of genetic exchange between cells. Nat Cell Biol. 2007;9(6):654–9.
36. Baglio SR, Pegtel DM, Baldini N. Mesenchymal stem cell secreted vesicles provide novel opportunities in (stem) cell-free therapy. Front Physiol. 2012;3:359.
37. Baglio SR, Rooijers K, Koppers-Lalic D, Verweij FJ, Perez LM, Zini N, et al. Human bone marrow- and adipose-mesenchymal stem cells secrete exosomes enriched in distinctive miRNA and tRNA species. Stem Cell Res Ther. 2015;6:127.
38. Boon RA, Vickers KC. Intercellular transport of microRNAs. Arterioscler Thromb Vasc Biol. 2013;33(2):186–92.
39. Ohshima K, Inoue K, Fujiwara A, Hatakeyama K, Kanto K, Watanabe Y, et al. Let-7 microRNA family is selectively secreted into the extracellular environment via exosomes in a metastatic gastric cancer cell line. PLoS One. 2010;5(10):e13247.
40. Montecalvo A, Larregina AT, Shufesky WJ, Stolz DB, Sullivan ML, Karlsson JM, et al. Mechanism of transfer of functional microRNAs between mouse dendritic cells via exosomes. Blood. 2012;119(3):756–66.
41. Macfarlane LA, Murphy PR. MicroRNA: biogenesis, function and role in cancer. Curr Genomics. 2010;11(7):537–61.
42. Ratajczak J, Wysoczynski M, Hayek F, Janowska-Wieczorek A, Ratajczak MZ. Membrane-derived microvesicles: important and underappreciated mediators of cell-to-cell communication. Leukemia. 2006;20(9):1487–95.
43. Palma CA, Tonna EJ, Ma DF, Lutherborrow MA. MicroRNA control of myelopoiesis and the differentiation block in acute myeloid leukaemia. J Cell Mol Med. 2012;16(5):978–87.
44. Bobrie A, Colombo M, Raposo G, Thery C. Exosome secretion: molecular mechanisms and roles in immune responses. Traffic. 2011;12(12):1659–68.
45. Hess C, Sadallah S, Hefti A, Landmann R, Schifferli JA. Ectosomes released by human neutrophils are specialized functional units. J Immunol. 1999;163(8):4564–73.
46. Wang J, Chen S, Ma X, Cheng C, Xiao X, Chen J, et al. Effects of endothelial progenitor cell-derived microvesicles on hypoxia/reoxygenation-induced endothelial dysfunction and apoptosis. Oxidative Med Cell Longev. 2013;2013:572729.
47. Guescini M, Genedani S, Stocchi V, Agnati LF. Astrocytes and glioblastoma cells release exosomes carrying mtDNA. J Neural Transm (Vienna). 2010;117(1):1–4.
48. Guescini M, Guidolin D, Vallorani L, Casadei L, Gioacchini AM, Tibollo P, et al. C2C12 myoblasts release micro-vesicles containing mtDNA and proteins involved in signal transduction. Exp Cell Res. 2010;316(12):1977–84.
49. Street JM, Barran PE, Mackay CL, Weidt S, Balmforth C, Walsh TS, et al. Identification and proteomic profiling of exosomes in human cerebrospinal fluid. J Transl Med. 2012;10:5.
50. Banigan MG, Kao PF, Kozubek JA, Winslow AR, Medina J, Costa J, et al. Differential expression of exosomal microRNAs in prefrontal cortices of schizophrenia and bipolar disorder patients. PLoS One. 2013;8(1):e48814.
51. Regehr WG, Carey MR, Best AR. Activity-dependent regulation of synapses by retrograde messengers. Neuron. 2009;63(2):154–70.
52. Korkut C, Li Y, Koles K, Brewer C, Ashley J, Yoshihara M, et al. Regulation of postsynaptic retrograde signaling by presynaptic exosome release. Neuron. 2013;77(6):1039–46.
53. Chivet M, Javalet C, Laulagnier K, Blot B, Hemming FJ, Sadoul R. Exosomes secreted by cortical neurons upon glutamatergic synapse activation specifically interact with neurons. J Extracell Vesicles. 2014;3:24722.
54. Nave KA, Trapp BD. Axon-glial signaling and the glial support of axon function. Annu Rev Neurosci. 2008;31:535–61.

55. Fruhbeis C, Frohlich D, Kuo WP, Kramer-Albers EM. Extracellular vesicles as mediators of neuron-glia communication. Front Cell Neurosci. 2013;7:182.
56. Bahrini I, Song JH, Diez D, Hanayama R. Neuronal exosomes facilitate synaptic pruning by up-regulating complement factors in microglia. Sci Rep. 2015;5:7989.
57. Fitzner D, Schnaars M, van Rossum D, Krishnamoorthy G, Dibaj P, Bakhti M, et al. Selective transfer of exosomes from oligodendrocytes to microglia by macropinocytosis. J Cell Sci. 2011;124(Pt 3):447–58.
58. Abbott NJ, Patabendige AA, Dolman DE, Yusof SR, Begley DJ. Structure and function of the blood-brain barrier. Neurobiol Dis. 2010;37(1):13–25.
59. Schiera G, Bono E, Raffa MP, Gallo A, Pitarresi GL, Di L, et al. Synergistic effects of neurons and astrocytes on the differentiation of brain capillary endothelial cells in culture. J Cell Mol Med. 2003;7(2):165–70.
60. Schiera G, Sala S, Gallo A, Raffa MP, Pitarresi GL, Savettieri G, et al. Permeability properties of a three-cell type in vitro model of blood-brain barrier. J Cell Mol Med. 2005;9(2):373–9.
61. Zhang Z, Chopp M. Neural stem cells and ischemic brain. J Stroke. 2016;18(3):267–72.
62. Li L, Xie T. Stem cell niche: structure and function. Annu Rev Cell Dev Biol. 2005;21:605–31.
63. Taupin P. Adult neural stem cells, neurogenic niches, and cellular therapy. Stem Cell Rev. 2006;2(3):213–9.
64. Chen J, Xiao X, Chen S, Zhang C, Chen J, Yi D, et al. Angiotensin-converting enzyme 2 priming enhances the function of endothelial progenitor cells and their therapeutic efficacy. Hypertension. 2013;61(3):681–9.
65. Spalding KL, Bergmann O, Alkass K, Bernard S, Salehpour M, Huttner HB, et al. Dynamics of hippocampal neurogenesis in adult humans. Cell. 2013;153(6):1219–27.
66. Ernst A, Alkass K, Bernard S, Salehpour M, Perl S, Tisdale J, et al. Neurogenesis in the striatum of the adult human brain. Cell. 2014;156(5):1072–83.
67. von Bohlen und HO. Immunohistological markers for proliferative events, gliogenesis, and neurogenesis within the adult hippocampus. Cell Tissue Res. 2011;345(1):1–19.
68. Batiz LF, Castro MA, Burgos PV, Velasquez ZD, Munoz RI, Lafourcade CA, et al. Exosomes as novel regulators of adult neurogenic niches. Front Cell Neurosci. 2015;9:501.
69. Agnati LF, Fuxe K. Extracellular-vesicle type of volume transmission and tunnelling-nanotube type of wiring transmission add a new dimension to brain neuro-glial networks. Philos Trans R Soc Lond Ser B Biol Sci. 2014;369(1652):pii:20130505.
70. Borroto-Escuela DO, Agnati LF, Bechter K, Jansson A, Tarakanov AO, Fuxe K. The role of transmitter diffusion and flow versus extracellular vesicles in volume transmission in the brain neural-glial networks. Philos Trans R Soc Lond Ser B Biol Sci. 2015;370(1672):20140183.
71. Chiasserini D, van Weering JR, Piersma SR, Pham TV, Malekzadeh A, Teunissen CE, et al. Proteomic analysis of cerebrospinal fluid extracellular vesicles: a comprehensive dataset. J Proteome. 2014;106:191–204.
72. Grapp M, Wrede A, Schweizer M, Huwel S, Galla HJ, Snaidero N, et al. Choroid plexus transcytosis and exosome shuttling deliver folate into brain parenchyma. Nat Commun. 2013;4:2123.
73. Pegtel DM, Peferoen L, Amor S. Extracellular vesicles as modulators of cell-to-cell communication in the healthy and diseased brain. Philos Trans R Soc Lond Ser B Biol Sci. 2014;369(1652):20130516.
74. Feliciano DM, Zhang S, Nasrallah CM, Lisgo SN, Bordey A. Embryonic cerebrospinal fluid nanovesicles carry evolutionarily conserved molecules and promote neural stem cell amplification. PLoS One. 2014;9(2):e88810.
75. Feigin VL, Lawes CM, Bennett DA, Barker-Collo SL, Parag V. Worldwide stroke incidence and early case fatality reported in 56 population-based studies: a systematic review. Lancet Neurol. 2009;8(4):355–69.
76. Elijovich L, Patel PV, Hemphill JC III. Intracerebral hemorrhage. Semin Neurol. 2008;28(5):657–67.

77. Sierra C, Coca A, Schiffrin EL. Vascular mechanisms in the pathogenesis of stroke. Curr Hypertens Rep. 2011;13(3):200–7.
78. Jeyaseelan K, Lim KY, Armugam A. MicroRNA expression in the blood and brain of rats subjected to transient focal ischemia by middle cerebral artery occlusion. Stroke. 2008;39(3):959–66.
79. Liu DZ, Tian Y, Ander BP, Xu H, Stamova BS, Zhan X, et al. Brain and blood microRNA expression profiling of ischemic stroke, intracerebral hemorrhage, and kainate seizures. J Cereb Blood Flow Metab. 2010;30(1):92–101.
80. Tan KS, Armugam A, Sepramaniam S, Lim KY, Setyowati KD, Wang CW, et al. Expression profile of MicroRNAs in young stroke patients. PLoS One. 2009;4(11):e7689.
81. Dharap A, Bowen K, Place R, Li LC, Vemuganti R. Transient focal ischemia induces extensive temporal changes in rat cerebral microRNAome. J Cereb Blood Flow Metab. 2009;29(4):675–87.
82. Yuan Y, Wang JY, Xu LY, Cai R, Chen Z, Luo BY. MicroRNA expression changes in the hippocampi of rats subjected to global ischemia. J Clin Neurosci. 2010;17(6):774–8.
83. Ouyang YB, Lu Y, Yue S, Giffard RG. miR-181 targets multiple Bcl-2 family members and influences apoptosis and mitochondrial function in astrocytes. Mitochondrion. 2012;12(2):213–9.
84. Yin KJ, Deng Z, Hamblin M, Xiang Y, Huang H, Zhang J, et al. Peroxisome proliferator-activated receptor delta regulation of miR-15a in ischemia-induced cerebral vascular endothelial injury. J Neurosci. 2010;30(18):6398–408.
85. Yin KJ, Deng Z, Huang H, Hamblin M, Xie C, Zhang J, et al. miR-497 regulates neuronal death in mouse brain after transient focal cerebral ischemia. Neurobiol Dis. 2010;38(1):17–26.
86. Lee ST, Chu K, Jung KH, Yoon HJ, Jeon D, Kang KM, et al. MicroRNAs induced during ischemic preconditioning. Stroke. 2010;41(8):1646–51.
87. Zeng L, Liu J, Wang Y, Wang L, Weng S, Tang Y, et al. MicroRNA-210 as a novel blood biomarker in acute cerebral ischemia. Front Biosci (Elite Ed). 2011;3:1265–72.
88. Buller B, Liu X, Wang X, Zhang RL, Zhang L, Hozeska-Solgot A, et al. MicroRNA-21 protects neurons from ischemic death. FEBS J. 2010;277(20):4299–307.
89. Liu L, Yu X, Guo X, Tian Z, Su M, Long Y, et al. miR-143 is downregulated in cervical cancer and promotes apoptosis and inhibits tumor formation by targeting Bcl-2. Mol Med Rep. 2012;5(3):753–60.
90. Wu K, Yang Y, Zhong Y, Ammar HM, Zhang P, Guo R, et al. The effects of microvesicles on endothelial progenitor cells are compromised in type 2 diabetic patients via downregulation of the miR-126/VEGFR2 pathway. Am J Physiol Endocrinol Metab. 2016;310(10):E828–37.
91. Gyorgy B, Hung ME, Breakefield XO, Leonard JN. Therapeutic applications of extracellular vesicles: clinical promise and open questions. Annu Rev Pharmacol Toxicol. 2015;55:439–64.
92. Ouyang YB, Stary CM, Yang GY, Giffard R. microRNAs: innovative targets for cerebral ischemia and stroke. Curr Drug Targets. 2013;14(1):90–101.
93. Yin KJ, Hamblin M, Chen YE. Angiogenesis-regulating microRNAs and ischemic stroke. Curr Vasc Pharmacol. 2015;13(3):352–65.
94. Teng H, Zhang ZG, Wang L, Zhang RL, Zhang L, Morris D, et al. Coupling of angiogenesis and neurogenesis in cultured endothelial cells and neural progenitor cells after stroke. J Cereb Blood Flow Metab. 2008;28(4):764–71.
95. Miyamoto N, Pham LD, Seo JH, Kim KW, Lo EH, Arai K. Crosstalk between cerebral endothelium and oligodendrocyte. Cell Mol Life Sci. 2014;71(6):1055–66.
96. Buller B, Chopp M, Ueno Y, Zhang L, Zhang RL, Morris D, et al. Regulation of serum response factor by miRNA-200 and miRNA-9 modulates oligodendrocyte progenitor cell differentiation. Glia. 2012;60(12):1906–14.
97. Gherardini L, Gennaro M, Pizzorusso T. Perilesional treatment with chondroitinase ABC and motor training promote functional recovery after stroke in rats. Cereb Cortex. 2015;25(1):202–12.

98. Zhang Y, Chopp M, Liu XS, Kassis H, Wang X, Li C, et al. MicroRNAs in the axon locally mediate the effects of chondroitin sulfate proteoglycans and cGMP on axonal growth. Dev Neurobiol. 2015;75(12):1402–19.
99. Zhang Y, Ueno Y, Liu XS, Buller B, Wang X, Chopp M, et al. The microRNA-17-92 cluster enhances axonal outgrowth in embryonic cortical neurons. J Neurosci. 2013;33(16):6885–94.
100. Park KK, Liu K, Hu Y, Smith PD, Wang C, Cai B, et al. Promoting axon regeneration in the adult CNS by modulation of the PTEN/mTOR pathway. Science. 2008;322(5903):963–6.
101. He X, Yu Y, Awatramani R, Lu QR. Unwrapping myelination by microRNAs. Neuroscientist. 2012;18(1):45–55.
102. Moubarik C, Guillet B, Youssef B, Codaccioni JL, Piercecchi MD, Sabatier F, et al. Transplanted late outgrowth endothelial progenitor cells as cell therapy product for stroke. Stem Cell Rev. 2011;7(1):208–20.
103. Thored P, Wood J, Arvidsson A, Cammenga J, Kokaia Z, Lindvall O. Long-term neuroblast migration along blood vessels in an area with transient angiogenesis and increased vascularization after stroke. Stroke. 2007;38(11):3032–9.
104. Ohab JJ, Fleming S, Blesch A, Carmichael ST. A neurovascular niche for neurogenesis after stroke. J Neurosci. 2006;26(50):13007–16.
105. Zhang ZG, Chopp M. Neurorestorative therapies for stroke: underlying mechanisms and translation to the clinic. Lancet Neurol. 2009;8(5):491–500.
106. Cantaluppi V, Biancone L, vliolini F, Beltramo S, Medica D, Deregibus MC, et al. Microvesicles derived from endothelial progenitor cells enhance neoangiogenesis of human pancreatic islets. Cell Transplant. 2012;21(6):1305–20.
107. Deregibus MC, Cantaluppi V, Calogero R, Lo IM, Tetta C, Biancone L, et al. Endothelial progenitor cell derived microvesicles activate an angiogenic program in endothelial cells by a horizontal transfer of mRNA. Blood. 2007;110(7):2440–8.
108. Skog J, Wurdinger T, van RS, Meijer DH, Gainche L, Sena-Esteves M, et al. Glioblastoma microvesicles transport RNA and proteins that promote tumour growth and provide diagnostic biomarkers. Nat Cell Biol. 2008;10(12):1470–6.
109. Haqqani AS, Delaney CE, Tremblay TL, Sodja C, Sandhu JK, Stanimirovic DB. Method for isolation and molecular characterization of extracellular microvesicles released from brain endothelial cells. Fluids Barriers CNS. 2013;10(1):4.
110. Yamamoto S, Niida S, Azuma E, Yanagibashi T, Muramatsu M, Huang TT, et al. Inflammation-induced endothelial cell-derived extracellular vesicles modulate the cellular status of pericytes. Sci Rep. 2015;5:8505.
111. Winkler EA, Bell RD, Zlokovic BV. Central nervous system pericytes in health and disease. Nat Neurosci. 2011;14(11):1398–405.
112. Schulz GB, Wieland E, Wustehube-Lausch J, Boulday G, Moll I, Tournier-Lasserve E, et al. Cerebral cavernous malformation-1 protein controls DLL4-notch3 signaling between the endothelium and pericytes. Stroke. 2015;46(5):1337–43.
113. Sharghi-Namini S, Tan E, Ong LL, Ge R, Asada HH. Dll4-containing exosomes induce capillary sprout retraction in a 3D microenvironment. Sci Rep. 2014;4:4031.
114. Sheldon H, Heikamp E, Turley H, Dragovic R, Thomas P, Oon CE, et al. New mechanism for Notch signaling to endothelium at a distance by delta-like 4 incorporation into exosomes. Blood. 2010;116(13):2385–94.
115. Tammela T, Zarkada G, Wallgard E, Murtomaki A, Suchting S, Wirzenius M, et al. Blocking VEGFR-3 suppresses angiogenic sprouting and vascular network formation. Nature. 2008;454(7204):656–60.
116. Taylor KL, Henderson AM, Hughes CC. Notch activation during endothelial cell network formation in vitro targets the basic HLH transcription factor HESR-1 and downregulates VEGFR-2/KDR expression. Microvasc Res. 2002;64(3):372–83.
117. Ihrie RA, Alvarez-Buylla A. Lake-front property: a unique germinal niche by the lateral ventricles of the adult brain. Neuron. 2011;70(4):674–86.

118. Cossetti C, Iraci N, Mercer TR, Leonardi T, Alpi E, Drago D, et al. Extracellular vesicles from neural stem cells transfer IFN-gamma via Ifngr1 to activate Stat1 signaling in target cells. Mol Cell. 2014;56(2):193–204.
119. Famakin BM. The immune response to acute focal cerebral ischemia and associated post-stroke immunodepression: a focused review. Aging Dis. 2014;5(5):307–26.
120. Zhang ZG, Chopp M. Exosomes in stroke pathogenesis and therapy. J Clin Invest. 2016;126(4):1190–7.
121. Higa GS, de SE, Walter LT, Kinjo ER, Resende RR, Kihara AH. MicroRNAs in neuronal communication. Mol Neurobiol. 2014;49(3):1309–26.
122. Kawikova I, Askenase PW. Diagnostic and therapeutic potentials of exosomes in CNS diseases. Brain Res. 2015;1617:63–71.
123. Faure J, Lachenal G, Court M, Hirrlinger J, Chatellard-Causse C, Blot B, et al. Exosomes are released by cultured cortical neurones. Mol Cell Neurosci. 2006;31(4):642–8.
124. Lachenal G, Pernet-Gallay K, Chivet M, Hemming FJ, Belly A, Bodon G, et al. Release of exosomes from differentiated neurons and its regulation by synaptic glutamatergic activity. Mol Cell Neurosci. 2011;46(2):409–18.
125. Goldie BJ, Dun MD, Lin M, Smith ND, Verrills NM, Dayas CV, et al. Activity-associated miRNA are packaged in Map1b-enriched exosomes released from depolarized neurons. Nucleic Acids Res. 2014;42(14):9195–208.
126. Goncalves MB, Malmqvist T, Clarke E, Hubens CJ, Grist J, Hobbs C, et al. Neuronal RARbeta signaling modulates PTEN activity directly in neurons and via exosome transfer in astrocytes to prevent glial scar formation and induce spinal cord regeneration. J Neurosci. 2015;35(47):15731–45.
127. Morel L, Regan M, Higashimori H, Ng SK, Esau C, Vidensky S, et al. Neuronal exosomal miRNA-dependent translational regulation of astroglial glutamate transporter GLT1. J Biol Chem. 2013;288(10):7105–16.
128. Dajas-Bailador F, Bonev B, Garcez P, Stanley P, Guillemot F, Papalopulu N. MicroRNA-9 regulates axon extension and branching by targeting Map 1b in mouse cortical neurons. Nat Neurosci. 2012. https://doi.org/10.1038/nn.3082.
129. Clarkson AN, Overman JJ, Zhong S, Mueller R, Lynch G, Carmichael ST. AMPA receptor-induced local brain-derived neurotrophic factor signaling mediates motor recovery after stroke. J Neurosci. 2011;31(10):3766–75.
130. Shen LH, Li Y, Gao Q, Savant-Bhonsale S, Chopp M. Down-regulation of neurocan expression in reactive astrocytes promotes axonal regeneration and facilitates the neurorestorative effects of bone marrow stromal cells in the ischemic rat brain. Glia. 2008;56(16):1747–54.
131. Edelstein L, Smythies J. The role of epigenetic-related codes in neurocomputation: dynamic hardware in the brain. Philos Trans R Soc Lond Ser B Biol Sci. 2014;369(1652):20130519.
132. Lausted C, Lee I, Zhou Y, Qin S, Sung J, Price ND, et al. Systems approach to neurodegenerative disease biomarker discovery. Annu Rev Pharmacol Toxicol. 2014;54:457–81.
133. Nedaeinia R, Manian M, Jazayeri MH, Ranjbar M, Salehi R, Sharifi M, et al. Circulating exosomes and exosomal microRNAs as biomarkers in gastrointestinal cancer. Cancer Gene Ther. 2017;24(2):48–56.
134. Perez-Gonzalez R, Gauthier SA, Kumar A, Saito M, Saito M, Levy E. A method for isolation of extracellular vesicles and characterization of exosomes from brain extracellular space. Methods Mol Biol. 2017;1545:139–51.
135. Wang Y, Sheng G, Juranek S, Tuschl T, Patel DJ. Structure of the guide-strand-containing argonaute silencing complex. Nature. 2008;456(7219):209–13.
136. Li M, Zeringer E, Barta T, Schageman J, Cheng A, Vlassov AV. Analysis of the RNA content of the exosomes derived from blood serum and urine and its potential as biomarkers. Philos Trans R Soc Lond Ser B Biol Sci. 2014;369(1652):20130502.
137. Chen Y, Song Y, Huang J, Qu M, Zhang Y, Geng J, et al. Increased circulating exosomal miRNA-223 is associated with acute ischemic stroke. Front Neurol. 2017;8:57.

138. Rice J, Roberts H, Burton J, Pan J, States V, Rai SN, et al. Assay reproducibility in clinical studies of plasma miRNA. PLoS One. 2015;10(4):e0121948.
139. Pritchard CC, Cheng HH, Tewari M. MicroRNA profiling: approaches and considerations. Nat Rev Genet. 2012;13(5):358–69.
140. Goodall EF, Heath PR, Bandmann O, Kirby J, Shaw PJ. Neuronal dark matter: the emerging role of microRNAs in neurodegeneration. Front Cell Neurosci. 2013;7:178.
141. Villarroya-Beltri C, Baixauli F, Gutierrez-Vazquez C, Sanchez-Madrid F, Mittelbrunn M. Sorting it out: regulation of exosome loading. Semin Cancer Biol. 2014;28:3–13.
142. Chaput N, Thery C. Exosomes: immune properties and potential clinical implementations. Semin Immunopathol. 2011;33(5):419–40.
143. Frohlich D, Kuo WP, Fruhbeis C, Sun JJ, Zehendner CM, Luhmann HJ, et al. Multifaceted effects of oligodendroglial exosomes on neurons: impact on neuronal firing rate, signal transduction and gene regulation. Philos Trans R Soc Lond Ser B Biol Sci. 2014;369(1652):20130510.
144. Kanninen KM, Bister N, Koistinaho J, Malm T. Exosomes as new diagnostic tools in CNS diseases. Biochim Biophys Acta. 2016;1862(3):403–10.
145. de Jong OG, Verhaar MC, Chen Y, Vader P, Gremmels H, Posthuma G, et al. Cellular stress conditions are reflected in the protein and RNA content of endothelial cell-derived exosomes. J Extracell Vesicles. 2012;1:eCollection.2012.
146. Ji Q, Ji Y, Peng J, Zhou X, Chen X, Zhao H, et al. Increased brain-specific MiR-9 and MiR-124 in the serum exosomes of acute ischemic stroke patients. PLoS One. 2016;11(9):e0163645.
147. Chimowitz MI, Lynn MJ, Derdeyn CP, Turan TN, Fiorella D, Lane BF, et al. Stenting versus aggressive medical therapy for intracranial arterial stenosis. N Engl J Med. 2011;365(11):993–1003.
148. Wei L, Wei ZZ, Jiang MQ, Mohamad O, Yu SP. Stem cell transplantation therapy for multifaceted therapeutic benefits after stroke. Prog Neurobiol. 2017. https://doi.org/10.1016/j.pneurobio.2017.03.003.
149. Cordeiro MF, Horn AP. Stem cell therapy in intracerebral hemorrhage rat model. World J Stem Cells. 2015;7(3):618–29.
150. Hu Y, Liu N, Zhang P, Pan C, Zhang Y, Tang Y, et al. Preclinical studies of stem cell transplantation in intracerebral hemorrhage: a systemic review and meta-analysis. Mol Neurobiol. 2016;53(8):5269–77.
151. Ma X, Qin J, Song B, Shi C, Zhang R, Liu X, et al. Stem cell-based therapies for intracerebral hemorrhage in animal model: a meta-analysis. Neurol Sci. 2015;36(8):1311–7.
152. Moskowitz MA, Lo EH, Iadecola C. The science of stroke: mechanisms in search of treatments. Neuron. 2010;67(2):181–98.
153. Diez-Tejedor E, Gutierrez-Fernandez M, Martinez-Sanchez P, Rodriguez-Frutos B, Ruiz-Ares G, Lara ML, et al. Reparative therapy for acute ischemic stroke with allogeneic mesenchymal stem cells from adipose tissue: a safety assessment: a phase II randomized, double-blind, placebo-controlled, single-center, pilot clinical trial. J Stroke Cerebrovasc Dis. 2014;23(10):2694–700.
154. Lee JS, Hong JM, Moon GJ, Lee PH, Ahn YH, Bang OY. A long-term follow-up study of intravenous autologous mesenchymal stem cell transplantation in patients with ischemic stroke. Stem Cells. 2010;28(6):1099–106.
155. Nakazaki M, Sasaki M, Kataoka-Sasaki Y, Oka S, Namioka T, Namioka A, et al. Intravenous infusion of mesenchymal stem cells inhibits intracranial hemorrhage after recombinant tissue plasminogen activator therapy for transient middle cerebral artery occlusion in rats. J Neurosurg. 2017;PMID:28059661:1–10.
156. Chen J, Chen J, Chen S, Zhang C, Zhang L, Xiao X, et al. Transfusion of CXCR4-primed endothelial progenitor cells reduces cerebral ischemic damage and promotes repair in db/db diabetic mice. PLoS One. 2012;7(11):e50105.
157. Doeppner TR, Kaltwasser B, Bahr M, Hermann DM. Effects of neural progenitor cells on post-stroke neurological impairment-a detailed and comprehensive analysis of behavioral tests. Front Cell Neurosci. 2014;8:338.

158. Zhang R, Zhang Z, Chopp M. Function of neural stem cells in ischemic brain repair processes. J Cereb Blood Flow Metab. 2016;36(12):2034–43.
159. Banerjee S, Bhat MA. Neuron-glial interactions in blood-brain barrier formation. Annu Rev Neurosci. 2007;30:235–58.
160. Sun D, Zhuang X, Xiang X, Liu Y, Zhang S, Liu C, et al. A novel nanoparticle drug delivery system: the anti-inflammatory activity of curcumin is enhanced when encapsulated in exosomes. Mol Ther. 2010;18(9):1606–14.
161. Dhillon N, Aggarwal BB, Newman RA, Wolff RA, Kunnumakkara AB, Abbruzzese JL, et al. Phase II trial of curcumin in patients with advanced pancreatic cancer. Clin Cancer Res. 2008;14(14):4491–9.
162. Pardridge WM. Drug transport across the blood-brain barrier. J Cereb Blood Flow Metab. 2012;32(11):1959–72.
163. Peng Q, Zhang S, Yang Q, Zhang T, Wei XQ, Jiang L, et al. Preformed albumin corona, a protective coating for nanoparticles based drug delivery system. Biomaterials. 2013;34(33):8521–30.
164. Veronese FM, Caliceti P, Schiavon O, Sergi M. Polyethylene glycol-superoxide dismutase, a conjugate in search of exploitation. Adv Drug Deliv Rev. 2002;54(4):587–606.
165. Yoshida K, Burton GF, McKinney JS, Young H, Ellis EF. Brain and tissue distribution of polyethylene glycol-conjugated superoxide dismutase in rats. Stroke. 1992;23(6):865–9.
166. Yang T, Martin P, Fogarty B, Brown A, Schurman K, Phipps R, et al. Exosome delivered anticancer drugs across the blood-brain barrier for brain cancer therapy in Danio rerio. Pharm Res. 2015;32(6):2003–14.
167. Zhuang X, Xiang X, Grizzle W, Sun D, Zhang S, Axtell RC, et al. Treatment of brain inflammatory diseases by delivering exosome encapsulated anti-inflammatory drugs from the nasal region to the brain. Mol Ther. 2011;19(10):1769–79.
168. Fruhbeis C, Frohlich D, Kuo WP, Amphornrat J, Thilemann S, Saab AS, et al. Neurotransmitter-triggered transfer of exosomes mediates oligodendrocyte-neuron communication. PLoS Biol. 2013;11(7):e1001604.
169. Alvarez-Erviti L, Seow Y, Yin H, Betts C, Lakhal S, Wood MJ. Delivery of siRNA to the mouse brain by systemic injection of targeted exosomes. Nat Biotechnol. 2011;29(4):341–5.
170. Liu Y, Li D, Liu Z, Zhou Y, Chu D, Li X, et al. Targeted exosome-mediated delivery of opioid receptor Mu siRNA for the treatment of morphine relapse. Sci Rep. 2015;5:17543.
171. Ohno S, Takanashi M, Sudo K, Ueda S, Ishikawa A, Matsuyama N, et al. Systemically injected exosomes targeted to EGFR deliver antitumor microRNA to breast cancer cells. Mol Ther. 2013;21(1):185–91.
172. Yeo RW, Lai RC, Zhang B, Tan SS, Yin Y, Teh BJ, et al. Mesenchymal stem cell: an efficient mass producer of exosomes for drug delivery. Adv Drug Deliv Rev. 2013;65(3):336–41.
173. Doeppner TR, Herz J, Gorgens A, Schlechter J, Ludwig AK, Radtke S, et al. Extracellular vesicles improve post-stroke neuroregeneration and prevent postischemic immunosuppression. Stem Cells Transl Med. 2015;4(10):1131–43.
174. Kim DK, Nishida H, An SY, Shetty AK, Bartosh TJ, Prockop DJ. Chromatographically isolated CD63+CD81+ extracellular vesicles from mesenchymal stromal cells rescue cognitive impairments after TBI. Proc Natl Acad Sci U S A. 2016;113(1):170–5.
175. Xin H, Li Y, Buller B, Katakowski M, Zhang Y, Wang X, et al. Exosome-mediated transfer of miR-133b from multipotent mesenchymal stromal cells to neural cells contributes to neurite outgrowth. Stem Cells. 2012;30(7):1556–64.
176. Xin H, Li Y, Cui Y, Yang JJ, Zhang ZG, Chopp M. Systemic administration of exosomes released from mesenchymal stromal cells promote functional recovery and neurovascular plasticity after stroke in rats. J Cereb Blood Flow Metab. 2013;33(11):1711–5.
177. Xin H, Li Y, Liu Z, Wang X, Shang X, Cui Y, et al. MiR-133b promotes neural plasticity and functional recovery after treatment of stroke with multipotent mesenchymal stromal cells in rats via transfer of exosome-enriched extracellular particles. Stem Cells. 2013;31(12):2737–46.

178. Zhang Y, Chopp M, Meng Y, Katakowski M, Xin H, Mahmood A, et al. Effect of exosomes derived from multipluripotent mesenchymal stromal cells on functional recovery and neurovascular plasticity in rats after traumatic brain injury. J Neurosurg. 2015;122(4):856–67.
179. Penfornis P, Vallabhaneni KC, Whitt J, Pochampally R. Extracellular vesicles as carriers of microRNA, proteins and lipids in tumor microenvironment. Int J Cancer. 2016;138(1):14–21.
180. Vallabhaneni KC, Penfornis P, Dhule S, Guillonneau F, Adams KV, Mo YY, et al. Extracellular vesicles from bone marrow mesenchymal stem/stromal cells transport tumor regulatory microRNA, proteins, and metabolites. Oncotarget. 2015;6(7):4953–67.
181. Jones EV, Bouvier DS. Astrocyte-secreted matricellular proteins in CNS remodelling during development and disease. Neural Plast. 2014;2014:321209.
182. Mackie AR, Klyachko E, Thorne T, Schultz KM, Millay M, Ito A, et al. Sonic hedgehog-modified human CD34+ cells preserve cardiac function after acute myocardial infarction. Circ Res. 2012;111(3):312–21.
183. Altaba A, Sanchez P, Dahmane N. Gli and hedgehog in cancer: tumours, embryos and stem cells. Nat Rev Cancer. 2002;2(5):361–72.
184. Goetz JA, Suber LM, Zeng X, Robbins DJ. Sonic Hedgehog as a mediator of long-range signaling. BioEssays. 2002;24(2):157–65.
185. Roberts DJ, Johnson RL, Burke AC, Nelson CE, Morgan BA, Tabin C. Sonic hedgehog is an endodermal signal inducing Bmp-4 and Hox genes during induction and regionalization of the chick hindgut. Development. 1995;121(10):3163–74.
186. Androutsellis-Theotokis A, Leker RR, Soldner F, Hoeppner DJ, Ravin R, Poser SW, et al. Notch signalling regulates stem cell numbers in vitro and in vivo. Nature. 2006;442(7104):823–6.
187. Liu XS, Chopp M, Wang XL, Zhang L, Hozeska-Solgot A, Tang T, et al. MicroRNA-17-92 cluster mediates the proliferation and survival of neural progenitor cells after stroke. J Biol Chem. 2013;288(18):12478–88.
188. Palma V, Lim DA, Dahmane N, Sanchez P, Brionne TC, Herzberg CD, et al. Sonic hedgehog controls stem cell behavior in the postnatal and adult brain. Development. 2005;132(2):335–44.
189. Wang L, Zhang ZG, Gregg SR, Zhang RL, Jiao Z, LeTourneau Y, et al. The Sonic hedgehog pathway mediates carbamylated erythropoietin-enhanced proliferation and differentiation of adult neural progenitor cells. J Biol Chem. 2007;282(44):32462–70.
190. Asahara T, Masuda H, Takahashi T, Kalka C, Pastore C, Silver M, et al. Bone marrow origin of endothelial progenitor cells responsible for postnatal vasculogenesis in physiological and pathological neovascularization. Circ Res. 1999;85(3):221–8.
191. Murayama T, Tepper OM, Silver M, Ma H, Losordo DW, Isner JM, et al. Determination of bone marrow-derived endothelial progenitor cell significance in angiogenic growth factor-induced neovascularization in vivo. Exp Hematol. 2002;30(8):967–72.
192. Zhang ZG, Zhang L, Jiang Q, Chopp M. Bone marrow-derived endothelial progenitor cells participate in cerebral neovascularization after focal cerebral ischemia in the adult mouse. Circ Res. 2002;90(3):284–8.
193. Yang Z, von Ballmoos MW, Faessler D, Voelzmann J, Ortmann J, Diehm N, et al. Paracrine factors secreted by endothelial progenitor cells prevent oxidative stress-induced apoptosis of mature endothelial cells. Atherosclerosis. 2010;211(1):103–9.
194. Gu S, Zhang W, Chen J, Ma R, Xiao X, Ma X, et al. EPC-derived microvesicles protect cardiomyocytes from Ang II-induced hypertrophy and apoptosis. PLoS One. 2014;9(1):e85396.
195. Cantaluppi V, Gatti S, Medica D, Figliolini F, Bruno S, Deregibus MC, et al. Microvesicles derived from endothelial progenitor cells protect the kidney from ischemia-reperfusion injury by microRNA-dependent reprogramming of resident renal cells. Kidney Int. 2012;82(4):412–27.
196. Ranghino A, Cantaluppi V, Grange C, Vitillo L, Fop F, Biancone L, et al. Endothelial progenitor cell-derived microvesicles improve neovascularization in a murine model of hindlimb ischemia. Int J Immunopathol Pharmacol. 2012;25(1):75–85.
197. Sluijter JP, Verhage V, Deddens JC, van den Akker F, Doevendans PA. Microvesicles and exosomes for intracardiac communication. Cardiovasc Res. 2014;102(2):302–11.

198. Xin H, Li Y, Chopp M. Exosomes/miRNAs as mediating cell-based therapy of stroke. Front Cell Neurosci. 2014;8:377.
199. Tavazoie M, Van d V, Silva-Vargas V, Louissaint M, Colonna L, Zaidi B, et al. A specialized vascular niche for adult neural stem cells. Cell Stem Cell. 2008;3(3):279–88.
200. Wang J, Chen Y, Yang Y, Xiao X, Chen S, Zhang C, et al. Endothelial progenitor cells and neural progenitor cells synergistically protect cerebral endothelial cells from Hypoxia/reoxygenation-induced injury via activating the PI3K/Akt pathway. Mol Brain. 2016;9:12.
201. Talaveron R, Matarredona ER, de la Cruz RR, Macias D, Galvez V, Pastor AM. Implanted neural progenitor cells regulate glial reaction to brain injury and establish gap junctions with host glial cells. Glia. 2014;62(4):623–38.

Part III
Neuronal Environment, Plasticity and Repair Mechanisms

Chapter 18
Integrating Molecular, Cellular, and Systems Approaches to Repairing the Brain After Stroke

Max O. Krucoff, Stephen C. Harward, Shervin Rahimpour, Keith Dombrowski, Erik F. Hauck, Shivanand P. Lad, and Dennis A. Turner

Abstract A stroke implies a sudden and spontaneous onset of neurological symptoms duc to a vascular insult. Dcspitc thc brain's inhcrcnt capacity for plasticity and spontaneous improvement, strokes still leave many patients with devastating deficits that can permanently affect independence and quality of life. This chapter focuses on ways to help restore the functionality of the central nervous system (CNS) after this type of injury. Understanding how neurons interact on both individual (i.e. cellular and molecular) and population (i.e. synapses and circuits) levels is crucial to developing successful restorative strategies, as is appreciating how these interactions change over the injury-recovery timeline. The CNS has several characteristics that make its restitution exceptionally difficult; beyond even its incredible intricacy, its parenchymal cells, or neurons, do not regenerate well after injury, and this damaged neuronal substrate embodies a consciousness system that must be engaged in its own recovery. In fact, there is now data suggesting that conscious intention, often invoked through goal-oriented rehabilitation, plays a crucial role in facilitating functional plasticity and long-range axonal sprouting. To capitalize on this principle, neural interfaces and electrical stimulation strategies are being integrated into rehabilitation paradigms to provide critically-timed feedback that

M.O. Krucoff, M.D. (✉) • S.C. Harward, M.D., Ph.D. • S. Rahimpour, M.D.
E.F. Hauck, M.D. • S.P. Lad, M.D., Ph.D.
Department of Neurosurgery, Duke University Medical Center,
Duke Neurosurgery Box 3807, Durham, NC 27710, USA
e-mail: max.krucoff@duke.edu

K. Dombrowski, M.D.
Department of Neurology, Duke University Medical Center, Durham, NC 27710, USA

D.A. Turner, M.D., M.A
Department of Neurosurgery, Duke University Medical Center,
Duke Neurosurgery Box 3807, Durham, NC 27710, USA

Department of Neurobiology, Duke University, Durham, NC 27710, USA

Research and Surgery Services, Durham Veterans Affairs Medical Center, Durham, NC, USA

P.A. Lapchak, J.H. Zhang (eds.), *Cellular and Molecular Approaches to Regeneration and Repair*, Springer Series in Translational Stroke Research,
https://doi.org/10.1007/978-3-319-66679-2_18

can reinvigorate injured circuits. Combining these approaches with interventions at the cellular and molecular level (e.g. immunological or genetic modulations aimed at promoting neuronal outgrowth, or stem cells that can replace damaged parenchyma) has the chance to improve neurological recovery to back toward baseline levels. Ultimately, because cells of the CNS do not regrow on their own, and because regrowth and synapse formation does not necessarily ensure restoration of function, harmonious application of synergistic approaches at both the micro- and macroscopic levels will be needed to establish long-lasting functional plasticity and meaningful recovery.

Keywords Neural repair • Neural regeneration • Stroke • Neurorehabilitation • Brain-machine interface • Brain-computer interface • Neural interface • Axonal regeneration • Neural restoration

Abbreviations

AMPA	α-Amino-3-hydroxy-5-methyl-4-isoxazolepropionic acid
AP	Action potential
BCI	Brain-computer interface
BCM	Bienenstock–Cooper–Munro
BMI	Brain-machine interface
BSDS	Brain state dependent stimulation
cAMP	Cyclic adenosine monophosphate
CIMT	Constraint-induced movement therapy
CNS	Central nervous system
CNTF	Ciliary neurotrophic factor
CPP	Cerebral perfusion pressure
CSPG	Chondroitin sulfate proteoglycans
DBS	Deep brain stimulation
DOC	Disorder of consciousness
DRG	Dorsal root ganglion
FDA	Federal Drug Administration
FES	Functional electrical stimulation
GABA	Gamma-aminobutyric acid
GAP43	Growth associated protein 43
GDF10	Growth and differentiation factor 10
ICA	Internal carotid artery
ICP	Intracranial pressure
IFG-1	Insulin-like growth factor 1
LTD	Long-term depression
LTP	Long-term potentiation

M1	Primary motor cortex
MAG	Myelin-associated
MAI	Myelin-associated inhibitory molecule
MCA	Middle cerebral artery
mTOR	Mechanistic target of rapamycin
NgR	Nogo receptor
NMDA	*N*-Methyl-D-aspartate
NSAID	Non-steroidal anti-inflammatory drug
OMgp	Oligodendrocyte-myelin glycoprotein
OPN	Osteopontin
PAS	Paired associative stimulation
PMC	Premotor cortices
PTEN	Phosphatase and tensin homolog
RGC	Retinal ganglion cell
rTMS	Repetitive transcranial magnetic stimulation
SCI	Spinal cord injury
SGZ	Subgranular zone
STDP	Spike-timing dependent plasticity
SVZ	Subventricular zone
TGF-β	Transforming growth factor beta
TGFβR	Transforming growth factor beta receptor
TMS	Transcranial magnetic stimulation

1 Introduction

The term "stroke" implies a sudden and spontaneous onset of neurological symptoms due to a vascular insult. Strokes can be ischemic from vascular occlusion or hemorrhagic from vessel rupture. Hemorrhagic strokes can be further divided by the location of extravasated blood into the subarachnoid, intraventricular, or intraparenchymal space. The mechanism and characteristics of neurological injury and resultant neurological deficits are specific to each type of stroke and its location in the brain. For example, large vessel occlusions (e.g. internal carotid artery [ICA] or middle cerebral artery [MCA]) result in cerebral infarcts with surrounding penumbras that can be devastating. On the other hand, small vessel occlusions tend to result in "lacunes," and their effects can range from asymptomatic to significant deficits (contingent on location). Hemorrhages also range from devastating to inconsequential depending on the location and size of the lesion. Because strokes occur suddenly, an inflammatory response results in edema and secondary injury over the next 2–10 days. After this process subsides, recovery begins, almost always resulting in some level of spontaneous functional improvement over the next 6–12 months [1].

Many strategies have evolved over time to either prevent (e.g. anti-hypertension therapy, diabetes control, lipid lowering agents, antiplatelet therapy) or rapidly treat (e.g. thrombolysis, thrombectomy, surgical intervention) strokes, as well as to minimize secondary injury through treatment of intracranial pressure (ICP), maintenance of cerebral perfusion pressure (CPP), and prevention of spreading depression [2]. While interventions undertaken before and during stroke have greatly improved outcomes, it has been more difficult to show clear benefit from pharmacological or other interventions undertaken to mitigate secondary injury [3]. After this phase has subsided and up until about 6–12 months, some spontaneous recovery typically occurs. Beyond this time period, the prospect for further improvement is greatly diminished [4], and the resultant neurological injury is generally considered chronic. Therefore, treatment efforts have tended to focus on the critical period of recovery (i.e. the first few months after injury). In this chapter, we focus on new discoveries in molecular, cellular, and systems neuroscience, and we discuss how their synergistic application may help further augment brain repair after stroke to improve functional outcomes in both the subacute and chronic phases of injury.

2 Molecular and Cellular Neural Repair

Although the brain demonstrates some spontaneous functional recovery after an injury via intrinsic plasticity mechanisms [1, 5–7], individual neurons of the mature central nervous system (CNS) do not self-repair. Rather, proximal axonal segments retract to form an end bulb [8–12], and distal segments initially remain electrically active before undergoing anterograde, or Wallerian, degeneration over the next 36–48 h [13]. The sparse attempts some CNS neurons make at growth cone formation and elongation after injury tend to fail because of a lack of guidance mechanisms [14–17] (although there is evidence that denervation can restore some developmental growth cues in the rat hippocampus [18]).

Though not restorative, neurogenesis does occur from niches of progenitor cells in the subgranular (SGZ) and subventricular (SVZ) zones of the healthy adult brain [19–23]. These processes support learning, memory, and olfaction. The SGZ supplies the dentate gyrus of the hippocampus, while the SVZ gives cells to the olfactory bulb and to CA1/CA3 regions of the hippocampus [24]. After a stroke, proliferation is stimulated in these areas and immature neurons are recruited to damaged sites in the striatum and cortex [25, 26]. Initially, tens of thousands of immature neurons can migrate; however, few of these cells mature and survive long-term [27, 28]. While this process is associated with some level of functional recovery, it is likely that most behavioral recovery is achieved through mechanisms other than neuronal replacement (e.g. growth factor production in local tissue and rewiring of existing connections) [22].

There are many known biological factors, both intra- and extracellular, that can alter the ability of CNS neurons to grow. Intrinsic mechanisms include transcription factors (e.g. c-Jun, Atf3, Klf family, Stat3, Sox11, and Smad1) and regeneration-associated genes (e.g. Gap43, Cap23, Arg1, Sprr1a, Hspb1, MARCKS, stathmin family, SCG10 L1, P21/waf1, and tubulins), both of which can enhance a neuron's ability to regenerate following an injury [29–32]. Phosphatase and tensin homolog (PTEN), a tumor suppressor, plays an important role in neural regeneration, as eliminating its gene both prevents apoptosis and induces axon extension in injured retinal ganglion cells (RGC) [33, 34]. Such regeneration is dependent on the mechanistic target of rapamycin (mTOR) pathways, as inhibiting mTOR eradicates the regenerative effect of PTEN deficiency [33]. Deletion of Socs3, a suppressor of signaling through the Jak-STAT pathway, also promotes regeneration by enhancing the efficacy of ciliary neurotrophic factor (CNTF), a protein known to prevent axonal degeneration and promote neuron survival and outgrowth following injury [35, 36]. The proto-oncogene bcl-2 (and expression of its anti-apoptotic protein) also plays a key role in preventing cell death after injury, enabling axonal regrowth in RGCs with the presence of trophic factors and physiologic electrical activity [37, 38].

Extrinsic factors preventing axonal regeneration include inhibitory proteins associated with myelin (e.g. NogoA, myelin-associated glycoprotein [MAG], and oligodendrocyte-myelin glycoprotein [OMgp]), proteoglycans in the perineuronal net and glial scar (e.g. chondroitin sulfate proteoglycans [CSPGs] like aggrecan, versican, brevican, neurocan, NG2, and phosphacan), and molecules that repel axon growth during development which continue to be expressed in the mature CNS (e.g. semaphorins, ephrins, slits, netrins, robos, and Wnts) [14, 39–41]. A summary of intrinsic and extrinsic factors now to affect neural growth and inhibition is provided in Fig. 18.1 [42].

Regrettably, removal or blockage of extracellular inhibitory factors alone so far has failed to achieve extensive axonal regeneration with a few exceptions [14, 39, 43, 44]. Interestingly, a strain of dorsal root ganglion (DRG) neurons grown from CAST/Ei knockout mice are less inhibited by the same extrinsic cues listed above [40]. Cells from these mice display increased axonal growth and enhanced regenerative responses following injury when compared to C57BL/6 mice, which are abilities largely dependent on the protein activin (a member of the transforming growth factor beta [TGF-β] family) [40]. Also, deletion of receptors that bind to myelin-associated inhibitory (MAI) molecules, or Nogo receptors (NgR), has been shown to increase regeneration potential in neurons [45]. As such, anti-Nogo immunotherapies are currently of great interest [46–49]. For example, in 2014, Wahl et al. demonstrated near full recovery of skilled forelimb function in rats with large strokes after intrathecal injection of an anti-NogoA antibody followed by intensive task-specific training [49]. Injection of the NogoA neutralizing agent was shown to promote growth of corticospinal fibers from the intact forebrain motor cortex across the midline of the cervical spinal cord to the hemicord that had lost its input from the motor cortex. This new fiber sprouting was then stabilized by a goal-directed, forced-use physical therapy regimen. Importantly, sequential application of drug then training was necessary to show benefit. When immunotherapy and training were combined simultaneously, functional outcome was *poorer* compared to no treatment at all or each treatment individually, likely due to aberrant fiber branching

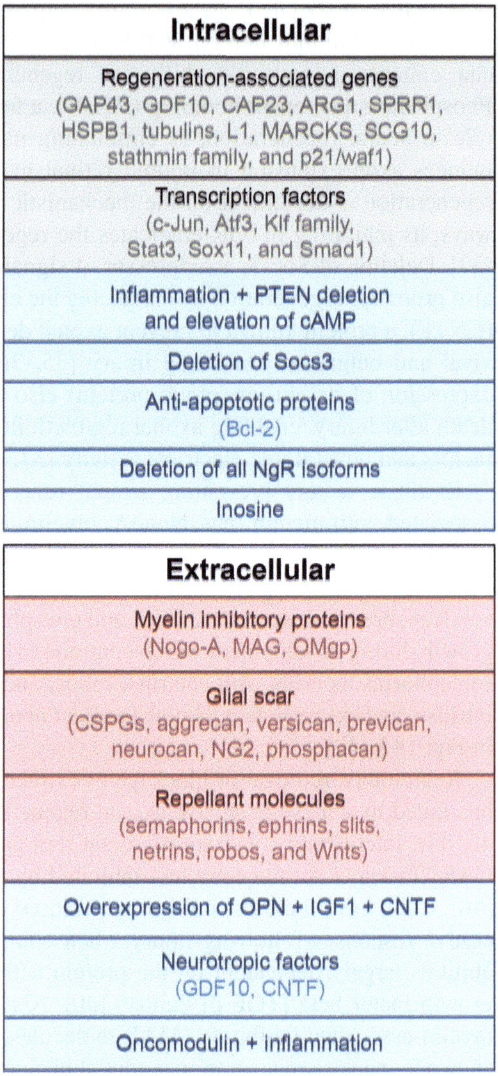

Fig. 18.1 Intra- and extracellular mechanisms of neuronal growth and inhibition. *Gray*—modulates both neuronal growth and inhibition. *Blue*—associated with neuronal growth. *Red*—associated with neuronal inhibition. Adapted with permission from Krucoff et al., 2016 [42]

(which was also seen in Maier et al. [48]). This example outlines the important distinction between regrowth and restoration of function, as well as the crucial interaction of micro- and macroscopic recovery systems.

Another example of axonal growth and synapse formation without restoration of function was demonstrated by Bei et al. when they induced adult mouse retinal axons to regrow and synapse in the superior colliculus via a PTEN/SOCS3 co-deletion and overexpression of osteopontin (OPN)/insulin-like growth factor 1 (IGF1)/CNTF [50]. However, visual function was not restored. In fact, these regenerated axons failed to conduct action potentials (AP) due to lack of myelination, and administration of voltage-gated potassium channel blockers was required to improve visual acuity.

In opposition to these inhibitory mechanisms, signals known to promote axonal growth are also present in the injured brain. For example, growth and differentiation factor 10 (GDF10) is induced in stroke and works through transforming growth factor beta receptors I and II (TGF-βRI and TGF-βRII) to promote axonal outgrowth [51]. Growth associated protein 43 (GAP43), a neuronal growth cone marker also called neuromodulin, is likewise induced in peri-infarct cortex after stroke, and it may contribute to a pro-growth environment following injury that allows for axonal sprouting and growth of dendritic trees [52–54]. Furthermore, the purine nucleoside inosine works through a direct intracellular mechanism to induce expression of genes associated with axonal growth (e.g. GAP43, L1, and α-1 tubulin) and has been shown to induce axonal reorganization and improve behavioral outcomes after spinal cord injury and stroke [55–57], as well as restore levels of GAP43 in the hippocampus in rats after stroke [34, 58, 59].

The role of inflammation in axonal regeneration is somewhat controversial. Some components of inflammation cause tissue damage and neuronal death, while others promote cell survival, axonal sprouting, and regeneration [18, 41, 60–63]. Both oncomodulin, a macrophage-derived growth factor for RGCs, and injury-induced cytokine release appear to play a role in inflammation-induced axonal regeneration [62, 64, 65]. Traditional anti-inflammatory therapies (e.g. non-steroidal anti-inflammatory drugs [NSAID]) may suppress beneficial as well as deleterious aspects of the immune response, and they can stimulate axonal regeneration via direct effects on neurons [39, 61, 66]. When combined with PTEN deletion and elevation of cyclic adenosine monophosphate (cAMP), intraocular inflammation enables some RGCs to regenerate axons from the retina to the brain and restore simple visual responses [34]. Timing of inflammation is also important, as it helps to prime the extracellular milieu for subsequent axonal entry and re-innervation for the first 10–14 days following injury [18]. This is the period when transplanted embryonic neurons have shown increased survival compared to normal brain or later transplantations [18]. For the next several months, the neural environment remains conducive to recovery due to its relatively loose extracellular space, enhanced neurotrophic factors, open synaptic sites, and probing axonal growth cones [29, 67, 68]. However, this window for recovery is limited because after 6–12 months there is less opportunity for further gain (although this doctrine is beginning to change) [69, 70]. A post-injury environment timeline is shown in Fig. 18.2.

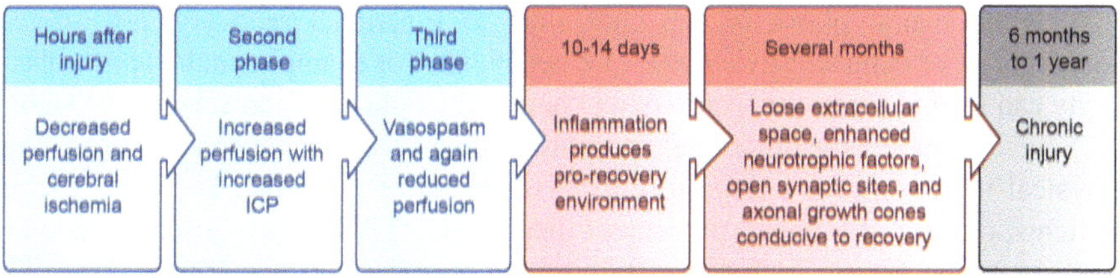

Fig. 18.2 Injury environment timeline. *Blue*—acute phase. *Red*—subacute phase. *Black*—chronic phase. Reproduced with permission from Krucoff et al., 2016 [42]

Attempts at pharmacotherapy (e.g. noradrenergic, dopaminergic, and growth factors) to limit secondary injury so far have shown limited evidence of improvement in outcomes beyond controls [71–73]. Some drugs, such as phenytoin, can even impede plasticity and recovery through suppression of growth cone-related physiological activity [3].

Cellular replacement therapy for stroke-related cell loss has the potential to significantly aid in reconstituting injured circuitry by replenishing some of the lost neuronal substrate [74]. Unfortunately, most stem cell transplants being considered in human trials are mesenchymal; in other words, they have non-neuronal origins and, thus, do not have defined mechanisms of action [75–77]. Some are even are intended to die out over time without facilitating circuit plasticity [78, 79]. Therefore, improving the stem cell derivations and defining expected mechanisms of action (e.g. facilitation of extracellular milieu, circuit reconstitution, axonal scaffolding, etc.) may lead to better results [80].

3 Neural Plasticity and Circuitry Restoration

Natural recovery after stroke occurs because the brain is inherently plastic [5, 7] and can adapt to the injury via the following mechanisms: collateral sprouting from neighboring neurons, strengthening or weakening existing synapses (i.e. long-term potentiation [LTP] or depression [LTD]), and altering concentrations of neurotransmitters, ions, gap junctions, and glial cells [81–84]. After a stroke, both hemispheres are known to assist with recovery depending on the size of the injury [85–87]. Following a small stroke within the primary motor cortex (M1), both ipsilesional dorsal and ventral premotor cortices (PMC) can reorganize themselves. However, when a lesion involves a larger portion of M1 and the dorsal PMC, the contralesional PMC appears to be critical for recovery-related reorganization [85, 87]. Initiation of post-infarct axonal sprouting from the intact cortical hemisphere to peri-infarct cortex and the contralateral dorsal striatum is signaled by synchronous neuronal activity [88]. In chronic stroke patients, activity in the ipsilesional M1 and medial-PMC is associated with good motor recovery, whereas increased cerebellar vermis activity signals poor recovery [89]. More recently, evidence has emerged for the possibility of long-range axonal sprouting in animal stroke models [49]. Context-dependent cortical activity paired with positive feedback seems to be critical for this type of axonal sprouting and lasting functional improvement. It should be noted that, while neural plasticity can contribute to functional recovery, not all re-innervations or connectivity changes are beneficial. For example, maladaptive plasticity can lead to spasticity, pathological pain, schizophrenia, and seizures [90–96].

Several avenues for functional restoration after stroke have been explored, and physical therapy plays a large role in virtually all of them [4]. For early stroke patients, constraint-induced movement therapy (CIMT) has been used to encourage use of the paretic limb by restraining the less affected one (Fig. 18.3a) [97,

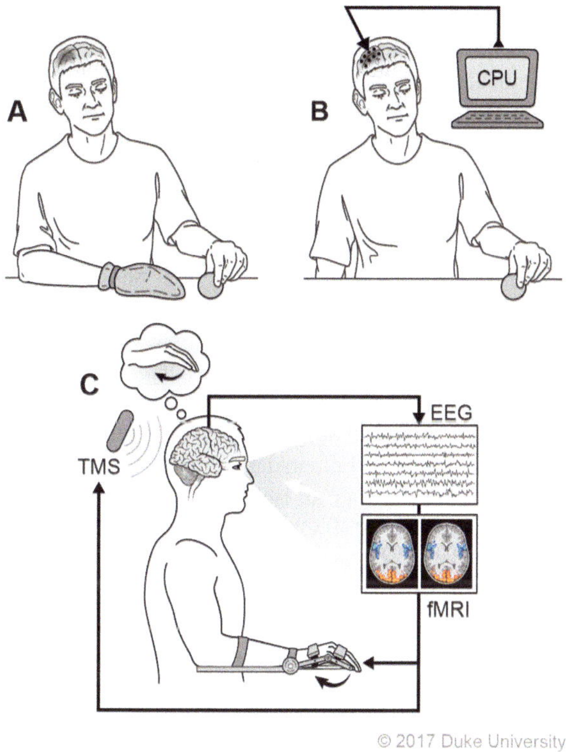

Fig. 18.3 Neurorehabilitation Modalities for Stroke. (**a**) Constraint-induced movement therapy (CIMT). (**b**) Cortical stimulation. (**c**) Closed-loop biofeedback, or brain state dependent stimulation (BSDS). Adapted with permission from Krucoff et al., 2016 [42]

98]. This is done to avoid learned non-use, as animal data has shown maladaptive changes and worse functional outcomes from allowing overcompensation with the less affected limb to dominate goal-directed tasks [99, 100]. Timing the application of CIMT appears to be crucial, however, as behavioral interventions employed too early after injury may be deleterious due to glutamate-*N*-methyl-D-aspartate (NMDA) receptor excitotoxicity in vulnerable tissue [101]. Several studies have examined the efficacy of CIMT for motor recovery in human stroke patients with mixed results [4, 102–104], and the optimal timing for its application in human stroke patients is yet to be determined [105]. Interestingly, early intensive training with immunotherapy in rat models of stroke has also been shown to be harmful due to hyperinnervation, aberrant growth, and wrong circuit connectivity, thus providing further evidence that timing of therapy is crucial yet poorly defined [49]. In yet another example, the application of γ-aminobutyric acid (GABA) antagonists or α-amino-3-hydroxy-5-methyl-4-isoxazolepropionic acid (AMPA) agents within a few days of stroke was found to increase the size of infarction; however, when the same agents were administered 3–5 days after a stroke, motor recovery improved [106].

Because most axonal plasticity depends on electrically active growth cones [5, 7], cortical electrical stimulation as an adjunct to physical therapy has generated significant interest (Fig. 18.3b) [107, 108]. This type of brain stimulation is different from Federal Drug Administration (FDA)-approved symptomatic treatments for movement disorders, for example, where the symptoms relapse when the stimulation is stopped. In stroke recovery, brain stimulation is intended to enhance plasticity for a short duration, resulting in stable improvement even after the stimulation period. Such stimulation has shown the ability to enhance plasticity and functional recovery after stroke in rats [109, 110], squirrel monkeys [111], and some humans [107, 108]. This improvement appears to coincide with both re-emergence of movement representation in peri-infarct areas as well as the emergence of new areas of representation [112, 113]. After repeated stimulation, areas of movement representation have been seen to shift several microns and increase in size with a corresponding increase in spine density in pyramidal cell layers III and V [114, 115]. While there is some human data supporting the use of subthreshold cortical stimulation for recovery after an ischemic infarct [107], recent phase III trials have been overall negative [108]. Other forms of stimulation include deep brain stimulation (DBS) and non-invasive external brain stimulation (e.g. transcranial magnetic stimulation [TMS]), both of which may be helpful particularly during the critical recovery period [116, 117].

Closed-loop TMS paradigms have also been developed for patients who cannot participate in traditional therapy (Fig. 18.3c). In 2014, Gharabaghi et al. published a brain state dependent stimulation (BSDS) protocol in which TMS of the motor cortex and haptic feedback to a paretic hand were controlled by sensorimotor desynchronization during motor-imagery [118]. Cortex-to-cortex interfaces have also been used to bridge damaged neural pathways directly. For example, Guggenmos et al. showed that a neural prosthetic could help reconnect premotor to somatosensory cortex in an injured rat brain to restore reach and grasp functions to pre-lesion levels [109]. Both TMS and direct cortical stimulation protocols remain in the early stages of development [116, 117].

In recent years, neural interfaces, also known as brain-machine (BMI) or brain-computer interfaces (BCI), have been developed to help engage cortical circuits and enhance native motor recovery by pairing motor action (real or imaginary) with real-time positive feedback [42]. These devices directly decode information from the nervous system to generate functional outputs based on the user's intent. The rehabilitative nature of newer designs comes from pairing goal-oriented tasks with expected outcomes and re-activating lesioned circuits, thus facilitating restorative functional plasticity. This is opposed to older bypass, or assistive, BMIs which simply circumvent the lesion to perform the intended action (Fig. 18.4). As discussed above, at a molecular and cellular level, restorative functional plasticity relies on activity-dependent modulation of synaptic transmission in the forms of both LTP and LTD. As their names imply, these synaptic changes have long-lasting impacts on functional connections, and thus provide a critical substrate for the rehabilitation of damaged circuitry. The temporal relationship between pre- and post-synaptic activity is critical in these cases, a concept commonly known as Hebbian,

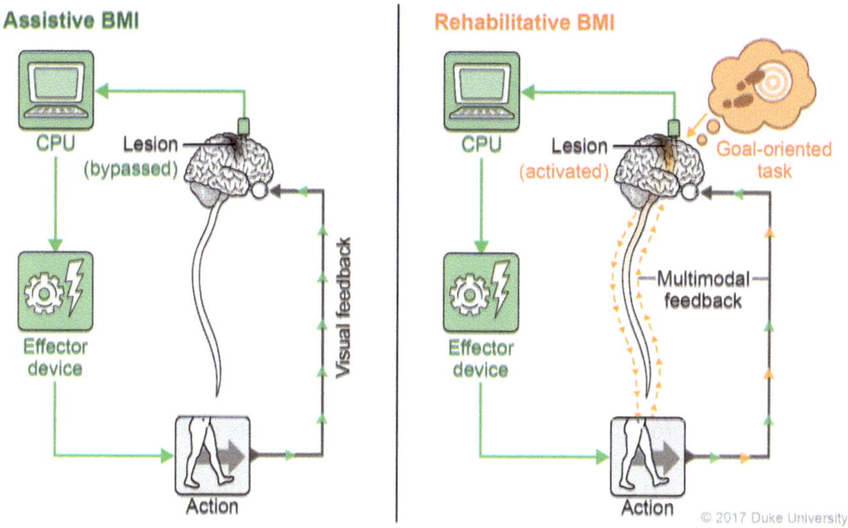

Fig. 18.4 Assistive vs. rehabilitative brain-machine interface (BMI) strategies. The assistive BMI uses nearby brain signals to bypass a neural lesion and generate an intended action. The rehabilitative BMI pairs goal-oriented tasks with positive feedback to re-activate lesioned circuits. Adapted with permission from Krucoff et al., 2016 [42]

or spike-timing dependent, plasticity (STDP) [119, 120]. Specifically, it has been shown that if pre-synaptic activity is preceded by post-synaptic activity, LTD will result. Conversely, if pre-synaptic activity is followed by post-synaptic activity, then LTP will result. This concept has been expressed in layman's terminology as, "neurons that fire together wire together," and captures the idea that synaptic strength is redistributed to favor functionally relevant pathways that are simultaneously active [121]. Mechanisms behind these observations remain largely unclear, but seem to involve the timing of NMDA receptor-mediated calcium influx into post-synaptic terminals [122]. For modeling of complex, larger scale circuits, the Bienenstock–Cooper–Munro (BCM) model can be used as a well-validated way to incorporate both pre- and postsynaptic firing rates into a circuitry stabilization metric [123, 124].

Some rehabilitative BMI approaches capitalize on STDP by using paired associative stimulation (PAS), or the act of pairing stimulation sites to promote plasticity [125, 126]. An example of a commonly used central stimulation strategy is TMS. TMS involves applying rapidly changing magnetic fields to the scalp via a magnetic stimulator. Continuous low frequency repetitive stimuli (≤1 Hz rTMS) decreases excitability of targets areas (similar to LTD which is maximally evoked at 1 Hz), while bursts of intermittent high frequency stimuli (≥5 Hz rTMS) enhance excitability (similar to LTP with high frequency bursts) [82, 113, 127]. These techniques have been used to induce modulation across cortico-subcortical and cortico-cortical networks through trans-synaptic spread, resulting in distant but specific changes along functional networks. Long term effects from TMS may be related to modulation of NMDA glutamatergic receptors, similar to induction of LTP/LTD, which also modulates number, location, and properties of synaptic NMDA receptors [84]. If timed correctly, corresponding sensory inputs can be potentiated [125]. In addition to TMS, functional electrical stimulation (FES) of paralyzed muscles or electrical stimulation of the nervous system distal to the injury timed with voluntary effort has also been shown to accelerate recovery after stroke [128, 129].

4 Future Directions and Conclusions

There are many reasons to be optimistic about the potential for improving neurological recovery after stroke. Despite the brain's inherent capacity for plasticity and spontaneous improvement, strokes still leave many patients with devastating deficits that can permanently affect independence and quality of life. Advances in modern neurobiological, neural engineering, and neurorehabilitation strategies have provided hope for better outcomes, and the synergistic potential of integrated protocols is only beginning to be realized. Because most strokes occur in older individuals whose inherent capacity for neural plasticity is diminished [12], microbiological interventions are needed to prime neurons for growth cone formation, goal-directed therapy paradigms are needed to engage the consciousness system, and neural interfaces are needed to provide real-time positive feedback to help stabilize functional neural circuits. How and when to employ each of these approaches requires further study [105].

In this chapter, we have discussed evidence that goal-directed therapy is a critical to *de novo* axonal sprouting, implying that consciousness plays a direct a role in neuronal regrowth, circuitry, rewiring, and guidance [49, 130]. On a practical level, this theory has two important corollaries: (1) that patients who suffer from disorders of consciousness (DOC) (i.e. comatose or vegetative patients) may need completely different therapeutic approaches, and (2) that experiments which have failed *in vitro* may still be viable therapies when integrated into a macroscopic neurorehabilitation framework that pairs conscious intention with real-time positive feedback. For patients with DOCs, perhaps deep brain or cortical stimulation will one day be able to substitute for active participation in rehabilitation, or at least prime the system to enhance awareness. For cellular or molecular approaches, integrating these techniques into models capable of fulfilling goal-directed rehabilitation paradigms will be necessary to test their true therapeutic potential, as a potentially viable approach may otherwise fail in isolation. This applies to stem cells [131], gene therapies [132], optogenetics [133], neuronal transplantation [134], and immunotherapies [43, 48]. It should be noted that task-based rehabilitation paradigms on their own do not always demonstrate improved functional outcomes when compared to controls in certain types of strokes [135], and exactly how consciousness modulates regrowth and connectivity mechanisms remains poorly understood [136–138].

While certain molecular approaches discussed in this chapter have shown promise in the lab (e.g. multiple NgR deletions, anti-NogoA antibodies, ionosine application, CAST/Ei knockouts), translation to bedside applications remains a challenge. This is true at least in part because replacing cells, generating plasticity, and enhancing axonal regrowth do not guarantee restoration of function by themselves [50]. The multi-year struggle to translate stem cell research into clinical benefit is an example of this challenge, as most stem cells are non-neural, do not have a defined or expected mechanism of action, and have not demonstrated the ability to alter CNS circuitry [74, 75, 77]. However, newer cells have been developed with improved neural capability that need to be tested in a strategic translational paradigm [80]. It is our hope that by combining microscopic neurobiological developments with macroscopic principles of neural circuitry and rehabilitation, success will be found where previous attempts at CNS restoration have failed [108, 135].

Conflict of Interest The authors declare they have no conflict of interest.

References

1. Kwakkel G, Kollen B, Lindeman E. Understanding the pattern of functional recovery after stroke: facts and theories. Restor Neurol Neurosci. 2004;22:281–99.
2. Christophe BR, Mehta SH, Garton ALA, Sisti J, Connolly ES. Current and future perspectives on the treatment of cerebral ischemia. Expert Opin Pharmacother. 2017;18:573–80.
3. Członkowska A, Leśniak M. Pharmacotherapy in stroke rehabilitation. Expert Opin Pharmacother. 2009;10:1249–59.
4. Winstein CJ, et al. AHA/ASA guideline guidelines for adult stroke rehabilitation and recovery. Stroke. 2016;47:e98–e169.
5. Alia C, et al. Neuroplastic changes following brain ischemia and their contribution to stroke recovery: novel approaches in neurorehabilitation. Front Cell Neurosci. 2017;11:1–22.
6. Hermann DM, Chopp M. Promoting neurological recovery in the post-acute stroke phase: Benefits and challenges. Eur Neurol. 2014;72:317–25.
7. Corbett D, Nguemeni C, Gomez-Smith M. How can you mend a broken brain? Neurorestorative approaches to stroke recovery. Cerebrovasc Dis. 2014;38:233–9.
8. Fishman HM, Bittner GD. Vesicle-mediated restoration of a plasmalemmal barrier in severed axons. News Physiol Sci. 2003;18:115–8.
9. Schlaepfer WW, Bunge RP. Effects of calcium ion concentration on the degeneration of amputated axons in tissue culture. J Cell Biol. 1973;59:456–70.
10. Hill CE, Beattie MS, Bresnahan JC. Degeneration and sprouting of identified descending supraspinal axons after contusive spinal cord injury in the rat. Exp Neurol. 2001;171:153–69.
11. Li Y, Raisman G. Sprouts from cut corticospinal axons persist in the presence of astrocytic scarring in long-term lesions of the adult rat spinal cord. Exp Neurol. 1995;134:102–11.
12. Shetty AK, Turner DA. Aging impairs axonal sprouting response of dentate granule cells following target loss and partial deafferentation. J Comp Neurol. 1999;414:238–54.
13. Conforti L, Gilley J, Coleman MP. Wallerian degeneration: an emerging axon death pathway linking injury and disease. Nat Rev Neurosci. 2014;15:394–409.
14. Benowitz LI, Yin Y. Combinatorial treatments for promoting axon regeneration in the CNS: strategies for overcoming inhibitory signals and activating neurons' intrinsic growth state. Dev Neurobiol. 2007;67:1148–65.
15. Bernstein DR, Stelzner DJ. Plasticity of the corticospinal tract following midthoracic spinal injury in the postnatal rat. J Comp Neurol. 1983;221:382–400.
16. Bulinski JC, et al. Changes in dendritic structure and function following hippocampal lesions: Correlations with developmental events? Prog Neurobiol. 1998;55:641–50.

17. Magavi SS, Leavitt BR, Macklis JD. Induction of neurogenesis in the neocortex of adult mice. Nature. 2000;405:951–5.
18. Shetty AK, Turner DA. Enhanced cell survival in fetal hippocampal suspension transplants grafted to adult rat hippocampus following kainate lesions: a three-dimensional graft reconstruction study. Neuroscience. 1995;67:561–82.
19. Alvarez-Buylla A, Lim D. For the long run: maintaining germinal niches in the adult brain. Neuron. 2004;41:683–6.
20. Doetsch F, Caillé I, Lim DA, García-Verdugo JM, Alvarez-Buylla A. Subventricular zone astrocytes are neural stem cells in the adult mammalian brain. Cell. 1999;97:703–16.
21. Laywell ED, Rakic P, Kukekov VG, Holland EC, Steindler DA. Identification of a multipotent astrocytic stem cell in the immature and adult mouse brain. Proc Natl Acad Sci U S A. 2000;97:13883–8.
22. Ohab JJ, Carmichael ST. Poststroke neurogenesis: emerging principles of migration and localization of immature neurons. Neuroscientist. 2008;14:369–80.
23. Seri B, García-Verdugo JM, McEwen BS, Alvarez-Buylla A. Astrocytes give rise to new neurons in the adult mammalian hippocampus. J Neurosci. 2001;21:7153–60.
24. Shetty A, Turner D. Fetal hippocampal cells grafted to kainate-lesioned CA3 region of adult hippocampus suppress aberrant supragranular sprouting of host mossy fibers. Exp Neurol. 1997;143:231–45.
25. Thored P, et al. Persistent production of neurons from adult brain stem cells during recovery after stroke. Stem Cells. 2006;24:739–47.
26. Macas J, Nern C, Plate KH, Momma S. Increased generation of neuronal progenitors after ischemic injury in the aged adult human forebrain. J Neurosci. 2006;26:13114–9.
27. Zhang RL, Zhang ZG, Zhang L, Chopp M. Proliferation and differentiation of progenitor cells in the cortex and the subventricular zone in the adult rat after focal cerebral ischemia. Neuroscience. 2001;105:33–41.
28. Arvidsson A, Collin T, Kirik D, Kokaia Z, Lindvall O. Neuronal replacement from endogenous precursors in the adult brain after stroke. Nat Med. 2002;8:963–70.
29. Carmichael ST, et al. Growth-associated gene expression after stroke: Evidence for a growth-promoting region in peri-infarct cortex. Exp Neurol. 2005;193:291–311.
30. Grenningloh G, Soehrman S, Bondallaz P, Ruchti E, Cadas H. Role of the microtubule destabilizing proteins SCG10 and stathmin in neuronal growth. J Neurobiol. 2004;58:60–9.
31. Sun F, He Z. Neuronal intrinsic barriers for axon regeneration in the adult CNS. Curr Opin Neurobiol. 2010;20:510–8.
32. Tedeschi A. Tuning the orchestra: transcriptional pathways controlling axon regeneration. Front Mol Neurosci. 2011;4:60.
33. Park KK, et al. Promoting axon regeneration in the adult CNS by modulation of the PTEN/mTOR pathway. Science. 2008;322:963–6.
34. de Lima S, et al. Full-length axon regeneration in the adult mouse optic nerve and partial recovery of simple visual behaviors. Proc Natl Acad Sci. 2012;109:9149–54.
35. Smith PD, et al. SOCS3 deletion promotes optic nerve regeneration in vivo. Neuron. 2009;64:617–23.
36. Xu B, Xie X. Neurotrophic factor control of satiety and body weight. Nat Rev Neurosci. 2016;17:282–92.
37. Chen D, Schneider G, Martinou J, Tonegawa S. Bcl-2 promotes regeneration of severed axons in mammalian CNS. Nature. 1997;385:434–9.
38. Goldberg JL, et al. Retinal ganglion cells do not extend axons by default. Neuron. 2002;33:689–702.
39. Benowitz LI, Carmichael ST. Promoting axonal rewiring to improve outcome after stroke. Neurobiol Dis. 2010;37:259–66.
40. Omura T, et al. Robust axonal regeneration occurs in the injured CAST/Ei mouse CNS. Neuron. 2015;86:1215–27.

41. de Lima S, Habboub G, Benowitz LI. Combinatorial therapy stimulates long-distance regeneration, target reinnervation, and partial recovery of vision after optic nerve injury in mice. Int Rev Neurobiol. 2012;106:153–72.
42. Krucoff MO, Rahimpour S, Slutzky MW, Edgerton VR, Turner DA. Enhancing nervous system recovery through neurobiologics, neural interface training, and neurorehabilitation. Front Neurosci. 2016;10
43. Alilain W, Horn KP, Hu H, Dick TE, Silver J. Functional regeneration of respiratory pathways after spinal cord injury. Nature. 2011;475:196–200.
44. Liu K, Tedeschi A, Park KK, He Z. Neuronal intrinsic mechanisms of axon regeneration. Annu Rev Neurosci. 2011;34:131–52.
45. Dickendesher TL, et al. NgR1 and NgR3 are receptors for chondroitin sulfate proteoglycans. Nat Neurosci. 2012;15:703–12.
46. Lee J-K, Kim J-E, Sivula M, Strittmatter SM. Nogo receptor antagonism promotes stroke recovery by enhancing axonal plasticity. J Neurosci. 2004;24:6209–17.
47. Freund P, et al. Nogo-A-specific antibody treatment enhances sprouting and functional recovery after cervical lesion in adult primates. Nat Med. 2006;12:790–2.
48. Maier IC, et al. Differential effects of anti-Nogo-A antibody treatment and treadmill training in rats with incomplete spinal cord injury. Brain. 2009;132:1426–40.
49. Wahl AS, et al. Asynchronous therapy restores motor control by rewiring of the rat corticospinal tract after stroke. Science. 2014;344:1250–5.
50. Bei F, et al. Restoration of visual function by enhancing conduction in regenerated axons. Cell. 2016;164:219–32.
51. Li S, et al. GDF10 is a signal for axonal sprouting and functional recovery after stroke. Nat Neurosci. 2015;18:1737–45.
52. Stroemer RP, Kent TA, Hulsebosch CE. Neocortical neural sprouting, synaptogenesis, and behavioral recovery after neocortical infarction in rats. Stroke. 1995;26:2135–44.
53. Schaechter JD, Moore CI, Connell BD, Rosen BR, Dijkhuizen RM. Structural and functional plasticity in the somatosensory cortex of chronic stroke patients. Brain. 2006;129:2722–33.
54. Murphy TH, Corbett D. Plasticity during stroke recovery: from synapse to behaviour. Nat Rev Neurosci. 2009;10:861–72.
55. Zai L, et al. Inosine alters gene expression and axonal projections in neurons contralateral to a cortical infarct and improves skilled use of the impaired limb. J Neurosci. 2009;29:8187–97.
56. Zai L, et al. Inosine augments the effects of a Nogo receptor blocker and of environmental enrichment to restore skilled forelimb use after stroke. J Neurosci. 2011;31:5977–88.
57. Kim D, et al. Inosine enhances axon sprouting and motor recovery after spinal cord injury. PLoS One. 2013;8:15–21.
58. Chen P, et al. Inosine induces axonal rewiring and behavioral outcome after stroke. PNAS. 2002;99:9031–6.
59. Dachir S, et al. Inosine improves functional recovery after experimental traumatic brain injury. Brain Res. 2014;1555:78–88.
60. Baldwin KT, Carbajal KS, Segal BM, Giger RJ. Neuroinflammation triggered by β-glucan/dectin-1 signaling enables CNS axon regeneration. Proc Natl Acad Sci U S A. 2015;112:2581–6.
61. Benowitz LI, Popovich PG. Inflammation and axon regeneration. Curr Opin Neurol. 2011;24:577–83.
62. Kurimoto T, et al. Neutrophils express oncomodulin and promote optic nerve regeneration. J Neurosci. 2013;33:14816–24.
63. Yin Y, et al. Macrophage-derived factors stimulate optic nerve regeneration. J Neurosci. 2003;23:2284–93.
64. Yin Y, et al. Oncomodulin is a macrophage-derived signal for axon regeneration in retinal ganglion cells. Nat Neurosci. 2006;9:843–52.
65. Yin Y, et al. Oncomodulin links inflammation to optic nerve regeneration. Proc Natl Acad Sci U S A. 2009;106:19587–92.

66. Benowitz L, Yin Y. Rewiring the injured CNS: lessons from the optic nerve. Exp Neurol. 2008;209:389–98.
67. Napieralski JA, Butler AK, Chesselet MF. Anatomical and functional evidence for lesion-specific sprouting of corticostriatal input in the adult rat. J Comp Neurol. 1996;373:484–97.
68. Nudo RJ. Recovery after brain injury: mechanisms and principles. Front Hum Neurosci. 2013;7:887.
69. Langhorne P, Bernhardt J, Kwakkel G. Stroke rehabilitation. Lancet. 2011;377:1693–702.
70. Teasell RW, Murie Fernandez M, McIntyre A, Mehta S. Rethinking the continuum of stroke rehabilitation. Arch Phys Med Rehabil. 2014;95:595–6.
71. DeFina PA, et al. Improving outcomes of severe disorders of consciousness. Restor Neurol Neurosci. 2010;28:769–80.
72. Breceda EY, Dromerick AW. Motor rehabilitation in stroke and traumatic brain injury: stimulating and intense. Curr Opin Neurol. 2013;26:595–601.
73. Krieger DW. Therapeutic drug approach to stimulate clinical recovery after brain injury. Front Neurol Neurosci. 2013;32:76–87.
74. Dobkin BH. Behavioral, temporal, and spatial targets for cellular transplants as adjuncts to rehabilitation for stroke. Stroke. 2007;38:832–9.
75. Cote DJ, et al. Ethical clinical translation of stem cell interventions for neurologic disease. Neurology. 2017;88(3):322–8. https://doi.org/10.1212/WNL.0000000000003506.
76. Smith EJ, et al. Implantation site and lesion topology determine efficacy of a human neural stem cell line in a rat model of chronic stroke. Stem Cells. 2012;30:785–96.
77. Wang Q, et al. Effect of stem cell-based therapy for ischemic stroke treatment: a meta-analysis. Clin Neurol Neurosurg. 2016;146:1–11.
78. Moniche F, et al. Intra-arterial bone marrow mononuclear cells (BM-MNCs) transplantation in acute ischemic stroke (IBIS trial): protocol of a phase II, randomized, dose-finding, controlled multicenter trial. Int J Stroke. 2015;10:1149–52.
79. Steinberg GK, et al. Clinical outcomes of transplanted modified bone marrow-derived mesenchymal stem cells in stroke: a phase 1/2a study. Stroke. 2016;47:1817–24.
80. Tornero D, et al. Synaptic inputs from stroke-injured brain to grafted human stem cell-derived neurons activated by sensory stimuli. Brain. 2017;140:692–706.
81. DeFina P, et al. The new neuroscience frontier: promoting neuroplasticity and brain repair in traumatic brain injury. Clin Neuropsychol. 2009;23:1391–9.
82. Demirtas-Tatlidede A, Vahabzadeh-Hagh AM, Bernabeu M, Tormos JM, Pascual-Leone A. Noninvasive brain stimulation in traumatic brain injury. J Head Trauma Rehabil. 2012;27:274–92.
83. Nahmani M, Turrigiano GG. Adult cortical plasticity following injury: recapitulation of critical period mechanisms? Neuroscience. 2014;283:4–16.
84. Villamar MF, Santos Portilla A, Fregni F, Zafonte R. Noninvasive brain stimulation to modulate neuroplasticity in traumatic brain injury. Neuromodulation. 2012;15:326–37.
85. Dancause N, Nudo R. Shaping plasticity to enhance recovery after injury. Prog Brain Res. 2011;192:273–95.
86. Dancause N. Extensive cortical rewiring after brain injury. J Neurosci. 2005;25:10167–79.
87. Kantak SS, Stinear JW, Buch ER, Cohen LG. Rewiring the brain: potential role of the premotor cortex in motor control, learning, and recovery of function following brain injury. Neurorehabil Neural Repair. 2012;26:282–92.
88. Carmichael ST, Chesselet M-F. Synchronous neuronal activity is a signal for axonal sprouting after cortical lesions in the adult. J Neurosci. 2002;22:6062–70.
89. Favre I, et al. Upper limb recovery after stroke is associated with ipsilesional primary motor cortical activity: a meta-analysis. Stroke. 2014;45:1077–83.
90. Kuner R. Central mechanisms of pathological pain. Nat Med. 2010;16:1258–66.
91. Thickbroom GW, Mastaglia FL. Plasticity in neurological disorders and challenges for noninvasive brain stimulation (NBS). J Neuroeng Rehabil. 2009;6:4.

92. Flor H, et al. Phantom-limb pain as a perceptual correlate of cortical reorganization following arm amputation. Nature. 1995;375:482–4.
93. Teyler TJ, Morgan SL, Russell RN, Woodside BL. Synaptic plasticity and secondary epileptogenesis. Int Rev Neurobiol. 2001;45:253–67.
94. Hasan A, et al. Dysfunctional long-term potentiation-like plasticity in schizophrenia revealed by transcranial direct current stimulation. Behav Brain Res. 2011;224:15–22.
95. Quartarone A, Rizzo V, Morgante F. Clinical features of dystonia: a pathophysiological revisitation. Curr Opin Neurol. 2008;21:484–90.
96. Dimitrijević MR, Nathan PW. Studies of spasticity in man. I Some features of spasticity. Brain. 1967;90:1–30.
97. Taub E, Uswatte G, Elbert T. New treatments in neurorehabilitation founded on basic research. Nat Rev Neurosci. 2002;3:228–36.
98. Taub E, Morris DM. Constraint-induced movement therapy to enhance recovery after stroke. Curr Atheroscler Rep. 2001;3:279–86.
99. Allred RP, Maldonado MA, Hsu And JE, Jones TA. Training the less-affected forelimb after unilateral cortical infarcts interferes with functional recovery of the impaired forelimb in rats. Restor Neurol Neurosci. 2005;23:297–302.
100. Allred RP, Jones TA. Maladaptive effects of learning with the less-affected forelimb after focal cortical infarcts in rats. Exp Neurol. 2008;210:172–81.
101. Kozlowski DA, James DC, Schallert T. Use-dependent exaggeration of neuronal injury after unilateral sensorimotor cortex lesions. J Neurosci. 1996;16:4776–86.
102. Wolf SL, et al. Effect of constraint-induced movement therapy on upper extremity function 3 to 9 months after stroke: the EXCITE randomized clinical trial. JAMA. 2006;296:2095–104.
103. Dromerick AW, et al. Very early constraint induced movement during stroke rehabilitation (VECTORS): a single-center RCT. Neurology. 2009;73:195–201.
104. McIntyre A, et al. Systematic review and meta-analysis of constraint-induced movement therapy in the hemiparetic upper extremity more than six months post stroke. Top Stroke Rehabil. 2012;19:499–513.
105. Lang CE, Lohse KR, Birkenmeier RL. Dose and timing in neurorehabilitation: prescribing motor therapy after stroke. Curr Opin Neurol. 2015;28:549–55.
106. Clarkson AN, et al. AMPA receptor-induced local brain-derived neurotrophic factor signaling mediates motor recovery after stroke. J Neurosci. 2011;31:3766–75.
107. Levy RM, et al. Cortical stimulation for the rehabilitation of patients with hemiparetic stroke: a multicenter feasibility study of safety and efficacy. J Neurosurg. 2008;108:707–14.
108. Levy RM, et al. Epidural electrical stimulation for stroke rehabilitation: results of the prospective, multicenter, randomized, single-blinded everest trial. Neurorehabil Neural Repair. 2016;30:107–19.
109. Guggenmos DJ, et al. Restoration of function after brain damage using a neural prosthesis. Proc Natl Acad Sci U S A. 2013;110:21177–82.
110. Kleim JA, et al. Motor cortex stimulation enhances motor recovery and reduces peri-infarct dysfunction following ischemic insult. Neurol Res. 2003;25:789–93.
111. Plautz EJ, et al. Post-infarct cortical plasticity and behavioral recovery using concurrent cortical stimulation and rehabilitative training: a feasibility study in primates. Neurol Res. 2003;25:801–10.
112. Nudo RJ, Milliken GW. Reorganization of movement representations in primary motor cortex following focal ischemic infarcts in adult squirrel monkeys. J Neurophysiol. 1996;75:2144–9.
113. Shin S, Dixon E, Okonkwo D, Richardson M. Neurostimulation for traumatic brain injury. J Neurosurg. 2014;121:1219–31.
114. Nudo RJ, Jenkins WM, Merzenich MM. Repetitive microstimulation alters the cortical representation of movements in adult rats. Somatosens Mot Res. 1990;7:463–83.
115. Monfils M-H, VandenBerg PM, Kleim JA, Teskey GC. Long-term potentiation induces expanded movement representations and dendritic hypertrophy in layer V of rat sensorimotor neocortex. Cereb Cortex. 2004;14:586–93.

116. Elsner B, Kugler J, Pohl M, Mehrholz J. Transcranial direct current stimulation (tDCS) for improving activities of daily living, and physical and cognitive functioning, in people after stroke. Cochrane Database Syst Rev. 2016;3:CD009645. https://doi.org/10.1002/14651858.CD009645.pub3.
117. Edwardson MA, Lucas TH, Carey JR, Fetz EE. New modalities of brain stimulation for stroke rehabilitation. Exp Brain Res. 2013;224:335–58.
118. Gharabaghi A, et al. Coupling brain-machine interfaces with cortical stimulation for brain-state dependent stimulation: enhancing motor cortex excitability for neurorehabilitation. Front Hum Neurosci. 2014;8:122.
119. Hebb DO. The organization of behavior: a neuropsychological theory. Hoboken, NJ: Wiley; 1949.
120. Cooper SJ. Donald O. Hebb's synapse and learning rule: a history and commentary. Neurosci Biobehav Rev. 2005;28:851–74.
121. Rebesco JM, Miller LE. Enhanced detection threshold for in vivo cortical stimulation produced by Hebbian conditioning. J Neural Eng. 2011;8:16011.
122. Purves D, et al. Neuroscience. In: Neuroscience. Sunderland, MA: Sinauer Associates; 2008.
123. Bienenstock EL, Cooper LN, Munro PW. Theory for the development of neuron selectivity: orientation specificity and binocular interaction in visual cortex. J Neurosci. 1982;2:32–48.
124. Cooper LN, Bear MF. The BCM theory of synapse modification at 30: interaction of theory with experiment. Nat Rev Neurosci. 2012;13:798–810.
125. Stefan K, Kunesch E, Cohen LG, Benecke R, Classen J. Induction of plasticity in the human motor cortex by paired associative stimulation. Brain. 2000;123(Pt 3):572–84.
126. Carson RG, Kennedy NC. Modulation of human corticospinal excitability by paired associative stimulation. Front Hum Neurosci. 2013;7:823.
127. Kobayashi M, Pascual-Leone A. Basic principles of magnetic stimulation. Lancet. 2003;2:145–56.
128. Daly JJ, et al. A randomized controlled trial of functional neuromuscular stimulation in chronic stroke subjects. Stroke. 2006;37:172–8.
129. Kafri M, Laufer Y. Therapeutic effects of functional electrical stimulation on gait in individuals post-stroke. Ann Biomed Eng. 2015;43:451–66.
130. van den Brand R, et al. Restoring voluntary control of locomotion after paralyzing spinal cord injury. Science. 2012;336:1182–5.
131. Mothe AJ, Tator CH. Advances in stem cell therapy for spinal cord injury. Thew J Clin Investig. 2012;122:3824–34.
132. Warren Olanow C, et al. Gene delivery of neurturin to putamen and substantia nigra in Parkinson disease: a double-blind, randomized, controlled trial. Ann Neurol. 2015;78:248–57.
133. Jarvis S, Schultz SR. Prospects for optogenetic augmentation of brain function. Front Syst Neurosci. 2015;9:157.
134. Furlanetti LL, et al. Continuous high-frequency stimulation of the subthalamic nucleus improves cell survival and functional recovery following dopaminergic cell transplantation in rodents. Neurorehabil Neural Repair. 2015;29:1001–12.
135. Winstein CJ, et al. Effect of a task-oriented rehabilitation program on upper extremity recovery following motor stroke: the ICARE randomized clinical trial. JAMA. 2016;315:571–81.
136. Brogaard B, Gatzia DE. What can neuroscience tell us about the hard problem of consciousness? Front Neurosci. 2016;10:1–4.
137. Koch C, Massimini M, Boly M, Tononi G. Neural correlates of consciousness: progress and problems. Nat Rev Neurosci. 2016;17:307–21.
138. Sandberg K, Frässle S, Pitts M. Future directions for identifying the neural correlates of consciousness. Nat Rev Neurosci. 2016. https://doi.org/10.1038/nrn.2016.104.

Chapter 19
Neural Network Regeneration After Stroke

Norihito Shimamura, Takeshi Katagai, Masato Naraoka, and Hiroki Ohkuma

Abstract Stroke remains a major cause of disability throughout the world: paralysis, cognitive impairment, aphasia, apraxia and so on. Surgical or medical intervention is curative in only a small number of cases. Stroke cases with morbidity require rehabilitation. Neurorehabilitation generally improves patient outcome, but the involved mechanisms have not been clarified. Recent advancements in technology are revealing the mechanisms of neurorehabilitation from the gene and up to neural network remodeling. Rehabilitation in clinical application, however, should be guided by convincing evidence. In this chapter we review the evidence for the regeneration of the neural network after stroke.

Keywords Cerebrovascular • Mechanism • Neural network • Rehabilitation • Reorganization • Stroke

Abbreviations

ADL	Activities of daily living
AHA	American Heart Association
AMP	Adenosine monophosphate
BDNF	Brain-derived neurotrophic factor
BOLD	Blood oxygen level dependent
CFA	Caudal forelimb area
CIMT	Constraint-induced movement therapy
DTI	Diffusion-tensor imaging
FA	Fractional anisotropy
f-MRI	Functional magnetic resonance imaging
GABA	γ-Aminobutyric acid
hBMSC	Human bone marrow stromal cells

N. Shimamura, M.D., Ph.D. (✉) • T. Katagai • M. Naraoka • H. Ohkuma
Department of Neurosurgery, Hirosaki University Graduate School of Medicine,
5-Zaihuchou, 036-8562 Hirosaki, Aomori Prefecture, Japan
e-mail: shimab@hirosaki-u.ac.jp

P.A. Lapchak, J.H. Zhang (eds.), *Cellular and Molecular Approaches to Regeneration and Repair*, Springer Series in Translational Stroke Research,
https://doi.org/10.1007/978-3-319-66679-2_19

IL	Interleukin
Muse	Multilineage differentiating stress enduring
RFA	Rostral forelimb area
TMS	Transcranial magnetic stimulation
USA	United State of America
VNS	Vagal nerve stimulation

1 Introduction

Stroke remains a major cause of disability throughout the world. It causes sudden neurological deterioration and sometimes death. From 2009 to 2012 an estimated 6.6 million Americans ≥20 years of age and 1.2 million Japanese suffered a stroke in 2014 [1, 2]. The stroke death rate has been decreasing for decades and the number of patients with a neurological deficit has thus been increasing.

The number of elderly stroke patients is also increasing, in particular in a society with progressively advancing aging [3–5]. The estimated, overall prevalence of stroke is 2.6% in the USA [2]. Previously, elderly patients were considered beyond the indication for intensive rehabilitation due to diminished treatment efficacy as compared to younger patients [6–8]. Aging is one of the key factors in stroke recovery, as older patients have less ability to fully recovered from a deficit [9]. But recent analyses suggest that inpatient neurorehabilitation provides benefits to all stroke patients, even patients over 80 years old [10]. We also found that early ambulation produces favorable outcomes and the non-demential state in elderly subarachnoid hemorrhage patients [11, 12]. Of course, not all young stroke patients will fully recover. Those who do not may still have a long life ahead. Young stroke patients also feel a greater burden of poor self-perception than do aged patients [13].

Patient neurological deficit depends on the severity and the region of stroke, not on the type of stoke. Some patients experience spontaneous recovery, but the degree of such cannot be estimated at the onset of stroke [14]. No one doubts the effectiveness of rehabilitation in counteracting a neurological deficit, but why are the outcomes of rehabilitation different in every patient? Patient-specific, stroke-type specific mechanisms of rehabilitation may well be at play, and the failure to understand the distinctions before the twenty-first century may have been the cause of therapeutic controversy.

The aim of this chapter is the clarification of the mechanisms of neural network regeneration after stroke.

1.1 Timing and Dosage of Rehabilitation in a Clinical Setting

The practice of rehabilitation started around the nineteenth century and the effectiveness of rehabilitation after stroke is common knowledge today. But the treatment strategy in rehabilitation relied on experience until a few decades ago, and the start and dosage of rehabilitation have been controversial. In 2013 the AHA guideline for ischemic stroke recommended early mobilization of less severely affected patients, and measures to prevent subacute complications of stroke were recommended (Class I; Level of Evidence C) [15]. In 2015 the AHA guideline for spontaneous intracranial hemorrhage recommended that rehabilitation can be beneficial when started as early as possible and continued in the community as part of a well-coordinated program of accelerated hospital discharge and home-based resettlement to promote ongoing recovery (Class IIa; Level of Evidence B). The definition of 'early' was controversial. "A very early rehabilitation trial for stroke" referred to early mobilization as within 24 h, while intensive rehabilitation for non-severe ischemic and spontaneous intracerebral hemorrhage patients actually decreased favorable outcome at 3 months by four-percent as compared to typical care (46% vs. 50%) [16]. We should be concerned with the time of mobilization and, in this randomized, controlled trial (RCT), the starting time for the usual care group was from 16.5 to 29.3 h after the stroke. Neurorehabilitation should generally be carried out within 1 or 2 days after stroke.

Dromerick et al. compared three rehabilitation protocols in a single-center RCT: traditional occupational therapy involving 1 h of activities of daily living (ADL) retraining, a group with 1 h of upper extremity bilateral training activities and standard constraint-induced movement therapy (CIMT), including 2 h of shaping therapy per day and wearing a padded constraint mitten for 6 h per day, and a high-intensity CIMT group, which underwent 3 h per day of shaping and wearing the mitten for 90% of waking hours [17]. All groups continued therapy for 2 weeks and the high dosage group had significantly less improvement on day 90 [17]. Another nonrandomized, parallel-group, dosage-controlled study of aphasia therapy with distributed therapy (6 h per week; 8 weeks) showed significantly greater improvement than with intensive therapy (16 h per week; 3 weeks) [18].

In 2016 the AHA guideline for adult stroke rehabilitation and recovery did not recommend high-dosage, because very early mobilization within 24 h of stroke onset can reduce the odds of a favorable outcome at 3 months. (ClassIII, Evidence A) [19]. We need to decide on the start of and the dosage of rehabilitation for patients from the combination of the evidence and the neurological condition of the patient. Our previous clinical studies revealed that early ambulation produces physically and psychologically favorable outcomes [11, 12, 20]. Elderly stroke patients lose their physical abilities abruptly during the acute phase of stroke. At the same time, full recovery of physical and psychological functioning is difficult for elderly patients; aging diminishes tissue stem cell- and cholinergic system functioning that regenerate the central nervous system, muscle and other organs [21, 22].

1.2 *Muscle Influence on Neural Networks*

Muscles and bone are organs not under the direct control of the brain, but they communicate with other organs, including the brain, exerting mutual influence [23, 24]. Muscle is not only a physical organ but also a secretary organ; it secretes several hundred peptides (e.g. interleukin (IL)-4, IL-6, IL-7, IL-15, myostatin, leukemia inhibitory factor 1, irisin, insulin-like growth factors 1, fibroblast growth factors 2, and brain-derived neurotrophic factor) [24]. Brain-derived neurotrophic factor (BDNF) is secreted by skeletal muscle, and muscle atrophy induces dementia, depression, diabetes and malignancies [21, 24–26]. BDNF protein is stored in human platelets and is released upon agonist stimulation [27]. Physical exercise training significantly increased resting concentrations of BDNF in peripheral blood, and resting concentrations of peripheral blood BDNF were significantly higher after intervention, especially in the aerobic state but not so with resistance training [28, 29]. BDNF is a contraction-inducible protein in skeletal muscle that is capable of enhancing lipid oxidation in skeletal muscle via activation of AMP-activated protein kinase [29]. Stranahan AM et al. revealed that a combination of exercise and caloric restriction increased levels of BDNF in the hippocampus, combined with an increase in dendritic spine density on the secondary and tertiary dendrites of dentate granule neurons in diabetic mice [30]. Dendrites transfected with the BDNF gene showed rapid turnover of spines, which may be a demonstration of the translation of activity patterns into specific morphological changes [31]. Rehabilitation increases circulating and intramuscular BDNF concentrations and produces neuro-reorganization in the central nervous system.

Therefore, loss of muscle strength is associated with physical disability and functional limitation [32]. In an analysis of females, reduction of both fat-free soft tissue and fat mass was associated with cognitive impairment [33]. Grip strength was also related to dementia in an aged population, especially in women [34, 35]. Careful clinical observational evaluation and treatment of muscle volume, however, is important for management. Tolea et al. showed that individuals with sarcopenia were six times more likely to experience combined cognitive impairment/physical impairment as compared to a normal control, and the effect of sarcopenia on cognition is related to low muscle strength rather than low muscle mass [21]. Sarcopenia, however, is not irreversible and we do not avoid starting rehabilitation in the chronic state. Intensive inpatient neurorehabilitation is beneficial for all stroke patients, even patients over 80 years of age [10].

1.3 Remodeling of Functional Networks Detected with MRI

Recent advanced magnetic resonance imaging (MRI) technology has revealed objective findings for rehabilitation. Rehme et al. reported functional MRI (f-MRI) scans showed robust and stronger blood-oxygen-level-dependent (BOLD) signal changes in post ischemic patients during movements of the affected or unaffected hand at various stages within 2 weeks [36]. They also revealed that the reinstatement of effective connectivity in the ipsilesional hemisphere is an important feature of motor recovery after stroke, but the supportive role of contralesional primary motor cortex into enhanced inhibitory coupling might indicate maladaptive processes that could be a target of non-invasive brain stimulation techniques [36].

In post-stroke aphasia patients, separate neuronal systems support each grammatical class—motor areas for verbs and perception areas for nouns—based on f-MRI

[37]. Semantic, feature-based rehabilitation consistently and significantly modulated the left inferior frontal gyrus in chronic post-stroke aphasia patients and in healthy volunteers [38]. Music therapy also represents a reduction in abnormal contralesional activity, an enhancement in f-MRI activity in the auditory and motor areas, and an increase in functional connectivity among several regions of the auditory-motor network in individuals suffering chronic stroke over several months [39]. Listening to music daily also significantly produced structural gray matter changes as compared to a group listening to an audiobook and a group without any listening material following middle cerebral artery stroke [40]. The gray matter reorganization in the frontal areas correlated with enhanced recovery of verbal memory, focused attention, and increased language skills, whereas gray matter reorganization in the subgenual anterior cingulate cortex correlated with reduced negative mood [40].

1.4 Transcranial Magnetic Stimulation for Stroke Patients

Transcranial magnetic stimulation (TMS) is available for the analysis of axonal integrity and to produce muscle contraction. Lefebvre et al. performed a double-blind, cross-over randomized, sham-controlled experiment with transcranial direct current stimulation [41]. Dual transcranial direct current stimulation applied during motor skill learning with a paretic upper limb resulted in prolonged shaping of brain activation, which supported behavioral enhancements in stroke patients [41]. Repetitive, continuous transcranial magnetic stimulation improved spatial neglect in stroke patients, and the effect can be evaluated by fractional anisotropy and mean diffusivity in MRI [42]. CIMT for 2 weeks yielded superior functional outcome for patients who started therapy within 9 months after stroke compared to patients who did so more than 12 months post-stroke [43]. Also, the TMS motor map shifted posteriorly in the late stroke group, and cortical reorganization did occur in those late treatment patients [43].

1.5 Neural Spine Formation in Healthy Rodents

In 2002, Trachtenberg et al. reported that new dendritic spines appeared daily in the barrel cortex of healthy mice and the elimination of some spines occurred, maintaining the density of spines [44]. Additionally, daily changes in dendritic spines in the barrel cortex *in vivo* in adult mice as well as spine sprouting and retraction are associated with synapse formation and elimination [44]. Yang et al. revealed that only 2 days of accelerated rotarod training induced new dendritic spines that were important for the reorganization of cortical circuits, and new experience led to pruning of existing synapses for the control of functional changes in cortical circuits [45]. Only 0.04% of total spines survived the first few weeks in synaptic circuits and novel experience produced pruning of a small fraction of existing spines, which promoted an integrated and stable structural basis for lifelong memory storage [45].

1.6 Animal Stroke Experiments

Animal experiments have been done to clarify among the possible mechanisms of rehabilitation. In rats with ischemic infarct in the endothelin-1 induced cortical focal ischemic model, early rehabilitation induced rapid improvement in motor function, and the forelimb motor maps were significantly enlarged compared with the no-rehab group 38 days after the stroke [46]. The authors concluded that early motor training after stroke can help shape the evolving post-stroke neural network [46]. Intensive rehabilitation significantly enhanced both the dendritic architecture and spine numbers in the adjoining rostral forelimb area in adult rats when the caudal forelimb region of the motor cortex was destroyed [22]. Those effects were diminished by cholinergic ablation [22]. Training in a forelimb reaching task leads to rapid (within an hour) formation of postsynaptic dendritic spines on the output pyramidal neurons in the contralateral motor cortex in a mouse model [47]. Forelimb skilled grasping rehabilitation did not influence the number of neurons in adult rats, but the spine density of neurons increased significantly by 22.5 ± 2.3% compared with untrained control subjects [48]. Grasping training also specifically induced spine plasticity, dendrite length and dendrite branches of C8 spine projecting corticospinal motor neurons without influencing the C4 spine situated in the same cortical layer V [48]. This synaptic plasticity contributes to neural circuits and neuro-rehabilitative outcome. Gulati et al. reported a brain-machine interface in a rat cortical focal injured model by which rats can control the firing of neurons in the perilesional cortex without motor functional recovery, and neurons form functional cell assemblies after training [49].

A time limitation for rehabilitation in various animals has been discussed. Xerri et al. reported early tactile stimulation (day 3–8) after the focal cortical ischemia reduced cortical tissue loss and digit representation loss as compared to delayed tactile stimulation (day 8–13) or without stimulation [50]. Early tactile stimulation also forced cortical re-mapping. The limitation to dendritic sprouting is 18 days after the neocortical damage [51, 52]. The immobilization of the ipsilateral forelimb (i.e. forced overuse impaired limb) after neocortical damage also prevents dendritic growth of the contralateral normal cortex and this deficiency extends beyond mobilization of the ipsilateral forelimb [53, 54]. Rehabilitation started at 5 days after the ischemic stroke achieved significant, favorable outcome as compared to starting the rehabilitation at 14 or 30 days after the stroke in the rat model [55]. An early start to rehabilitation increased the number of branches and the complexity of layer-V neurons in the undamaged motor cortex [55]. High-dosage rehabilitation treatment started at 15 days after middle cerebral artery occlusion induced improving forelimb motor function and enhanced dendritic growth of neurons at layer-V in the undamaged motor cortex [56]. Rehabilitative training for caudal forelimb area (CFA) ischemic stroke induced motor map reorganization of the rat rostral forelimb area (RFA) significantly after 2 weeks, while rehabilitative training increased the number of neurons projecting from the RFA to both the upper and lower cervical cord [57]. Four weeks of rehabilitative training after CFA ischemic stroke also increased functional recovery and expansion of RFA due to endogenous neurogenesis in a rat model [58]. Winship et al. reported single neuron rewiring occurred after S1 cortical infarction within one month and remapping of cortical function had occurred naturally [59].

In primate focal cortical ischemic injury, 12-week rehabilitation starting 2 weeks after the stroke led to recovery to near-baseline levels, and this recovery relied on a small change in the contralesional primary motor cortex and significantly increasing motor outputs from the ipsilateral pontomedullary reticular formation [60]. Evidence from animal experiments confirms the starting time and intensity of rehabilitation in individual clinical practices.

1.7 Concept of the Cortical Network via Fasciculi

Before the twenty-first century, neurological deficit was discussed in terms of the concept of the 'eloquent' or 'non-eloquent' nucleus and specific fasciculi (i.e. the arcuate fasciculus, corpus callosum). Microsurgical tractographic anatomy and advanced magnetic resonance imaging came to represent three-dimensional white matter networks of fasciculi throughout the brain [61–67]. These findings suggested that a reconsideration of pathophysiology and the mechanism of neurological deficit was in order. Several researchers made full use of state-of-the-art diffusion-tensor imaging (DTI) technology to determine the potential neural network connectivity not only by fractional anisotropy (FA), but also using several other parameters [37, 38, 68, 69]. Steiner et al. were the first to report on the complementary nature of TMS and MRI [70]. They showed that stroke patients with a muscle evoked potentials (MEPs) reaction achieved meaningful functional recovery, but patients without MEPs combined with fractional anisotropy asymmetry that exceeded a value of 0.25 did not achieve functional recovery [70]. DTI correlated with functional recovery, and brain-computer interface training also influenced the contralesional corticospinal tract [71]. Stewart et al. compared chronic, left hemisphere stroke in ten individuals with 16 age-matched controls with respect to brain activation during right hand motor tasks by DTI [72]. The control group showed increased activation in the left inferior parietal lobule (IPL), while the stroke group showed increased activation in several right/contralesional regions, including the right IPL [72]. DTI permits an estimation of the functional recovery of patients and an evaluation of the outcome of rehabilitation [42, 71].

The semantic anatomical networks can be evaluated with MRI [73]. White matter integrity in the frontoparietal network was significantly related to haptic performance in chronic stroke, and thus the thalamus to the primary motor cortex and the precuneus interhemispheric tracts were appropriate targets for rehabilitation [74]. In an electroencephalogram experiment, weaker cortical connectivity in thalamic ischemic stroke patients suggested functional impairment of information transmission in cortical connections [75]. Both the inter-cortical network and the thalamus to the primary motor cortex may be targets for sensory rehabilitation training.

1.8 Innovative Approach to Significant Recovery

Low-dose $GABA_A$ antagonism treatment at 7 days after focal cortical ischemic injury produced 25% better functional outcome and a significant reduction in the stroke volume as compared to placebo in the rat model [76]. Left vagal nerve stimulation (VNS) paired with an isometric force task produced significant forelimb functioning in an endothelin-induced focal motor area ischemic, aged rodent model [77]. The authors speculated that the mechanism of this result is that the VNS drives neural activity in proplasticity neuromodulatory centers of the brain, including the noradrenergic locus coeruleus and the cholinergic basal forebrain [77]. VNS increases BDNF and fibroblast growth factor in the hippocampus and cerebral cortex, and activates the BDNF binding receptor (TrkB) in the hippocampus [78, 79].

Stem cell transplantation is one of the candidates for the fundamental treatment of stroke [80–86]. Human bone marrow stromal cells (hBMSC) cultured with animal protein-free medium show the same surface markers and trophic factors as BMSC cultured with animal proteins. Those cells cultured with animal protein-free hBMSC are safely available for neurotransplantation [87]. Intra carotid arterial injection of neural stem cells produced a significantly large amount of cells that migrated into brain compared to intravenous application in rat ischemic stroke [88, 89]. But the intraarterial injection of stem cells risks arterial occlusion. Animal protein-free medium cultured hBMSC survived, migrated and differentiated within infarcted brain tissue until 8 weeks after the stereotaxic implantation [90]. Uchida et al. reported early and delayed functional recovery by multilineage differentiating stress enduring (Muse) cell stereotaxic injection into an ischemic lesion in the rat model [91, 92]. We also reported Muse cell treatment in the intracerebral hemorrhage mouse model; Muse cells significantly improved neurological functioning after the early state [93]. Transplanted stem cells sprouting new axons into the cortex and spine, combined in the future with the transplantation of stem cells and neurorehabilitation, will become the fundamental treatment for stroke. Of course, we do not rush into stem cell therapy without a one-by-one evaluation of clinical trials [94].

2 Conclusion

We cannot achieve the regeneration state after stroke through one treatment alone. To rescue the brain, physicians perform thrombolysis, retrieval of the thrombus, revascularization, removal of a mass lesion or the hemostasis of a bleeding point as soon as possible after stroke (Fig. 19.1). In a next step, we medicate using several drugs to relieve edema, inflammation and free radicals. From the acute stage to the chronic stage, neurorehabilitation is done to preserve muscle volume, prevent contracture and for axonal sprouting. Consecutive control of axonal sprouting, functional compensation through the reorganization of intercortical connections,

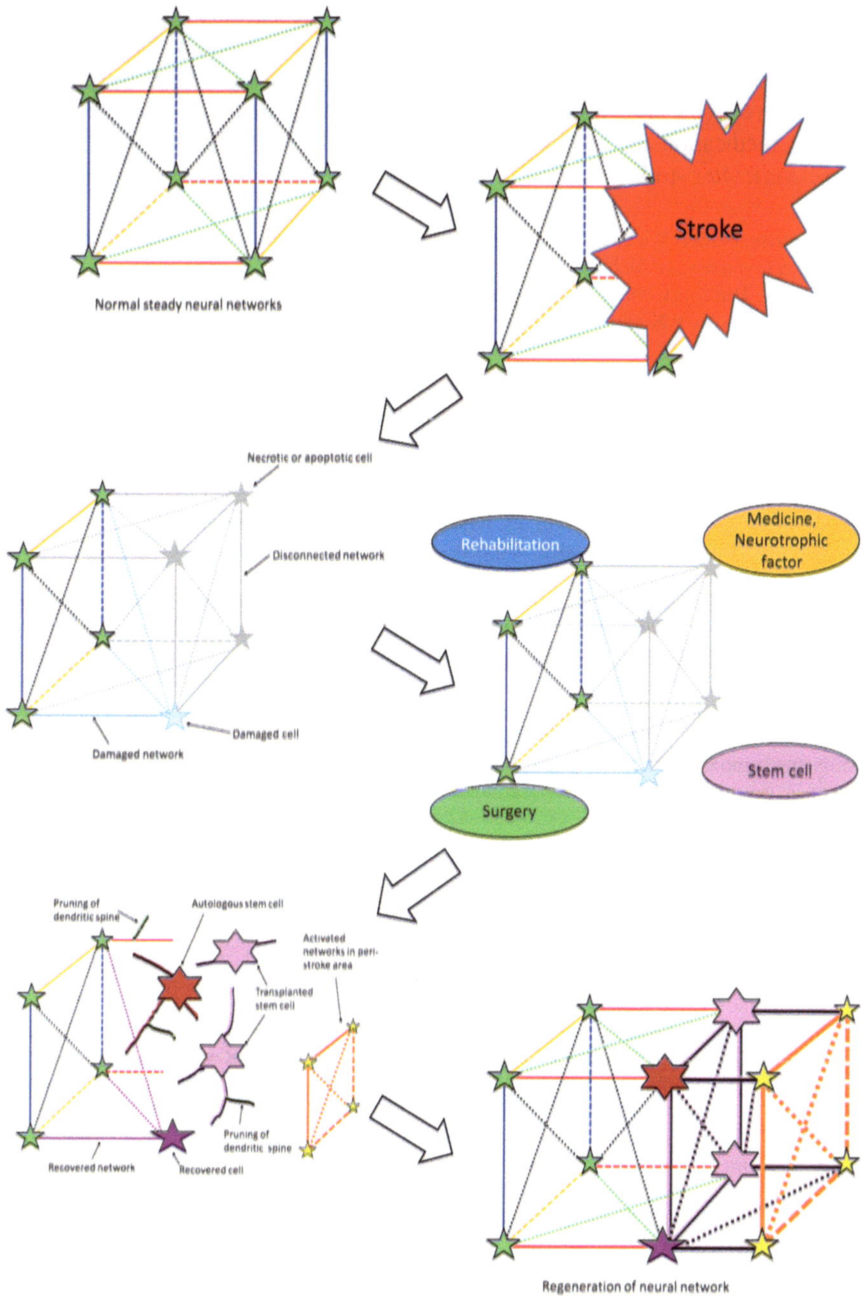

Fig. 19.1 Concept of reorganization of neural network after stroke. Several cascades are activated after the stroke and artificial influence is added. Finally, reorganization of neural networks is established. A *star* represents a neuron; *line* or *dotted line*, axonal network

reorganization through the influence of other organs, especially muscle, are indispensible processes for favorable outcome after stroke. Additionally, stem cell application to stroke provides the cellular constituents and trophic factors necessary for favorable recovery after stroke. Comprehensive and continuous treatments through surgery, medication and rehabilitation are essential for the regeneration of the neural network after stroke.

Acknowledgements We thank Mark Inglin (University of Basel) for his editorial assistance.

Competing interests We have no conflict of interest.

Funding This work was supported by Grants-in-Aid for Scientific Research of Japan, grant number 26462149 for Norihito Shimamura.

References

1. Minister of Health LaW. Survey of patient. Tokyo: Minister of Health LaW; 2016.
2. Mozaffarian D, Benjamin EJ, Go AS, Arnett DK, Blaha MJ, Cushman M, et al. Heart disease and stroke statistics--2015 update: a report from the American Heart Association. Circulation. 2015;131(4):e29–322.
3. Brinjikji W, Lanzino G, Rabinstein AA, Kallmes DF, Cloft HJ. Age-related trends in the treatment and outcomes of ruptured cerebral aneurysms: a study of the nationwide inpatient sample 2001–2009. AmJ Neuroradiol. 2013;34(5):1022–7.
4. Nieuwkamp DJ, Algra A, Blomqvist P, Adami J, Buskens E, Koffijberg H, et al. Excess mortality and cardiovascular events in patients surviving subarachnoid hemorrhage: a nationwide study in Sweden. Stroke. 2011;42(4):902–7.
5. Nieuwkamp DJ, Setz LE, Algra A, Linn FH, de Rooij NK, Rinkel GJ. Changes in case fatality of aneurysmal subarachnoid haemorrhage over time, according to age, sex, and region: a meta-analysis. Lancet Neurol. 2009;8(7):635–42.
6. Forti P, Maioli F, Procaccianti G, Nativio V, Lega MV, Coveri M, et al. Independent predictors of ischemic stroke in the elderly: prospective data from a stroke unit. Neurology. 2013;80(1):29–38.
7. Kammersgaard LP, Jorgensen HS, Reith J, Nakayama H, Pedersen PM, Olsen TS, et al. Short- and long-term prognosis for very old stroke patients. The Copenhagen Stroke Study. Age Ageing. 2004;33(2):149–54.
8. Nakayama H, Jorgensen HS, Raaschou HO, Olsen TS. The influence of age on stroke outcome. The Copenhagen Stroke Study. Stroke. 1994;25(4):808–13.
9. Koh SH, Park HH. Neurogenesis in stroke recovery. Transl Stroke Res. 2017;8(1):3–13.
10. Knecht S, Rossmuller J, Unrath M, Stephan KM, Berger K, Studer B. Old benefit as much as young patients with stroke from high-intensity neurorehabilitation: cohort analysis. J Neurol Neurosurg Psychiatry. 2016;87(5):526–30.
11. Shimamura N, Matsuda N, Satou J, Nakano T, Ohkuma H. Early ambulation produces favorable outcome and nondemential state in aneurysmal subarachnoid hemorrhage patients older than 70 years of age. World Neurosurg. 2014;81(2):330–4.
12. Shimamura N, Naraoka M, Katagai T, Katayama K, Kakuta K, Matsuda N, et al. Analysis of factors that influence long-term independent living for elderly subarachnoid hemorrhage patients. World Neurosurg. 2016;90:504–10.
13. Ren H, Liu C, Li J, Yang R, Ma F, Zhang M, et al. Self-perceived burden in the young and middle-aged inpatients with stroke: a cross-sectional survey. Rehabil Nurs. 2016;41(2):101–11.
14. Kwakkel G, Kollen BJ, van der Grond J, Prevo AJ. Probability of regaining dexterity in the flaccid upper limb: impact of severity of paresis and time since onset in acute stroke. Stroke. 2003;34(9):2181–6.

15. Jauch EC, Saver JL, Adams HP Jr, Bruno A, Connors JJ, Demaerschalk BM, et al. Guidelines for the early management of patients with acute ischemic stroke: a guideline for healthcare professionals from the American Heart Association/American Stroke Association. Stroke. 2013;44(3):870–947.
16. Group ATC, Bernhardt J, Langhorne P, Lindley RI, Thrift AG, Ellery F, et al. Efficacy and safety of very early mobilisation within 24 h of stroke onset (AVERT): a randomised controlled trial. Lancet. 2015;386(9988):46–55.
17. Dromerick AW, Lang CE, Birkenmeier RL, Wagner JM, Miller JP, Videen TO, et al. Very early constraint-induced movement during stroke rehabilitation (VECTORS): a single-center RCT. Neurology. 2009;73(3):195–201.
18. Dignam J, Copland D, McKinnon E, Burfein P, O'Brien K, Farrell A, et al. Intensive versus distributed aphasia therapy: a nonrandomized, parallel-group, dosage-controlled study. Stroke. 2015;46(8):2206–11.
19. Winstein CJ, Stein J, Arena R, Bates B, Cherney LR, Cramer SC, et al. Guidelines for adult stroke rehabilitation and recovery: a guideline for healthcare professionals from the American Heart Association/American Stroke Association. Stroke. 2016;47(6):e98–e169.
20. Shimamura N, Munakata A, Ohkuma H. Current management of subarachnoid hemorrhage in advanced age. Acta Neurochir Suppl. 2011;110(Pt 2):151–5.
21. Tolea MI, Galvin JE. Sarcopenia and impairment in cognitive and physical performance. Clin Interv Aging. 2015;10:663–71.
22. Wang L, Conner JM, Nagahara AH, Tuszynski MH. Rehabilitation drives enhancement of neuronal structure in functionally relevant neuronal subsets. Proc Natl Acad Sci U S A. 2016;113(10):2750–5.
23. Karsenty G, Olson Eric N. Bone and muscle endocrine functions: unexpected paradigms of inter-organ communication. Cell. 2016;164(6):1248–56.
24. Pedersen BK, Febbraio MA. Muscles, exercise and obesity: skeletal muscle as a secretory organ. Nat Rev Endocrinol. 2012;8(8):457–65.
25. Pedersen BK. Exercise-induced myokines and their role in chronic diseases. Brain Behav Immun. 2011;25(5):811–6.
26. Pedersen BK, Pedersen M, Krabbe KS, Bruunsgaard H, Matthews VB, Febbraio MA. Role of exercise-induced brain-derived neurotrophic factor production in the regulation of energy homeostasis in mammals. Exp Physiol. 2009;94(12):1153–60.
27. Fujimura H, Altar CA, Chen R, Nakamura T, Nakahashi T, Kambayashi J, et al. Brain-derived neurotrophic factor is stored in human platelets and released by agonist stimulation. Thromb Haemost. 2002;87(4):728–34.
28. Dinoff A, Herrmann N, Swardfager W, Liu CS, Sherman C, Chan S, et al. The effect of exercise training on resting concentrations of peripheral brain-derived neurotrophic factor (BDNF): a meta-analysis. PLoS One. 2016;11(9):e0163037.
29. Matthews VB, Astrom MB, Chan MH, Bruce CR, Krabbe KS, Prelovsek O, et al. Brain-derived neurotrophic factor is produced by skeletal muscle cells in response to contraction and enhances fat oxidation via activation of AMP-activated protein kinase. Diabetologia. 2009;52(7):1409–18.
30. Stranahan AM, Lee K, Martin B, Maudsley S, Golden E, Cutler RG, et al. Voluntary exercise and caloric restriction enhance hippocampal dendritic spine density and BDNF levels in diabetic mice. Hippocampus. 2009;19(10):951–61.
31. Horch HW, Kruttgen A, Portbury SD, Katz LC. Destabilization of cortical dendrites and spines by BDNF. Neuron. 1999;23(2):353–64.
32. Hairi NN, Cumming RG, Naganathan V, Handelsman DJ, Le Couteur DG, Creasey H, et al. Loss of muscle strength, mass (sarcopenia), and quality (specific force) and its relationship with functional limitation and physical disability: the Concord Health and Ageing in Men Project. J Am Geriatr Soc. 2010;58(11):2055–62.
33. Nourhashemi F, Andrieu S, Gillette-Guyonnet S, Reynish E, Albarede JL, Grandjean H, et al. Is there a relationship between fat-free soft tissue mass and low cognitive function? Results from a study of 7,105 women. J Am Geriatr Soc. 2002;50(11):1796–801.

34. Shin HY, Kim SW, Kim JM, Shin IS, Yoon JS. Association of grip strength with dementia in a Korean older population. Int J Geriatr Psychiatry. 2012;27(5):500–5.
35. Sternang O, Reynolds CA, Finkel D, Ernsth-Bravell M, Pedersen NL, Dahl Aslan AK. Factors associated with grip strength decline in older adults. Age Ageing. 2015;44(2):269–74.
36. Rehme AK, Eickhoff SB, Wang LE, Fink GR, Grefkes C. Dynamic causal modeling of cortical activity from the acute to the chronic stage after stroke. NeuroImage. 2011;55(3):1147–58.
37. Gleichgerrcht E, Fridriksson J, Rorden C, Nesland T, Desai R, Bonilha L. Separate neural systems support representations for actions and objects during narrative speech in post-stroke aphasia. Neuroimage Clin. 2016;10:140–5.
38. Kiran S, Meier EL, Kapse KJ, Glynn PA. Changes in task-based effective connectivity in language networks following rehabilitation in post-stroke patients with aphasia. Front Hum Neurosci. 2015;9:316.
39. Ripolles P, Rojo N, Grau-Sanchez J, Amengual JL, Camara E, Marco-Pallares J, et al. Music supported therapy promotes motor plasticity in individuals with chronic stroke. Brain Imaging Behav. 2015;10(4):1289–307.
40. Sarkamo T, Ripolles P, Vepsalainen H, Autti T, Silvennoinen HM, Salli E, et al. Structural changes induced by daily music listening in the recovering brain after middle cerebral artery stroke: a voxel-based morphometry study. Front Hum Neurosci. 2014;8:245.
41. Lefebvre S, Dricot L, Laloux P, Gradkowski W, Desfontaines P, Evrard F, et al. Neural substrates underlying stimulation-enhanced motor skill learning after stroke. Brain. 2015;138(Pt 1):149–63.
42. Yang W, Liu TT, Song XB, Zhang Y, Li ZH, Cui ZH, et al. Comparison of different stimulation parameters of repetitive transcranial magnetic stimulation for unilateral spatial neglect in stroke patients. J Neurol Sci. 2015;359(1–2):219–25.
43. Sawaki L, Butler AJ, Leng X, Wassenaar PA, Mohammad YM, Blanton S, et al. Differential patterns of cortical reorganization following constraint-induced movement therapy during early and late period after stroke: a preliminary study. NeuroRehabilitation. 2014;35(3):415–26.
44. Trachtenberg JT, Chen BE, Knott GW, Feng G, Sanes JR, Welker E, et al. Long-term in vivo imaging of experience-dependent synaptic plasticity in adult cortex. Nature. 2002;420(6917):788–94.
45. Yang G, Pan F, Gan WB. Stably maintained dendritic spines are associated with lifelong memories. Nature. 2009;462(7275):920–4.
46. Nishibe M, Urban ET III, Barbay S, Nudo RJ. Rehabilitative training promotes rapid motor recovery but delayed motor map reorganization in a rat cortical ischemic infarct model. Neurorehabil Neural Repair. 2015;29(5):472–82.
47. Xu T, Yu X, Perlik AJ, Tobin WF, Zweig JA, Tennant K, et al. Rapid formation and selective stabilization of synapses for enduring motor memories. Nature. 2009;462(7275):915–9.
48. Wang L, Conner JM, Rickert J, Tuszynski MH. Structural plasticity within highly specific neuronal populations identifies a unique parcellation of motor learning in the adult brain. Proc Natl Acad Sci U S A. 2011;108(6):2545–50.
49. Gulati T, Won SJ, Ramanathan DS, Wong CC, Bodepudi A, Swanson RA, et al. Robust neuroprosthetic control from the stroke perilesional cortex. J Neurosci. 2015;35(22):8653–61.
50. Xerri C, Zennou-Azogui Y. Early and moderate sensory stimulation exerts a protective effect on perilesion representations of somatosensory cortex after focal ischemic damage. PLoS One. 2014;9(6):e99767.
51. Jones TA, Schallert T. Overgrowth and pruning of dendrites in adult rats recovering from neocortical damage. Brain Res. 1992;581(1):156–60.
52. Jones TA, Schallert T. Use-dependent growth of pyramidal neurons after neocortical damage. J Neurosci. 1994;14(4):2140–52.
53. Jones TA, Kleim JA, Greenough WT. Synaptogenesis and dendritic growth in the cortex opposite unilateral sensorimotor cortex damage in adult rats: a quantitative electron microscopic examination. Brain Res. 1996;733(1):142–8.
54. Kozlowski DA, James DC, Schallert T. Use-dependent exaggeration of neuronal injury after unilateral sensorimotor cortex lesions. J Neurosci. 1996;16(15):4776–86.
55. Biernaskie J, Chernenko G, Corbett D. Efficacy of rehabilitative experience declines with time after focal ischemic brain injury. J Neurosci. 2004;24(5):1245–54.

56. Biernaskie J, Corbett D. Enriched rehabilitative training promotes improved forelimb motor function and enhanced dendritic growth after focal ischemic injury. J Neurosci. 2001;21(14):5272–80.
57. Okabe N, Shiromoto T, Himi N, Lu F, Maruyama-Nakamura E, Narita K, et al. Neural network remodeling underlying motor map reorganization induced by rehabilitative training after ischemic stroke. Neuroscience. 2016;339:338–62.
58. Shiromoto T, Okabe N, Lu F, Maruyama-Nakamura E, Himi N, Narita K, et al. The role of endogenous neurogenesis in functional recovery and motor map reorganization induced by rehabilitative therapy after stroke in rats. J Stroke Cerebrovasc Dis. 2017;26(2):260–72.
59. Winship IR, Murphy TH. Remapping the somatosensory cortex after stroke: insight from imaging the synapse to network. Neuroscientist. 2009;15(5):507–24.
60. Herbert WJ, Powell K, Buford JA. Evidence for a role of the reticulospinal system in recovery of skilled reaching after cortical stroke: initial results from a model of ischemic cortical injury. Exp Brain Res. 2015;233(11):3231–51.
61. Baydin S, Gungor A, Tanriover N, Baran O, Middlebrooks EH, Rhoton AL Jr. Fiber tracts of the medial and inferior surfaces of the cerebrum. World Neurosurg. 2017;98:34–49.
62. Fernandez-Miranda JC, Rhoton AL Jr, Alvarez-Linera J, Kakizawa Y, Choi C, de Oliveira EP. Three-dimensional microsurgical and tractographic anatomy of the white matter of the human brain. Neurosurgery. 2008;62(6 Suppl 3):989–1026. discussion 8
63. Kucukyuruk B, Yagmurlu K, Tanriover N, Uzan M, Rhoton AL Jr. Microsurgical anatomy of the white matter tracts in hemispherotomy. Neurosurgery. 2014;10(Suppl 2):305–24. discussion 24
64. Rubino PA, Rhoton AL Jr, Tong X, Oliveira E. Three-dimensional relationships of the optic radiation. Neurosurgery. 2005;57(4 Suppl):219–27. discussion 27
65. Kraft E, Schaal MC, Lule D, Konig E, Scheidtmann K. The functional anatomy of motor imagery after sub-acute stroke. NeuroRehabilitation. 2015;36(3):329–37.
66. Catani M, Ffytche DH. The rises and falls of disconnection syndromes. Brain. 2005;128(Pt 10):2224–39.
67. Catani M, Howard RJ, Pajevic S, Jones DK. Virtual in vivo interactive dissection of white matter fasciculi in the human brain. NeuroImage. 2002;17(1):77–94.
68. Lee MH, Shin YI, Lee SH, Cha YJ, Kim DY, Han BS, et al. Diffusion tensor imaging to determine the potential motor network connectivity between the involved and non-involved hemispheres in stroke. Biomed Mater Eng. 2015;26(Suppl 1):S1447–53.
69. Feng W, Wang J, Chhatbar PY, Doughty C, Landsittel D, Lioutas VA, et al. Corticospinal tract lesion load: an imaging biomarker for stroke motor outcomes. Ann Neurol. 2015;78(6):860–70.
70. Stinear CM, Barber PA, Smale PR, Coxon JP, Fleming MK, Byblow WD. Functional potential in chronic stroke patients depends on corticospinal tract integrity. Brain. 2007;130(Pt 1):170–80.
71. Young BM, Stamm JM, Song J, Remsik AB, Nair VA, Tyler ME, et al. Brain-computer interface training after stroke affects patterns of brain-behavior relationships in corticospinal motor fibers. Front Hum Neurosci. 2016;10:457.
72. Stewart JC, Dewanjee P, Shariff U, Cramer SC. Dorsal premotor activity and connectivity relate to action selection performance after stroke. Hum Brain Mapp. 2016;37(5):1816–30.
73. Fang Y, Han Z, Zhong S, Gong G, Song L, Liu F, et al. The semantic anatomical network: evidence from healthy and brain-damaged patient populations. Hum Brain Mapp. 2015;36(9):3499–515.
74. Borstad AL, Choi S, Schmalbrock P, Nichols-Larsen DS. Frontoparietal white matter integrity predicts haptic performance in chronic stroke. Neuroimage Clin. 2016;10:129–39.
75. Liu S, Guo J, Meng J, Wang Z, Yao Y, Yang J, et al. Abnormal EEG complexity and functional connectivity of brain in patients with acute thalamic ischemic stroke. Comput Math Methods Med. 2016;2016:2582478.
76. Lake EM, Chaudhuri J, Thomason L, Janik R, Ganguly M, Brown M, et al. The effects of delayed reduction of tonic inhibition on ischemic lesion and sensorimotor function. J Cereb Blood Flow Metab. 2015;35(10):1601–9.

77. Hays SA, Ruiz A, Bethea T, Khodaparast N, Carmel JB, Rennaker RL II, et al. Vagus nerve stimulation during rehabilitative training enhances recovery of forelimb function after ischemic stroke in aged rats. Neurobiol Aging. 2016;43:111–8.
78. Follesa P, Biggio F, Gorini G, Caria S, Talani G, Dazzi L, et al. Vagus nerve stimulation increases norepinephrine concentration and the gene expression of BDNF and bFGF in the rat brain. Brain Res. 2007;1179:28–34.
79. Furmaga H, Carreno FR, Frazer A. Vagal nerve stimulation rapidly activates brain-derived neurotrophic factor receptor TrkB in rat brain. PLoS One. 2012;7(5):e34844.
80. Chen J, Tang YX, Liu YM, Chen J, Hu XQ, Liu N, et al. Transplantation of adipose-derived stem cells is associated with neural differentiation and functional improvement in a rat model of intracerebral hemorrhage. CNS Neurosci Ther. 2012;18(10):847–54.
81. Honmou O, Houkin K, Matsunaga T, Niitsu Y, Ishiai S, Onodera R, et al. Intravenous administration of auto serum-expanded autologous mesenchymal stem cells in stroke. Brain. 2011;134(Pt 6):1790–807.
82. Liang H, Yin Y, Lin T, Guan D, Ma B, Li C, et al. Transplantation of bone marrow stromal cells enhances nerve regeneration of the corticospinal tract and improves recovery of neurological functions in a collagenase-induced rat model of intracerebral hemorrhage. Mol Cells. 2013;36(1):17–24.
83. Parr AM, Tator CH, Keating A. Bone marrow-derived mesenchymal stromal cells for the repair of central nervous system injury. Bone Marrow Transplant. 2007;40(7):609–19.
84. Qin J, Gong G, Sun S, Qi J, Zhang H, Wang Y, et al. Functional recovery after transplantation of induced pluripotent stem cells in a rat hemorrhagic stroke model. Neurosci Lett. 2013;554(0):70–5.
85. Qin J, Song B, Zhang H, Wang Y, Wang N, Ji Y, et al. Transplantation of human neuro-epithelial-like stem cells derived from induced pluripotent stem cells improves neurological function in rats with experimental intracerebral hemorrhage. Neurosci Lett. 2013;548(0):95–100.
86. Yamauchi T, Kuroda Y, Morita T, Shichinohe H, Houkin K, Dezawa M, et al. Therapeutic effects of human multilineage-differentiating stress enduring (MUSE) cell transplantation into infarct brain of mice. PLoS One. 2015;10(3):e0116009.
87. Shichinohe H, Kuroda S, Sugiyama T, Ito M, Kawabori M, Nishio M, et al. Biological features of human bone marrow stromal cells (hBMSC) cultured with animal protein-free medium-safety and efficacy of clinical use for neurotransplantation. Transl Stroke Res. 2011;2(3):307–15.
88. Auriat AM, Rosenblum S, Smith TN, Guzman R. Intravascular stem cell transplantation for stroke. Transl Stroke Res. 2011;2(3):250–65.
89. Pendharkar AV, Chua JY, Andres RH, Wang N, Gaeta X, Wang H, et al. Biodistribution of neural stem cells after intravascular therapy for hypoxic-ischemia. Stroke. 2010;41(9):2064–70.
90. Ito M, Kuroda S, Sugiyama T, Shichinohe H, Takeda Y, Nishio M, et al. Validity of bone marrow stromal cell expansion by animal serum-free medium for cell transplantation therapy of cerebral infarct in rats-a serial MRI study. Transl Stroke Res. 2011;2(3):294–306.
91. Uchida H, Morita T, Niizuma K, Kushida Y, Kuroda Y, Wakao S, et al. Transplantation of unique subpopulation of fibroblasts, muse cells, ameliorates experimental stroke possibly via robust neuronal differentiation. Stem Cells. 2016;34(1):160–73.
92. Uchida H, Niizuma K, Kushida Y, Wakao S, Tominaga T, Borlongan CV, et al. Human muse cells reconstruct neuronal circuitry in subacute lacunar stroke model. Stroke. 2017;48(2):428–35.
93. Shimamura N, Kakuta K, Wang L, Naraoka M, Uchida H, Wakao S, et al. Neuro-regeneration therapy using human muse cells is highly effective in a mouse intracerebral hemorrhage model. Exp Brain Res. 2017;235(2):565–72.
94. Kokaia Z, Darsalia V. Neural stem cell-based therapy for ischemic stroke. Transl Stroke Res. 2011;2(3):272–8.

Chapter 20
The Role of Matricellular Proteins in Experimental Subarachnoid Hemorrhage-Induced Early Brain Injury

Lei Liu and Hidenori Suzuki

Abstract Subarachnoid hemorrhage (SAH) is a serious life-threatening type of stroke caused by bleeding into the subarachnoid space surrounding the brain. It elicits a wide range of stress responses in brain tissues and results in brain injury. The term early brain injury (EBI) is a concept to explain pathophysiological changes that occur in brain within 72 h of SAH. Matricellular proteins (MCPs) are a class of nonstructural extracellular matrix proteins that exert diverse functions through binding to cell surface receptors, growth factors, cytokines and other MCPs. Until now, some of MCPs have been investigated in clinical SAH settings and laboratory studies. Here, we review the role of MCPs in post-SAH EBI by focusing on osteopontin, tenascin-C, and periostin.

Keywords Subarachnoid hemorrhage • Early brain injury • Matricellular proteins • Osteopontin • Tenascin-C • Periostin

Abbreviations

Ang	Angiopoietin
BBB	Blood-brain barrier
CCN	Cyr61/CTGF/NOV
CICES	Complement inhibiting component of *Ephedra sinica*

L. Liu, M.D.
Department of Neurosurgery, Mie University Graduate School of Medicine,
2-174 Edobashi, Tsu, Mie 514-8507, Japan

H. Suzuki, M.D., Ph.D. (✉)
Department of Neurosurgery, Mie University Graduate School of Medicine,
2-174 Edobashi, Tsu, Mie 514-8507, Japan

Research Center for Matrix Biology, Mie University Graduate School of Medicine,
Tsu, Japan
e-mail: suzuki02@clin.medic.mie-u.ac.jp

P.A. Lapchak, J.H. Zhang (eds.), *Cellular and Molecular Approaches to Regeneration and Repair*, Springer Series in Translational Stroke Research,
https://doi.org/10.1007/978-3-319-66679-2_20

CSF	Cerebrospinal fluid
EBI	Early brain injury
ECM	Extracellular matrix
EGFR	Epidermal growth factor receptor
ERK	Extracellular signal-regulated kinase
FAK	Focal adhesion kinase
IL	Interleukin
ILK	Integrin-linked kinase
JNK	c-Jun N-terminal kinase
MAPK	Mitogen-activated protein kinase
MCP	Matricellular protein
MKP	Mitogen-activated protein kinase phosphatase
MMP	Matrix metalloproteinase
NF	Nuclear factor
OPN	Osteopontin
PDGFR	Platelet-derived growth factor receptor
P-gp	P-Glycoprotein
PI3K	Phosphatidylinositol 3-kinase
RGD	L-Arginyl-glycyl-L-aspartate
r-OPN	Recombinant osteopontin
SAH	Subarachnoid hemorrhage
siRNA	Short-interfering ribonucleic acid
SMA	Smooth muscle actin
SMemb	Embryonic smooth muscle myosin heavy chain
TLR	Toll-like receptor
TNC	Tenascin-C
TNKO	Tenascin-C knockout
VEGF	Vascular endothelial growth factor
VEGFR	Vascular endothelial growth factor receptor
VSMC	Vascular smooth muscle cell
ZO	Zona occludens

1 Introduction

Aneurysmal subarachnoid hemorrhage (SAH) is one of the most life-threatening diseases in the central nervous system [1]. It elicits a wide range of acute inflammatory responses in brain parenchyma and results in brain injury. The detailed mechanisms are still undefined even through there are many clinical and laboratory studies focusing on post-SAH brain injury. Recently, accumulating evidences have suggested that early brain injury (EBI), which is a concept to explain pathophysiological changes occurring in brain within 72 h of SAH, may be the primary cause of poor outcome after SAH [2].

The extracellular matrix (ECM) is a dynamic structural network and a crucial component of tissue microenvironment. Matricellular proteins (MCPs) are nonstructural ECM proteins that exert diverse functions through binding to cell surface receptors, growth factors, cytokines, and other MCPs [3]. This group of proteins includes thrombospondin, osteonectin, tenascin, osteopontin (OPN), periostin, galectin, Cyr61/CTGF/NOV (CCN) and others [3], and the members of MCPs are still increasing. During central nervous system development, the MCPs provide a microenvironment that regulates cell migration, axonal guidance, and synaptogenesis, However, upregulation of MCPs leads to blood-brain barrier (BBB) disruption and activation of inflammatory responses in pathological conditions [4].

Among MCPs, OPN, tenascin-C (TNC) and periostin have been reported to be involved in brain injury after SAH. In this chapter, we summarize the evidences regarding the role of MCPs in post-SAH brain injury.

2 Osteopontin (OPN)

OPN is a highly modified ECM glycophosphoprotein and modulates intracellular signaling pathways by binding various integrins and CD44 variants [5]. OPN plays an important role in the initiation of inflammation by affecting cell adhesion, chemotaxis, immune regulation, and apoptosis [5].

2.1 OPN Expression in SAH Brain

It has been reported that the serum level of OPN was significantly increased in intracerebral hemorrhage patients [6], but the expression level of OPN has never been investigated in aneurysmal SAH patients. However, in laboratory researches, it was reported that OPN plays a protective role in post-SAH EBI [7]. The expression level of OPN was very low in healthy adult rat brain. Conversely, OPN was obviously induced in brain after 24 h and peaked at 72 h of SAH. The trend of post-SAH OPN expression in brain was consistent with the rat's recovery of body weight losses and neurobehavior impairments as well as alleviation of BBB disruption [7]. Immunohistochemical staining showed that the expression of OPN was located at astrocytes and capillary endothelial cells after SAH in brain [7].

2.2 OPN Ameliorates EBI by Preventing BBB Disruption and Neuronal Apoptosis

The effect of endogenous OPN on EBI was investigated by suppressing and inducing OPN expression in a filament punctured SAH model. Specific OPN short-interfering ribonucleic acid (siRNA) treatment obviously aggravated neurobehavioral impairments and resulted in significantly increased BBB permeability [7]. The mechanism study demonstrated that the blockage of endogenous OPN expression suppressed mitogen-activated protein kinase (MAPK) phosphatase (MKP)-1 induction, a kind of endogenous MAPKs inhibitor, and activated c-Jun N-terminal kinase (JNK), p38, and extracellular signal-regulated kinase (ERK) 1/2 [7]. Vascular endothelial growth factor (VEGF)-A is a pivotal regulator of angiogenesis that controls vascular growth under physiological and pathological conditions and promotes vascular permeability [8]. MAPKs not only mediated the effect of VEGF-A on BBB permeability but also induced VEGF-A expression [9]. Meanwhile, inactivating endogenous OPN decreased angiopoietin (Ang)-1 expression [7] which has a potent anti-vascular permeability property and inhibits the activation of VEGF-A [10]. These evidences indicated that endogenous OPN plays an important role in preserving BBB integrity after SAH (Fig. 20.1).

Thus, it is hopeful to prevent EBI by inducing endogenous OPN upregulation after SAH. Zuo et al. found that Complement Inhibiting Component of Ephedra Sinica (CICES), a kind of Chinese herb which has ability to inhibit the activity of the classical and alternative pathways of complement, alleviated neurological deficits, cortex cell apoptosis, BBB disruption and brain edema after 24 and 72 h of experimental SAH [11]. The neuroprotective effects of CICES on post-SAH EBI was associated with complement C3 inhibition and upregulated OPN expression as well as reduced MMP-9 expression [11]. In another post-SAH EBI study, Enkhjargal et al. demonstrated that intranasal administration of vitamin D3 improved neurological impairment and brain edema, and was associated with upregulation of endogenous OPN within 72 h after experimental SAH [12]. Advanced study found that vitamin D3 upregulates expression of endogenous brain OPN in astrocytes and that it protects BBB via CD44 splicing and P-glycoprotein in the vascular endothelial cells after SAH in rats. Blockage of vitamin D receptor and OPN exacerbated neurobehavioral impairment and BBB disruption and was associated with a decrease of mature/full-length glycosylated P-glycoprotein in rat brain vascular endothelial cells [12].

On the other hand, recombinant OPN (r-OPN) pretreatment prevented the loss of body weight, neurobehavioral impairments, BBB disruption, and brain edema formation after 24–72 h of experimental SAH [13]. The mechanism studies demonstrated that the neuroprotective effects of r-OPN were associated with inactivation of nuclear factor (NF)-κB and matrix metalloproteinase (MMP)-9, and preservation of inter-endothelial tight junction protein zona occludens (ZO)-1. Simultaneously, the expression of tissue inhibitor of MMP-1 was significantly upregulated to maintain the matrix stability [13]. MMP-9 is a key factor to cause BBB disruption by degrading the ECM of cerebral microvessel basal lamina, including collagen IV, laminin, fibronectin and ZO-1 [14–16]. Interestingly, r-OPN pretreatment did not suppress interleukin (IL)-1β induction after SAH [13] even though IL-1β has been reported to cause brain injury after SAH and IL-1β-induced MMP-9 overexpression was dependent on NF-κB transduction pathway in astrocytes [17, 18]. One possible speculation is IL-1β activation via NK-κB independent signaling pathways after SAH, such as sympathetic activation or catecholamine release [19, 20]. Thus, more researches would be needed in future.

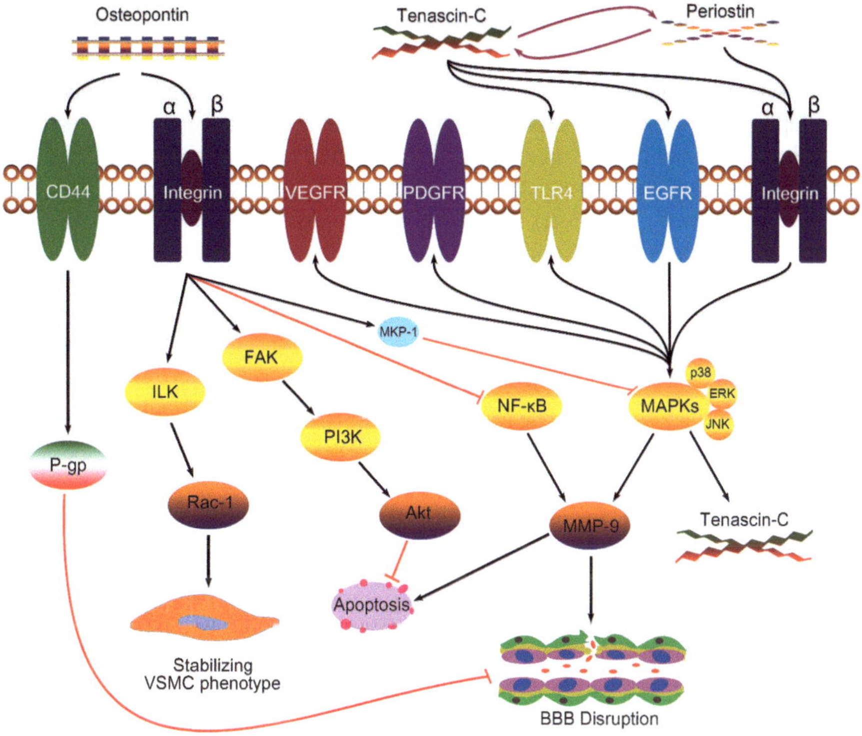

Fig. 20.1 The mechanisms for matricellular proteins to cause early brain injury (EBI) after experimental subarachnoid hemorrhage (SAH). Osteopontin prevents post-SAH EBI by (1) stabilizing vascular smooth muscle cell (VSMC) phenotype via integrin-linked kinase (ILK)/Rac-1 signaling pathways; (2) inhibiting neuronal apoptosis via focal adhesion kinase (FAK)/phosphatidylinositol 3-kinase (PI3K)/Akt signaling pathway activation; and (3) protecting blood-brain barrier (BBB) integrity via inactivating nuclear factor (NF)-κB/matrix metalloproteinase (MMP)-9 signaling pathway, via activating mitogen-activated protein kinase (MAPK) phosphatase (MKP)-1 that inhibits MAPKs/MMP-9 signaling pathway or via increasing mature/full-length glycosylated P-glycoprotein (P-gp) in brain vascular endothelial cells. Tenascin-C (TNC) induces post-SAH EBI by causing neuronal apoptosis and BBB disruption via MAPKs/MMP-9 signaling pathways. Activated MAPKs also induce TNC expression by relevant receptors, including vascular endothelial growth factor receptor (VEGFR), platelet-derived growth factor receptor (PDGFR), toll-like receptor (TLR) 4, epidermal growth factor receptor (EGFR) and integrins. Periostin induces post-SAH BBB disruption, which is mediated by MAPK signaling pathways. The interaction between periostin and TNC also plays an important role in post-SAH EBI by modulating downstream signaling pathways. *ERK* extracellular signal-regulated kinase, *JNK* c-Jun N-terminal kinase

Except for preventing BBB disruption, the noninvasive nasal application of r-OPN suggested that intranasal r-OPN treatment attenuated brain edema and neuronal apoptosis, which ultimately improved neurobehavioral impairments in rats after SAH [21]. The antiapoptotic mechanisms of r-OPN treatment possibly involved focal adhesion kinase-phosphatidylinositol 3-kinase-Akt signaling pathways activation, which inhibited caspase-3 cleavage in cortex and hippocampus (Fig. 20.1) [21].

2.3 OPN Stabilizes Vascular Smooth Muscle Cell Phenotype

The vascular neural network is a physiological unit to be considered for therapeutic development in stroke. The vascular smooth muscle is an important part of vascular neural network and may be an alternative therapeutic target for post-SAH EBI [22, 23]. The vascular smooth muscle of cerebral arteries typically switches from contractile to synthetic type after injury or hemorrhagic stroke [24]. The vascular smooth muscle cells (VSMCs) can be stained with α-smooth muscle actin (α-SMA), a marker of contractile VSMCs, and embryonic smooth muscle myosin heavy chain (SMemb)/non-muscle myosin heavy chain isoform, an accurate marker of synthetic VSMCs [25]. It is reported that the expression of α-SMA was significantly decreased and the expression of SMemb was significantly increased at 24 and 72 h after experimental SAH [26]. Unbalanced contractile/synthetic vascular smooth muscle phenotype may result in decreasing auto-regulatory capacity and aggravating brain edema [27]. r-OPN alleviated neurological impairments and protected the vascular smooth muscle phenotypic transformation, and the mechanisms involved integrin/integrin-linked kinase (ILK)/Rac-1 signaling pathways activation [26]. A L-Arginyl-glycyl-L-aspartate (RGD)-dependent integrin receptor antagonist, RGD motif-containing hexapeptide (GRGDSP), reduced r-OPN-induced ILK upregulation and Rac-1 activation, meanwhile, knockdown of ILK by siRNA or selective Rac-1 inhibition by NSC23766 abolished r-OPN-induced preservation of vascular smooth muscle phenotypic transformation [26].

In summary, OPN plays a protective role in SAH-induced EBI by protecting BBB integrity, preventing neuronal apoptosis, and stabilizing VSMC phenotype (Fig. 20.1).

3 Tenascin-C (TNC)

TNC belongs to the MCP family that is downregulated in healthy adult tissues and transiently induced during inflammatory responses [28]. It exerts various functions through binding to cell surface receptors, other MCPs, and growth factors [29].

3.1 TNC Expression in SAH Brain

In aneurysmal SAH patients, TNC expression levels were markedly increased in serum and cerebrospinal fluid (CSF). CSF TNC level peaked immediately after SAH, and the highest level in CSF occurred in the first 3 days followed by a decrease over time, whereas serum TNC level increased transiently and peaked on days 4–6 [30]. Higher TNC expression levels in CSF were observed in patients with worse admission clinical grade, more severe SAH on admission computed tomography, acute obstructive hydrocephalus, subsequent angiographic vasospasm, delayed cerebral ischemia, chronic shunt-dependent hydrocephalus, and a worse outcome [30, 31]. In *in-vivo* studies, expression of TNC was weakly detected in normal adult animals, however, overexpressed TNC was observed in the cerebral VSMC layers and brain parenchyma after experimental SAH in rats [32, 33].

3.2 Overexpressed TNC Induces BBB Disruption After SAH

It is well known that TNC expression was obviously upregulated during acute or chronic inflammatory diseases [28]. In an experimental SAH mice model, overexpressed TNC was associated with severe neurobehavioral impairment, brain edema, and BBB disruption [34]. Endogenous TNC induction activates downstream signaling pathways, such as platelet-derived growth factor receptors (PDGFRs), MAPKs and MMP-9 [32]. It was reported that imatinib mesylate, a kind of PDGFR inhibitor, prevented cerebral vasospasm by inhibiting PDGFR activation, TNC induction, and MAPK activation [32]. Another study demonstrated that imatinib mesylate preserved BBB integrity by inhibiting JNK-mediated MMP-9 activation after experimental SAH in rats [35]. TNC knockout (TNKO) mice was used to determine the effect of TNC on post-SAH BBB disruption [34]. It was demonstrated that TNKO alleviated neurobehavioral impairments, brain edema, and BBB disruption after SAH. The following mechanism studies found that the effect of TNKO on post-SAH BBB protection was associated with MMP-9 inhibition and tight junction protein ZO-1 preservation, as well as inactivation of MAPKs in cerebral capillary endothelial cells and brain parenchyma [34]. Conversely, exogenous TNC treatment re-aggravated neurological impairments, brain edema, and BBB disruption compared with the vehicle treatment in TNKO SAH mice [34]. These findings indicated that TNC plays an important role in post-SAH BBB disruption (Fig. 20.1).

3.3 Overexpressed TNC Is Responsible for Neuronal Apoptosis After SAH

Apoptosis is one of the most significant pathological processes in EBI after SAH [36]. It is believed that there are a number of apoptotic pathways playing a role in SAH: the death receptor pathway, caspase-dependent and caspase-independent pathways, as well as the mitochondrial pathway [36]. It was reported that TNC upregulated MMPs expression in cultured VSMCs, which lead to TNC fragmentation and cell apoptosis [37]. In an experimental SAH study, it was demonstrated that TNC is a mediator of post-SAH neuronal apoptosis through the mechanism of PDGF and MAPKs, conversely, imatinib mesylate, a PDGFR inhibitor, prevented TNC induction, MAPKs activation, and neuronal apoptosis. Meanwhile, recombinant TNC administration reactivated MAPKs pathways and induced neurons apoptosis, as well as endogenous TNC expression in SAH brain [33]. These indirect evidences suggested that the positive feedback mechanism of TNC-induced cascades reaction plays an important role in post-SAH neurons apoptosis (Fig. 20.1).

4 Periostin

Periostin is a secreted MCP that has been reported to be a critical player in the pathobiology of various diseases, including inflammatory disease [38]. Periostin plays a regulatory role rather than structural functions: binding of periostin to integrin receptors activates many intracellular signaling pathways and modulates multiple downstream proteins [39]. Previous studies suggested that periostin alternates the pathological processes by modulating MMPs expression in inflammatory diseases through different signaling pathways [40, 41].

4.1 Periostin Expression in SAH Brain

The expression level of periostin has never been investigated in aneurysmal SAH patients. In a laboratory study, the expression level of periostin was very low in sham operated mice brain, but periostin was obviously induced in brain after 24 h of SAH [42]. Immunohistochemical staining showed that periostin was weakly detected in neurons in sham mice, and was intensively expressed in brain capillary endothelial cells as well as neurons after SAH [42].

4.2 Overexpressed Periostin Causes Post-SAH EBI

To assess the role of periostin in post-SAH EBI, recombinant periostin protein and monoclonal anti-periostin antibody were administrated by an intracerebroventricular injection in a SAH mice model [42]. Recombinant periostin protein treatment obviously aggravated neurobehavioral impairments and brain edema after 24 h of SAH. However, monoclonal anti-periostin antibody treatment significantly improved neurobehavioral impairments and obviously ameliorated brain edema after 24–48 h of SAH. Immunohistochemical staining showed that periostin was effectively neutralized in brain capillary endothelial cells and neurons by anti-periostin antibody, meanwhile, BBB disruption, which was evaluated by immunoglobulin G extravasation, was significantly suppressed by periostin neutralization at 24 h of SAH [42]. The mechanism study demonstrated that overexpressed periostin activated the downstream p38/ERK/MMP-9 signaling pathways and caused ZO-1 degradation (Fig. 20.1). Thus, these evidences indicated that periostin plays an important role in post-SAH EBI.

5 Interaction Between Periostin and TNC Aggravates Post-SAH EBI

It has been reported that periostin interacts with MCPs, such as TNC and bone morphogenetic protein-1, and regulates tissue homeostasis [43, 44]. In post-SAH EBI study, administrating of recombinant periostin not only aggravated neurobehavioral impairments and brain edema but also induced TNC expression [42]. Conversely, periostin neutralization suppressed post-SAH TNC induction and EBI [42]. Furthermore, deficiency of TNC inhibited periostin induction and improved neurobehavioral impairments although periostin expression in TNKO sham mice was identical to that of wild-type sham mice [42]. These results indicated that periostin and TNC regulate each other and alter downstream signaling pathways in post-SAH EBI. Moreover, periostin and TNC may form a positive feedback mechanism to aggravate BBB disruption (Fig. 20.1).

6 Conclusions

This review summarizes the findings that MCPs, which include OPN, TNC and periostin, play important roles in SAH-induced EBI by regulating downstream signaling pathways and modulating BBB integrity and cell apoptosis. The current evidences suggest that MCPs can be therapeutic targets for preventing post-SAH EBI. However, there are many other known MCPs that have never been investigated in the context of EBI after SAH. In addition, future studies are required to determine how MCPs are involved in the pathophysiological process of SAH-induced EBI in more details.

References

1. Suarez JI, Tarr RW, Selman WR. Aneurysmal subarachnoid hemorrhage. N Engl J Med. 2006;354(4):387–96.
2. Cahill J, Zhang JH. Subarachnoid hemorrhage: is it time for a new direction? Stroke. 2009;40(3 Suppl):S86–7.
3. Chiodoni C, Colombo MP, Sangaletti S. Matricellular proteins: from homeostasis to inflammation, cancer, and metastasis. Cancer Metastasis Rev. 2010;29(2):295–307.
4. Benarroch EE. Extracellular matrix in the CNS: dynamic structure and clinical correlations. Neurology. 2015;85(16):1417–27.

5. Wang KX, Denhardt DT. Osteopontin: role in immune regulation and stress responses. Cytokine Growth Factor Rev. 2008;19(5-6):333–45.
6. Acar A, Cevik MU, Arikanoglu A, Evliyaoglu O, Basarili MK, Uzar E, et al. Serum levels of calcification inhibitors in patients with intracerebral hemorrhage. Int J Neurosci. 2012;122(5):227–32.
7. Suzuki H, Hasegawa Y, Kanamaru K, Zhang JH. Mechanisms of osteopontin-induced stabilization of blood-brain barrier disruption after subarachnoid hemorrhage in rats. Stroke. 2010;41(8):1783–90.
8. Ferrara N, Gerber HP, LeCouter J. The biology of VEGF and its receptors. Nat Med. 2003;9(6):669–76.
9. Kusaka G, Ishikawa M, Nanda A, Granger DN, Zhang JH. Signaling pathways for early brain injury after subarachnoid hemorrhage. J Cereb Blood Flow Metab. 2004;24(8):916–25.
10. Gavard J, Patel V, Gutkind JS. Angiopoietin-1 prevents VEGF-induced endothelial permeability by sequestering Src through mDia. Dev Cell. 2008;14(1):25–36.
11. Zuo S, Li W, Li Q, Zhao H, Tang J, Chen Q, et al. Protective effects of Ephedra sinica extract on blood-brain barrier integrity and neurological function correlate with complement C3 reduction after subarachnoid hemorrhage in rats. Neurosci Lett. 2015;609:216–22.
12. Enkhjargal B, McBride DW, Manaenko A, Reis C, Sakai Y, Tang J, et al. Intranasal administration of vitamin D attenuates blood-brain barrier disruption through endogenous upregulation of osteopontin and activation of CD44/P-gp glycosylation signaling after subarachnoid hemorrhage in rats. J Cereb Blood Flow Metab. 2016;1:271678X16671147.
13. Suzuki H, Ayer R, Sugawara T, Chen W, Sozen T, Hasegawa Y, et al. Protective effects of recombinant osteopontin on early brain injury after subarachnoid hemorrhage in rats. Crit Care Med. 2010;38(2):612–8.
14. Sehba FA, Mostafa G, Knopman J, Friedrich V Jr, Bederson JB. Acute alterations in microvascular basal lamina after subarachnoid hemorrhage. J Neurosurg. 2004;101(4):633–40.
15. Guo Z, Sun X, He Z, Jiang Y, Zhang X, Zhang JH. Matrix metalloproteinase-9 potentiates early brain injury after subarachnoid hemorrhage. Neurol Res. 2010;32(7):715–20.
16. Mahajan SD, Aalinkeel R, Reynolds JL, Nair B, Sykes DE, Bonoiu A, et al. Suppression of MMP-9 expression in brain microvascular endothelial cells (BMVEC) using a gold nanorod (GNR)-siRNA nanoplex. Immunol Investig. 2012;41(4):337–55.
17. Wu CY, Hsieh HL, Jou MJ, Yang CM. Involvement of p42/p44 MAPK, p38 MAPK, JNK and nuclear factor-kappa B in interleukin-1beta-induced matrix metalloproteinase-9 expression in rat brain astrocytes. J Neurochem. 2004;90(6):1477–88.
18. Sozen T, Tsuchiyama R, Hasegawa Y, Suzuki H, Jadhav V, Nishizawa S, et al. Role of interleukin-1beta in early brain injury after subarachnoid hemorrhage in mice. Stroke. 2009;40(7):2519–25.
19. Naredi S, Lambert G, Friberg P, Zall S, Eden E, Rydenhag B, et al. Sympathetic activation and inflammatory response in patients with subarachnoid haemorrhage. Intensive Care Med. 2006;32(12):1955–61.
20. Tan KS, Nackley AG, Satterfield K, Maixner W, Diatchenko L, Flood PM. Beta2 adrenergic receptor activation stimulates pro-inflammatory cytokine production in macrophages via PKA- and NF-kappaB-independent mechanisms. Cell Signal. 2007;19(2):251–60.
21. Topkoru BC, Altay O, Duris K, Krafft PR, Yan J, Zhang JH. Nasal administration of recombinant osteopontin attenuates early brain injury after subarachnoid hemorrhage. Stroke. 2013;44(11):3189–94.
22. Zhang JH, Badaut J, Tang J, Obenaus A, Hartman R, Pearce WJ. The vascular neural network—a new paradigm in stroke pathophysiology. Nat Rev Neurol. 2012;8(12):711–6.
23. Zhang JH. Vascular neural network in subarachnoid hemorrhage. Transl Stroke Res. 2014;5(4):423–8.
24. Edvinsson LI, Povlsen GK. Vascular plasticity in cerebrovascular disorders. J Cereb Blood Flow Metab. 2011;31(7):1554–71.
25. Rensen SS, Doevendans PA, van Eys GJ. Regulation and characteristics of vascular smooth muscle cell phenotypic diversity. Neth Heart J. 2007;15(3):100–8.

26. Wu J, Zhang Y, Yang P, Enkhjargal B, Manaenko A, Tang J, et al. Recombinant osteopontin stabilizes smooth muscle cell phenotype via integrin receptor/integrin-linked kinase/Rac-1 pathway after subarachnoid hemorrhage in rats. Stroke. 2016;47(5):1319–27.
27. Shimamura N, Ohkuma H. Phenotypic transformation of smooth muscle in vasospasm after aneurysmal subarachnoid hemorrhage. Transl Stroke Res. 2014;5(3):357–64.
28. Tucker RP, Chiquet-Ehrismann R. The regulation of tenascin expression by tissue microenvironments. Biochim Biophys Acta. 2009;1793(5):888–92.
29. Udalova IA, Ruhmann M, Thomson SJ, Midwood KS. Expression and immune function of tenascin-C. Crit Rev Immunol. 2011;31(2):115–45.
30. Suzuki H, Kinoshita N, Imanaka-Yoshida K, Yoshida T, Taki W. Cerebrospinal fluid tenascin-C increases preceding the development of chronic shunt-dependent hydrocephalus after subarachnoid hemorrhage. Stroke. 2008;39(5):1610–2.
31. Suzuki H, Kanamaru K, Shiba M, Fujimoto M, Imanaka-Yoshida K, Yoshida T, et al. Cerebrospinal fluid tenascin-C in cerebral vasospasm after aneurysmal subarachnoid hemorrhage. J Neurosurg Anesthesiol. 2011;23(4):310–7.
32. Shiba M, Suzuki H, Fujimoto M, Shimojo N, Imanaka-Yoshida K, Yoshida T, et al. Imatinib mesylate prevents cerebral vasospasm after subarachnoid hemorrhage via inhibiting tenascin-C expression in rats. Neurobiol Dis. 2012;46(1):172–9.
33. Shiba M, Fujimoto M, Imanaka-Yoshida K, Yoshida T, Taki W, Suzuki H. Tenascin-C causes neuronal apoptosis after subarachnoid hemorrhage in rats. Transl Stroke Res. 2014;5(2):238–47.
34. Fujimoto M, Shiba M, Kawakita F, Liu L, Shimojo N, Imanaka-Yoshida K, et al. Deficiency of tenascin-C and attenuation of blood-brain barrier disruption following experimental subarachnoid hemorrhage in mice. J Neurosurg. 2015;124(6):1693–702.
35. Zhan Y, Krafft PR, Lekic T, Ma Q, Souvenir R, Zhang JH, et al. Imatinib preserves blood-brain barrier integrity following experimental subarachnoid hemorrhage in rats. J Neurosci Res. 2015;93(1):94–103.
36. Cahill J, Calvert JW, Zhang JH. Mechanisms of early brain injury after subarachnoid hemorrhage. J Cereb Blood Flow Metab. 2006;26(11):1341–53.
37. Wallner K, Li C, Shah PK, Wu KJ, Schwartz SM, Sharifi BG. EGF-Like domain of tenascin-C is proapoptotic for cultured smooth muscle cells. Arterioscler Thromb Vasc Biol. 2004;24(8):1416–21.
38. Kudo A. Periostin in fibrillogenesis for tissue regeneration: periostin actions inside and outside the cell. Cell Mol Life Sci. 2011;68(19):3201–7.
39. Conway SJ, Izuhara K, Kudo Y, Litvin J, Markwald R, Ouyang G, et al. The role of periostin in tissue remodeling across health and disease. Cell Mol Life Sci. 2014;71(7):1279–88.
40. Lv S, Liu H, Cui J, Hasegawa T, Hongo H, Feng W, et al. Histochemical examination of cathepsin K, MMP1 and MMP2 in compressed periodontal ligament during orthodontic tooth movement in periostin deficient mice. J Mol Histol. 2014;45(3):303–9.
41. Attur M, Yang Q, Shimada K, Tachida Y, Nagase H, Mignatti P, et al. Elevated expression of periostin in human osteoarthritic cartilage and its potential role in matrix degradation via matrix metalloproteinase-13. FASEB J. 2015;29(10):4107–21.
42. Liu L, Kawakita F, Fujimoto M, Nakano F, Imanaka-Yoshida K, Yoshida T, et al. Role of periostin in early brain injury after subarachnoid hemorrhage in mice. Stroke. 2017;48(4):1108–11.
43. Kii I, Nishiyama T, Li M, Matsumoto K, Saito M, Amizuka N, et al. Incorporation of tenascin-C into the extracellular matrix by periostin underlies an extracellular meshwork architecture. J Biol Chem. 2010;285(3):2028–39.
44. Maruhashi T, Kii I, Saito M, Kudo A. Interaction between periostin and BMP-1 promotes proteolytic activation of lysyl oxidase. J Biol Chem. 2010;285(17):13294–303.

Chapter 21
Chemokines and Proteolysis: Implications for Stem Cell Dynamics in Ischemic Stroke

Umadevi V. Wesley and Robert J. Dempsey

Abstract Stroke still remains a significant clinical challenge, with only a small proportion of the ischemic patients benefiting from current treatments which are limited by a narrow therapeutic time window. Cerebral ischemic stroke results in severe neurological deficits due to massive loss of neurons and disruption of vasculature. Although our understanding of the stroke pathology has remarkably increased, further insight into the cellular and molecular mechanisms involved in the post-stroke brain repair is still required to identify more effective drug targets with wider time window. Cerebral ischemia and reperfusion injury alters the brain microenvironment including dysregulation of cytokines, chemokines and abnormal release of proteases leading to neuronal cell death, endothelial cell and stem/progenitor cell dysfunction, disruption of blood brain barrier and the vascular unit. Thus, delineating the timely and balanced regulation of proteases, cytokines, chemokines, and stem/progenitor cells is critical for enhancing post-stroke brain protection and repair, and neurological functional recovery. In this chapter, we will present the facts about interactions of chemokines, proteases and stem cells in the context of pathophysiology of stroke.

Keywords Stroke • Ischemic brain injury • Chemokines • Proteases • Stem cells • Matrix metalloproteases • Dipeptidyl peptidase IV • Stromal derived factor

Abbreviations

CVD	Cerebro-vascular diseases
DPPIV	Dipeptidyl peptidase 4
EPC	Endothelial progenitor cells

U.V. Wesley, Ph.D. (✉) • R.J. Dempsey, M.D., F.A.C.S.
Department of Neurosurgery, School of Medicine and Public Health,
University of Wisconsin, Madison, WI 53792, USA
e-mail: wesley@neurosurgery.wisc.edu; dempsey@neurosurgery.wisc.edu

P.A. Lapchak, J.H. Zhang (eds.), *Cellular and Molecular Approaches to Regeneration and Repair*, Springer Series in Translational Stroke Research,
https://doi.org/10.1007/978-3-319-66679-2_21

HIF-1a	Hypoxia-inducible factor-1α (HIF-1α)
IL	Interleukin
MCAO	Middle cerebral artery occlusion
MCP-1	Monocyte chemoattractant protein-1
MMP	Matrix metalloproteases
MSC	Mesenchymal stem cells
NPC	Neural progenitor cells
SDF1	Stromal derived factor

1 Introduction

Stroke is a leading cause of mortality and long-term disability worldwide. Stroke still remains a significant clinical challenge, with only a small proportion of the ischemic patients benefiting from current treatments which are limited by a narrow therapeutic time window [1–8]. Two major types of strokes have been identified: hemorrhagic and ischemic stroke. Ischemic stroke accounts for more than 80% of all strokes [4–9]. Cerebral ischemic stroke results in severe neurological deficits due to massive loss of neurons and disruption of vasculature. Restoration of both neuro and vascular units is thus important for functional recovery [4, 6–15]. Although our understanding of the stroke pathology has remarkably increased, further insight into the cellular and molecular mechanisms involved in the post-stroke brain repair is still required to identify more effective drug targets with wider time window.

Cerebral ischemia and reperfusion injury alters the brain microenvironment including dysregulation of cytokines, chemokines and abnormal release of proteases leading to neuronal cell death, endothelial cell dysfunction, disruption of blood brain barrier and the vascular unit [4, 6, 12–18]. Thus, delineating the timely and balanced regulation of proteases, cytokines, and chemokines is critical for enhancing endothelial and neural stem cell recruitment that contributes to post-stroke brain protection and repair. As a self-defense mechanism, ischemic-stroke upregulates chemokines that promote stem and progenitor cell migration, survival, and angiogenesis. However, this attempt to self-repair the post-stroke brain is short lived and has limited success, presumably due to disruption of chemokine activity by proteases. Indeed, through mutual balance and interactions, chemokines and proteases regulate angiogenesis and neurogenesis, and have gained much attention due to their therapeutic potential for post-stroke recovery [4, 6, 8–23]. Thus interplay between proteases, chemokines and stem cells represents important molecular-cellular cross-talk for stroke outcome. In this chapter, we will present the cross-talk between chemokines, proteases and stem cells in the context of pathophysiology of stroke.

2 Chemokines and Ischemic Stroke/Cerebral Ischemia

Chemokines are members of cytokine superfamily with distinct functional roles. Chemokines are low-molecular-weight proteins that are well established for their capability to accelerate chemotaxis or gradient directed migration for recruitment of cells including leukocytes, progenitor and stem cells. Chemokines, in addition to chemotactic activities, are also involved in autocrine and paracrine signaling and thus critically modulate cellular migration, survival, angiogenesis and neurogenesis [22–26]. Currently there are more than 50 chemokines and 20 chemokine receptors. Chemokines have been classified into four main subfamilies: CXC, CC, CX3C and XC, on the basis of the number and location of the cysteine residues at the N-terminus of the molecule. These chemokines exert their biological effects through the binding to selective G protein-coupled transmembrane receptors including CXCR, CCR, and CX3CR that are present on the surfaces of their target cells. Among these, the CC and CXC chemokines are the major mediators of cerebral injury and protection following ischemic stroke. The expression and activities of these chemokines are modulated by local and systemic inflammation. Indeed, during acute phase, ischemic-stroke induces inflammation not only in the brain, but also in the blood stream which in turn influences critical steps of the post-stroke recovery processes. The CXC and CC chemokines further exacerbate the inflammatory response in infarcted brain, as they promote recruitment and activation of inflammatory cells [25–35]. On the other hand, they also enhance stem cell homing to site of injury, neuronal and glial cell survival, and neo-angiogenesis, thus favoring brain repair [35–39]. Given these abilities chemokines represent potential therapeutic targets, and drug reagents for stroke treatment. Overall, well-orchestrated balance between the accumulation and degradation of these chemokines is integral to brain injury and repair.

Under normal physiological conditions, chemokines are expressed in brain by microglia, astrocytes and neurons, and endothelial cells at low levels, but rapidly upregulated in response to ischemic damage and re-perfusion injury. While some cytokines exacerbate cerebral injury, others are shown to provide neuroprotection [20–36]. Chemokines levels are also increased in response to signals from pro and anti-inflammatory cytokines. Higher concentration of chemokines create ligand gradient and play an important role in selectively recruiting stem cells, monocytes, neutrophils, and lymphocytes that express respective chemokine receptors. Due to their chemotactic and signaling properties, chemokines have gained much attention for investigation in many neurological disorders including ischemic stroke. Among these chemokines, Stromal cell-derived factor-1a (SDF1), Monocyte chemoattractant protein-1 (MCP-1), fractalkine (CX3CL1), and Glucagon like peptide-1 (GLP-1) macrophage inflammatory protein-1 (MIP-1), RANTES, and GRO-alpha (CXCL1) have shown to be involved in ischemic brain injury and repair [39–70]. The involvement of these chemokines in multiple biological functions that are crucial for tissue re-modeling makes them potentially useful for improving the clinical application of stem cell therapy. Of note, the functions of many of these chemokines are tightly controlled by proteolytic events [28, 33]. Here we discuss three most widely studied chemokines, SDF1, MCP-1, and fractalkine.

2.1 *Stromal Derived Factor 1 (SDF1)*

The chemokine stromal-derived factor 1 (SDF1) also known as CXCL12 belongs to the CXC group of chemokines. SDF1 elicits major control over hematopoietic cells, tissue/organ-committed progenitor/stem cells, and neural stem cell migration, proliferation and survival, and angiogenesis by binding and stimulating its receptors CXCR4 and CXCR7. Indeed SDF1-CXCR4 knockout mice display defects in colonization of bone marrow stem cells, and impaired development of heart, brain, and large vessels. Thus, the SDF1-CXCR4 axis plays important roles during various biological activities including organogenesis [39]. Of particular interest, in the ischemic brain SDF1 is released at the site of injury and is shown to enhance stem/progenitor cell recruitment and survival. It has been shown that CXCR4 is expressed on neural stem cells (NSC) and neuroblasts after stroke, and SDF1 promotes directional migration of these newly formed neuroblasts to ischemic damaged areas. SDF1 expression is increased primarily in the ischemic penumbra, particularly in perivascular astrocytes. This chemokine has been suggested to provide neuroprotection also by increasing the homing of bone marrow-derived stromal stem cells to sites of injury [40–50].

Research in recent decade have demonstrated that angiogenesis is critical in improving post-stroke neurological functional recovery. Thus elucidating the underlying mechanisms of this interplay between stroke-induced neurogenesis and angiogenesis is of great importance for neuro-restorative therapy. Angiogenesis is formation of new micro-vessels from pre-existing vessels. It is a multi-step biological process, including proliferation and sprouting of endothelial cells, formation of tube-like vascular structures, and branching. SDF1 has also been demonstrated to play an important role in angiogenesis and improving local cerebral blood flow. Following ischemic stroke, the majority of SDF1 expression is shown to be associated with activated microglia in the perivascular region. In addition, in animal models of stroke, SDF1 is expressed and released, predominantly by activated astrocytes that are associated with blood vessels. SDF1 produced in brain may act as a chemoattractant for peripheral blood derived cells. SDF1 contributes to angiogenesis by recruiting endothelial progenitor cells (EPCs) from the bone marrow, and by increasing endothelial cell proliferation through a CXCR4 dependent mechanism [51–54]. These findings have linked SDF1 to the restoration of both neuro and vascular unit to areas of ischemic injury as the microvascular endothelial cells are greatly involved in this process. Thus, SDF1 is a critical player in post-stroke brain repair.

2.2 Monocyte Chemoattractant Protein-1 (MCP-1)

MCP-1 also known as CCL2 is the first identified CC chemokine of the C-C chemokine family. It signals through its cognate receptor CCR2. MCP-1 is ubiquitously expressed by many cell types including endothelial, fibroblasts, epithelial, smooth muscle, astrocytes, and microglial cells. The major source of MCP-1 are monocyte/macrophages. It regulates the migration and infiltration of monocytes, T lymphocytes, natural killer (NK) cells, and neuroblasts. Using middle cerebral artery occlusion (MCAO) model of focal ischemic stroke, we and others have demonstrated upregulation of MCP-1 expression in the rat brain. MCP-1 mRNA and protein levels increase significantly within a few hours of ischemia-reperfusion injury in the brain of the rat and it remains high for several days. Double immune-histochemical analysis in MCAO mice has revealed the increased levels of MCP-1 in neurons as early as 12 h after focal brain ischemia, but astrocytes and microglia show MCP-1 expression at a later stage following the ischemia/reperfusion. The MCP-1 levels are also shown to increase in the serum and CSF of ischemic stroke patients during the early stages of ischemic stroke. MCP-1 has been suggested to contribute to tissue damage through recruitment of inflammatory cells. These data indicate that higher levels of MCP-1 may worsen stroke outcome. In support of this idea, mice deficient in MCP-1 show significant decrease in infarct volume as a consequence of focal brain ischemia and a decrease in macrophage accumulation in the infarct area one or two weeks after the stroke. Furthermore, mice lacking the gene for the MCP-1 receptor, CCR2, exhibit reduced infarct size, edema, leukocyte infiltration and expression of inflammatory mediators following focal ischemia, indicating its role in brain injury. Detrimental effects of MCP-1 is further demonstrated in which blockade of MCP-1or its receptor CCR2 decreases the permeability of blood brain barrier (BBB) after reperfusion. Paradoxically, MCP-1 may also contribute to post-stroke brain restoration. In rats and mice, MCP-1 may promote ischemic stroke induced migration of neuroblasts derived from neural progenitors in the sub ventricular zone (SVZ-derived neuroblasts) towards the ischemic striatum and cortex. Thus, in addition to attracting leukocytes, CCL2 plays an important role in the migration of newly formed neuroblasts from neurogenic regions to the injured regions of the brain after a stroke [55–58].

2.3 Fractalkine

Fractalkine is also known as chemokine (C-X3-C motif) ligand 1. It is the only member of the CX3C chemokine family. Interestingly, CX3CL1 is widely expressed in the brains of young, but decreased in those of aged mice and rat with reduced hippocampal neurogenesis. The genetic deletion or pharmacological inhibition of CX3CR1 leads to reduced neurogenesis in the dentate gyrus of mouse hippocampus. Fractalkine is implicated in ischemic pathophysiology and its expression is increased in neurons and in endothelial cells after a focal ischemic insult. Interestingly, its receptor CX3CR1, is expressed only in microglia/macrophages, indicating that fractalkine is involved in neuron–microglia interaction and signaling. These results suggest a possible role for fractalkine in protection against neurotoxicity produced by activated microglia. Fractalkine is shown to display neuroprotective and anti-inflammatory activities in several animal models of disease, and its expression correlates with positive outcomes in human neuro-pathologies. Fractalkine is upregulated by hypoxia and/or inflammation-induced inflammatory cytokines and it is shown to promote angiogenesis and microvasculature remodeling. On the other hand, this chemokine is shown to be involved in the activation and chemo-attraction of leukocytes and microglia into the infracted tissue. Indeed, fractalkine-deficient mice exhibit a smaller infarct size and lower mortality after transient focal cerebral ischemia. It is evident that CX3CL1 may have either beneficial or destructive potential in the CNS, depending on the activation state of the microglia cell population [59–68]. These studies indicate fractalkine (CX3CL1) and its receptor CX3CR1 as potential targets for future therapeutics for post-stroke repair.

3 Proteolytic Enzymes in Focal Ischemic Brain Damage and Repair

Proteases are generally known to be involved in the degradative processes. However, now it has become evident that proteolytic enzymes regulate key molecular and cellular signaling pathways, thus contributing to homeostasis and pathological conditions. By initiating intracellular signals through both catalytically and non-catalytically, proteases play important roles in many biological processes including injury and tissue remodeling [69–71].

Posttranslational modifications of chemokine functions have gained major attention due to their impact on physio-pathological status. Clearly, the best known type of posttranslational modification of chemokines is proteolysis by specific enzymes. Both the NH_2-terminal and the COOH-terminal end of a chemokine can be subjected to proteolysis, and internal cleavage is also carried out by endopeptidases. All forms of proteases including intracellular, membrane bound, and secreted forms of proteases are involved in truncation or degradation of proteins that modulate several critical physiological processes such as cell proliferation, cell death, injury repair, tissue remodeling, homeostasis, and immune responses [71–73].

Stroke induces significant alterations in the proteolytic events in the ischemic brain micro-environment. In particular, serine proteases and matrix metallic-proteinases (MMP) are aberrantly expressed in response to ischemic injury in brain causing time dependent effects on post-stroke brain repair and neurological functional outcome. Proteolytic enzymes have gained much attention as they are involved in the catalytic processing of peptides and small proteins including chemokines in normal brain and during pathogenesis of neurodegenerative disorders. Particularly, dysregulation of these proteolytic enzymes play an important role in post-stroke brain injury and repair, and are thus potential target for therapeutic intervention in stroke [72–77]. In this chapter, we will review two key proteases involved in truncation and regulation of chemokines in ischemic stroke. The role of the matrix metalloproteases (MMPs), and dipeptidyl peptidase (DPPIV) will be discussed.

3.1 Matrix Metalloproteases (MMP)

MMPs are very well studied protease enzymes whose catalytic mechanism involves a metal. MMPs are established to have an important role in extracellular matrix (ECM) degradation and cell migration/invasion. MMPs cleave components of the ECM, such as collagen, proteoglycan and laminin, but also process a number of cell-surface and soluble proteins, including receptors, inflammatory cytokines and chemokines. Thus, in addition to their physiological roles, such as tissue remodeling, MMPs contribute to the regulation of neuro-inflammatory responses to injury [78–84].

Numerous studies have demonstrated the involvement of MMPs in ischemic stroke patho-physiology. Two members of this class of proteases, the gelatinases MMP-2 and MMP-9, have been strongly implicated in ischemic brain pathology because they contribute to the disruption of the blood brain barrier (BBB) and hemorrhagic transformation following injury both in animal models and in stroke patients. Previous studies have described increased expression and activity of MMPs in the brain following transient focal ischemia. Their protease activity was shown to increase immediately after the start of reperfusion in the regions of injured brain in a rat model of transient Middle Cerebral Artery Occlusion (MCAO). MMP enzyme activity was mainly detected in neuronal nuclei during the early stages of ischemia/reperfusion, but was detected in the cytosolic compartment and in non-neuronal glial cells at later reperfusion times. MMPs are also expressed by stromal cells and leukocytes making it a potential target for neuroprotection and repair. Indeed treatment with MMP inhibitors or MMP neutralizing antibodies showed decreased infarct volume and prevented BBB disruption after permanent or transient MCAO in rodents. During the early acute phase of stroke, MMP9 contributes to inflammation and neuronal cell death. Paradoxically, MMP9 is demonstrated to be beneficial during late repair phase as it promotes neuro-vascular remodeling through enhancing angiogenesis and neuroblast migration to the injured brain region [78–86]. Thus, timely regulation of expression and activity of these proteases is critical for stroke outcome.

Molecular mechanisms of MMPs are also well studied. Studies have shown that MMP-3 and 9 guide sub ventricular neuroblasts migration via the activation of PI3K/Akt and ERK1/2 signaling pathways. It has been postulated that neuroblasts express and secret MMPs that support their migration to ischemic lesion. Further studies are required to conclusively determine the role of MMPs in post-stroke brain. MMPs truncate or cleave several CC and CXC chemokines, thus altering their functions. Amino terminal proteolysis of CXCL8 (interlukin-8 (IL-8) by MMP-9 generates CXCL8 that in turn stimulates MMP-9 secretion and enhances mobilization of intracellular calcium and migration of leukocytes. MMP-8-, MMP-9-, or MMP-12 also mediate proteolysis of CXCL11and CXCL12 also called as Interferon-inducible T-cell alpha (I-Tac) and stromal derived factor (SDF) leading to decrease in their chemotactic activity. Among CC chemokines, MCPs are well known substrates for MMPs cleavage and inactivation. MMPs not only cleave chemokines but also promote degradation of the extracellular matrix [79, 85–88] that may impact post-stroke brain injury and repair.

3.2 *Dipeptidyl Peptidase IV (DPPIV)*

Dipeptidyl peptidase (DPP) IV was identified for the first time by Hopsu-Havu and Glenner in 1966 in livers of rats [89]. This serine protease is also called as adenosine complexing protein 2 or cluster of differentiation 26 (CD26). The DPP enzymes have been associated with a wide variety of physiological and pathophysiological processes of the immune system [90–92]. Here, we will discuss the role of DPPIV in chemokine and stem cell regulation implicated in stroke pathology.

The human DPPIV gene contains 26 exons and is located on chromosome 2q.24.3. It is a 110 kDa multi-functional protein. It is a type II integral membrane protein with a six amino acids short amino terminal domain in cytoplasm, 22 amino acids membrane span domain and a long 732 amino acids extracellular domain. It functions as a receptor, as an adhesion molecule for ECM collagen and fibronectin, as a costimulatory signal for T lymphocytes, and is involved in apoptosis [91–111]. Low oxygen levels (ischemia/hypoxia) induce DPPIV through hypoxia-inducible Factor-1α [112]. In addition, several cytokines including interferons (IFNs) and interleukins (IL-1β), differentiating factor retinoic acid, and HNF-1 can also upregulate DPPIV expression in various cell types including fibroblasts, epithelial cells, endothelial cells, and leukocytes [113, 114]. DPPIV exists in both soluble and membrane-bound form. Membrane-bound DPPIV forms homodimer and active heterodimers with fibroblast activating protein (FAPα). DPPIV also binds to adenosine deaminase (ADA), an enzyme that converts 2′-deoxyadenosine into inosine and 2′-deoxyinosine which plays a role in the development and functioning of lymphoid tissues, the activation and proliferation of T and B cells, and release of inflammatory cytokines such as IFN-γ, TNF-α, and IL-6. DPPIV is shown to bind the zymogen plasminogen that promotes its conversion to active plasmin which in turn leads to degradation of the ECM, thus regulating cell migration and invasion. The serine protease enzyme activity of DPPIV is well studied. It selectively cleaves or truncates small peptides whose penultimate position in the NH_2-terminal amino acid sequence is occupied by a proline or an alanine residue. Several in vitro and in vivo studies have demonstrated that many cytokines and chemokines as substrates for DPPIV truncation, as many of them contain a proline or an alanine residue in the penultimate position of their NH_2-terminal sequence [92–105].

Proteolytic cleavage of N-terminal amino acids disrupts the ability of chemokines to bind their receptors. Since chemokines are involved in a number of critical cellular functions that impacts physiological and pathological status, this protease-chemokine interactions have significant impact on various metabolic and regenerative diseases including diabetes and stroke [114–117]. Due to these reasons DPPIV inhibitors have been studied both in vitro and in vivo studies.

In contrast to other proteases, much less is known about DPPIV in cerebral ischemia. DPPIV is a major regulator of cytokines and neuropeptides involved in inflammation, immunity, vascular, and stem cell function. Dysregulated DPPIV expression is associated with tumor development. We and others have shown that DPPIV inhibits tumor cell migration and angiogenic potential through blocking the SDF1-CXCR4 signaling pathways. Recent studies have suggested a role for DPPIV in organ-ischemia including lung, renal, cerebral, and limb ischemic injury [105–109]. Emerging studies have shown that inhibition of DPPIV improves cardiovascular outcomes after cerebral and myocardial ischemia-reperfusion injury [117, 118]. However, little is known about its role in cerebral-ischemia and in chemokine mediated neuro-angiogenesis.

4 Stem Cell Regulation by Proteases in Ischemic Brain

The potential contribution of stem cells in cerebro-vascular diseases has gained enormous attention within neuro-regenerative research. Endogenous progenitor/stem cells recruited from angio-neurogenic niches, or recruited from the peripheral circulation, protects from the detrimental consequences of cerebrovascular events, especially ischemia. This has led to many investigations into the role of many stem cell niches for post-stroke therapy. Indeed, emerging studies in the field of brain injury support the use of endogenous and exogenous stem cells to remodel and repair brain tissue [118–121]. Three major types of progenitor/stem cells including endothelial progenitor cells (EPC), neuronal progenitor/stem cells and mesenchymal stem cells have been largely studied in the context of stroke.

EPC are bone marrow-derived progenitor cells characterized by the expression of CD34, a stem cell marker, and the endothelial marker such as vascular endothelial growth factor receptor-2 or CD309. Endothelial progenitor cells (EPCs) through maturation into endothelial cells, mainly contribute to neo-angiogenesis, formation of new blood vessels, and thus supply required nutrients and oxygen to injured brain. The contribution of EPCs in re-establishment of vascular unit in the brain after ischemic stroke has been well demonstrated. Increased circulating EPCs levels are shown to be correlated with reduced infarct volume, and enhanced neurological functional recovery in patients following acute ischemic stroke [122–126]. A number of studies are also exploring the role of neural progenitor/stem cells (NSCs) in ischemic stroke and have reported that NSCs protect the brain against ischemic injury, promote neurovascular repair, and improve long-term neurobehavioral outcomes. Neural progenitor stem cells differentiate into mature neurons, and thus replace the injured or dead neurons. The migration of neural stem cells is not only required for brain protection in the context of neurodegeneration but plays an important role in brain repair. Our studies have clearly demonstrated to migration of neural progenitor cells/neuroblasts to the site of injured brain following focal ischemic stroke [127–129]. Mesenchymal Stem Cells (MSCs) have gained significant attention due to their potential use in regeneration and remodeling of brain tissue. Endogenous or exogenous MSCs can differentiate into neurons, glial cells and astrocytes that support recovery of lost neurons. In addition, MSCs release various growth factors and chemokines including SDF1, brain derived neurotrophic and nerve growth factors which recruit resident neural stem cells to infarcted area, and establish network between neurons, which in turn increases the repair and recovery processes. In addition, all these stem cells regulate the immunological responses that may reduce inflammation and apoptotic cell death, which significantly reduces further neurological damage [130–132].

Physiological or pathological factors stimulate the production of bone marrow-derived EPCs and MSC, and NSCs that migrate to infarct region and attempt to repair of damaged blood vessels and neurons in ischemic tissues. However, beneficial effects of endogenous or exogenous stem cells is not fully realized in the brain likely due to their poor recruitment and survival in the injured site. The proteases MMPs and DPPIV that are induced following ischemic injury, truncate or degrade the chemotactic chemokines, thus disrupting chemokine-receptor signaling including CXCL12–CXCR4 signaling [133, 134]. This phenomenon leads to poor recruitment and proliferation of stem cells. Thus, improving the chemokine gradient from the peripheral blood to the damaged tissue may play a critical role in the recruitment of stem/progenitor cells to the ischemic brain. Taken together, inhibiting or blocking protease function rescues chemokine activities and is of benefit to promote stem cell mediated brain repair and recovery of neurological functions following ischemic stroke.

5 Concluding Remarks

Proteolytic enzymes are highly pertinent to the processes of stroke severity and functional outcome. Their interaction with other signaling molecules, particularly cytokines and chemokines plays a critical role in regulation of cross-talk between

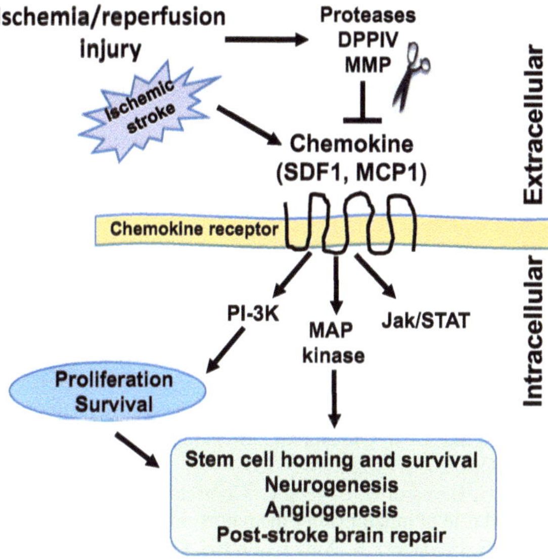

Fig. 21.1 Schematic illustration of interactions between proteases, chemokines and stem cells in response to ischemia/reperfusion injury. Ischemia/reperfusion injury is associated with upregulation of proteases and chemokines levels. Higher concentration of chemokines creates ligand gradient, and play an important role in selectively recruiting stem cells, monocytes, neutrophils, and lymphocytes that express respective chemokine receptors. This leads to activation of proliferation and survival signaling pathways. Excessive production of proteases in the ischemic microenvironment inactivates these chemokines by proteolytic events, thus disrupting chemokine-receptor signaling pathways leading to impaired stem cell homing, apoptosis, inefficient angio-neurogenesis, and poor outcome

post-stroke brain injury and ischemic microenvironment. Ischemic stroke disrupts balance between proteases and chemokines levels in the brain microenvironment. The posttranslational modifications of chemokines substantially modulate their chemotactic potential affecting various biological activities including stem cell recruitment, neurogenesis and angiogenesis. Several decades of work has established that chemokines are particularly subjected to proteolytic processing by DPPIV and MMPs. This disrupts the ability of chemokines to interact with their specific receptors, and thus impair recruitment of stem cells to injured area. In addition, truncation of chemokines disrupts survival and proliferation signaling pathways potentially leading to impaired post-stroke brain repair and recovery as illustrated in Fig. 21.1. Overall, continued research in the area of proteases, chemokines, and stem cell regulation is needed for our complete understanding of the post-stroke brain repair and recovery.

Acknowledgement Supported in part by National Institute of Health (R.J.D.) and American Heart Association (U.V.W.), and the department of Neurosurgery, UW Madison, WI 53792.

References

1. Nagy Z, Nardai S. Cerebral ischemia/repefusion injury: from bench space to bedside. Brain Res Bull. 2017;pii:S0361-9230(16)30227-1.
2. Lo EH, Ning M. Mechanisms and challenges in translational stroke research. J Investig Med. 2016;64(4):827–9.
3. Lo EH, Dalkara T, Moskowitz MA. Mechanisms, challenges and opportunities in stroke. Nat Rev Neurosci. 2003;4:2123–6.
4. Moskowitz MA, Lo EH, Iadecola C. The science of stroke: mechanisms in search of treatments. Neuron. 2010;67:181–98.
5. Hossmann KA. Pathophysiology and therapy of experimental stroke. Cell Mol Neurobiol. 2006;26:1057–84.
6. Dirnagl U, Iadecola C, Moskowitz MA. Pathobiology of ischaemic stroke: an integrated view. Trends Neurosci. 1999;22:391–7.
7. Pappachan J, Kirkham FJ. Cerebrovascular disease and stroke. Arch Dis Child. 2008;93:890–8.
8. Borlongan CV, Rodrigues AA Jr, Oliveira MC. Breaking the barrier in stroke: what should we know? Curr Pharm Des. 2012;18(25):3615–23.
9. Amantea D, Nappi G, Bernardi G, Bagetta G, Corasaniti MT. Post-ischemic brain damage: pathophysiology and role of inflammatory mediators. FEBS J. 2009;276:13–26.
10. Arai K, Lok J, Guo S, Hayakawa K, Xing C, Lo EH. Cellular mechanisms of neurovascular damage and repair after stroke. J Child Neurol. 2011;26:1193–8.
11. Lo EH. A new penumbra: transitioning from injury into repair after stroke. Nat Med. 2008;14:497–500.
12. Xiong Y, Mahmood A, Chopp M. Angiogenesis, neurogenesis and brain recovery of function following injury. Curr Opin Investig Drugs. 2010;11:298–308.
13. Love S. Apoptosis and brain ischaemia. Prog Neuro-Psychopharmacol Biol Psychiatry. 2003;27:267–82.
14. Kalluri HS, Dempsey RJ. Growth factors, stem cells, and stroke. Neurosurg Focus. 2008;24:E14.
15. Iadecola C, Ross ME. Molecular pathology of cerebral ischemia: delayed gene expression and strategies for neuroprotection. Ann N Y Acad Sci. 1997;835:203–17.
16. Mergenthaler P, Dirnagl U, Meisel A. Pathophysiology of stroke: lessons from animal models. Metab Brain Dis. 2004;19(3–4):151–67.
17. Arvidsson A, Collin T, Kirik D, Kokaia Z, Lindvall O. Neuronal replacement from endogenous precursors in the adult brain after stroke. Nat Med. 2002;8:963–70.
18. Yamashita T, Abe K. Mechanisms of endogenous endothelial repair in stroke. Curr Pharm Des. 2012;18:3649–52.
19. Del Zoppo GJ, Becker KJ, Hallenbeck JM. Inflammation after stroke: is it harmful? Arch Neurol. 2001;58:669–72.
20. Kriz J. Inflammation in ischemic brain injury: timing is important. Crit Rev Neurobiol. 2006;18:145–57.
21. Bosisio D, Salvi V, Gagliostro V, Sozzani S. Angiogenic and antiangiogenic chemokines. Chem Immunol Allergy. 2014;99:89–104.
22. Ahmed S, Malemud CJ, Koch AE, Athar M, Taub DD. Cytokines and chemokines: disease models, mechanisms, and therapies. Mediat Inflamm. 2014;2014:296356. https://doi.org/10.1155/2014/296356.
23. Goazigo AR-L. Current status of chemokines in the adult CNS. Prog Neurobiol. 2013;104:67–92.
24. Stone MJ, Hayward JA, Huang C, Huma Z, Sanchez J. Mechanisms of regulation of the chemokine-receptor network. Int J Mol Sci. 2017;18(2):pii:E342. https://doi.org/10.3390/ijms18020342.
25. Minami M, Satoh M. Chemokines and their receptors in the brain: pathophysiological roles in ischemic brain injury. Life Sci. 2003;74(2–3):321–7.

26. Bajetto A, Bonavia R, Barbero S, Florio T, Schettini G. Chemokines and their receptors in the central nervous system. Front Neuroendocrinol. 2001;22(3):147–84.
27. Du Y, Deng W, Wang Z, Ning M, Zhang W, Zhou Y, Lo EH, Xing C. Differential subnetwork of chemokines/cytokines in human, mouse, and rat brain cells after oxygen-glucose deprivation. J Cereb Blood Flow Metab. 2017;37(4):1425–34.
28. Mortier A, Van Damme J, Proost P. Regulation of chemokine activity by posttranslational modifcation. Pharmacol Ther. 2008;120:197–217.
29. Newton RC, Vaddi K. Biological responses to C-C chemokines. Methods Enzymol. 1997;287:174–86.
30. Mirabelli-Badenier M, Braunersreuther V, Viviani GL, Dallegri F, Quercioli A, Veneselli E, Mach F, Montecucco F. CC and CXC chemokines are pivotal mediators of cerebral injury in ischaemic stroke. Thromb Haemost. 2011;105(3):409–20.
31. Réaux-Le Goazigo A, Van Steenwinckel J, Rostène W, Mélik Parsadaniantz S. Current status of chemokines in the adult CNS. Prog Neurobiol. 2013;104:67–92.
32. Graves DT, Jiang Y. Chemokines, a family of chemotactic cytokines. Crit Rev Oral Biol Med. 1995;6(2):109–18.
33. Metzemaekers M, Van Damme J, Mortier A, Proost P. Regulation of chemokine activity – a focus on the role of dipeptidyl peptidase IV/CD26. Front Immunol. 2016;7(483):1–23.
34. Zlotnik A, Yoshie O. The chemokine superfamily revisited. Immunity. 2012;36(5):705–16.
35. Ahuja SK, Gao JL, Murphy PM. Chemokine receptors and molecular mimicry. Immunol Today. 1994;15(6):281–7.
36. Motaln H, Turnsek TL. Cytokines play a key role in communication between mesenchymal stem cells and brain cancer cells. Protein Pept Lett. 2015;22(4):322–31.
37. Sullivan R, Duncan K, Dailey T, Kaneko Y, Tajiri N, Borlongan CV. A possible new focus for stroke treatment – migrating stem cells. Expert Opin Biol Ther. 2015;15(7):949–58.
38. Pelus LM, Fukuda S. Chemokine-mobilized adult stem cells; defining a better hematopoietic graft. Leukemia. 2008;22(3):466–73.
39. Nagasawa T. CXCL12/SDF-1 and CXCR4. Front Immunol. 2015;6:301. https://doi.org/10.3389/fimmu.2015.00301.
40. Wang Y, Huang J, Li Y, Yang GY. Roles of chemokine CXCL12 and its receptors in ischemic stroke. Curr Drug Targets. 2012;13(2):166–72.
41. Richter R, Jochheim-Richter A, Ciuculescu F, et al. Identification and characterization of circulating variants of CXCL12 from human plasma: effects on chemotaxis and mobilization of hematopoietic stem and progenitor cells. Stem Cells Dev. 2014;23(16):1959–74.
42. Hill WD, Hess DC, Martin-Studdard A, et al. SDF-1 (CXCL12) is upregulated in the ischemic penumbra following stroke: association with bone marrow cell homing to injury. J Neuropathol Exp Neurol. 2004;63(1):84–96.
43. Cui L, Qu H, Xiao T, Zhao M, Jolkkonen J, Zhao C. Stromal cell-derived factor-1 and its receptor CXCR4 in adult neurogenesis after cerebral ischemia. Restor Neurol Neurosci. 2013;31(3):239–51.
44. Robin AM, Zhang ZG, Wang L, et al. Stromal cell-derived factor 1alpha mediates neural progenitor cell motility after focal cerebral ischemia. J Cereb Blood Flow Metab. 2006;26(1):125–34.
45. Bakondi B, Shimada IS, Peterson BM, Spees JL. SDF-1alpha secreted by human CD133-derived multipotent stromal cells promotes neural progenitor cell survival through CXCR7. Stem Cells Dev. 2011;20:1021–9.
46. Ceradini DJ, Kulkarni AR, Callaghan MJ, et al. Progenitor cell trafficking is regulated by hypoxic gradients through HIF-1 induction of SDF-1. Nat Med. 2004;10:858–64.
47. Filippo TR, Galindo LT, Barnabe GF, et al. CXCL12 N-terminal end is sufficient to induce chemotaxis and proliferation of neural stem/progenitor cells. Stem Cell Res. 2013;11:913–25.
48. Imitola J, Raddassi K, Park KI, et al. Directed migration of neural stem cells to sites of CNS injury by the stromal cell-derived factor 1alpha/CXC chemokine receptor 4 pathway. Proc Natl Acad Sci. 2004;101:18117–22.

49. Merino JJ, Bellver-Landete V, Oset-Gasque MJ, Cubelos B. CXCR4/CXCR7 molecular involvement in neuronal and neural progenitor migration: focus in CNS repair. J Cell Physiol. 2015;230:27–42.
50. Yin W, Ma L, Zhang J, et al. The migration of neural progenitor cell mediated by SDF-1 is NF-kappaB/HIF-1alpha dependent upon hypoxia. CNS Neurosci Ther. 2013;19:145–53.
51. Zheng H, Fu G, Dai T, Huang H. Migration of endothelial progenitor cells mediated by stromal cell-derived factor-1alpha/CXCR4 via PI3K/Akt/eNOS signal transduction pathway. J Cardiovasc Pharmacol. 2007;50:274–80.
52. Kuhlmann CR, Schaefer CA, Reinhold L, Tillmanns H, Erdogan A. Signalling mechanisms of SDF-induced endothelial cell proliferation and migration. Biochem Biophys Res Commun. 2005;335(4):1107–14.
53. Salcedo R, Wasserman K, Young HA, et al. Vascular endothelial growth factor and basic fibroblast growth factor induce expression of CXCR4 on human endothelial cells: In vivo neovascularization induced by stromal-derived factor-1alpha. Am J Pathol. 1999;154(4):1125–35.
54. Bajetto A, Barbero S, Bonavia R, et al. Stromal cell-derived factor-1alpha induces astrocyte proliferation through the activation of extracellular signal-regulated kinases 1/2 pathway. J Neurochem. 2001;77:1226–36.
55. Deshmane SL, Kremlev S, Amini S, Bassel E, Sawaya BE. Monocyte chemoattractant protein-1 (MCP-1): an overview. J Interf Cytokine Res. 2009;29(6):313–26.
56. Chen Y, Hallenbeck JM, Ruetzler C, et al. Overexpression of monocyte chemoattractant protein 1 in the brain exacerbates ischemic brain injury and is associated with recruitment of inflammatory cells. J Cereb Blood Flow Metab. 2003;23:748–55.
57. Yan YP, Sailor KA, Lang BT, Park SW, Vemuganti R, Dempsey RJ. Monocyte chemoattractant protein-1 plays a critical role in neuroblast migration after focal cerebral ischemia. J Cereb Blood Flow Metab. 2007;27(6):1213–24.
58. He X, Li DR, Cui C, Wen LJ. Clinical significance of serum MCP-1 and VE-cadherin levels in patients with acute cerebral infarction. Eur Rev Med Pharmacol Sci. 2017;21(4):804–8.
59. Lauro C, Catalano M, Di Paolo E, et al. Fractalkine/CX3CL1 engages different neuroprotective responses upon selective glutamate receptor overactivation. Front Cell Neurosci. 2015;21(8):472. https://doi.org/10.3389/fncel.2014.00472.
60. Lauro C, Catalano M, Trettel F, Limatola C. Fractalkine in the nervous system: neuroprotective or neurotoxic molecule? Ann N Y Acad Sci. 2015;1351:141–8.
61. Mizuno T, Kawanokuchi J, Numata K, Suzumura A. Production and neuroprotective functions of fractalkine in the central nervous system. Brain Res. 2003;979(1–2):65–70.
62. Zhang Y, Zheng J, Zhou Z, et al. Fractalkine promotes chemotaxis of bone marrow-derived mesenchymal stem cells towards ischemic brain lesions through Jak2 signaling and cytoskeletal reorganization. FEBS J. 2015;282(5):891–903.
63. Qin W, Li Z, Luo S, Wu R, Pei Z, Huang R. Exogenous fractalkine enhances proliferation of endothelial cells, promotes migration of endothelial progenitor cells and improves neurological deficits in a rat model of ischemic stroke. Neurosci Lett. 2014;569:80–4.
64. R C, Villa P, Chece G, et al. CX3CL1 is neuroprotective in permanent focal cerebral ischemia in rodents. J Neurosci. 2011;31(45):16327–35.
65. Liu Y, Wu XM, Luo QQ, et al. CX3CL1/CX3CR1-mediated microglia activation plays a detrimental role in ischemic mice brain via p38MAPK/PKC pathway. J Cereb Blood Flow Metab. 2015;35(10):1623–31.
66. Tang Z, Gan Y, Liu Q, et al. CX3CR1 deficiency suppresses activation and neurotoxicity of microglia/macrophage in experimental ischemic stroke. J Neuroinflammation. 2014;11:26. https://doi.org/10.1186/1742-2094-11-26.
67. D'Haese JG, Friess H, Ceyhan GO. Therapeutic potential of the chemokine-receptor duo Fractalkine/CX3CR1: an update. Expert Opin Ther Targets. 2012;16(6):613–8.
68. Chapman GA, Moores K, Harrison D, Campbell CA, Stewart BR, Strijbos PJ. Fractalkine cleavage from neuronal membranes represents an acute event in the inflammatory response to excitotoxic brain damage. J Neurosci. 2000;20:RC87(1–5).
69. Davis M, Mantle D, Mendelow AD. The role of proteolytic enzymes in focal ischaemic brain damage. Acta Neurochir Suppl. 2000;76:261–4.

70. Lee SR, Wang X, Tsuji K, Lo EH. Extracellular proteolytic pathophysiology in the neurovascular unit after stroke. Neurol Res. 2004;26:854–61.
71. Zhao BQ, Tejima E, Lo EH. Neurovascular proteases in brain injury, hemorrhage and remodeling after stroke. Stroke. 2007;38(2 Suppl):748–52.
72. Wolf M, Albrecht S, Marki C. Proteolytic processing of chemokines: implications in physiological and pathological conditions. Int J Biochem Cell Biol. 2008;40:1185–98.
73. Kryczka J, Boncela J. Proteases revisited: roles and therapeutic implications in fibrosis. Mediat Inflamm. 2017;2017:2570154. https://doi.org/10.1155/2017/2570154.
74. Vivien D, Buisson A. Serine protease inhibitors: novel therapeutic targets for stroke? J Cereb Blood Flow Metab. 2000;20:755–64.
75. Wang X, Li X, Xu L, et al. Up-regulation of secretory leukocyte protease inhibitor (SLPI) in the brain after ischemic stroke: adenoviral expression of SLPI protects brain from ischemic injury. Mol Pharmacol. 2003;64:833–40.
76. Apte SS, Parks WC. Metalloproteinases: a parade of functions in matrix biology and an outlook for the future. Matrix Biol. 2015;44-46:1–6.
77. Yang Y, Rosenberg GA. Matrix metalloproteinases as therapeutic targets for stroke. Brain Res. 2015;1623:30–8.
78. Kurzepa J, Kurzepa J, Golab P, Czerska S, Bielewicz J. The significance of matrix metalloproteinase (MMP)-2 and MMP-9 in the ischemic stroke. Int J Neurosci. 2014;124:707–16.
79. Song J, Wu C, Korpos E, et al. Focal MMP-2 and MMP-9 activity at the blood-brain barrier promotes chemokine-induced leukocyte migration. Cell Rep. 2015;10(7):1040–54.
80. Zhao BQ, Wang S, Kim HY, et al. Role of matrix metalloproteinases in delayed cortical responses after stroke. Nat Med. 2006;12:441–5.
81. Siwetz M, Blaschitz A, Kremshofer J, et al. Metalloprotease dependent release of placenta derived fractalkine. Mediat Inflamm. 2014;2014:839290. https://doi.org/10.1155/2014/839290.
82. Lucivero V, Prontera M, Mezzapesa DM, et al. Different roles of matrix metalloproteinases-2 and -9 after human ischaemic stroke. Neurol Sci. 2007;28:165–70.
83. Rosenberg GA, Cunningham LA, Wallace J, et al. Immunohistochemistry of matrix metalloproteinases in reperfusion injury to rat brain: activation of MMP-9 linked to stromelysin-1 and microglia in cell cultures. Brain Res. 2001;893(1–2):104–12.
84. Seo JH, Guo S, Lok J, et al. Neurovascular matrix metalloproteinases and the blood-brain barrier. Curr Pharm Des. 2012;18(25):3645–8.
85. Park KP, Rosell A, Foerch C, et al. Plasma and brain matrix metalloproteinase-9 after acute focal cerebral ischemia in rats. Stroke. 2009;40(8):2836–42.
86. Lenglet S, Montecucco F, Mach F. Role of matrix metalloproteinases in animal models of ischemic stroke. Curr Vasc Pharmacol. 2015;13(2):161–6.
87. McQuibban GA, Gong JH, Wong JP, Wallace JL, Clark-Lewis I, Overall CM. Matrix metalloproteinase processing of monocyte chemoattractant proteins generates CC chemokine receptor antagonists with anti-inflammatory properties in vivo. Blood. 2002;100:1160–7.
88. Le NT, Xue M, Castelnoble LA, Jackson CJ. The dual personalities of matrix metalloproteinases in inflammation. Front Biosci. 2007;12:1475–87.
89. Hopsu-Havu VK, Glenner GG. A new dipeptide naphthylamidase hydrolyzing glycyl-prolyl-β-naphthylamide. Histochemie. 1966;7:197–201.
90. Koivisto V. Discovery of dipeptidyl-peptidase IV – a 40 year journey from bench to patient. Diabetologia. 2008;51:1088–9.
91. Waumans Y, Baerts L, Kehoe K, Lambeir AM, De Meester I. The dipeptidyl peptidase family, prolyl oligopeptidase, and prolyl carboxypeptidase in the immune system and inflammatory disease, including atherosclerosis. Front Immunol. 2015;6:387. https://doi.org/10.3389/fimmu.2015.00387.
92. Mortier A, Gouwy M, Van Damme J, Proost P, Struyf S. CD26/dipeptidylpeptidase IV-chemokine interactions: double-edged regulation of inflammation and tumor biology. J Leukoc Biol. 2016;99(6):955–69.
93. Klemann C, Wagner L, Stephan M, von Hörsten S. Cut to the chase: a review of CD26/dipeptidyl peptidase-4's (DPP4) entanglement in the immune system. Clin Exp Immunol. 2016;185:1–21.

94. Scharpé S, De Meester I. Peptide truncation by dipeptidyl peptidase IV: a new pathway for drug discovery? Verh K Acad Geneeskd Belg. 2001;63(1):5–32.
95. Proost P, De Meester I, Schols D, et al. Amino-terminal truncation of chemokines by CD26/dipeptidyl-peptidase IV. Conversion of RANTES into a potent inhibitor of monocyte chemotaxis and HIV-1-infection. J Biol Chem. 1998;273(13):7222–7.
96. Lambeir AM, Proost P, Durinx C, et al. Kinetic investigation of chemokine truncation by CD26/dipeptidyl peptidase IV reveals a striking selectivity within the chemokine family. J Biol Chem. 2001;276(32):29839–45.
97. Shibuya-Saruta H, Kasahara Y, Hashimoto Y. Human serum dipeptidyl peptidase IV (DPPIV) and its unique properties. J Clin Lab Anal. 1996;10(6):435–40.
98. Wesley UV, Hatcher JF, Ayvaci ER, Klemp A, Dempsey RJ. Regulation of dipeptidyl peptidase IV in the post-stroke rat brain and in vitro ischemia: Implications for chemokine-mediated neural progenitor cell migration and angiogenesis. Mol Neurobiol. 2016. https://doi.org/10.1007/s12035-016-0039-4.
99. Arscott WT, LaBauve AE, May V, Wesley UV. Suppression of neuroblastoma growth by dipeptidyl peptidase IV: relevance of chemokine regulation and caspase activation. Oncogene. 2009;28(4):479–91.
100. Christopherson KW II, Hangoc G, Broxmeyer HE. Cell surface peptidase CD26/dipeptidyl-peptidase IV regulates CXCL12/stromal cell-derived factor-1 alpha-mediated chemotaxis of human cord blood CD34+ progenitor cells. J Immunol. 2002;169:7000–8.
101. Wesley UV, Albino AP, Tiwari S, Houghton AN. A role for dipeptidyl peptidase IV in suppressing the malignant phenotype of melanocytic cells. J Exp Med. 1999;190:311–22.
102. Wesley UV, McGroarty M, Homoyouni A. Dipeptidyl peptidase inhibits malignant phenotype of prostate cancer cells by blocking basic fibroblast growth factor signaling pathway. Cancer Res. 2005;65:1325–34.
103. Wesley UV, Tiwari S, Houghton AN. Role for dipeptidyl peptidase IV in tumor suppression of human non small cell lung carcinoma cells. Int J Cancer. 2004;1099(6):855–66.
104. Sun YX, Pedersen EA, Shiozawa Y, et al. CD26/dipeptidyl peptidase IV regulates prostate cancer metastasis by degrading SDF-1/CXCL12. Clin Exp Metastasis. 2008;25:765–76.
105. Jungraithmayr W, De Meester I, Matheeussen V, Baerts L, Arni S, Weder W. CD26/DPP-4 inhibition recruits regenerative stem cells via stromal cell-derived factor-1 and beneficially influences ischaemia-reperfusion injury in mouse lung transplantation. Eur J Cardiothorac Surg. 2012;41:1166–73.
106. Rohnert P, Schmidt W, Emmerlich P, et al. Dipeptidyl peptidase IV, aminopeptidase N and DPIV/APN-like proteases in cerebral ischemia. J Neuroinflammation. 2012;9:44. https://doi.org/10.1186/1742-2094-9-44.
107. Chua S, Sheu JJ, Chen YL, et al. Sitagliptin therapy enhances the number of circulating angiogenic cells and angiogenesis-evaluations in vitro and in the rat critical limb ischemia model. Cytotherapy. 2013;15:1148–63.
108. Vaghasiya J, Sheth N, Bhalodia Y, Manek R. Sitagliptin protects renal ischemia reperfusion induced renal damage in diabetes. Regul Pept. 2011;166:48–54.
109. Chua S, Lee FY, Tsai TH, et al. Inhibition of dipeptidyl peptidase-IV enzyme activity protects against myocardial ischemia-reperfusion injury in rats. J Transl Med. 2014;12:357. https://doi.org/10.1186/s12967-014-0357-0.
110. Boonacker E, Van Noordan CJ. The multifunctional or moonlighting protein CD26/DPPIV. Eur J Cell Biol. 2003;82:53–73.
111. Havre PA, Abe M, Urasaki Y, Ohnuma K, Morimoto C, Dang NH. The role of CD26/dipeptidyl peptidase IV in cancer. Front Biosci. 2008;13:1634–45.
112. Dang DT, Chun SY, Burkitt K, et al. Hypoxia- inducible factor-1 target genes as indicators of tumor vessel response to vascular endothelial growth factor inhibition. Cancer Res. 2008;68:1872–80.
113. Bauvois B, Djavaheri-Mergny M, Rouillard D, Dumont J, Wietzerbin J. Regulation of CD26/DPPIV gene expression by interferons and retinoic acid in tumor B cells. Oncogene. 2000;19:265–72.

114. Gu N, Tsuda M, Matsunaga T, et al. Glucose regulation of dipeptidyl peptidase IV gene expression is mediated by hepato- cyte nuclear factor-1alpha in epithelial intestinal cells. Clin Exp Pharmacol Physiol. 2008;35:1433–9.
115. Proost P, Struyf S, Loos T, et al. Coexpression and interaction of CXCL10 and CD26 in mesenchymal cells by synergising inflammatory cytokines: CXCL8 and CXCL10 are discriminative markers for autoimmune arthropathies. Arthritis Res Ther. 2006;8:1–14.
116. Kim NH, Yu T, Lee DH. The nonglycemic actions of dipeptidyl peptidase-4 inhibitors. Biomed Res Int. 2014;2014:368703. https://doi.org/10.1155/2014/368703.
117. El-Sahar AE, Safar MM, Zaki HF, Attia AS, Ain-Shoka AA. Sitagliptin attenuates transient cerebral ischemia/reperfusion injury in diabetic rats: implication of the oxidative-inflammatory-apoptotic pathway. Life Sci. 2015;126:81–6.
118. Inthachai T, Lekawanvijit S, Kumfu S, Apaijai N, Pongkan W, Chattipakorn SC, Chattipakorn N. Dipeptidyl peptidase-4 inhibitor improves cardiac function by attenuating adverse cardiac remodelling in rats with chronic myocardial infarction. Exp Physiol. 2015;100(6):667–79.
119. Tang YH, Ma YY, Zhang ZJ, Wang YT, Yang GY. Opportunities and challenges: stem cell-based therapy for the treatment of ischemic stroke. CNS Neurosci Ther. 2015;21(4):337–47.
120. Burns TC, Steinberg GK. Stem cells and stroke: opportunities, challenges and strategies. Expert Opin Biol Ther. 2011;11(4):447–61.
121. Chang YC, Shyu WC, Lin SZ, Li H. Regenerative therapy for stroke. Cell Transplant. 2007;16:171–81.
122. Shyu WC, Lee YJ, Liu DD, Lin SZ, Li H. Homing genes, cell therapy and stroke. Front Biosci. 2006;11:899–907.
123. Meamar R, Nikyar H, Dehghani L, et al. The role of endothelial progenitor cells in transient ischemic attack patients for future cerebrovascular events. J Res Med Sci. 2016;21:47. https://doi.org/10.4103/1735-1995.183995.
124. Li Y-F, Ren L-N, Guo G, et al. Endothelial progenitor cells in ischemic stroke: an exploration from hypothesis to therapy. J Hematol Oncol. 2015;8:33. https://doi.org/10.1186/s13045-015-0130-8.
125. Farag SS, Srivastava S, Messina-Graham S, et al. In vivo DPP-4 inhibition to enhance engraftment of single-unit cord blood transplants in adults with hematological malignancies. Stem Cells Dev. 2013;22:1007–15.
126. Urbich C, De Souza AI, Rossig L, et al. Proteomic characterization of human early pro-angiogenic cells. J Mol Cell Cardiol. 2011;50(2):333–6.
127. Jin K, Sun Y, Xie L, et al. Directed migration of neuronal precursors into the ischemic cerebral cortex and striatum. Mol Cell Neurosci. 2003;24:171–89.
128. Yamashita T, Ninomiya M, Hernandez Acosta P, et al. Subventricular zone-derived neuroblasts migrate and differentiate into mature neurons in the post-stroke adult striatum. J Neurosci. 2006;26:6627–36.
129. Dempsey RJ, Sailor KA, Bowen KK, Tureyen K, Vemuganti R. Stroke-induced progenitor cell proliferation in adult spontaneously hypertensive rat brain: effect of exogenous IGF-1 and GDNF. J Neurochem. 2003;87:586–97.
130. Maria Ferri AL, Bersano A, Lisini D, Boncoraglio G, Frigerio S, Parati E. Mesenchymal stem cells for ischemic stroke: progress and possibilities. Curr Med Chem. 2016;23(16):1598–608.
131. Bang OY, Kim EH, Cha JM, Moon GJ. Adult stem cell therapy for stroke: challenges and progress. J Stroke. 2016;18(3):256–66.
132. Nurkovic J, Dolicanin Z, Mustafic F, Mujanovic R, Memic M, Grbovic V, Skevin AJ, Nurkovic S. Mesenchymal stem cells in regenerative rehabilitation. J Phys Ther Sci. 2016;28(6):1943–8.
133. Ou X, O'Leary HA, Broxmeyer HE. Implications of DPP4 modification of proteins that regulate stem/progenitor and more mature cell types. Blood. 2013;122:161–9.
134. Fadini GP, Avogaro A. Dipeptidyl peptidase-4 inhibition and vascular repair by mobilization of endogenous stem cells in diabetes and beyond. Atherosclerosis. 2013;229:23–9.

Chapter 22
The NLRP3 Inflammasome: A Possible Therapeutic Target for Treatment of Stroke

Tauheed Ishrat and Sanaz Nasoohi

Abstract Ischemic stroke is a complex systemic disease causing severe long-term disability and death worldwide. Experimental and clinical data have demonstrated that inflammation is a major component of ischemic stroke pathobiology. The post-ischemic neuroinflammatory response is characterized by microglial and astro-glial activation and increased expression of inflammatory mediators. Recent findings have provided insight into a newly discovered inflammatory mechanism that contributes to neuronal and glial cell death in neurodegenerative diseases and stroke mediated by inflammasomes. Interestingly, of inflammasomes described to date, NLRP3 (nucleotide-binding domain (NOD)-like receptor protein 3) inflammasome is the best characterized multi-protein complexes and most strongly associated with sterile inflammation. In this chapter, we discuss in detail the prominent contribution and regulation of NLRP3-inflammasome activation in the pathophysiology of ischemic stroke. Furthermore, provide recent developments on the potential of NLRP3 inhibitors in the therapeutic management of stroke outcomes. The significant contribution of regulatory mechanisms of NLRP3 inflammasome with the development of stroke, may improve our understanding of NLRP3 inhibition for developing future therapies and novel drug targets for stroke.

Keywords NLRP3 • Inflammasome • Ischemic stroke • Inflammation • NOD-like receptor protein • Neuronal death • Inhibitors

---

Tauheed Ishrat and Sanaz Nasoohi contributed equally to this work.

T. Ishrat, Ph.D. (✉)
Department of Anatomy and Neurobiology, College of Medicine, University of Tennessee Health Science Center, 855 Monroe Avenue, Wittenborg Building, Room-231, Memphis, TN 38163, USA
e-mail: tishrat@uthsc.edu

S. Nasoohi, Ph.D.
Neuroscience Research Center, Shahid Beheshti University of Medical Sciences, Tehran, Iran

P.A. Lapchak, J.H. Zhang (eds.), *Cellular and Molecular Approaches to Regeneration and Repair*, Springer Series in Translational Stroke Research,
https://doi.org/10.1007/978-3-319-66679-2_22

Abbreviations

Aβ	Amyloid beta
ADP	Adenosine di phosphate
AIM2	Absence in melanoma
AMD	Age-related macular degeneration
AMPK	AMP-activated protein kinase
ASC	Apoptosis-associated speck like
ASK-1	Apoptosis signal-regulating kinase
ASICS	Acid-sensing ion channels
ATP	Adenosine di phosphate
BBB	Blood brain barrier
BBG	Brilliant blue G
BHB	β-hydroxybutyrate
BRCC3	BRCA1/BRCA2-containing complex, subunit 3
BTK	Bruton's tyrosine kinase
CaSR	Calcium sensing receptor
CB2R	Cannabinoid receptor 2
CLR	C-type lectin receptor
CM	Carbon monoxide
Cox2	Cyclooxygenase-2
CNS	Central nervous system
CRID	Cytokine release inhibitory drugs
CXCL1	Chemokine (C-X-C motif) ligand-1
DAMP	Damage-associated molecular patterns
DPI	Diphenylene iodonium
DUBs	Deubiquitinating enzymes
eMCAO	Embolic middle cerebral artery occlusion
EAE	Experimental autoimmune encephalomyelitis
EGCG	Epigallocatechin gallate
EP-4	Prostaglandin E2 receptor 4
ER-stress	Endoplasmic reticulum stress
FFA	Free fatty acid
FBXL2	SCF complex subunit F-box L2
GFAP	Glial fibrillary acidic protein
GLUT1	Glucose transporter-1
GPR6CA	G protein-coupled receptor family C group 6 member A
GSDMD	Gasdermin D
HCA	Hydroxy-carboxylic acid receptor 2
HIF-α	Hypoxia inducible factor-1 alpha
HRECs	Human retinal endothelial cells
HSP90	Heat shock protein 90
ICH	Intracerebral hemorrhage
IFN	Interferon

IFNAR	Interferon-alpha/beta receptor
IL-1β	Interleukin 1 beta
IP3	Inositol trisphosphate
IER-1	Inositol-requiring 1
iNOS	Inducible nitric oxide synthase
IRAK	Interleukin-1 receptor-associated kinase 1
JAC	Janus kinase
JNK	c-jun-N-terminal kinase
KO	Knock-out
LPS	Lipopolysaccharide
LRR	Leucine rich repeat domain
MAP	Mitogen activated protein
MAVS	Mitochondrial anti-viral signaling protein
MCAO	Middle cerebral artery occlusion
MFGE8	Milk fat globule-EGF 8
MI	Myocardial infarction
MAPK	Mitogen-activated protein kinase
miRNA	Micro ribonucleic acid
mtROS	Mitochondrial Reactive oxygen species
MCT	Monocarboxylate transporter 1
MMP-9	Matrix metalloproteinase 9
MNS	3,4-Methylenedioxy-β-nitrostyrene
MRI	Magnetic resonance imaging
MyD-88	Myeloid differentiation primary response 88
NADPH	Nicotinamide adenine dinucleotide phosphate-oxidase
NAD	Nicotinamide adenine dinucleotide phosphate
NADH	Nicotinamide adenine dinucleotide phosphate
NBD	Nucleotide-binding domain
NFκB	Nuclear factor kappa-B
NLR	Nucleotide-binding oligomerization domain like receptor
NLRP	NOD-like receptor proteins
NMDA	N-methyl-D-Aspartate
NOD	Nucleotide-binding oligomerization domain
NOX	Nicotinamide adenine dinucleotide phosphate oxidase
OGD	Oxygen glucose deprivation
Pal-BSA	Palmitate coupled to bovine serum albumin
PAMP	Pathogen-associated molecular patterns
PERK	dsRNA-activated protein kinase-like ER kinase
PGE2	Prostaglandin E2
PKA	Protein kinase A
PLC	Phospholipase c
PPA2	Pyrophosphatase 2
PPAR-γ	Peroxisome proliferator-activated receptor-γ
PRR	Pattern recognition receptor
PTP	Phospho-tyrosine phosphatases

P2X7R	P2X purinoceptor 7
RKIP	Raf-1 kinase inhibitory protein
ROS	Reactive oxygen species
SAR	Structure activity relationship
Ser	Serine
SGT1	Suppressor of g2 allele of skp1
SLC	Solute carriers
SCF	SKP1-cullin-F-box protein
SUR1	Sulfonylurea receptor 1
STAT	Signal transducers and activators of transcription
TAK1	Transforming growth factor beta-activated kinase 1
tMCAO	Transient middle cerebral artery occlusion
Thr	Thioredoxin
TLR	Toll-like receptor
TNF-α	Tumor necrotizing factor α
TRIF	TIR-domain-containing adapter-inducing interferon-β
Trpm4	Transient receptor potential melastatin 4
Trx	Thioredoxin
TXNIP	Thioredoxin interacting protein
WT	Wild type

1 Introduction

According to the latest statistics, stroke when considered separately from other cerebrovascular disorders is standing as the fifth cause of death, behind diseases of the heart, cancer, chronic lower respiratory disease, and unintentional injuries/accidents [1]. Around two thirds of stroke patients either die or are left disabled [2]. Of all strokes, 87% are ischemic while the rest 13% is kind of hemorrhagic. Roughly 40% of hemorrhagic cases are fatal while of those, <40% of the surviving patients may live independently in the first year after. Thrombolytic therapy is the only existing therapeutic approach, being applicable for patients with no hemorrhagic events it is advantageous only for those admitted to clinical care centers not later than 6 h after symptoms onset. Given the symptoms of either type of stroke is not quite specific, it is rational that <5% of stroke patients can get to efficient therapeutic measures [1, 3]. Therefore increasing efforts are being focused to unravel the detailed pathological process of stroke insult proving more promising therapeutic targets. Neuroinflammation has been recognized as of early implicated pathways starting from the very early beginning from stroke-associated endothelial damage and lasting for several weeks in parallel with functional shekels. For that reason it is intricately investigated in terms of its critical effectors which inherently parallel with neural loss [4, 5]. During the acute phase, referring to minutes to hours after stroke, ischemic cell death leads to the release of a variety intracellular contents working as danger signals. Theses agent including Ca^{2+}, enzymes and cytokines (i.e., IL-1 and TNF-α) induce an initial inflammatory response which in turn, propagate the neuro-inflammatory response through activation of resident immune cells (i.e., microglia) and recruitment of inflammatory cells. Neutrophils are the first circulating cells to migrate into the brain after stroke, followed by gradual appearance of macrophages and lymphocytes in the site of injury, including core and peri-infarct regions [6, 7]. Formation of large multiprotein complexes called inflammasomes, are of the early established events efficiently amplifying the primary immune responses. Working as of main innate immune system receptors and sensors, inflammasomes are increasingly defined to act as the key mediator in detecting cellular fait and inflammatory responses after stroke [8]. Conspicuously, inflammasomes modulation could remarkably prevent neural cell death and attenuate ischemia/reperfusion (I/R) injury in *in-vivo* and *in-vitro* stroke models [9, 10], the observation that might be explained with the widespread events downstream to inflammasomes. Indeed inflammasomes activation not only instigate a variety of deteriorating and death signals, but also it is not confined to immune cells i.e. cerebral microglia. In this line, recent researches have revealed that almost all CNS territory residents including neurons, astrocytes, granulocytes and even endothelial cells also express functional inflammasomes which contribute to either CNS disorders and systemic inflammation [8]. Furthermore, upon activation, the inflammasomes induce two caspase-1-dependent inflammatory cascades: necrotic cell death (pyroptosis) and processing of IL-1β and IL-18 to active forms, both of which produce substantial consequences. Pyroptosis is the lytic cell death occurring through cleavage of gasdermin D (GSDMD) by active caspase-1. The N-terminus of GSDMD then may form pores in the plasma membrane which permits passage of fluids disturbing ion hemostasis in intracellular space besides bringing pro-inflammatory cytokines. IL-1β and IL-18 are the only cytokines produced as inactive precursors which are also processed by inflammasomes. Mature IL-1β and IL-18 cytokines activate IL-1 receptor/toll-like receptor (IL-1R/TLR) signaling. Pivotal for activation of innate immunity and inflammation, this signaling stimulate the transcription of many pro-inflammatory and antiviral genes, all contributing to amplify the background inflammatory response. IL-1β is a critical cytokine in CNS pathologies which shows true co-localization within areas of early focal neuronal injury implies that it might be the major form of IL-1 contributing to inflammation following stroke [11, 12].

In essence, this chapter would provide basic information required for understanding inflammasome function from sensing danger signals to contributing to stroke pathology. With the central role of NLRP3 inflammasome and its major implication in ischemic insult, the main effort is to demonstrate the details for its regulation during pathological processes of tissue injury in cerebral ischemia. Finally, an overview of recent advances on developing inflammasome inhibitors are provided, with a particular view on searching for promising therapeutic tools in stroke research.

2 NLRPs as Tissue Injury Sensors

NLRPs (NOD-like receptor proteins) are one of pattern recognition receptors (PRRs) in macrophages which play essential role to upregulate pro-inflammatory gene expression [13, 14]. Based on cellular location and major structural features, PRR may be classified into (a) transmembranal TLRs, (b) transmembranal C-type lectin receptors (CLRs); (c) cytosolicnucleotide-binding oligomerization domain-like receptors or NOD-like receptor proteins (NLRPs); (d) intracellular retinoic acid-inducible gene-1 (RIG-1)-like receptors (RLRs), which are primarily involved in antiviral responses; and (e) absent in melanoma 2 (AIM-2)-like receptors (or non-NLRs).

All these receptors may induce major pro-inflammatory and pro-apoptotic pathways upon activation with their cognate ligands. Depending on initial source, PPR ligands are classified to two classes named pathogen associated molecular patterns (PAMPs) or damage associated molecular patterns (DAMPs). PAMPs are motifs commonly carried by pathogens, e.g. bacterial endotoxin (or LPS) on gram-negative bacteria. Monosodium urate or cholesterol crystals and β-amyloid plaques or endogenous molecules released by necrosis like ATP are among well characterized DAMPs. The inflammation that ensues DAMPs-induced PPRs stimulation is called the sterile inflammation and either resolves the initial insult or leads to disease [15].

3 NLRP3 as the Predominant Inflammasome

Unlike many other PRRs, NLRs have not been restricted to a specific ligand or a typical cognate molecular pattern (PAMP or DAMP) [14, 16]. The NLR family is one the most extensively studied PRRs due to its major role in several pathological inflammation. NLRs are generally composed of three separate domains: 1) the N-terminal domain: which contains a pyrin domain, a caspase recruitment domain, or a baculovirus inhibitory repeat domain and has been used as a structural subclassification for the NLR family, 2) The central NBD or nucleotide-binding domain, which is responsible for dNTPase activity and oligomerization in the presence of nucleotides, primarily ATP, 3) A Leucine rich repeat domain (LRRs) at the C terminus of NLR proteins [17]. Upon activation, the NLR protein oligomerizes with the ASC (apoptosis-associated speck like) adaptor protein which then recruits procaspase-1, allowing its autocleavage and activation. Activated caspase-1 enzyme in turn cleaves upregulated premature proinflammatory cytokines: interleukin-1 (IL-1) and interleukin-18 (IL-18) and causes their release [18, 19]. Several NLRs have the capability to activate the inflammasome in vitro, including: NLRP1, NLRP2, NLRP3, NLRP6, NLRP12, NLRC4 and NOD-2, however, only a handful of prominent NLRPs including NLRP1, NLRP3, NLRC4 and NAIP5 have been recognized as functional inflammasome activators [17]. Comparatively, in contrast to most innate immune receptors [20], NLRP3 might be activated by a variety of pathogen- and host-derived "danger" signals including whole RNA, RNA/DNA hybrids and proteins from gram-positive and gram-negative bacteria, viruses, fungi, protozoa (as PAMPs), as well as ATP, monosodium urate crystals and calcium pyrophosphate dihydrate crystals (as DAMPs) [21, 22]. Given a diversity of stimuli may engage NLRP3, NLRP3 activation is very likely to happen downstream to many PRRs converging on a common pathway. Interestingly, of inflammasomes identified to date, NLRP3 is most strongly associated with sterile inflammation [15]. In fact the NLRP3 inflammasome is established as one of the most critical multi-protein responsible for instigating metabolic, cardiovascular, and neurodegenerative disease-associated inflammation, and is therefore named as a sensor for metabolic danger [23–25].

4 NLP3 Activation and Regulation

According to the pictorial description in Fig. 22.1, based on the most addressed models; NLRP3 activation necessitates two main signals "Priming" and "Activation" both of which initiating following sensing tissue damage through RRPs. Priming is best defined as preparing inflammasomes at transcriptional and post-transcriptional level, to get instigated by the activating signal to induce IL-β secretion from immune cells. The well-known canonical pathway for NLRP3 inflammasome activation illustrate the main dynamic interplay between tissue injury/infection, innate immune system and pathological inflammatory responses. However recent years investigations has unraveled a non-canonical pathway that influence NLRP3 inflammasome priming and activation particularly in gram negative infections. The general principles are so common though, the discriminative attributes of these two terms are detailed in the principal steps in NLRP3 priming, activation and regulation as described below.

4.1 Step 1: Priming; Canonical and Non-canonical

4.1.1 Transcriptional Priming

In physiological conditions, NLRP3 and pro-IL-1β are not expressed in sufficient amounts. That is signaling through transcriptionally active receptors (e.g.; TLRs, NOD2, TNFR1 and TNFR2) seem constitutionally necessary for NLRP3-inflammasome activation [26, 27]. Given that critical mass of NLRP3 is required for inflammasome activation, it has been shown upregulation of NLRP3 is enough for priming-independent activation [28]. Indeed, NLRP3 substantial expression lowers the threshold for NLRP3-inflammasome activation and enhances caspase-1 cleavage as well [29]. Nuclear factor kappa-B (NFκB) and mitogen-activated protein kinases (MAPKs) pathways are central for driving the gene expression and denovo

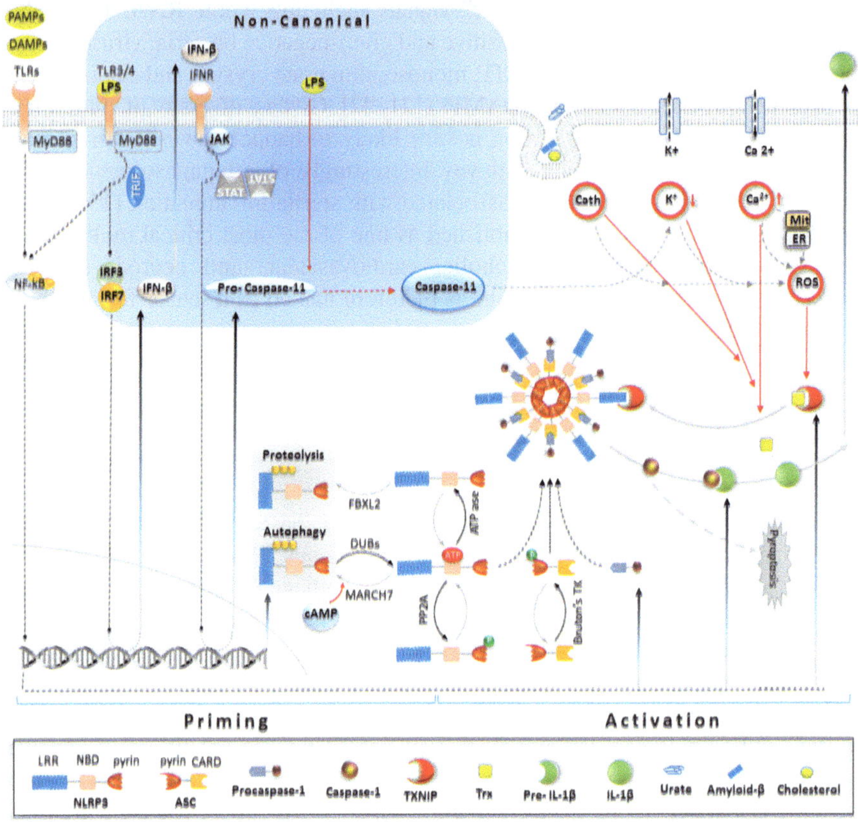

Fig. 22.1 Schematic representation of mechanisms involved in NLRP3 priming and activation. Priming step contributes to preparing NLRP3 inflammasome constituents and Pro-IL-1b (or IL-6) in transcriptional and post-translational levels. Basically TLRs stimulation by DAMPs or PAMPs (e.g. LPS) coupled with MyD 88-NFkB pathway enhances the due transcripts. Following construction in cytosolic mileu, the readily ubiquinated NLRP3 might be subject to degradation by autophagy or DUBs action. The deubiquinated NLRP3 then may recruit ASC and then pro-caspase 1 if is in a non-phosphorylated state devoid of electrostatic repulsion of pyrin domains with the ASC phosphorylated pyrin. After binding to ASC, the NLP3 ATPase activity brings ATP molecules to attach to NLRP3, the step is required for NLRP3 oligomerization. A wide range of stimuli are known to activate the NLP3 inflammasome. As shown briefly in the *up-right panel*, any factor that causes lysosomal cathepsin release following phagocytosis of non-functional particles (e.g. cholesterol crystals) and any stimuli that ends with intracellular K^+ depletion, Ca^{2+} over load and ROS generation may result in NLRP3 activation. The direct mechanism through which this effectors interface with NLRP3 to boost its caspase-1 cleavage activity is un-known. However ROS either is generated by external stimuli or downstream to other effectors has been known to explain NLRP3 inflammasome activation. Under high levels of intracellar ROS, TXNIP which is normally controlling Trx activity is translocated to NRLP3 inflammasome and instigate its activity. Following activation, NLRP3 inflammasome leads to caspase-1 activation, IL-1b release and pyroptosis. NLRP3 inflammasomes might be exported to ECF and subsequent proteolysis. Along with the basic pathways in priming and activation, there are some alternate mechanisms specific to gram negative bacteria called non-canonical as determined in light blue background. Typically LPS in gram negative

protein synthesis of NLRP3 and pro-IL-1β. Several DAMPs and PAMPs bind to pro-inflammatory receptors, including TLRs, NLRs, and tumor necrosis factor receptors 1 and 2 (TNFR1 and TNFR2), for which NF-κB or MAPKs work as transcriptional effectors. The signal is basically mediated by toll-like receptor (TLR)-adaptor molecules MyD88 (myeloid differentiation primary response 88) working along with IRAK1 and 4 (IL-1R associated kinase family members) and does not merely result in transcriptional priming but also may get involved in transcription-independent priming pathways [30, 31].

While priming step is somewhat common between canonical and non-canonical pathway, it may be differentiated based on particular transcripts might be specifically induced in the later phase. That is with the initiating stimuli to be gram negative bacteria (endotoxin), non-canonical pathway is not limited to MyD88 recruitment and IL-1β, IL-18, and NLRP3 transcription, but as well; engages toll/IL-1 receptor homology-domain-containing adapter-inducing interferon-β (TRIF) pathway. Through enhancing IRF3–IRF7 complex formation [32, 33], this elicits the expression of interferon (IFN)-α/β, and then JAK/STAT pathway activation which eventually elevates transcription of mouse caspase-11 (or human caspase 4-5). This caspase subtype acts as a prominent adjuvant activating stimulus specific to non-canonical NLRP3-inflammasome activation [34–36].

In terms of preparing licensed inflammasomes for activation, it is important to note that, the transcription of the due substrate pro-IL-1β simultaneously starts to rise taking hours to peak, but may not be temporally related to inflammasome activation. In contrast, pro-IL-18 is constitutively expressed in sufficient levels and available for inflammasome processing leading to IL-18 Release within minutes of appropriate stimulation [26, 37]. Apparently while transcriptional priming seems the essential step for inflammasome activation, transcriptional upregulation of NLRP3 and pro-IL-1β may also amplify inflammasome activity.

4.1.2 Non-transcriptional Priming

Based on numerous evidences NLRP3 inflammasome priming might take place independent of new protein synthesis, either in canonical or non-canonical pathway. That is inhibition of transcription or translation may not totally block

Fig. 22.1 (continued) bacteria activate the TLR4–MyD88 and TRIF pathways, with subsequent IRF3–IRF7 complex formation. This elicits the expression of IFN-β that binds the IFNR. The consequent JAK/STAT activation leads to elevated transcripts of caspase-11 which is then cleaved and activated by the intracellular signals of LPS11 that in turn promotes the activation of NLRP3-ASC-caspase-1 pathway. Abbreviations: *NLRP3* nucleotide-binding oligomerization domain, leucine rich repeat and pyrin domain-containing protein 3, *TLRs* toll-like receptors, *MyD88* adaptor molecules myeloid differentiation primary response 88, *TNFR* tumor necrosis factor receptor, *NF-κB* nuclear factor-κB, *ATP* adenosine triphosphate, *ROS* reactive oxygen species, *TRIF* toll/IL-1 receptor homology (TIR)-domain-containing adapter-inducing interferon-β, *IRF* interferon regulatory factor, *IFN* interferon, *IFNAR* interferon-α/β receptor, *IL* interleukin, *JAK/STAT* janus kinase/signal transducers and activators of transcription, *ER* endoplasmic reticulum, *Mit* mitochondria

inflammasome priming and activation, suggesting an important regulatory role for post translational modifications in NLRP3 inflammasome. Supporting evidence highlighting post-translational modification, it has been shown IL-18 release takes place within minutes of monocyte stimulation, which is much faster than the discernible increase in NLRP3 protein abundance [26, 29, 38]. NLRP3 receptors are constitutionally ubiquitinated following expression providing a critical regulatory mechanism for their stability and function for an immediate switch between inflammatory and silence immune responses. In fact as far as ubiquitinated, NLRP3 reside inactive in cytoplasmic milieu not capable of self-oligomerization. Upon sufficient stimuli, the main deubiquitinase enzyme BRCC3 (BRCA1/BRCA2-containing complex, subunit 3) removes the ubiquitin leading to dissociation of HSP90 and SGT1 from NLRP3 which is critical for relieving from its auto-inhibition.

Such a regulation might be more complex regarding to the fact that multiple agents may control NLRP3 ubiquitination which takes place in both NLRP3 K48- and K63 regions. (1) As mentioned above, the DUB enzyme BRCC3 works as a prominent regulatory point [39], but does not work as the only check point. (2) Instantly, SCF complex subunit F-box L2 (FBXL2) may constitutively ubiquitinate NLRP3 for degradation at the proteasome. The interaction between FBXL2 and NLRP3 might be abolished by bacterial endotoxin preserving inflammasome activity [40], (3) E3 Ubiquitin-Protein ligase, MARCH7; which also mediates anti-inflammatory function of the neurotransmitter Dopamine, ubiquitinates NLRP3 and targets it for degradation by autophagy [41].

Few other post-transcriptional alterations have been also characterized for NLRP3 activation. Intracellular kinases and phosphatases play a key role in balanced phosphorylation forms in ASC and NLRP3 for appropriate interaction. Bruton's tyrosine kinase (BTK) has been recently identified as an essential component of the NLRP3 inflammasome activation in unprimed cells, in which BTK physically needs to interacts with ASC and NLRP3 and contribute to ASC phosphorylation at pyrin domain [42]. On the contrary side, in unprimed cells, NLRP3 is phosphorylated at S5, resulting in electrostatic repulsion between pyrin domain with that of ASC. Thus PP2A induces dephosphorylation licenses NLRP3 for activation and is required before inflammasome assembly [43].

4.2 Step 2: Activation; Canonical and Non-canonical

NRP3 inflammasome activation might be defined as NRP3 oligomerization rendering it capable of pro-Caspase 1 cleavage. Following essential priming, NLRP3 may recruit the adaptor protein ASC, which has a pivotal role in the activation of procaspase 1 following the formation of the multiprotein complex consisting of NLRP3, ASC, and pro-caspase-1 [44–46]. Nevertheless, an efficient activating stimulus is required to initiate the assembly and promote NLRP3 inflammasome oligomerization. In specific feature of non-canonical pathway binding of lipopolysaccharide (LPS) to inactive mouse caspase-11 (or human caspase 4-5) produced through the earlier priming, has been shown to activate the effector functions of caspase-11 to activate NLRP3 inflammasome oligomerization [47]. Nevertheless, variety of stimuli namely ATP, osmolarity or pH alterations may activate NLRP3 and ASC recruitment, might be still involved in NLRP3 activation by gram negative infections. Major classes of NLRP3 activators including several DAMPs and PAMPs end with a few alterations in intracellular territory namely potassium depletion, which in turn lead to eventual NLRP3 inflammasome-induced caspase-1 cleavage. Given most of the diverse stimuli converge on at least one of these downstream events, in the following section the discussion continues based on the type of the effectors relying downstream to one or more DAMPs or PAMPs which are addressed in this classified context.

4.2.1 Intracellular Potassium Depletion

Long before the discovery of inflammasomes, studies on LPS-primed macrophages confirmed IL-1β proteolytic processing and secretion depends on cytosolic K^+ depletion, as a common mechanism in response to different stimuli like ATP or nigericin [48, 49]. Conversely, increased extracellular K^+ concentrations was demonstrated to block NLRP3 inflammasome assembly in response to most of the identified NLRP3 triggers. Now known to be stimulated by a variety of initiating stimuli, K^+ efflux remains the best-characterized responsible stimulus for NLRP3 inflammasome activation [46]. However the exact mechanism leading to NLRP3 activation is not fully understood, recent findings has determined the Ser/Thr kinase involved in mitotic cell division, NEK7, is specifically required for NLRP3 inflammasome activation downstream to K^+ efflux, which is recruited to NLRP3 upstream of inflammasome formation [50, 51]. The following sub-sections brief the main NLRP3 activating stimuli converging on reducing the cytosolic level of K^+ ion, trough different mechanisms.

4.2.1.1 Extracellular ATP

Extracellular ATP which might increase following detrimental tissue damage and release of intracellular contents, has been established as an agonist of a ligand-gated cation channel called P2X7 receptor (P2X7R). Upon activation this channels opens and allows for an exchange of intracellular K^+ ions for extracellular Na^+ or Ca^{2+} ions. Producing a net K^+ efflux; this results in NLRP3 inflammasome activation. Activation of P2X7R may also follow opening of pannexin-1 channels, mediating caspase 11 (or human caspase 4-5) induced NLRP3 activation. That is following recognition of LPS caspase-11 cleaves intracellular domain of pannexin-1, the intramembranous receptors in monocytes, leading to opening of the channel and K^+ and ATP efflux into the extracellular space. Consequently, the leakage of K^+ ions activates the NLRP3 inflammasome and ATP acts as an agonist for the P2X7R to amplify the K^+ depletion and NLRP3 inflammasome activation. Intriguingly, it has been demonstrated the acquired levels of ATP through this process, are much lower (nanomolar concentrations) than the amounts of ATP typically proposed to activate P2X7R receptors [52].

4.2.1.2 K^+ Ionophores

K^+ ionophores have been shown to induce NLRP3 inflammasome activation when they produce a net K^+ efflux. A variety of cytotoxic agents like nigericin, gramicidin and valinomycin are amongst the well-studied K^+ ionophores capable of NLP3 inflammasome stimulation. Nigericin, is a lipophilic ionophore existing in a free membrane-impermeant anionic form or a neutral membrane-permeant complex. In its anionic form, nigericin binds to H^+ on the outside of the cell releasing the proton on the intracellular side, leading to acidification of cytosol. Then in intracellular side, nigericin anion binds to K^+ and releases it on the outside of the cell, leading to K^+ efflux. Valinomycin, also forms equimolar complexes with K^+, however; unlike to that of the neutral nigericin-K complexes, it forms a single positive charge complex (valinomycin-K^+). Therefore, valinomycin mediated K^+ efflux is electrogenic and thus is limited to the balanced gradients between both sides of cell membrane. Gramicidin as a peptide ionophore, allows for K^+ efflux balanced by Na^+ influx in a manner electrochemically similar to the P2X7R [53, 54].

4.2.1.3 Extracellular Hypo-tonicity

Extracellular Hypo-tonicity may result from pathological edema or any event disturbing massive ion hemostasis and is shown to contribute to NLRP3 inflammasome activation mainly through intracellular K^+ depletion. Two main mechanisms may underlie the following K^+ deregulation. Firstly extracellular tonicity leads to K^+ and Cl^- channels opening, driving an efflux of K^+ and Cl^- ions to balance the intracellular and extracellular osmolarity values. Secondly Na^+/K^+-ATPase pump dysfunction may lead to increased net influx of Na^+ ions which in turn will promote an osmotic movement of water through aquaporins into the cell. Besides diluting intracellular K^+ ions concentrations and NLRP3 inflammasome activation, a strong osmosis would be deteriorating enough to end with cell death [22, 45].

4.2.2 Intracellular Ca^{2+} Overload

Some of NLRP3 inflammasome activating stimuli seem to act through increased intracellular levels of Ca^{2+}. Extracellular Ca^{2+} may influx trough different plasma membrane-resident Ca^{2+} channels namely ASICs (Acid-sensing ion channels) [55, 56]. Importantly, acidosis as well as mitochondrial and endoplasmic reticulum (ER) stress are among the best characterized pathological features activating NLRP3 inflammasomes probably via Ca^{2+} overload. ER stress occurring following accumulation of unfolded proteins in ER compartment or disruption of the ER-Ca^{2+} homoeostasis, may lead to increased cytosolic Ca^{2+} concentration through phospho lipase C (PLC)/inositol triphosphate pathway i.e. following ATP stimulation. Mitochondria may also absorb or release Ca^{2+} under different conditions.

Pathological Ca^{2+} overload can lead to NLRP3 activation trough two different pathways: (1) it can increase mitochondrial ROS (mtROS) production, besides collapsing of the mitochondrial membrane potential [57], (2) Excessive intracellular Ca^{2+} also activate kinase TAK1 which has been shown to be involved in NLRP3 activation induced by lysosomal damage or hypotonic stimuli [45, 58, 59]. First demonstrated by Brough et al. Ca^{2+} chelator BAPTA-AM was shown to reduce IL-1β production in ATP-treated macrophages [60]. Later, it was discovered that both incubation with Ca^{2+}-free media and inhibition of intracellular ER Ca^{2+} stores by thapsigargin would prevent caspase-1 activation and IL-1β secretion by ATP, nigericin and alum [61], implying Ca^{2+} acts as a pivotal mediator for some NLRP3 inflammasome activators. However, there are still some uncertainties for significance of the hypothesis, in particular as in some studies; required concentrations of small-molecule compounds required to inhibit NLRP3 inflammasomes, significantly exceeds their IC50 values reported for other processes [62, 63].

4.2.3 Lysosomal Rupture and Pyroptosis

Insoluble particulates may manifest in extracellular milieu following exposure to environmental irritants or as a consequence of pathological tissue injury. This might include adjuvants like alum, the air pollutants such as silica, or the disease associated particles like monosodium urate and calcium pyrophosphate dehydrate are all well characterized as the NLRP3 inflammasome activators. It has been established that endocytosis of these crystalline particles triggers the disintegration of endolysosomal organelles which may lead to lysosomes rupture or pyroptosis. Based on evidences on lysosomal rupture theory, release of lysosomal contents into the cytosol, in particular for cathepsin B, could also trigger NLRP3 signaling [64–66]. In this line inhibitors of lysosomal cathepsins have been shown to block caspase-1 activation by all tested NLRP3 inflammasome inducers [67]. In the event that pyroptosis inducers rather than lysosomal deregulators, are engulfed by phagosomes, distinct cellular responses are believed to induce NLRP3 activation, involving catepsin-C instead of catepsin-1. In fact pyroptosis inducing agents are shown to induce more profound NLRP3 activation in line with caspase-B-mediated cell death, cascade of events starting with the release of proteolytic enzymes into the cytosol and leading to a loss of plasma membrane integrity and systemic immune response [68].

4.2.4 Reactive Oxygen Species

According to a pile of concrete evidences ROS is an essential signaling component required for NLRP3-inflammasome activation, besides affecting the prerequisite priming step. Xanthine oxidase, peroxisome oxidases, uncoupling of cytochrome p450 and nitric oxide synthases; in conjunction with NADPH oxidases and mitochondrial are among the multiple subcellular sources for ROS production. Of these, NADPH oxidases and mitochondrial ROS (mROS) are extensively studied and identified as sources of ROS involved in NLRP3-inflammasome activation [18, 30, 69]. Indeed empirical evidences indicating ROS is produced by a variety of NLRP3 inflammasomes activators, and works as a critical underlying mechanism in triggering NLRP3 inflammasome formation, suggest the hypothesis that ROS formation is a major mediator of a diversity of NLRP3 inflammasome activating DAMPs and PMPs [70]. Nevertheless NLRP3 is originally assumed as a cytosolic receptor with broad range of activating stimuli, it seems outlandish that NLRP3 acts as a receptor capable of directly binding to the effectors of all these remarkably different stimuli. Apparently ROS production offers a link to many different stimuli, including a variety of DAMPs and PAMPs stimulating canonical and non-canonical pathways to NLRP3 inflammasome activation [70].

In spite of the pile of evidences implicative of pivotal role of ROS in NLRP3 inflammasome's activation, little is known about the exact mechanisms by which NLRP3 senses oxidative stress. So far thioredoxin-interacting protein (TXNIP) and mitochondrial anti-viral signaling protein (MAVS) are characterized as to be associated with NLRP3 inflammasome stimulation by ROS. Demonstrated by Subramanian et al., MAVS is a mitochondrial adaptor protein to mediate the relocalization and association of NLRP3 inflammasome to mitochondria in stress conditions [71]. However, apparently MAVS mediating effect is specific to ATP and nigericin stimulation as well as few viral challenges [72].

TXNIP, the endogenous inhibitor of the antioxidant thioredoxin (Trx), may time dependently dissociate from Trx, bind to NLRP3 inflammasome to stimulate caspase-1 cleavage activity. Association between these two proteins was defined by the study of Zhou et al. as a necessary step in NLRP3 inflammasome activation in pancreatic islet cells in response to high glucose [73]. Ever since, several investigations have confirmed the necessity of TXNIP for NLRP3 inflammasome activation [74–76]. To confirm the specific role of TXNIP, shRNA transfection targeting TXNIP [77], has empirically prevented TXNIP-NLRP3 binding and subsequent homocysteine-induced glomerular injury. It might be of note, according to the existing evidences a cell-type specific manner exists for TXNIP, limiting its effects to mediate the proinflammatory effects of ROS signaling molecules, in particular cell types [78].

4.2.4.1 TXNIP and Thioredoxin System

The thioredoxin (Trx) is a thiol-reducing system consisting of NADPH (nicotinamide adenine dinucleotide phosphate-oxidase) and homodimeric seleno-protein thioredoxin reductase besides Trx [79]. There are two main critical roles for Trx system both implicated in the pathophysiology of diseases such as diabetes, arthritis and neurodegenerative disorders. First is maintaining the reducing environment which is critical to mask proteins disulfide to protect against reactive oxygen species. Next Trx exerts antiapoptotic effects which manifests trough binding to the pro-apoptotic protein apoptosis signal-regulating kinase 1 (ASK-1) and blocking its activity. TXNIP is a critical protein for regulating Trx function. As a stress sensor, TXNIP expression can be induced by neumerous exogenous detrimental stimuli, including metabolic stress and changes in calcium and oxygen levels [75, 80, 81]. Both expression and activity of Trx is strictly controlled by TXNIP which tightly limiting Trx's diverse roles in redox-dependent processes including protein folding, regulation of apoptosis, and protection from oxidative stress [79, 82]. TXNIP provides two discrete Trx related effects: (1) Affecting Trx redox dependent signaling by limiting the availability of free sulfhydryl (thiol) group of Trx [83, 84]. (2) Translocating Trx as a member of the alpha arrestin protein family, the later highlighting a major role of TXNIP as a scaffolding protein [85]. In terms of shuttling and targeting proteins into different subcellular compartments, under increased ROS levels TXNIP may translocate Trx to the plasma membrane [86], or may shuttle from the nucleus to bind the mitochondrial Trx and mediate the inhibition of glucose uptake in response to increased ROS [87, 88].

4.2.4.2 TXNIP and NLRP3-Inflammasome Activation

As a link between increased cellular ROS levels and the proinflammatory genes expression, TXNIP has attracted many investigations for developing promising therapeutics. Through a direct interaction with NLRP3 this protein seems to convey the message from a wide variety of stimuli to NLRP3 and subsequently to caspase and Interleukines activation. The phenomena first identified in 2010 by Zhou et al. in cultured macrophages, different NLRP3-inflammasome activators including ATP, monosodium urate crystals and silica were shown to enable TXNIP to dissociate from Trx and bind to the inflammasome receptor, NLRP3; following increases in cellular ROS levels. Using pharmacological and genetic tools they provided concrete evidences implying TXNIP binding is the essential prerequisite for in vitro and in vivo NLRP3 inflammasome activation [73]. Pursued by other scientists it was further understood, more diverse type of agents could engage TXNIP translocation subsequent to ROS production. High fat diet in experimental animals [89], high glucose-exposed human retinal microvasculature [90] and thrombin-exposed BV2 cells [91] are among the most recent models have been closely inspected for involvement of TXNIP/NLRP3. Taken together with other evidences on LPS, viral and pathological stimuli the obtained findings from these investigations conclude that TXNIP works as a wide road for deteriorating insults to reach out to NLRP3 inflammasome as the governing inflammasome in inflammation and apoptosis [92–94].

4.2.4.3 Modes of NLRP3 Activation Trough TXNIP

According to several studies ROS generating stimuli may augment TXNIP/NLRP3 through enhancing TXNIP transcription and expression. Besides several data about enhanced TXNIP protein expression following exposure to NLRP3 activators [95, 96], recent findings indicate that TXNIP gene transcript are induced simultaneously with NLRP3. As part of inflammasome priming, Feng et al. showed TXNIP transcripts also raises in parallel with NLRP3 mRNA when they exposed mesangial cells to LPS [97], leading to the due proteins over expression. Thereafter in the study of potential disease modifying agents, using carbon monoxide (CM) to reduce lung inflammatory response to LPS, Jiang et al. represented in vivo evidences for elevated TXNIP mRNAs, concurrent with that of NLRP3 and IL-1β following LPS exposure which was reversed by CM [98]. A recent work also demonstrated, berberine's anti-inflammatory effects on macrophages stimulated by monosodium urate crystal is associated with reversed transcript levels of TXNIP as well as NLRP3 inflammasome [99].

Enhanced shuttering activity and TXNIP translocation to the mitochondria milieu is the other well documented behavior of TXNIP following ROS generating agents. Early after their first demonstration, Zhou et al. found ROS induced activation of the NLRP3 inflammasome is associated with increased translocation of TXNIP into the mitochondria [100]. It is well confirmed by a recent work in which intracellular shuttling of TXNIP was assessed by immunofluorescent staining using MitoTracker Red, showing enhanced translocation to mitochondria in high glucose exposure which was efficiently attenuated by antioxidants [101].

4.2.4.4 TXNIP/NLRP3 Inhibition in Inflammation and Disease

Given that TXNIP/NLRP3 pathway have been shown to be functional in several cell types and organ systems, its pharmacological inhibition trough different specific or pleiotropic agents has been shown to contribute to significant ameliorating effects. Genetic deletion of TXNIP, as a precisely specific method has established TXNIP as an incomparable effector for NLRP3 stimulation. Validated TXNIP silencing by continuous delivery of TXNIP deoxy ribozyme (DNA zyme) in a period of 12 weeks in an animal model of renal injury [102], TXNIP targeted shRNA treatment in high glucose-induced NADPH oxidase activation [94], or siRNA interference against TXNIP in rat hepatic inflammatin [103] have demonstrated successful TXNIP ablation which has resulted in substantial prevention of NLRP3 formation and disease progress.

Additionally, pharmacological inhibition of TXNIP utilizing natural or pleiotropic agents has provided putative potential therapeutic targeting TXNIP/NLRP3 pathway. Instantly quercetin (a natural antioxidant of flavonoid origin), rutin (a flavonol quercetin) or ascorbic acid all provide ameliorating effect on disease model of high glucose/fructose stress [101, 104, 105]. Xanthohumol, a principal prenyl-flavonoid is among the most recently introduced agents with substantial efficacy in TXNIP/NLRP3 blockage in acute lung injury model [92].

4.2.4.5 TXNIP as an Experimental Target in Ischemic/Reperfusion Injury

Regarding the profound generation of ROS during ischemic/reperfusion injury, there is remarkable emerging interest toward investigating novel therapeutics targeting TXNIP. Cardiac NLRP3 inflammasome activation has been shown to be associated with over-production of ROS and TXNIP in MI dogs [106]. Conspicuously, co-over expression of TXNIP and NLRP3 in animal model of cerebral stroke has been shown to be suppressed by ameliorating agent like ruscogenin, an important steroid sapogenin [107]. The ameliorating effects of umbelliferone, a natural antioxidant belonging to coumarin derivatives [108] as well as curcumin, a natural polyphenolic compound [109] have been also ascribed to attenuation of TXNIP/NLRP3 pathway, while providing neuroprotective in experimental model of cerebral ischemia and ER-stress, respectively. In a recent effort to precisely address TXNIP role in NLRP3 activation in human stroke, we utilized mice embolic stroke model, and we found TXNIP knock down or pharmacological inhibition by resveratrol leads to substabstantial protection against stroke with close implication of NLRP3 and PPARs in conjunction with inflammatory cytokines IL-1β and TNF-α [110].

4.3 Endogenous Regulators of NLRP3

4.3.1 Autophagy

The substantial intracellular degradation system autophagy, plays a key role in inflammasome inactivation. The autophagosome surrounds a portion of the cytoplasm forming the autolysosome, which leads to degradation of the contents. Engulfing several cytoplasmic molecules, autophagy may result in broad effects on inflammasomes function as a multi-compartment assembly. Autophagy have been defined to probably block inflammasomes, by its degrading activity at least at three different levels (1) activating noncanonical pathway (2) priming ASC and pro-IL-1β, (3) mature IL-1β secretion [111]. In brief, autophagy deficiency may lead to accumulation of damaged bacteria residues and enhance activation of the noncanonical inflammasome. Also it has been shown that Pro-IL-1β and IL-1β can be degraded in autophagosomes, leading to decreased inflammatory responses [112]. Interestingly, it is also proposed that autophagy preferentially controls NLRP3 activation rather than others inflammasomes. For all these reasons it is pretty rationale that autophagy impairment may contributes to caspase-1 over-activation [113] as well as NLRP3 response to various infections [114]. Some endogenous agents like cannabinoids have been proved to contribute to modulate NLRP3 inflammasome activity via autophagy induction. Cannabinoid receptor 2 (CB2R) demonstrated as a therapeutic target in inflammation-related diseases, recently has been shown to induce autophagy which may explain why activation of the anti-inflammatory CB2R attenuates NLRP3 inflammasome in mouse BV2 microglia as well as in a mouse model of EAE [115, 116]. Thus CB2R agonists (i.e., HU-308) may provide effective therapy for treating NLRP3 inflammasome-related diseases. Some naturally occurring compounds also act as autophagy inducers, and would be discussed later.

4.3.2 cAMP Enhancers

There are concrete evidences implying cAMP may inhibit NRP3 inflammasome. It is believed to bind to NLRP3, and recruits the ubiquitin ligases which polyubiquitinates NLRP3 and prepares it for autophagosomal degradation. In this line it has been shown that pharmacological activators of adenylyl cyclases [117], or agonists of GPCRs enhancing adenylyl cyclase activity [117, 118], lead to a decrease in typical NLRP3 activation. Prostaglandin E2 (PGE2) having a broad range of effects, acts as a vasodilator and facilitates tissue influx of neutrophils [119] and macrophages [120] in early phases of inflammation, however; PGE2 also has many potent immunosuppressive properties [121, 122]. NLRP3 inflammasome activation is shown to be blocked by PGE2 in human macrophages, mediated through prostaglandin E receptor 4 (EP4) and an increase in intracellular cAMP, apparently independently of protein kinase A [123]. In a minor pathway cAMP enhancers may also improve protein kinase A (PKA) which directly phosphorylate the cytoplasmic receptor NLRP3 and attenuate its ATPase function [124]. Intriguingly, dopamine may also modulate NLRP3 activation specifically, which can be ascribed to D1 receptor to produce the second messenger cAMP which in turn enhances E3 ligase induced ubiquitination of NLRP3 inflammasome and the subsequent autophagy [41, 125].

4.3.3 Micro RNAs

Micro RNAs may provide another way for regulating inflammasomes activity. MicroRNAs are 22 nt non-coding RNAs that bind to the 3′ untranslated region (3′-UTR) of protein-coding mRNAs to regulate their translation [126]. On inflammasome priming and activation, micro RNAs may interfere in several levels. Instantly, mRNA-223 binds to a conserved site in the 3′ UTR of the NLRP3 transcript, suppressing NLRP3 expression and priming and thus IL-1β production [127, 128]. By targeting TXNIP, MicroRNA-20a negatively regulates expression of NLRP3-inflammasome [129]. Reportedly, several other microRNAs are involved in the activation of the NLRP3 inflammasome, namely microRNA-133a-1 [130], microRNA-155, and microRNA-377 [131]. Repressing the levels of these factors may be advantageous in inflammasome-related disease.

4.3.4 Type 1 Interferons

Type I IFN receptor (IFNAR) is a member of the TLR family with several downstream effector proteins like Janus kinases and several kind of signal transducers and activators of transcriptions (STATs). Type 1 interferons (IFNs), including IFN-α and IFN-β are non-specific inflammasome inhibitors, produced by specialized immune cells such as macrophages, microglia and astrocytes in response to extra cellular stimuli and irritants [132]. As a practical therapeutic tool these IFNs have been used in various auto-immune and auto-inflammatory diseases, namely multiple sclerosis and rheumatic diseases [133, 134]. However despite several investigations the underlying mechanism of inflammasome inhibition by type IFN 1 is largely unclear. A few plausible mechanisms have been suggested by Guarda et al. experiments on bone marrow-derived macrophages concluding that IFN-β may inhibit IL-1β production through at least two pathways [135]. First, NLRP1 and NLRP3 inflammasomes might be repressed following phosphorylation of STAT1 transcription. In the second alternate cascade, IL-10 levels increase in a STAT-dependent mechanism which as an immune response modulator reduces the levels of pro-IL-1α and pro-IL-1β. Conspicuously, while both IFN-α and IFN-β are non-specific inflammasome inhibitors, there are concrete evidences implicative of NLRP3 involvement in type 1 IFNs therapeutic advantages. That is IFN-β therapy might be effective in EAE mice only when the NLRP3 inflammasome contributes directly to the disease process [133].

5 Picture of Inflammasome Activation in Stroke

Based on several empirical investigations stroke induced insult has been discovered to be associated with inflammasomes activation. Based on early studies it was assumed that NLRP3 inflammasome is mainly expressed in immune cells. Nevertheless soon NLRP3, ASC, and caspase-1 expression were observed in microglia, similar to bone marrow-derived macrophages, but not in astrocytes [136]. Dietrich and co-workers in 2008 found remarkable association between expression of NLRP1, ASC, caspase-1 with IL-1β and IL-18 activation in in neurons, astrocytes, and microglia/macrophages after ischemic stroke in mice according to immunofluorescence and cellular localization experiments [137]. The co-localization of IL-1β within areas of early focal neuronal injury implies that it might be the major form of IL-1 contributing to inflammation following stroke [11]. The continuous research works by Fann et al. providing informative data in the subject of inflammasomes involvement in stroke, described an evident increase in levels of NLRP1 and NLRP3-inflammasome proteins, and IL-1β and IL-18 in stroke patient's brain in consistent with cellular and animal models of stroke [138]. This was not however in full consistency with Yang et al. findings implying that NLRP3 was expressed in microglia and endothelial cells but not in neurons in MCAO model in mice [10].

The ground breaking data from a cohort prospective 6-years long study, indicating patients carrying the NLRP3-Q705K minor allele are more likely to get affected by stroke/transient ischemic attack (TIA), suggests a link between NLRP3 inflammasome with stroke incidence [18]. This is well supported with later works implicative of a positive correlation between NLRP3 inflammasome levels with severity of coronary atherosclerosis [139]. Nevertheless NLRP3 inhibitors have been shown efficiently ameliorating stroke induced injury, there is little consistency among findings about NLRP3 expression profiles. Instantly the NLRC4 (NLR family, CARD domain containing 4) and AIM2 inflammasomes have been demonstrated to contribute to acute ischemic brain injury, without NLRP3 inflammasome involvement [140]. Apparently the controversy mainly concerns NLRP3, where gene expression does not always collate with the protein expression at the same time point [141] and might be more confined to microglial and endothelial cells as major resources of NLRP3 inflammasomes in regions vulnerable to ischemic insult [10]. The different ischemia models and interventions, in terms of duration and severity may explain such variations. Nonetheless, regional expression of NLRP3 inflammasomes might be litteraly different in the brain. Interestingly, there are evidences implying NLRP3, ASC, and caspase-1 were abundant in the wall of human cerebral aneurysms, highlighting NLRP3 inflammasome critical involvement in the development of cerebral vascular diseases like stroke [142]. Consistently further works indicated alcohol-induced accumulation and crystallization of cholesterol activates NLRP3/caspase-1 in the cerebral vessel leading to early development of atherosclerosis [143].

The precise mechanism through which NLRP3 contributes to neurovascular damage in ischemic stroke is yet to be concluded. Fundamentally it is believed NLRP3-mediated release of IL-1β may escalate brain microvessels' endothelial cell permeability and microglia-mediated neurotoxicity. IL-1β existing at low levels in the healthy brain modulating several physiological functions; may aggravate glutamate excitotoxicity and oxidative stress in pathological levels, leading to ROS generation, ER stress and TXNIP/NLRP3 inflammasome activation [142, 144].

6 Plausible Stimulus Involved in NLRP3 Receptor Activation in Cerebral Ischemia

While there are several potential factors posited as molecular and cellular stimuli for NLRP1 and NLRP3 receptor activation during cerebral ischemia, the exact mechanism is not clear. The comprehensive view on potential mechanism of inflammasomes activation has been reviewed elsewhere [145]. As of the main interest here, the principal potential mechanisms are illustrated in Fig. 22.2 and briefly

Fig. 22.2 (continued) and ER stress which in conjunction with the preexisting excitotoxicity dramatically enhances intracellular Ca^{2+} levels (**b**). Dring the eventual blood reperfusion massive oxygen flow to the injury site is much more than the capacity of living cells to controll oxidative mitochondrial phosphorylation. For the uncoupling oxidative phosphorylation the excess O_2 would turn to superoxide mitochondria. Similarly being a substrate for NADPH oxidase the excessive O_2 would be consumed for more ROS generation (**c**). Abbreviations: *GPR6CA* G protein-coupled receptor family C group 6 member A, *ROS* reactive oxygen species, *ECF* extracellular fluids, *DAG* diacylglycerol, *IP3* inositol triphosphate, *ATP* adenosine three phosphate

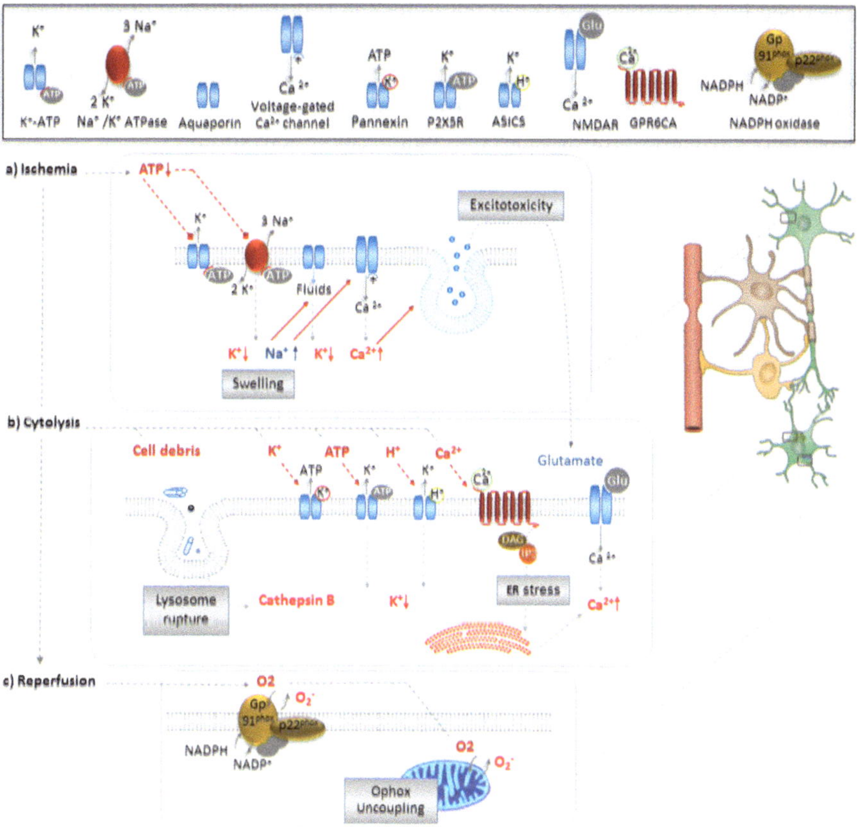

Fig. 22.2 Simplified illustration of plausible effectors involved in NLRP3 activation in acute ischemic stroke. The involved complex pathways may be described in terms of blood flow occlusion leading to ischemia (**a**) which is followed by ischemic cell death (**b**) and the spontaneous or therapeutically induced reperfusion (**c**). *Red colored letters* present the principal effectors may explain NLRP3 activation following stroke. Intracellular ATP deficiency as the major earliest manifestation results in the opening of K^+-ATP channels and leads to the channel opening and K^+ efflux. The falling ATP also leads to profound repression of Na^+/K^+ ATPase pump which in turn may trigger several NLRP3 activating pathways. First the disturbed ion exchange would result in an inward flow of ECF for the increased intracellular osmotic pressure. Besides the induced cellular edema, diluting all the cytosolic elements amplifies the intracellular K^+ depletion. Secondly, the impaired ion balance would result in a partial hyperpolarization and subsequent opening of voltage dependent Ca^{2+} channels. Ca^{2+} inward flow would instantly work to massive release of excitatory neurotransmitters affecting adjacent CNS cells (**a**). If the initial ischemic injury is deteriorating enough to end with CNS cells death the intracellular constitutional compounds and ions are released in to extracellular fluid. Non-functional particulate materials are engulfed by phagosomes primarily by glial cells leading to eventual lysosomal cathepsin B release. Cytolytic events also enriches ECF from K^+ which may activate pannexin channels exporting the intracellular ATP residues which together with the cytolytic ATP release leads to remarkable activation of P2X7 channels and massive K^+ efflux. The induced K^+ depletion is then exacerbated with ischemic induced acidosis either through lactic acidosis or cytolytic H^+ release, may activate ASICs channels permitting for more K^+ efflux. The elevated ECF Ca^{2+} levels would also activate G-protein/IP3/DAG coupled receptors

described in a classified manner in the following text. In the view of pathological start point, cerebral stroke initiates with sudden shortage of oxygen and glucose subsequent to blood flow occlusion. This result in three main specific features of stroke: ATP deficiency, cellular necrosis and reperfusion injury, each of which seemingly engage several cascades to inflammasome activation:

6.1 ATP Shortage

As the very early feature in ischemic insult, a dramatic fall in ATP generation follows the occlusive stroke. This may initiate inflammatory cascades involving inflammasome activation, at least through two interrelated pathway as follows.

6.1.1 Dysfunction of the Na^{+}/K^{+} ATPase

The NLRP1 and NLRP3 receptors can be activated by a decrease in K^{+} levels in the cytoplasm caused by dysfunction of the Na^{+}/K^{+} ATPase pump due to a decreased production of ATP. The increased influx of Na^{+} promotes the osmotic movement of water through aquaporins into the cell diluting the concentration of K^{+} in the cytoplasm; amplifying the initial K^{+} depletion subsequent to Na^{+}/K^{+} ATPase pumps dysfunction and leads to inflammasome activation. The inward movement of water through aquaporins into the cells may also end with sever cellular swelling rather than inflammasome activation.

6.1.2 Excitotoxicity

The partial hyperpolarization of cell membrane is an immediate consequent of Na^{+}/K^{+} ATPase pumps dysfunction directly affect voltage gated Ca^{2+} channels. The subsequent Ca^{2+} inward flow would instigate NLRP3 inflammasome activation. In the nervous system this intracellular Ca^{2+} overload would result in profound release of terminal glutamate vesicles leading to excitotoxicity. Trough postsynaptic receptors the excitatory neurotransmitter may augment the influx of Ca^{2+} to the cells and escalate inflammatory responses.

6.2 Necrotic Cells

Necrotic cells in the ischemic core releasing their content in surrounding space or secretions from metabolically active leukocytes may lead to enhanced extracellular concentrations of K^{+}, H^{+}, Ca^{2+}, ATP and particulate crystals all capable of stimulating NLRP3 inflammasome.

6.2.1 Acidosis

Extracellular H^+ binding onto ASIC1a on neurons and glial cells result in the influx of Ca^{2+}. Enhanced levels of Ca^{2+} ions in the intracellular environment, may activate the NLRP3 receptor through specific mechanisms. Reduction in oxygen availability also ends up with intracellular acidosis, as a result of accumulation of lactic acid following minimal anaerobic glycolysis; which may activate the NLRP3 receptor in synergy with a decreased intracellular K^+ concentration.

6.2.2 Ca^{2+} Increase

Extracellular Ca^{2+} may activate CaSRs (calcium-sensing receptor) and the orphan G-protein coupled receptor GPR6CA both leading to activate PLC/DAG/InsP3 pathway, which in turn ends with release of Ca^{2+} from endoplasmic reticulum to activate NLRP3 in different ways.

6.2.3 Extracellular ATP Increase

ATP elevation in extracellular fluid enhances the chance of binding to the plasma membrane P2X4 receptors on neurons, astrocytes or microglia, leading to K^+ efflux along with its associated K^+ decrease in the cytoplasm.

6.2.4 Extracellular K^+ Increase

P2X4R associated K^+ efflux amplifying the release of K^+ by necrotic cells, produces hyperkalemic extracellular environment and activate Pannexin 1 channels on the plasma membrane. As described before Pannexin 1 opening will lead to the release of more ATP, which can further activate more P2X4 creating a positive feedback loop for extracellular K^+ and ATP.

6.3 *Lysosomal Membrane Destabilization and Rupture*

As a consequence of release of damage-associated particulate materials and excessive phagocytosis, lysosomal permeation and cathepsin release is one of the expected stimuli to induce NLRP3 inflammasome in ischemic injury.

6.4 Reperfusion Injury

Reperfusion injury, the specific feature to temporary arterial occlusion, is well characterized with its associated superoxide generation. In fact blood reperfusion while rescues starving neurons works as a switch point for highly injured cells firing them by abundant ROS generation and extensive inflammatory responses. Reperfusion of the blood to the injury site brings massive oxygen flow much more than the capacity of living cells to controll oxidative mitochondrial phosphorylation and thus would be substrate for intracellular sources of ROS generation particularly mitochondria and NADPH oxidase. Accordingly oxidative stress works as one of the enormous events during ischemic injury, providing a particular pathway for ROS generation during cerebral ischemia which in turn translocates TXNIP to bind with the NLRP3 receptor leading to its activation.

7 Small Molecule NLRP3 Inhibitors and the Therapeutic Potential

Several NLRP3 inflammasome inhibitors have been designed, synthetized or discovered and evaluated in different pathological context in search for promising therapeutics. The growing number of the molecules might be classified based on different paradigms though, here we would summarize them according to their mode of action on different steps of inflammasome activation. Nevertheless given that many of these agents does not act merely trough one mechanism, it is important to note the provided classification is utterly based on the main defined mechanism concluded by recent findings which for very recent compounds are highly subject to further revisions with later investigations. Keeping in mind the main purpose is discerning on promising therapies for cerebral ischemia, in Table 22.1 a concise look has been provided to overview the reports about experimental therapeutics dealing with inflammasomes in cerebral ischemia.

7.1 Compounds Blocking NLRP3 Priming

7.1.1 Auranofin

Commonly used for the treatment of rheumatoid arthritis (RA), auranofin later was determined to antagonize NLRP3 activity at different levels. This compound suppresses pro-IL-1β mRNA expression and secretion by activated onocytes and macrophages in the synovial fluid that leads to the pathophysiological changes associated with RA [146, 147]. Auranofin has been also shown to suppress LPS-induced gene expression of NLRP3 and of IL-1 in macrophages [148], besides of other

Table 22.1 Evidences on inflammasome inhibition by experimental therapeutics against stroke

Inflammasome inhibitor	Characteristic	Disease model	Effect	Reference	Citation
In-house-made antibody	Neutralizing antibody against NLRP1	eMCAO	Reduced IL-1β/I8 levels	[137]	166
5Z-7-oxozeaenol	NF-κB inhibitor	eMCAO	Associated with a reduction in JNK and c-Jun signaling	[160]	16
		tMCAO	Improved Infarction and neurological scores	[161]	6
TAK-242	TLR4 antagonist	tMCAO	Inhibited Phosphorylation of downstream TLR4 signaling pathway, inflammatory cytokines Improved infarction and neurological scores	[259]	35
IU1	Specific USP14 inhibitor	tMCAO	Associated with reduced Protein aggregates and improved proteasome functionality Infarction and neurological scores	[164]	1
miR-124	DUB (Usp14) Silencing	tMCAO OGD (primary neurons)	Improved Resistance of neurons to in vitro ischemia, Infarction, neurological scores and neurovascular remodeling in vivo	[163]	62
ibrutinib (PCI-32765)	Bruton's tyrosine kinase inhibitors	tMCAO	Reduce IL-1β, IL-6, IL-23A And infiltrating microglia	[42]	45
Parthenolide	NF-κB inhibitor NLRP3 ATPase blocker	tMCAO	Reduced NF-κB, phosho-p38MAPK, and caspase-1 expressions improved BBB permeability and stroke outcomes	[169]	28

(continued)

Table 22.1 (continued)

Inflammasome inhibitor	Characteristic	Disease model	Effect	Reference	Citation
BAY 11-7082	NF-κB inhibitor NLRP3 ATPase inhibitor	OGD (pc12 cells)	Protects cells against ischemic injury	[172]	1
		OGD (Microglial cells)	Improved cells survival rate	[173]	
		tMCAO OGD (primary neurons)	Inhibits NF-κB and MAPK signaling, expression and activation of NLRP1 and NLRP3	[260]	2
Nicorandil	K^+_{ATP} Opener	OGD (BV-2 cells)	Reduced IL-1β, Caspase-1 and NLRP3	[195]	14
A438079	Selective P2X7R inhibitor	intracerebral hemorrhage	Inhibits RhoA activation	[192]	6
Brilliant Blue G	P2X7 receptor antagonist	OGD (brain section) (BV2 cells)	Protects microglial against OGD-Induced cell death	[191]	22
		tMCAO	Reduce caspase-3 dependent neuronal apoptosis	[193]	
Probenecid	Pannexin 1 inhibitor	OGD (Primary astrocytes)	Reduces NLRP3, caspase-1, and Aquporins	[261]	4
Ca-074ME	cathepsin B specific inhibitor	dMCAO	Reduced up-regulation of endosomes and apoptosis in peri-infarct neurons.	[204]	14
MCC950	ASC oligomerization inhibitor	tMCAO	Reduce caspase-3 dependent neuronal apoptosis	[193]	
		Platelet aggregation	Inhibits platelet activation/aggregation and in vitro thrombus formation	[222]	3

BHB	ASC oligomerization inhibitor K^+ efflux inhibitor	tMCAO	Upregulates HIF-1α	[234]	110
		MCAO electrocoagulation	Depends on hydroxy-carboxylic acid receptor 2	[235]	59
Apocynin	NOX inhibitor	tMCAO	Reduced levels of NOX2, NOX4 and ROS. inhibited degradation of IκBα, NF-κBp65 nuclear localization and the expression of its target gene (COX2 and iNOS), suppressed the expression of NLRP3, ASC, caspase-1, interleukin (IL)-1β and IL-18 in	[262]	
Ac-YVAD.CMK	Caspase-1 inhibitor	tMCAO OGD (primary neuron)	Reduced inflammasome proteins, IL-1β, IL-18, and cleaved caspase-1/3 in vivo/vitro	[138]	98
Resveratrol	polyphenolic compound (Natural)	eMCAO	Inhibits TXNIP Attenuate PARP activity caspase-1/3cleavage NLRP3 activation and IL-1β release	[110]	22
Paeoniflorin	bioactive monoterpene glucoside (Natural)	OGD (hippocampal slices)	Reduce NLRP1 and NLRP3 inflammasomes, as well as IL-18, IL-1β, and caspase-3.	[263]	3

(continued)

Table 22.1 (continued)

Inflammasome inhibitor	Characteristic	Disease model	Effect	Reference	Citation
Sinomenine	Alkaloid compound (Natural)	tMCAO OGD (astrocytes/microglia)	Associates with AMPK activation Attenuates astrocytic and microglial activation in vivo Inhibits Caspase-1/3, ASC, IL-1β and NLRP3 expression in vivo/vitro	[264]	1
Ruscogenin	steroid sapogenin (Natural)	tMCAO OGD (bEnd.3 cells)	Associates with MAPK suppression Inhibits Caspase-1, IL-1β, TXNIP and NLRP3 expression in vivo/vitro Reduced ROS generation in vitro	[107]	1
Umbelliferon	coumarin derivatives (Natural)	tMCAO	Associates with PPAR-γ upregulation inhibited TXNIP/NLRP3 as well as IL-1β-18	[108]	18
Curcumin	Polyphenolic compound (Natural)	tMCAO OGD (Mice hippocampus) Glutamate Toxicity (Mice hippocampus)	Associated with IRE1α and PERK suppression Reduced ROS generation via AMPK activation attenuated glutamate neurotoxicity, TXNIP/NLRP3/Caspase-1/IL-1β in vivo/vitro	[109]	42
Chrysophanol	Anthraquinone derivative (Natural)	tMCAO	Inhibits Caspase-1/3, ASC, IL-1β and NLRP3 expression	[265]	38

17β-estradiol progesterone	Steroid hormone (Natural)	tMCAO	Inflammasomes NLRC4, AIM2 and ASC, and decreased ASC and NLRP3 proteins and transcripts of IL-1β (IL1β), IL18 and TNFα	[141]	15
Telmisartan	Angiotensin receptor blocker Insulin sensitiver	tMCAO SH Rats	Reduces NLRP3, MMP-9, GFAP positive cells	[266]	11
Ruscogenin	Steroid sapogenin	tMCAO OGD (microvascular endothelial cells)	Reduced ROS and MAPK activity, IL-1β and caspase-1, NLRP3 and TXNIP expression	[107]	1
Minocycline	Antibiotic Immunosuppressive	tMCAO OGD (BV2 microglial cells)	Attenuated NLRP3 and cerebral edema and activation of microglia	[267]	1
Cordycepin	Adenosine derivative	Intracerebral hemorrhage	Suppresses of NLRP3 inflammasome activation alleviated neurological deficits, brain edema, and peri-hematomal tissue damage	[268]	

proinflammatory cytokines, namely IL-6 and TNF-α [146]. The underlying mechanism by which auranofin blocks NLRP3 inflammasome could be explained on different targets. Firstly it has been consistently shown to block IKK phosphorylation inhibiting IκB ubiquitinylation and proteasomal degradation which prevents nuclear translocation of free NF-κB [28, 149]. At transcriptional levels, auranofin also inhibits AP-1 and IRF3, as two other major factors important for cell signaling in non-canonical pathway. Conspicuously, auranofin also has been demonstrated to prevent the homodimerization of TLR4 which is required for NF-κB and IRF3 activation [150, 151]. Besides modulating TLR4-MyD88 pathway, the drug was discovered to interact with thioredoxin reductase, a redox enzyme responsible in controlling macrophage activation, to block LPS-induced pro-IL-1β and NLRP3 gene expression.

7.1.2 TAK-242, Bromoxone, and 5Z-7-Oxozeaenol

TAK-242 the specific TLR4 antagonist [152] as well as bromoxone and 5Z-7-oxozeaenol which inhibit nuclear translocation of NF-κB p65 [153] attenuate inflammasomes activity in priming step. These are all effective NLRP3 inflammasome inhibitors that suppress LPS+ATP-induced IL-1β release upstream of IKK. For TAK-242 it takes place through binding to the intracellular domain of human TLR4 leading to its irreversible modification [154, 155] and repressed processing of pro-inflammatory cytokines like IL-1, IL-6, IL-8, and TNF-α [156, 157]. Bromoxone is a non-specific irreversible inflammasome inhibitor (NLRP1, NLRP3, AIM2). However the target of bromoxone is not well known, there are evidences showing it resides upstream of IKK. Indeed, Gong and colleagues demonstrated that bromoxone abolishes NF-κB nuclear translocation without directly affecting NF-κB or by suppressing IKKβ kinase activity. That is while 5Z-7-Oxozeaenol is known to be a TAK1 inhibitor which specifically inhibits NLRP3 inflammasome activation in response to LPS, nigericin, and alum [158].

7.1.2.1 Experimental Effects on Stroke

TAK-242 has been shown to be able to cross blood-brain barrier, block TLR4 signaling and attenuate the expression of inflammatory cytokines in animal models of stroke [159], however little is known about the importance of plausible inhibition of inflammasomes priming. 5Z-7-oxozeaenol therapeutic potential has been also evaluate in animal models of cerebral ischemia. In early studies it was shown delayed treatment, even several hours after stroke onset, was beneficiary following either intraperitoneal or intracerebroventricular administration. The TAK1 inhibitor reduced infarct size and improved behavioral, apparently independent of AMPK activation while associated with a reduction in JNK and c-Jun signaling [160]. It was later confirmed that TNF Receptor Associated Factor 3 (TRAF3) detrimental effects through TAK1, could be blocked by 5Z-7-oxozeaenol and contributes to improved stroke outcomes [161].

7.1.3 DUB Inhibitors

While posttranslational regulation of NLRP3 is poorly characterized, inflammasome activation via deubiquitinylation discovered by Juliana and co-workers, is an established intriguing target to develop specific NLRP3 inhibitors. Among all identified deubiquitinating enzymes (DUBs), BRCC3 has been determined as the major DUB responsible for the specific deubiquitination of NLRP3 [39]. PR-619, WP1130 and eyarestatin-I are among the synthetized DUB inhibitors to be considered as promising therapeutic agents to combat diseases involving NLRP3 activation [29, 39, 162].

7.1.3.1 Experimental Effects on Stroke

Regarding the potential benefits in stroke therapy, there are few evidences about these small molecules. MicroRNAs (miRNAs) decreasing the expression of the deubiquitinating enzyme Usp14, has been shown to reduce neural injury either in cultured oxygen-glucose-deprived cortical neurons in vitro or in mice subjected to middle cerebral artery occlusion [163]. Such findings were confirmed by a recent remarkable work by Min et al. who demonstrated IU1 a specific small molecule inhibitor of USP14, attenuated ischemic stroke-caused neuronal injury, which was reflected by increased survival rate, reduced infarct volume, as well as decreased neuronal loss, while it was associated with reduced protein aggregates and enhanced proteasome functionality [164].

7.2 Compounds Blocking NBD ATPase Activity

7.2.1 Parthenolide and BAY 11-7082

The herbal sesquiterpene lactone parthenolide and BAY 11-7082 are structurally related compounds that inhibit the NF-κB pathway. However these compounds have been shown to selectively inhibit NLRP3 inflammasome activity at multiple levels independent of their inhibitory effect on NF-κB activity in macrophages [165]. Nevertheless, both compounds could also inhibit NF-κB-induced NLRP3 and pro-IL-1β gene expression. That is parthenolide directly inhibits IKK81 and NF-κB [166, 167], while BAY 11-7082 can inhibit IKK [168]. Basically, parthenolide and Bay 11-7082 inhibit ATP as well as nigericin induced NLRP3 inflammasomes with parthenolide may also directly target caspase-1.

7.2.1.1 Experimental Effects on Stroke

In the view of a potential pharmacological tool, stroke animals treated with parthenolide have been demonstrated to manifest dramatically improved neurological deficit and infarct volume, down-regulated phospho-p38 MAPK and caspase-1 expressions, and up-regulated claudin-5 expression in ischemic brain tissue. However the potential involvement of NLRP3 inhibition was not considered, NF-κB and caspase-1 modulation were identified as the potential underlying effectors for decreased stroke outcomes and BBB permeability [169]. Comparing to BAY 11-7082, the original parthenolide compound showing poor bioavailability is not a suitable compound for clinical trials and thus water soluble derivatives are being evaluated [170, 171]. This may somehow explain more recent investigations focused on BAY 11-7082 which permeate cell membranes relatively easily [165]. In a recent work by Sue et al. investigating Raf-1 kinase inhibitory protein (RKIP) neuroprotective effects against oxygen-glucose deprivation, BAY 11-7082 was shown to mimic RKIP ameliorating effects in PC12 cells, presumptively through NF-κB inhibition [172]. Further works in their lab confirmed BAY 11-7082 as a NF-κB inhibitor regulates microglial activity after oxygen-glucose deprivation (OGD) [173].

7.2.2 3,4-Methylenedioxy-β-Nitrostyrene, MNS

3,4-methylenedioxy-β-nitrostyrene, MNS was recently defined to inhibit NLRP3 ATPase activity required for ATP-dependent oligomerization of the NLRP3 inflammasome [174]. In early biological evaluation, MNS significantly inhibited NLRP3 inflammasome activation and inflammatory cytokine production in burn wounds attenuating neutrophil infiltration and accelerating wound healing [175]. As an NLRP3 inflammasome inhibitor MNS, was later shown to prevent burn sepsis by attenuating the inflammatory response in vital organs and alleviating subsequent inflammatory infiltrations [176].

7.3 Compounds Interfering with NLRP3 Activating Stimuli

7.3.1 P2X7R Antagonists

Design of P2X7R antagonists is of significant interest among medicinal chemists hence many of the due compounds have been empirically recognized as promising therapeutics. So that (AZD9056) [177], (CE-224,535) [178] and (GSK1482160) [179] are all of P2X7R blockers entering clinical trials. Nevertheless, given that there is large genetic variation in the P2X7R gene, a large variability in human responsiveness to the drugs is predictable [180, 181]. Basically by blocking activation of the P2X7R these antagonists prevent ATP-induced K^+ efflux in vitro [182] and in vivo [183, 184] attenuating NLRP3 inflammasome-mediated IL-1β maturation and release. In the absence of these blockers, extracellular ATP causes pannexin-1 recruitment to the plasma membrane which interacts with P2X7R leading to K^+ efflux [185–187] which is sufficient for NLRP3 inflammasome activation [46].

7.3.1.1 Experimental Effects on Stroke

P2X7R has been shown to play a predominant role in anoxic depolarization after stroke and neuroinflammation [188–190]. In this line, brilliant blue G (BBG), a potent P2X7R antagonist, has been shown to protect microglial OGD-induced cell death. BBG exposure in tissue slices from P2X7R null mice also confirmed deteriorating role of P2X7R [191]. Intriguingly P2X7R suppression was also shown to protects blood-brain barrier (BBB) after intracerebral hemorrhage (ICH) through inhibiting RhoA activation [192]. Given the emerged link between P2X7R and inflammasomes, involvement of P2X7R/NLRP3 pathway has been studied and remarkable expressions of P2X7R and NLRP3 inflammasome components has been reported following stroke. Treatment of stroke animals with P2X7R antagonist BBG or NLRP3 inhibitor (MCC950); both reducing the cerebral injury neurological impairment support the hypothesis that P2X7R/NLRP3 pathway plays a vital role in caspase-3 dependent neuronal apoptosis following stroke [193].

7.3.2 K^+_{ATP} Channel Openers

ATP-sensitive potassium (K^+_{ATP}) channels are membrane associated channels that are gated by adenosine and ATP. During resting state with high ATP/ADP ratio this channels are open and contribute to inward K^+ current. In stress and ATP deficiency conditions this channel go to closing status leading to less K^+ influx. Accordingly K^+_{ATP} channel openers are investigated as efficient tools to affect inflammasome activation in various disease models. Instantly iptakalim is a K^+_{ATP} channel opener that can readily cross the blood-brain barrier. It has shown to improve neuroinflammation and neurogenesis. The beneficiary effects of iptakalim are in parallel with attenuating microglial activity as well as NLRP3-inflammasome/caspase-1/interleukin 1β axis in the hippocampus [194]. Interestingly nicorandil another K^+_{ATP} channel opener, has been demonstrated to reverse OGD-induced IL-1β production, either interfering with TLR4 or glial activation in mice [195].

7.3.3 Cathepsin B Inhibitors

The lysosomal rupture and the cytosolic release of cysteine protease cathepsin B may at least partly, mediate particle induced NLRP3 inflammasome activation. The cathepsin B inhibitor, Ca-074Me significantly suppresses NLRP3 and NLRP1 inflammasome activation [196, 197] and may represent a promising tool to prevent excessive IL-1β release. The cathepsin inhibitor Ca074Me may affect multiple cathepsins for which it has been utilized in several biochemical and cellular examinations. That is while Ca074Me is cited as a cathepsin B-specific inhibitor in many studies and used to implicate cathepsin B in NLRP3 activation [64, 198, 199]. In fact Ca074Me, being a pro-drug methyl ester is processed in lysosomes into Ca-074, a highly cathepsin B-selective free acid. However, the very low rate of this processing provides time for Ca074Me to inhibit multiple cathepsins [200, 201]. K777 (N-methyl-piperazine-phenylalanyl-homophenylalanyl-vinylsulfone-phenyl), is a newly developed broad cathepsin inhibitor, affecting cathepsins B, L, S, C, V and K in cell-free assays [202]. Surprising recent data indicate that K777 or Ca074Me suppress IL-1β or TNF-α secretion similarly in both WT and cathepsin B-deficient PMs. Taken together with some previous evidences, these data imply the individual examined cathepsins including cathepsin B, are not sufficient for the activation of particle-induced NLRP3 inflammasome activation and IL-1β secretion [203]. However this does not rule out the potential effect of cathepsin B inhibitors, particularly regarding the recent reports on their protective effects. According to the findings of Zeng and co-workers, endosomes up-regulation following stroke is attenuated by Ca-074ME treatment in parallel with reduced apoptosis in peri-infarct neurons [204].

7.3.4 ROS Inhibitors

As emphasized in previous sections ROS play a key role in NLRP3 inflammasome activation. Almost all NLRP3 agonists induce and require ROS [205] [206]. Mitochondrial and lysosomal NADPH oxidase appears to be a good candidate for the source of ROS which can increase enormously in stress conditions [207]. The first due report in 2008 demonstrated specific knockdown of NADPH oxidase subunit p22phox as well as exposure of general ROS scavengers such as N-acetylcysteine and antioxidant ammonium pyrrolidine dithiocarbamate in THP-1 cells ends with less caspase-1 activation and IL-1b release in response to particulate stimuli like asbestos and silica [207]. The knowledge on NLRP3 involvement was later provided by Latz and co-workers in 2013, identifying NADPH oxidase (NOX) contribute to intracellular ROS generation which in turn is responsible for NLRP3 activation in response to particulate matter [208]. In line with this, several NOX inhibitors has been shown to attenuate IL-1β and/or NLRP3 inflammasome activation. Instantly NADPH oxidase inhibition either with diphenylene iodonium (DPI); a broad spectrum NADPH oxidase inhibitor, or using NOX2ds-tat blocking gp91phox subunit of NOX which is responsible for one-electron reduction of oxygen and superoxide generation; can prevent NLRP3 inflammasome activation [209–211]. Interestingly, mtROS seems to play a critical role not only in NLRP3 inflammasome activation but also in the prerequisite priming process. Studies on mouse macrophages have provided evidences indicating that TLR4 with Myd88 can prime NLRP3 through its deubiquitination. Utilizing antioxidants may efficiently block many mtROS-dependent processes [29].

7.4 Compounds Blocking ASC Oligomerization

The cytokine release inhibitory drugs (CRIDs) 1 and 2 developed by Gabel and co-workers are sort of diaryl-sufonylurea analogues showing irreversible inhibitory activity against LPS plus ATP and hypotonic stress-induced IL-1β maturation in human monocytes in vitro [212]. Continued work by Coll et al. demonstrated the related new compound MCC950. However CRIDs 1 and 2 and in particular MCC950 are all promising candidates to further develop novel therapeutics for NLRP3-dependent pathologies, no definite mechanism has been yet identified. It has been suggested that nucleophilic attack on NLRP3 cysteine residues on epoxide functional groups might explain the inhibitory effects [213]. The ketone metabolite β-hydroxybutyrate and glybenclamid the anti-diabetic sulfonylurea compound were later discovered to act as ASC oligomerization inhibitor for which yet little has been discovered as the main underlying mechanisms.

7.4.1 MCC950

MCC950 (also named as CP-456,773) was recently introduced by Coll et al. [214] as a selective NLRP3 inhibitor, blocking both canonical and non-canonical activation of the NLRP3 inflammasome. In their cutting-edge study MCC950 was shown to efficiently reduce NLRP3-induced ASC oligomerization, associated with remarkable IL-1β secretion probably through inhibiting caspase-1-dependent pathway in mouse and human macrophages. Nevertheless the exact mechanism of the compound was not ultimately specified in their experiments, several critical mechanisms were considered. Accordingly the ATP-sensitive K^+ channels as well as ATP-induced Ca^{2+} flux both of the main requirement for activation of NLRP3, was ruled out to be involved in MCC950 pharmacodynamics. NLRP3 trimerization as the requisite for efficient IL-1β production [215], was not also disturbed by MCC950. The drugs pharmacokinetic studies showed it is relatively stable following incubation with human or mouse liver microsomes, with a half-life of 3.27 h and oral bioavailability of 68% in in vivo experiments in mice. Emerging efforts to unravel the full characteristics of the drug are now on and one of the most recent works gives first insight into the SAR of MCC950 highlighting hexahydroindacene moiety as key part of pharmacophore [216].

In the earliest biological evaluations of MCC950 by Coll and colleagues, they opt for central nervous system disorders. EAE is a multiple sclerosis pathological model with typical T cell-mediated inflammation and demyelination in which IL-1 signaling and NLRP3 have been recently shown to play crucial roles. To confirm MCC950 activity in vivo, Coll and colleagues first indicated MCC950 pretreatment significantly abolished serum levels of IL-1β and IL-6 in LPS treated mice. Then in EAE mice treated with MCC950, they showed the severity of EAE was reduced and the onset was delayed [214]. The emerging interests on MCC950 as a therapeutic tool in other CNS disorders was then pursued by other studies on Alzheimers' disease, intracerebral hemorrhage and stroke. According to such investigation it was shown MCC950 could efficiently improve cognitive function in in mice through attenuating Amyloid-β (Aβ) induced NLRP3 inflammasome activation in mice [140, 217], preventing IL-1β release from microglia and at the same time, promoting Aβ phagocytosis. In BV2 cells, the murine microglia-like cell line, MCC950 treatment was also shown to attenuate thrombin-induced cell apoptosis and expression of apoptotic proteins [91]. Aside central territory, NLRP3 inflammasome is a multi-protein complex involved in instigating inflammation in peripheral nervous system and thus MCC950 could attenuate inflammatory hyperalgesia [218].

7.4.1.1 Experimental Effects on Stroke

Sterile inflammatory responses have been long shown to be implicated in the development of myocardial ischemia, with IL-1β as an early prominent mediator [219, 220]. Recent evidence have also indicated the NLRP1 and NLRP3 inflammasome are involved in neurovascular diseases namely neurodegenerative disorders and stroke [221].

Consistently, administration of MCC950 has been shown to significantly attenuate the expression of NLRP3 components, and cleaved caspase-3 as well as neuronal apoptosis and brain infarction volume following stroke [193], with the optimal effect on 3 days after stroke. The effects of the drug have been also evaluated in thrombotic events closely associated with stroke incidence and injury. In ICH induced injury as a common event following severe stroke, MCC950 may provide protection. Interestingly, in the view of thrombus formation and incidence of occlusive strokes, it has been shown both platelet activation/aggregation and in vitro thrombus formation could be substantially inhibited by MCC950 [222]. In this line, Bruton's tyrosine kinase governing production of bioactive IL-1β and other proinflammatory cytokines; has been defined as the regulator of these NLRP3-dependent platelet effects.

7.4.2 β-Hydroxybutirate

β-Hydroxybutyrate (BHB), a ketone metabolite, has been long appeared to possess several intriguing aspects making it a molecule of interest for developing new therapies. The advantages of BHB to the body can be presumptively explained through diverse pathways. Basically, BHB could be consumed as an efficient alternative energy source by the vital organs like the heart and brain during high-intensity exercise or caloric deficiency [223, 224]. In this connection it can reduce production of mitochondrial reactive oxygen species by changing the $NAD^+/NADH$ ratio, besides upregulation of genes involved in protection against oxidative stress via blocking histone deacetylases. Conspicuously, as a signaling molecule BHB may also influence opening of K^+ channels and regulation of Ca^{2+} channels resulting in drastic impact on many cellular functions. Indeed, the characteristic blockage of K^+ efflux from macrophages ascribes the alleviating impact of BHB on immune responses, specifically in terms of inflammasome modulation. Comparable to MCC950, BHB also appears to block inflammasome activation by inhibiting NLRP3-induced ASC oligomerization, while BHB affects only canonical activation [225]. In fact, Youm et al. [226] revealed that the ketone metabolite BHB reduce NLRP3 inflammasome activity as well as IL-1β and IL-18 production in human monocytes. According to their in vivo experiments, BHB as well as ketogenic diet alleviate caspase-1-mediated IL-1β production and secretion without affecting the NLRC4 or AIM2 inflammasomes. To investigate about the pharmacodynamics of BHB in specific NLRP3 inhibition, they concluded BHB inhibits NLRP3 by preventing K^+ efflux and reducing ASC oligomerization and speck formation, independently of uncoupling protein-2 (UCP2) and classical starvation regulated mechanisms like AMPK and sirtuins, with no correlation with magnitude of histone acetylation in macrophages [226]. However involvement of AMPK is still under debate, as instantly in a recent study by Bae et al. in hepatoma HepG2 cells, AMPK activation was shown to be required for NLRP3 inhibition by BHB [227].

The use of BHB as a therapeutic is currently limited though, metabolism of ketone bodies including BHB has been demonstrated in full details elsewhere [224]. Low-carbohydrate, high fat ketogenic diet is a more widely used approach to alleviate disorders like seizures [228]. BHB as a monocarboxylate passes BBB through monocarboxylate transporters. Members of the SLC16 family, MCT1 (SLC16A1) and MCT2 (SLC16A7) are respectively low affinity transporters mainly in BBB and high affinity transporters mostly located in neurons [229, 230]. After entry to the CNS, all cell types use BHB in respiration which in neurons and oligodendrocytes is three times more efficient than that in astrocytes [231].

7.4.2.1 Experimental Effects on Stroke

Upon the described beneficiary effects of BHB, probably there might be a benefit to the brain being in a mildly ketotic body [232]. BHB consumption may meet baseline energy needs in the brain but may not support synaptic activity during excitotoxicity. In this connection higher on-admission BHB values have been shown to be associated with poorer stroke outcomes in stroke patients [233]. In early studies, neuroprotective properties of BHB in ischemic brain injury were primarily ascribed to the upregulation of hypoxia inducible factor (HIF)-1alpha, through a fourfold elevation of intracellular succinate a known inhibitor of HIF-1α degradation [234]. In an outstanding investigation of immune responses to ketogenic diet in stroke animals, the hydroxy-carboxylic acid receptor 2 (HCA2) on monocytes was shown to be required for the neuroprotective effect of BHB, inducing a neuroprotective phenotype of monocytes and/or macrophages [235]. These novel understanding was confirmed by another study confirming HCA2 role in neuroinflammatory diseases like multiple sclerosis [236].

7.4.3 Glyburide

Basically described as an anti-diabetes mellitus type II drug, glyburide stimulates insulin release by blocking ATP-sensitive potassium (K_{ATP}) channels in pancreatic β cells and islet chloride channels [237, 238]. Later it was shown by Gabel and colleagues that the drug may also inhibit IL-1β release in LPS-activated human monocytes [239], the findings predating the discovery of inflammasomes [240]. According to the primary SAR studies both the benzamido and sulfonyl groups of the compound contribute to inhibition of LPS+ATP-stimulated caspase-1 activation and IL-1β secretion [241]. As IL-1β secretion was later shown not significantly affected in the presence of the NLRC4 or AIM2 activators [242], glyburide was later recognized as a specific inhibitor of the NLRP3 inflammasome [241]. However the effect was confined to concentrations much higher than what can be clinically achievable. Given that K^+ efflux is required for NLRP3 inflammasome activation and K^+ channels are the main target of glyburide, [243, 244] obstruction of ATP-dependent K^+ efflux was of early assumed mechanisms for NLRP3 inhibition [245]. However, a later study showed that inhibitory activity of glyburide was maintained even when K^+_{ATP} channel subunits $Kir_{6.1}$, $Kir_{6.2}$, and SUR2 were knocked out, implying that K^+_{ATP} channels are dispensable for the effect [61]. Interestingly, not all sulfonylurea drugs inhibit inflammasome activation [241], neither sulfonylurea compounds may prevent K^+ efflux caused by NLRP3 activators [214], collectively concluding these inhibitors act in downstream of K^+ depletion and may prevent ASC oligomerization. This might occur during or upstream to NLRP3 inflammasome assembly.

7.4.3.1 Experimental Effects on Stroke

Glibenclamide (glyburide) as a NLRP3 oligomerization inhibitor, has received renewed attention in the last decade for its pleiotropic protective effects in acute CNS injury. According to several clinical studies, glybseverideis is effective in preventing edema and improving outcome after focal ischemia [234–236]. Instantly glyburide has been shown to be associated with T2 fluid-attenuated inversion recovery signal intensity ratio on brain MRI, diminished tissue water and reduced blood MMP-9 level [237]. In preclinical studies, even a very low dosage of glyburide has been shown to improve functional outcomes following stroke besides improving NeuN-positive neurons in the cortex and hippocampus, and enhanced angiogenesis in the hippocampus [238]. Corroborated with evidences implicative of trivial role of blocking K^+_{ATP} channels by glyburide in the CNS, glibenclamide seems to exert its effects primarily via inhibition of the characterized Sur1-Trpm4 (sulfonylurea receptor 1-transient receptor potential melastatin 4) channel. Nevertheless, blockade of K^+_{ATP} channels have been determined to be required for neuroprotective role of microglia in the early stages of stroke [239] as well as glyburide neuroprotection against stroke induces oxidative stress and inflammatory responses [240, 241]. In fact, there are pile of concrete evidences demonstrating glibenclamide ameliorating effects in rodent models of various CNS pathologies, importantly ischemic [242–244] and hemorrhagic stroke [245, 246], is mediated through Sur1 inhibition while there is no examination on potential implication of inflammasomes.

7.5 Compounds Blocking the Activity of NLRP3 Assembly

In 2013 Liu et al. developed Fc11a-2 a novel benzoimidazole compound, which could specifically inhibit NLRP3. Mechanistically it interferes with the proximity-induced autocleavage of procaspase-1 in NLRP3 inflammasome complex which leads to less release of activated caspase-1 [246]. Directly targeting NLRP3 complex, Fc11a-2 was shown to be highly effective in suppressing IL-1β/18 production. This promising results encouraged them to work on more benzoimidazole derivatives of which their recent publication introduced TBZ-09 and TBZ-21 as new lead compounds with comparable efficacies with Fc11a-2 [247]. All their suggested molecules provided remarkable ameliorating effects in dextran sulfate sodium (DSS)-induced experimental colitis in mice beside significant reduction in IL-1β/18 production in LPS+ATP-stimulated THP-1 cells. These encouraging effects warrant further studies to evaluate efficacy in other NLRP3 inflammasome relevant disease models. In this view caspase 1 inhibitors may be somehow regarded as NLRP3 inhibitors.

8 Naturally Occurring Inhibitors

Plant-derived natural compounds have been long considered in traditional medicine as complementary supplements in the treatment of chronic inflammatory diseases or included in preventive approaches. However according to the complexity of the effects and lack of the detailed mechanistic explanation little could be implemented as a pharmacological tool, still they provide unique promising lead compound to investigate for appropriate therapeutics. Interestingly, more than 8000 natural polyphenols have been described [248] as potential inflammasome inhibitors. Many of them existing in the form of polyphenols like resveratrol, curcumin, arglabin, EGCG, and quercetin are potent inhibitors, typically acting at more than one element in NLRP3 inflammasome regulating pathways. The comprehensive study of such naturally occurring inflammasme inhibitors have been reviewed elsewhere [249]. The exemplary well known compound, resveratrol, provides a concise instance acting at multiple levels reducing expression of NLRP3, ASC and P2X7R [250]; increasing cAMP content [251]; induce autophagy and AMPK stimulation [252]. However, as they influence a variety of pathways, it should be noted that these polyphenols have a broad biological effect all may not be explained upon inflammasome inactivation [253]. In fact, polyphenols are known to provide antioxidant effect, contribute to activate glutathione S transferases, and inhibit COX enzymes [254, 255]. To explain the complex effects of these kinds of compounds all together in a unique feature, an interesting theory termed as "Xenohormesis" have been posited. Based on this theory, many of these polyphenols like resveratrol and curcumin are produced in stressed plants, triggering effectors of caloric restriction. Receiving such signals by consuming these compounds, might be interpreted as a potential risk of future limitation of food availability and augmentation of adapting and protecting signals [256, 257]. Nevertheless the theory needs to be validated by empirical evidences, there are some backing evidences. Interestingly, resveratrol and curcumin which are structurally unrelated, share this caloric restriction mimicking feature by activating AMPK and Sirt1 [258]. Moreover aspirin as of the most frequently used COX inhibitors, is a derivative of another stress-induced phytochemical, salicylic acid.

9 Summary and Future Directions

In summary, in light of above discussed studies, this chapter lends further credit for the central role of NLRP3-inflammasome that is a vital player in both acute and chronic diseases of the brain including stroke. These findings further suggest a potential clinical benefit of therapeutic interventions that target NLRP3 inflammasome assembly and activity. On the other hand, TXNIP-NLRP3 inflammasome activation plays a major role in mediating the pro-inflammatory response involved in the pathophysiology of the stroke. Taken together, NLPP3 represents a logical therapeutic targets for stroke, but it have never been clinically evaluated due to a lack of ideal drug candidates. Recent innovations in NLP3 inhibition and drug delivery highlight new potentials for stroke therapy. Further investigation into the mechanisms of NLRP3 components as well as pre-clinical and clinical studies of targeted inhibitors are needed for development of novel drug for stroke therapy.

Acknowledgements This chapter was supported by R01NS097800-01 to Tauheed Ishrat.
Disclosure of conflict of interest: None

References

1. Benjamin EJ, Blaha MJ, Chiuve SE, Cushman M, Das SR, Deo R, et al. Heart disease and stroke statistics—2017 update: a report from the American Heart Association. Circulation. 2017;135(10):e146–603.
2. Corbyn Z. A growing global burden. Nature. 2014;510(7506):S2.
3. Barrington J, Lemarchand E, Allan SM. A brain in flame; do inflammasomes and pyroptosis influence stroke pathology? Brain Pathol. 2017;27:205.
4. Iadecola C, Anrather J. The immunology of stroke: from mechanisms to translation. Nat Med. 2011;17(7):796–808.
5. Macrez R, Ali C, Toutirais O, Le Mauff B, Defer G, Dirnagl U, et al. Stroke and the immune system: from pathophysiology to new therapeutic strategies. Lancet Neurol. 2011;10(5):471–80.

6. Gelderblom M, Leypoldt F, Steinbach K, Behrens D, Choe C-U, Siler DA, et al. Temporal and spatial dynamics of cerebral immune cell accumulation in stroke. Stroke. 2009;40(5):1849–57.
7. Lakhan SE, Kirchgessner A, Hofer M. Inflammatory mechanisms in ischemic stroke: therapeutic approaches. J Transl Med. 2009;7(1):97.
8. Lénárt N, Brough D, Dénes Á. Inflammasomes link vascular disease with neuroinflammation and brain disorders. J Cereb Blood Flow Metab. 2016;36(10):1668–85.
9. Deroide N, Li X, Lerouet D, Van Vré E, Baker L, Harrison J, et al. MFGE8 inhibits inflammasome-induced IL-1β production and limits postischemic cerebral injury. J Clin Invest. 2013;123(3):1176–81.
10. Yang F, Wang Z, Wei X, Han H, Meng X, Zhang Y, et al. NLRP3 deficiency ameliorates neurovascular damage in experimental ischemic stroke. J Cereb Blood Flow Metab. 2014;34(4):660–7.
11. Luheshi NM, Kovács KJ, Lopez-Castejon G, Brough D, Denes A. Interleukin-1α expression precedes IL-1β after ischemic brain injury and is localised to areas of focal neuronal loss and penumbral tissues. J Neuroinflammation. 2011;8(1):186.
12. Vezzani A, Maroso M, Balosso S, Sanchez M-A, Bartfai T. IL-1 receptor/Toll-like receptor signaling in infection, inflammation, stress and neurodegeneration couples hyperexcitability and seizures. Brain Behav Immun. 2011;25(7):1281–9.
13. Chen GY, Nuñez G. Sterile inflammation: sensing and reacting to damage. Nat Rev Immunol. 2010;10(12):826–37.
14. Takeuchi O, Akira S. Pattern recognition receptors and inflammation. Cell. 2010;140(6):805–20.
15. Cassel SL, Sutterwala FS. Sterile inflammatory responses mediated by the NLRP3 inflammasome. Eur J Immunol. 2010;40(3):607–11.
16. Suresh R, Mosser DM. Pattern recognition receptors in innate immunity, host defense, and immunopathology. Adv Physiol Educ. 2013;37(4):284–91.
17. Davis BK, Wen H, Ting JP. The inflammasome NLRs in immunity, inflammation, and associated diseases. Annu Rev Immunol. 2011;29:707–35.
18. Gross O, Thomas CJ, Guarda G, Tschopp J. The inflammasome: an integrated view. Immunol Rev. 2011;243(1):136–51.
19. Tsuchiya K, Hara H. The inflammasome and its regulation. Crit Rev Immunol. 2014;34(1):41–80.
20. Akira S, Uematsu S, Takeuchi O. Pathogen recognition and innate immunity. Cell. 2006;124(4):783–801.
21. Man SM, Kanneganti TD. Regulation of inflammasome activation. Immunol Rev. 2015;265(1):6–21.
22. Próchnicki T, Mangan MS, Latz E. Recent insights into the molecular mechanisms of the NLRP3 inflammasome activation. F1000Res. 2016;5:F1000 Faculty Rev-1469.
23. Strowig T, Henao-Mejia J, Elinav E, Flavell R. Inflammasomes in health and disease. Nature. 2012;481(7381):278–86.
24. Schroder K, Zhou R, Tschopp J. The NLRP3 inflammasome: a sensor for metabolic danger? Science. 2010;327(5963):296–300.
25. Lamkanfi M, Dixit VM. Inflammasomes and their roles in health and disease. Annu Rev Cell Dev Biol. 2012;28:137–61.
26. Ghonime MG, Shamaa OR, Das S, Eldomany RA, Fernandes-Alnemri T, Alnemri ES, et al. Inflammasome priming by lipopolysaccharide is dependent upon ERK signaling and proteasome function. J Immunol. 2014;192(8):3881–8.
27. Embry CA, Franchi L, Nuñez G, Mitchell TC. Mechanism of impaired NLRP3 inflammasome priming by monophosphoryl lipid A. Sci Signal. 2011;4(171):ra28.
28. Bauernfeind FG, Horvath G, Stutz A, Alnemri ES, MacDonald K, Speert D, et al. Cutting edge: NF-κB activating pattern recognition and cytokine receptors license NLRP3 inflammasome activation by regulating NLRP3 expression. J Immunol. 2009;183(2):787–91.
29. Juliana C, Fernandes-Alnemri T, Kang S, Farias A, Qin F, Alnemri ES. Non-transcriptional priming and deubiquitination regulate NLRP3 inflammasome activation. J Biol Chem. 2012;287(43):36617–22.

30. Jin C, Flavell RA. Molecular mechanism of NLRP3 inflammasome activation. J Clin Immunol. 2010;30(5):628–31.
31. Wen H, Miao EA, Ting JP-Y. Mechanisms of NOD-like receptor-associated inflammasome activation. Immunity. 2013;39(3):432–41.
32. Sander LE, Davis MJ, Boekschoten MV, Amsen D, Dascher CC, Ryffel B, et al. Detection of prokaryotic mRNA signifies microbial viability and promotes immunity. Nature. 2011;474(7351):385–9.
33. Gurung P, Malireddi RS, Anand PK, Demon D, Walle LV, Liu Z, et al. Toll or interleukin-1 receptor (TIR) domain-containing adaptor inducing interferon-β (TRIF)-mediated caspase-11 protease production integrates Toll-like receptor 4 (TLR4) protein-and Nlrp3 inflammasome-mediated host defense against enteropathogens. J Biol Chem. 2012;287(41):34474–83.
34. Aachoui Y, Sagulenko V, Miao EA, Stacey KJ. Inflammasome-mediated pyroptotic and apoptotic cell death, and defense against infection. Curr Opin Microbiol. 2013;16(3):319–26.
35. Broz P, Ruby T, Belhocine K, Bouley DM, Kayagaki N, Dixit VM, et al. Caspase-11 increases susceptibility to Salmonella infection in the absence of caspase-1. Nature. 2012;490(7419):288–91.
36. Rathinam VA, Vanaja SK, Waggoner L, Sokolovska A, Becker C, Stuart LM, et al. TRIF licenses caspase-11-dependent NLRP3 inflammasome activation by gram-negative bacteria. Cell. 2012;150(3):606–19.
37. Schroder K, Sagulenko V, Zamoshnikova A, Richards AA, Cridland JA, Irvine KM, et al. Acute lipopolysaccharide priming boosts inflammasome activation independently of inflammasome sensor induction. Immunobiology. 2012;217(12):1325–9.
38. Fernandes-Alnemri T, Kang S, Anderson C, Sagara J, Fitzgerald KA, Alnemri ES. Cutting edge: TLR signaling licenses IRAK1 for rapid activation of the NLRP3 inflammasome. J Immunol. 2013;191(8):3995–9.
39. Py BF, Kim M-S, Vakifahmetoglu-Norberg H, Yuan J. Deubiquitination of NLRP3 by BRCC3 critically regulates inflammasome activity. Mol Cell. 2013;49(2):331–8.
40. Han S, Lear TB, Jerome JA, Rajbhandari S, Snavely CA, Gulick DL, et al. Lipopolysaccharide primes the NALP3 inflammasome by inhibiting its ubiquitination and degradation mediated by the SCFFBXL2 E3 ligase. J Biol Chem. 2015;290(29):18124–33.
41. Yan Y, Jiang W, Liu L, Wang X, Ding C, Tian Z, et al. Dopamine controls systemic inflammation through inhibition of NLRP3 inflammasome. Cell. 2015;160(1):62–73.
42. Ito M, Shichita T, Okada M, Komine R, Noguchi Y, Yoshimura A, et al. Bruton's tyrosine kinase is essential for NLRP3 inflammasome activation and contributes to ischaemic brain injury. Nat Commun. 2015;6:7360.
43. Stutz A, Kolbe C-C, Stahl R, Horvath GL, Franklin BS, van Ray O, et al. NLRP3 inflammasome assembly is regulated by phosphorylation of the pyrin domain. J Exp Med. 2017.; doi:jem. 20160933
44. Mariathasan S, Weiss DS, Newton K, McBride J, O'rourke K, Roose-Girma M, et al. Cryopyrin activates the inflammasome in response to toxins and ATP. Nature. 2006;440(7081):228–32.
45. Compan V, Baroja-Mazo A, López-Castejón G, Gomez AI, Martínez CM, Angosto D, et al. Cell volume regulation modulates NLRP3 inflammasome activation. Immunity. 2012;37(3):487–500.
46. Muñoz-Planillo R, Kuffa P, Martínez-Colón G, Smith BL, Rajendiran TM, Núñez G. K+ efflux is the common trigger of NLRP3 inflammasome activation by bacterial toxins and particulate matter. Immunity. 2013;38(6):1142–53.
47. Viganò E, Mortellaro A. Caspase-11: The driving factor for noncanonical inflammasomes. Eur J Immunol. 2013;43(9):2240–5.
48. Walev I, Reske K, Palmer M, Valeva A, Bhakdi S. Potassium-inhibited processing of IL-1 beta in human monocytes. EMBO J. 1995;14(8):1607.
49. Perregaux D, Barberia J, Lanzetti AJ, Geoghegan KF, Carty T, Gabel C. IL-1 beta maturation: evidence that mature cytokine formation can be induced specifically by nigericin. J Immunol. 1992;149(4):1294–303.
50. He Y, Zeng MY, Yang D, Motro B, Núñez G. NEK7 is an essential mediator of NLRP3 activation downstream of potassium efflux. Nature. 2016;530:354.

51. Shi H, Wang Y, Li X, Zhan X, Tang M, Fina M, et al. NLRP3 activation and mitosis are mutually exclusive events coordinated by NEK7, a new inflammasome component. Nat Immunol. 2016;17:250.
52. Bartlett R, Stokes L, Sluyter R. The P2X7 receptor channel: recent developments and the use of P2X7 antagonists in models of disease. Pharmacol Rev. 2014;66(3):638–75.
53. Kelkar DA, Chattopadhyay A. The gramicidin ion channel: a model membrane protein. Biochim Biophys Acta. 2007;1768(9):2011–25.
54. Pressman BC. Biological applications of ionophores. Annu Rev Biochem. 1976;45(1):501–30.
55. Clapham DE. Calcium signaling. Cell. 1995;80(2):259–68.
56. Walsh C, Barrow S, Voronina S, Chvanov M, Petersen OH, Tepikin A. Modulation of calcium signalling by mitochondria. Biochim Biophys Acta. 2009;1787(11):1374–82.
57. Duchen MR. Mitochondria and calcium: from cell signalling to cell death. J Physiol. 2000;529(1):57–68.
58. Okada M, Matsuzawa A, Yoshimura A, Ichijo H. The lysosome rupture-activated TAK1-JNK pathway regulates NLRP3 inflammasome activation. J Biol Chem. 2014;289(47):32926–36.
59. Yaron J, Gangaraju S, Rao M, Kong X, Zhang L, Su F, et al. K+ regulates Ca2+ to drive inflammasome signaling: dynamic visualization of ion flux in live cells. Cell Death Dis. 2015;6(10):e1954.
60. Brough D, Le Feuvre RA, Wheeler RD, Solovyova N, Hilfiker S, Rothwell NJ, et al. Ca2+ stores and Ca2+ entry differentially contribute to the release of IL-1β and IL-1α from murine macrophages. J Immunol. 2003;170(6):3029–36.
61. Murakami T, Ockinger J, Yu J, Byles V, McColl A, Hofer AM, et al. Critical role for calcium mobilization in activation of the NLRP3 inflammasome. Proc Natl Acad Sci. 2012;109(28):11282–7.
62. Maruyama T, Kanaji T, Nakade S, Kanno T, Mikoshiba K. 2APB, 2-aminoethoxydiphenyl borate, a membrane-penetrable modulator of Ins (1, 4, 5) P3-induced Ca2+ release. The. J Biochem. 1997;122(3):498–505.
63. Bleasdale JE, Thakur NR, Gremban RS, Bundy GL, Fitzpatrick FA, Smith RJ, et al. Selective inhibition of receptor-coupled phospholipase C-dependent processes in human platelets and polymorphonuclear neutrophils. J Pharmacol Exp Ther. 1990;255(2):756–68.
64. Hornung V, Bauernfeind F, Halle A, Samstad EO, Kono H, Rock KL, et al. Silica crystals and aluminum salts activate the NALP3 inflammasome through phagosomal destabilization. Nat Immunol. 2008;9(8):847–56.
65. Eisenbarth SC, Colegio OR, O'Connor W, Sutterwala FS, Flavell RA. Crucial role for the Nalp3 inflammasome in the immunostimulatory properties of aluminium adjuvants. Nature. 2008;453(7198):1122–6.
66. Hentze H, Lin X, Choi M, Porter A. Critical role for cathepsin B in mediating caspase-1-dependent interleukin-18 maturation and caspase-1-independent necrosis triggered by the microbial toxin nigericin. Cell Death Differ. 2003;10(9):956–68.
67. Deng D, Jiang N, Hao S-J, Sun H, G j Z. Loss of membrane cholesterol influences lysosomal permeability to potassium ions and protons. Biochim Biophys Acta. 2009;1788(2):470–6.
68. Lima H Jr, Jacobson L, Goldberg M, Chandran K, Diaz-Griffero F, Lisanti MP, et al. Role of lysosome rupture in controlling Nlrp3 signaling and necrotic cell death. Cell Cycle. 2013;12(12):1868–78.
69. Martinon F. Signaling by ROS drives inflammasome activation. Eur J Immunol. 2010;40(3):616–9.
70. Tschopp J, Schroder K. NLRP3 inflammasome activation: the convergence of multiple signalling pathways on ROS production? Nat Rev Immunol. 2010;10(3):210–5.
71. Subramanian N, Natarajan K, Clatworthy MR, Wang Z, Germain RN. The adaptor MAVS promotes NLRP3 mitochondrial localization and inflammasome activation. Cell. 2013;153(2):348–61.
72. Park S, Juliana C, Hong S, Datta P, Hwang I, Fernandes-Alnemri T, et al. The mitochondrial antiviral protein MAVS associates with NLRP3 and regulates its inflammasome activity. J Immunol. 2013;191(8):4358–66.

73. Zhou R, Tardivel A, Thorens B, Choi I, Tschopp J. Thioredoxin-interacting protein links oxidative stress to inflammasome activation. Nat Immunol. 2010;11(2):136–40.
74. Wang W, Wang C, Ding XQ, Pan Y, TT G, Wang MX, et al. Quercetin and allopurinol reduce liver thioredoxin-interacting protein to alleviate inflammation and lipid accumulation in diabetic rats. Br J Pharmacol. 2013;169(6):1352–71.
75. El-Azab M, Baldowski B, Mysona B, Shanab A, Mohamed I, Abdelsaid M, et al. Deletion of thioredoxin-interacting protein preserves retinal neuronal function by preventing inflammation and vascular injury. Br J Pharmacol. 2014;171(5):1299–313.
76. Mohamed IN, Hafez SS, Fairaq A, Ergul A, Imig JD, El-Remessy AB. Thioredoxin-interacting protein is required for endothelial NLRP3 inflammasome activation and cell death in a rat model of high-fat diet. Diabetologia. 2014;57(2):413–23.
77. Chen J, Cha-Molstad H, Szabo A, Shalev A. Diabetes induces and calcium channel blockers prevent cardiac expression of proapoptotic thioredoxin-interacting protein. Am J Physiol Endocrinol Metab. 2009;296(5):E1133–E9.
78. Masters SL, Dunne A, Subramanian SL, Hull RL, Tannahill GM, Sharp FA, et al. Activation of the NLRP3 inflammasome by islet amyloid polypeptide provides a mechanism for enhanced IL-1 [beta] in type 2 diabetes. Nat Immunol. 2010;11(10):897–904.
79. Meyer Y, Buchanan BB, Vignols F, Reichheld J-P. Thioredoxins and glutaredoxins: unifying elements in redox biology. Annu Rev Genet. 2009;43:335–67.
80. Patwari P, Higgins LJ, Chutkow WA, Yoshioka J, Lee RT. The interaction of thioredoxin with Txnip evidence for formation of a mixed disulfide by disulfide exchange. J Biol Chem. 2006;281(31):21884–91.
81. Singh LP. Thioredoxin interacting protein (TXNIP) and pathogenesis of diabetic retinopathy. J Clin Exp Ophthalmol. 2013;4
82. Pejnovic NN, Pantic JM, Jovanovic IP, Radosavljevic GD, Milovanovic MZ, Nikolic IG, et al. Galectin-3 deficiency accelerates high-fat diet–induced obesity and amplifies inflammation in adipose tissue and pancreatic islets. Diabetes. 2013;62(6):1932–44.
83. Chung JW, JH JEON, SR YOON, Choi I. Vitamin D3 upregulated protein 1 (VDUP1) is a regulator for redox signaling and stress-mediated diseases. J Dermatol. 2006;33(10):662–9.
84. Junn E, Han SH, Im JY, Yang Y, Cho EW, Um HD, et al. Vitamin D3 up-regulated protein 1 mediates oxidative stress via suppressing the thioredoxin function. J Immunol. 2000;164(12):6287–95.
85. Polekhina G, Ascher DB, Kok SF, Waltham M. Crystallization and preliminary X-ray analysis of the N-terminal domain of human thioredoxin-interacting protein. Acta Crystallogr Sect F Struct Biol Cryst Commun. 2011;67(5):613–7.
86. Spindel ON, Berk BC. Thioredoxin-Interacting Protein Mediates TRX1 Translocation to the Plasma Membrane in Response to Tumor Necrosis Factor-α. Arterioscler Thromb Vasc Biol. 2011;31(8):1890–7.
87. Saxena G, Chen J, Shalev A. Intracellular shuttling and mitochondrial function of thioredoxin-interacting protein. J Biol Chem. 2010;285(6):3997–4005.
88. Wu N, Zheng B, Shaywitz A, Dagon Y, Tower C, Bellinger G, et al. AMPK-dependent degradation of TXNIP upon energy stress leads to enhanced glucose uptake via GLUT1. Mol Cell. 2013;49(6):1167–75.
89. Ding C, Zhao Y, Shi X, Zhang N, Zu G, Li Z, et al. New insights into salvianolic acid A action: regulation of the TXNIP/NLRP3 and TXNIP/ChREBP pathways ameliorates HFD-induced NAFLD in rats. Sci Rep. 2016;6:28734.
90. Chen W, Zhao M, Zhao S, Lu Q, Ni L, Zou C, et al. Activation of the TXNIP/NLRP3 inflammasome pathway contributes to inflammation in diabetic retinopathy: a novel inhibitory effect of minocycline. Inflamm Res. 2017;66:157–66.
91. Ye X, Zuo D, Yu L, Zhang L, Tang J, Cui C, et al. ROS/TXNIP pathway contributes to thrombin induced NLRP3 inflammasome activation and cell apoptosis in microglia. Biochem Biophys Res Commun. 2017;485(2):499–505.

92. Lv H, Liu Q, Wen Z, Feng H, Deng X, Ci X. Xanthohumol ameliorates lipopolysaccharide (LPS)-induced acute lung injury via induction of AMPK/GSK3β-Nrf2 signal axis. Redox Biol. 2017;12:311–24.
93. Yin Y, Zhou Z, Liu W, Chang Q, Sun G, Dai Y. Vascular endothelial cells senescence is associated with NOD-like receptor family pyrin domain-containing 3 (NLRP3) inflammasome activation via reactive oxygen species (ROS)/thioredoxin-interacting protein (TXNIP) pathway. Int J Biochem Cell Biol. 2017;84:22.
94. Gao P, He F-F, Tang H, Lei C-T, Chen S, Meng X-F, et al. NADPH oxidase-induced NALP3 inflammasome activation is driven by thioredoxin-interacting protein which contributes to podocyte injury in hyperglycemia. J Diabetes Res. 2015;2015:504761.
95. Liu W, Gu J, Qi J, Zeng XN, Ji J, Chen ZZ, et al. Lentinan exerts synergistic apoptotic effects with paclitaxel in A549 cells via activating ROS-TXNIP-NLRP3 inflammasome. J Cell Mol Med. 2015;19(8):1949–55.
96. Xiao J, Liu Y, Xing F, Leung TM, Liong EC, Tipoe GL. Bee's honey attenuates non-alcoholic steatohepatitis-induced hepatic injury through the regulation of thioredoxin-interacting protein–NLRP3 inflammasome pathway. Eur J Nutr. 2016;55(4):1465–77.
97. Feng H, Gu J, Gou F, Huang W, Gao C, Chen G, et al. High glucose and lipopolysaccharide prime NLRP3 inflammasome via ROS/TXNIP pathway in mesangial cells. J Diabet Res. 2016;2016:6973175.
98. Jiang L, Fei D, Gong R, Yang W, Yu W, Pan S, et al. CORM-2 inhibits TXNIP/NLRP3 inflammasome pathway in LPS-induced acute lung injury. Inflamm Res. 2016;65(11):905–15.
99. Dinesh P, Rasool M. Berberine, an isoquinoline alkaloid suppresses TXNIP mediated NLRP3 inflammasome activation in MSU crystal stimulated RAW 264.7 macrophages through the upregulation of Nrf2 transcription factor and alleviates MSU crystal induced inflammation in rats. Int Immunopharmacol. 2017;44:26–37.
100. Zhou R, Yazdi AS, Menu P, Tschopp J. A role for mitochondria in NLRP3 inflammasome activation. Nature. 2011;469(7329):221–5.
101. Choe J-Y, Kim S-K. Quercetin and ascorbic acid suppress fructose-induced NLRP3 inflammasome activation by blocking intracellular shuttling of TXNIP in human macrophage cell lines. Inflammation. 2017;40:980–94.
102. Tan CY, Weier Q, Zhang Y, Cox AJ, Kelly DJ, Langham RG. Thioredoxin-interacting protein: a potential therapeutic target for treatment of progressive fibrosis in diabetic nephropathy. Nephron. 2015;129(2):109–27.
103. Zhang X, Zhang J-H, Chen X-Y, Q-H H, Wang M-X, Jin R, et al. Reactive oxygen species-induced TXNIP drives fructose-mediated hepatic inflammation and lipid accumulation through NLRP3 inflammasome activation. Antioxid Redox Signal. 2015;22(10):848–70.
104. Wang W, Wu Q-h, Sui Y, Wang Y, Qiu X. Rutin protects endothelial dysfunction by disturbing Nox4 and ROS-sensitive NLRP3 inflammasome. Biomed Pharmacother. 2017;86:32–40.
105. Zhang Q-Y, Pan Y, Wang R, Kang L-L, Xue Q-C, Wang X-N, et al. Quercetin inhibits AMPK/TXNIP activation and reduces inflammatory lesions to improve insulin signaling defect in the hypothalamus of high fructose-fed rats. J Nutr Biochem. 2014;25(4):420–8.
106. Hu Q, Wei B, Wei L, Hua K, Yu X, Li H, et al. Sodium tanshinone IIA sulfonate ameliorates ischemia-induced myocardial inflammation and lipid accumulation in Beagle dogs through NLRP3 inflammasome. Int J Cardiol. 2015;196:183–92.
107. Cao G, Jiang N, Hu Y, Zhang Y, Wang G, Yin M, et al. Ruscogenin attenuates cerebral ischemia-induced blood-brain barrier dysfunction by suppressing TXNIP/NLRP3 inflammasome activation and the MAPK pathway. Int J Mol Sci. 2016;17(9):1418.
108. Wang X, Li R, Wang X, Fu Q, Ma S. Umbelliferone ameliorates cerebral ischemia–reperfusion injury via upregulating the PPAR gamma expression and suppressing TXNIP/NLRP3 inflammasome. Neurosci Lett. 2015;600:182–7.
109. Li Y, Li J, Li S, Li Y, Wang X, Liu B, et al. Curcumin attenuates glutamate neurotoxicity in the hippocampus by suppression of ER stress-associated TXNIP/NLRP3 inflammasome activation in a manner dependent on AMPK. Toxicol Appl Pharmacol. 2015;286(1):53–63.

110. Ishrat T, Mohamed IN, Pillai B, Soliman S, Fouda AY, Ergul A, et al. Thioredoxin-interacting protein: a novel target for neuroprotection in experimental thromboembolic stroke in mice. Mol Neurobiol. 2015;51(2):766–78.
111. Saitoh T, Akira S. Regulation of inflammasomes by autophagy. J Allergy Clin Immunol. 2016;138(1):28–36.
112. Harris J, Hartman M, Roche C, Zeng SG, O'Shea A, Sharp FA, et al. Autophagy controls IL-1β secretion by targeting pro-IL-1β for degradation. J Biol Chem. 2011;286(11):9587–97.
113. Nakahira K, Haspel JA, Rathinam VA, Lee S-J, Dolinay T, Lam HC, et al. Autophagy proteins regulate innate immune responses by inhibiting the release of mitochondrial DNA mediated by the NALP3 inflammasome. Nat Immunol. 2011;12(3):222–30.
114. Lupfer C, Thomas PG, Anand PK, Vogel P, Milasta S, Martinez J, et al. Receptor interacting protein kinase 2-mediated mitophagy regulates inflammasome activation during virus infection. Nat Immunol. 2013;14(5):480–8.
115. Smoum R, Baraghithy S, Chourasia M, Breuer A, Mussai N, Attar-Namdar M, et al. CB2 cannabinoid receptor agonist enantiomers HU-433 and HU-308: an inverse relationship between binding affinity and biological potency. Proc Natl Acad Sci. 2015;112(28):8774–9.
116. Shao BZ, Wei W, Ke P, ZQ X, Zhou JX, Liu C. Activating cannabinoid receptor 2 alleviates pathogenesis of experimental autoimmune encephalomyelitis via activation of autophagy and inhibiting NLRP3 inflammasome. CNS Neurosci Ther. 2014;20(12):1021–8.
117. Lee G-S, Subramanian N, Kim AI, Aksentijevich I, Goldbach-Mansky R, Sacks DB, et al. The calcium-sensing receptor regulates the NLRP3 inflammasome through Ca2+ and cAMP. Nature. 2012;492(7427):123–7.
118. Yan Y, Jiang W, Spinetti T, Tardivel A, Castillo R, Bourquin C, et al. Omega-3 fatty acids prevent inflammation and metabolic disorder through inhibition of NLRP3 inflammasome activation. Immunity. 2013;38(6):1154–63.
119. Desouza IA, CF F-P, Camargo EA, Lima CS, Teixeira SA, Muscará MN, et al. Inflammatory mechanisms underlying the rat pulmonary neutrophil influx induced by airway exposure to staphylococcal enterotoxin type A. Br J Pharmacol. 2005;146(6):781–91.
120. Tajima T, Murata T, Aritake K, Urade Y, Hirai H, Nakamura M, et al. Lipopolysaccharide induces macrophage migration via prostaglandin D2 and prostaglandin E2. J Pharmacol Exp Ther. 2008;326(2):493–501.
121. Kvirkvelia N, McMenamin M, Chaudhary K, Bartoli M, Madaio MP. Prostaglandin E2 promotes cellular recovery from established nephrotoxic serum nephritis in mice, prosurvival, and regenerative effects on glomerular cells. Am J Physiol Renal Physiol. 2013;304(5):F463–F70.
122. MacKenzie KF, Clark K, Naqvi S, McGuire VA, Nöehren G, Kristariyanto Y, et al. PGE2 induces macrophage IL-10 production and a regulatory-like phenotype via a protein kinase A–SIK–CRTC3 pathway. J Immunol. 2013;190(2):565–77.
123. Sokolowska M, Chen L-Y, Liu Y, Martinez-Anton A, Qi H-Y, Logun C, et al. Prostaglandin E2 inhibits NLRP3 inflammasome activation through EP4 receptor and intracellular cyclic AMP in human macrophages. J Immunol. 2015;194(11):5472–87.
124. Mortimer L, Moreau F, MacDonald JA, Chadee K. NLRP3 inflammasome inhibition is disrupted in a group of auto-inflammatory disease CAPS mutations. Nat Immunol. 2017;17:1176.
125. Swanson KV, Ting JP. Reining in uncontrolled inflammasome with PKA. Nat Immunol. 2016;17(10):1137–8.
126. Chen S, Sun B. Negative regulation of NLRP3 inflammasome signaling. Protein Cell. 2013;4(4):251.
127. Haneklaus M, Gerlic M, Kurowska-Stolarska M, Rainey A-A, Pich D, McInnes IB, et al. Cutting edge: miR-223 and EBV miR-BART15 regulate the NLRP3 inflammasome and IL-1β production. J Immunol. 2012;189(8):3795–9.
128. Bauernfeind F, Rieger A, Schildberg FA, Knolle PA, Schmid-Burgk JL, Hornung V. NLRP3 inflammasome activity is negatively controlled by miR-223. J Immunol. 2012;189(8):4175–81.

129. Li X-F, Shen W-W, Sun Y-Y, Li W-X, Sun Z-H, Liu Y-H, et al. MicroRNA-20a negatively regulates expression of NLRP3-inflammasome by targeting TXNIP in adjuvant-induced arthritis fibroblast-like synoviocytes. Joint Bone Spine. 2016;83(6):695–700.
130. Bandyopadhyay S, Lane T, Venugopal R, Parthasarathy PT, Cho Y, Galam L, et al. MicroRNA-133a-1 regulates inflammasome activation through uncoupling protein-2. Biochem Biophys Res Commun. 2013;439(3):407–12.
131. Wang W, Ding X-Q, T-T G, Song L, Li J-M, Xue Q-C, et al. Pterostilbene and allopurinol reduce fructose-induced podocyte oxidative stress and inflammation via microRNA-377. Free Radic Biol Med. 2015;83:214–26.
132. Meylan E, Tschopp J, Karin M. Intracellular pattern recognition receptors in the host response. Nature. 2006;442(7098):39–44.
133. Inoue M, Williams KL, Oliver T, Vandenabeele P, Rajan JV, Miao EA, et al. IFNβ therapy against EAE is effective only when development of the disease depends on the NLRP3 inflammasome. Sci Signal. 2011;5(225):ra38.
134. Inoue M, Shinohara ML. The role of interferon-β in the treatment of multiple sclerosis and experimental autoimmune encephalomyelitis–in the perspective of inflammasomes. Immunology. 2013;139(1):11–8.
135. Guarda G, Braun M, Staehli F, Tardivel A, Mattmann C, Förster I, et al. Type I interferon inhibits interleukin-1 production and inflammasome activation. Immunity. 2011;34(2):213–23.
136. Gustin A, Kirchmeyer M, Koncina E, Felten P, Losciuto S, Heurtaux T, et al. NLRP3 inflammasome is expressed and functional in mouse brain microglia but not in astrocytes. PLoS One. 2015;10(6):e0130624.
137. Abulafia DP, de Rivero Vaccari JP, Lozano JD, Lotocki G, Keane RW, Dietrich WD. Inhibition of the inflammasome complex reduces the inflammatory response after thromboembolic stroke in mice. J Cereb Blood Flow Metab. 2009;29(3):534–44.
138. Fann DY-W, Lee S, Manzanero S, Tang S-C, Gelderblom M, Chunduri P, et al. Intravenous immunoglobulin suppresses NLRP1 and NLRP3 inflammasome-mediated neuronal death in ischemic stroke. Cell Death Dis. 2013;4(9):e790.
139. Afrasyab A, Qu P, Zhao Y, Peng K, Wang H, Lou D, et al. Correlation of NLRP3 with severity and prognosis of coronary atherosclerosis in acute coronary syndrome patients. Heart Vessels. 2016;31(8):1218–29.
140. Halle A, Hornung V, Petzold GC, Stewart CR, Monks BG, Reinheckel T, et al. The NALP3 inflammasome is involved in the innate immune response to amyloid-β. Nat Immunol. 2008;9(8):857–65.
141. Lammerding L, Slowik A, Johann S, Beyer C, Zendedel A. Poststroke inflammasome expression and regulation in the peri-infarct area by gonadal steroids after transient focal ischemia in the rat brain. Neuroendocrinology. 2016;103(5):460–75.
142. Zhang D, Yan H, Hu Y, Zhuang Z, Yu Z, Hang C. Increased expression of NLRP3 inflammasome in wall of ruptured and unruptured human cerebral aneurysms: preliminary results. J Stroke Cerebrovasc Dis. 2015;24(5):972–9.
143. Abdul-Muneer P, Alikunju S, Mishra V, Schuetz H, Szlachetka AM, Burnham EL, et al. Activation of NLRP3 inflammasome by cholesterol crystals in alcohol consumption induces atherosclerotic lesions. Brain Behav Immun 2017: 62: 291.
144. Soria FN, Pérez-Samartín A, Martin A, Gona KB, Llop J, Szczupak B, et al. Extrasynaptic glutamate release through cystine/glutamate antiporter contributes to ischemic damage. J Clin Invest. 2014;124(8):3645–55.
145. Fann DY-W, Lee S-Y, Manzanero S, Chunduri P, Sobey CG, Arumugam TV. Pathogenesis of acute stroke and the role of inflammasomes. Ageing Res Rev. 2013;12(4):941–66.
146. Han S, Kim K, Kim H, Kwon J, Lee Y-H, Lee C-K, et al. Auranofin inhibits overproduction of pro-inflammatory cytokines, cyclooxygenase expression and PGE 2 production in macrophages. Arch Pharm Res. 2008;31(1):67–74.
147. Yamada R, Sano H, Hla T, Hashiramoto A, Fukui W, Miyazaki S, et al. Auranofin inhibits interleukin-1β-induced transcript of cyclooxygenase-2 on cultured human synoviocytes. Eur J Pharmacol. 1999;385(1):71–9.

148. Cox AG, Brown KK, Arner ES, Hampton MB. The thioredoxin reductase inhibitor auranofin triggers apoptosis through a Bax/Bak-dependent process that involves peroxiredoxin 3 oxidation. Biochem Pharmacol. 2008;76(9):1097–109.
149. Jeon K-I, Jeong J-Y, Jue D-M. Thiol-reactive metal compounds inhibit NF-κB activation by blocking IκB kinase. J Immunol. 2000;164(11):5981–9.
150. Kataoka K, Handa H, Nishizawa M. Induction of cellular antioxidative stress genes through heterodimeric transcription factor Nrf2/small Maf by antirheumatic gold (I) compounds. J Biol Chem. 2001;276(36):34074–81.
151. Youn HS, Lee JY, Saitoh SI, Miyake K, Hwang DH. Auranofin, as an anti-rheumatic gold compound, suppresses LPS-induced homodimerization of TLR4. Biochem Biophys Res Commun. 2006;350(4):866–71.
152. Sha T, Sunamoto M, Kitazaki T, Sato J, Ii M, Iizawa Y. Therapeutic effects of TAK-242, a novel selective Toll-like receptor 4 signal transduction inhibitor, in mouse endotoxin shock model. Eur J Pharmacol. 2007;571(2):231–9.
153. Zhang D, Yan H, Li H, Hao S, Zhuang Z, Liu M, et al. TGFβ-activated kinase 1 (TAK1) inhibition by 5Z-7-oxozeaenol attenuates early brain injury after experimental subarachnoid hemorrhage. J Biol Chem. 2015;290(32):19900–9.
154. Chang CA, Kanak MA, Yoshimatsu G, Lawrence MC, Kane RR, Naziruddin B. A small molecule inhibitor of toll-like receptor-4 (tlr-4) effectively protects islets from Ibmir. Xenotransplantation. 2015;22:S156.
155. Takashima K, Matsunaga N, Yoshimatsu M, Hazeki K, Kaisho T, Uekata M, et al. Analysis of binding site for the novel small-molecule TLR4 signal transduction inhibitor TAK-242 and its therapeutic effect on mouse sepsis model. Br J Pharmacol. 2009;157(7):1250–62.
156. de Seny D, Cobraiville G, Charlier E, Neuville S, Esser N, Malaise D, et al. Acute-phase serum amyloid a in osteoarthritis: regulatory mechanism and proinflammatory properties. PLoS One. 2013;8(6):e66769.
157. Glushkova OV, Parfenyuk SB, Khrenov MO, Novoselova TV, Lunin SM, Fesenko EE, et al. Inhibitors of TLR-4, NF-κ B, and SAPK/JNK signaling reduce the toxic effect of lipopolysaccharide on RAW 264.7 cells. J Immunotoxicol. 2013;10(2):133–40.
158. Gong Y-N, Wang X, Wang J, Yang Z, Li S, Yang J, et al. Chemical probing reveals insights into the signaling mechanism of inflammasome activation. Cell Res. 2010;20(12):1289–305.
159. Hua F, Tang H, Wang J, Prunty MC, Hua X, Sayeed I, et al. TAK-242, an antagonist for Toll-like receptor 4, protects against acute cerebral ischemia/reperfusion injury in mice. J Cereb Blood Flow Metab. 2015;35(4):536–42.
160. White BJ, Tarabishy S, Venna VR, Manwani B, Benashski S, McCullough LD, et al. Protection from cerebral ischemia by inhibition of TGFβ-activated kinase. Exp Neurol. 2012;237(1):238–45.
161. Gong J, Li Z-Z, Guo S, Zhang X-J, Zhang P, Zhao G-N, et al. Neuron-specific tumor necrosis factor receptor–associated factor 3 is a central regulator of neuronal death in acute ischemic strokenovelty and significance. Hypertension. 2015;66(3):604–16.
162. Lopez-Castejon G, Luheshi NM, Compan V, High S, Whitehead RC, Flitsch S, et al. Deubiquitinases regulate the activity of caspase-1 and interleukin-1β secretion via assembly of the inflammasome. J Biol Chem. 2013;288(4):2721–33.
163. Doeppner TR, Doehring M, Bretschneider E, Zechariah A, Kaltwasser B, Müller B, et al. MicroRNA-124 protects against focal cerebral ischemia via mechanisms involving Usp14-dependent REST degradation. Acta Neuropathol. 2013;126(2):251–65.
164. Min JW, Lü L, Freeling J, Martin D, Wang H. USP14 inhibitor attenuates cerebral ischemia/reperfusion-induced neuronal injury in mice. J Neurochem. 2017;140:826.
165. Juliana C, Fernandes-Alnemri T, Wu J, Datta P, Solorzano L, J-W Y, et al. Anti-inflammatory compounds parthenolide and Bay 11-7082 are direct inhibitors of the inflammasome. J Biol Chem. 2010;285(13):9792–802.
166. Saadane A, Masters S, DiDonato J, Li J, Berger M. Parthenolide inhibits IκB kinase, NF-κB activation, and inflammatory response in cystic fibrosis cells and mice. Am J Respir Cell Mol Biol. 2007;36(6):728–36.

167. García-Piñeres AJ, Vc C, Mora G, Schmidt TJ, Strunck E, Pahl HL, et al. Cysteine 38 in p65/NF-κB plays a crucial role in DNA binding inhibition by sesquiterpene lactones. J Biol Chem. 2001;276(43):39713–20.
168. Meng X, Martinez MA, MA R-S, Winter SS, Wilson BS. IKK inhibitor bay 11-7082 induces necroptotic cell death in precursor-B acute lymphoblastic leukaemic blasts. Br J Haematol. 2010;148(3):487–90.
169. Dong L, Qiao H, Zhang X, Zhang X, Wang C, Wang L, et al. Parthenolide is neuroprotective in rat experimental stroke model: downregulating NF-B, phospho-p38MAPK, and caspase-1 and ameliorating BBB permeability. Mediators Inflamm. 2013;2013:370804.
170. Guzman ML, Rossi RM, Neelakantan S, Li X, Corbett CA, Hassane DC, et al. An orally bioavailable parthenolide analog selectively eradicates acute myelogenous leukemia stem and progenitor cells. Blood. 2007;110(13):4427–35.
171. D'anneo A, Carlisi D, Lauricella M, Puleio R, Martinez R, Di Bella S, et al. Parthenolide generates reactive oxygen species and autophagy in MDA-MB231 cells. A soluble parthenolide analogue inhibits tumour growth and metastasis in a xenograft model of breast cancer. Cell Death Dis. 2013;4(10):e891.
172. Su L, Du H, Dong X, Zhang X, Lou Z. Raf kinase inhibitor protein regulates oxygen-glucose deprivation-induced PC12 cells apoptosis through the NF-κB and ERK pathways. J Clin Biochem Nutr. 2016;59(2):86–92.
173. Su L, Zhang R, Chen Y, Ma C, Zhu Z. Raf kinase inhibitor protein attenuates ischemic-induced microglia cell apoptosis and activation through NF-κB pathway. Cell Physiol Biochem. 2017;41(3):1125–34.
174. He Y, Varadarajan S, Muñoz-Planillo R, Burberry A, Nakamura Y, Núñez G. 3, 4-methylenedioxy-β-nitrostyrene inhibits NLRP3 inflammasome activation by blocking assembly of the inflammasome. J Biol Chem. 2014;289(2):1142–50.
175. Xiao M, Li L, Li C, Liu L, Yu Y, Ma L. 3, 4-methylenedioxy-β-nitrostyrene ameliorates experimental burn wound progression by inhibiting the NLRP3 inflammasome activation. Plast Reconstr Surg. 2016;137(3):566e–75e.
176. Long H, Xu B, Luo Y, Luo K. Artemisinin protects mice against burn sepsis through inhibiting NLRP3 inflammasome activation. Am J Emerg Med. 2016;34(5):772–7.
177. Keystone EC, Wang MM, Layton M, Hollis S, McInnes IB, Team DS. Clinical evaluation of the efficacy of the P2X7 purinergic receptor antagonist AZD9056 on the signs and symptoms of rheumatoid arthritis in patients with active disease despite treatment with methotrexate or sulphasalazine. Annal Rheum Dis. 2012;71:1630.
178. Stock TC, Bloom BJ, Wei N, Ishaq S, Park W, Wang X, et al. Efficacy and safety of CE-224,535, an antagonist of P2X7 receptor, in treatment of patients with rheumatoid arthritis inadequately controlled by methotrexate. J Rheumatol. 2012;39(4):720–7.
179. Ali Z, Laurijssens B, Ostenfeld T, McHugh S, Stylianou A, Scott-Stevens P, et al. Pharmacokinetic and pharmacodynamic profiling of a P2X7 receptor allosteric modulator GSK1482160 in healthy human subjects. Br J Clin Pharmacol. 2013;75(1):197–207.
180. Ferrari D, Pizzirani C, Adinolfi E, Lemoli RM, Curti A, Idzko M, et al. The P2X7 receptor: a key player in IL-1 processing and release. J Immunol. 2006;176(7):3877–83.
181. Sorge RE, Trang T, Dorfman R, Smith SB, Beggs S, Ritchie J, et al. Genetically determined P2X7 receptor pore formation regulates variability in chronic pain sensitivity. Nat Med. 2012;18(4):595–9.
182. Ferrari D, Chiozzi P, Falzoni S, Dal Susino M, Melchiorri L, Baricordi OR, et al. Extracellular ATP triggers IL-1 beta release by activating the purinergic P2Z receptor of human macrophages. J Immunol. 1997;159(3):1451–8.
183. Solle M, Labasi J, Perregaux DG, Stam E, Petrushova N, Koller BH, et al. Altered cytokine production in mice lacking P2X7Receptors. J Biol Chem. 2001;276(1):125–32.
184. Labasi JM, Petrushova N, Donovan C, McCurdy S, Lira P, Payette MM, et al. Absence of the P2X7 receptor alters leukocyte function and attenuates an inflammatory response. J Immunol. 2002;168(12):6436–45.

185. Pelegrin P, Surprenant A. Pannexin-1 mediates large pore formation and interleukin-1β release by the ATP-gated P2X7 receptor. EMBO J. 2006;25(21):5071–82.
186. Locovei S, Scemes E, Qiu F, Spray DC, Dahl G. Pannexin1 is part of the pore forming unit of the P2X 7 receptor death complex. FEBS Lett. 2007;581(3):483–8.
187. Kanneganti T-D, Lamkanfi M, Kim Y-G, Chen G, Park J-H, Franchi L, et al. Pannexin-1-mediated recognition of bacterial molecules activates the cryopyrin inflammasome independent of Toll-like receptor signaling. Immunity. 2007;26(4):433–43.
188. Bravo D, Maturana C, Pelissier T, Hernández A, Constandil L. Interactions of pannexin 1 with NMDA and P2X7 receptors in central nervous system pathologies: possible role on chronic pain. Pharmacol Res. 2015;101:86–93.
189. Brough D, Le Feuvre RA, Iwakura Y, Rothwell NJ. Purinergic (P2X7) receptor activation of microglia induces cell death via an interleukin-1-independent mechanism. Mol Cell Neurosci. 2002;19(2):272–80.
190. Arbeloa J, Pérez-Samartín A, Gottlieb M, Matute C. P2X7 receptor blockade prevents ATP excitotoxicity in neurons and reduces brain damage after ischemia. Neurobiol Dis. 2012;45(3):954–61.
191. Eyo UB, Miner SA, Ahlers KE, L-J W, Dailey ME. P2X7 receptor activation regulates microglial cell death during oxygen-glucose deprivation. Neuropharmacology. 2013;73:311–9.
192. Zhao H, Zhang X, Dai Z, Feng Y, Li Q, Zhang JH, et al. P2X7 receptor suppression preserves blood-brain barrier through inhibiting RhoA activation after experimental intracerebral hemorrhage in rats. Sci Rep. 2016;6:23286.
193. Ye X, Shen T, Hu J, Zhang L, Zhang Y, Bao L, et al. Purinergic 2X7 receptor/NLRP3 pathway triggers neuronal apoptosis after ischemic stroke in the mouse. Exp Neurol. 2017;292:46–55.
194. Lu M, Yang J-Z, Geng F, Ding J-H, Hu G. Iptakalim confers an antidepressant effect in a chronic mild stress model of depression through regulating neuro-inflammation and neurogenesis. Int J Neuropsychopharmacol. 2014;17(9):1501–10.
195. Zhao AP, Dong YF, Liu W, Gu J, Sun XL. Nicorandil inhibits inflammasome activation and toll-like receptor-4 signal transduction to protect against oxygen–glucose deprivation-induced inflammation in BV-2 cells. CNS Neurosci Ther. 2014;20(2):147–53.
196. Heid ME, Keyel PA, Kamga C, Shiva S, Watkins SC, Salter RD. Mitochondrial reactive oxygen species induces NLRP3-dependent lysosomal damage and inflammasome activation. J Immunol. 2013;191(10):5230–8.
197. Biswas R, Hamilton RF, Holian A. Role of lysosomes in silica-induced inflammasome activation and inflammation in absence of MARCO. J Immunol Res. 2014;2014:304180.
198. Morishige T, Yoshioka Y, Tanabe A, Yao X, Tsunoda S-i, Tsutsumi Y, et al. Titanium dioxide induces different levels of IL-1β production dependent on its particle characteristics through caspase-1 activation mediated by reactive oxygen species and cathepsin B. Biochem Biophys Res Commun. 2010;392(2):160–5.
199. Bruchard M, Mignot G, Derangère V, Chalmin F, Chevriaux A, Végran F, et al. Chemotherapy-triggered cathepsin B release in myeloid-derived suppressor cells activates the Nlrp3 inflammasome and promotes tumor growth. Nat Med. 2013;19(1):57–64.
200. Jacobson LS, Lima H, Goldberg MF, Gocheva V, Tsiperson V, Sutterwala FS, et al. Cathepsin-mediated necrosis controls the adaptive immune response by Th2 (T helper type 2)-associated adjuvants. J Biol Chem. 2013;288(11):7481–91.
201. Montaser M, Lalmanach G, Mach L. CA-074, but not its methyl ester CA-074Me, is a selective inhibitor of cathepsin B within living cells. Biol Chem. 2002;383(7-8):1305–8.
202. Chen YT, Brinen LS, Kerr ID, Hansell E, Doyle PS, McKerrow JH, et al. In vitro and in vivo studies of the trypanocidal properties of WRR-483 against Trypanosoma cruzi. PLoS Negl Trop Dis. 2010;4(9):e825.
203. Orlowski GM, Colbert JD, Sharma S, Bogyo M, Robertson SA, Rock KL. Multiple cathepsins promote pro–IL-1β synthesis and NLRP3-mediated IL-1β activation. J Immunol. 2015;195(4):1685–97.

204. Hou Q, Ling L, Wang F, Xing S, Pei Z, Zeng J. Endostatin expression in neurons during the early stage of cerebral ischemia is associated with neuronal apoptotic cell death in adult hypertensive rat model of stroke. Brain Res. 2010;1311:182–8.
205. Cruz CM, Rinna A, Forman HJ, Ventura AL, Persechini PM, Ojcius DM. ATP activates a reactive oxygen species-dependent oxidative stress response and secretion of proinflammatory cytokines in macrophages. J Biol Chem. 2007;282(5):2871–9.
206. Petrilli V, Papin S, Dostert C, Mayor A, Martinon F, Tschopp J. Activation of the NALP3 inflammasome is triggered by low intracellular potassium concentration. Cell Death Differ. 2007;14(9):1583–9.
207. Dostert C, Pétrilli V, Van Bruggen R, Steele C, Mossman BT, Tschopp J. Innate immune activation through Nalp3 inflammasome sensing of asbestos and silica. Science. 2008;320(5876):674–7.
208. Latz E, Xiao TS, Stutz A. Activation and regulation of the inflammasomes. Nat Rev Immunol. 2013;13(6):397–411.
209. Álvarez S, Muñoz-Fernández MÁ. TNF-α may mediate inflammasome activation in the absence of bacterial infection in more than one way. PLoS One. 2013;8(8):e71477.
210. Wen H, Gris D, Lei Y, Jha S, Zhang L, Huang MT-H, et al. Fatty acid-induced NLRP3-ASC inflammasome activation interferes with insulin signaling. Nat Immunol. 2011;12(5):408–15.
211. Xiao H, Lu M, Lin TY, Chen Z, Chen G, Wang W-C, et al. SREBP2 activation of NLRP3 inflammasome in endothelium mediates hemodynamic-induced atherosclerosis susceptibility. Circulation. 2013;128:632.
212. Sreerama L, Sladek NE. Identification and characterization of a novel class 3 aldehyde dehydrogenase overexpressed in a human breast adenocarcinoma cell line exhibiting oxazaphosphorine-specific acquired resistance. Biochem Pharmacol. 1993;45(12):2487–505.
213. Laliberte RE, Perregaux DG, Hoth LR, Rosner PJ, Jordan CK, Peese KM, et al. Glutathione S-transferase omega 1-1 is a target of cytokine release inhibitory drugs and may be responsible for their effect on interleukin-1β posttranslational processing. J Biol Chem. 2003;278(19):16567–78.
214. Coll RC, Robertson AA, Chae JJ, Higgins SC, Muñoz-Planillo R, Inserra MC, et al. A small-molecule inhibitor of the NLRP3 inflammasome for the treatment of inflammatory diseases. Nat Med. 2015;21(3):248–55.
215. Sušjan P, Roškar S, Hafner-Bratkovič I. The mechanism of NLRP3 inflammasome initiation: trimerization but not dimerization of the NLRP3 pyrin domain induces robust activation of IL-1β. Biochem Biophys Res Commun. 2017;483:823.
216. Salla M, Butler MS, Pelingon R, Kaeslin G, Croker DE, Reid JC, et al. Identification, synthesis, and biological evaluation of the major human metabolite of NLRP3 inflammasome inhibitor MCC950. ACS Med Chem Lett. 2016;7(12):1034–8.
217. Dempsey C, Araiz AR, Bryson K, Finucane O, Larkin C, Mills E, et al. Inhibiting the NLRP3 inflammasome with MCC950 promotes non-phlogistic clearance of amyloid-β and cognitive function in APP/PS1 mice. Brain Behav Immun. 2017;61:306–16.
218. Dolunay A, Senol SP, Temiz-Resitoglu M, Guden DS, Sari AN, Sahan-Firat S, et al. Inhibition of NLRP3 inflammasome prevents LPS-induced inflammatory hyperalgesia in mice: contribution of NF-κB, caspase-1/11, ASC, NOX, and NOS isoforms. Inflammation. 2017;40:366–86.
219. Takahashi M. NLRP3 inflammasome as a novel player in myocardial infarction. Int Heart J. 2014;55(2):101–5.
220. van Hout GP, Bosch L, Ellenbroek GH, de Haan JJ, van Solinge WW, Cooper MA, et al. The selective NLRP3-inflammasome inhibitor MCC950 reduces infarct size and preserves cardiac function in a pig model of myocardial infarction. Eur Heart J. 2017;38:828.
221. Mohamed IN, Ishrat T, Fagan SC, El-Remessy AB. Role of inflammasome activation in the pathophysiology of vascular diseases of the neurovascular unit. Antioxid Redox Signal. 2015;22(13):1188–206.
222. Murthy P, Durco F, Miller-Ocuin JL, Takedai T, Shankar S, Liang X, et al. The NLRP3 inflammasome and bruton's tyrosine kinase in platelets co-regulate platelet activation, aggregation, and in vitro thrombus formation. Biochem Biophys Res Commun. 2017;483(1):230–6.

223. Mercken EM, Crosby SD, Lamming DW, JeBailey L, Krzysik-Walker S, Villareal DT, et al. Calorie restriction in humans inhibits the PI3K/AKT pathway and induces a younger transcription profile. Aging Cell. 2013;12(4):645–51.
224. Newman JC, Verdin E. Ketone bodies as signaling metabolites. Trends Endocrinol Metab. 2014;25(1):42–52.
225. Netea MG, Joosten LA. Inflammasome inhibition: putting out the fire. Cell Metab. 2015;21(4):513–4.
226. Youm Y-H, Nguyen KY, Grant RW, Goldberg EL, Bodogai M, Kim D, et al. The ketone metabolite [beta]-hydroxybutyrate blocks NLRP3 inflammasome-mediated inflammatory disease. Nat Med. 2015;21(3):263–9.
227. Bae HR, Kim DH, Park MH, Lee B, Kim MJ, Lee EK, et al. β-Hydroxybutyrate suppresses inflammasome formation by ameliorating endoplasmic reticulum stress via AMPK activation. Oncotarget. 2016;7(41):66444–54.
228. Cotter DG, Schugar RC, Crawford PA. Ketone body metabolism and cardiovascular disease. Am J Physiol Heart Circ Physiol. 2013;304(8):H1060–H76.
229. Halestrap AP. Monocarboxylic acid transport. Compr Physiol. 2013;3:1611.
230. Bergersen LH, Magistretti PJ, Pellerin L. Selective postsynaptic co-localization of MCT2 with AMPA receptor GluR2/3 subunits at excitatory synapses exhibiting AMPA receptor trafficking. Cereb Cortex. 2005;15(4):361–70.
231. Edmond J, Robbins R, Bergstrom J, Cole R, De Vellis J. Capacity for substrate utilization in oxidative metabolism by neurons, astrocytes, and oligodendrocytes from developing brain in primary culture. J Neurosci Res. 1987;18(4):551–61.
232. Cahill GF Jr, Veech RL. Ketoacids? Good medicine? Trans Am Clin Climatol Assoc. 2003;114:149.
233. Pikija S, Trkulja V, Simundic A-M, Vrcek E, Boskovic K, Bacani S. Is on-admission capillary blood beta-hydroxybutyrate concentration associated with the acute stroke severity and short-term functional outcome? Neurol Res. 2013;35(9):959–67.
234. Puchowicz MA, Zechel JL, Valerio J, Emancipator DS, Xu K, Pundik S, et al. Neuroprotection in diet-induced ketotic rat brain after focal ischemia. J Cereb Blood Flow Metab. 2008;28(12):1907–16.
235. Rahman M, Muhammad S, Khan MA, Chen H, Ridder DA, Müller-Fielitz H, et al. The β-hydroxybutyrate receptor HCA2 activates a neuroprotective subset of macrophages. Nat Commun. 2014;5:3944.
236. Offermanns S, Schwaninger M. Nutritional or pharmacological activation of HCA 2 ameliorates neuroinflammation. Trends Mol Med. 2015;21(4):245–55.
237. Kinard TA, Satin LS. An ATP-sensitive Cl− channel current that is activated by cell swelling, cAMP, and glyburide in insulin-secreting cells. Diabetes. 1995;44(12):1461–6.
238. Ashcroft FM. ATP-sensitive potassium channelopathies: focus on insulin secretion. J Clin Invest. 2005;115(8):2047–58.
239. Watanabe T, Takeda A, Tsukiyama T, Mise K, Okuno T, Sasaki H, et al. Identification and characterization of two novel classes of small RNAs in the mouse germline: retrotransposon-derived siRNAs in oocytes and germline small RNAs in testes. Genes Dev. 2006;20(13):1732–43.
240. Perregaux DG, McNiff P, Laliberte R, Hawryluk N, Peurano H, Stam E, et al. Identification and characterization of a novel class of interleukin-1 post-translational processing inhibitors. J Pharmacol Exp Ther. 2001;299(1):187–97.
241. Lamkanfi M, Mueller JL, Vitari AC, Misaghi S, Fedorova A, Deshayes K, et al. Glyburide inhibits the Cryopyrin/Nalp3 inflammasome. J Cell Biol. 2009;187(1):61–70.
242. Coll RC, O'Neill LA. The cytokine release inhibitory drug CRID3 targets ASC oligomerisation in the NLRP3 and AIM2 inflammasomes. PLoS One. 2011;6(12):e29539.
243. Sturgess N, Cook D, Ashford MJ, Hales CN. The sulphonylurea receptor may be an ATP-sensitive potassium channel. Lancet. 1985;326(8453):474–5.
244. Henquin J-C. Tolbutamide stimulation and inhibition of insulin release: studies of the underlying ionic mechanisms in isolated rat islets. Diabetologia. 1980;18(2):151–60.

245. Gaidt MM, Ebert TS, Chauhan D, Schmidt T, Schmid-Burgk JL, Rapino F, et al. Human monocytes engage an alternative inflammasome pathway. Immunity. 2016;44(4):833–46.
246. Liu W, Guo W, Wu J, Luo Q, Tao F, Gu Y, et al. A novel benzo [d] imidazole derivate prevents the development of dextran sulfate sodium-induced murine experimental colitis via inhibition of NLRP3 inflammasome. Biochem Pharmacol. 2013;85(10):1504–12.
247. Pan L, Hang N, Zhang C, Chen Y, Li S, Sun Y, et al. Synthesis and biological evaluation of novel benzimidazole derivatives and analogs targeting the NLRP3 inflammasome. Molecules. 2017;22(2):213.
248. Bravo L. Polyphenols: chemistry, dietary sources, metabolism, and nutritional significance. Nutr Rev. 1998;56(11):317–33.
249. Cerella C, Radogna F, Dicato M, Diederich M. Natural compounds as regulators of the cancer cell metabolism. Int J Cell Biol. 2013;2013:639401.
250. Huang T-T, Lai H-C, Chen Y-B, Chen L-G, Y-H W, Ko Y-F, et al. cis-Resveratrol produces anti-inflammatory effects by inhibiting canonical and non-canonical inflammasomes in macrophages. Innate Immun. 2014;20(7):735–50.
251. Chang YP, Ka SM, Hsu WH, Chen A, Chao LK, Lin CC, et al. Resveratrol inhibits NLRP3 inflammasome activation by preserving mitochondrial integrity and augmenting autophagy. J Cell Physiol. 2015;230(7):1567–79.
252. Wu J, Li X, Zhu G, Zhang Y, He M, Zhang J. The role of resveratrol-induced mitophagy/autophagy in peritoneal mesothelial cells inflammatory injury via NLRP3 inflammasome activation triggered by mitochondrial ROS. Exp Cell Res. 2016;341(1):42–53.
253. Barrajón-Catalán E, Herranz-López M, Joven J, Segura-Carretero A, Alonso-Villaverde C, Menéndez JA, et al. Molecular promiscuity of plant polyphenols in the management of age-related diseases: far beyond their antioxidant properties. In: Oxidative stress and inflammation in non-communicable diseases-molecular mechanisms and perspectives in therapeutics. New York, NY: Springer; 2014. p. 141–59.
254. Baur JA, Sinclair DA. Therapeutic potential of resveratrol: the in vivo evidence. Nat Rev Drug Discov. 2006;5(6):493–506.
255. Sharma R, Gescher A, Steward W. Curcumin: the story so far. Eur J Cancer. 2005;41(13):1955–68.
256. Howitz KT, Sinclair DA. Xenohormesis: sensing the chemical cues of other species. Cell. 2008;133(3):387–91.
257. Hooper PL, Hooper PL, Tytell M, Vígh L. Xenohormesis: health benefits from an eon of plant stress response evolution. Cell Stress Chaperones. 2010;15(6):761–70.
258. Queen BL, Tollefsbol TO. Polyphenols and aging. Curr Aging Sci. 2010;3(1):34–42.
259. Hua F, Tang H, Wang J, Prunty MC, Hua X, Sayeed I, et al. TAK-242, an antagonist for toll-like receptor 4, protects against acute cerebral ischemia/reperfusion injury in mice. J Cereb Blood Flow Metab. 2015;1:7.
260. Fann D, Lim Y, Cheng Y, Lok K, Chunduri P, Baik S, et al. Evidence that NF-κB and MAPK signaling promotes NLRP inflammasome activation in neurons following ischemic stroke. Mol Neurobiol. 2017;
261. Jian Z, Ding S, Deng H, Wang J, Yi W, Wang L, et al. Probenecid protects against oxygen–glucose deprivation injury in primary astrocytes by regulating inflammasome activity. Brain Res. 2016;1643:123–9.
262. Qin Y-Y, Li M, Feng X, Wang J, Cao L, Shen X-K, et al. Combined NADPH and the NOX inhibitor apocynin provides greater anti-inflammatory and neuroprotective effects in a mouse model of stroke. Free Radic Biol Med. 2017;104:333–45.
263. He Y-B, Nan L-H, Huang M, Zheng Y-F, Yang L, Xu W, et al. Paeoniflorin down-regulates the expression of NLRP1 and NLRP3 inflammasomes in rat hippocampal slices after oxygen-glucose deprivation. Int J Clin Exp Med. 2016;9(6):10907–14.
264. Qiu J, Wang M, Zhang J, Cai Q, Lu D, Li Y, et al. The neuroprotection of Sinomenine against ischemic stroke in mice by suppressing NLRP3 inflammasome via AMPK signaling. Int Immunopharmacol. 2016;40:492–500.

265. Zhang N, Zhang X, Liu X, Wang H, Xue J, Yu J, et al. Chrysophanol inhibits NALP3 inflammasome activation and ameliorates cerebral ischemia/reperfusion in mice. Mediators Inflamm. 2014;2014:370530.
266. Kono S, Kurata T, Sato K, Omote Y, Hishikawa N, Yamashita T, et al. Neurovascular protection by telmisartan via reducing neuroinflammation in stroke-resistant spontaneously hypertensive rat brain after ischemic stroke. J Stroke Cerebrovasc Dis. 2015;24(3):537–47.
267. Lu Y, Xiao G, Luo W. Minocycline suppresses NLRP3 inflammasome activation in experimental ischemic stroke. Neuroimmunomodulation. 2016;23(4):230–8.
268. Cheng Y, Wei Y, Yang W, Song Y, Shang H, Cai Y, et al. Cordycepin confers neuroprotection in mice models of intracerebral hemorrhage via suppressing NLRP3 inflammasome activation. Metab Brain Dis. 2017;32:1133–45.

Chapter 23
Microglial Function in Intracerebral Hemorrhage Injury and Recovery

A-Hyun Cho, Neethu Michael, David H. Cribbs, and Mark J. Fisher

Abstract Intracerebral hemorrhage (ICH) accounts for 10–15% of all strokes and is a major cause of disability and mortality. Introduction of blood components (e.g., thrombin, heme, and platelets) following ICH initiates neuroinflammatory responses mainly mediated by microglia, which are the resident immune cells in the central nervous system. Microglia have been shown to have dual roles in ICH, both beneficial and detrimental. The beneficial role involves phagocytosis of cellular debris and red blood cells after the hemorrhagic incident, while the detrimental role involves the production of pro-inflammatory cytokines and chemokines resulting in neuroinflammation. These dual and contradictory roles of microglia are thought to be implemented by two distinct phenotypes: classically-activated microglia and alternatively-activated microglia. We discuss herein the role of microglia in ICH with particular emphasis on its role in brain injury and recovery after ICH.

A.-H. Cho, M.D., Ph.D.
Department of Neurology, Yeoudio St. Mary's Hospital, Catholic University of Korea, Seoul, South Korea

Department of Neurology, University of California, Irvine, CA, USA
e-mail: ahyun@catholic.ac.kr

N. Michael, Ph.D.
Department of Neurology, University of California, Irvine, CA, USA
e-mail: neethum@uci.edu

D.H. Cribbs, Ph.D.
UCI MIND, University of California, Irvine, CA, USA
e-mail: cribbs@uci.edu

M.J. Fisher, M.D. (✉)
Department of Neurology, University of California, Irvine, CA, USA

Department of Anatomy & Neurobiology, University of California, Irvine, CA, USA

Department of Pathology & Laboratory Medicine, University of California, Irvine, CA, USA

Department of Neurology, UC Irvine Medical Center, 101 The City Drive South, Shanbrom Hall, Room 121, Orange, CA 92868, USA
e-mail: mfisher@uci.edu

P.A. Lapchak, J.H. Zhang (eds.), *Cellular and Molecular Approaches to Regeneration and Repair*, Springer Series in Translational Stroke Research,
https://doi.org/10.1007/978-3-319-66679-2_23

Keywords Microglia • Intracerebral hemorrhage • Brain injury • Brain recovery • Neuroinflammation

Abbreviations

BBB	Blood-brain barrier
Bcl-2	B-cell lymphoma-2
Bcl-xl	B-cell lymphoma-extra large
CD36	Cluster of differentiation 36
CD47	Cluster of differentiation 47
CEBP α	CCAAT/enhancer-binding protein alpha
CNS	Central nervous system
CX3CR-1	CX3C chemokine receptor-1
CXCL2	Chemokine (C-X-C motif) ligand 2
HO	Heme oxygenase
ICH	Intracerebral hemorrhage
IL	Interleukin
KO	Knock-out
MHCII	Major histocompatibility complex II
mTOR	Mechanistic target of rapamycin
NF-κB	Nuclear factor kappa-light-chain-enhancer of activated B cells
Nrf2	Nuclear factor (erythroid-derived 2)-like 2
PAR-1	Protease activated receptor-1
PI3K	Phosphoinositide 3-kinase
PPAR-γ	Peroxisome proliferator-activated receptor gamma
ROS	Reactive oxygen species
SIRPα	Signal-regulatory protein α
TGF-β1	Transforming growth factor-beta 1
TLR	Toll-like receptors
TNF-α	Tumor necrosis factor-α

1 Introduction

Intracerebral hemorrhage (ICH) accounts for ~10–15% of all strokes and is associated with high morbidity and mortality [1, 2]. Prognosis after ICH is often poor and clinical interventions are extremely limited. Mechanisms of brain injury and recovery after ICH have been described in detail [3, 4]. Primary brain injury at the time of ICH occurs within minutes-to-hours after hemorrhage development. It includes tissue disruption, mass effect due to increased intracranial pressure, and ischemia. This induces inflammation-driven breakdown of the blood-brain barrier (BBB) and, consequently, blood components, e.g., cells (erythrocytes, leukocytes and macrophages) and plasma proteins (thrombin, complement factors, etc.), enter the brain. Thrombin contributes to early blood-brain-barrier disruption and edema formation. This in turn initiates a cascade of inflammatory reactions, comprising of both cellular and molecular components. Importantly, there is inflammatory cell migration and glial activation, followed by tissue repair responses restoring some neurologic functions [4–7].

Microglia are the critical innate immune cells in the central nervous system. They are considered to be the earliest inflammatory cells to react to ICH and the primary cell type responsible for secondary injury after ICH. Microglia in the resting state serve an immune surveillance function, and they can sense even subtle changes in the microenvironment through a variety of surface receptors and by extending and retracting their processes [8]. They can respond to these changes by producing pro- or anti-inflammatory cytokines and neurotrophic factors. In general, microglia have two different activated polarization states: a classically-activated phenotype which is pro-inflammatory, and an alternatively-activated phenotype which is interpreted to be healing or reparative. The pro-inflammatory phenotype release inflammatory factors and damage neurons by releasing oxidative metabolites and proteases. The second phenotype has anti-inflammatory and phagocytic activity and release various protective and trophic factors repairing nerve cells [4]. The involvement of microglia and its polarization in ICH has been demonstrated [4]. There is a substantial need for improvement of prognosis after ICH, and new therapeutic targets are thus of high importance. Microglia can serve as one such highly effective target in ICH treatment, owing to its critical role in the brain immune reactions. In this chapter, we will discuss the role of microglia in brain injury and repair mechanism after ICH.

2 Microglial Role in Brain Injury After ICH

2.1 Microglial Activation During Brain Injury

Microglia are the initially activated immune cells in the brain in response to ICH. There are also infiltrating macrophages entering the perihematomal tissue [9]. Microglia respond to pro-inflammatory triggers in the surrounding brain and change to an activated phenotype. Activated microglia are detected within 1 h, and peak at ~3–7 days, decrease after 7 days, and return to basal levels by 21 days [10]. They are regarded to be the primary cell type responsible for secondary injury after ICH via release of cytokines, chemokines, prostaglandins, proteases, ferrous iron, and other immunoactive molecules [11]. Microglia also create crosstalk with T cells through antigen presentation via MHCII expression [12].

Activation of TLR4, widely expressed by microglia, leads to neuroinflammation [13] which in turn causes the upregulation of pro-inflammatory genes via NF-kB signaling [14]. Microglia are also activated by blood plasma components such as thrombin, fibrin, and heme through TLR/NF-kB pathway. TLR4 contributes to neuronal damage in ICH, and blocking of TLR4 reduces inflammatory injury and neurological deficits [14, 15]. TLR4-induced autophagy contributes to microglial activation [16]. Inhibition of the TLR4 signaling pathway may thus be a therapeutic target for ICH.

Protease activated receptor-1 (PAR-1) is involved in pro-inflammatory process, which can be detrimental to neuronal survival and are related to brain injuries including ICH. PAR-1 expression increases after ICH and PAR-1 KO mice show less brain edema and neuronal death. Thrombin-induced PAR-1 activation contributes to microglial activation/polarization toward pro-inflammatory phenotype, amplifying the release of pro-inflammatory cytokines after ICH [17].

Heme oxygenase (HO) is the rate limiting enzyme of physiological heme degradation that catalyzes heme to biliverdin, carbon monoxide, and iron. HO-1, one of the two active enzymes of HO, is found to increase after ICH, predominantly in microglia and endothelial cells. A comparison study with HO-1 KO and wild type mice showed that KO mice exhibited lesser brain injury volume and neuronal deficits, thus suggesting that HO-1 is a potent therapeutic target for ICH treatment [18]. Iron, a byproduct of the action of HO, is neurotoxic as it can produce reactive free radicals by the Fenton reaction [19]. Clearance of the iron overload in the hemorrhagic brain can thus serve as yet another target in ICH treatment.

ICH also promotes miRNA144 expression, which leads to inflammation that can affect brain function. miRNA144 is a regulator of microglial autophagy and inflammation by regulating the mTOR (mechanistic target of rapamycin) pathway, which in turn regulates the inflammatory immune response. Inhibition of miRNA144 promotes mTOR expression and decreases IL-6, IL-1β, TNF-α after ICH [20]. Chemokines such as CXCL2 produced by microglia have chemotactic activity for neutrophils and also exacerbate the inflammatory reaction [21].

2.2 Microglial Polarization

Microglia have two broad polarization states: a classically-activated pro-inflammatory phenotype and an alternatively-activated anti-inflammatory phenotype. Some have labelled these two phenotypes as M1 and M2, respectively, similar to that of macrophages. Note, however, that the use of M1 and M2 nomenclature has received criticism [22]. The classically-activated phenotype releases pro-inflammatory factors such as TNF-α, interleukin-1β, IL-6 and reactive oxygen species. It injures neurons by releasing oxidative metabolites and proteases. In contrast, the alternatively-activated microglia have anti-inflammatory and phagocytic activity, and release various protective and trophic factors repairing nerve cells. The anti-inflammatory phenotype plays an important role in brain recovery involving neurogenesis and angiogenesis [11]. However, mechanisms of phenotype shift have not yet been clearly identified. There is evidence that IL-4 is essential for anti-inflammatory polarization after ischemia, with loss of IL-4 inducing classically-activated microglia while inhibiting alternatively-activated microglia [23, 24]. Phenotype shifts can also be triggered by extra- and intra-cellular signaling [25, 26]. There is also evidence of a novel microglial phenotype ('dark microglia') which is rarely present in the steady state condition, but present in abundance during aging, stress, and Alzheimer's disease pathology [27].

3 Microglial Role in Brain Recovery After ICH

3.1 Microglial Role in Recovery Mechanism After ICH or Other Neuronal Injury

Microglia and macrophages are among the most potent modulators of CNS repair and regeneration [28]. These cells become polarized with different phenotypes at various stages after brain injuries and have contradictory functions. The status of these functions *in vivo* is complex.

Alternatively-activated microglia absorb hematoma and cell debris by phagocytic activity. CD36, a scavenger receptor of microglia/macrophages, has been reported to mediate phagocytosis of erythrocytes. Induction of CD36 by peroxisome proliferator-activated receptor gamma (PPARγ) promotes hematoma absorption in mice [29]. Inflammatory factors that are induced by TLR signaling can regulate the expression of CD36 in macrophages [30, 31] with TLR signaling inhibiting CD36 expression and hematoma absorption [32, 33]. CD47 is an integrin-associated protein expressed on erythrocytes and other cells, which blocks phagocytosis via its interaction with signal-regulatory protein-α (SIRPα), expressed by microglia/macrophages. There is evidence that CD47 has a key role in hematoma clearance by microglia after ICH [34]. More rapid hematoma clearance after ICH is induced by intracerebral injection of blood from mice lacking CD47. Erythrocyte CD47 inhibits microglial erythrophagocytosis, and CD47 may be a potential target for its regulation [34].

Neurogenesis is induced in response to ICH and this may contribute to brain repair and functional recovery [35, 36]. Microglia have positive and negative effects on neurogenesis after brain injury. Pro-inflammatory microglia impair basal neurogenesis and secrete destructive factors that hinder neurogenesis and aggravate long-term neurological deficits after injury [37-39]. Microglia with anti-inflammatory phenotype have the potential to promote the proliferation and migration of neural progenitor cells via secretion of trophic factors [38, 40–42]. Whether both these phenotypes are pro-neurogenic or anti-neurogenic is not yet clearly demonstrated and they may have overlapping functions. The direction of their function may depend on both phenotype and the specific time after injury.

Axonal regeneration after CNS injury occurs to a limited extent. Classically-activated microglia hinder axonal regeneration [43, 44]. However, after spinal cord injury, monocyte-derived macrophages have been shown to be critical for recovery [45].

Brain angiogenesis is yet another recovery process after ICH [46]. It is well known that tissue macrophages have proangiogenic function during wound healing [47, 48] and produce proangiogenic factors such as vascular endothelial growth factor (VEGF) and IL-8 [48, 49]. Activated microglia regulate endothelial cell proliferation [50]. However, the exact role of resident microglia in angiogenesis after ICH is still unclear.

Microglia have been shown to regulate synaptic plasticity after CNS injuries and activated microglia have an adaptive role in the regulation of synaptic homeostasis and plasticity [51]. Oligodendrocyte differentiation is promoted by the anti-inflammatory phenotype and impaired by its blockage [52]. Most of the reports describing microglia/macrophage function in repair mechanisms have been performed in ischemic brain injury or spinal cord injury models rather than in ICH models. More studies are thus needed for a more complete understanding of microglial role in ICH recovery.

3.2 *Mediators of Microglia-Facilitated ICH Recovery*

Identification of specific mediators of microglial activation is particularly important to determine specific therapeutic targets for attenuating microglia-mediated brain injury. Nrf2 and PPAR-γ promote transcription of antioxidant genes. Nrf2- deficient mice have severe neurological deficits with increased leukocyte infiltration, ROS production, DNA damage, and cytochrome c release [53]. Nrf2 is present in neurons, astrocytes, and microglia, and its expression is neuroprotective against early inflammatory brain injury in hemorrhagic stroke models. In addition, PPAR-γ activators reduce the expression of pro-inflammatory genes and prevent neuronal damage [29, 54]. PPAR-γ thus appears to play a protective role after hemorrhagic stroke. The chemokine receptor CX3CR-1 is required for anti-inflammatory polarization of microglia, which facilitates recovery after ICH [13]. miRNA124 is another factor which enhances alternatively-activated anti-inflammatory polarization and ameliorates ICH-induced inflammation via CEBP α pathway [55].

During the repair process, IL-4 production in the CNS controls autoimmune inflammation by inducing the anti-inflammatory phenotype [4, 56]. IL-10 is an anti-inflammatory cytokine also produced by glial cells, which inhibits pro-inflammatory cytokines of microglia. IL-10 has also been shown to prevent apoptosis by activating PI3K, enhancing apoptotic factors such as Bcl-2 and Bcl-xl, and depleting caspase-3 [57]. Intracerebral injection of IL-4 inhibits activation of the pro-inflammatory phenotype while enhancing the anti-inflammatory phenotype, along with improving neurobehavioral recovery from deficits after ICH [23]. TGF-β1 has been shown to reduce microglial inflammation, and improve functional recovery after intracerebral hemorrhage [58]. Regulatory T cells also contribute to altering microglial polarization and reducing inflammation in ICH [59]. ST2 is a member of interleukin-1 receptor family while IL-33 acts as its ligand, and both have critical roles in inflammatory responses. Work by Yang et al. showed that ST2 receptor deficiency aggravates brain infarction and that ST2/IL-33 interaction stimulates microglial IL-10 production, which is also a marker of the anti-inflammatory microglial phenotype [60].

4 Conclusion

Microglia play a vital role in both brain injury and recovery after ICH. With its detrimental and beneficial roles (pro-inflammatory and anti-inflammatory, respectively), microglia can act as a double-edged sword. Studies of mediators of inflammatory response implicate microglia as a potential therapeutic target. Future studies emphasizing *in vivo* models and utilizing advanced imaging methods [61, 62] represent a promising approach to improve clinical outcome following ICH by targetting microglia.

Acknowledgement This work is supported by NIH R01 NS20989

References

1. Qureshi AI, Mendelow AD, Hanley DF. Intracerebral haemorrhage. Lancet. 2009; 373(9675):1632–44.
2. Qureshi AI, Suri MF, Nasar A, Kirmani JF, Ezzeddine MA, Divani AA, et al. Changes in cost and outcome among US patients with stroke hospitalized in 1990 to 1991 and those hospitalized in 2000 to 2001. Stroke. 2007;38(7):2180–4.
3. Egashira Y, Hua Y, Keep RF, Xi G. Intercellular cross-talk in intracerebral hemorrhage. Brain Res. 2015;1623:97–109.
4. Hu X, Leak RK, Shi Y, Suenaga J, Gao Y, Zheng P, et al. Microglial and macrophage polarization-new prospects for brain repair. Nat Rev Neurol. 2015;11(1):56–64.
5. Wang J, Tsirka SE. Contribution of extracellular proteolysis and microglia to intracerebral hemorrhage. Neurocrit Care. 2005;3(1):77–85.
6. Wang J, Rogove AD, Tsirka AE, Tsirka SE. Protective role of tuftsin fragment 1-3 in an animal model of intracerebral hemorrhage. Ann Neurol. 2003;54(5):655–64.
7. Askenase MH, Sansing LH. Stages of the inflammatory response in pathology and tissue repair after intracerebral hemorrhage. Semin Neurol. 2016;36(3):288–97.
8. Nimmerjahn A, Kirchhoff F, Helmchen F. Resting microglial cells are highly dynamic surveillants of brain parenchyma in vivo. Science. 2005;308(5726):1314–8.
9. Hammond MD, Taylor RA, Mullen MT, Ai Y, Aguila HL, Mack M, et al. CCR2+ Ly6C(hi) inflammatory monocyte recruitment exacerbates acute disability following intracerebral hemorrhage. J Neurosci. 2014;34(11):3901–9.
10. Keep RF, Hua Y, Xi G. Intracerebral haemorrhage: mechanisms of injury and therapeutic targets. Lancet Neurol. 2012;11(8):720–31.
11. Zhang Z, Zhang Z, Lu H, Yang Q, Wu H, Wang J. Microglial Polarization and Inflammatory Mediators After Intracerebral Hemorrhage. Mol Neurobiol. 2017;54(3):1874–86.

12. Starossom SC, Mascanfroni ID, Imitola J, Cao L, Raddassi K, Hernandez SF, et al. Galectin-1 deactivates classically activated microglia and protects from inflammation-induced neurodegeneration. Immunity. 2012;37(2):249–63.
13. Taylor RA, Sansing LH. Microglial responses after ischemic stroke and intracerebral hemorrhage. Clin Dev Immunol. 2013;2013:746068.
14. Lin S, Yin Q, Zhong Q, Lv FL, Zhou Y, Li JQ, et al. Heme activates TLR4-mediated inflammatory injury via MyD88/TRIF signaling pathway in intracerebral hemorrhage. J Neuroinflammation. 2012;9:46.
15. Wang YC, Wang PF, Fang H, Chen J, Xiong XY, Yang QW. Toll-like receptor 4 antagonist attenuates intracerebral hemorrhage-induced brain injury. Stroke. 2013;44(9):2545–52.
16. Yang Z, Liu B, Zhong L, Shen H, Lin C, Lin L, et al. Toll-like receptor-4-mediated autophagy contributes to microglial activation and inflammatory injury in mouse models of intracerebral haemorrhage. Neuropathol Appl Neurobiol. 2015;41(4):e95–106.
17. Wan S, Cheng Y, Jin H, Guo D, Hua Y, Keep RF, et al. Microglia Activation and Polarization After Intracerebral Hemorrhage in Mice: the Role of Protease-Activated Receptor-1. Transl Stroke Res. 2016;7(6):478–87.
18. Wang J, Doré S. Heme oxygenase-1 exacerbates early brain injury after intracerebral haemorrhage. Brain : a journal of neurology. 2007;130(Pt 6):1643–1652.
19. Zecca L, Youdim MB, Riederer P, Connor JR, Crichton RR. Iron, brain ageing and neurodegenerative disorders. Nat Rev Neurosci. 2004;5(11):863–73.
20. Yu A, Zhang T, Zhong W, Duan H, Wang S, Ye P, et al. miRNA-144 induces microglial autophagy and inflammation following intracerebral hemorrhage. Immunol Lett. 2017;182:18–23.
21. Shiratori M, Tozaki-Saitoh H, Yoshitake M, Tsuda M, Inoue K. P2X7 receptor activation induces CXCL2 production in microglia through NFAT and PKC/MAPK pathways. J Neurochem. 2010;114(3):810–9.
22. Ransohoff RM. A polarizing question: do M1 and M2 microglia exist? Nat Neurosci. 2016;19(8):987–91.
23. Yang J, Ding S, Huang W, Hu J, Huang S, Zhang Y, et al. Interleukin-4 ameliorates the functional recovery of intracerebral hemorrhage through the alternative activation of microglia/macrophage. Front Neurosci. 2016;10:61.
24. Liu X, Liu J, Zhao S, Zhang H, Cai W, Cai M, et al. Interleukin-4 is essential for microglia/macrophage M2 polarization and long-term recovery after cerebral ischemia. Stroke. 2016;47(2):498–504.
25. Hu X, Li P, Guo Y, Wang H, Leak RK, Chen S, et al. Microglia/macrophage polarization dynamics reveal novel mechanism of injury expansion after focal cerebral ischemia. Stroke. 2012;43(11):3063–70.
26. Chhor V, Le Charpentier T, Lebon S, Ore MV, Celador IL, Josserand J, et al. Characterization of phenotype markers and neuronotoxic potential of polarised primary microglia in vitro. Brain Behav Immun. 2013;32:70–85.
27. Bisht K, Sharma KP, Lecours C, Sanchez MG, El Hajj H, Milior G, et al. Dark microglia: a new phenotype predominantly associated with pathological states. Glia. 2016;64(5):826–39.
28. Hanisch UK, Kettenmann H. Microglia: active sensor and versatile effector cells in the normal and pathologic brain. Nat Neurosci. 2007;10(11):1387–94.
29. Zhao X, Sun G, Zhang J, Strong R, Song W, Gonzales N, et al. Hematoma resolution as a target for intracerebral hemorrhage treatment: role for peroxisome proliferator-activated receptor gamma in microglia/macrophages. Ann Neurol. 2007;61(4):352–62.
30. Zamora C, Canto E, Nieto JC, Angels Ortiz M, Juarez C, Vidal S. Functional consequences of CD36 downregulation by TLR signals. Cytokine. 2012;60(1):257–65.
31. Fang H, Chen J, Lin S, Wang P, Wang Y, Xiong X, et al. CD36-mediated hematoma absorption following intracerebral hemorrhage: negative regulation by TLR4 signaling. J Immunol. 2014;192(12):5984–92.
32. Fang H, Wang PF, Zhou Y, Wang YC, Yang QW. Toll-like receptor 4 signaling in intracerebral hemorrhage-induced inflammation and injury. J Neuroinflammation. 2013;10:27.

33. Rodriguez-Yanez M, Brea D, Arias S, Blanco M, Pumar JM, Castillo J, et al. Increased expression of Toll-like receptors 2 and 4 is associated with poor outcome in intracerebral hemorrhage. J Neuroimmunol. 2012;247(1-2):75–80.
34. Ni W, Mao S, Xi G, Keep RF, Hua Y. Role of erythrocyte CD47 in intracerebral hematoma clearance. Stroke. 2016;47(2):505–11.
35. Masuda T, Isobe Y, Aihara N, Furuyama F, Misumi S, Kim TS, et al. Increase in neurogenesis and neuroblast migration after a small intracerebral hemorrhage in rats. Neurosci Lett. 2007;425(2):114–9.
36. Shen J, Xie L, Mao X, Zhou Y, Zhan R, Greenberg DA, et al. Neurogenesis after primary intracerebral hemorrhage in adult human brain. J Cereb Blood Flow Metab. 2008;28(8):1460–8.
37. Ekdahl CT, Claasen JH, Bonde S, Kokaia Z, Lindvall O. Inflammation is detrimental for neurogenesis in adult brain. Proc Natl Acad Sci U S A. 2003;100(23):13632–7.
38. Butovsky O, Ziv Y, Schwartz A, Landa G, Talpalar AE, Pluchino S, et al. Microglia activated by IL-4 or IFN-gamma differentially induce neurogenesis and oligodendrogenesis from adult stem/progenitor cells. Mol Cell Neurosci. 2006;31(1):149–60.
39. Kim BJ, Kim MJ, Park JM, Lee SH, Kim YJ, Ryu S, et al. Reduced neurogenesis after suppressed inflammation by minocycline in transient cerebral ischemia in rat. J Neurol Sci. 2009;279(1-2):70–5.
40. Nikolakopoulou AM, Dutta R, Chen Z, Miller RH, Trapp BD. Activated microglia enhance neurogenesis via trypsinogen secretion. Proc Natl Acad Sci U S A. 2013;110(21):8714–9.
41. Yan YP, Lang BT, Vemuganti R, Dempsey RJ. Galectin-3 mediates post-ischemic tissue remodeling. Brain Res. 2009;1288:116–24.
42. Choi YS, Cho HY, Hoyt KR, Naegele JR, Obrietan K. IGF-1 receptor-mediated ERK/MAPK signaling couples status epilepticus to progenitor cell proliferation in the subgranular layer of the dentate gyrus. Glia. 2008;56(7):791–800.
43. Kitayama M, Ueno M, Itakura T, Yamashita T. Activated microglia inhibit axonal growth through RGMa. PLoS One. 2011;6(9):e25234.
44. Horn KP, Busch SA, Hawthorne AL, van Rooijen N, Silver J. Another barrier to regeneration in the CNS: activated macrophages induce extensive retraction of dystrophic axons through direct physical interactions. J Neurosci. 2008;28(38):9330–41.
45. Shechter R, London A, Varol C, Raposo C, Cusimano M, Yovel G, et al. Infiltrating blood-derived macrophages are vital cells playing an anti-inflammatory role in recovery from spinal cord injury in mice. PLoS Med. 2009;6(7):e1000113.
46. Tang T, Liu XJ, Zhang ZQ, Zhou HJ, Luo JK, Huang JF, et al. Cerebral angiogenesis after collagenase-induced intracerebral hemorrhage in rats. Brain Res. 2007;1175:134–42.
47. Zajac E, Schweighofer B, Kupriyanova TA, Juncker-Jensen A, Minder P, Quigley JP, et al. Angiogenic capacity of M1- and M2-polarized macrophages is determined by the levels of TIMP-1 complexed with their secreted proMMP-9. Blood. 2013;122(25):4054–67.
48. Medina RJ, O'Neill CL, O'Doherty TM, Knott H, Guduric-Fuchs J, Gardiner TA, et al. Myeloid angiogenic cells act as alternative M2 macrophages and modulate angiogenesis through interleukin-8. Mol Med. 2011;17(9-10):1045–55.
49. Willenborg S, Lucas T, van Loo G, Knipper JA, Krieg T, Haase I, et al. CCR2 recruits an inflammatory macrophage subpopulation critical for angiogenesis in tissue repair. Blood. 2012;120(3):613–25.
50. Welser JV, Li L, Milner R. Microglial activation state exerts a biphasic influence on brain endothelial cell proliferation by regulating the balance of TNF and TGF-beta1. J Neuroinflammation. 2010;7:89.
51. Stellwagen D, Malenka RC. Synaptic scaling mediated by glial TNF-alpha. Nature. 2006;440(7087):1054–9.
52. Miron VE, Boyd A, Zhao JW, Yuen TJ, Ruckh JM, Shadrach JL, et al. M2 microglia and macrophages drive oligodendrocyte differentiation during CNS remyelination. Nat Neurosci. 2013;16(9):1211–8.
53. Wang J, Fields J, Zhao C, Langer J, Thimmulappa RK, Kensler TW, et al. Role of Nrf2 in protection against intracerebral hemorrhage injury in mice. Free Radic Biol Med. 2007;43(3):408–14.

54. Zhao X, Zhang Y, Strong R, Grotta JC, Aronowski J. 15d-Prostaglandin J2 activates peroxisome proliferator-activated receptor-gamma, promotes expression of catalase, and reduces inflammation, behavioral dysfunction, and neuronal loss after intracerebral hemorrhage in rats. J Cereb Blood Flow Metab. 2006;26(6):811–20.
55. Yu A, Zhang T, Duan H, Pan Y, Zhang X, Yang G, et al. MiR-124 contributes to M2 polarization of microglia and confers brain inflammatory protection via the C/EBP-alpha pathway in intracerebral hemorrhage. Immunol Lett. 2017;182:1–11.
56. Ponomarev ED, Maresz K, Tan Y, Dittel BN. CNS-derived interleukin-4 is essential for the regulation of autoimmune inflammation and induces a state of alternative activation in microglial cells. J Neurosci. 2007;27(40):10714–21.
57. Moore KW, de Waal Malefyt R, Coffman RL, O'Garra A. Interleukin-10 and the interleukin-10 receptor. Annu Rev Immunol. 2001;19:683–765.
58. Taylor RA, Chang CF, Goods BA, Hammond MD, Mac Grory B, Ai Y, et al. TGF-beta1 modulates microglial phenotype and promotes recovery after intracerebral hemorrhage. J Clin Invest. 2017;127(1):280–92.
59. Zhou K, Zhong Q, Wang YC, Xiong XY, Meng ZY, Zhao T, et al. Regulatory T cells ameliorate intracerebral hemorrhage-induced inflammatory injury by modulating microglia/macrophage polarization through the IL-10/GSK3beta/PTEN axis. J Cereb Blood Flow Metab. 2017;37(3):967–79.
60. Yang Y, Liu H, Zhang H, Ye Q, Wang J, Yang B, et al. ST2/IL-33-dependent microglial response limits acute ischemic brain injury. J Neurosci. 2017;37(18):4692–704.
61. Venneti S, Lopresti BJ, Wiley CA. Molecular imaging of microglia/macrophages in the brain. Glia. 2013;61(1):10–23.
62. Flogel U, Ding Z, Hardung H, Jander S, Reichmann G, Jacoby C, et al. In vivo monitoring of inflammation after cardiac and cerebral ischemia by fluorine magnetic resonance imaging. Circulation. 2008;118(2):140–8.

Chapter 24
The Role of T Cells in Post-stroke Regeneration

Julia V. Cramer and Arthur Liesz

Abstract The interaction of the immune system with the brain is necessary for development and surveillance of the healthy brain. The influence of the adaptive immune system on several brain diseases has been described in great detail. In ischemic stroke, a growing body of evidence has demonstrated a key role for T cells in the acute phase after stroke. Pro- and anti-inflammatory T cell subpopulations impact in this early phase the inflammatory milieu and directly affect secondary lesion progression and neuronal injury. Recently, a functional role for T cells has also become more evident also in delayed neuronal (dys-)function and late-phase recovery after stroke. Here, T cells may also affect various non-immunological pathways involved in tissue repair, neuronal plasticity and functional recovery. These pleiotropic effects of T cells on mechanisms such as neurogenesis and angiogenesis suggest T cells as potential therapeutic target to modulate post-stroke regeneration. This chapter will provide a comprehensive overview of the current knowledge about the role of T cells in stroke with a particular focus on regenerative processes in the chronic phase.

Keywords Stroke • Brain ischemia • T cell • Lymphocyte • Inflammation • Neuroimmunology • Regeneration • Cytokines

Abbreviations

APC	Antigen presenting cell
BBB	Blood brain barrier
BDNF	Brain derived neurotrophic factor
CCR	Chemokine receptor
CNS	Central nervous system
CD	Cluster of Differentiation

J.V. Cramer • A. Liesz (✉)
Institute for Stroke and Dementia Research (ISD), Klinikum der Universität München, Feodor-Lynen-Straße 17, 81377 Munich, Germany
e-mail: julia.cramer@med.uni-muenchen.de; arthur.liesz@med.uni-muenchen.de

P.A. Lapchak, J.H. Zhang (eds.), *Cellular and Molecular Approaches to Regeneration and Repair*, Springer Series in Translational Stroke Research,
https://doi.org/10.1007/978-3-319-66679-2_24

CTLA-4	Cytotoxic T-lymphocyte-associated Protein 4
DAMPs	Damage-Associated Molecular Patterns
GITR	Glucocorticoid-induced TNF receptor
IFNγ	Interferon gamma
IgE	Immunoglobulin E
IGF-1	Insulin like growth factor 1
IL	Interleukin
MCAO	Middle Cerebral Artery Occlusion
MHC	Major Histocompatibility Complex
MMP	Metalloproteinase
NPC	Neural Precursor Cell
RAGE	Receptor for advanced glycosylation endproducts
SGZ	Subgranular zone
SVZ	Subventricular zone
TGF-β	Transforming growth factor beta
Th cell	T helper cell
TLRs	Toll-like receptors
TNF-α	Tumor necrosis factor alpha
Treg	Regulatory T cell
VEGF	Vascular Endothelial Growth Factor

1 An Introduction to T Cells in Brain Function

1.1 T Cell Subsets and Their Physiological Role in the Body

T lymphocytes are part of the adaptive immune system and are divided in several subtypes depending on their functional properties, which are determined by transcription factors, surface markers and the cytokine secretion profile. There are two major groups, the CD8+ cytotoxic T cells and the CD4+ T helper (Th) cells. The former can act directly via cell-cell interaction dependent cytolytic pathways, involving the perforin-granzyme effector mechanism [1]. Cytotoxic T cells recognize antigens presented via MHC-I and thereby play a critical role in microbial defense [1] and primary autoimmune diseases [2, 3] (Fig. 24.1).

CD4+ Th cells have mainly indirect effector function and are able to induce proliferation of other immune cells. Despite recognizing a specific antigen, naïve CD4+ T helper cells possess still a vast plasticity. Hence, they are able to polarize into different effector and regulatory subtypes. The differentiation depends on three signals which they receive both from innate and adaptive immune cells: antigen-presentation, co-receptor stimulation and cytokine signaling. Classically it is reported, that Interferon-gamma (IFNγ) and Interleukin (IL) -12 signal can induce the pro-inflammatory T helper 1 (Th1) subtype [4, 5]. Th1 cells help resolve bacterial infection and release pro-inflammatory cytokines such as IFNγ and IL-12. In contrast, T helper 2 (Th2)

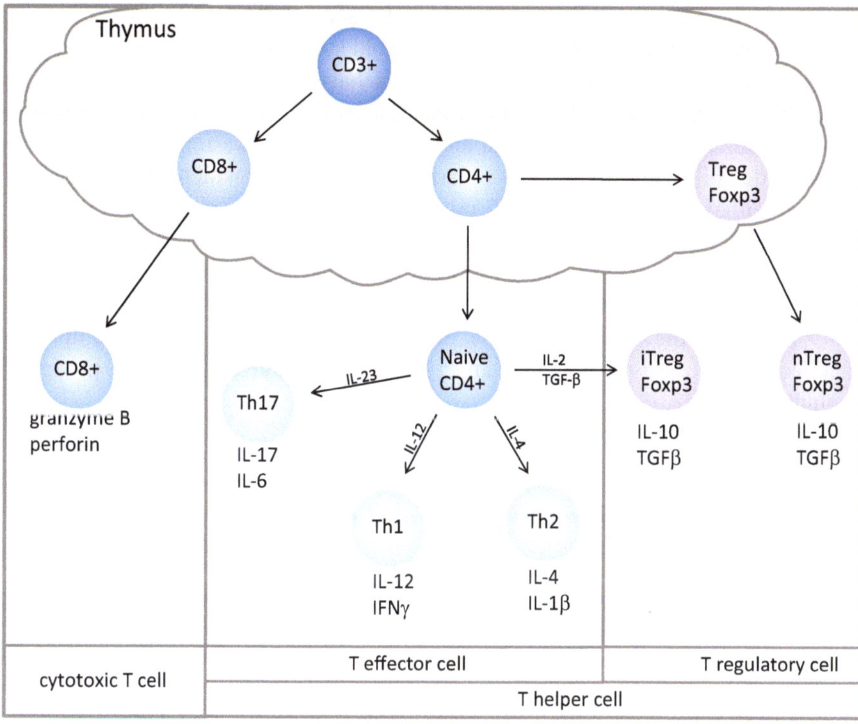

Fig. 24.1 Simplified overview of the main T cell subpopulations. Cytokines which promote polarization to the individual subsets are added to the *arrows*, cytokines released by the different cells are indicated below

polarization is mainly driven by IL-4 [6, 7]. Th2 cells are able to enhance B cell responses and initiate in synergy with IL-4 an IgE class-switch. IgE is a key mediator in mast cell activation and therefore, Th2 cells have a critical role in the pathophysiology of allergy; additionally, they provide protection against parasitosis [8]. Over the past decade a substantial variety of other CD4+ subtypes has been described, from which pro-inflammatory Th17 cells and regulatory T cells (Tregs) are the best characterized and with direct implications for neuroinflammatory diseases and modulation of the post-stroke immune response. Developing in presence of IL-23, Th17 cells produce cytokines associated with chronic inflammation such as IL-17, IL-6 and Tumor necrosis factor alpha (TNF-α) [9, 10]. In contrast, CD4+CD25+Foxp3+ Tregs play a key role in self-tolerance by suppressing autoreactive T cells under physiological conditions. Furthermore, these cells are able to dampen over-activation by attenuating T cell responses mainly by secretion of the anti-inflammatory molecules IL-10, CTLA-4 and TGF-β, in addition to other cell-cell contact dependent mechanisms [11–13].

Another subtype distinct from the classical T cells are γδ T cells. The composition of their T cell receptor by γ and δ subunits separates them from conventional

T cells in which the T cell receptor consists of α and β subunits. They are belonging to the adaptive immune cell family, but also hold some characteristics of the innate immune system in addition [14]. γδ T cells have different antigen-recognition requirements, most likely do not need antigen-processing and MHC-presentation, and can additionally recognize lipid antigens. Their adaptive immune characteristics include ability to differentially polarize. Depending on which stimuli received, they can either polarize to an IL-17 producing pro-inflammatory type or a regulatory phenotype [15, 16]. The activation of this cell population is very rapid similar to innate immune cells [15]. γδ T cells are preactivated after their development in the thymus and there is no need for clonal expansion [14].

Considering the diversity of the different T cell subsets, their complexity enables them to play a role in many physiological and pathological processes including stroke.

1.2 T Cells in the Healthy Brain and Its Homeostasis

In contrast to the previously prevailing concept of the brain as an immune privileged organ, several more recent studies have unequivocally demonstrated constant surveillance of the brain by circulating leukocytes. Patrolling leukocytes have been shown to be present in certain parts of the brain under healthy conditions and to be critical in maintaining physiological function and plasticity [17–19]. Under physiological conditions, T cells are able to migrate across the endothelium, preferentially in post-capillary venules and enter the subarachnoid and perivascular space as well as the CSF [20]. It is most likely that T cell activation is facilitating the cerebral extravasation [21]. However, lymphocytes are rarely found in the brain parenchyma itself. Highly specialized cells and membranes regulate T cell access from the perivascular space to the brain [22]. Astrocyte ensheathment of the vasculature [23] and the glia limitans, which encompass the microvessels in brain parenchyma and surface [22], restrict the perivascular space and impede immune cell infiltration [24].

Cerebral immunosurveillance by circulating lymphocytes is restricted to certain parts of the brain: In the perivascular and subarachnoid space T cells encounter resident antigen presenting cells (APCs) such as perivascular macrophages [20, 25]. Via their MHC-II molecules these APCs present antigens which are drained from the brain parenchyma [22]. Interacting with them, T cells can become reactivated, if their cognate antigen is presented [26]. Missing antigen recognition has been reported to terminate cerebral T cell patrolling and inducing T cell egress exit via the CSF to cervical lymph nodes [27] without entering the parenchyma [20]. By this mode of antigen-specific brain surveillance, the lymphocyte population presented in the brain and surrounding structures is turned over almost twice a day [20]. Some T cells, mainly CD4+ effector memory T cells, can reside in the choroid plexus with receptors specific for CNS antigens [28] and provide a cell pool for immediate immune reactions. Furthermore, the same T cell subset can be found in healthy human CSF samples [29].

Immune cells and especially T cells are known to be involved in many physiological processes in brain homeostasis. Their presence positively influences hippocampal neurogenesis and hippocampal brain-derived neurotrophic factor (BDNF) expression [17, 30]. These changes were not only observed on a biochemical or histological level. Notably, the connection between the immune system and neurogenesis has been clearly demonstrated in animal behavior studies. Behavioral deficits, especially in spatial memory tasks, are apparent in animals devoid of a functional immune system [18, 30, 31]. Interestingly, these memory deficits can at least be partially restored by T cell transfer to immunodeficient animals [31]. This finding was further verified by following studies which were able to identify that these characteristics were specifically attributable to the subpopulation of CD4+ Th cells [32]. Moreover, lymphocyte deficiency has been associated with an impact on social behavior in mice [33]. In addition to this, it was found that blocking T cell migration to the meninges was also able to evoke behavioral abnormalities in mice [30, 33, 34]. Taken together, these findings highlight the importance of T cells for physiological brain function and behavior.

Yet, the precise mechanism by which T cells cause this impact is insufficiently known. Previous reports have associated T cell cytokines such as IFNγ, IL-6 and IL-4 to the above-mentioned behavioral and structural differences between T cell bearing and deficient animals [33–36]. Further research for the analysis of T cell mechanisms in brain homeostasis under physiological conditions will be warranted for the identification of potentially novel targets of T cell-mediated pathologies in a variety of brain disorders.

2 T Cells in Stroke

2.1 Dynamics of Leukocyte Infiltration After Cerebral Ischemia

Acute brain ischemia leads to sterile neuroinflammation attracting both local and systemic immune cells to the lesion site. In the ischemic core, destruction of the cellular integrity in necrotic cells allows intracellular content to be released into the extracellular space. These soluble components are termed damage-associated molecular patterns (DAMPs) [37]. DAMPs initiate the inflammatory cascade both locally and systemically. Pattern recognition receptors, including Toll-like receptors (TLRs) [38] and the receptor for advanced glycosylation endproducts (RAGE) [39] sense DAMPs. Interaction of cerebral DAMPs with TLRs found on microglia—the resident immune cells of the brain—induces their activation and migration to the infarct lesion [40]. Consequently, microglia generate a pro-inflammatory milieu by secretion of cytokines and chemoattractant molecules [41, 42]. Microglia-derived cytokines, such as IL-6, IL-1β and TNF-α induce two processes involved in the recruitment of circulating lymphocytes to the damaged brain [43]. Cytokines promote the activation of peri-infarct endothelial cells and their upregulation of adhesion molecule expression. This facilitates leukocyte rolling, an initial step in extravasation of circulating leukocytes [43]. Simultaneously, microglia-derived cytokines pass on the signal to circulating peripheral immune cells which in turn become activated as well. As a consequence, activated immune cells enhance expression of proteins such as integrins, which are necessary for firm adhesion to the endothelium which is then followed by transmigration [44]. Additionally, activated T cells secrete cytokines which are also able to induce endothelial activation and thereby amplify the inflammatory response [45]. Another mechanism enabling increased leukocyte entry to the post-ischemic brain parenchyma is the degradation of the glia limitans. Under healthy conditions this membrane—which is part of the outer layer of the blood brain barrier (BBB)—is limiting the access for leukocytes to the parenchyma. After its degradation by metalloproteinases (MMPs), especially MMP-2 and MMP-9 [46], T cell infiltration to the lesion site is eased.

The dynamics of leukocyte recruitment to the brain differs substantially for different cell types of circulating leukocytes (Fig. 24.2): After stroke, invading macrophages appear within minutes to hours after ischemia at the lesion site [47, 48]. Thereafter, the number of invading neutrophils peaks at 3 days post infarct [48]. Their presence persists at least until day 7 and declines afterwards [48]. However, one report by Enzmann et al. has suggested that the majority of granulocytes attracted to the ischemic brain might not enter the brain parenchyma but are found in one murine stroke model only in the perivascular space [49]. T lymphocytes in general migrate preferentially to the lesion borders, increase in cell number of several days after ischemia [50] and can be detected throughout at least 30 days post infarct in the brain parenchyma [51]. Looking more detailed at the individual

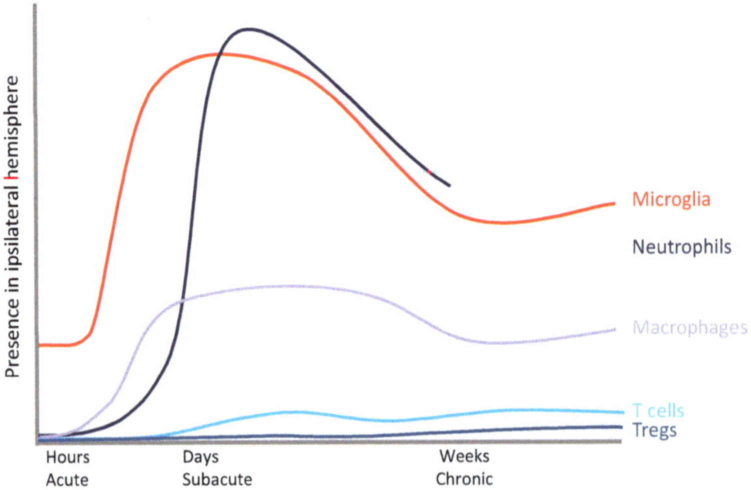

Fig. 24.2 Dynamics of immune cell infiltration into the ipsilateral hemisphere. Modified from [48, 51]

subsets of the T cells, initially, CD8+ cytotoxic T cells are recruited as early as 3 h after onset of ischemia [52]. CD4+ and Natural Killer T cells are following at around 24 h after ischemia [52]. However, Treg aggregation culminates later and this subtype is still significantly present at day 30 post lesion [51].

2.2 *The Impact of T Cell Subsets in the Acute Phase*

T cells play a crucial role in secondary stroke progression [41, 53, 54] albeit representing only a very small subpopulation of brain-invading leukocytes—about 3000 T cells compared to 65,000 neutrophils per hemisphere [48]. Transgenic mice deficient of lymphocytes have been consistently demonstrated to have smaller lesions after transient and permanent middle cerebral artery occlusion (MCAO) compared to immunocompetent control animals [55–60]. Further, restoring the lymphocyte population in such immunodeficient animals by adoptive cell transfer reversed the protective effect, resulting in infarct volumes comparable to WT animals [58]. Finally, antibody mediated depletion of single T cell subsets, namely CD8+, CD4+ and γδ T cells, was also able to attenuate ischemic injury [56, 61–63]. But not only depletion of leukocytes or different subsets but also changing the ratio of pro inflammatory and anti-inflammatory subsets alleviates post-stroke neuroinflammation. Pro-inflammatory Th1 and Th17 cells constitute the majority of T cells entering the brain after stroke. Shifting the pro-inflammatory T cell polarization towards Th2 cell activation was shown to reduce secondary neuronal degeneration after stroke [64]. The processes leading to this protection are incompletely understood, but in focus of current research.

Amelioration of stroke outcome due to T cell-targeted immunomodulation was detectable as early as some hours after ischemia. However, antigen-specific clonal expansion of T cells requires approximately a week for efficient generation of antigen-specific immune response. Therefore, it is rather unlikely that antigen-specific mechanisms play a major role in post-stroke T cell immunity. It is more likely that cytokines secreted by T cells after invading the ischemic brain or already in the circulation are key contributors to T cell-mediated effects on stroke outcome. Consequently, since the individual subtypes of T cells play different physiological roles and produce characteristic cytokines, it is important to examine the functional role of individual T cell subpopulations separately.

Cytotoxic T cells have detrimental effects after stroke. Antigen-dependent activation of this cell population was associated with their invasion to the lesion site [61]. Secretion of perforin and granzymes by cytotoxic T cells leads to secondary stroke progression (Fig. 24.3a). Mracsko et al. reported that CD8+ T cells were interacting with neurons in the subacute phase after stroke and contributed to neuronal cell death [61]. Additionally, the release of IFNγ leads to upregulation of MHC-I molecules in neurons [65] potentially contributing to secondary immunological cell death by promoting interaction of neurons with cytotoxic T cells.

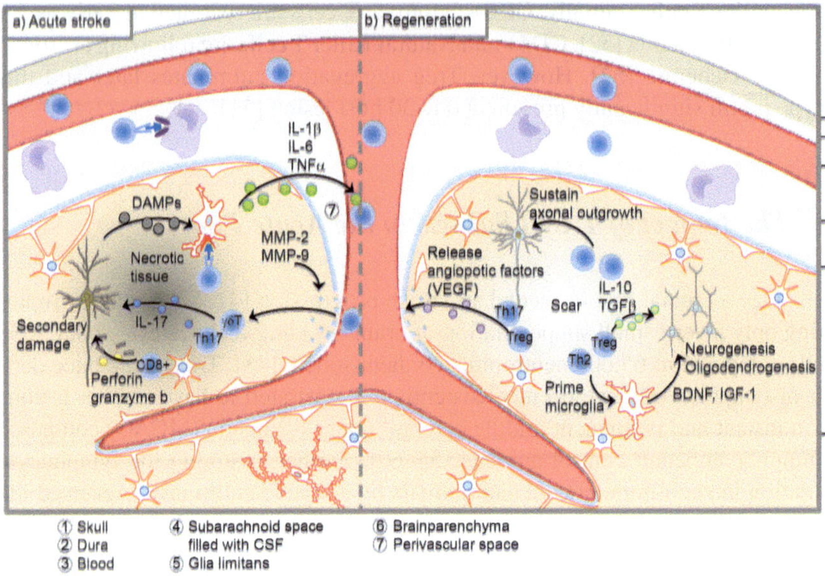

Fig. 24.3 Functional role of T cells in the acute phase and regeneration phase after stroke. (**a**) In the acute phase after stroke, necrotic cells release DAMPs which activate local microglia. Activated microglia secrete cytokines and chemoattractants facilitating T cell entry to the lesion site. T cells promote exacerbation of neuroinflammation and increase secondary cell death. (**b**) Possible roles of T cells in recovery after stroke. T cells are involved in several processes involved in tissue regeneration such as neurogenesis, angiogenesis and axonal outgrowth

γδ T cells and to a lesser extent Th17 harm the damaged tissue by secretion of IL-17 [62] which peaks at day 3 post lesion [66]. γδ T cell deficient mice—in genetic and antibody-mediated depletion models—have smaller infarct volumes compared to respective control mice [62]. Moreover, neutralization of IL-17 using specific antibodies also significantly reduced infarct size [63]. IL-17 binds to the IL-17 receptor which is upregulated in astrocytes, microglia and neurons after stroke [67]. Downstream signaling of the IL-17 receptor enhances expression of NF-κB and GSK3beta which were associated with induction of neuronal apoptosis in different neurodegenerative diseases [68–70]. NF-κB is critical in several neuroinflammatory processes after stroke including microglial activation, leukocyte trafficking and secretion of pro-inflammatory cytokines [68]. These manifold functions of IL-17 suggest a key role of the cytokine and its main producers, gamma delta and Th17 cells, in post-stroke lesion progression [71] (Fig. 24.3a). Interestingly, a second peak in IL-17 expression has been described to occur in the late phase around day 28 after stroke [66]. At this late time point mainly reactive astrocytes secreted the cytokine [66]. Lin et al. showed that at this late stage, IL-17 might improve neurogenesis in contrast to the rather detrimental function during the acute phase [66]. This example of a potentially opposing function of a specific cytokine—pro-inflammatory functions of IL-17 in the acute phase and potentially restorative capacities in the later stages—highlights the complexity of the adaptive immune response after stroke (Fig. 24.3). While the previously discussed T cell subpopulations secrete pro-inflammatory mediators and contribute mainly to propagation of the neuroinflammatory milieu, Foxp3+ Treg have in contrary been identified to limit an overshooting immune response after stroke [55]. Although the function of Treg cells in vascular inflammation after stroke is still under debate, the vast majority of studies by now have verified a protective function of this cell type on parenchymal post-stroke neuroinflammation [72]. Several studies showed exacerbation of neuronal degeneration in models of Treg depletion including genetic or antibody-mediated depletion or selective adoptive cell transfer models [55, 73, 74]. IL-10 has been identified as the key mediator of Treg in facilitating their neuroprotective function [73, 75]. Accordingly, intraventricular IL-10 injection reversed the effect of Treg depletion [55]. In transgenic mice which overexpress IL-10, lower levels of pro-apoptotic caspase 3 were detected in the injured hemisphere on day 4 post stroke [76]. *In vitro* microglia stimulated by IL-10 enhanced neuronal survival [77]. In addition, both overexpressing IL-10 and IL-10 gene transfer lead to reduced infarct sizes and attenuated leukocyte infiltration [76, 78]. Tregs might also affect BBB integrity during acute stroke. Metalloproteinases (MMPs), in particular MMP-2 and MMP-9, are critical in degradation of extracellular matrix and contributing to vascular leakage [46]. Adoptive Treg transfer has been associated with inhibition of MMP-9 activity and as a consequence protection of the neurovascular unit integrity [79]. Taken together, Tregs play a central role in limiting the detrimental impact of neuroinflammation after stroke.

3 T Cells in Regeneration

T cells have multiple effects on other cell populations which are known to play a major role in regeneration after stroke. In the following, we will focus on the direct effect of T cells on different processes involved in post-stroke regeneration (Fig. 24.3b).

3.1 T Cells and Neurogenesis

Neurogenesis in the adult brain is limited to only few specialized localizations, namely the subventricular zone (SVZ) and subgranular zone (SGZ) of the dentate gyrus [80–82]. After stroke, neural precursor cells (NPC) have the ability to migrate to the lesion site [83, 84]. Here, NPC might not only replace dead cells [85, 86] but also—and maybe even more importantly—contribute to a pro-regenerative milieu and thereby support neuronal plasticity by secreting growth factors such as BDNF [87, 88]. Proliferation, migration and differentiation of NPC are substantially influenced by immune cells, including T cells. Vice versa, NPC might play a role in modifying the post-ischemic immune response [89, 90]. Under physiological conditions, T cells have an impact on cerebral cell renewal [91]. Without entering the brain, lymphocytes facilitate neurogenesis [34, 92]. After brain injury, the interaction of the adaptive immune system and neuronal precursor cells (NPC) is also evident [93]. Similar to the acute phase after stroke, different T cell subsets seem to have opposing functions during recovery, in particular on NPC proliferation and differentiation. T helper cells impair regeneration after stroke and promote ineffective functional recovery. T cell deficiency, both in transgenic mice and by CD4-specific antibody-mediated cell depletion, was associated with reduced apoptosis and increased proliferation rates of NPC in a cortical lesion model [94]. A specific subset of CD4+ T helper cells expressing the glucocorticoid-induced TNF receptor (GITR) has been identified to mediate this substantial impact of T cells on post-stroke neurogenesis [95]. In contrast, Tregs have been demonstrated to support neurogenesis. Treg depletion by CD25-specific antibodies resulted in a reduced number of NPC after experimental stroke [94]. Correspondingly, the higher number of Tregs in the ventricle of the ischemic hemisphere was associated with increased proliferation of NPC and attributable to the secretion of IL-10 by Treg [96]. Taking these opposite effects of T cells on neurogenesis depending on their subtype, it reveals again the intricacy of neuroimmunological processes after cerebral ischemia.

Besides direct effects on NPC proliferation, T cells might also indirectly affect NPC via their impact on microglial cells. Cytokines released from the different T cell subsets can activate and differentially polarize microglia depending on the secreted cytokine profile [97]. In turn, activated microglia have been shown to have two main routes to influence neurogenesis: Microglia secrete IGF-1, a growth hormone which reinforces NPC proliferation in general [98]. Additionally, they are in control of balancing oligodendrogensis and neurogensis. Activated via IL-4, microglia facilitate oligodendrogenesis albeit IFNγ shifts microglia towards a neurogenesis supporting polarization type [99]. Whether these findings also affect regeneration after stroke is yet to be determined.

3.2 T Cells and Axonal Sprouting

Another essential process for neuronal repair is axonal sprouting. In spinal cord injury and transection of the perforant pathway it was shown that T cells can affect axonal outgrowth [100, 101]. *In vitro* experiments underscore their influence on sprouting, but again different subsets play distinct roles: CD4+ Th cells can enhance axonal growth, whereas CD8+ cytotoxic T cells might impair this process [102]. So far, the impact of T cells on axonal outgrowth has not been investigated in animal stroke models. Yet, better understanding of T cell-driven mechanisms in neuronal plasticity and their impact on regeneration after stroke might open new therapeutic targets for stroke patients.

3.3 T Cells and Angiogenesis

Angiogenesis is a critical mechanism in the restoration of injured tissue. After stroke, endothelial cells proliferate and immature vessels begin to outgrow to regions with stimulatory signal [103]. There is a wide variety of factors sustaining vascular growth which can be secreted by different cell types including T cells. The influence of T cell on angiogenesis was studied in different experimental conditions. Recently, it was shown in a model of hind limb ischemia that the presence of CCR7+ T cells is beneficial for effective arteriogenesis [104]. Accordingly, in transgenic mice deficient of total CD4+ cells collateral vascularization after hindlimb ischemia was reduced and associated with impaired limb function [105]. The concept of a differential impact of the individual T cell subsets described above applies also for angiogenesis. It was previously shown that pro-inflammatory CD4+ Th1 cells and their characteristic cytokine IFNγ dampen vascular growth, in models of tumor and lung ischemia [106, 107]. In contrast, anti-inflammatory Tregs promote angiogenesis not only by suppressing the impact of effector T cells [107] but also by secreting chemokines, the proangiogenic vascular endothelial growth factor (VEGF) and TGF-β [108, 109]. TGF-β signaling is known to be essential for vascular development during embryogenesis [110]. In experimental cancer research, IL-6, a classical lymphocyte derived cytokine was shown to enhance angiogenesis also via upregulation of VEGF [111]. Furthermore, T cell-derived IL-17 was shown to facilitate neovascularization of the rat cornea and promote chemotactic function in tumor vascularization by upregulation of proangiogenic factors [112]. The named mechanisms of T cells regulating angiogenesis in a variety of disease models has so far not been investigated in ischemic stroke. However, in light of the important function of angiogenesis in restoration of tissue function, further investigations on the specific role of T cells in this processes will be of high relevance.

4 Conclusion

The presence of T cells is critical in physiological brain function. A large body of evidence over the past decade has highlighted a key role for T cells in the pathophysiology of stroke in the acute phase. Within the first days after stroke, pro-inflammatory T cell subpopulations play a detrimental role and contribute to the exacerbation of neuronal damage. In contrast, immunosuppressive Treg cells are beneficial for stroke outcome by limiting the inflammatory collateral damage. While the role of T cells in the acute and subacute phase after stroke has been extensively investigated, the contribution of T cells to the chronic regenerative phase is so far barely understood. Nevertheless, studies in other research field indicated a role for T cells for several reparative processes such as neurogenesis, angiogenesis and axonal growth. Unfortunately, until now only very few studies have specifically investigated the function of T cells in post-stroke regeneration and their therapeutic potential to improve regenerative capabilities after stroke. Yet, T cells are ideal therapeutic targets due to their excellent druggability as circulating cells and numerous T cell-targeted therapies readily available. Therefore, further studies particularly testing the function of T cells in post-stroke recovery are urgently needed for the development of novel therapeutics for stroke patients.

References

1. Stenger S, Hanson DA, Teitelbaum R, Dewan P, Niazi KR, Froelich CJ, et al. An antimicrobial activity of cytolytic T cells mediated by granulysin. Science. 1998;282:121–5.
2. Kägi D, Odermatt B, Seiler P, Zinkernagel RM, Mak TW, Hengartner H. Reduced incidence and delayed onset of diabetes in perforin-deficient nonobese diabetic mice. J Exp Med. 1997;186(7):989–97.
3. Rieux-Laucat F, Le Deist F, De Saint Basile G. Autoimmune lymphoproliferative syndrome and perforin. N Engl J Med. 2005;352:306–7.
4. Hsieh C, Macatonia S, Tripp C, Wolf S, O'Garra A, Murphy K. Development of TH1 CD4+ T cells through IL-12 produced by Listeria-induced macrophages. Science. 1993;260:547–9.
5. Scharton TM, Scott P. Natural killer cells are a source of interferon gamma that drives differentiation of CD4 + T cell subsets and induces early resistance to leishmania major in mice. J Exp Med. 1993;178:567–77.
6. Min B, Prout M, Hu-Li J, Zhu J, Jankovic D, Morgan ES, et al. Basophils produce IL-4 and accumulate in tissues after infection with a Th2-inducing parasite. J Exp Med. 2004;200(4):507–17.
7. Shinkai K, Mohrs M, Locksley RM. Helper T cells regulate type-2 innate immunity in vivo. Nature. 2002;420:825–9.
8. Mosmann TR, Sad S. The expanding universe of T-cell subsets: Th1, Th2 and more. Immunol Today. 1996;17(3):138–46.
9. Langrish CL, Chen Y, Blumenschein WM, Mattson J, Basham B, Sedgwick JD, et al. IL-23 drives a pathogenic T cell population that induces autoimmune inflammation. J Exp Med. 2005;0(2):233–40.
10. Weaver CT, Harrington LE, Mangan PR, Gavrieli M, Murphy KM. Th17: an effector CD4 T cell lineage with regulatory T cell ties. Immunity. 2006;24:677–88.
11. Sakaguchi S, Regulatory T. cells: minireview key controllers of immunologic self-tolerance. Cell. 2000;101:455–8.
12. Sakaguchi S, Yamaguchi T, Nomura T, Ono M. Regulatory T cells and immune tolerance. Cell. 2008;133:775–87.
13. Fontenot JD, Gavin MA, Rudensky AY. Foxp3 programs the development and function of CD4+ CD25+ regulatory T cells. Nat Immunol. 2003;4:330–6.
14. Vantourout P, Hayday A. Six-of-the-best: unique contributions of γδ T cells to immunology. Nat Rev Immunol. 2013;13:88–100.
15. Bonneville M, O'Brien RL, Born WK. γδ T cell effector functions: a blend of innate programming and acquired plasticity. Nat Rev Immunol. 2010;10:469–78.
16. Vermijlen D, Ellis P, Langford C, Klein A, Engel R, Willimann K, et al. Distinct cytokine-driven responses of activated blood γδ T cells: insights into unconventional T cell pleiotropy1. J Immunol. 2007;178(7):4304–14.
17. Ziv Y, Ron N, Butovsky O, Landa G, Sudai E, Greenberg N, et al. Immune cells contribute to the maintenance of neurogenesis and spatial learning abilities in adulthood. Nat Neurosci. 2006;9(2):268–75.
18. Kipnis J, Cohen H, Cardon M, Ziv Y, Schwartz M. T cell deficiency leads to cognitive dysfunction: implications for therapeutic vaccination for schizophrenia and other psychiatric conditions. Proc Natl Acad Sci U S A. 2004;101(21):8180–95.

19. Lewitus GM, Wilf-Yarkoni A, Ziv Y, Shabat-Simon M, Gersner R, Zangen A, et al. Vaccination as a novel approach for treating depressive behavior. Biol Psychiatry. 2009;65:283–8.
20. Engelhardt B, Ransohoff RM. The ins and outs of T-lymphocyte trafficking to the CNS: anatomical sites and molecular mechanisms. Trends Immunol. 2005;26(9):485–95.
21. Hickey WF. Migration of hematogenous cells through the blood-brain barrier and the initiation of CNS inflammation. Brain Pathol. 1991 Jan;1(2):97–105.
22. Engelhardt B, Carare RO, Bechmann I, Flügel A, Laman JD, Weller RO. Vascular, glial, and lymphatic immune gateways of the central nervous system. Acta Neuropathol. 2016;132:317–38.
23. Allt G, Lawrenson JG. Is the pial microvessel a good model for blood-brain barrier studies? Brain Res Rev. 1997;24:67–76.
24. Bechmann I, Galea I, Perry VH. What is the blood-brain barrier (not)? Trends Immunol. 2007;28(1):5–11.
25. Bartholomäus I, Kawakami N, Odoardi F, Schläger C, Miljkovic D, Ellwart JW, et al. Effector T cell interactions with meningeal vascular structures in nascent autoimmune CNS lesions. Nature. 2009;462:94–8.
26. Kivisakk P, Imitola J, Rasmussen S, Elyaman W, Zhu B, Ransohoff RM, et al. Localizing central nervous system immune surveillance: meningeal antigen-presenting cells activate T cells during experimental autoimmune encephalomyelitis. Ann Neurol. 2009;65(4):457–69.
27. Goldmann J, Kwidzinski E, Brandt C, Mahlo J, Richter D. T cells traffic from brain to cervical lymph nodes via the cribroid plate and the nasal mucosa. J Leukoc Biol. 2006;80(October):797–801.
28. Baruch K, Ron-Harel N, Gal H, Deczkowska A, Shifrut E, Ndifon W, et al. CNS-specific immunity at the choroid plexus shifts toward destructive Th2 inflammation in brain aging. Proc Natl Acad Sci U S A. 2013;110(6):2264–9.
29. Kivisäkk P, Mahad DJ, Callahan MK, Trebst C, Tucky B, Wei T, et al. Human cerebrospinal fluid central memory CD4 T cells: evidence for trafficking through choroid plexus and meninges via P-selectin. Proc Natl Acad Sci U S A. 2003;100(14):8389–94.
30. Wolf SA, Steiner B, Akpinarli A, Kammertoens T, Nassenstein C, Braun A, et al. CD4-positive T lymphocytes provide a neuroimmunological link in the control of adult hippocampal neurogenesis. J Immunol. 2009;182:3979–84.
31. Brynskikh A, Warren T, Zhu J, Kipnis J. Adaptive immunity affects learning behavior in mice. Brain Behav Immun. 2008;22(6):861–9.
32. Radjavi A, Smirnov I, Kipnis J. Brain antigen-reactive CD4 + T cells are sufficient to support learning behavior in mice with limited T cell repertoire. Brain Behav Immun. 2014;35:58–63.
33. Filiano AJ, Xu Y, Tustison NJ, Marsh RL, Baker W, Smirnov I, et al. Unexpected role of interferon-γ in regulating neuronal connectivity and social behaviour. Nature. 2016;535:425–9.
34. Derecki NC, Cardani AN, Yang CH, Quinnies KM, Crihfield A, Lynch KR, et al. Regulation of learning and memory by meningeal immunity: a key role for IL-4. J Exp Med. 2010;207(5):1067–80.
35. Balschun D, Wetzel W, Del Rey A, Pitossi F, Schneider H, Zuschratter W, et al. Interleukin-6: a cytokine to forget. FASEB J. 2004;18:1788–91.
36. Filiano AJ, Gadani SP, Kipnis J. How and why do T cells and their derived cytokines affect the injured and healthy brain? Nat Rev Neurosci. 2017;18(6):375–84.
37. Chen GY, Nuñez G. Sterile inflammation: sensing and reacting to damage. Nat Rev Immunol. 2010;10:826–37.
38. Kono H, Rock KL. How dying cells alert the immune system to danger. Nat Rev Immunol. 2008;8(4):279–89.
39. Lotze MT, Tracey KJ. High - mobility group box 1 protein (HMGB1): nuclear weapon in the immune arsenal. Nat Rev Immunol. 2005;5:331–42.
40. Denes A, Vidyasagar R, Feng J, Narvainen J, McColl BW, Kauppinen RA, et al. Proliferating resident microglia after focal cerebral ischaemia in mice. J Cereb Blood Flow Metab. 2007;27(12):1941–53.

41. Iadecola C, Anrather J. The immunology of stroke: from mechanisms to translation. Nat Med. 2011;17:796–808.
42. Kuric E, Ruscher K. Dynamics of major histocompatibility complex class II-positive cells in the postischemic brain - influence of levodopa treatment. J Neuroimmunol. 2014;11(145):1–12.
43. Yilmaz G, Granger DN. Cell adhesion molecules and ischemic stroke. Neurol Res. 2008;30(8):783–93.
44. Huang J, Upadhyay UM, Vascular TRJ. Inflammation in stroke and focal cerebral ischemia. Surg Neurol. 2006;66:323–245.
45. Doll DN, Barr TL, Simpkins JW. Cytokines: their role in stroke and potential use as biomarkers and therapeutic targets. Aging Dis. 2014;5(5):294–306.
46. Bauer AT, Bürgers HF, Rabie T, Marti HH. Matrix metalloproteinase-9 mediates hypoxia-induced vascular leakage in the brain via tight junction rearrangement. J Cereb Blood Flow Metab. 2010;30:837–48.
47. Schilling M, Besselmann M, Leonhard C, Mueller M, Ringelstein EB, Kiefer R. Microglial activation precedes and predominates over macrophage infiltration in transient focal cerebral ischemia: a study in green fluorescent protein transgenic bone marrow chimeric mice. Exp Neurol. 2003;183:25–33.
48. Gelderblom M, Leypoldt F, Steinbach K, Behrens D, Choe CU, Siler DA, et al. Temporal and spatial dynamics of cerebral immune cell accumulation in stroke. Stroke. 2009;40:1849–57.
49. Enzmann G, Mysiorek C, Roser G, Cheng Y-J, Sharang G, Hannocks M-J, et al. The neurovascular unit as a selective barrier to polymorphonuclear granulocyte (PMN) infiltration into the brain after ischemic injury. Acta Neuropathol. 2013;125:395–412.
50. Jander S, Kraemer M, Schroeter M, Witte OW, Stoll G. Lymphocytic infiltration and expression of intercellular adhesion molecule-l in photochemically induced ischemia of the rat cortex. J Cereb Blood Flow Metab. 1995;15:42–51.
51. Stubbe T, Ebner F, Richter D, Engel OR, Klehmet J, Royl G, et al. Regulatory T cells accumulate and proliferate in the ischemic hemisphere for up to 30 days after MCAO. J Cereb Blood Flow Metab. 2013;33(10):37–47.
52. Chu HX, Kim HA, Lee S, Moore JP, Chan CT, Vinh A, et al. Immune cell infiltration in malignant middle cerebral artery infarction: comparison with transient cerebral ischemia. J Cereb Blood Flow Metab. 2013;34(10):450–9.
53. Gill D, Veltkamp R. Dynamics of T cell responses after stroke. Curr Opin Pharmacol. 2016;26:26–32.
54. Chamorro Á, Meisel A, Planas AM, Urra X, van de Beek D, Veltkamp R. The immunology of acute stroke. Nat Rev Neurol. 2012;8(7):401–10.
55. Liesz A, Suri-Payer E, Veltkamp C, Doerr H, Sommer C, Rivest S, et al. Regulatory T cells are key cerebroprotective immunomodulators in acute experimental stroke. Nat Med. 2009;15(2):192–9.
56. Liesz A, Zhou W, Mracskó É, Karcher S, Bauer H, Schwarting S, et al. Inhibition of lymphocyte trafficking shields the brain against deleterious neuroinflammation after stroke. Brain. 2011;134:704–20.
57. Kleinschnitz C, Schwab N, Kraft P, Hagedorn I, Dreykluft A, Schwarz T, et al. Early detrimental T-cell effects in experimental cerebral ischemia are neither related to adaptive immunity nor thrombus formation. Blood. 2010;115:3835–42.
58. Yilmaz G, Arumugam TV, Stokes KY, Granger DN. Role of T lymphocytes and interferon-gamma in ischemic stroke. Circulation. 2006;113:2105–12.
59. Hurn PD, Subramanian S, Parker SM, Afentoulis ME, Kaler LJ, Vandenbark AA, et al. T-and B-cell-deficient mice with experimental stroke have reduced lesion size and inflammation. J Cereb Blood Flow Metab. 2007;27:1798–805.
60. Subramanian S, Zhang B, Kosaka Y, Burrows GG, Grafe MR, Vandenbark AA, et al. Recombinant T cell receptor ligand (RTL) treats experimental stroke. Stroke. 2009;40(7):2539–45.

61. Mracsko E, Liesz A, Stojanovic A, Pak-Kin Lou W, Osswald M, Zhou W, et al. Antigen dependently activated cluster of differentiation 8-positive T cells cause perforin-mediated neurotoxicity in experimental stroke. J Neurosci. 2014;34:16784–95.
62. Shichita T, Sugiyama Y, Ooboshi H, Sugimori H, Nakagawa R, Takada I, et al. Pivotal role of cerebral interleukin-17–producing gdT cells in the delayed phase of ischemic brain injury. Nat Med. 2009;15:946–50.
63. Gelderblom M, Weymar A, Bernreuther C, Velden J, Arunachalam P, Steinbach K, et al. Neutralization of the IL-17 axis diminishes neutrophil invasion and protects from ischemic stroke. Blood. 2012;120:3793–802.
64. Luo Y, Zhou Y, Xiao W, Liang Z, Dai J, Weng X, et al. Interleukin-33 ameliorates ischemic brain injury in experimental stroke through promoting Th2 response and suppressing Th17 response. Brain Res. 2015;1597:86–94.
65. Neumann H, Cavalie A, Jenne DE, Wekerle H. Induction of MHC class I genes in neurons. Science. 1995;269:549–52.
66. Lin Y, Zhang J-C, Yao C-Y, Wu Y, Abdelgawad A, Yao S-L, et al. Critical role of astrocytic interleukin-17 A in post-stroke survival and neuronal differentiation of neural precursor cells in adult mice. Cell Death Dis. 2016;7:1–14.
67. Wang D-D, Zhao Y-F, Wang G-Y, Sun B, Kong Q-F, Zhao K, et al. IL-17 potentiates neuronal injury induced by oxygen–glucose deprivation and affects neuronal IL-17 receptor expression. J Neuroimmunol. 2009;212:17–25.
68. Zepp J, Wu L, Li X. IL-17 receptor signaling and Th17-mediated autoimmune demyelinating disease Pathogenic Th17 cells and Autoimmune Diseases. Trends Immunol. 2011;32(5):232–9.
69. Hetman M, Cavanaugh JE, Kimelman D, Xia Z. Role of glycogen synthase kinase-3beta in neuronal apoptosis induced by trophic withdrawal. J Neurosci. 2000;20(7):2567–74.
70. Pap M, Cooper GM. Role of translation initiation factor 2B in control of cell survival by the phosphatidylinositol 3-kinase/Akt/glycogen synthase kinase 3beta signaling pathway. Mol Cell Biol. 2002;22(2):578–86.
71. Swardfager W, Winer DA, Herrmann N, Winer S, Lanctôt KL. Interleukin-17 in post-stroke neurodegeneration. Neurosci Biobehav Rev. 2013;37:436–47.
72. Liesz A, Hu X, Kleinschnitz C, Offner H. Functional role of regulatory lymphocytes in stroke: facts and controversies. Stroke. 2015 May;46(5):1422–30.
73. Liesz A, Zhou W, Na S-Y, Hämmerling GJ, Garbi N, Karcher S, et al. Boosting regulatory T cells limits neuroinflammation in permanent cortical stroke. J Neurosci. 2013;33(44):17350–62.
74. Xie L, Sun F, Wang J, Mao X, Xie L, Yang S-H, et al. mTOR signaling inhibition modulates macrophage/microglia-mediated neuroinflammation and secondary injury via regulatory T cells after focal ischemia. J Immunol. 2014;192(2):6009–19.
75. Na SY, Mracsko E, Liesz A, Hünig T, Veltkamp R. Amplification of regulatory T cells using a CD28 superagonist reduces brain damage after ischemic stroke in mice. Stroke. 2015;46(1):212–20.
76. De Bilbao F, Arsenijevic D, Moll T, Garcia-Gabay I, Vallet P, Langhans W, et al. In vivo over-expression of interleukin-10 increases resistance to focal brain ischemia in mice. J Neurochem. 2009;110(1):12–22.
77. Kipnis J, Avidan H, Caspi RR, Schwartz M. Dual effect of CD4 regulatory T cells in neurodegeneration: a dialogue with microglia. Proc Natl Acad Sci U S A. 2004;101:14663–9.
78. Ooboshi H, Ibayashi S, Shichita T, Kumai Y, Takada J, Ago T, et al. Postischemic gene transfer of interleukin-10 protects against both focal and global brain ischemia. Circulation. 2005;111(7):913–9.
79. Li P, Mao L, Liu X, Gan Y, Zheng J, Thomson AW, et al. Essential role of PD-L1 in regulatory T cell-afforded protection against blood-brain barrier damage after stroke. Stroke. 2014;45(3):857–64.
80. Gage FH. Mammalian neural stem cells. Science. 2000;287:1433–8.
81. Lois C, Alvarez-Buylla A. Long-distance neuronal migration in the adult mammalian brain. Science. 1994;264:1145–7.

82. Kriegstein A, Alvarez-Buylla A. The glial nature of embryonic and adult neural stem cells. Annu Rev Neurosci. 2009;32:149–84.
83. Parent JM, Vexler ZS, Gong C, Derugin N, Ferriero DM. Rat forebrain neurogenesis and striatal neuron replacement after focal stroke. Ann Neurol. 2002;52:802–13.
84. Arvidsson A, Collin T, Kirik D, Kokaia Z, Lindvall O. Neuronal replacement from endogenous precursors in the adult brain after stroke. Nat Med. 2002;8:963–70.
85. Hou SW, Wang YQ, Xu M, Shen DH, Wang JJ, Huang F, et al. Functional integration of newly generated neurons into striatum after cerebral ischemia in the adult rat brain. Stroke. 2008;39(10):2837–44.
86. Jin K, Wang X, Xie L, Mao XO, Zhu W, Wang Y, et al. Evidence for stroke-induced neurogenesis in the human brain. Proc Natl Acad Sci U S A. 2006;103(35):13198–202.
87. Lu P, Jones LL, Snyder EY, Tuszynski MH. Neural stem cells constitutively secrete neurotrophic factors and promote extensive host axonal growth after spinal cord injury. Exp Neurol. 2003;181:115–29.
88. Lindvall O, Kokaia Z. Stem cell research in stroke: how far from the clinic? Stroke. 2011;42(8):2369–75.
89. Einstein O, Karussis D, Grigoriadis N, Mizrachi-Kol R, Reinhartz E, Abramsky O, et al. Intraventricular transplantation of neural precursor cell spheres attenuates acute experimental allergic encephalomyelitis. Mol Cell Neurosci. 2003;24:1074–82.
90. Martino G, Pluchino S. The therapeutic potential of neural stem cells. Nat Rev Neurosci. 2006;7:395–406.
91. Schwartz M, Shechter R. Protective autoimmunity functions by intracranial immunosurveillance to support the mind: the missing link between health and disease. Mol Psychiatry. 2010;15:342–54.
92. Ron-Harel N, Cardon M, Schwartz M. Brain homeostasis is maintained by "danger" signals stimulating a supportive immune response within the brain's borders. Brain Behav Immun. 2011;25:1036–43.
93. Kokaia Z, Martino G, Schwartz M, Lindvall O. Cross-talk between neural stem cells and immune cells: the key to better brain repair? Nat Neurosci. 2012;15:1078–87.
94. Saino O, Taguchi A, Nakagomi T, Nakano-Doi A, Kashiwamura SI, Doe N, et al. Immunodeficiency reduces neural stem/progenitor cell apoptosis and enhances neurogenesis in the cerebral cortex after stroke. J Neurosci Res. 2010;88:2385–97.
95. Takata M, Nakagomi T, Kashiwamura S, Nakano-Doi A, Saino O, Nakagomi N, et al. Glucocorticoid-induced TNF receptor-triggered T cells are key modulators for survival/death of neural stem/progenitor cells induced by ischemic stroke. Cell Death Differ. 2011;19(10):756–67.
96. Wang J, Xie L, Yang C, Ren C, Zhou K, Wang B, et al. Activated regulatory T cell regulates neural stem cell proliferation in the subventricular zone of normal and ischemic mouse brain through interleukin 10. Front Cell Neurosci. 2015;9(361):1–11.
97. Butovsky O, Talpalar AE, Ben-Yaakov K, Schwartz M. Activation of microglia by aggregated B-amyloid or lipopolysaccharide impairs MHC-II expression and renders them cytotoxic whereas IFN-gamma and IL-4 render them protective. Mol Cell Biol. 2005;29:381–93.
98. Gudi V, Kuljec JŠ, Yildiz Z, Frichert K, Skripuletz T, Moharregh-Khiabani D, et al. Spatial and temporal profiles of growth factor expression during CNS demyelination reveal the dynamics of repair priming. PLoS One. 2011;6(7):e22623.
99. Butovsky O, Ziv Y, Schwartz A, Landa G, Talpalar AE, Pluchino S, et al. Microglia activated by IL-4 or IFN-γ differentially induce neurogenesis and oligodendrogenesis from adult stem/progenitor cells. Mol Cell Neurosci. 2006;31(1):149–60.
100. Nielsen HH, Toft-Hansen H, Lykke Lambertsen K, Owens T, Finsen B. Stimulation of adult oligodendrogenesis by myelin-specific T cells. Am J Pathol. 2011;179:2028–41.
101. Wu B, Matic D, Djogo N, Szpotowicz E, Schachner M, Jakovcevski I. Improved regeneration after spinal cord injury in mice lacking functional T- and B-lymphocytes. Exp Neurol. 2012;237:274–85.

102. Pool M, Rambaldi I, Darlington PJ, Wright MC, Fournier AE, Bar-Or A. Neurite outgrowth is differentially impacted by distinct immune cell subsets. Mol Cell Neurosci. 2012;49:68–76.
103. An C, Shi Y, Li P, Hu X, Gan Y, Stetler RA, et al. Molecular dialogs between the ischemic brain and the peripheral immune system: dualistic roles in injury and repair. Prog Neurobiol. 2014;115:6):6–24.
104. Nossent AY, Bastiaansen AJNM, Peters EAB, de Vries MR, Aref Z, Welten SMJ, et al. CCR7-CCL19/CCL21 axis is essential for effective arteriogenesis in a murine model of hindlimb ischemia. J Am Heart Assoc. 2017;6:e005281.
105. Stabile E, Susan Burnett M, Watkins C, Kinnaird T, Bachis A, La Sala A, et al. Impaired arteriogenic response to acute hindlimb ischemia in CD4-knockout mice. Circulation. 2003;108(2):205–10.
106. Albini A, Marchisone C, Del GF, Benelli R, Masiello L, Tacchetti C, et al. Inhibition of angiogenesis and vascular tumor growth by interferon-producing cells. Am J Pathol. 2000;156(4):1381–93.
107. Zhong Q, Jenkins J, Moldobaeva A, D'Alessio F, Wagner EM. Effector T cells and ischemia-induced systemic angiogenesis in the lung. Am J Respir Cell Mol Biol. 2016 Mar;54(3):394–401.
108. Facciabene A, Peng X, Hagemann IS, Balint K, Barchetti A, Wang L-P, et al. Tumour hypoxia promotes tolerance and angiogenesis via CCL28 and T reg cells. Nature. 2011;475:226–30.
109. D'Alessio FR, Zhong Q, Jenkins J, Moldobaeva A, Wagner EM. Lung angiogenesis requires CD4(+) forkhead homeobox protein-3(+) regulatory T cells. Am J Respir Cell Mol Biol. 2015 May;52(5):603–10.
110. Larsson J, Goumans MJ, Sjöstrand LJ, Van Rooijen MA, Ward D, Levéen P, et al. Abnormal angiogenesis but intact hematopoietic potential in TGF-β type I receptor-deficient mice. EMBO J. 2001;20(7):1663–73.
111. Wei L-H, Kuo M-L, Chen C-A, Chou C-H, Lai K-B, Lee C-N, et al. Interleukin-6 promotes cervical tumor growth by VEGF-dependent angiogenesis via a STAT3 pathway. Oncogene. 2003;22:1517–27.
112. Numasaki M, Fukushi J-I, Ono M, Narula SK, Zavodny PJ, Kudo T, et al. Interleukin-17 promotes angiogenesis and tumor growth. Blood. 2003;101(7):2620.

Chapter 25
The Inflammatory Response and Its Effect on Rehabilitation-Induced Repair Processes After Stroke

Ali Alawieh, Farris Langley, and Stephen Tomlinson

Abstract Post-stroke inflammation is associated with a significant exacerbation of acute injury, and at the same time promotes an unfavorable environment for regeneration and recovery. Sustained inflammation after stroke is associated with poor motor and cognitive recovery and limits the ability of the brain to engage in and benefit from rehabilitation paradigms. Stroke comorbidities such as aging, diabetes, and smoking are all associated with a more robust neuroinflammatory response after stroke and poor outcomes. Preclinical and clinical studies have not yet investigated the role of post-stroke neuroinflammation in predicting the response to rehabilitation therapy. Here, we review the interaction between post-stroke neuroinflammation and determinants of response to rehabilitation therapy, we discuss the few studies that used anti-inflammatory therapy to boost the response to rehabilitative interventions, and we emphasize the need for combining anti-inflammatory therapy and rehabilitation in both therapeutic and mechanistic studies of experimental stroke.

Keywords Inflammation • Post-stroke plasticity • Stroke • Neurogenesis • Rehabilitation • Combination therapy

---

A. Alawieh • F. Langley
Department of Microbiology and Immunology, Medical University of South Carolina, Charleston, SC 29425, USA

Medical Scientist Training Program, Medical University of South Carolina, Charleston, SC 29425, USA
e-mail: alawieh@musc.edu; langleel@musc.edu

S. Tomlinson, Ph.D. (✉)
Department of Microbiology and Immunology, Medical University of South Carolina, Charleston, SC 29425, USA

Ralph H. Johnson VA Medical Center, Charleston, SC, USA

173 Ashley Avenue, BSB 201, MSC 504, Charleston, SC 29425, USA
e-mail: tomlinss@musc.edu

P.A. Lapchak, J.H. Zhang (eds.), *Cellular and Molecular Approaches to Regeneration and Repair*, Springer Series in Translational Stroke Research,
https://doi.org/10.1007/978-3-319-66679-2_25

Abbreviations

NSAID	Nonsteroidal anti-inflammatory drug
CRP	C-reactive protein
SVZ	Sub-ventricular zone
SGZ	Sub-granular zone
CR2-fH	Complement receptor 2—Factor H
RV	Resveratrol
sTNFαR1	soluble Tumor Necrosis Factor α receptor 1
TNFα	Tumor Necrosis Factor α

1 Introduction

Despite the fact that the extent of initial deficit and the size of injury are major predictors of outcome after stroke, repair and regeneration that occurs in the brain following ischemic injury is key to limit the evolution of acute injury and promote recovery of function [1–3]. In fact, the current clinical strategy to improve recovery of stroke patients beyond the acute phase is motor and cognitive rehabilitation that targets mechanisms of repair and regeneration after stroke [4–9]. However, the detailed pathophysiological processes involved in the response to rehabilitation therapy are still largely unknown, and a standard paradigm for optimizing the response to rehabilitation in clinical settings has not yet been established.

Rehabilitation therapy has been found to promote favorable pathophysiological responses such as increased growth factor levels, increased axonal and dendritic sprouting and increased neurogenesis and neuronal migration ([10–15], reviewed in [4, 8]). However, the same processes are also expected to be under a tonic inhibitory drive by pathological inflammatory responses that persist in the infarcted brain after stroke [16–18]. Inflammatory responses sustained in the brain after the initial insult are associated with degenerative changes, and with an anti-regenerative environment that impede the response to rehabilitation therapy. Therefore, it may be hypothesized that the extent of neuroinflammation propagating beyond the acute phase of stroke may determine the response to rehabilitation therapy, and that suppression of inflammation may augment the response to rehabilitation therapy. Studies from stroke patients have shown that increased markers of inflammation, including acute phase reactants (IL6, CRP), can predict outcomes after ischemic stroke [19–21]; however, the impact of inflammation on response to rehabilitation has not yet been investigated at the clinical level. The aim of this work is to provide an overview of the interaction between inflammation and rehabilitation therapy during chronic recovery after stroke and to emphasize the need for investigating combinatory strategies of anti-inflammatory therapy and rehabilitation paradigms at preclinical levels.

1.1 Determinants of Response to Rehabilitation Therapy

In absence of treatment, stroke patients tend to show spontaneous recovery in both motor and cognitive functions beyond the acute phase. Rehabilitation therapy aims to speed up or augment the recovery trajectory and promote better outcomes. It is natural that the extent of initial injury is a key determinant of response to rehabilitation since it defines the extent of damage, the ability of patients to engage with rehabilitation paradigms, and the availability of neuronal reserve able to engage in regenerative and remodeling processes [3, 4, 22]. Reducing the extent of initial injury is a major focus in acute ischemic stroke treatment and neuroprotection therapy but beyond the acute phase when a stable infarct is defined, reducing the extent of lesion is not a major pathophysiological target of interventions such as rehabilitation therapy. Additional determinants of the response to rehabilitation include the ability to re-wire existing connections taking advantage of increased post-stroke plasticity, and the ability to form and incorporate new neurons into existing circuitry. These two components will be further discussed in the following sections to assess how inflammation contributes to the ability of the recovering brain to augment and re-organize existing resources to optimize overall outcome.

2 Pathological Inflammation After Stroke

Following cerebral ischemia and reperfusion injury leading to neuronal apoptosis and necrosis, a neuroinflammatory response is triggered that includes the expression of danger-associated molecular patterns, the release of reactive oxygen species, and the subsequent activation of cells of the innate and adaptive immune systems [23]. A prominent feature of neuroinflammatory responses is their ability to self-amplify and maintain a sustained activation, at a pathological level, that extends beyond the acute phase [24]. Post-acute inflammation is characterized by the presence of a vicious cycle that prevents the resolution of this response to allow for recovery (Fig. 25.1). Inflammatory stress to neurons kindles the growing inflammatory response, which in turn promotes more stress to recovering neurons ultimately leading to a degenerative response. Although activation of immune cells and activation products are implicated in both injury and recovery after stroke (reviewed in [1, 25]), pathologic immune activation after stroke, characterized by a delay in resolution of inflammation and a significant feed-forward mechanisms, is majorly associated with significant delay in recovery and increased degenerative responses. The fact that the same immune cells and components contribute to both reparatory and degenerative inflammation is a major hurdle for therapeutic strategies, especially those that aim at depletion of specific immune components or processes. Lessons learned from several decades of research in anti-inflammatory and neuroprotective therapies for ischemic stroke have demonstrated that efforts should be focused on strategies that promote a faster resolution of inflammation and interrupt the feed-forward amplification

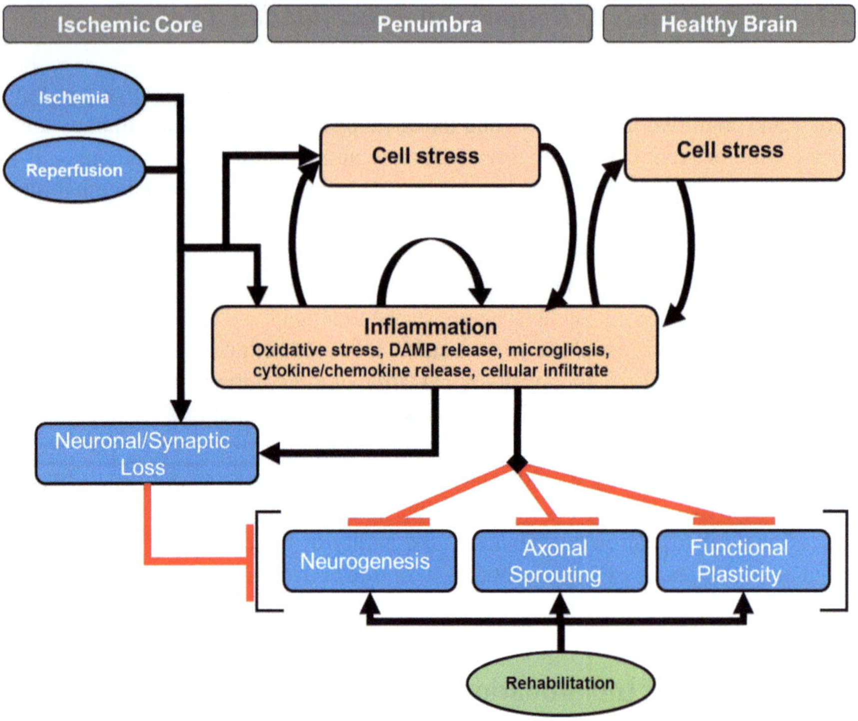

Fig. 25.1 Ischemia and reperfusion injury to the ischemic core after stroke contributes to neuronal and synaptic loss and initiates an inflammatory response that propagates to the ischemic penumbra. Inflammatory mechanisms induced after stroke have the propensity to self-amplify and elicit cellular stress on recovering neurons in the penumbra. This initiates a vicious cycle where more inflammation triggers more cellular stress and subsequent death of neurons, that in turn fosters a more pronounced inflammatory response. This cycle results in an expansion in the size of the penumbra, reducing the reserve of neurons and synapses available to engage in recovery, and promoting an anti-regenerative environment that limits the response to rehabilitative interventions

mechanisms, without eliminating pathways within the inflammatory casade that may serve important reparatory roles. The dual role of immune activation in injury and recovery after stroke has been extensively reviewed [1, 2, 25]; however, this work will review the interaction between neuroinflammation and regenerative mechanisms induced or exploited by rehabilitation interventions.

2.1 *Inflammation and Post-stroke Neurogenesis*

Neuronal progenitor cells in the sub-ventricular zone (SVZ) or subgranular zone (SGZ) proliferate after stroke and migrate, in the absence of inhibitory signals, to perilesional brain and populate cortical, hippocampal and striatal targets [26–29].

Newly generated neurons are capable of incorporating into existing circuitry and constitute a substrate for axonal and dendritic remodeling. Formation and incorporation of newly formed neurons after stroke leads to improvements in both cognitive and motor recovery [30–32]. However, pathologic inflammation has been consistently associated with significant reduction in basal and ischemia-induced neurogenesis, and inhibition of migration of newly formed neuroblasts [16, 33–43]. Interruption of the propagation of neuroinflammation has resulted in a significant increase in neuroblast proliferation, migration and subsequent functional outcomes [16, 33–43]. Specific inflammatory mediators that have been implicated in suppressing neurogenesis include IL-1β, IL-6, TNF-α, C3, [1, 44–46]. A few studies have implicated neuroimmune interactions such as activation of T-cells and microglia in promoting basal neurogenesis [47, 48]. This data may apply in the context of mild inflammatory responses; however, post-stroke inflammation characterized by robust activation of microglia and infiltration of immune cells leads to a detrimental effect on neurogenesis after stroke [46]. In addition, depletion of immune system components may not be an optimal strategy to investigate the role of post-stroke inflammation in neurogenesis due to potentially inhibiting beneficial homeostatic interactions. In a previous work, we demonstrated that when the ability of the complement system to self-amplify is inhibited after MCAO, there is a reduction in pathological inflammation, resulting in significantly improved neuroblasts migration from the SVZ with improved cognitive performance [43]. Complement activation is amplified by the alternative complement pathway, and we have shown that site-targeted inhibition of this pathway with CR2-fH interrupts this feed-forward mechanisms fostering uncontrolled inflammation, and is optimal to facilitate the resolution of inflammation while allowing for a regenerative environment [43]. Supporting our hypothesis, complete inhibition of complement by blocking all pathways of complement activation interrupted the migration of neuroblasts and resulted in poor recovery [43]. Although rehabilitation is implicated in promoting neurogenesis after stroke, the interaction between inflammation, rehabilitation, and neurogenesis has not been investigated. Notably, advanced age, a major determinant of clinical response to rehabilitation, is associated with a significant reduction in post-stroke neurogenesis in rodent models [49].

2.2 *Inflammation and Post-stroke Plasticity*

Neuronal remodeling beyond the acute phase of ischemic stroke exploits a window of neuroplasticity and allows peri-lesional neurons, or even neurons in the contralateral hemisphere, to establish new synaptic connections that can compensate for functions previously performed within the infarcted brain [50–53]. The process of remodeling has been documented in both humans using functional imaging and brain stimulation studies and in animal models [3], and has been shown to enhance recovery and restore function, at least partially. The pathophysiological mechanisms underlying neuronal re-wiring involve axonal sprouting and subsequent synaptogenesis. In fact, extensive preclinical research has been devoted to investigate strategies to enhance axonal sprouting to promote recovery after stroke. For instance, growth inhibitory factors such as Nogo-A has been shown to suppress axonal sprouting after CNS injury, and inhibition or antagonism of Nogo-A activity has been found to enhance neuronal plasticity and improve recovery in young and adult rodents after stroke [54–59]. In parallel, rehabilitation strategies aim to exploit the window of neuronal plasticity in the brain after stroke to promote beneficial re-wiring [3], and anti-inflammatory therapies aim to lift the brakes off the growth promoting factors to extend the window and magnitude of post-stroke plasticity and potentiate the effects of rehabilitation. Chronic inflammation after stroke is considered a plasticity-impeding factor, an effect that is more prominent in aged animals [4]. Infiltration of immune cells, including CD45+, CD3+ and CD4+ cells [60, 61], increased TNF-alpha signaling [62], and increased microglial activation and release of cytokine and chemokines [60, 63], have all been implicated in suppressing plasticity and worsening recovery after stroke [4, 63]. Supporting these findings, inhibition of inflammatory pathways such as TNF-alpha, CXCL12 signaling, microglial activation, or complement activation leads to enhanced post-stroke plasticity and improved recovery [1, 61–63]. In addition, several interventions that are shown to enhance post-stroke plasticity such as erythropoietin and VEGF administration also mediate their effects, at least in part, by suppression of neuroinflammatory responses [60, 64–66]. These studies as well as others (reviewed in [67]; [4]) demonstrate that excessive inflammation after stroke is associated with depression of post-stroke plasticity, an outcome that may limit the response to rehabilitation therapy (discussed in the following section). It is also worth noting that immune activation products may also facilitate post-stroke plasticity, as seen in neurogenesis; however, this is not the hallmark of potent neuroinflammatory responses after stroke, and may be relevant for conditions with a mild inflammatory response.

2.3 Inflammation and Response to Rehabilitation in Preclinical Models

Despite the significance of assessing the interaction between inflammatory mechanisms and outcomes of rehabilitation, only a few studies have investigated the combination of anti-inflammatory therapy with rehabilitation to assess for the presence of potential cooperative, synergistic or antagonistic effects. In addition, from a therapeutic standpoint, and since most stroke patients receive some form of rehabilitation therapy after stroke, it is essential to investigate the efficacy of an anti-inflammatory or neuroprotective agent in the context of rehabilitation prior to further translation to human trials.

In one of the first studies combining anti-inflammatory therapy with rehabilitative training in animal models, Liebigt et al. [68] used a combination of minocycline or indomethacin, and skilled reaching training during recovery from photothrombotic infarct in rats. Indomethacin is a nonsteroidal anti-Inflammatory drug (NSAID) that inhibits cyclo-oxygenases, whereas minocycline inhibits the activation of microglia after stroke. Following 4 weeks of rehabilitative training combined with minocycline or indomethacin treatment, both pharmacological treatments significantly improved motor performance on reaching and ladder tasks compared to rehabilitation alone. Notably, the effects seen with minocycline treatment were observed in the absence of a significant reduction in lesion volume at 14 or 42 days after injury. Anti-inflammatory therapy with minocycline or indomethacin was associated with a significant increase in the number of BrdU+ neurons and a significant decrease in proliferating microglia in the sensorimotor cortex compared to rehabilitation alone [68]. These findings demonstrate that anti-inflammatory therapy favorably interacts with rehabilitation training to improve outcomes after stroke, even in absence of a significant decrease in infarct size with anti-inflammatory therapy.

The interaction between NSAIDs and experience-dependent plasticity after stroke was further investigated in a rat model of photothrombotic stroke with whisker deprivation. Whisker deprivation, by clipping all but one row of vibrissae, is associated with a significant increase in cortical representation of the spared row, thus evoking a strong neuroplastic response [69]. Stroke was associated with a significant decrease in cortical plasticity after deprivation, an effect that was reversed when the NSAID, Ibuprofen, was administered following stroke. Ibuprofen showed a dose-dependent decrease in cyclo-oxygenase 2 levels in the brain, and a dose-dependent increase in the cortical representation of the spared vibrissae indicating increased cortical plasticity [69]. A similar approach was used to assess whether inhibition of TNF-alpha can also preserve cortical plasticity after stroke. Following stroke and whisker deprivation, soluble TNF-alpha receptor 1 (sTNFαR1) was osmotically infused into the murine brain without any influence on lesion volume [62]. Anti-inflammatory therapy with sTNFαR1 inhibited signaling downstream of TNFαR1, decreased TNFα levels, and restored cortical plasticity that was inhibited after stroke [62].

In a recent work, Huaifang et al. [70] used environmental enrichment and resveratrol (RV) treatment to assess the combinatory effects of both therapies on overall recovery after stroke. Enriched environment is one of the commonly used models of motor and cognitive rehabilitation that has been associated with increased neurogenesis and neuronal plasticity. Following middle cerebral artery occlusion in rats, administration of RV combined with enriched environment resulted in a more prominent functional recovery compared to either intervention alone. Combination therapy also significantly reduced oxidative stress and ERK1/2 signaling involved in exacerbating post-stroke pathology [70].

In addition to these studies investigating the combination of anti-inflammatory therapies with rehabilitation paradigms, some reports have also demonstrated a cooperative or synergistic effect of combining neuroprotective therapy and rehabilitation after stroke [71–73].

3 Conclusion

The role of neuroinflammation in modulating the response to rehabilitative interventions after stroke is still largely under-investigated. Collectively, current studies provide preliminary evidence that anti-inflammatory therapy may facilitate the response to rehabilitation paradigm; however, further and more robust behavioral and molecular characterization is still needed. Equally important is the careful distinction between pathologic inflammation, a self-amplified response with slow resolution, and mild inflammation or activation of immune cells in the brain after injury. Studies depleting specific immune cells and components showing rebound injury are not sufficient to challenge the pathologic role of excessive inflammation that occurs after stroke and constitutes a major target for intervention. Alternatively, efforts should be devoted to breaking the loop of inflammation and stimulating faster resolution of the inflammatory response by masking the triggers of sustained activation or breaking the feed-forward mechanisms along this pathway.

In addition, subsequent studies should investigate whether rehabilitation and anti-inflammatory therapies modulate different pathophysiological processes, and provide a more detailed analysis of cognitive and motor recovery measures to assess whether this interaction is consistent across the two aspects of recovery. Finally, the significance of investigating the efficacy of anti-inflammatory therapies, and neuroprotective therapies in general, in the context of rehabilitation should be further emphasized as a requirement for preclinical stroke research to help bridge the translational roadblock for many novel interventions.

Acknowledgements Authors acknowledge support from the NIH (1P20GM109040, U54 GM10494) to ST, the Department of Veterans Affairs (Merit Award 1I01RX001141 and 1BX001218) to ST, and an American Heart Association Pre-doctoral Fellowship to AA (15PRE25250009).

References

1. Alawieh A, Elvington A, Tomlinson S. Complement in the homeostatic and ischemic brain. Front Immunol. 2015;6:417.
2. Alawieh A, Narang A, Tomlinson S. Complementing regeneration. Oncotarget. 2015;6(26):21769–70.
3. Alawieh A, Zhao J, Feng W. Factors affecting post-stroke motor recovery: implications on neurotherapy after brain injury. Behav Brain Res. 2016;
4. Murphy TH, Corbett D. Plasticity during stroke recovery: from synapse to behaviour. Nat Rev Neurosci. 2009;10(12):861–72.
5. Langhorne P, Coupar F, Pollock A. Motor recovery after stroke: a systematic review. Lancet Neurol. 2009;8(8):741–54.
6. Di Pino G, Pellegrino G, Assenza G, Capone F, Ferreri F, Formica D, et al. Modulation of brain plasticity in stroke: a novel model for neurorehabilitation. Nat Rev Neurol. 2014;10(10):597–608.

7. Bowden MG, Woodbury ML, Duncan PW. Promoting neuroplasticity and recovery after stroke: future directions for rehabilitation clinical trials. Curr Opin Neurol. 2013;26(1):37–42.
8. Dimyan MA, Cohen LG. Neuroplasticity in the context of motor rehabilitation after stroke. Nat Rev Neurol. 2011;7(2):76–85.
9. Krakauer JW, Carmichael ST, Corbett D, Wittenberg GF. Getting neurorehabilitation right: what can be learned from animal models? Neurorehabil Neural Repair. 2012;26(8):923–31.
10. Zhao LR, Risedal A, Wojcik A, Hejzlar J, Johansson BB, Kokaia Z. Enriched environment influences brain-derived neurotrophic factor levels in rat forebrain after focal stroke. Neurosci Lett. 2001;305(3):169–72.
11. Hicks AU, Hewlett K, Windle V, Chernenko G, Ploughman M, Jolkkonen J, et al. Enriched environment enhances transplanted subventricular zone stem cell migration and functional recovery after stroke. Neuroscience. 2007;146(1):31–40.
12. Luo CX, Jiang J, Zhou QG, Zhu XJ, Wang W, Zhang ZJ, et al. Voluntary exercise-induced neurogenesis in the postischemic dentate gyrus is associated with spatial memory recovery from stroke. J Neurosci Res. 2007;85(8):1637–46.
13. Kim MW, Bang MS, Han TR, Ko YJ, Yoon BW, Kim JH, et al. Exercise increased BDNF and trkB in the contralateral hemisphere of the ischemic rat brain. Brain Res. 2005;1052(1):16–21.
14. Stroemer RP, Kent TA, Hulsebosch CE. Enhanced neocortical neural sprouting, synaptogenesis, and behavioral recovery with D-amphetamine therapy after neocortical infarction in rats. Stroke. 1998;29(11):2381–93. discussion 93–5.
15. Stroemer RP, Kent TA, Hulsebosch CE. Neocortical neural sprouting, synaptogenesis, and behavioral recovery after neocortical infarction in rats. Stroke. 1995;26(11):2135–44.
16. Monje ML, Toda H, Palmer TD. Inflammatory blockade restores adult hippocampal neurogenesis. Science. 2003;302(5651):1760–5.
17. Greifzu F, Schmidt S, Schmidt KF, Kreikemeier K, Witte OW, Lowel S. Global impairment and therapeutic restoration of visual plasticity mechanisms after a localized cortical stroke. Proc Natl Acad Sci U S A. 2011;108(37):15450–5.
18. Marquardt L, Ruf A, Mansmann U, Winter R, Buggle F, Kallenberg K, et al. Inflammatory response after acute ischemic stroke. J Neurol Sci. 2005;236(1-2):65–71.
19. Winovich DT, Longstreth WT, Jr., Arnold AM, Varadhan R, Zeki Al Hazzouri A, Cushman M, et al. Factors associated with ischemic stroke survival and recovery in older adults. Stroke 2017.
20. Kuo HK, Yen CJ, Chang CH, Kuo CK, Chen JH, Sorond F. Relation of C-reactive protein to stroke, cognitive disorders, and depression in the general population: systematic review and meta-analysis. Lancet Neurol. 2005;4(6):371–80.
21. Rost NS, Wolf PA, Kase CS, Kelly-Hayes M, Silbershatz H, Massaro JM, et al. Plasma concentration of C-reactive protein and risk of ischemic stroke and transient ischemic attack: the Framingham study. Stroke. 2001;32(11):2575–9.
22. Kelly PJ, Furie KL, Shafqat S, Rallis N, Chang Y, Stein J. Functional recovery following rehabilitation after hemorrhagic and ischemic stroke. Arch Phys Med Rehabil. 2003;84(7):968–72.
23. Chamorro A, Meisel A, Planas AM, Urra X, van de Beek D, Veltkamp R. The immunology of acute stroke. Nat Rev Neurol. 2012;8(7):401–10.
24. Iadecola C, Anrather J. The immunology of stroke: from mechanisms to translation. Nat Med. 2011;17(7):796–808.
25. Ekdahl CT, Kokaia Z, Lindvall O. Brain inflammation and adult neurogenesis: the dual role of microglia. Neuroscience. 2009;158(3):1021–9.
26. Adamczak J, Aswendt M, Kreutzer C, Rotheneichner P, Riou A, Selt M, et al. Neurogenesis upregulation on the healthy hemisphere after stroke enhances compensation for age-dependent decrease of basal neurogenesis. Neurobiol Dis. 2016;99:47–57.
27. Koh SH, Park HH. Neurogenesis in stroke recovery. Transl Stroke Res. 2017;8(1):3–13.
28. Shiromoto T, Okabe N, Lu F, Maruyama-Nakamura E, Himi N, Narita K, et al. The role of endogenous neurogenesis in functional recovery and motor map reorganization induced by rehabilitative therapy after stroke in rats. J Stroke Cerebrovasc Dis. 2017;26(2):260–72.

29. Xiong XY, Liu L, Yang QW. Functions and mechanisms of microglia/macrophages in neuroinflammation and neurogenesis after stroke. Prog Neurobiol. 2016;142:23–44.
30. Thored P, Arvidsson A, Cacci E, Ahlenius H, Kallur T, Darsalia V, et al. Persistent production of neurons from adult brain stem cells during recovery after stroke. Stem Cells. 2006;24(3):739–47.
31. Kokaia Z, Lindvall O. Neurogenesis after ischaemic brain insults. Curr Opin Neurobiol. 2003;13(1):127–32.
32. Lakhan SE, Kirchgessner A, Hofer M. Inflammatory mechanisms in ischemic stroke: therapeutic approaches. J Transl Med. 2009;7:97.
33. Chien MY, Chuang CH, Chern CM, Liou KT, Liu DZ, Hou YC, et al. Salvianolic acid A alleviates ischemic brain injury through the inhibition of inflammation and apoptosis and the promotion of neurogenesis in mice. Free Radic Biol Med. 2016;99:508–19.
34. Kim H, Wei Y, Lee JY, Wu Y, Zheng Y, Moskowitz MA, et al. Myeloperoxidase inhibition increases neurogenesis after ischemic stroke. J Pharmacol Exp Ther. 2016;359(2):262–72.
35. Ahmed ME, Tucker D, Dong Y, Lu Y, Zhao N, Wang R, et al. Methylene Blue promotes cortical neurogenesis and ameliorates behavioral deficit after photothrombotic stroke in rats. Neuroscience. 2016;336:39–48.
36. Xia CF, Yin H, Yao YY, Borlongan CV, Chao L, Chao J. Kallikrein protects against ischemic stroke by inhibiting apoptosis and inflammation and promoting angiogenesis and neurogenesis. Hum Gene Ther. 2006;17(2):206–19.
37. Tobin MK, Bonds JA, Minshall RD, Pelligrino DA, Testai FD, Lazarov O. Neurogenesis and inflammation after ischemic stroke: what is known and where we go from here. J Cereb Blood Flow Metab. 2014;34(10):1573–84.
38. Saino O, Taguchi A, Nakagomi T, Nakano-Doi A, Kashiwamura S, Doe N, et al. Immunodeficiency reduces neural stem/progenitor cell apoptosis and enhances neurogenesis in the cerebral cortex after stroke. J Neurosci Res. 2010;88(11):2385–97.
39. Pluchino S, Muzio L, Imitola J, Deleidi M, Alfaro-Cervello C, Salani G, et al. Persistent inflammation alters the function of the endogenous brain stem cell compartment. Brain. 2008;131(Pt 10):2564–78.
40. Kang SS, Keasey MP, Arnold SA, Reid R, Geralds J, Hagg T. Endogenous CNTF mediates stroke-induced adult CNS neurogenesis in mice. Neurobiol Dis. 2013;49:68–78.
41. Strassburger M, Braun H, Reymann KG. Anti-inflammatory treatment with the p38 mitogen-activated protein kinase inhibitor SB239063 is neuroprotective, decreases the number of activated microglia and facilitates neurogenesis in oxygen-glucose-deprived hippocampal slice cultures. Eur J Pharmacol. 2008;592(1-3):55–61.
42. Zhou J, Cheng G, Kong R, Gao DK, Zhang X. The selective ablation of inflammation in an acute stage of ischemic stroke may be a new strategy to promote neurogenesis. Med Hypotheses. 2011;76(1):1–3.
43. Alawieh A, Elvington A, Zhu H, Yu J, Kindy MS, Atkinson C, et al. Modulation of post-stroke degenerative and regenerative processes and subacute protection by site-targeted inhibition of the alternative pathway of complement. J Neuroinflammation. 2015;12:247.
44. Ben-Hur T, Ben-Menachem O, Furer V, Einstein O, Mizrachi-Kol R, Grigoriadis N. Effects of proinflammatory cytokines on the growth, fate, and motility of multipotential neural precursor cells. Mol Cell Neurosci. 2003;24(3):623–31.
45. Peng H, Whitney N, Wu Y, Tian C, Dou H, Zhou Y, et al. HIV-1-infected and/or immune-activated macrophage-secreted TNF-alpha affects human fetal cortical neural progenitor cell proliferation and differentiation. Glia. 2008;56(8):903–16.
46. Whitney NP, Eidem TM, Peng H, Huang Y, Zheng JC. Inflammation mediates varying effects in neurogenesis: relevance to the pathogenesis of brain injury and neurodegenerative disorders. J Neurochem. 2009;108(6):1343–59.
47. Ziv Y, Ron N, Butovsky O, Landa G, Sudai E, Greenberg N, et al. Immune cells contribute to the maintenance of neurogenesis and spatial learning abilities in adulthood. Nat Neurosci. 2006;9(2):268–75.

48. Ziv Y, Avidan H, Pluchino S, Martino G, Schwartz M. Synergy between immune cells and adult neural stem/progenitor cells promotes functional recovery from spinal cord injury. Proc Natl Acad Sci U S A. 2006;103(35):13174–9.
49. Darsalia V, Heldmann U, Lindvall O, Kokaia Z. Stroke-induced neurogenesis in aged brain. Stroke. 2005;36(8):1790–5.
50. Carmichael ST. Themes and strategies for studying the biology of stroke recovery in the post-stroke epoch. Stroke. 2008;39(4):1380–8.
51. Brown CE, Li P, Boyd JD, Delaney KR, Murphy TH. Extensive turnover of dendritic spines and vascular remodeling in cortical tissues recovering from stroke. J Neurosci. 2007;27(15):4101–9.
52. Liu Z, Li Y, Zhang X, Savant-Bhonsale S, Chopp M. Contralesional axonal remodeling of the corticospinal system in adult rats after stroke and bone marrow stromal cell treatment. Stroke. 2008;39(9):2571–7.
53. Liu Z, Zhang RL, Li Y, Cui Y, Chopp M. Remodeling of the corticospinal innervation and spontaneous behavioral recovery after ischemic stroke in adult mice. Stroke. 2009;40(7):2546–51.
54. Sozmen EG, Rosenzweig S, Llorente IL, DiTullio DJ, Machnicki M, Vinters HV, et al. Nogo receptor blockade overcomes remyelination failure after white matter stroke and stimulates functional recovery in aged mice. Proc Natl Acad Sci U S A. 2016;113(52):E8453–E62.
55. Schmidt A, Minnerup J. Promoting recovery from ischemic stroke. Expert Rev Neurother. 2016;16(2):173–86.
56. Lee JK, Kim JE, Sivula M, Strittmatter SM. Nogo receptor antagonism promotes stroke recovery by enhancing axonal plasticity. J Neurosci. 2004;24(27):6209–17.
57. Wiessner C, Bareyre FM, Allegrini PR, Mir AK, Frentzel S, Zurini M, et al. Anti-Nogo-A antibody infusion 24 hours after experimental stroke improved behavioral outcome and corticospinal plasticity in normotensive and spontaneously hypertensive rats. J Cereb Blood Flow Metab. 2003;23(2):154–65.
58. Tsai SY, Papadopoulos CM, Schwab ME, Kartje GL. Delayed anti-nogo-a therapy improves function after chronic stroke in adult rats. Stroke. 2011;42(1):186–90.
59. Lindau NT, Banninger BJ, Gullo M, Good NA, Bachmann LC, Starkey ML, et al. Rewiring of the corticospinal tract in the adult rat after unilateral stroke and anti-Nogo-A therapy. Brain. 2014;137(Pt 3):739–56.
60. Herz J, Reitmeir R, Hagen SI, Reinboth BS, Guo Z, Zechariah A, et al. Intracerebroventricularly delivered VEGF promotes contralesional corticorubral plasticity after focal cerebral ischemia via mechanisms involving anti-inflammatory actions. Neurobiol Dis. 2012;45(3):1077–85.
61. Ruscher K, Kuric E, Liu Y, Walter HL, Issazadeh-Navikas S, Englund E, et al. Inhibition of CXCL12 signaling attenuates the postischemic immune response and improves functional recovery after stroke. J Cereb Blood Flow Metab. 2013;33(8):1225–34.
62. Liguz-Lecznar M, Zakrzewska R, Kossut M. Inhibition of Tnf-alpha R1 signaling can rescue functional cortical plasticity impaired in early post-stroke period. Neurobiol Aging. 2015;36(10):2877–84.
63. Kriz J, Lalancette-Hebert M. Inflammation, plasticity and real-time imaging after cerebral ischemia. Acta Neuropathol. 2009;117(5):497–509.
64. Chen J, Zhang C, Jiang H, Li Y, Zhang L, Robin A, et al. Atorvastatin induction of VEGF and BDNF promotes brain plasticity after stroke in mice. J Cereb Blood Flow Metab. 2005;25(2):281–90.
65. Reitmeir R, Kilic E, Kilic U, Bacigaluppi M, ElAli A, Salani G, et al. Post-acute delivery of erythropoietin induces stroke recovery by promoting perilesional tissue remodelling and contralesional pyramidal tract plasticity. Brain. 2011;134(Pt 1):84–99.
66. Mengozzi M, Cervellini I, Villa P, Erbayraktar Z, Gokmen N, Yilmaz O, et al. Erythropoietin-induced changes in brain gene expression reveal induction of synaptic plasticity genes in experimental stroke. Proc Natl Acad Sci U S A. 2012;109(24):9617–22.
67. Liguz-Lecznar M, Kossut M. Influence of inflammation on poststroke plasticity. Neural Plast. 2013;2013:258582.

68. Liebigt S, Schlegel N, Oberland J, Witte OW, Redecker C, Keiner S. Effects of rehabilitative training and anti-inflammatory treatment on functional recovery and cellular reorganization following stroke. Exp Neurol. 2012;233(2):776–82.
69. Jablonka JA, Kossut M, Witte OW, Liguz-Lecznar M. Experience-dependent brain plasticity after stroke: effect of ibuprofen and poststroke delay. Eur J Neurosci. 2012;36(5):2632–9.
70. Su Q, Pu H, Hu C. Neuroprotection by combination of resveratrol and enriched environment against ischemic brain injury in rats. Neurol Res. 2016;38(1):60–8.
71. Wang J, Feng X, Du Y, Wang L, Zhang S. Combination treatment with progesterone and rehabilitation training further promotes behavioral recovery after acute ischemic stroke in mice. Restor Neurol Neurosci. 2013;31(4):487–99.
72. Chang HC, Yang YR, Wang PS, Wang RY. Quercetin enhances exercise-mediated neuroprotective effects in brain ischemic rats. Med Sci Sports Exerc. 2014;46(10):1908–16.
73. Griva M, Lagoudaki R, Touloumi O, Nousiopoulou E, Karalis F, Georgiou T, et al. Long-term effects of enriched environment following neonatal hypoxia-ischemia on behavior, BDNF and synaptophysin levels in rat hippocampus: effect of combined treatment with G-CSF. Brain Res. 2017;1667:55.

Chapter 26
Complement C3a: Shaping the Plasticity of the Post-stroke Brain

Anna Stokowska and Marcela Pekna

Abstract Complement is part of the innate immune system that plays a major role in the initiation of inflammation and host defence against pathogenic bacteria. Complement activation is also a contributor to tissue damage in a range of autoimmune conditions. For those reasons, the activation of the complement system in the central nervous system (CNS) was for long considered deleterious. Based on the evidence accumulated during the past decade, this view has been dramatically changing and complement is gaining recognition for its non-immune surveillance related functions, including regulation of morphogenesis, and adult tissue regeneration. C3a is a 77 amino acid, 9 kDa peptide generated through the proteolytic activation of the central molecule of the complement system, the third complement component, C3. C3a exerts most of its functions through its canonical G-protein coupled receptor C3aR that is expressed by many cell types including neurons and glia. This chapter considers recent insights into the novel roles of the complement system, in particular C3a, in the CNS with focus on brain plasticity and recovery after ischemic brain injury.

Keywords C3a • C3a receptor • The complement system • Ischemic stroke • Neural plasticity • Recovery

---

A. Stokowska, Ph.D. • M. Pekna, M.D., Ph.D. (✉)
Laboratory of Regenerative Neuroimmunology, Department of Clinical Neuroscience, Institute of Neuroscience and Physiology, Sahlgrenska Academy at University of Gothenburg, Box 440, 405 30, Gothenburg, Sweden
e-mail: Marcela.Pekna@neuro.gu.se

P.A. Lapchak, J.H. Zhang (eds.), *Cellular and Molecular Approaches to Regeneration and Repair*, Springer Series in Translational Stroke Research,
https://doi.org/10.1007/978-3-319-66679-2_26

Abbreviations

AMPAR	α-Amino-3-hydroxy-5-methyl-4-isoxazolepropionic acid receptor
C1q	Complement component 1q
C3	The third complement component
C3a	The smaller of the two fragments generated by proteolytic activation of C3
C3a-desArg	C3a after removal of C-terminal arginine residue
C3aR	C3a receptor
C3b	The larger of the two fragments generated by proteolytic activation of C3
C5a	The smaller of the two fragments generated by proteolytic activation of the fifth complement component
C5aR2	Second receptor for C5a
CD	Cluster of differentiation
CNS	Central nervous system
CR3	Complement receptor 3
ERK	Extracellular signal-regulated kinase
GAP-43	Growth associated protein 43
GFAP	Glial fibrillary acidic protein
GFAP-C3a	Glial fibrillary acidic protein promoter driven expression of C3a
IgG	Immunoglobulin G
IgM	Immunoglobulin M
IL	Interleukin
IXa	Activated coagulation factor IX
MBL	Mannose binding lectin
NGF	Nerve growth factor
XIa	Activated coagulation factor XI
Xa	Activated coagulation factor X
VGLUT1	Vesicular glutamate transporter 1

1 Introduction

The complement system is a critical constituent of the humoral innate immune response and its fundamental importance in the elimination of pathogenic bacteria, inflammation and clearance of immune complexes, as well as its role in tissue damage in various autoimmune conditions have been well recognized for decades. Research findings in the last years provide growing evidence for the involvement of immune cells and molecules, including complement proteins and their activation products, in tissue repair and regeneration. In amphibians, complement proteins or their orthologs are expressed in the regenerating limbs and lens and have been implicated in adult limb regeneration [1–3]. By controlling cell migration, complement activation products seem to play a role in early vertebrate development and morphogenesis [4–6]. In rodents, complement is critical for hepatocyte proliferation and liver regeneration [7–9], and promotes homing [10], chemotaxis [11] and retention of hematopoietic stem and progenitor cells in the bone marrow [12]. Through its angiogenic activity, complement can support wound healing [13] but also plays a role in choroidal neovascularization [14–17], which is a typical feature of age-related macular degeneration [18]. Recent evidence points to the complement system as an important modulator of regenerative responses in the mammalian spinal cord [19–21], and implicates complement in the regulation of brain plasticity in the injured as well as healthy brain. Detailed understanding of the multiple immune as well as non-immune functions of the complement system in the CNS will be necessary for the design of complement-targeting strategies for the treatment of neurological disease conditions. The type of injury and its temporal aspects are also likely to be of critical importance, as will be the route of administration of the complement-targeting therapeutics. The potential effects on CNS of interventions targeting complement for the treatment of non-neurological disorders will also need to be carefully considered.

1.1 The Complement System

The complement system is a major effector of innate immune response that serves as a universally distributed rapid and effective first line of defense against pathogens that breach the mechanical and chemical barriers of the body as well as elimination of body's own dead cells. The complement system is a general term for a group of more than 50 soluble proteins, cell receptors and control proteins found in the blood and tissues. Their roles in innate immunity include the opsonisation and lysis of pathogens, elimination of soluble antigen–antibody complexes, stimulation of leukocyte chemotaxis and initiation of inflammation. Complement affects also adaptive immunity by regulating B and T lymphocyte function [22]. Hepatocytes are the predominant source of complement proteins is liver, however other cell types such as macrophages, lymphocytes, fibroblasts and endothelial also contribute to systemic pool of complement factors. Complement factors are also produced locally in the brain and spinal cord [22].

1.2 The Complement Cascade

Activation of the circulating complement proteins in response to injury or challenge by pathogens results in a cascade of reactions involving structural rearrangements of proteins, proteolytic cleavages, and the assembly of terminal lytic complexes on the activating/target surface. The cascade is initiated through three major pathways: the classical, lectin or alternative pathway [22]. The classical pathway of complement activation is triggered primarily by binding of the recognition protein C1q to antigen–antibody complexes through the Fc region of IgGs and IgMs, or by antibody-independent mechanisms involving direct binding of C1q to viral envelopes, cell walls of Gram-negative bacteria, C-reactive protein, intermediate filaments and myelin. The lectin pathway is initiated by the binding of mannose-binding lectin (MBL) and ficolins to carbohydrate moieties on the surface of bacterial and yeast cells or parasite envelopes or cells. MBL can also bind to and opsonize apoptotic and necrotic cells [23]. C1q, MBL and ficolins are typical pattern-recognition molecules, which allow anchoring of C1q- or MBL-associated serine proteases that propagate the proteolytic cascade leading to the formation of C3-convertase, an enzymatic complex that cleaves and activates the central molecule of the cascade, the third complement component (C3). Alternative pathway is initiated at a very low rate in the fluid phase by a spontaneous conformational change of C3, which upon recruitment of factor B and a proteolytically-active factor D leads to the formation of an alternative C3-convertase complex on the activating surface [22]. In addition, C3 can be activated directly by MBL-associated serine protease 1 [24], non-complement proteases such as neutrophil elastase, cathepsins [25, 26], granulocyte neutral proteases [27], lysosomal enzymes, kallikrein, as well as coagulation factors XIa, Xa, IXa, thrombin, and plasmin [28, 29]. Cleavage of C3 generates a small C3a fragment, and a larger fragment, C3b, that binds to bacterial cell wall or altered mammalian cell membrane, where it participates in the amplification of the alternative pathway and triggers the terminal part of the complement cascade that results in cell lysis. As a ligand of the complement receptor 3 (CR3, CD11b/CD18)), C3b also facilitates the phagocytosis of C3b tagged target molecules or structures, including neuronal synapses [30] (Fig. 26.1).

2 Complement System Activation in the Ischemic Brain

In the CNS, complement proteins are produced by astrocytes, microglia and neurons [31–34]. Brain ischemia not only triggers systemic as well as local activation of the complement system but there is a growing body of evidence for multiple immune and non-immune functions exerted by complement proteins and their activation products in the ischemic brain parenchyma with very distinct and even opposing effects on outcome.

Ischemic stroke triggers pronounced activation of complement in the systemic circulation [35–39] and immunohistochemical evidence from human post-mortem brain tissue shows local complement activation in brain parenchyma and markedly reduced expression of membrane bound complement regulatory proteins after ischemic stroke [40, 41]. Experimental studies have been instrumental in elucidating the mechanisms involved. Mocco and co-workers demonstrated that C3 activation is the key constituent in ischemia-induced brain tissue injury and pointed to C3a as the main mediator [42]. While C1q, the complement protein initiating the classical pathway of complement activation, does not seem to be involved in complement activation in the ischemic brain parenchyma [42, 43], genetic deficiency of mannose

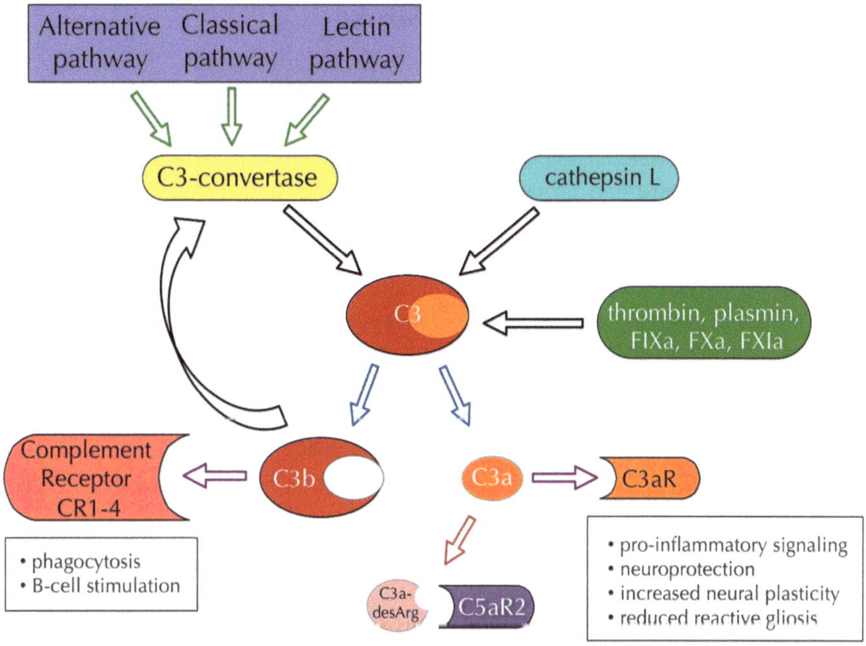

Fig. 26.1 C3a is released through the proteolytic activation of C3. C3a exerts its effects through binding to the C3a receptor (C3aR), the activation of which has been shown to induce inflammation, have neuroprotective effects and increase neural plasticity after ischemic stroke

binding lectin (MBL) abrogated C3 cleavage as well as the sub-acute accumulation of mononuclear cells in the ischemic region and improved outcome at 24 h post-stroke [44]. However, the neuroprotective effect of MBL deficiency was not sustained in the post-acute phase [44]. These findings, together with the report on lasting protection against acute brain injury conferred by acute pharmacological lectin pathway inhibition [45] imply that complement activation by the lectin pathway acutely contributes to tissue damage, but has beneficial functions in the post-acute repair phase. Although the initial steps leading to lectin pathway activation after ischemia are not fully understood, ischemia-induced appearance in brain parenchyma of neoepitopes that bind naturally occurring IgM [46] and binding of MBL to the neoepitope-bound IgM [47] seem to play an important role in this process. Indeed, IgM deposits were detected in the necrotic zones of post-mortem human brain 5–7 days after ischemic stroke [41]. How persistent is the neoepitope expression and what are the mechanisms of complement activation in the peri-infarct region and more remote brain regions in the post-acute and chronic phase are some of the outstanding questions that merit further investigation. On its own, the alternative pathway is not sufficient to initiate complement activation in the ischemic brain, but seems to propagate brain tissue injury via amplification of the cascade. Notably, deficiency of factor B as well as targeted pharmacological inhibition of the alternative pathway had sustained neuroprotective effect [48].

3 The Roles of Complement in Stroke

Given that complement activation products are powerful mediators of inflammation and neutrophil infiltration, complement activation, has been regarded as a contributor to secondary tissue damage after injury including brain ischemia [49] or trauma [50]. However, there is a growing body of evidence showing that the protective effect of complement inhibition in the acute phase after stroke is not always maintained in the subacute and chronic phases (for review see e.g. [51]). These results point to the critical role of complement in both tissue damage and repair processes after ischemic brain injury. The complement peptide C3a and its canonical receptor C3aR in particular appear to elicit multiple responses in immune and brain cells after ischemia; often with opposing effects on outcome. These consequences of C3aR activation on the long-term outcome depend on the specific cell type and the timing of the response in relation to the ischemia onset.

3.1 C3a and Its Receptor C3aR in the Healthy and Ischemic Brain

C3a is a 9 kDa, 77 amino acid peptide generated by proteolytic activation of C3 (Fig. 26.1). C3a exerts its functions through its canonical/cognate receptor, C3aR and through the considerably less studied second receptor for C5a, C5aR2 (previously known as C5L2) [52]. After its generation, C3a is rapidly cleaved by carboxypeptidases that remove the C-terminal arginine [53, 54], which ensures that the actions of C3a are tightly controlled and highly localized. While the C3a-desArg peptide does not bind to C3aR [55], it was shown to bind C5aR2 [56, 57]. C3aR is a member of the rhodopsin family of seven transmembrane G-protein-coupled receptors [58] and is widely expressed in many tissues including the brain. C3aR activation leads to increased vascular permeability, smooth muscle contraction, activation of myeloid cells such as neutrophils, monocytes/macrophages, basophils, and platelets, as well as directed migration of inflammatory cells such as eosinophilic leucocytes and mast cells [59]. C3aR expressed by endothelial cells [60, 61] regulates their expression of cytokines such as IL-8 and IL-1beta [62], and plays a critical role in endothelial activation and leukocyte recruitment into the brain by regulating the endothelial cell expression of intercellular cell adhesion molecule 1, and vascular cell adhesion molecule 1 [63]. Furthermore, stimulation of C3aR on epithelial cells of the choroid plexus has been shown to cause disruption of blood-cerebrospinal fluid barrier [64].

The high constitutive expression of C3aR on rodent as well as human neurons [60, 65–67], points to the involvement of C3a signaling in the brain beyond its established immune-related functions. Further, adult neural progenitor cells express C3aR [68], and C3a directly regulates their differentiation and migration through the extracellular signal-regulated kinase (ERK)1/2 signaling pathway [69]. C3a was

Table 26.1 Cellular expression and functions of C3aR in the CNS

Cell type	C3aR functions	References
Neural stem/progenitor cells	Neuronal differentiation Migration	[69] [69]
Neurons	Migration Neurite outgrowth Modulation of synaptic strength Modulation of dendritic morphology	[70, 71] [69] [112] [112]
Astrocytes	Intracellular signaling Cytokine expression Survival	[73] [74–76] [72]
Microglia	Intracellular calcium release NGF upregulation Regulation of phagocytosis	[77] [78] [79]
Endothelial cells	Cytokine expression Expression of cell adhesion molecules	[62] [63]
Epithelial cells of choroid plexus	Disorganization of tight junctions	[64]

also reported to accelerate the migration of granule cells of the developing cerebellum [70] and regulate neuronal migration during cortical development [71]. The expression of C3aR on astrocytes is upregulated by ischemia [60, 65, 72]. C3a affects astrocyte intracellular signaling [73] and the expression of cytokines such as interleukin (IL)-6, IL-8 and nerve growth factor (NGF) [74–76]. Through its inhibitory effect on ERK signaling-mediated apoptotic pathway and caspase-3 cleavage, C3a promotes astrocyte survival after ischemia [72]. Microglia express functional C3aR and respond to C3a stimulation with an increase in intracellular calcium concentration [77] and upregulation of NGF [78]. C3a exerts dual effect on microglial phagocytic functions such that acute C3a activation promotes, whereas chronic C3a treatment attenuates, microglial phagocytosis [79], Table 26.1.

Although unable to couple to G-proteins, C5aR2 can function as a positive modulator for both C5a- and C3a-induced responses [80]. C5aR2 was shown to be expressed on neurons and glial cells, exert anti-inflammatory functions [81], and provide neuroprotection after spinal cord injury [122].

3.2 *The Dual Role of C3a in the Acute Phase After Ischemic Injury*

In the acute phase after stroke, C3a can contribute to brain tissue loss as it plays a critical role in endothelial activation and leukocyte recruitment into the brain [63], and C3 deficiency as well as pre-treatment with systemically administered C3aR antagonist resulted in reduced granulocyte infiltration, reduced infarct volume and reduced neurological deficit scores in mice 24 h after transient cerebral ischemia [42]. Similarly, acute systemic administration of a low dose of a C3aR antagonist starting prior to transient ischemia induction resulted in smaller subcortical infarcts 7 days post-ischemia [83]. These results point to the therapeutic benefit of systemic inhibition of C3aR signaling in the acute phase after stroke. Indeed, systemic C3a levels are elevated in the first days after ischemic as well as haemorhagic stroke and in some stroke subtypes show association with unfavourable outcome [36, 38, 39, 84]. Genetic studies provide further support for the involvement of C3 in the pathogenesis of ischemic stroke [85]. On the other hand, C3a was shown to be protective against excitotoxicity-induced neuronal death, an effect that appears to be mediated by astrocytes [86], to support survival of astrocytes after ischemic stress [72] and to induce neuroprotective phenotype in microglial cells as indicated by their increased production of NGF [78]. Mice expressing biologically active C3a controlled by the glial fibrillary acidic protein promoter (*GFAP-C3a*), *i.e.* expressing C3a in reactive astrocytes, were protected against shock induced by lipopolysaccharide injection [87]. In further support of the positive effects of C3a in the injured brain, *GFAP-C3a* mice showed also reduced brain tissue loss in a model of neonatal hypoxic-ischemic brain injury and single dose intraventricular treatment with C3a ameliorated neonatal hypoxia-ischemia-induced memory impairment in wild type control mice but not in mice lacking C3aR ($C3aR^{-/-}$) [88]. Notably, daily intranasal treatment with C3a for 3 days starting 1 h after hypoxia-ischemia prevented cognitive impairment in wild-type mice tested 42 days after hypoxia-ischemia and reduced injury-induced reactive gliosis in the hippocampus [89].

Jointly, these findings support the conclusion that in the acute phase after ischemic injury, C3a can contribute to tissue loss by exerting its pro-inflammatory effects on endothelial cells and recruitment of inflammatory cells from the systemic circulation. However, through its direct effects on astrocytes and microglia, C3a can increase the survival of brain cells, including neurons, thus limiting tissue damage and promoting recovery.

3.3 C3a and Post-stroke Neural Plasticity

Brain ischemia induces endogenous plasticity and repair processes that range from proliferation, differentiation and migration of neural stem and progenitor cells [90–92], to sprouting of axonal projections and establishing new synaptic contacts that form the basis for extensive rewiring of the existing neuronal connections and rearrangement of cortical maps [93, 94]. These constituents of neural plasticity play an important role in normal learning, are enhanced by the milieu created following the injury and jointly contribute to recovery of function after stroke and other CNS injuries [95]. Recent findings provide novel insights into the regulation of the various structural constituents of post-stroke neural plasticity, implicate C3a-C3aR signaling in promoting some of its important constituents and point to C3aR as a target to facilitate functional recovery after ischemic brain injury, Table 26.2.

Table 26.2 The functions of C3a in the regulation of neural plasticity and in neuroprotection

Function	References
Basal neurogenesis	[68]
Post-stroke neurogenesis	[68]
Post-stroke synaptogenesis	[113]
Post-stroke expression of GAP-43, marker of axonal and glial plasticity	[113]
Modulation of reactive gliosis	[89]
Retina regeneration	[121]
Neuroprotection, survival of astrocytes after ischemic stress	[72, 88]

3.3.1 C3a and Post-stroke Neurogenesis

There is a mounting body of evidence for the role for C3a in the regulation of adult mammalian neurogenesis. Hippocampal neural stem cells in vitro as well as migrating neuroblasts in vivo express C3aR [68] and in vitro studies show that C3a directly regulates the differentiation and migration of adult neural progenitor cells [69]. Findings that C3aR-deficient, C3aR antagonist-treated as well as C3-deficient mice had impaired basal neurogenesis, support the contention that signaling through C3aR acts as a positive regulator of adult neurogenesis [68]. In a model of permanent focal cerebral ischemia, mice deficient in C3 showed reduced ischemia-induced neurogenesis and larger infarct volume at both 7 and 21 days after ischemia [68]. These studies indicate that the products of C3 activation positively regulate post-stroke neurogenesis and protect the brain tissue after ischemia. New findings from our laboratory underscore the importance of C3aR signaling for the neurogenic response to ischemic brain injury. Using C3aR-deficient mice, *GFAP-C3a* mice expressing C3a in reactive astrocytes, and their respective wild type controls, we observed that C3a overexpression increased whereas C3aR deficiency decreased the number of newly born neurons in the peri-infarct region, implying that the stroke-induced neurogenic response is at least partially regulated by C3a-C3aR signaling (Fig. 26.2). While low levels of *GFAP* promoter activity in unchallenged adult mice likely explain unchanged basal neurogenesis in *GFAP-C3a* mice [96], pronounced and persistent reactive astrogliosis in the peri-infarct tissue [97] conceivably leads to transgene-derived C3a levels that are sufficiently high to affect post-stroke neurogenesis in this region.

It is noteworthy that daily systemic administration of a low dose of a C3aR antagonist SB 290157 [98] starting already prior to ischemia induction increased neuronal precursor cell proliferation in the ipsilesional subventricular zone of the lateral ventricle 7 days after transient middle cerebral artery occlusion in mice [83]. Given that this treatment did not affect basal neurogenesis, the positive effect of low dose C3aR antagonist treatment on post-stroke neurogenesis could be attributed to its inhibitory effect on inflammatory response and striatal infiltration by activated

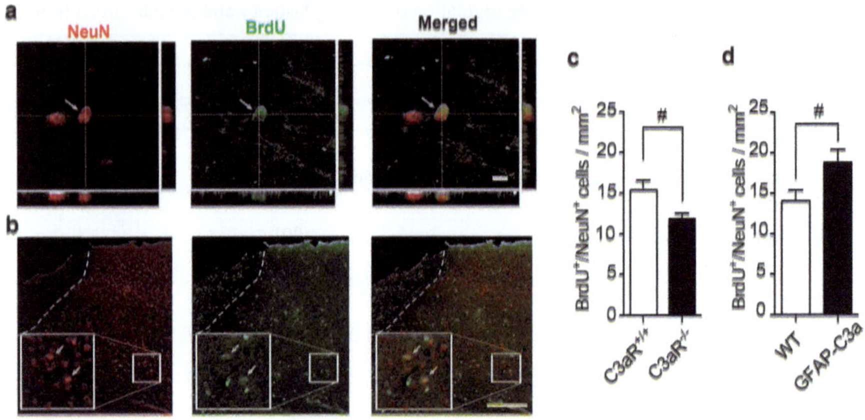

Fig. 26.2 C3a-C3aR signaling positively regulates post-stroke neurogenesis in the peri-infarct region. High (**a**) and low (**b**) magnification images of BrdU$^+$/NeuN$^+$ neurons. Lack of C3aR (*C3aR*$^{-/-}$) reduced (**c**), whereas over-expression of C3a in reactive astrocytes (*GFAP-C3a*) increased (**d**), the number of newly formed neurons (BrdU$^+$/NeuN$^+$) in the peri-infarct region. Mean ± SEM; *C3aR*$^{+/+}$ n = 10, *C3aR*$^{-/-}$ n = 14, WT n = 13, *GFAP-C3a* n = 12; #$P < 0.05$ determined by t test). Scale bar 10 μm (**a**) and 200 μm (**b**).

T-lymphocytes, that reduce neurogenesis, in the C3aR antagonist-treated animals rather than to the direct effect of the drug on progenitor cells [83]. In line with our previous report [68], higher dose of the C3aR antagonist, which is more likely to reach neural stem cell niche after systemic delivery, reduced the number of proliferating doublecortin-positive neural progenitor cells in the subventricular zone of unchallenged mice [83]. Further investigations need to determine the mechanism for C3a generation in an unchallenged neurogenic niche and the mechanism for C3a generation in the post-acute and chronic phase after stroke.

3.3.2 C3a and Post-stroke Synaptic Plasticity

Although the specific mechanism of complement activation in normal CNS is currently unknown, complement has a clear role in regulating the number of synapses in the developing brain. Transforming growth factor β secreted by immature astrocytes triggers the neuronal expression of complement component C1q in the developing brain [99, 100]. C1q targets an as yet unknown synapse-associated target and the synapse is tagged for elimination through the activation of the classical complement cascade, deposition of C3b and its subsequent recognition by microglial CR3 [30, 99]. Excessive classical complement pathway activity and synapse elimination have been implicated in the development of schizophrenia [101], and the C3b-CR3-mediated elimination of synapses seems to be re-activated in neurodegenerative diseases such as glaucoma [102] and Alzheimer's disease [103]. It is noteworthy that in the hippocampus of C3-deficient mice, the excessive number of glutamatergic synapses (due to the impaired developmental synapse elimination) is compensated by their reduced release probability, which is conceivably the reason for the absence of any signs of spontaneous epileptiform activity in these mice [104]. Notably, the C3-deficient mice exhibit better hippocampus-dependent learning and memory functions [104], and are protected from age-related region-specific loss of neurons and synapses in the hippocampus, and age-related cognitive decline [105]. Similarly, adult C3-deficient mice showed faster recovery of motor functions lost after sciatic nerve transection and were protected from axotomy-induced inhibitory synapse removal [106].

Stroke leads to neuronal death, gliosis, and axonal degeneration that occur also in non-ischemic remote brain regions that have synaptic connections with the primary lesion site [107]. This secondary degeneration has been linked to neurological deficits such as depression and cognitive impairment [108, 109], and can predict motor outcome after stroke [107] but the underlying molecular mechanism are only poorly understood. Given that synaptic dysfunction and loss may be the initial step in neurodegeneration [110, 111], studies on the involvement of C3b-CR3 signaling in the elimination of synapses in the regions affected by secondary degeneration in the post-stroke brain are warranted.

While the C3b-CR3 interaction constitutes a major mechanism for synapse removal, neuronal C3aR signaling promotes α-amino-3-hydroxy-5-methyl-4-isoxazolepropionic acid receptor (AMPAR) membrane localization to increase synaptic strength, and treatment with a C3aR antagonist or C3aR deficiency in neurons reduced dendritic complexity [112]. Excessive activation of neuronal C3aR can, on the other hand, negatively alter dendritic morphology and synaptic function [112]. A recent study investigating the role of C3a-C3aR signaling in structural synaptic responses to cerebral ischemia demonstrated that C3a-C3aR signaling increases the density and size of glutamatergic pre-synaptic terminals, and presumably synapses, in the peri-infarct region as well as in the contralesional hemisphere (Fig. 26.3), but does not affect any of these parameters in an unchallenged brain, and this C3a-C3aR-mediated response to ischemia is cortical region and layer specific [113]. Daily intranasal treatment with C3a starting 7 days after ischemia induction led to robust increase in synaptic responses and was associated with faster and sustained functional recovery [113], Fig. 26.4. Intranasal administration enables rapid and non-invasive delivery of peptides to the brain. The transport of peptides occurs mainly via peri-vascular bulk flow along the olfactory and trigeminal nerves and thus permits peptides to bypass the periphery and the blood-brain barrier, reaching the brain and entering the cerebrospinal fluid within minutes [114]. The findings of the plasticity and recovery promoting effects of C3a given intranasally are particularly intriguing not only because they show that C3a can be delivered to the brain through this clinically feasible route but also that such treatment, conceivably in combination with relevant rehabilitative interventions, could provide therapeutic benefit to stroke survivors in the post-acute or even chronic phase.

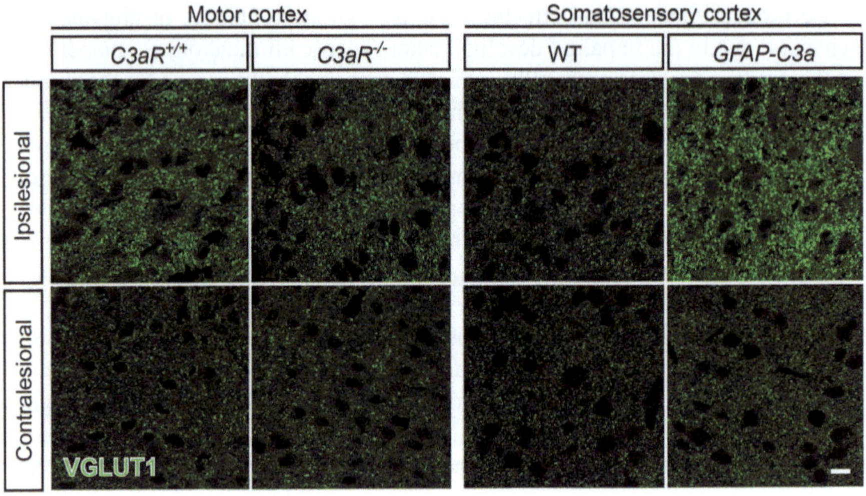

Fig. 26.3 After ischemic stroke, C3a-C3aR signaling positively regulates the density of excitatory synapses in the peri-infarct region. The expression of vesicular glutamate transporter 1 (VGLUT1), a marker of excitatory synapses, was reduced in the peri-infarct motor and somatosensory cortex in mice lacking C3aR (*C3aR*$^{-/-}$) and increased in mice over-expressing of C3a in reactive astrocytes (*GFAP-C3a*) compared with their respective controls (*C3aR*$^{+/+}$, WT). Scale bar 10 μm. Reproduced with permission from [113]

3.3.3 C3a and Post-stroke Axonal Plasticity

Axonal plasticity is a hallmark of CNS regeneration that is associated with reactivation of the intrinsic neuronal growth program and upregulation of the membrane phosphoprotein growth associated protein (GAP)-43 [115]. GAP-43 associates with axonal growth cones and is used as marker of axonal sprouting and plasticity [116, 117]. GAP-43 is also upregulated during reactive synaptogenesis [116, 118] and astrocyte-derived GAP-43 promotes neuronal survival and plasticity [119]. We have recently shown that after cerebral ischemia, the expression of GAP-43 in the peri-infarct regions is reduced in the absence of C3aR and increased when C3a is expressed in reactive astrocytes or administered intranasally [113], Fig. 26.5. As ischemia leads to upregulation of C3 in sprouting neurons [120] and C3a promotes neurite outgrowth *in vitro* [69], these findings support the conclusion that C3a signaling through C3aR plays a positive role in post-stroke axonal plasticity, possibly including axonal sprouting.

4 Concluding Remarks

Whereas complement activation and C3a in particular can contribute to tissue injury in the acute phase after cerebral ischemia, there is a mounting evidence to support the role of C3a-C3aR signaling in tissue repair and recovery by stimulating

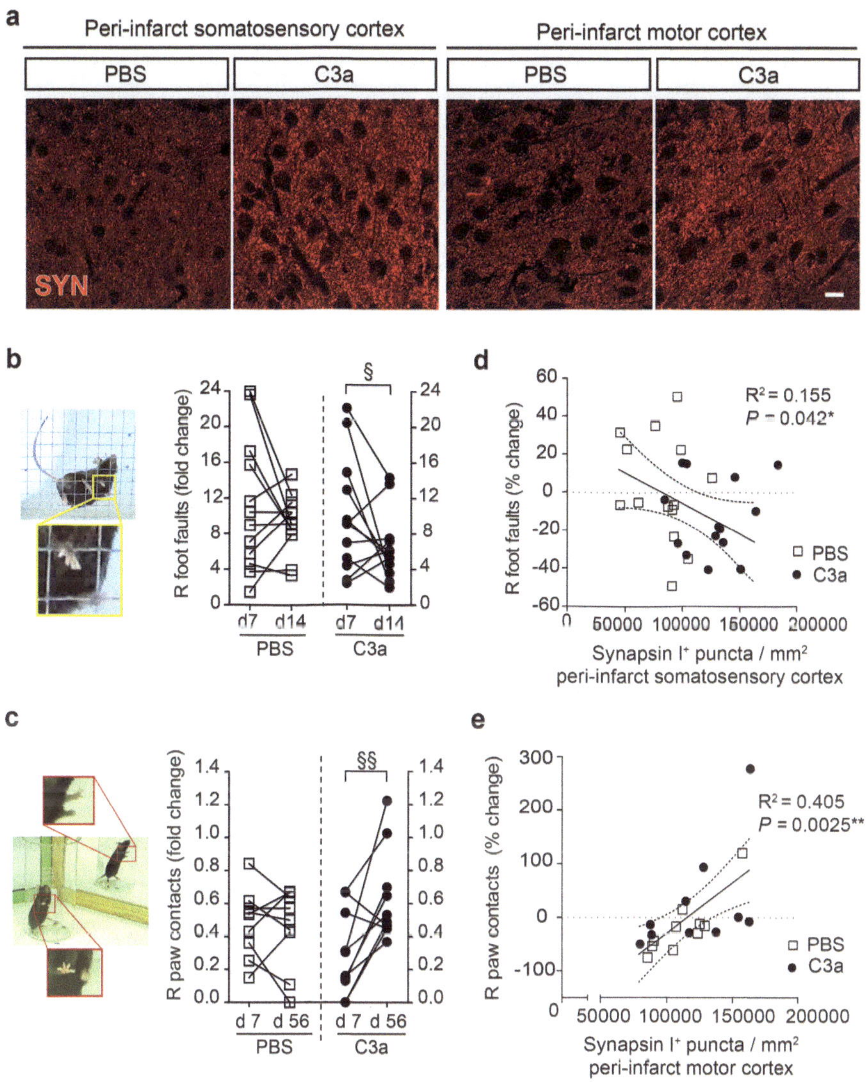

Fig. 26.4 Intranasal treatment with C3a increases pre-synaptic terminal density in the cortex and stimulates recovery of forepaw function after ischemic stroke. (**a**) Representative images of peri-infarct somatosensory and motor cortex stained with antibody against a pan-synaptic marker synapsin I on day 21 after stroke (scale bar: 10 μm). (**b**) *Left*: A typical foot fault during grid walking task; *Right*: Change in *right* (R, affected) paw foot faults in the grid walking task of individual mice between days 7 and 14 after stroke. (**c**) *Left*: An example of behavior scored in the cylinder test. *Right*: change in the performance of individual mice between days 7 and 56 post-stroke. (**d**, **e**) Scatter plots and linear regression fit of association between the density of synapsin I$^+$ puncta and change in performance between days 7 and 21 post-stroke in (**c**) grid walking test and (**e**) cylinder test. Intranasal treatment was performed between day 7 and 28 after stroke. §$P < 0.05$; §§$P < 0.01$. Reproduced with permission from [113]

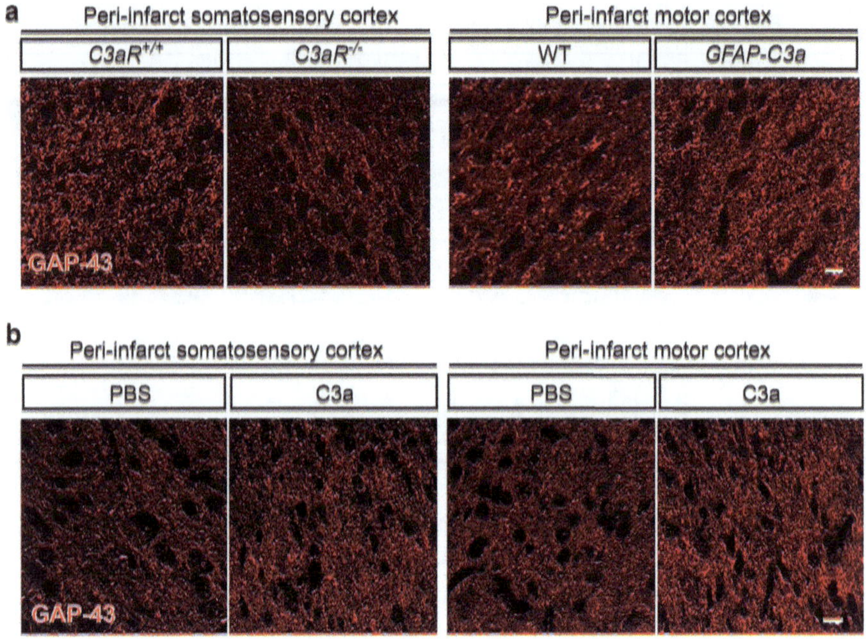

Fig. 26.5 C3a-C3aR signaling positively regulates peri-infarct axonal density after ischemic stroke. The expression of growth-associated protein 43 (GAP-43), a marker of axonal plasticity, in the peri-infarct cortex was reduced in mice lacking C3aR ($C3aR^{-/-}$), and increased in mice over-expressing of C3a in reactive astrocytes (*GFAP-C3a*) (**a**) as well as mice treated with intranasal C3a (**b**) compared to their respective controls ($C3aR^{+/+}$, WT, and PBS-treated mice). Scale bar 10 μm. Reproduced with permission from [113]

post-stroke neural plasticity including cell replacement, reorganization of axonal circuitry, and consequently, regulation of synaptic input. In light of the role of neuronal C3aR in modulation of synaptic strength and dendritic morphology [112], and the effects of C3a on neural progenitor cell differentiation and migration [69], the C3a-C3aR-mediated increase in peri-infarct neurogenesis, upregulation of expression of GAP-43 and increased number of pre-synaptic terminals, particularly glutamatergic terminals [113], are conceivably due at least in part to a direct effect of C3a on neurons. However, given the broad expression of C3aR in the brain, C3a can also exert its effects on post-stroke plasticity indirectly by modulating the functions of glial, endothelial, immune cells, stem/progenitor cells, and epithelial cells in the choroid plexus. Although the elucidation of the underlying cellular and molecular mechanism warrants additional experimental studies, the available data point to intranasal delivery of C3aR agonists in the post-acute phase as an attractive approach to improve functional recovery after ischemic stroke. While this non-invasive and directed route of administration presents clear advantages over systemic modes of delivery, further preclinical studies will need to determine the optimal therapeutic window and dose as well as the potential benefits of this type of treatment for other types of brain injuries.

References

1. Del Rio-Tsonis K, Tsonis PA, Zarkadis IK, Tsagas AG, Lambris JD. Expression of the third component of complement, C3, in regenerating limb blastema cells of urodeles. J Immunol. 1998;161:6819–24.
2. Kimura Y, Madhavan M, Call MK, Santiago W, Tsonis PA, Lambris JD, et al. Expression of complement 3 and complement 5 in newt limb and lens regeneration. J Immunol. 2003;170:2331–9.
3. Morais da Silva S, Gates PB, Brockers JP. The newt ortholog of CD59 is implicated in proximodistal identity during amphibian limb regeneration. Dev Cell. 2002;3:547–55.
4. Broders-Bondon F, Paul-Gilloteaux P, Gazquez E, Heysch J, Piel M, Mayor R, et al. Control of the collective migration of enteric neural crest cells by the Complement anaphylatoxin C3a and N-cadherin. Dev Biol. 2016;414:85–99. https://doi.org/10.1016/j.ydbio.2016.03.022.
5. Leslie JD, Mayor R. Complement in animal development: unexpected roles of a highly conserved pathway. Semin Immunol. 2013;25:39–46. https://doi.org/10.1016/j.smim.2013.04.005.
6. Szabo A, Cobo I, Omara S, McLachlan S, Keller R, Mayor R. The molecular basis of radial intercalation during tissue spreading in early development. Dev Cell. 2016;37:213–25. https://doi.org/10.1016/j.devcel.2016.04.008.
7. Mastellos D, Papadimitriou JC, Franchini S, Tsonis PA, Lambris JD. A novel role of complement: mice deficient in the fifth component of complement (C5) exhibit impaired liver regeneration. J Immunol. 2001;166:2479–86.
8. Strey CW, Markiewski M, Mastellos D, Tudoran R, Spruce LA, Greenbaum LE, et al. The proinflammatory mediators C3a and C5a are essential for liver regeneration. J Exp Med. 2003;198:913–23. https://doi.org/10.1084/jem.20030374.
9. Daveau M, Benard M, Scotte M, Schouft M-T, Hiron M, Francois A, et al. Expression of a functional C5a receptor in regenerating hepatocytes and its involvement in a proliferative signalling pathway in rat. J Immunol. 2004;173:3418–24.
10. Reca R, Mastellos D, Majka M, Marquez L, Ratajczak J, Franchini S, et al. Functional receptor for C3a anaphylatoxin is expressed by normal hematopoietic stem/progenitor cells, and C3a enhances their homing.related responses to SDF-1. Blood. 2003;101:3784–93. https://doi.org/10.1182/blood-2002-10-3233.
11. Honczarenko M, Ratajczak MZ, Nicholson-Weller A, Silberstein LE. Complement C3a enhances CXCL12 (SDF-1)-mediated chemotaxis of bone marrow hematopoietic cells independently of C3a receptor. J Immunol. 2005;175:3698–706.
12. Ratajczak J, Reca R, Kucia M, Majka M, Allendorf DJ, Baran JT, et al. Mobilization studies in mice deficient in either C3 or C3a receptor (C3aR) reveal a novel role for complement in retention of hematopoietic stem/progenitor cells in bone marrow. Blood. 2004;103:2071–8. https://doi.org/10.1182/blood-2003-06-2099.
13. Bossi F, Tripodo C, Rizzi L, Bulla R, Agostinis C, Guarnotta C, et al. C1q as a unique player in angiogenesis with therapeutic implication in wound healing. Proc Natl Acad Sci U S A. 2014;111:4209–14. https://doi.org/10.1073/pnas.1311968111.
14. Bora NS, Kaliappan S, Jha P, Xu Q, Sivasankar B, Harris CL, et al. CD59, a complement regulatory protein, controls choroidal neovascularization in a mouse model of wet-type age-related macular degeneration. J Immunol. 2007;178:1783–90.
15. Bora NS, Kaliappan S, Jha P, Xu Q, Sohn JH, Dhaulakhandi DB, et al. Complement activation via alternative pathway is critical in the development of laser-induced choroidal neovascularization: role of factor B and factor H. J Immunol. 2006;177:1872–8.
16. Bora PS, Sohn JH, Cruz JM, Jha P, Nishihori H, Wang Y, et al. Role of complement and complement membrane attack complex in laser-induced choroidal neovascularization. J Immunol. 2005;174:491–7.
17. Lyzogubov V, Wu X, Jha P, Tytarenko R, Triebwasser M, Kolar G, et al. Complement regulatory protein CD46 protects against choroidal neovascularization in mice. Am J Pathol. 2014;184:2537–48. https://doi.org/10.1016/j.ajpath.2014.06.001.

18. Anderson DH, Radeke MJ, Gallo NB, Chapin EA, Johnson PT, Curletti CR, et al. The pivotal role of the complement system in aging and age-related macular degeneration: hypothesis re-visited. Prog Retin Eye Res. 2010;29:95–112. https://doi.org/10.1016/j.preteyeres.2009.11.003.
19. Peterson SL, Anderson AJ. Complement and spinal cord injury: traditional and non-traditional aspects of complement cascade function in the injured spinal cord microenvironment. Exp Neurol. 2014;258:35–47. https://doi.org/10.1016/j.expneurol.2014.04.028.
20. Brennan FH, Gordon R, Lao HW, Biggins PJ, Taylor SM, Franklin RJ, et al. The complement receptor C5aR controls acute inflammation and astrogliosis following spinal cord injury. J Neurosci. 2015;35:6517–31. https://doi.org/10.1523/jneurosci.5218-14.2015.
21. Peterson SL, Nguyen HX, Mendez OA, Anderson AJ. Complement protein C1q modulates neurite outgrowth in vitro and spinal cord axon regeneration in vivo. J Neurosci. 2015;35:4332–49. https://doi.org/10.1523/jneurosci.4473-12.2015.
22. Bajic G, Degn SE, Thiel S, Andersen GR. Complement activation, regulation, and molecular basis for complement-related diseases. EMBO J. 2015;34:2735–57. 10.15252/embj.201591881.
23. Nauta AJ, Raaschou-Jensen N, Roos A, Daha MR, Madsen HO, Borrias-Essers MC, et al. Mannose-binding lectin engagement with late apoptotic and necrotic cells. Eur J Immunol. 2003;33:2853–63. https://doi.org/10.1002/eji.200323888.
24. Matsushita M, Fujita T. Cleavage of the third component of complement (C3) by mannose-binding protein-associated serine protease (MASP) with subsequent complement activation. Immunobiology. 1995;194:443–8. https://doi.org/10.1016/s0171-2985(11)80110-5.
25. Liszewski MK, Kolev M, Le Friec G, Leung M, Bertram PG, Fara AF, et al. Intracellular complement activation sustains T cell homeostasis and mediates effector differentiation. Immunity. 2013;39:1143–57. https://doi.org/10.1016/j.immuni.2013.10.018.
26. Yuan X, Shan M, You R, Frazier MV, Hong MJ, Wetsel RA, et al. Activation of C3a receptor is required in cigarette smoke-mediated emphysema. Mucosal Immunol. 2015;8:874–85. https://doi.org/10.1038/mi.2014.118.
27. Johnson U, Ohlsson K, Olsson I. Effects of granulocyte neutral proteases on complement components. Scand J Immunol. 1976;5:421–6.
28. Markiewski MM, Lambris JD. The role of complement in inflammatory diseases from behind the scenes into the spotlight. Am J Pathol. 2007;171:715–27. https://doi.org/10.2353/ajpath.2007.070166.
29. Amara U, Flierl MA, Rittirsch D, Klos A, Chen H, Acker B, et al. Molecular intercommunication between the complement and coagulation systems. J Immunol. 2010;185:5628–36. https://doi.org/10.4049/jimmunol.0903678.
30. Schafer DP, Lehrman EK, Kautzman AG, Koyama R, Mardinly AR, Yamasak R, et al. Microglia sculpt postnatal neural circuits in an activity and complement-dependent manner. Neuron. 2012;74:691–705. https://doi.org/10.1016/j.neuron.2012.03.026.
31. Gasque P, Fontaine M, Morgan BP. Complement expression in human brain. Biosynthesis of terminal pathway components and regulators in human glial cells and cell lines. J Immunol. 1995;154:4726–33.
32. Gasque P, Ischenko A, Legoedec J, Mauger C, Schouft MT, Fontaine M. Expression of the complement classical pathway by human glioma in culture. A model for complement expression by nerve cells. J Biol Chem. 1993;268:25068–74.
33. Gasque P, Julen N, Ischenko AM, Picot C, Mauger C, Chauzy C, et al. Expression of complement components of the alternative pathway by glioma cell lines. J Immunol. 1992;149:1381–7.
34. Thomas A, Gasque P, Vaudry D, Gonzalez B, Fontaine M. Expression of a complete and functional complement system by human neuronal cells in vitro. Int Immunol. 2000;12:1015–23.
35. Pedersen ED, Waje-Andreassen U, Vedeler CA, Aamodt G, Mollnes TE. Systemic complement activation following human acute ischaemic stroke. Clin Exp Immunol. 2004;137:117–22. https://doi.org/10.1111/j.1365-2249.2004.02489.x.

36. Mocco J, Wilson DA, Komotar RJ, Sughrue ME, Coates K, Sacco RL, et al. Alterations in plasma complement levels after human ischemic stroke. Neurosurgery. 2006;59:28–33. https://doi.org/10.1227/01.neu.0000219221.14280.65.
37. Széplaki G, Szegedi R, Hirschberg K, Gombos T, Varga L, Karádi I, et al. Strong complement activation after acute ischemic stroke is associated with unfavorable outcomes. Atherosclerosis. 2009;204:315–20. https://doi.org/10.1016/j.atherosclerosis.2008.07.044.
38. Stokowska A, Olsson S, Holmegaard L, Jood K, Blomstrand C, Jern C, et al. Plasma C3 and C3a levels in cryptogenic and large vessel disease stroke: associations with outcome. Cerebrovasc Dis. 2011;32:114–22. https://doi.org/10.1159/000328238.
39. Stokowska A, Olsson S, Holmegaard L, Jood K, Blomstrand C, Jern C, et al. Cardioembolic and small vessel disease stroke show differences in associations between systemic C3 levels and outcome. PLoS One. 2013;8:e72133. https://doi.org/10.1371/journal.pone.0072133.
40. Lindsberg PJ, Ohman J, Lehto T, Karjalainen-Lindsberg ML, Paetau A, Wuorimaa T, et al. Complement activation in the central nervous system following blood-brain barrier damage in man. Ann Neurol. 1996;4:587–96.
41. Pedersen ED, Løberg EM, Vege E, Daha MR, Maehlen J, Mollnes TE. In situ deposition of complement in human acute brain ischaemia. Scand J Immunol. 2009;69:555–62. https://doi.org/10.1111/j.1365-3083.2009.02253.x.
42. Mocco J, Mack WJ, Ducruet AF, Sosunov AA, Sughrue ME, Hassid BG, et al. Complement component C3 mediates inflammatory injury following focal cerebral ischemia. Circ Res. 2006;99:209–17. https://doi.org/10.1161/01.res.0000232544.90675.42.
43. De Simoni MG, Storini C, Barba M, Catapano L, Arabia AM, Rossi E, et al. Neuroprotection by complement (C1) inhibitor in mouse transient brain ischemia. J Cereb Blood Flow Metab. 2003;23:232–9. https://doi.org/10.1097/01.wcb.0000046146.31247.a1.
44. Ducruet AF, Sosunov SA, Zacharia BE, Gorski J, Yeh ML, Derosa P, et al. The neuroprotective effect of genetic mannose-binding lectin deficiency is not sustained in the sub-acute phase of stroke. Transl Stroke Res. 2011;2:588–99. https://doi.org/10.1007/s12975-011-0104-2.
45. Fumagalli S, De Simoni MG. Lectin complement pathway and its bloody interactions in brain ischemia. Stroke. 2016;47:3067–73. https://doi.org/10.1161/strokeaha.116.012407.
46. Elvington A, Atkinson C, Kulik L, Zhu H, Yu J, Kindy MS, et al. Pathogenic natural antibodies propagate cerebral injury following ischemic stroke in mice. J Immunol. 2012;188:1460–8. https://doi.org/10.4049/jimmunol.1102132.
47. Zhang M, Takahashi K, Alicot EM, Vorup-Jensen T, Kessler B, Thiel S, et al. Activation of the lectin pathway by natural IgM in a model of ischemia/reperfusion injury. J Immunol. 2006;177:4727–34.
48. Elvington A, Atkinson C, Zhu H, Yu J, Takahashi K, Stahl GL, et al. The alternative complement pathway propagates inflammation and injury in murine ischemic stroke. J Immunol. 2012;189:4640–7. https://doi.org/10.4049/jimmunol.1201904.
49. D'Ambrosio AL, Pinsky DJ, Connolly ES. The role of the complement cascade in ischemia/reperfusion injury: implications for neuroprotection. Mol. Medicine. 2001;7:367–82.
50. Stahel PF, Morganti-Kossmann MC, Kossmann T. The role of the complement system in traumatic brain injury. Brain Res Rev. 1998;27:243–56.
51. Alawieh A, Elvington A, Tomlinson S. Complement in the homeostatic and ischemic brain. Front Immunol. 2015;6:417. https://doi.org/10.3389/fimmu.2015.00417.
52. Coulthard LG, Woodruff TM. Is the complement activation product C3a a proinflammatory molecule? Re-evaluating the evidence and the myth. J Immunol. 2015;194(8):3542. https://doi.org/10.4049/jimmunol.1403068.
53. Huey R, Bloor CM, Kawahara MS, Hugli TE. Potentiation of the anaphylatoxins in vivo using an inhibitor of serum carboxypeptidase N (SCPN). I. Lethality and pathologic effects on pulmonary tissue. Am J Pathol. 1983;112:48–60.
54. Campbell WD, Lazoura E, Okada N, Okada H. Inactivation of C3a and C5a octapeptides by carboxypeptidase R and carboxypeptidase N. Microbiol Immunol. 2002;46:131–4.
55. Wilken HC, Gotze O, Werfel T, Zwirner J. C3a(desArg) does not bind to and signal through the human C3a receptor. Immunol Lett. 1999;67:141–5.

56. Kalant D, Cain SA, Maslowska M, Sniderman AD, Cianflone K, Monk PN. The chemoattractant receptor-like protein C5L2 binds the C3a des-Arg77/acylation stimulating protein. J Biol Chem. 2003;278:11123–9. https://doi.org/10.1074/jbc.M206169200.
57. Kalant D, Maclaren R, Cui W, Samanta R, Mon PN, Laporte SA, et al. C5L2 is a functional receptor for acylation stimulating protein. J Biol Chem. 2005;280:23936–44. https://doi.org/10.1074/jbc.M406921200.
58. Ember JA, Jagels MA, Hugli T. Characterization of complement anaphylatoxins and biological responses. In: Volanakis JE, Frank MM, editors. The human complement system in health and disease. New York, NY: Marcel Dekker; 1998. p. 241–84.
59. Klos A, Tenner AJ, Johswich KO, Ager RR, Reis ES, Kohl J. The role of the anaphylatoxins in health and disease. Mol Immunol. 2009;46:2753–66. https://doi.org/10.1016/j.molimm.2009.04.027.
60. van Beek J, Bernaudin M, Petit E, Gasque P, Nouvelot A, MacKenzie ET, et al. Expression of receptors for complement anaphylatoxins C3a and C5a following permanent focal cerebral ischemia in the mouse. Exp Neurol. 2000;161:373–82. https://doi.org/10.1006/exnr.1999.7273.
61. Schraufstatter IU, Trieu K, Sikora L, Sriramarao P, DiScipio R. Complement C3a and C5a Induce different signal transduction cascades in endothelial cells. J Immunol. 2002;169:2102–10.
62. Monsinjon T, Gasque P, Chan P, Ischenko A, Brady JJ, Fontaine MC. Regulation by complement C3a and C5a anaphylatoxins of cytokine production in human umbilical vein endothelial cells. FASEB J. 2003;17:1003–14. https://doi.org/10.1096/fj.02-0737com.
63. Wu F, Zou Q, Ding X, Shi D, Zhu X, Hu W, et al. Complement component C3a plays a critical role in endothelial activation and leukocyte recruitment into the brain. J Neuroinflammation. 2016;13:23. https://doi.org/10.1186/s12974-016-0485-y.
64. Boire A, Zou Y, Shieh J, Macalinao DG, Pentsova E, Massague J. Complement component 3 adapts the cerebrospinal fluid for leptomeningeal metastasis. Cell. 2017;168:1101–13.e13. https://doi.org/10.1016/j.cell.2017.02.025.
65. Davoust N, Jones J, Stahel PF, Ames RS, Barnum SR. Receptor for the C3a anaphylatoxin is expressed by neurons and glial cells. Glia. 1999;26:201–11.
66. Benard M, Gonzalez BJ, Schouft M-T, Falluel-Morel A, Chan P, Vaudry H, et al. Characterization of C3a and C5a Receptors in rat cerebellar granule neurons during maturation. Neuroprotective effect of C5a against apoptotic cell death. J Biol Chem. 2004;279:43487–96. https://doi.org/10.1074/jbc.M404124200.
67. Pedersen ED, Froyland E, Kvissel AK, Pharo AM, Skalhegg BS, Rootwelt T, et al. Expression of complement regulators and receptors on human NT2-N neurons--effect of hypoxia and reoxygenation. Mol Immunol. 2007;44:2459–68. https://doi.org/10.1016/j.molimm.2006.10.022.
68. Rahpeymai Y, Hietala MA, Wilhelmsson U, Fotheringham A, Davies I, Nilsson AK, et al. Complement: a novel factor in basal and ischemia-induced neurogenesis. EMBO J. 2006;25:1364–74. https://doi.org/10.1038/sj.emboj.7601004. 7601004 [pii].
69. Shinjyo N, Ståhlberg A, Dragunow M, Pekny M, Pekna M. Complement-derived anaphylatoxin C3a regulates in vitro differentiation and migration of neural progenitor cells in vitro. Stem Cells. 2009;27:2824–32. https://doi.org/10.1002/stem.225.
70. Benard M, Raoult E, Vaudry D, Leprince J, Falluel-Morel A, Gonzalez BJ, et al. Role of complement anaphylatoxin receptors (C3aR, C5aR) in the development of the rat cerebellum. Mol Immunol. 2008;45:3767–74. https://doi.org/10.1016/j.molimm.2008.05.027.
71. Gorelik A, Sapir T, Haffner-Krausz R, Olender T, Woodruff TM, Reiner O. Developmental activities of the complement pathway in migrating neurons. Nat Commun. 2017;8:15096. https://doi.org/10.1038/ncomms15096.
72. Shinjyo N, de Pablo Y, Pekny M, Pekna M. Complement peptide C3a promotes astrocyte survival in response to ischemic stress. Mol Neurobiol. 2016;53:3076–87. https://doi.org/10.1007/s12035-015-9204-4.
73. Sayah S, Jauneau AC, Patte C, Tonon MC, Vaudry H, Fontaine M. Two different transduction pathways are activated by C3a and C5a anaphylatoxins on astrocytes. Mol Brain Res. 2003;112:53–60.

74. Sayah S, Ischenko A, Zhakhov A, Bonnard AS, Fontaine M. Expression of cytokines by human astrocytomas following stimulation by C3a and C5a anaphylatoxins: specific increase in interleukin-6 mRNA expression. J Neurochem. 1999;72:2426–36.
75. Jauneau AC, Ischenko A, Chan P, Fontaine M. Complement component anaphylatoxins upregulate chemokine expression by human astrocytes. FEBS Lett. 2003;537:17–22.
76. Jauneau A-C, Ischenko A, Chatagner A, Benard M, Chan P, Schouft M-T, et al. Interleukin-1β and anaphylatoxins exert a synergistic effect on NGF expression by astrocytes. J Neuroinflammation. 2006;3:8. https://doi.org/10.1186/1742-2094-3-8.
77. Möller T, Nolte C, Burger R, Verkhratsky A, Kettermann H. Mechanisms of C5a and C3a complement fragment-induced [Ca2+]i signaling in mouse microglia. J Neurosci. 1997;17:615–24.
78. Heese K, Hock C, Otten U. Inflammatory signals induce neurotropin expression in human microglial cells. J Neurochem. 1998;70:699–707.
79. Lian H, Litvinchuk A, Chiang AC, Aithmitti N, Jankowsky JL, Zheng H. Astrocyte-microglia cross talk through complement activation modulates amyloid pathology in mouse models of Alzheimer's disease. J Neurosci. 2016;36:577–89. https://doi.org/10.1523/jneurosci.2117-15.2016.
80. Chen NJ, Mirtsos C, Suh D, YC L, Lin WJ, McKerlie C, et al. C5L2 is critical for the biological activities of the anaphylatoxins C5a and C3a. Nature. 2007;446:203–7. https://doi.org/10.1038/nature05559.
81. Gavrilyuk V, Kalinin S, Hilbush BS, Middlecamp A, McGuire S, Pelligrino D, et al. Identification of complement 5a-like receptor (C5L2) from astrocytes: characterization of anti-inflammatory properties. J Neurochem. 2005;92:1140–9. https://doi.org/10.1111/j.1471-4159.2004.02942.x.
82. Woodruff TM, Ager RR, Tenner AJ, Noakes PG, Taylor SM. The role of the complement system and the activation fragment C5a in the central nervous system. Neuromolecular Med. 2010;12:179–92. https://doi.org/10.1007/s12017-009-8085-y.
83. Ducruet AF, Zacharia BE, Sosunov SA, Gigante PR, Yeh ML, Gorski JW, et al. Complement inhibition promotes endogenous neurogenesis and sustained anti-inflammatory neuroprotection following reperfused stroke. PLoS One. 2012;7:e38664. https://doi.org/10.1371/journal.pone.0038664.
84. Mack WJ, Ducruet AF, Hickman ZL, Garrett MC, Albert EJ, Kellner CP, et al. Early plasma complement C3a levels correlate with functional outcome after aneurysmal subarachnoid hemorrhage. Neurosurgery. 2007;61:255–260 . discussion 60–1. https://doi.org/10.1227/01neu0000255518968 37.8e.
85. Olsson S, Stokowska A, Holmegaard L, Jood K, Blomstrand C, Pekna M, et al. Genetic variation in complement component C3 shows association with ischaemic stroke. Eur J Neurol. 2011;18:1272–4. https://doi.org/10.1111/j.1468-1331.2011.03377.x.
86. van Beek J, Nicole O, Ali C, Ischenko A, MacKenzie ET, Buisson A, et al. Complement anaphylatoxin C3a is selectively protective against NMDA-induced neuronal cell death. Neuroreport. 2001;12:289–93.
87. Boos L, Szalai AJ, Barnum SR. C3a expressed in the central nervous system protects against LPS-induced shock. Neurosci Lett. 2005;387:68–71. https://doi.org/10.1016/j.neulet.2005.07.015.
88. Järlestedt K, Rousset CI, Ståhlberg A, Sourkova H, Atkins AL, Thornton C, et al. Receptor for complement peptide C3a: a therapeutic target for neonatal hypoxic-ischemic brain injury. FASEB J. 2013;27:3797–804. https://doi.org/10.1096/fj.13-230011.
89. Moran J, Stokowska A, Walker FR, Mallard C, Hagberg H, Pekna M. Intranasal C3a treatment ameliorates cognitive impairment in a mouse model of neonatal hypoxic-ischemic brain injury. Exp Neurol. 2017;290:74–84. https://doi.org/10.1016/j.expneurol.2017.01.001.
90. Arvidsson A, Collin T, Kirik D, Kokaia Z, Lindvall O. Neuronal replacement from endogenous precursors in the adult brain after stroke. Nat Med. 2002;8:963–70. https://doi.org/10.1038/nm747.
91. Gu W, Brannstrom T, Wester P. Cortical neurogenesis in adult rats after reversible photothrombotic stroke. J Cereb Blood Flow Metab. 2000;20:1166–73. https://doi.org/10.1097/00004647-200008000-00002.

92. Osman AM, Porritt MJ, Nilsson M, Kuhn HG. Long-term stimulation of neural progenitor cell migration after cortical ischemia in mice. Stroke. 2011;42:3559–65. https://doi.org/10.1161/strokeaha.111.627802.
93. Carmichael ST. Plasticity of cortical projections after stroke. Neuroscientist. 2003;9:64–75. https://doi.org/10.1177/1073858402239592.
94. Winship IR, Murphy TH. Remapping the somatosensory cortex after stroke: insight from imaging the synapse to network. Neuroscientist. 2009;15:507–24. https://doi.org/10.1177/1073858409333076.
95. Pekna M, Pekny M, Nilsson M. Modulation of neural plasticity as a basis for stroke rehabilitation. Stroke. 2012;43:2819–28. https://doi.org/10.1161/strokeaha.112.654228.
96. Bogestål RY, Barnum SR, Smith PL, Mattisson V, Pekny M, Pekna M. Signaling through C5aR is not involved in basal neurogenesis. J Neurosci Res. 2007;85:2892–7. https://doi.org/10.1002/jnr.21401.
97. Nowicka D, Rogozinska K, Aleksy M, Witte OW, Skangiel-Kramska J. Spatiotemporal dynamics of astroglial and microglial responses after photothrombotic stroke in the rat brain. Acta Neurobiol Exp. 2008;68:155–68.
98. Ames RS, Lee D, Foley JJ, Jurewicz AJ, Tornetta MA, Bautsch W, et al. Identification of a selective nonpeptide antagonist of the anaphylatoxin C3a receptor that demonstrates antiinflammatory activity in animal models. J Immunol. 2001;166:6341–8.
99. Stevens B, Allen NJ, Vazquez LE, Howell GR, Christopherson KS, Nouri N, et al. The classical complement cascade mediates CNS synapse elimination. Cell. 2007;131:1164–78. https://doi.org/10.1016/j.cell.2007.10.036.
100. Bialas AR, Stevens B. TGF-β signaling regulates neuronal C1q expression and developmental synaptic refinement. Nat Neurosci. 2013;16:1773–82. https://doi.org/10.1038/nn.3560.
101. Sekar A, Bialas AR, de Rivera H, Davis A, Hammond TR, Kamitaki N, et al. Schizophrenia risk from complex variation of complement component 4. Nature. 2016;530:177–83. https://doi.org/10.1038/nature16549.
102. Howell GR, Macalinao DG, Sousa GL, Walden M, Soto I, Kneeland SC, et al. Molecular clustering identifies complement and endothelin induction as early events in a mouse model of glaucoma. J Clin Invest. 2011;121:1429–44. https://doi.org/10.1172/jci44646.
103. Hong S, Beja-Glasser VF, Nfonoyim BM, Frouin A, Li S, Ramakrishnan S, et al. Complement and microglia mediate early synapse loss in Alzheimer mouse models. Science. 2016;352:712–6. https://doi.org/10.1126/science.aad8373.
104. Perez-Alcazar M, Daborg J, Stokowska A, Wasling P, Björefeldt A, Kalm M, et al. Altered cognitive performance and synaptic function in the hippocampus of mice lacking C3. Exp Neurol. 2014;253:154–64. https://doi.org/10.1016/j.expneurol.2013.12.013.
105. Shi Q, Colodner KJ, Matousek SB, Merry K, Hong S, Kenison JE, et al. Complement C3-deficient mice fail to display age-related hippocampal decline. J Neurosci. 2015;35:13029–42. https://doi.org/10.1523/jneurosci.1698-15.2015.
106. Berg A, Zelano J, Stephan A, Thams S, Barres BA, Pekny M, et al. Reduced removal of synaptic terminals from axotomized spinal motoneurons in the absence of complement C3. Exp Neurol. 2012;237:8–17. https://doi.org/10.1016/j.expneurol.2012.06.008.
107. Zhang J, Zhang Y, Xing S, Liang Z, Zeng J. Secondary neurodegeneration in remote regions after focal cerebral infarction: a new target for stroke management? Stroke. 2012;43:1700–5. https://doi.org/10.1161/strokeaha.111.632448.
108. Kronenberg G, Balkaya M, Prinz V, Gertz K, Ji S, Kirste I, et al. Exofocal dopaminergic degeneration as antidepressant target in mouse model of poststroke depression. Biol Psychiatry. 2012;72:273–81. https://doi.org/10.1016/j.biopsych.2012.02.026.
109. Jones KA, Zouikr I, Patience M, Clarkson AN, Isgaard J, Johnson SJ, et al. Chronic stress exacerbates neuronal loss associated with secondary neurodegeneration and suppresses microglial-like cells following focal motor cortex ischemia in the mouse. Brain Behav Immun. 2015;48:57–67. https://doi.org/10.1016/j.bbi.2015.02.014.
110. Cunningham C, Deacon R, Wells H, Boche D, Waters S, Diniz CP, et al. Synaptic changes characterize early behavioural signs in the ME7 model of murine prion disease. Eur J Neurosci. 2003;17:2147–55.

111. Conforti L, Adalbert R, Coleman MP. Neuronal death: where does the end begin? Trends Neurosci. 2007;30:159–69. https://doi.org/10.1016/j.tins.2007.02.004.
112. Lian H, Yang L, Cole A, Sun L, Chiang AC, Fowler SW, et al. NFkappaB-activated astroglial release of complement C3 compromises neuronal morphology and function associated with Alzheimer's disease. Neuron. 2015;85:101–15. https://doi.org/10.1016/j.neuron.2014.11.018.
113. Stokowska A, Atkins AL, Moran J, Pekny T, Bulmer L, Pascoe MC, et al. Complement peptide C3a stimulates neural plasticity after experimental brain ischemia. Brain. 2017;140:353–69. https://doi.org/10.1093/brain/aww314.
114. Lochhead JJ, Thorne RG. Intranasal delivery of biologics to the central nervous system. Adv Drug Deliv Rev. 2012;64:614–28. https://doi.org/10.1016/j.addr.2011.11.002.
115. Carmichael ST, Archibeque I, Luke L, Nolan T, Momiy J, Li S. Growth-associated gene expression after stroke: evidence for a growth-promoting region in peri-infarct cortex. Exp Neurol. 2005;193:291–311. https://doi.org/10.1016/j.expneurol.2005.01.004.
116. Benowitz LI, Rodriguez WR, Neve RL. The pattern of GAP-43 immunostaining changes in the rat hippocampal formation during reactive synaptogenesis. Brain Res Mol Brain Res. 1990;8:17–23.
117. Benowitz LI, Routtenberg A. GAP-43: an intrinsic determinant of neuronal development and plasticity. Trends Neurosci. 1997;20:84–91.
118. Lin LH, Bock S, Carpenter K, Rose M, Norden JJ. Synthesis and transport of GAP-43 in entorhinal cortex neurons and perforant pathway during lesion-induced sprouting and reactive synaptogenesis. Brain Res Mol Brain Res. 1992;14:147–53.
119. Hung CC, Lin CH, Chang H, Wang CY, Lin SH, Hsu PC, et al. Astrocytic GAP43 Induced by the TLR4/NF-kappaB/STAT3 Axis Attenuates Astrogliosis-Mediated Microglial Activation and Neurotoxicity. J Neurosci. 2016;36:2027–43. https://doi.org/10.1523/jneurosci.3457-15.2016.
120. Li S, Overman JJ, Katsman D, Kozlov SV. C.J. D, Twiss JL, et al. An age-related sprouting transcriptome provides molecular control of axonal sprouting after stroke. Nat Neurosci. 2010;13:1496–504. https://doi.org/10.1038/nn.2674.
121. Haynes T, Luz-Madrigal A, Reis ES, Echeverri Ruiz NP, Grajales-Esquivel E, Tzekou A, et al. Complement anaphylatoxin C3a is a potent inducer of embryonic chick retina regeneration. Nat Commun. 2013;4:2312. https://doi.org/10.1038/ncomms3312.
122. Biggins PJC, Brennan FH, Taylor SM, Woodruff TM, Ruitenberg MJ. The alternative receptor for complement component 5a, C5aR2, conveys neuroprotection in traumatic spinal cord injury. Journal of Neurotrauma 2017;34(12):2075–85.

MIX
Papier aus verantwortungsvollen Quellen
Paper from responsible sources
FSC® C105338
www.fsc.org

If you have any concerns about our products,
you can contact us on
ProductSafety@springernature.com

In case Publisher is established outside the EU,
the EU authorized representative is:
Springer Nature Customer Service Center GmbH
Europaplatz 3, 69115 Heidelberg, Germany

Printed by Libri Plureos GmbH
in Hamburg, Germany